FLUIDS, COLLOIDS AND SOFT MATERIALS: AN INTRODUCTION TO SOFT MATTER PHYSICS

FLUIDS, COLLOIDS AND SOFT MATERIALS: AN INTRODUCTION TO SOFT MATTER PHYSICS

Edited by

ALBERTO FERNANDEZ-NIEVES

ANTONIO MANUEL PUERTAS

Published by John Wiley & Sons, Inc., Hoboken, New Jersey
Published simultaneously in Canada

Library of Congress Cataloging-in-Publication Data:

Names: Fernandez-Nieves, Alberto. | Puertas, Antonio Manuel, 1973-
Title: Fluids, colloids, and soft materials : an introduction to soft matter
physics / edited by Alberto Fernandez-Nieves, Antonio Manuel Puertas.
Description: Hoboken, New Jersey : John Wiley & Sons, Inc., [2016] | Includes
bibliographical references and index.
Identifiers: LCCN 2015047736 (print) | LCCN 2015048370 (ebook) | ISBN
9781118065624 (cloth) | ISBN 9781119220527 (pdf) | ISBN 9781119220534
(epub)
Subjects: LCSH: Soft condensed matter. | Colloids.
Classification: LCC QC173.458.S62 F59 2016 (print) | LCC QC173.458.S62
(ebook) | DDC 532–dc23
LC record available at http://lccn.loc.gov/2015047736

Printed in the United States of America

10 9 8 7 6 5 4 3 2 1

CONTENTS

PREFACE

Soft matter is a highly multidisciplinary field that encompasses the study of a wide range of very different materials, which nevertheless share some commonalities. In the eyes of P. G. de Gennes (Nobel Prize in physics, 1991), soft materials are materials with large response functions. As a result, mild external influences can cause very large effects. Ultimately, the origin of this behavior arises from their internal characteristic energies, which are comparable to room-temperature thermal energy, k_BT, and internal characteristic structures, which typically have characteristic length scales, l, in the mesoscopic scale. It is for these reasons that soft materials have low elastic moduli and are easy to deform. Indeed, to a first approximation, one finds that the scale of the elastic modulus is $G \sim k_B T/l^3$, which, for typical values of l, is between six and nine orders of magnitude smaller than the modulus of atomic or molecular solids.

Soft materials are, in many cases, mixtures of phases; they are structured fluids consisting of a solute that adds structure to a background solvent. They can also be single-phase materials, but with constituent molecules that are anisotropic in shape and that, as a result, are able to acquire a degree of order in between that characteristic of the isotropic liquid and crystalline states (see Fig. 1a). Polymer solutions, surfactants, liquid crystals, colloidal suspensions, aerosols, emulsions, and foams are examples of soft materials. Granular materials are also soft, albeit in a broader sense, as they are athermal and have characteristic energies related to interparticle normal and friction forces.

Interestingly, soft materials exhibit emergent behavior, whereby collections of particles develop properties and behavior that is absent without interactions and/or correlations, and that is typically hard to anticipate in view of the properties of the individual building blocks. This is common in traditional condensed matter; superconductivity, superfluidity, and Bose–Einstein condensation are examples,

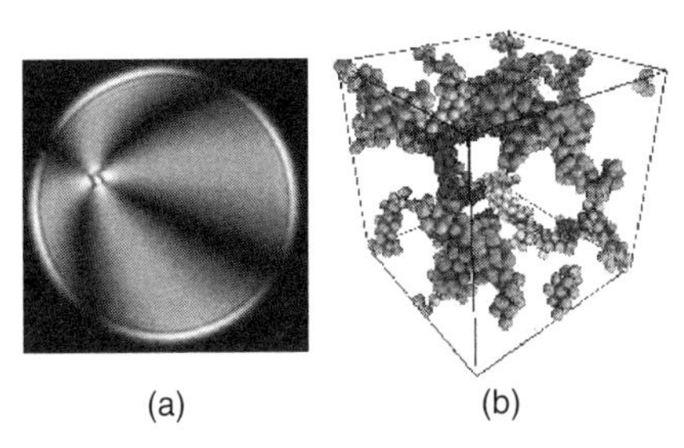

Figure 1 Soft matter examples: (a) nematic liquid crystal droplet seen under crossed polarizers and (b) simulation snapshot of a colloidal gel.

where collections of electrons or atoms all cooperate to result in remarkable behavior. Notably, these material states exist in equilibrium and are a result of quantum mechanics, whereas in soft matter quantum effects are negligible.

Soft materials in equilibrium also exist in intricate and remarkable states. Diblock copolymer melts, for instance, are known to adopt a bicontinuous gyroid structure. The physics leading to this intricate organization has a classical origin. It is the balance between the energetic cost for having a large contact area between the two immiscible blocks making the polymers, which drives phase separation, and the entropic penalty associated with stretching the polymer, which decreases its available configurations, what results in the formation of the gyroid structure. In soft condensed matter, all under the realm of classical mechanics, the equilibrium ground state of the material can consist of intricate structures and have properties that emerge as a result of interparticle interactions and correlations. Of course, this can also be achieved by bringing the material out of equilibrium. In this case, even simple materials can achieve remarkable steady states; for instance, a simple fluid, when heated from below, can self-organize into convection cells arranged in a hexagonal lattice. What is remarkable is that many soft materials exhibit emergent behavior even at equilibrium.

Another interesting and differentiating aspect of soft materials compared to atomic and molecular systems is that their dynamics are typically slower; this ultimately results from the larger length scales characteristic of soft matter. Furthermore, since these materials are commonly multicomponent, there are different relevant timescales involved enabling the structural relaxation of the system after the application of an external stress or the occurrence of a sudden fluctuation. In suspensions, the atomic or molecular solvent has a characteristic timescale of 10^{-12} s, while the particles can have a characteristic relaxation time of the order of microseconds and miliseconds. In many situations, only the latter dynamics is of interest, allowing the underlying solvent to be thought of as an equilibrium fluid. From the experimental point of view, this implies that ultrafast relaxations need not be studied, and in some cases, direct observation with an optical microscope is enough to characterize the relevant dynamics of the system.

Since it is common for soft materials to be comprised of macromolecules, such as polyelectrolytes or biomolecules, the relaxation of the internal degrees of freedom can also be relevant. But even in the absence of internal motions, the slow microscopic dynamics of the system can result in nonergodic transitions, such as vitrification and gelation. The slowing down of the dynamics by several decades can be monitored, for example, with light scattering, allowing thorough comparisons with theoretical models and predictions. Some of these dynamical processes are common in atomic and molecular systems, whereas others are specific of soft matter; gelation (see Fig. 1b) is a prominent example of this fact. This aspect, in addition to the capability to tailor the interactions among the macromolecules or colloidal particles, has made soft materials an excellent bench for testing theories of dynamical processes.

One further implication of the different timescales for structural relaxation is the nonlinear response of the system. This can be probed by, for example, applying mechanical stresses and studying the material response. The different timescales in the system imply nontrivial frequency dependencies, where mild stresses can drive the system deep into the nonlinear regime. Shear thinning, where the viscosity decreases with applied stress, and shear thickening, where it increases, are common examples of these nonlinear effects. Also remarkably, small stresses can fluidize a solid soft material, such as toothpaste, and cause it to flow in fluid-like manner.

There is a large community of scientists and engineers examining these types of materials from many different perspectives. Some investigators use soft materials as model systems to address questions in condensed matter. For others, their focus is to understand and describe the fundamental and unique properties and phenomena observed in soft materials. There is also research being done related to synthesizing novel building blocks that when put together result in materials with interesting optical, electrical, and mechanical properties; these are often designed for specific applications, for example, in biomedical engineering and optics. The aim of this book is to introduce the reader to the physics of soft materials, emphasizing the physics and leaving behind detailed mathematical treatments. Each of the chapters starts from the essential physics and brings the subject to the point where related research can then be discussed.

We have structured the book in sections or blocks dedicated to a general theme within soft matter. We start with a first block dedicated to fluid flows; these are often encountered in microfluidics, which has become an important tool nowadays for controllably generating drops and particles. The second block discusses colloids in the presence of gravitational, electric, shear, and optical fields. The third block discusses three major experimental techniques employed to accessing and quantifying the properties of suspensions: scattering, rheology, and optical microscopy. We then move to covering many-particle behavior and discuss in the fourth block, colloidal fluids, crystals, glasses, and gels. The fifth block is then dedicated to other types of soft materials. We focus on emulsions, liquid crystals, granular media, and foams. We leave polymer and polyelectrolyte solutions out as theseare arguably the most studied soft material; there are excellent textbooks already published dedicated to this type and related systems. The last block of the book is dedicated to ordered materials in curved spaces, where curving space introduces geometric frustration forcing the presence of topological defects in the ground state.

We believe that the collection of chapters provides an overview of the richness, peculiarities, and behavior of soft materials. Since each chapter starts from the basic physics and then brings the reader to the level of current research, we think that the book is ideal for teaching at the graduate level, with appropriate more advanced complements depending on the course. It was with this hope that we undertook this endeavor. Additionally, the book can help introduce researchers to specific types of soft materials and to some of the techniques currently used to study and characterize them. Certainly, soft materials will remain a rich topic in the future. Thus, this book does not offer a complete description of soft matter; instead, the contributions aim to provide the background for researchers and engineers working with these fascinating materials, with the hope of inspiring future work and applications.

Alberto Fernandez-Nieves and Antonio M. Puertas
September 2015

LIST OF CONTRIBUTORS

Martin Bates, Department of Chemistry, University of York, York, United Kingdom

Yuri Belotti, SUPA, School of Science and Engineering, University of Dundee, Dundee, United Kingdom

Daniel L. Blair, Department of Physics, The Institute for Soft Matter Synthesis and Metrology, Georgetown University, Washington DC, USA

Elena Castro-Hernandez, Departmento de Ingenieria Aereoespacial y Mecanica de Fluidos, University of Sevilla, Sevilla, Spain

Zhengdong Cheng, Artie McFerrin Department of Chemical Engineering, Materials Science and Engineering, The Professional Program in Biotechnology, Texas A&M University, College Station, TX, USA

Luca Cipelletti, Laboratoire Charles Coulomb (L2C), UMR 5221 CNRS-Université de Montpellier 2, Montpellier, France

Sylvie Cohen-Addad, Institut des NanoSciences de Paris, CNRS-UMR 7588, UPMC Univ Paris 06, Paris, France; Université Paris-Est, Champs-sur-Marne, France

Emanuela Del Gado, Department of Physics and Institute for Soft Matter Synthesis and Metrology, Georgetown University, Washington, DC, USA

Sudeep K. Dutta, Department of Physics, The Institute for Soft Matter Synthesis and Metrology, Georgetown University, Washington, DC, USA

Alberto Fernandez-Nieves, School of Physics, Georgia Institute of Technology, Atlanta, GA, USA

Davide Fiocco, Institute of Theoretical Physics, EPFL, Lausanne, Switzerland

Giuseppe Foffi, Laboratoire de Physique des Solides, CNRS UMR 8502, Université Paris-Sud XI, Orsay, France

Daniel I. Goldman, School of Physics, Georgia Institute of Technology, Atlanta, GA, USA

Jose M. Gordillo, Departmento de Ingenieria Aereoespacial y Mecanica de Fluidos, University of Sevilla, Sevilla, Spain

Nick Gravish, School of Physics, Georgia Institute of Technology, Atlanta, GA, USA

Josefa Guerrero, School of Physics, Georgia Institute of Technology, Atlanta, GA, USA

Francisco J. Higuera, E. T. S. Ingenieros Aeron_auticos, UPM, Madrid, Spain

Reinhard Höhler, Institut des NanoSciences de Paris, CNRS-UMR 7588, UPMC Univ Paris 06, Paris, France; Université Paris-Est, Champs-sur-Marne, France

Elizabeth Knowlton, Department of Physics, The Institute for Soft Matter Synthesis and Metrology, Georgetown University, Washington, DC, USA

Vinzenz Koning, Instituut-Lorentz, Universiteit Leiden, 2300 RA Leiden, The Netherlands

Jan P. F. Lagerwall, Physics and Materials Science Research Unit, University of Luxembourg, Luxembourg

Taewoo Lee, Department of Physics and Liquid Crystal Materials Research Center, University of Colorado, Boulder, CO, USA

Minne P. Lettinga, Forschungszentrum Jülich, Institute of Complex Systems (ICS-3), 52425 Jülich, Germany and Laboratory for Soft Matter and Biophysics, KU Leuven, Celestijnenlaan 200D, B-3001 Leuven, Belgium

Ignacio G. Loscertales, Escuela Técnica Superior de Ingenieros Industriales, Universidad de Málaga, Málaga, Spain

José Luis Arauz-Lara, Instituto de Física Manuel Sandoval Vallarta, Universidad Autónoma de San Luis Potosí, San Luis, Potosí, S.L.P., Mexico

Suliana Manley, Laboratory of Physics of Biological Systems, EPFL, Lausanne, Switzerland

Johan Mattsson, School of Physics and Astronomy, University of Leeds, Leeds, United Kingdom

Craig McDonald, SUPA, School of Science and Engineering, University of Dundee, Dundee, United Kingdom

David McGloin, SUPA, School of Science and Engineering, University of Dundee, Dundee, United Kingdom

David J. Pine, Center for Soft Matter Research, New York University, New York, NY, USA

Phil N. Segrè, Department of Physics, Oxford College of Emory University, Oxford, GA, USA

Bohdan Senyuk, Department of Physics and Liquid Crystal Materials Research Center, University of Colorado, Boulder, CO, USA

Ivan I. Smalyukh, Department of Physics and Liquid Crystal Materials Research Center, University of Colorado, Boulder, CO, USA; Renewable and Sustainable Energy Institute, National Renewable Energy Laboratory and University of Colorado, Boulder, CO, USA

Todd M. Squires, Department of Chemical Engineering, University of California, Santa Barbara, CA, USA

Véronique Trappe, Département de Physique, Université de Fribourg, Fribourg, Switzerland

Rahul P. Trivedi, Department of Physics and Liquid Crystal Materials Research Center, University of Colorado, Boulder, CO, USA

Vincenzo Vitelli, Instituut-Lorentz, Universiteit Leiden, RA Leiden, The Netherlands

Hans M. Wyss, Department of Mechanical Engineering and ICMS, Eindhoven University of Technology, 5612AJ Eindhoven, the Netherlands

Alessio Zaccone, Department of Chemical Engineering and Biotechnology, University of Cambridge, Cambridge, United Kingdom

SECTION I

FLUID FLOWS

1

DROP GENERATION IN CONTROLLED FLUID FLOWS

ELENA CASTRO HERNANDEZ[1], JOSEFA GUERRERO[2], ALBERTO FERNANDEZ-NIEVES[2], & JOSE M. GORDILLO[1]

[1]*Departamento de Ingenieria Aeroespacial y Mecanica de Fluidos, University of Sevilla, Sevillam 41092, Spain*
[2]*School of Physics, Georgia Institute of Technology, Atlanta, GA 30332, USA*

1.1 INTRODUCTION

Drops are present in our everyday life in kitchens and showers and are also used in fountains for aesthetic reasons. In addition, drops are of fundamental importance in many industrial processes [1–6]. Chemical and metallurgical engineers rely on drop formation for operations as varied as distillation, absorption, flotation, and spray drying [7]. Mechanical engineers have studied droplet behavior in connection with combustion operations [6]. In the food industry they are used to mask flavors and change textures [8, 9], and in the pharmaceutical sector they are involved in the production of creams and syrups [10–12].

The most common devices for mass production of drops are mixers and ultrasound emulsificators. In the case of mixers, the breakup of the dispersed phase results from the turbulent motion induced by the mobile parts of the mixer [13], whereas the operation of an ultrasound emulsificator relies on the collapse of cavitation bubbles, which induces velocity gradients in the continuous phase that cause formation of jets and subsequent breakup [14]. These methods are simple, robust, and of low cost, but they typically result in a wide size distribution and a poor control on drop size. These drawbacks can be overcome by filtering the droplets [15, 16]. However, the need of this second step complicates the process and increases costs.

In the last decades, great efforts have been made to improve current production methods in order to obtain monodisperse micron-sized droplets at high production rates. Recent fabrication methods rely on microfluidics as this technology provides great control over fluid flow and mixing of components.

In general, two different regimes can be experimentally observed depending on the operating conditions: dripping and jetting. The dripping regime operates at low flow rates and drop formation occurs right at the exit of the injection tube. In this regime, the resulting droplets are very monodisperse, but the production frequency is low.

Fluids, Colloids and Soft Materials: An Introduction to Soft Matter Physics, First Edition. Edited by Alberto Fernandez Nieves and Antonio Manuel Puertas.

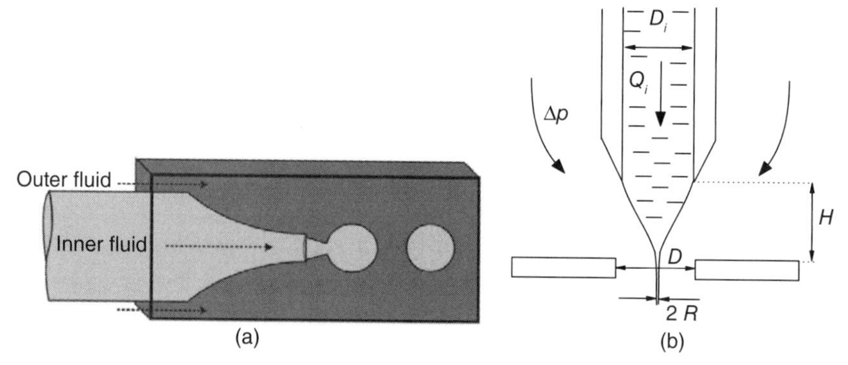

Figure 1.1 (a) Schematic of the coflowing configuration. Source: Reprinted Figure 2a with permission from Ref. [20]. (b) Schematic of the axisymmetric flow focusing configuration. Source: Reprinted Figure 1 from Ref. [52].

In addition, the droplet size is proportional to the diameter of the injection tube, and as a consequence, in order to obtain droplets that are smaller than ~10 μm, needles with such a small diameter are required. However, in this case, clogging becomes an issue. The jetting regime operates at higher flow rates than the dripping regime. In this case, the drop diameter is proportional to the diameter of the jet, which under the right conditions can be much smaller than the diameter of the injection tube. Jetting can only be achieved under the action of a force field. If only hydrodynamic forces act on the liquid, jetting can be obtained in either the coflow or the flow focusing configurations. Another alternative is the use of electric forces, with techniques such as electrospray (see Chapter 2).

The coflow configuration is characterized by the coaxial flow of two immiscible fluids as shown in Figure 1.1a. In this case, jetting occurs when the tangential stresses exerted by the continuous phase on the dispersed phase overcome surface tension stresses. In flow focusing, two fluids are forced to flow through a small orifice located in front of an injection tube, as shown in Figure 1.1b. In this configuration, the outer pressure gradient is favorable, and, as a consequence, not only the outer tangential stresses but also the outer pressure gradient imposed by the geometry accelerates the inner fluid through the orifice inducing formation of a jet that can be much smaller than the injection tube.

Other emulsification schemes besides coflow and flow focusing, which is the focus of this chapter, have also been explored. An important example is the T-junction geometry introduced by Garstecki *et al.* [19], where an inlet channel containing the dispersed phase perpendicularly intersects the main channel hosting the continuous phase [20] (see Figure 1.2a). Both phases form an interface at the junction, and as the fluid flow continues, the tip of the dispersed phase enters the main channel. The shear stresses of the continuous phase and the subsequent pressure gradient cause the head of the dispersed phase to elongate into the main channel until breakup occurs and a drop is formed, as shown in Figure 1.2b. The size of the drop can be changed by altering the flow rates, the channel dimensions, and the viscosity ratio of the two liquids. This geometry is very popular due to its simplicity and flexibility [21]. The main disadvantage is that the minimum size of the drops is limited by the size of the channel. Despite the production rate is not very high, parallel production can reduce this limitation.

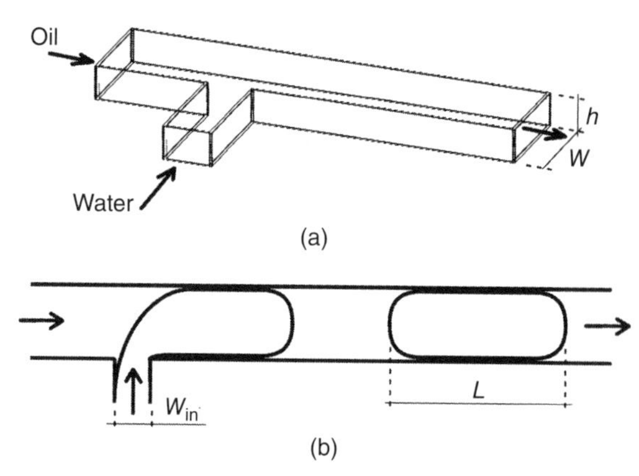

Figure 1.2 (a) Schematic illustration of the microfluidic T-junction composed of rectangular channels. (b) Top view of the same schematic in a two-dimensional representation. Source: Reprinted Figure 1 from Ref. [17].

1.2 COFLOW

1.2.1 Problem and Dimensionless Numbers

One of the simplest designs of a coflow device consists in coaxially aligning two capillary tubes. Typically, the inner one is cylindrical, with a tapered tip, and the outer tube has a square cross section [17]. Coaxial alignment is achieved by matching the outer diameter of the untapered portion of the inner capillary to the inner dimension of the square cross section of the outer tube (Figure 1.1a). As the length scales are below the capillary length, the effects of gravity are negligible. In addition, even if the flow in the outer capillary is not axisymmetric, since the tip is centered in the square cross section and $D_i \ll D_o$, with D_i the inner diameter of the tip and D_o the inner length of the square cross section of the outer capillary, the local flow around the tip is approximately axisymmetric. The inner fluid is injected through the cylindrical capillary tube at a flow rate Q_i and the outer fluid is supplied through the voids between the cross sections of both tubes at a flow rate Q_o.

Drop formation in a coflow device is characterized by nine physical parameters: the densities of both fluids, ρ_i and ρ_o, their viscosities, μ_i and μ_o, their flow rates, the surface tension between the two fluids, σ, the inner diameter of the injection tube, and the inner length of the square cross section of the outer tube. Since there are three fundamental quantities (mass, length, and time), there are $9 - 3 = 6$ independent dimensionless groups [22, 23]. Assuming that the outer viscous stresses are the most relevant stresses in the problem, the selection of μ_o, D_i, and $Q_o = U_o D_o^2$, where U_o is the mean velocity of the outer stream, as characteristic quantities leads to the following six dimensionless parameters:

$$\frac{D_o}{D_i},\quad q = \frac{Q_i}{Q_o},\quad \frac{\mu_i}{\mu_o},\quad Ca_o = \frac{\mu_o U_o}{\sigma},\quad Re_o = \frac{\rho_o U_o D_i}{\mu_o},$$
$$Re_i = \frac{\rho_i U_i D_i}{\mu_i}$$

where Ca_o is the capillary number of the outer fluid, Re_i is the Reynolds number of the inner fluid and Re_o that of the outer fluid in the scale of the tip diameter.

Generally, for low values of the inner to outer flow rate ratio, the dripping regime is observed, whereas for higher values of q, jetting occurs. The inner to outer viscosity ratio, μ_i/μ_o, must play a significant role in the jet formation process, since gas ligaments have not been reported in a coflow configuration at low Reynolds numbers. The inner to outer length ratio, D_i/D_o, is fixed for each device and should be sufficiently smaller than one to avoid wall effects. The rest of the parameters measure the relative importance between different forces in the problem: capillary forces, viscous forces, and inertial forces. The outer capillary number characterizes the relative importance of the outer viscous stress compared to the surface tension stress, while the Reynolds numbers express the ratio between inertial and viscous forces.

Other dimensionless parameters that will appear in the discussion are the inner capillary number, $Ca_i = \mu_i U_i/\sigma$ and the inner Weber number, $We_i = \rho_i U_i^2 D_i/\sigma$; the latter determines the relative importance between inertial and surface tension forces for the inner fluid, with $U_i = 4Q_i/(\pi D_i^2)$ the mean velocity of the inner stream. The inner Ohnesorge number, $Oh_i = \mu_i/(\rho_i v_c D_i)$ will also play a significant role; it measures the importance of inner viscous forces in the development of capillary instabilities, where $v_c = [\sigma/(\rho_i D_i)]^{1/2}$ is the capillary velocity. Note that these numbers can be obtained from combinations of the above-mentioned six parameters.

For given values of the physical parameters in the problem, the magnitude of the dimensionless groups allows identification of the most relevant forces and hence of the expected operating regime of the device.

1.2.2 Dripping and Jetting

When a liquid is forced through an orifice in the presence of a coflowing, immiscible fluid, it can drip or form a jet, depending on the flow rates. In dripping, the growing droplet experiences a force due to the viscous drag exerted by the coflowing fluid and a force due to surface tension, which keeps the drop at the tip of the capillary tube (Figure 1.3a). As the outer flow rate increases, the drop size concomitantly decreases until a critical value of $Ca_o \sim O(1)$ is reached [24]. At this point, the dripping regime transitions into a jetting regime, where a long jet, which narrows in the downstream direction, is formed. This jet ultimately breaks into drops due to the Rayleigh–Plateau instability (Figure 1.3b) [25, 26].

Jetting can also result if the kinetic energy due to the flow of the inner stream at the jet interface overcomes the surface tension energy. In this case, a widening jet results, as shown in Figure 1.3c. Dripping can thus transition into jetting when $We_i > 1$, provided Re_i is also larger than one. These jets are very different from the narrowing jets not only in their shape. The breakup is also very different. In fact, drop formation from these widening jets is reminiscent of dripping at the tip, even if the process takes place a distance downstream of it.

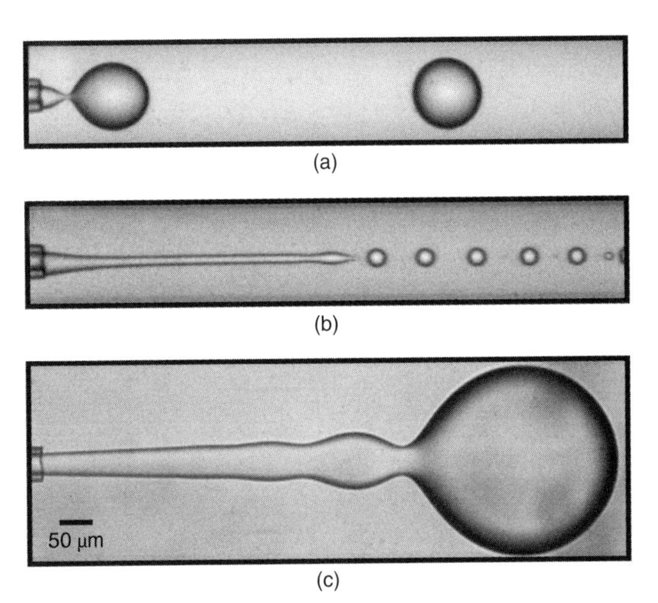

Figure 1.3 Images in the (a) dripping, (b) narrowing jet, and (c) widening jet regimes. Source: Reprinted Figure 1 from Ref. [23]. Copyright 2007 by the American Physical Society.

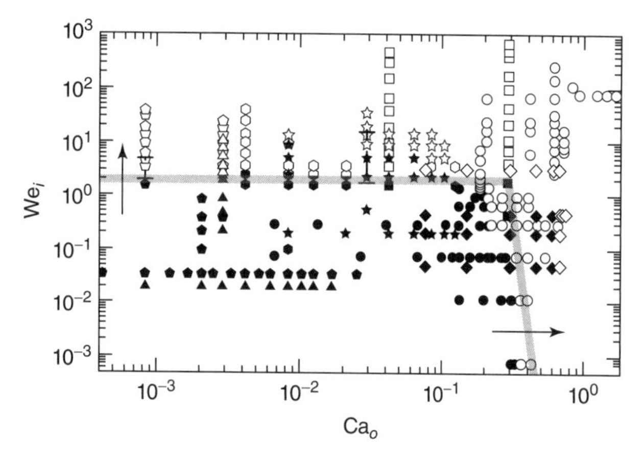

Figure 1.4 State diagram of the dripping-to-jetting transition for coflowing streams as a function of Ca_o and We_i. Filled symbols represent dripping while open symbols represent jetting. Each shape is a different viscosity ratio, surface tension, or geometry. Surface tension is $\sigma = 40$ mN/m unless otherwise stated. Square: $\mu_i/\mu_o = 0.01$. Diamond: $\mu_i/\mu_o = 0.01$, with the extra capillary tube to increase U_o. Hexagon: $\mu_i/\mu_o = 0.1$. Circle: $\mu_i/\mu_o = 0.1$. Pentagon: $\mu_i/\mu_o = 1$. Triangle: $\mu_i/\mu_o = 10$. Star: $\mu_i/\mu_o = 10$ and $\sigma = 4$ mN/m. Source: Reprinted Figure 4 from Ref. [23]. Copyright 2007 by the American Physical Society.

For $Re_o \ll 1$ and $Re_i > 1$, Utada *et al.* [24] proposed a state diagram for the dripping to jetting transition in terms of We_i and Ca_o. For $We_i < 1$, jetting occurs if the outer viscous forces overcome capillary forces, $Ca_o \gg O(1)$, whereas for $Ca_o < 1$, jetting is observed when the inertial forces of the inner liquid dominate over surface tension forces, $We_i > 1$ (Figure 1.4).

More recently, Castro-Hernández *et al.* [27] reported that when $Re_i < 1$, the inner Weber number no longer predicts the transition from dripping to jetting. In this case, the appropriate dimensionless group is the capillary number of the inner fluid; jetting occurs when $Ca_i > 1$.

1.2.3 Narrowing Jets

In this regime, the jet thins downstream until, eventually, its diameter reaches a nearly constant value, as shown in Figure 1.3b. For these jets, when the outer flow rate is increased for a fixed value of the inner flow rate, the jet diameter decreases, whereas when the inner flow rate is increased, for a fixed value of the outer flow rate, the jet diameter increases. The drop size mimics this behavior, as shown in Figure 1.5. Utada *et al.* [28] proposed a simple model to predict the jet diameter for these jets in the axisymmetric coflow configuration. Solving the motion of two coaxial liquids in Stokes flow, $\nabla P = \mu\nabla^2 u$, and relating the mean velocities of both fluids to the flow rates, one obtains

$$\frac{Q_i}{Q_o} = \frac{\mu_o}{\mu_i}\frac{\epsilon^4}{(1-\epsilon^2)^2} + 2\frac{\epsilon^2}{1-\epsilon^2}, \qquad (1.1)$$

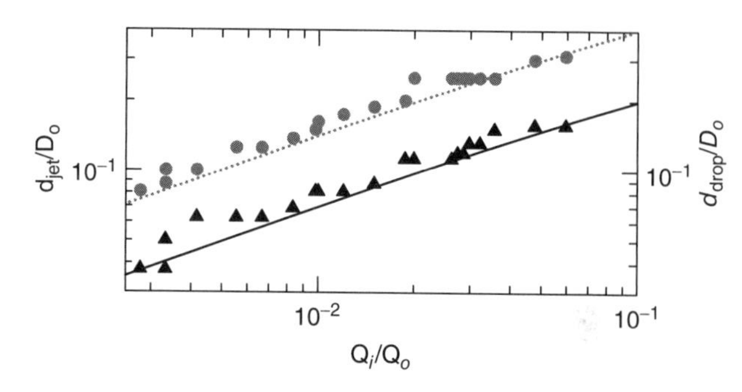

Figure 1.5 Experimentally measured jet diameter (triangles) and droplet diameter (circles) scaled by D_o as a function of the inner to outer flow rate ratio for a viscosity ratio $\mu_i/\mu_o = 0.1$ ($Oh_i = 0.11$). The solid line is the prediction from the model for d_{jet} with no fitting parameters. The dashed line is the predicted result assuming $d_{drop} \approx 2d_{jet}$ from the Rayleigh–Plateau instability. Source: Reprinted Figure 2 from Ref. [23]. Copyright 2007 by the American Physical Society.

where $\epsilon = d_{jet}/D_o$ is the ratio between the diameter of the jet and the inner radius of the outer tube. When $\epsilon \ll 1$ the leading term in Equation 1.1 results in a normalized jet diameter, $d_{jet}/D_o = \sqrt{Q_i/2Q_o}$, which correctly accounts for the measurements, as shown in Figure 1.5.

The observed proportionality between d_{jet} and the drop size can be understood by considering the Rayleigh–Plateau breakup of a jet, which predicts that the wavelength of the fastest unstable mode, λ^*, is proportional to the jet diameter. The proportionality constant is only a function of the viscosity ratio for sufficiently large Oh_i [29]. Considering that this mode causes the breakup of the jet, we equate the volume of a cylinder of length λ^*, to the volume of the resulting spherical droplet: $\pi d_{drop}^3/6 = \pi d_{jet}^2 \lambda^*/4$, with d_{drop} the drop diameter. This suggests that the drop size should be proportional to the jet diameter, consistent with the experimental results. In addition, since $\mu_i/\mu_o = 0.1$ and $\lambda^* = 5.48 d_{jet}$, $d_{drop} \approx 2d_{jet}$, consistent also with the experimental results (Figure 1.5).

Interestingly, Suryo and Basaran [30] found out numerically that in this coflow geometry, d_{jet} can be much smaller than D_i. Soon after, Marín *et al.* [31] showed that under the right operating conditions, droplets below the micron could be obtained from needles with $D_i = 100$ μm. The requirements for the observation of this regime are that (i) $Re_o \ll 1$, so that the outer flow remains attached at the jet interface, (ii) $Re_i \ll 1$, so that the momentum of the outer fluid effectively diffuses to the inner stream, and (iii) $Ca_o \gtrsim O(1)$. In this case, if $q \ll 1$, a cone-jet transition is observed and the jet diameter can be much smaller than D_i.

Recently, Castro-Hernández *et al.* [32] have studied the role of μ_i/μ_o, Ca_o and q in the coflow configuration when operated under the narrowing-jet regime. They experimentally observed that when the inner to outer flow rate decreases, for fixed values of the viscosity ratio and the outer capillary number, the diameter of the jet and of the droplets

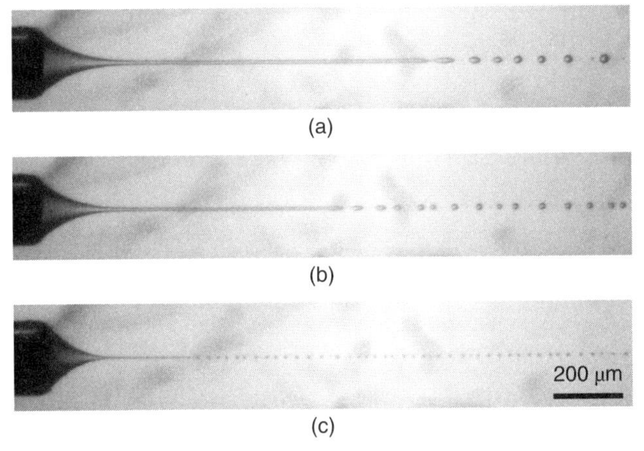

Figure 1.6 Experimental images showing jet formation and breakup when silicone oil of 10 cP is used as inner fluid and glycerin as the outer fluid. $\mu_i/\mu_o = 10^{-2}$ and $Ca_o = 5$. (a) $U_i/U_o = 10^{-2}$, $Q_i = 5 \times 10^{-1}$ μl/min, $Q_o = 7$ ml/min; (b) $U_i/U_o = 6 \times 10^{-3}$, $Q_i = 3 \times 10^{-1}$ μl/min, $Q_o = 7$ ml/min; (c) $U_i/U_o = 5 \times 10^{-4}$, $Q_i = 2 \times 10^{-2}$ μl/min, $Q_o = 7$ ml/min. Source: Reprinted Figure 2 from Ref. [31].

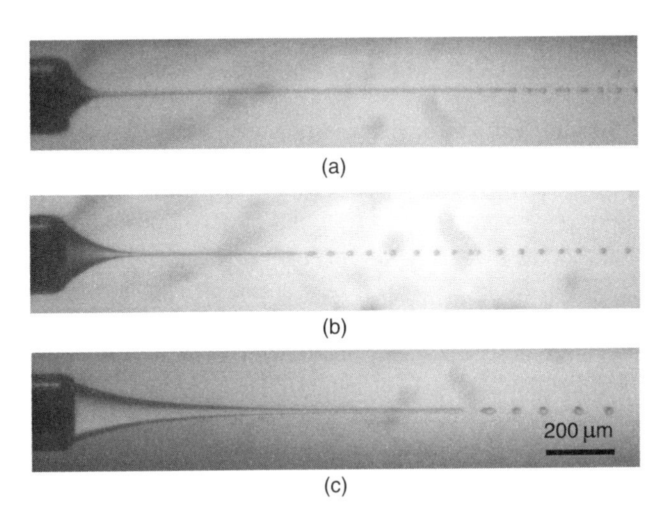

Figure 1.7 Experimental images showing jet formation and breakup for $Ca_o = 5$, $U_i/U_o = 4 \times 10^{-3}$ and different values of the viscosity ratio: (a) silicone oil of 100 cP/glycerin, $\mu_i/\mu_o = 10^{-1}$; (b) silicone oil of 10 cP/glycerin, $\mu_i/\mu_o = 10^{-2}$; (c) water/silicone oil of 1000 cP, $\mu_i/\mu_o = 10^{-3}$. $Q_i = 2 \times 10^{-2}$ μl/min, $Q_o = 7$ ml/min. Source: Reprinted Figure 6 from Ref. [31].

that result from its breakup both decrease (see Figure 1.6), consistent with the results shown in Fig. 1.5. In addition, when Ca_o reaches a value above the threshold at which the transition between dripping and jetting occurs, they reported that the length of the jet before breakup is proportional to Ca_o, whereas the jet diameter and, as a consequence, the droplet size remain constant. Lastly, they could verify that if the viscosity ratio decreases, the cone-jet structure becomes more elongated (see Figure 1.7), eventually resulting in

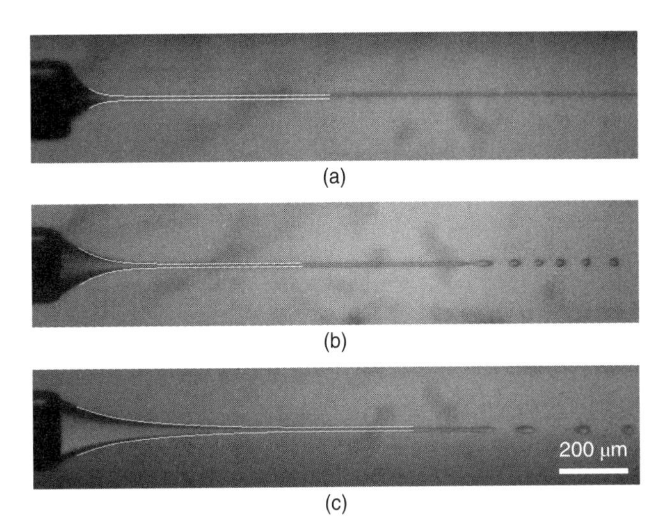

Figure 1.8 Experimental images obtained using different silicone oils as the inner fluid and glycerin as the outer fluid for $Ca_o = 5$. The continuous white line corresponds to the theoretical jet shape. The values of the control parameters in each of the three cases shown are (a) $\mu_i/\mu_o = 10^{-1}$, $U_i/U_o = 6 \times 10^{-3}$, $Q_i = 3 \times 10^{-1}$ μl/min, $Q_o = 7$ ml/min; (b) $\mu_i/\mu_o = 10^{-2}$, $U_i/U_o = 10^{-2}$, $Q_i = 5 \times 10^{-1}$ μl/min, $Q_o = 7$ ml/min; (c) $\mu_i/\mu_o = 10^{-3}$ $U_i/U_o = 10^{-2}$, $Q_i = 8 \times 10^{-1}$ μl/min, $Q_o = 12$ ml/min. Source: Reprinted Figures 18–20 from Ref. [32].

aperiodic jet breakup and a drop size that is no longer uniform.

Making use of the continuity equation, the axial momentum equation, the normal stress balance across the interface and the kinematic boundary condition at the free interface, they obtained a parameter-free theoretical prediction for the jet shape, finding good agreement with the experimental results (Figure 1.8).

1.2.4 Unified Scaling of the Drop Size in Both Narrowing and Widening Regimes

The drop size that results from the breakup of a widening jet decreases with Q_o for a fixed Q_i, consistent with the results for narrowing jets. In contrast, when Q_i is increased for a fixed value of Q_o, two different situations are observed: If the viscosity of the outer fluid is $\mu_o \simeq 10$ cP, the behavior of d_{drop} is consistent with the situation encountered with narrowing jets; d_{drop} increases with Q_i. However, when $\mu_o \simeq 1$ cP, the opposite behavior is observed and the drop size decreases with Q_i, as shown in Figure 1.9.

To understand this dependence, let us revisit the behavior of narrowing jets. Since for these jets, $Re_o < 1$, there is an effective diffusion of momentum across the whole cross section of the jet. As a result, the inner and outer velocities become equal at some distance downstream of the injection

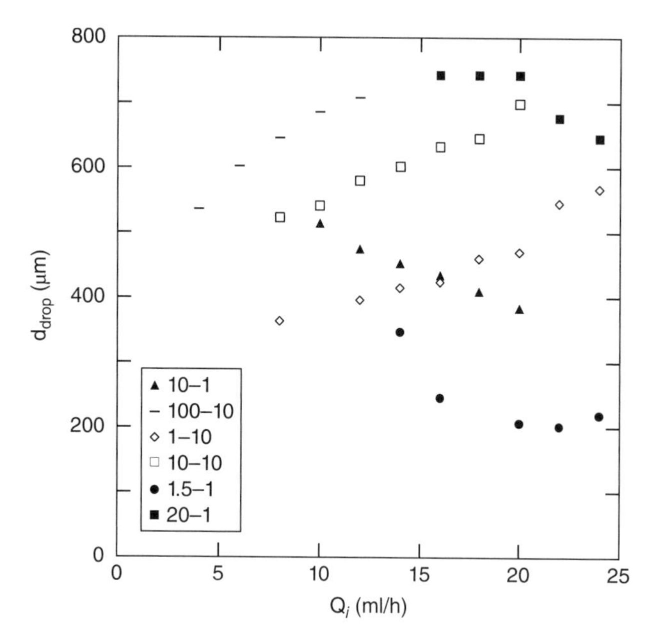

Figure 1.9 Dependence of the drop diameter on Q_i for a fixed value of $Q_o = 200$ ml/h and various inner/outer viscosities. Observe that the trends are different depending on the values of the inner and outer viscosities. Numbers in the caption indicate inner/outer viscosities in centipoises. Source: Reprinted Figure 7c from Ref. [26].

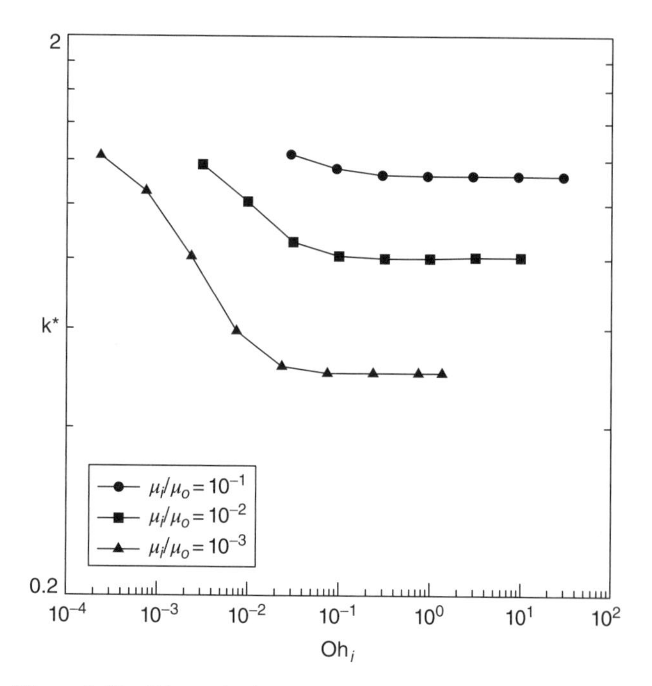

Figure 1.10 Dimensionless wave number versus inner Ohnesorge number for different viscosity ratios.

tube and the jet diameter simply results from

$$\frac{\pi d_{\text{jet}}^2}{4} U_o = Q_i \to d_{\text{jet}} = \left(\frac{4Q_i}{\pi U_o}\right)^{1/2}, \tag{1.2}$$

which is consistent with Equation 1.1 for $d_{\text{jet}} \ll D_o$, except for a numerical prefactor related to the details of the velocity profile. In addition, since these jets are convectively unstable (see Section 2.5), the size of the drops obtained from their breakup can be determined from the mass balance: $\pi\, d_{\text{drop}}^3/6 = \pi^2 d_{\text{jet}}^3/(4k^*)$, where $k^* = k^*(\mu_i/\mu_o, Oh_i)$ is the dimensionless wave number corresponding to the maximum growth rate of sinusoidal capillary perturbations and $\lambda^* = \pi d_{\text{jet}}/k^*$ is its corresponding wavelength. Since k^* depends weakly on Oh_i for relatively large values of this parameter, as shown in Figure 1.10, it is sensible to write $k^* = k_t^*$, with $k_t^* = k_t^*(\mu_i/\mu_o)$ the dimensionless wave number corresponding to the maximum growth rate in the limit, first considered by Tomotika [29], $Oh_i \to \infty$. With these considerations and using Equation 1.2 , we obtain

$$d_{\text{drop}} = \left(\frac{144}{\pi}\right)^{1/6} (k_t^*)^{-1/3} \left(\frac{Q_i}{U_o}\right)^{1/2}. \tag{1.3}$$

Castro-Hernández *et al.* [27] extended these ideas to describe the behavior of d_{drop} for both narrowing and widening jets. The main quantity in this approach is the time for drop formation, T, which is the sum of the convective time, t_{conv}, and the pinch-off time, t_{pinch}: $T = t_{\text{conv}} + t_{\text{pinch}}$ [33]. The convective time, t_{conv}, is the time required to convect the inner fluid a distance λ at a velocity U_p, where λ is the distance traveled by the downstream location of the jet within two consecutive pinch-off events and U_p is the velocity of the tip of the jet, as shown in Figure 1.11. The pinch-off time is the time needed to break the liquid thread. Since $t_{\text{pinch}} \ll t_{\text{conv}}$ in most experimental situations, breakup can be considered to take place almost instantaneously. However, for breakup to happen, a length equal to $\lambda^* = \pi d_{\text{jet}}/k^*$ is required. This means that the downstream location of the jet would need to travel a distance λ^* before breakup can occur. As a result,

$$T = \frac{\lambda}{U_p} = \frac{\pi\, d_{\text{jet}}}{k^*\, U_p}, \tag{1.4}$$

and since continuity demands that $\pi d_{\text{drop}}^3/6 = Q_i T$, then

$$\frac{d_{\text{drop}}}{D_i} = \frac{1}{D_i}\left(\frac{6\,Q_i\, d_{\text{jet}}}{k^*\, U_p}\right)^{1/3}. \tag{1.5}$$

The drop diameter is then determined by d_{jet}, k^*, and U_p. For the narrowing jets, $U_p \simeq U_o$ [24] and Equation 1.5 corresponds to Equation 1.3 . However, for the widening jets, this is not always the case. If the outer fluid viscosity is large compared to the viscosity of water, $U_p \simeq U_o$ and Equation 1.5

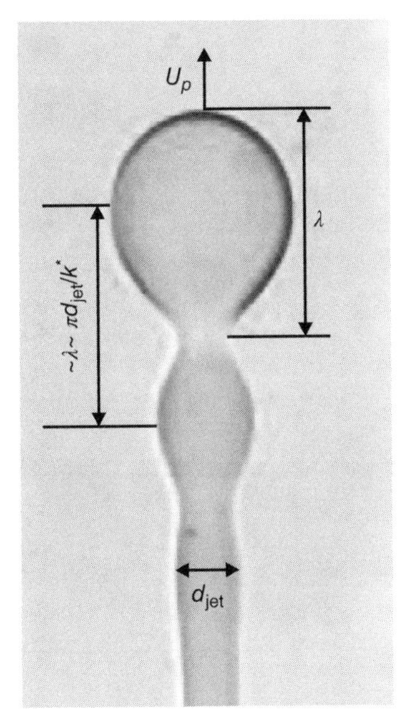

Figure 1.11 Image showing the velocity at the more downstream position of the jet, U_p, the jet diameter, d_{jet}, and the axial distance traveled by the tip of the jet between two consecutive pinch-off events, λ. The value of λ can be approximated by the wavelength corresponding to the maximum growth rate of capillary perturbations, $\pi d_{\text{jet}}/k^*$. Source: Reprinted Figure 8 from Ref. [26].

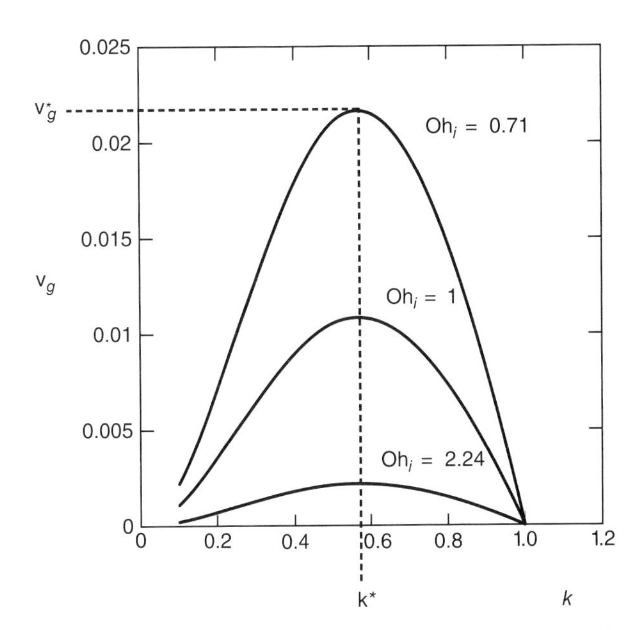

Figure 1.12 Growth rate, v_g, versus dimensionless wave number for different values of the inner Ohnesorge number and $\mu_i/\mu_o = 0.1$. Source: Reprinted Figure 11 from Ref. [26].

captures the increase of d_{drop} with Q_i observed experimentally. In contrast, when the outer viscosity is similar to that of water, $U_p \neq U_o$ since the inner fluid can drag the outer fluid and affect the velocity of the jet interface, which will then be larger than the outer velocity. As a result, $U_p = f(Q_i)$ and could result in a decreasing d_{drop} with Q_i. See [27] for further details.

This is not the only difference between narrowing and widening jets. As we have seen, the jet diameter for the case of the narrowing jets is simply $d_{\text{jet}} = [4Q_i/(\pi U_o)]^{1/2}$ [24, 31]. However, for the widening jets this equality is generally not correct since the inner liquid velocity can be different from U_o, as pointed out in the previous paragraph. Hence, a different way to estimate d_{jet} is needed. To do this, we recall that upstream the breakup point, this inner liquid velocity is larger than the speed of capillary perturbations; this explains why the jets do not break in this region. As the jet widens downstream, the inner liquid velocity decreases and at some axial location, it becomes comparable to the speed of capillary disturbances. At this place, the jet breaks. We emphasize that this can happen before the inner velocity of the jet equals the outer fluid velocity. Based on this, we estimate d_{jet} from the condition $t_{\text{pinch}} \simeq t_{\text{prop}}$, where t_{prop} is a characteristic time for the propagation of capillary perturbations, $t_{\text{prop}} = d_{\text{jet}}/U_i$, with $U_i = 4Q_i/(\pi d_{\text{jet}}^2)$. The characteristic pinch-off time is $t_{\text{pinch}} = 1/(v_g^*)(\rho_i/\mu_i)(d_{\text{jet}}/2)^2$, with v_g^* the maximum dimensionless growth rate associated to k^* (see Figure 1.12). With all these facts, we obtain

$$d_{\text{jet}} = \frac{1}{v_g^*}\frac{\rho_i Q_i}{\pi \mu_i}, \tag{1.6}$$

Taking all these aspects into account, we consider the data in Figure 1.9 together with the drop size data of narrowing jets and plot d_{drop}/D_i versus $[6Q_i d_{\text{jet}}/(k^* U_p)]^{1/3}/D_i$, where d_{jet} is obtained from Equation 1.6 and U_p is taken equal to U_o when $\mu_o \geq 5$ cP or it is directly measured experimentally. For k^*, we solve Tomotika's complete equations for any value of Oh_i; see Ref [29]. We find that there is a linear relation between the two quantities, as shown in Figure 1.13 and consistent with the expectations from Equation 1.5 . Furthermore, the slope of the best fit is 0.9, which is close to 1, and the intercept is 0.75, which is small compared to the values of d_{drop}/D_i. Thus, the proposed model correctly describes the drop size that results from the breakup of both widening and narrowing jets.

1.2.5 Convective Versus Absolute Instabilities

Dripping is a common example of an absolute instability, where the perturbations that induce breakup grow in time at a fixed location in space, at a frequency that is intrinsic to the system. As a result, dripping is insensitive to external noise and results in extremely monodisperse droplets. By contrast, jetting is often the result of a convective instability, where

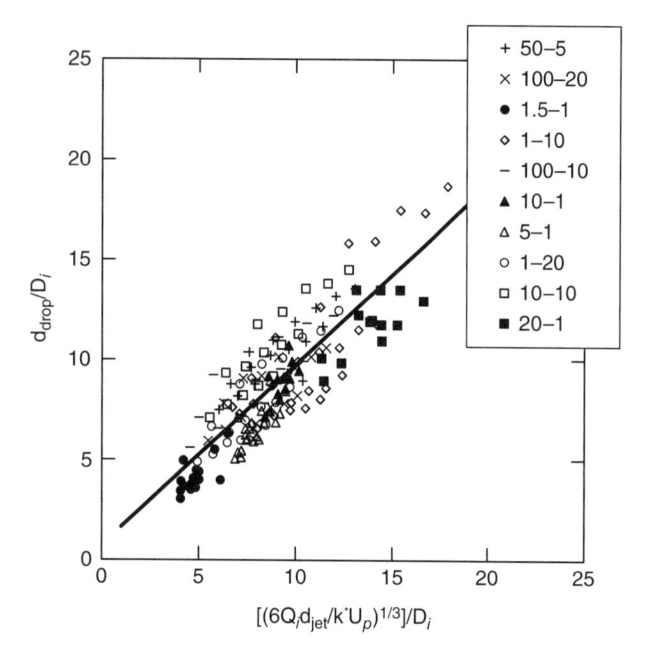

Figure 1.13 Experimentally measured drop diameters d_{drop}/D_i as a function of the parameter $(6Q_i\ d_{\text{jet}}/(D_i^3\ k^*\ U_p))^{1/3}$. The slope of the linear regression fit to the experimental data is very close to 1, consistent with the theoretical prediction given by Equation (1.5). The relative errors, however, are ±30%. The maximum experimental error associated to the measurement of the tip velocity is of the order of ∼ 10%. Hence, the dispersion in the data is attributable to necessary simplifications in the way the wavelength of maximum growth rate and the tip velocity, U_p, are calculated. Numbers in the caption indicate inner/outer viscosities in centipoises. Source: Reprinted Figure 13b from Ref. [26].

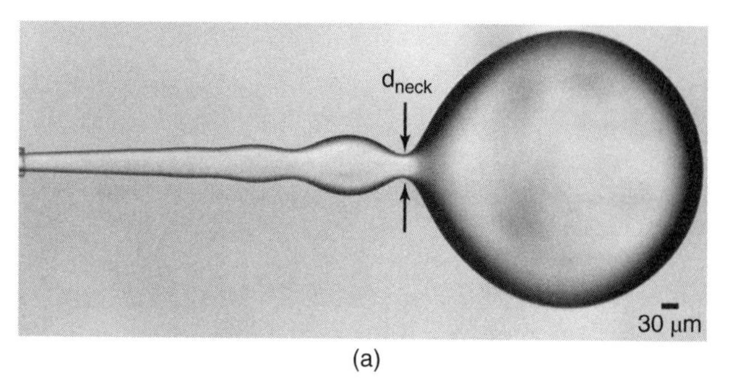

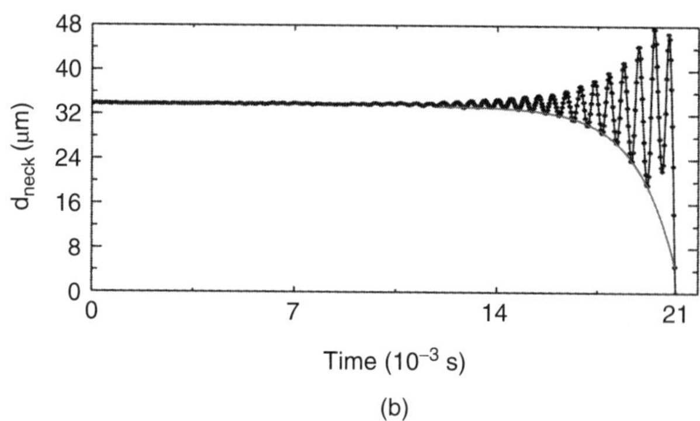

Figure 1.14 (a) High-speed image of a typical widening jet. The small neck between the jet and the bulb has a diameter, d_{neck}. The outer diameter of the tip is ∼ 40 μm, while $d_{\text{tip}} \simeq 30$ μm and the inner diameter of the surrounding cylindrical capillary is ∼ 600 μm. Here $We_i = 5.5$. (b) Neck diameter as a function of time. The line is an exponential fit to the envelope. The associated growth rate is ∼ 40 Hz. The frequency of oscillation is ∼ 2000 Hz. The flow rates of the outer and inner fluids are 9×10^4μl/h and 6×10^3μl/h, respectively. Source: Reprinted Figure 1 from Ref. [32]. Copyright 2008 by the American Physical Society.

the perturbation that leads to breakup amplifies the external noise as it is advected downstream, usually leading to less uniform droplets.

Absolute instabilities in the jetting regime are rarely observed. However, Utada *et al.* [34] reported that drop formation in the widening jet regime happens via absolute, rather than convective instabilities. Experimentally, this is supported by the following facts: (i) Despite the widening jets are generated by injecting the inner fluid at $We_i > O(1)$, drop pinch-off from the end of the jet occurs only after the jet diameter has widened sufficiently such that We_i decreases to order unity as in the dripping regime, which only occurs when $We_i \lesssim O(1)$ and results in drop formation via absolute instabilities [35, 36]. (ii) The large difference in velocity between U_i and the neck of the widening jet, highlighted with arrows in Figure 1.14(a), coupled with the spatially stationary oscillations throughout the entire pinch-off process suggests that the superposition of the perturbations produces the condition of zero group velocity at a fixed location. (iii) The envelope associated to the oscillatory motion of the neck grows exponentially in time, as shown in Figure 1.14(b) for the case of $\mu_i/\mu_o = 0.1$, implying that the growth rate of the instability is positive.

Narrowing jets essentially break via Rayleigh-Plateau instabilities that are convected by the flow. Nevertheless, assessing whether widening jets indeed result from absolute instabilities requires performing a linear stability analysis. Using the classical quasi-parallel approximation, any parameter associated with the flow, such as the velocity and the pressure, is assumed to be proportional to exp $(\mathrm{i}(kz - wt))$, where z is the axial position measured with respect to the injection needle and t is the time. In general, the frequency, $w = w_r + \mathrm{i}w_i$, and the wave number, $k = k_r + \mathrm{i}k_i$, are complex and are related through a dispersion relation $D(\omega, k) = 0$. Furthermore, the superposition of all possible modes generates wave packets that travel both up- and downstream along the interface of the jet with group velocity $v_g = \partial w_r/\partial k_r$. Typically, a temporal stability analysis is used to determine whether or not a system is stable.

This corresponds to examining the behavior of the perturbations over time. In this analysis, the wave number is assumed to be real and the frequency is a complex quantity. If the growth rate of the instability is negative, $\omega_i < 0$, the perturbations decay in time, the jet is temporally stable and breakup will not occur. In contrast, if $\omega_i > 0$, the perturbations grow exponentially in time, the jet is temporally unstable and breakup can occur. The breakup time of the jet is calculated from the growth rate, $t_b = 1/\omega_i$ and the drop size can be determined from $Q_i t_b = \pi d_{\mathrm{drop}}^3/6$.

For the unstable situations in the temporal analysis, we perform an additional spatiotemporal study to distinguish absolute from convective instabilities. This analysis is based on the Briggs–Bers criterion [37, 38], which is employed to determine whether perturbations introduced at a fixed spatial location in the flow are amplified or decay at that spatial location. For that purpose, we look for frequencies and wave numbers, ω_0 and k_0, that satisfy the dispersion relation, $D(\omega_0, k_0) = 0$, and result in a zero group velocity, $v_g|_{\omega=\omega_0,k=k_0} = 0$ at some specific spatial location, in the laboratory frame of reference. If $\mathrm{Im}(\omega_0) < 0$, the instability is convective since the perturbations decay in time at that specific spatial location. In contrast, if $\mathrm{Im}(\omega_0) > 0$, the instability is absolute since the perturbations grow exponentially in time at that spatial location. When the instability is convective, unstable waves are convected in the direction of the flow. In this case, what one visually sees is the result of an unperturbed flow with superimposed waves that grow and propagate in the downstream direction. If, however, the instability is absolute, the stability analysis predicts that tiny perturbations will exponentially grow in time right at the place where the noise is introduced, preventing the unperturbed flow to be experimentally observable. In this way, whether the instability is convective or absolute can be detected experimentally.

The simplest stability analysis is the parallel stability analysis, which consists in solving the stability problem at every spatial location of the base flow, once it is assumed that, as far as the stability problem is concerned, the velocity profile at such axial location remains unchanged in the downstream and upstream directions up to $\pm\infty$, respectively. Interestingly, this approach correctly captures the differences observed experimentally in the resulting flow [34, 39]. For this reason, despite a global stability analysis [40, 41] would be conceptually more appropriate to predict what is experimentally observed, in this section we only present results obtained with the local parallel flow assumption.

To perform the stability analysis under the parallel flow assumption, we need to know the steady base flow. Hence, we need to calculate the downstream evolution of the velocity profiles of both inner and outer fluids. These can be obtained from the continuity and Navier–Stokes equations, which in the slender jet approximation can be written as [42]

$$\frac{\partial u^{i,o}}{\partial z} + \frac{1}{r}\frac{\partial (r v^{i,o})}{\partial r} = 0, \tag{1.7}$$

$$u^{i,o}\frac{\partial u^{i,o}}{\partial z} + v^{i,o}\frac{\partial u^{i,o}}{\partial r} = \nu^{i,o}\frac{1}{r}\frac{\partial}{\partial r}\left(r\frac{\partial u^{i,o}}{\partial r}\right) - \frac{1}{\rho^{i,o}}\frac{\partial P^{i,o}}{\partial z}, \tag{1.8}$$

where u and v are the axial and radial velocities, respectively, and the superscripts refer to outer and inner fluids, respectively. Once the system of equations above is solved subjected to the boundary conditions (see [42]), we find that the parabolic velocity profile of the inner fluid is very pronounced near the tip, as shown by the curve for $z = 36$ μm in Figure 1.15, and progressively flattens downstream, as shown by the curves for $z = 160$ μm, $z = 336$ μm, and $z = 513$ μm in Figure 1.15. In contrast, the velocity profile of the outer liquid is essentially flat on the length scale of the jet. We confirm that the local Weber number is of order unity at the experimental distance from the tip where jet breakup happens.

This base flow is then perturbed at each z to obtain the sign of the imaginary part of ω_0. The axial distance, z^*, where the instability transitions between convective and absolute, can be located studying the behavior of the solutions of the dispersion relation in the complex k-plane. These solutions can be seen as propagating wave packets with an amplitude that increases or decreases depending on whether they correspond to unstable or stable modes, respectively. The way to distinguish if these solutions represent upstream or downstream wave packets is by noting the sign of the group velocity. If $v_g > 0$, the wave packet travels downstream, while if $v_g < 0$, the wave packet travels

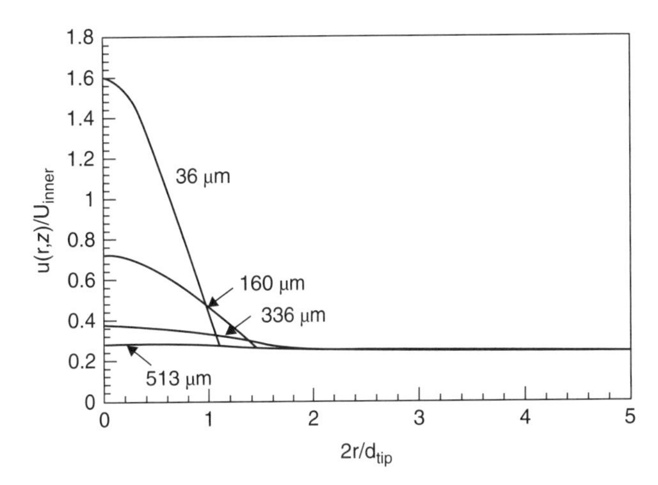

Figure 1.15 Velocity profile for different axial positions. The parameters used in this case are $\mu_i/\mu_o = 0.1$, $We_i = 1.66$ and $Ca_o = 0.67$.

upstream. Now, since we are only interested in a local stability analysis, we can perform a Taylor expansion of $D(\omega, k)$ around (ω_0, k_0) and keep terms up to leading order in ω and k: $D(k, \omega) = D(k_0, \omega_0) + [\partial D(k, \omega)/\partial \omega]_{k_0,\omega_0} (\omega - \omega_0) + [\partial^2 D(k, \omega)/\partial k^2]_{k_0,\omega_0} (k - k_0)^2/2$, where we have used that $[\partial D(k, \omega)/\partial k]_{k_0,\omega_0} = [\partial D(k, \omega)/\partial \omega]_{k_0,\omega_0} [\partial \omega/\partial k]_{k_0,\omega_0} = 0$, given that the group velocity at (ω_0, k_0) is zero. In this case, the dependence between ω and k is quadratic, implying that, in the vicinity of (ω_0, k_0), there are two solutions; these correspond to wave packets propagating in opposite directions and appear as branches in the complex k-plane, as shown in Figure 1.16 for representative values of z. The value of z^* can then be located as the axial distance where the two branches coalesce. This corresponds to the existence of a saddle point in the complex k-plane. Mathematically, this means that at z^*: $D(\omega_0, k_0) = 0$, $[\partial D(\omega, k)/\partial k]_{\omega_0,k_0} = 0$, $\text{Im}(\omega_0) = 0$ and $[\partial^2 D(\omega, k)/\partial k^2]_{k_0,\omega_0} \neq 0$.

The way to find the saddle point is equivalent to solving the so- called signaling problem [43] at every spatial location, for situations where the stable perturbations are convected upstream. Note first that the signaling problem consists in finding the response of the unperturbed flow to a periodic forcing of small amplitude. In the case of a stable flow, the amplitude of the perturbations will decay both in the upstream and downstream directions. In contrast, the flow will be convectively unstable if stable/unstable perturbations propagate in the upstream/downstream direction. When the conditions for a saddle point are fulfilled, the group velocity of unstable perturbations is zero, implying that the energy seeded in the flow by the forcing cannot be evacuated away from the location where it is introduced.

For the case of $\mu_i/\mu_o = 0.1$, $Ca_o = 0.67$, and $We_i = 1.66$, Utada et al. [34] find that this point is located at $z^* \approx 275$ μm. For $z < z^*$, $\text{Im}(\omega_0) > 0$ and the instability is absolute, while for $z > z^*$, $\text{Im}(\omega_0) < 0$ and the instability is convective. When the region of the jet located right at the exit of the injection tube that is absolutely unstable is much larger than the characteristic wavelength of the absolute mode, $\lambda_0 = 2\pi/k_0$, the jet breaks via an absolute instability. The transition to a convective instability takes place for the values of the Capillary and Weber numbers for which the extent of the region adjacent to the injector where the instability is absolute, either is zero, or possesses a length much smaller than that of the wavelength of the absolute mode.

As a further test to the interpretation of the experimental results, Utada et al. [34] did additional experiments to induce a transition from an absolute to a convective jet instability. The idea was to start with a jet that breaks up via absolute instabilities and increase Ca_o sufficiently to induce the formation of a narrowing jet, which breaks up convectively. To achieve larger values of the outer capillary number, they used a more viscous outer fluid and set $\mu_i/\mu_o = 0.01$. They

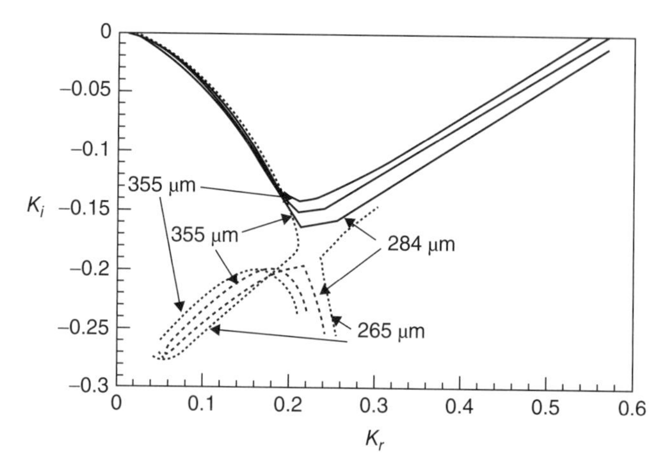

Figure 1.16 Saddle point in the complex k-plane indicating the axial location at which the instability transitions to absolute. The parameters used are $\mu_i/\mu_o = 0.1$, $We_i = 1.66$, and $Ca_o = 0.67$.

started with an absolutely breaking jet at $We_i = 3.1$ and $Ca_o < 1$ (see Figure 1.17a) and gradually increased Q_o and thus Ca_o. Interestingly, at a critical $Ca_o^* \simeq 0.65$, they observed an abrupt and dramatic increase in the length of the jet, as shown in Figure 1.17b–d. This large increase in the jet length also coincided with the suppression of the spatial oscillations that were observed initially, as also shown in Figure 1.17b and c. Remarkably, the stability analysis predicts that at $Ca_o^* = 0.69$ the instability transitions from absolute to convective, consistent with the experimental results. Furthermore, the theoretical analysis also predicts that Ca_o^* should not change significantly with We_i; this implies that the shear from the outer liquid determines the transition from an absolute to convective instability irrespective of the inertia of the inner liquid. This is remarkable because when $We_i < 1$ and $Ca_o < Ca_o^*$, the system is in a dripping regime; above Ca_o^*, however, the system transitions from an absolutely unstable dripping regime [36, 43–46] to a convectively unstable jetting one.

The experiments and their theoretical analysis then suggest that the widening jet regime results in drop formation via absolute instabilities, consistent with the dripping regime. In contrast, the narrowing jet regime results in drop formation via convective instabilities. The presence of absolute instabilities in jets then enables a route that is alternative to dripping for the generation of uniform emulsions. The key advantage here is that absolute instabilities are, in principle, not affected by noise.

1.3 FLOW FOCUSING

The flow focusing configuration can be implemented in an axisymmetric device [28, 48] or in a two- dimensional device that can be generated by soft-lithography techniques [49].

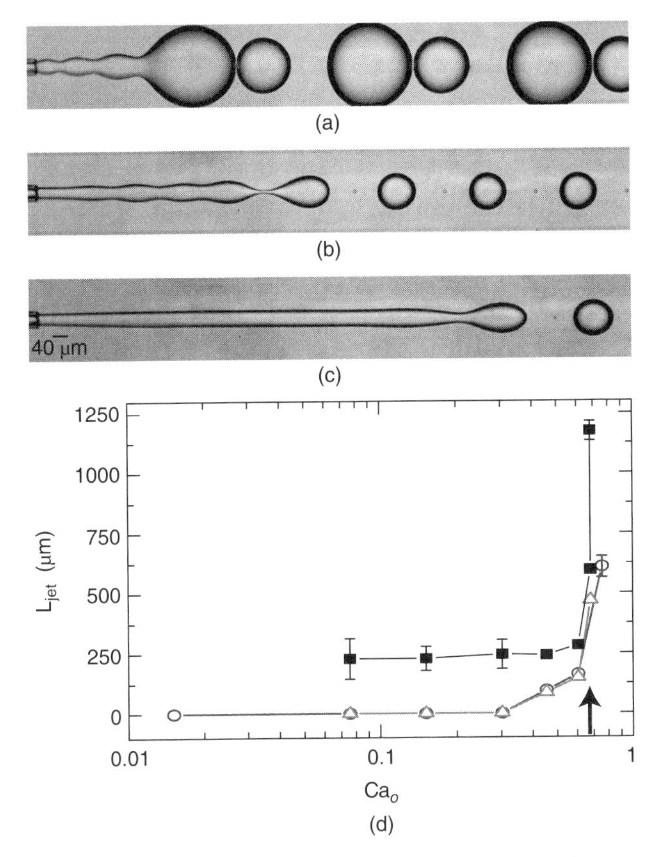

Figure 1.17 (a) Image of the jet generated at $We_i = 3.1$ and $Ca_o = 0.07$. (b) Transient image of the jet generated at $Ca_o = 0.69$. The oscillations on the jet gradually die out as the length increases at the critical value, Ca_o^*. (c) Image of the jet at $Ca_o = 0.69$ after the lengthening. (d) Plot of the jet length as a function of Ca_o. The filled squares are the measured jet lengths. The open circles and triangles are dripping. The We_i for the circles and triangles are 0.19 and 0.05, respectively. For these experiments, $\mu_i/\mu_o = 0.01$. Source: Reprinted Figure 3 from Ref. [32]. Copyright 2008 by the American Physical Society.

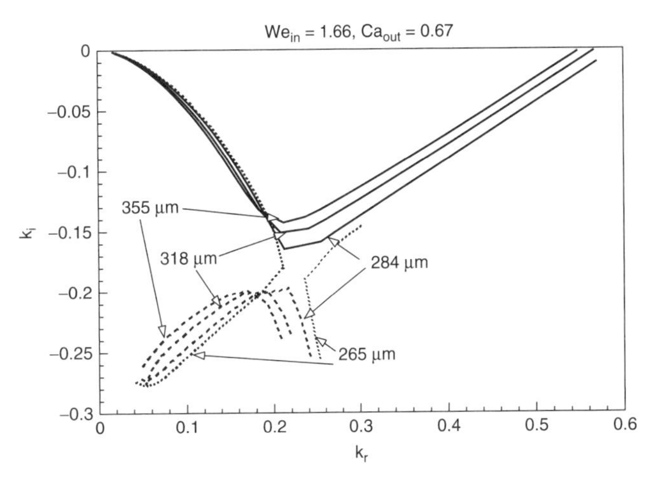

Figure 1.18 Different configurations in an axisymmetric flow focusing device. (a) Simple gas–liquid. (b) Simple liquid–liquid. (c) Concentric. Source: Reprinted Figure 21 from L. Martín-Banderas's PhD Thesis Reference [52].

Different configurations are possible with an axisymmetric flow focusing device, as shown in Figure 1.18. Using a gas as focusing fluid and a liquid as focused stream in a simple configuration, the generation of a spray is observed. If the fluids are both liquids, this technique allows the generation of emulsions that may be used as templates to produce particles, for example, by means of an evaporative solvent method [50]. Using a concentric device where the immiscible fluids are focused by an outer stream, the production of capsules has also been reported [51].

In the simplest axisymmetric version, a fluid is injected through a needle of inner diameter D_i, which is in front of an orifice of diameter D, located at a distance H from the needle, surrounded by an outer immiscible fluid that focuses the inner stream through the orifice. As in the coflow configuration, a "dripping" mode and a jetting regime can be obtained. The "dripping" regime is characterized by the generation of droplets close to the orifice, within a distance of about one orifice diameter. By contrast, when jetting occurs, the droplets are produced at the end of a jet that extends at least three orifice diameters.

When operated under adequate conditions, the inner stream can develop a cusp-like shape that is stable and that results in a thin jet of diameter d_{jet} that eventually breaks into droplets due to capillary instabilities. In these devices, the Weber and Reynolds numbers are generally much larger than one, which means that the process is essentially controlled by inertia.

The simple flow focusing configuration when the outer fluid is a gas and the focused stream is a liquid has been extensively described by Gañán-Calvo [48] when operated in the jetting regime. The physical properties involved in this problem are the density of the inner stream, ρ_i, its viscosity, μ_i, the inner flow rate, Q_i, the outer pressure drop, ΔP_o, the surface tension between both phases, σ, the diameter of the injection tube, the diameter of the orifice, and the distance between the capillary tube and the orifice. As the characteristic lengths are below the capillary length, gravitational forces are negligible.

Perhaps the most important quantity to control is the jet diameter, which is almost constant along the jet, or alternatively the jet velocity $U_{\text{jet}} = 4Q_i/(\pi d_{\text{jet}}^2)$. Using d_{jet} and U_{jet} as characteristic length and velocity, the Weber and Reynolds numbers based on the properties of the inner fluid are $We_i = \rho_i U_{\text{jet}}^2 d_{\text{jet}}/\sigma$ and $Re_i = \rho_i U_{\text{jet}} d_{\text{jet}}/\mu_i$. For large values of these numbers, surface tension and viscous effects can be neglected and it is possible to assume that the energy injected in the system is transformed into kinetic energy of

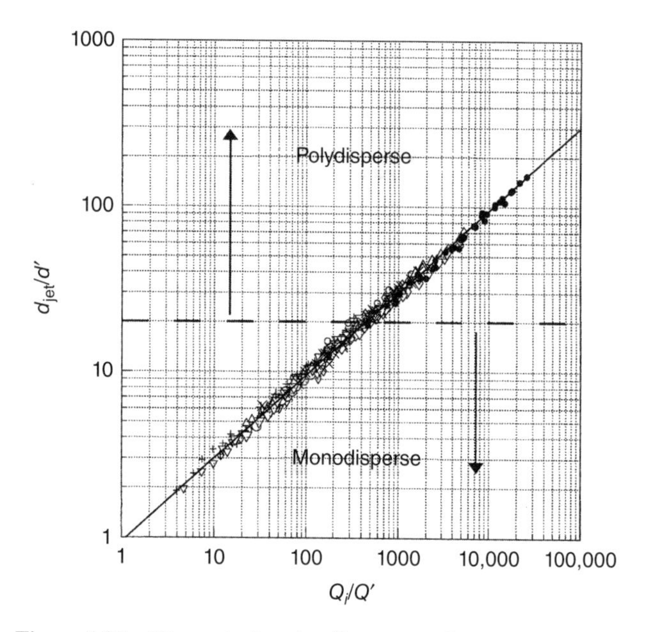

Figure 1.19 Dimensionless jet diameter at the constriction as a function of the nondimensional inner flow rate for different liquids and geometrical configurations. Source: Reprinted Figure 5 from Ref. [47]. Copyright 1998 by the American Physical Society.

the jet. This allows the estimation of the jet diameter [52]:

$$\Delta P_o \approx \frac{1}{2}\rho_i U_{\text{jet}}^2 \quad \Rightarrow \quad d_{\text{jet}} = \left(\frac{8\rho_i}{\pi^2 \Delta P_o}\right)^{\frac{1}{4}} Q_i^{1/2}. \tag{1.9}$$

Another important quantity is the minimum liquid flow rate, for a given device and hence a given ΔP_o, resulting in a stable jet. This results from the requirement that $We_i = 1$, as otherwise jet formation would not be achieved. The result is $Q_{\min} = (\sigma d_{\text{jet}}^3/\rho_i)^{1/2}$.

Measurements of d_{jet} for different liquids and geometrical configurations all collapse together in a single mastercurve when scaled by a reference length, $d' = \sigma/\Delta P_o$, and plotted versus a dimensionless inner-fluid flow rate, Q_i/Q', with $Q' = (\sigma^4/(\rho_i \Delta P_o)^3)^{1/2}$ (see Figure 1.19). This implies that

$$d_{\text{jet}}/d' = (8/\pi^2)^{1/4}(Q_i/Q')^{1/2}. \tag{1.10}$$

Since μ_i is not a relevant quantity, there are seven variables and hence four independent dimensionless groups. Two of them have already been identified as d' and Q'. The other two can be chosen as D_i/D and H/D and involve geometrical details of the device. These are indeed important since for a chosen D_i, there is an optimum value of H for which $Q_{\min}$ is minimum. Moreover, this value significantly depends on D. As expected, then, if D_i is increased, H and D should increase accordingly to ensure similarity in the working conditions.

In flow focusing, both global and local instabilities are important. We say that flow focusing is globally unstable if a steady meniscus cannot be formed [18]. Three mechanisms responsible for global instability have been identified. (i) If d_{jet} is too small, the injected energy is invested in surface energy resulting in the lack of steady-state emission. (ii) When the applied pressure is not big enough, the stresses exerted by the outer stream on the jet interface do not overcome surface tension forces and the meniscus never forms. (iii) For small enough inner flow rates, recirculation cells are formed inside the meniscus preventing jet formation and ultimately interrupting the flow.

Importantly, global stability is a necessary but not a sufficient condition to obtain a stable jet. Flow focusing must also be locally stable, which means that the jet must be convectively unstable. As a consequence, the growing perturbations must be convected downstream and result in a steady liquid ligament. If the jet is absolutely unstable, "dripping" kicks in. Vega *et al.* [18] performed a stability analysis showing that the "dripping" to jetting transition can be described as a transition from an absolute to a convective instability. Hence, the "dripping" mode is associated to an absolutely unstable jet, while the jetting mode is associated to a convectively unstable jet.

Figure 1.20 shows the stability regions in the We_i, Re_i representation when water is used as focused fluid and air as focusing fluid with $H = D_i = 200$ μm. Three different regimes are observed: (i) the steady jetting regime, where the meniscus is stable and the jet is convectively unstable (jetting), (ii) the local instability regime, where the meniscus is stable and the jet is absolutely unstable (dripping), and (iii) the global instability regime, where the meniscus is unstable. The experimental data is plotted in the upper graph, while the lower graph reproduces the boundary lines seen experimentally and identifies the regions where these regimes are observed. The open symbols show transitions from steady states to local instability situations, while the solid symbols corresponds to transitions from local to absolutely unstable situations. The dotted line shows the prediction obtained from the linear stability analysis of the basic flow. The dashed-dotted line corresponds to $Q_{\min} = 2.9$ ml/h for the fluids and device used in this particular experiment. Note that there is a turning point in the two experimental transition lines; above this point, as We_i increases, both lines almost coincide with the curve corresponding to a constant $Q_{\min}$. Hence, there is a minimum value of the inner flow rate below which flow focusing becomes globally unstable independently of the outer pressure drop.

Vega *et al.* [52] perform the same stability analysis for different H/D and reported that the transition from the steady jetting mode to the locally unstable "dripping" mode was not affected by the geometry. The same conclusion was reached for the transition from local to global instability

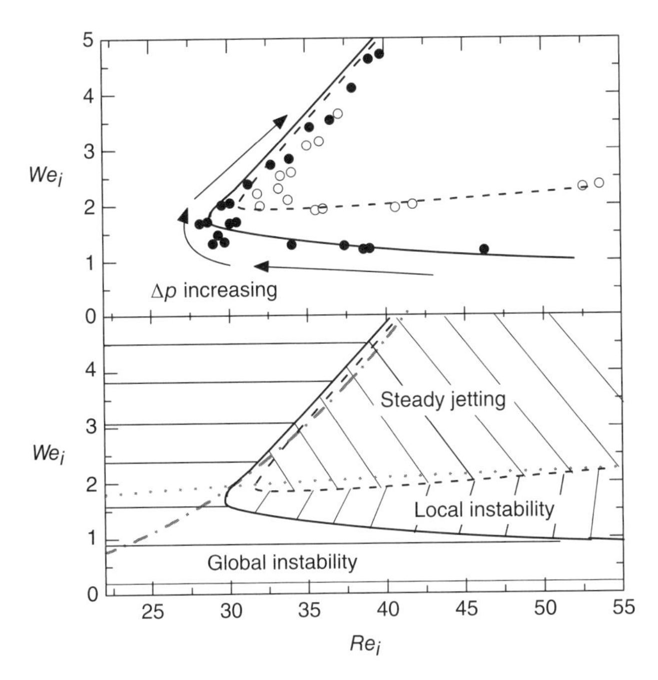

Figure 1.20 The lower graph shows the stability regions plotted from the experimental data shown in the upper graph. The dashed-dotted line is the curve of constant minimum flow rate, while the dotted line is the Leib and Goldstein prediction for the convective/absolute transition. Source: Reprinted Figure 6 from Ref. [52].

at small Weber numbers. In contrast, the geometry and the properties of the focused fluid have an effect on $Q_{\min}$ and on the pressure drop corresponding to the turning point of the transition curves. In addition, the relative importance of the global instability region is smaller when μ_i increases; in this case, an additional dimensionless group accounting for the relative influence of the viscous force must be considered.

1.4 SUMMARY AND OUTLOOK

We have discussed the generation of droplets with a narrow size distribution using coflowing fluids and flow focusing:

(1) In the coflow configuration, a fluid is injected through an injection tube in the presence of an immiscible fluid that flows in parallel. Depending on the operating conditions, two different regimes are obtained: dripping and jetting. In the "dripping" regime, extremely monodisperse droplets are generated at the tip of the capillary tube with a low production rate. In contrast, in the jetting regime less monodisperse droplets are obtained at the end of a long liquid ligament or jet, but at much higher production rates. Within the jetting regime, we find that the jet can either widen or narrow in the downstream direction. Despite these jets break into droplets via either absolute or convective instabilities, respectively, the drop size can be unified by a single scaling law.

(2) In the flow focusing configuration, a focused fluid is forced by an immiscible focusing fluid through a constriction. This configuration can be implemented in an axisymmetric or in a two-dimensional device. In the "dripping" regime, the generation of the droplets occurs close to the orifice while in the jetting regime it happens within a distance at least three orifice diameters. We have reviewed the relevant scaling laws for the jet size and hence for the resulting drop size. In addition, we have discussed the operating regimes in terms of global and local instabilities, as well as in terms of absolute and convective instabilities.

The advantage of these techniques and the many related techniques that are currently in use or under development compared to the more traditional methods for emulsion generation rests on the exquisite fluid flow control offered by microfluidics. The high throughout is still the major hurdle for these techniques to completely dominate the field of emulsion generation.

REFERENCES

[1] Basaran OA. Small-scale free surface flows with break-up: drop formation and emerging applications. AIChE J 2002;48:1842.

[2] Squires TM, Quake SR. Microfluidics: fluid physics at the nanoliter scale. Rev Mod Phys 2005;77:977.

[3] Stone HA, Adjari A. Engineering flows in small devices: microfluidics toward a lab-on-a-chip. Annu Rev Fluid Mech 2004;36:381.

[4] Gunther A, Jensen KF. Multiphase microfluidics: from flow characteristics to chemical and materials synthesis. Lab Chip 2006;6:1487.

[5] Barrero A, Loscertales IG. Micro- and nanoparticles via capillary flows. Annu Rev Fluid Mech 2007;39:89.

[6] Clift R, Grace JR, Weber ME. *Bubbles, Drops, and Particles*. Mineola (NY): Dover Publications, Inc.; 2005.

[7] Liu H. *Science and Engineering of Droplets: Fundamentals and Applications*. Norwich (NY): William Andrew; 1999.

[8] Zúñiga RN, Aguilera JM. Aerated food gels: fabrication and potential applications. Trends Food Sci Technol 2008;19:176.

[9] Williams PA, Phillips GO, editors. *Gums and Stabilisers for the Food Industry 17: The Changing Face of Food Manufacture: The Role of Hydrocolloids*. Volume 346. Cambridge: The Royal Society of Chemistry; 2014.

[10] Martin-Banderas L. New trends in micro-and nano-systems development for pharmaceutical actives release. Med Chem (Shariqah (United Arab Emirates)) 2012;8(4):515.

[11] Michalet XEA. Quantum dots for live cells, in vivo imaging, and diagnostics. Science 2005;307:538.

[12] Gouin S. Microencapsulation: industrial appraisal of existing technologies and trends. Trends Food Sci Technol 2004;15:330.

[13] Friberg S, Jones S, Kroschwitz S. *Kirk-Othmer Encyclopedia of Chemical Technology*. 4th ed., Volume 9. New York: John Wiley and Sons; 1994.

[14] Kentish S, Simons L. The use of ultrasonics for nanoemulsion preparation. Innovative Food Sci Emerg Technol 2008;9(2):170.

[15] Yamada M, Seki M. Hydrodynamic filtration for on-chip particle concentration and classification utilizing microfluidics. Lab Chip 2005;5(11):1233–1239.

[16] Bhagat AAS, Kuntaegowdanahalli SS, Papautsky I. Inertial microfluidics for continuous particle filtration and extraction. Microfluid Nanofluid 2009;7(2):217–226.

[17] Shah RK, Shum HC, Rowat AC, Lee D, Agresti JJ, Utada AS, Chu L, Kim WK, Fernández-Nieves A, Martínez CJ, Weitz DA. Designer emulsions using microfluidics. Mater Today 2008;11(4):18–27.

[18] Vega EJ, Montanero JM, Herrada MA, Ga nán Calvo AM. Global and local instability of flow focusing: the influence of the geometry. Phys Fluids 2010;22(6):064105.

[19] Garstecki P, Fuerstman MJ, Stone HA, Whitesides GM. Formation of droplets and bubbles in a microfluidic T-junctions scaling and mechanism of break-up. Lab Chip 2006;6:437–446.

[20] Teh S, Lin R, Hung L, Lee AP. Droplet microfluidics. Lab Chip 2007;8:198–220.

[21] Christopher GF, Anna SL. Microfluidic methods for generating continuous droplets streams. J Phys D Appl Phys 2007;40:319–336.

[22] Vaschy A. Sur les lois de similitude en physique. Ann Télégraphiques 1892;19:25.

[23] Buckingham E. On physically similar systems. Illustrations of the use of dimensional equations. Phys Rev 1914;4:345.

[24] Utada AS, Fernández-Nieves A, Stone HA, Weitz D. Dripping to jetting transitions in co-flowing liquid streams. Phys Rev Lett 2007;99:094502.

[25] Rayleigh WS. On the capillary phenomena of jets. Proc Lond Math Soc 1879;29:71–97.

[26] Plateau J. Statique experimentale et theorique des liquides soumis aux seules forces moleculaires. Acad Sci Bruxelles Mem 1849;23:5.

[27] Castro-Hernández E, Gundabala V, Fernández-Nieves A, Gordillo JM. Scaling the drop size in coflow experiments. New J Phys 2009;11(7):075021.

[28] Utada AS, Lorenceau E, Link DR, Kaplan PD, Stone HA, Weitz D. Monodisperse double emulsions generated from a microcapillary device. Science 2005;308:537–541.

[29] Tomotika S. On the instability of a cylindrical thread of a viscous liquid surrounded by another viscous fluid. Proc R Soc A 1935;150:322–337.

[30] Suryo R, Basaran O. Tip streaming from a liquid drop forming from a tube in a co-flowing outer fluid. Phys Fluids 2006;18:082102.

[31] Marín AG, Campo-Cortés F, Gordillo JM. Generation of micron-sized drops and bubbles through viscous coflows. Colloids Surf A Physicochem Eng Aspects 2009;344:2–7.

[32] Castro-Hernández E, Campo-Cortés F, Gordillo JM. Slender-body theory for the generation of micrometre-sized emulsions through tip streaming. J Fluid Mech 2012; 698:423.

[33] Gordillo JM, Sevilla A, Martínez-Bazán C. Bubbling in a coflow at high Reynolds numbers. Phys Fluids 2007;19:077102.

[34] Utada AS, Fernández-Nieves A, Gordillo JM, Weitz D. Absolute instability of a liquid jet in a coflowing stream. Phys Rev Lett 2008;100:014502.

[35] Clanet C, Lasheras JC. Transition from dripping to jetting. J Fluid Mech 1999;383:307.

[36] Ambravaneswaran B, Subramani HJ, Philips SD, Basaran OA. Dripping-jetting transitions in a dripping faucet. Phys Rev Lett 2004;93:034501.

[37] Briggs RJ. *Electron Stream Interaction with Plasmas*. Cambridge (MA): MIT Press; 1964.

[38] Bers A. *Space Time Evolution of Plasma Instabilities*. Amsterdam: North Holland; 1983. p 451–517.

[39] Sevilla A, Gordillo JM, Martínez-Bazán C. Bubble formation in a coflowing air–water stream. J Fluid Mech 2005;530:181.

[40] Rubio-Rubio M, Sevilla A, Gordillo JM. On the thinnest steady threads obtained by gravitational stretching of capillary jets. J Fluid Mech 2013;729:471–483.

[41] Gordillo JM, Sevilla A, Campo-Cortes F. Global stability of stretched jets: conditions for the generation of monodisperse micro-emulsions using coflows. J Fluid Mech 2014;738:335–357.

[42] Gordillo JM, Pérez-Saborid M, Gañán Calvo AM. Linear stability of co-flowing liquid-gas jets. J Fluid Mech 2001; 448:23.

[43] Gordillo JM, Perez-Saborid M. Transient effects in the signaling problem. Phys Fluids 2002;14:4329.

[44] Gordillo JM, Gañán Calvo AM, Pérez-Saborid M. Monodisperse microbubbling: absolute instabilities in coflowing gas–liquid jets. Phys Fluids 2001;13:3839.

[45] Sevilla A, Gordillo JM, Martínez-Bazán C. Transition from bubbling to jetting in a coaxial air–water jet. Phys Fluids 2005;17:018105.

[46] Gañán Calvo AM, Riesco-Chueca P. Jetting–dripping transition of a liquid jet in a lower viscosity co-flowing immiscible liquid: the minimum flow rate in flow focusing. J Fluid Mech 2006;553:75.

[47] Guillot P, Colin A, Utada AS, Ajdari A. Stability of a jet in confined pressure-driven biphasic flows at low Reynolds numbers. Phys Rev Lett 2007;99:104502.

[48] Gañán Calvo AM. Generation of steady liquid microthreads and micron-size monodisperse sprays in gas streams. Phys Rev Lett 1998;80(2):285–288.

[49] Anna SL, Bontoux N, Stone HA. Formation of dispersions using Flow Focusing in microchannels. Appl Phys Lett 2003;82:364–366.

[50] Martín-Banderas L, Rodríguez-Gil A, Cebolla A, Chávez S, Berdún-Álvarez T, Flores-Mosquera M, Gañán Calvo AM. Towards high-throughput production of uniformly encoded microparticles. Adv Mater 2006;18(5):559–564.

[51] Gañán Calvo AM, Chávez S, Cebolla A, Flores-Mosquera M, Castro-Hernández E. Method of preparing micro and nanometric particles with labile products. PCT/ES2006/000212. 2009.

[52] Martín-Banderas L. Microencapsulación mediante la tecnología Flow-Focusing para aplicaciones biotecnológicas y biomédicas [PhD thesis]. Sevilla: CRAI Antonio Ulloa. Facultad de Farmacia, Universidad de Sevilla; 2007.

2

ELECTRIC FIELD EFFECTS

FRANCISCO J. HIGUERA
UPM Madrid, E. T. S. Ingenieros Aeronáuticos, Spain

2.1 INTRODUCTION

Electrospray is a technique for generating nearly monodisperse sprays of small electrically charged drops of an electrically conducting liquid using electric forces to disrupt the liquid surface; see Fernández de la Mora [1] for a recent review. A meniscus of the liquid to be sprayed is subject to an electric field by applying a high voltage between the liquid, which is in contact with an electrode, and another electrode at some distance from the meniscus. The electric field induces an electric current in the liquid that accumulates electric charge at its surface and causes an electric stress that elongates the meniscus in the direction of the field. Under different conditions, the meniscus may either shed charged drops or emit one or several jets that in turn break into drops. The different functioning modes of an electrospray and the transitions between them have been classified by Cloupeau and Prunet-Foch [2, 3] and Jaworek and Krupa [4]. The cone-jet mode [5], in which the meniscus takes a conical shape with a single stationary jet issuing from its tip and breaking into drops at some distance downstream, is the most suitable for generating small monodisperse drops and has been extensively studied. In his pioneering work with a meniscus formed at the end of a capillary, Zeleny [6, 7] first showed that the balance of electric and surface tension stresses in the cone-jet mode requires a voltage of order $(\gamma a/\epsilon_0)^{1/2}$, where γ is the surface tension of the liquid, a is the radius of the base of the meniscus, and ϵ_0 is the permittivity of the surrounding medium, be this a vacuum or a passive dielectric fluid. Taylor [8] carried out an analysis of the balance of stresses using a spheroidal approximation for the shape of the meniscus in order to determine the cone-jet onset voltage and found that a conical meniscus of semiangle $\alpha = 49.29°$ is an exact hydrostatic solution (at least in the vicinity of the tip) for a special value of the voltage. Smith [9] and Cloupeau and Prunet-Foch [5] found hysteresis, whereby the cone-jet can be maintained when the voltage is decreased below the onset voltage at which the cone-jet first appears when the voltage is gradually increased. This behavior can be traced to the intensification

Fluids, Colloids and Soft Materials: An Introduction to Soft Matter Physics, First Edition. Edited by Alberto Fernandez Nieves and Antonio Manuel Puertas.

of the electric field around a pointed meniscus, which leads to electric stresses that may oppose surface tension and keep the meniscus pointed even when the electric field away from the tip would only cause a moderate stretching of a rounded hydrostatic meniscus. Once a cone-jet is formed, the main parameters determining the size of the cone-to-jet transition region and the drops are the conductivity of the liquid and the flow rate injected into the meniscus. The size of the drops decreases when the flow rate decreases or the conductivity increases; see Fernández de la Mora and Loscertales [10], Chen and Pui [11], and Gañán-Calvo *et al.* [12], among others, for scaling laws of the size of the drops and the electric current, and Barrero *et al.* [13] for extensions of these results to cone-jets in baths of dielectric liquids. A stable cone-jet exists only within a range of values of the flow rate that shifts toward lower flow rates when the conductivity increases [9].

Fernández de la Mora and Loscertales [10] claim that charge relaxation effects are always important in a certain region around the tip of the meniscus of characteristic size $R_e = (\epsilon_0 \epsilon Q/K)^{1/3}$, which is also the order of the size of the spray drops. Here K and ϵ are the electrical conductivity of the liquid and its dielectric constant (the ratio of its permittivity to the permittivity of the surrounding medium), and Q is the flow rate. The residence time of the liquid in this region, $t_{\mathrm{re}} = R_e/v_e$ with $v_e = Q/R_e^2$, is of the order of the electric relaxation time $t_e = \epsilon\epsilon_0/K$, so that the electric charge of a material element of the liquid surface cannot increase at the same pace as the electric field acting on the surface element during its transit across the relaxation region. This condition determines the order of the electric current as a function of the flow rate as $I \sim (\gamma K Q/\epsilon)^{1/2}$. It also implies that the electric field due to the applied voltage enters the liquid in the relaxation region and leads to an electric shear stress that can drive the liquid into the jet. Gañán-Calvo *et al.* [12] proposed that charge relaxation effects occur in a certain region of the jet, rather than around the tip of the meniscus. Higuera [14] estimated the characteristic time of charging of the liquid surface in the region of the jet where the conduction current is transferred to the surface, showing that this time may be large compared to the electric relaxation time of the liquid.

Cloupeau and Prunet-Foch [5] experimentally determined the domain of operation or region of the voltage/flow-rate plane where a cone-jet of a given liquid can be established. This region is bounded by a minimum voltage that depends on the flow rate and at which the system jumps to an oscillatory mode, and a maximum voltage above which instabilities, or a multiple-jet mode, or electrical discharges in the surrounding gas appear. As was already mentioned, the flow rate can be varied at constant voltage in the domain of operation between certain minimum and maximum values. Fernández de la Mora [1] noted that the existence of a maximum flow rate may reflect a reduction of the electric field acting on the meniscus due to the charged spray drops, whose density increases with the flow rate, especially in the presence of a surrounding medium that opposes a resistance to the motion of the drops. The minimum flow rate determines the thinnest jet and the highest surface electric field that can be attained with a given liquid [10]. The breakup of the jet into drops is most regular near this minimum flow rate, leading to the smallest drops and the narrowest size distributions [15]. The domain of operation of an electrospray of a very viscous liquid of small conductivity has been described numerically using a simplified model [16].

The origin of the minimum flow rate is not yet clear, despite the large amount of work devoted to ascertain it. Fernández de la Mora and Loscertales [10] experimentally found that the minimum flow rate is of order ϵQ_0 in many cases, where $Q_0 = \epsilon_0 \gamma/\rho K$ with ρ the density of the liquid. Fernández de la Mora [1] pointed out that the scaling laws for the length and radius of the jet derived by Cherney [17] on the basis of Fernández de la Mora and Loscertales [10] model suggest that the jet could effectively disappear, or break up before the transfer of charge to its surface is complete, when the flow rate is of order ϵQ_0. As an alternative explanation, Guerrero *et al.* [18] noted that the dynamics of the relaxation region is largely independent of the rest of the meniscus. This region is able to suck a flow rate of order $\epsilon\ Q_0$ or larger independently of the flow rate injected through the capillary, which may make a stationary configuration impossible when the latter flow rate is small compared to ϵQ_0. On the other hand, Gañán-Calvo *et al.* [12] find a minimum flow rate of order $\epsilon^{1/2} Q_0$.

In this chapter, the basic mechanisms of the electrospray are reviewed and some applications and extensions are briefly discussed.

2.2 MATHEMATICAL FORMULATION AND ESTIMATES

The flow, the electric field, and the transport of electric charge in an electrospray can be described in the framework of the leaky dielectric model [19, 20]. Let $\boldsymbol{v}$ and p denote the velocity and pressure distributions of the liquid, which satisfy the Navier–Stokes equations. The dielectric medium surrounding the liquid is assumed to play no role in the dynamics, which is justified when this medium is a gas or a vacuum. In the absence of a magnetic field, the electric fields in the liquid and the surrounding medium are of the form $\boldsymbol{E}^l = -\nabla \varphi^l$ and $\boldsymbol{E} = -\nabla \varphi$, where φ^l and φ are the electric potentials in the liquid and outside. The electric field outside the liquid satisfies $\nabla \cdot \boldsymbol{E} = 0$ in the absence of space charge. The electric field in the liquid leads to a conduction current whose density is assumed to be given by Ohm's law $\boldsymbol{j} = K\boldsymbol{E}^l$ with a constant conductivity K, and conservation of charge requires $\nabla \cdot \boldsymbol{j} = 0$ in the absence of space charge in the liquid. Therefore, the electric potentials satisfy Laplace's equations $\nabla^2\varphi^l = \nabla^2\varphi = 0$ in the liquid and in the surrounding medium.

At the surface of the liquid, say $f(\boldsymbol{x}, t) = 0$ with $f < 0$ in the liquid, which is a free surface to be found as part of the solution, the electric conditions $\varphi^l = \varphi$ and $\epsilon_0(\boldsymbol{E} - \epsilon\boldsymbol{E}^l) \cdot \boldsymbol{n} = \sigma$ must be satisfied. Here $\boldsymbol{n} = \nabla f/|\nabla f|$ is the unit normal to the surface pointing away from the liquid and σ is the density of free surface charge.

Electric charge reaches the surface by conduction in the liquid and is convected by the flow. The density of surface charge satisfies the conservation equation $D\sigma/Dt = K\boldsymbol{E}^l \cdot \boldsymbol{n} + \sigma\boldsymbol{n} \cdot \nabla\boldsymbol{v} \cdot \boldsymbol{n}$, where $D\sigma/Dt = \partial\sigma/\partial t + \boldsymbol{v} \cdot \nabla\sigma$ is the material derivative of σ at the surface. The first term on the right-hand side of the equation for σ is the rate at which conduction brings charge to the unit area of the surface, and $\boldsymbol{n} \cdot \nabla\boldsymbol{v} \cdot \boldsymbol{n}$ in the second term is the negative of the strain rate of a material element of the surface; see, for example, Batchelor [21]. Thus, if δA is an elementary material element of the surface, $D(\delta A)/Dt = -\delta A\, \boldsymbol{n} \cdot \nabla\boldsymbol{v} \cdot \boldsymbol{n}$ and the charge conservation equation for the surface element reads $D(\sigma\, \delta A)/Dt = K\boldsymbol{E}^l \cdot \boldsymbol{n}\, \delta A$. Conduction under the action of the applied field tends to increase the density of surface charge, but this charge reduces the normal electric field $\boldsymbol{E}^l \cdot \boldsymbol{n}$ that drives conduction (because $\boldsymbol{E}^l \cdot \boldsymbol{n} = (\boldsymbol{E} \cdot \boldsymbol{n} - \sigma/\epsilon_0)/\epsilon$ from the second electric condition at the surface). The characteristic time for conduction to screen the liquid from the applied field is the electric relaxation time $t_e = \epsilon_0\epsilon/K$, which follows from the balance $\sigma/t_e \sim K\boldsymbol{E}^l \cdot \boldsymbol{n}$ with $\sigma \sim \epsilon_0\boldsymbol{E} \cdot \boldsymbol{n} \sim \epsilon_0\epsilon\boldsymbol{E}^l \cdot \boldsymbol{n}$.

The electric field and electric charge at the surface of the liquid lead to an electric stress whose components normal and tangent to the surface are [22, 20]

$$\tau_n^e = \frac{1}{2}\epsilon_0\left[(\boldsymbol{E} \cdot \boldsymbol{n})^2 - \epsilon(\boldsymbol{E}^l \cdot \boldsymbol{n})^2\right] + \frac{1}{2}\epsilon_0(\epsilon - 1)|\boldsymbol{E} \times \boldsymbol{n}|^2,$$

$$\boldsymbol{\tau}_t^e = \sigma(\boldsymbol{E} - \boldsymbol{n}\, \boldsymbol{E} \cdot \boldsymbol{n}).$$

In summary, the governing equations are

$$\nabla \cdot \boldsymbol{v} = 0, \frac{\partial\boldsymbol{v}}{\partial t} + \boldsymbol{v} \cdot \nabla\boldsymbol{v} = -\frac{1}{\rho}\nabla p + \nu\nabla^2\boldsymbol{v}, \quad \nabla^2\varphi^l = 0 \tag{2.1}$$

in the liquid, for $f(\boldsymbol{x}, t) < 0$,

$$\nabla^2\varphi = 0 \tag{2.2}$$

in the surrounding dielectric, for $f(\boldsymbol{x}, t) > 0$, and

$$\frac{Df}{Dt} = 0, -p + \boldsymbol{n} \cdot \tau' \cdot \boldsymbol{n} + \gamma\nabla \cdot \boldsymbol{n} = \tau_n^e, \boldsymbol{t} \cdot \tau' \cdot \boldsymbol{n} = \boldsymbol{t} \cdot \boldsymbol{\tau}_t^e,$$

$$\varphi^l = \varphi, \epsilon_0(\boldsymbol{E} - \epsilon\boldsymbol{E}^l) \cdot \boldsymbol{n} = \sigma,$$

$$\frac{D\sigma}{Dt} = K\boldsymbol{E}^l \cdot \boldsymbol{n} + \sigma\boldsymbol{n} \cdot \nabla\boldsymbol{v} \cdot \boldsymbol{n} \tag{2.3}$$

at the surface, $f(\boldsymbol{x}, t) = 0$. Here ρ, ν, and γ are the density, kinematic viscosity, and surface tension of the liquid, the pressure p is referred to the uniform pressure of the surrounding medium, $\tau' = \mu[\nabla\boldsymbol{v} + (\nabla\boldsymbol{v})^T]$ with $\mu = \rho\nu$ is the viscous stress tensor, and $\boldsymbol{t}$ is a unit vector tangent to the surface.

These equations must be supplemented with boundary conditions at the electrodes and other solid surfaces bounding the system. The electrode in contact with the liquid is often a metallic tube through which the liquid is injected. At the surface of this tube: $\varphi^l = \varphi = 0$; the liquid surface meets the tube at a contact line whose location depends on the wetting conditions of the liquid and the solid (e.g., the contact line may coincide with the edge of the tube); and the flow rate of liquid issuing from the tube is a given constant Q. At the far electrode: $\varphi^l = \varphi = -V$, where V is the applied voltage.

In many cases, the radius of the electrically charged jet emerging from the meniscus at the end of the tube is small compared to the radius of the tube and the length of the jet. Fernández de la Mora and Loscertales [10] noted that this condition allows a simplified problem to be posited for a region around the tip of the meniscus comprising the meniscus-to-jet transition and, as will be seen below, a certain leading stretch of the jet. This region, which is sketched in Figure 2.1, plays a crucial role, as it determines the electric current of the electrospray. It is small compared to any other length of the system, be this the radius of the tube, the distance to jet breakup, or the interelectrode distance. All such lengths are irrelevant to the local problem. The voltage applied between the electrodes is the cause of the electric field seen by the region of interest, but this field is modified by the charge of the spray in the space between the jet and the far electrode and, to a lesser extent, by the adjustment of the bulk of the meniscus. In addition, the field

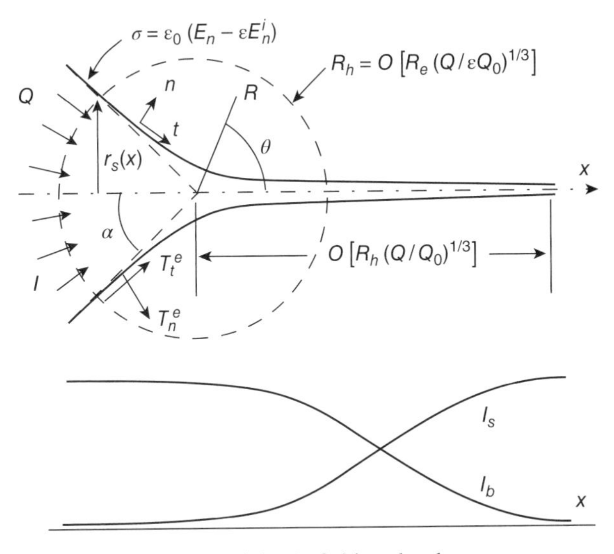

Figure 2.1 Definition sketch.

is intensified locally by the elongated meniscus (see Section 2.2.1). All these effects make the electric field in the small region of interest largely independent of the details of any particular experimental configuration. It is this separation of scales that underlies the search for universal scaling laws for the cone-jet mode. Such laws should not be expected to hold when the radius of the jet is not small compared to the radius of the tube.

2.2.1 Conical Meniscus

Consider first an intermediate region upstream of the cone-to-jet transition region and much larger than it but still small compared to the radius of the tube. The velocity of the liquid in this region is very small compared to its velocity in the jet, so that the pressure variations and viscous stresses due to the flow can be neglected in first approximation. The residence time of the liquid in this region is much larger than the electric relaxation time, so that the electric displacement at the liquid side of the surface ($\epsilon_0 \epsilon \boldsymbol{E}^l$) is very small and the surface is nearly equipotential. In these conditions, the shape of the liquid surface is determined by the balance of surface tension and electric stresses. If, in addition, the electric field induced at the meniscus by the charge of the jet is neglected, then the meniscus is a Taylor cone in the intermediate region [8]. In terms of the semiangle of the cone, α, and the distance R from its apparent vertex (see Figure 2.1), the balance of stresses reads $\gamma/(R \tan\alpha) = \frac{1}{2}\epsilon_0 E^2$, determining the electric field at the equipotential surface as $E = [2\gamma/(\epsilon_0 R \tan\alpha)]^{1/2}$. Since $\boldsymbol{E} = -\boldsymbol{\nabla}\varphi$ and $\nabla^2\varphi = 0$ outside the liquid, this surface field amounts to $\varphi = \varphi_T = AR^{1/2}P_{1/2}(\cos\theta)$, where θ is the angle around the vertex measured from the prolongation of the cone, which is the surface $\theta = \pi - \alpha$; $P_{1/2}(\cos\theta)$ is the Legendre function of degree 1/2 regular at $\theta = 0$; and $A = (2\gamma)^{1/2}/[-\epsilon_0 P'_{1/2}(-\cos\alpha)\sin\alpha \tan^{1/2}\alpha]^{1/2}$. The condition that $\varphi = 0$ at the surface requires that $\cos(\pi - \alpha)$ be the first zero of $P_{1/2}$; that is, that $\alpha \approx 49.29°$, for which $A \approx 1.3459\ (\gamma/\epsilon_0)^{1/2}$.

The hydrostatic balance in Taylor's solution determines the angle of the cone and the strength of the electric field around the cone, measured by A. This means that, for a given configuration of the electrodes creating the electric field and a given pressure at the inlet of the tube holding the meniscus (or a given volume of liquid), the solution will be realized only for a special value of the voltage applied between the electrodes, V_T say. This voltage and the shape of the whole meniscus have been computed by Pantano *et al.* [23] for a needle-plate geometry. However, this hydrostatic solution is unstable [8, 24, 25], and therefore it can approach the stable meniscus of an electrospray only to a certain degree. Probably the difference is in the electric field induced by the charge of the jet or the drops, which may have a stabilizing effect not accounted for in the previous computation. Menisci that are not strictly conical and have apparent angles at the tip different from that of a Taylor cone are often observed in experiments.

The field induced by the charge of the jet far downstream of the conical meniscus is a small perturbation to the field of a Taylor cone. This perturbation can be estimated noticing that, if the jet does not break into drops, it acts as a line of charge of stretch (Γ, charge per unit length) equal to the product of the density of surface charge and the perimeter of the jet cross section. Using the estimates of these magnitudes worked out in the following section (Eq. 2.7 with x replaced by R) we find $\Gamma = O(\epsilon_0^{5/4}\rho^2 Q^{1/2} I^{1/2}\gamma^{-1/4}R^{-1/4})$, where I is the electric current transported by the jet. The field induced by this line of charge at the meniscus is of order $\Gamma/\epsilon_0 R$ up to logarithms (see, e.g., [26] or [27]), which amounts to a perturbation $\varphi_1 = O(\epsilon_0^{1/4}\rho^2 Q^{1/2} I^{1/2}\gamma^{-1/4}R^{-1/4})$ to the electric potential φ_T of a Taylor cone. The condition that $\varphi_T + \varphi_1 = 0$ at the surface of the meniscus, $\theta = \pi - \alpha + \delta(R)$ say, determines the small departure from the surface of a Taylor cone as $\delta(R) = O(\epsilon_0^{3/4}\rho^2 Q^{1/2} I^{1/2}\gamma^{-3/4}R^{-3/4})$. Since $\varphi_1 < 0$ and $\partial\varphi_T/\partial\theta > 0$ at the meniscus, the perturbation $\delta > 0$; that is, the surface is more pointed than a Taylor cone.

Another cause for the lack of observations of a pure Taylor cone is that the meniscus is not strictly hydrostatic. The flow rate Q must reach the jet through the meniscus, leading to a radial sink flow with velocity $v_R = -Q/[2\pi(1-\cos\alpha)R^2]$ in the spherical coordinates used to describe Taylor's solution above. Similarly, the electric current I reaches the jet by conduction in the meniscus, which requires a radial electric field $E_R = -I/[2\pi(1-\cos\alpha)KR^2]$, whose electric potential is $\varphi^l = I/[2\pi(1-\cos\alpha)KR]$, so that the surface of the meniscus is not strictly equipotential. In combination with the surface charge of Taylor's solution, $\sigma = (2\epsilon_0\gamma/R\tan\alpha)^{1/2}$, the field E_R causes an electric shear stress $\tau_R = \sigma E_R$ directed toward the apparent vertex of the meniscus, which induces a recirculating flow additional to the sink flow mentioned above. This flow has been analyzed by Barrero *et al.* [28]. Viscous forces dominate the flow in the meniscus far from the apparent vertex, where the sink and recirculating flow contributions to the stream function can be added to give

$$\psi = \frac{Q(1+\cos\theta)}{2\pi(1-\cos\alpha)} - \frac{\epsilon_0^{1/2}\gamma^{1/2}I/\mu}{2^{1/2}\pi\tan^{1/2}\alpha(1-\cos\alpha)\sin\alpha} R^{1/2} f_B(-\cos\theta), \tag{2.4}$$

where $f_B(\xi)$ is the solution of [28]

$$(1-\xi^2)\left(f_B^{iv} - 4\xi f_B''' + \frac{3}{2}f_B''\right) - \frac{15}{16}f_B = 0,$$

$$f_B(1) = f_B(\cos\alpha) = f_B''(\cos\alpha) - 1 = 0, \quad f_B'(1) < \infty.$$

The far electric field of the solution described here and in the rest of Section 2.2 tends to that of a Taylor cone.

Even accounting for the effect of the space charge due to the charged drops of the spray, and for the adjustment of the meniscus between the exit of the tube and the apparent vertex of the cone, this condition can be expected to be fulfilled only in a narrow range of applied voltages for any realistic configuration of the electrodes. No cone-jet should exist outside of this range, which is in line with experimental observations (e.g., [5]). It is also for this reason that the applied voltage does not appear in the results of Section 2.2.

2.2.2 Cone-to-Jet Transition Region and Beyond

If liquid is continuously injected through the meniscus, then, in the range of voltages and flow rates corresponding to the cone-jet regime, this liquid leaves the meniscus as a jet that issues from an axisymmetric cone-to-jet transition region around the vertex of the cone. The jet may extend to a large distance downstream of the transition region before it breaks up into charged drops. The convection and conduction contributions to the electric current in a stationary axisymmetric flow are

$$I_s(x) = 2\pi\sigma v_s r_s \quad \text{and} \quad I_b = 2\pi K \int_0^{r_s} E_x^l r \, dr, \qquad (2.5)$$

where x and r are distances along the symmetry axis and normal to it, $r_s(x)$ and $v_s(x)$ are the radius of the surface cross section and the velocity of the liquid at the surface, and E_x^l is the axial component of $\boldsymbol{E}^l$. The conservation equation for the free surface charge, $\sigma(x)$, becomes

$$\frac{dI_s}{dx} = 2\pi r_s K E_n^l (1 + r_s'^2)^{1/2}, \qquad (2.6)$$

which is an integrated form of the last condition 2.3 . Hereafter, $E_n^l = \boldsymbol{E}^l \cdot \boldsymbol{n}$ and $E_n = \boldsymbol{E} \cdot \boldsymbol{n}$.

As can be seen using Equations 2.4 and 2.5, and Poisson's law $\nabla \cdot \boldsymbol{E}^l = 0$ integrated over the jet cross section, the total current $I = I_b + I_s$ is a constant independent of x.

The balance of electric and surface tension stresses characteristic of the conical meniscus is upset in the transition region, both because the pressure variations due to the motion of the liquid come into play [29] and because this motion changes the electric field and the charge distribution at the surface, and therefore the electric stress [10]. Let R_h denote the characteristic size of the transition region, and $v_h = Q/R_h^2$ and E_h the characteristic velocity of the liquid and the characteristic electric field around the liquid in this region. Assuming that the inertia of the liquid is important, so that $\Delta p = O(\rho v_h^2)$, the order-of-magnitude balances $\rho v_h^2 \sim \gamma/R_h \sim \epsilon_0 E_h^2$ give $R_h = (\rho Q^2/\gamma)^{1/3}$ and $E_h = (\gamma/\epsilon_0 R_h)^{1/2}$.

The importance of charge relaxation effects in the transition region is measured by the ratio of the residence time $t_r = R_h/v_h$ to the electric relaxation time $t_e = \epsilon_0\epsilon/K$; that is, $t_r/t_e = (R_h/R_e)^3$, where $R_e = (\epsilon_0\epsilon Q/K)^{1/3}$ is the characteristic size of the charge relaxation region of Fernández de la Mora and Loscertales [10]. The condition $t_r \sim t_e$ amounts to $R_h \sim R_e$ and is realized for $Q \sim \epsilon Q_0$, with $Q_0 = \epsilon_0\gamma/\rho K$. Then the surface charge has no time to reach its equilibrium and cannot screen the liquid in the transition region from the outer field. This field is of order $E_e = (\gamma/\epsilon_0 R_e)^{1/2}$, so that the field in the liquid is $E^l \sim E_e/\epsilon$ and the electric current is $I \sim K R_e^2 E^l = I_0/\epsilon^{1/2}$ with $I_0 = \epsilon_0^{1/2}\gamma/\rho^{1/2}$. The values of Q and I for which $t_r \sim t_e$ are also the orders of the minimum flow rate and the minimum current experimentally attainable in the cone-jet mode, which means that the residence time is never much smaller than the electric relaxation time (but see Ref. [30] for further analysis).

The ratio t_r/t_e is large in the high flow rate regime $Q \gg \epsilon Q_0$, which implies that the surface charge has time to reach equilibrium in the transition region of size R_h and therefore $\epsilon_0\epsilon E_n^l$ is negligible in the last but one condition 2.3. There is no relaxation region around the tip of the meniscus in these conditions. The flow becomes a quasi-unidirectional jet downstream of the transition region, and the surface charge is able to reach equilibrium ($\sigma \approx \epsilon_0 E_n$) and screen the liquid also in a certain initial region of the jet. The evolution of the flow in this region of the jet is determined by the following conditions (hereafter, x is the axial distance from the apparent vertex of the conical meniscus):

- The jet acts electrically as a line of charge of strength $2\pi r_s \epsilon_0 E_n$, which induces an axial electric field $E_{x_{jet}} \sim E_n r_s/x$ at its surface (up to a logarithmic factor). In order to keep the surface nearly equipotential, this field must balance the axial field due to the conical meniscus, which is $E_T \sim [\gamma/(\epsilon_0 x)]^{1/2}$. This condition reads $E_{x_{jet}} \sim E_T$.
- The electric shear stress $\sigma \boldsymbol{E} \cdot \boldsymbol{t}$, where $\boldsymbol{t}$ is a unit vector tangent to the surface in its meridional plane, is small in the region where the surface of the jet is nearly equipotential. The normal electric stress, $\frac{1}{2}\epsilon_0 E_n^2$, points outward and causes a depression in the liquid that accelerates the jet, while surface tension stresses are small. Therefore, $\rho v^2 \sim \epsilon_0 E_n^2$ with $v \sim Q/r_s^2$.

These two conditions give

$$r_s \sim \frac{\rho^{1/2} Q}{\gamma^{1/2} x^{1/2}} \quad \text{and} \quad E_n \sim \frac{\gamma x}{\epsilon_0^{1/2} \rho^{1/2} Q}. \qquad (2.7)$$

In the region where these estimates hold, the surface convection current is $I_s \sim 2\pi\sigma v r_s \sim \epsilon_0^{1/2}\gamma^{3/2} x^{3/2}/\rho Q$, which increases with x at the expense of the conduction current; see Figure 2.1. The normal electric field required at the liquid side of the surface to make for the variation of I_s is

$E_n^l \sim I_s/Kr_s x$, from Equation 2.5, and the associated axial field in the liquid is $E_x^l \sim E_n^l x/r_s \sim \epsilon_0^{1/2}\gamma^{5/2}x^{5/2}/\rho^2 KQ^3$, from Poisson's law $\nabla \cdot \boldsymbol{E}^l = 0$. This field increases with x and becomes of the order of the outer field E_T for $x \sim \rho^{2/3}K^{1/3}Q/\epsilon_0^{1/3}\gamma^{2/3} = R_h(Q/Q_0)^{1/3}$, where $r_s \sim \epsilon_0^{1/6}\rho^{1/6}Q^{1/2}/\gamma^{1/6}K^{1/6} = R_h/(Q/Q_0)^{1/6}$. These conditions define the current transfer region where the surface of the liquid ceases to be equipotential, the outer field E_T enters the liquid, and the bulk conduction current becomes surface convection current. It is also in this region where the electric shear stress begins to play a role in the evolution of the jet; see Ref. [31] for details.

The asymptotic evolution of the jet downstream of the current transfer region is determined by the balance of inertia and electric shear stress ($\rho r_s^2 v^2/x \sim \sigma E_x r_s$ in the momentum equation in 2.1 integrated over the jet cross section), and the condition that most of the current is due to convection, $2\pi\sigma v r_s \approx I$. Together with the conditions of mass conservation, $\pi v r_s^2 \approx Q$, and $E_x \sim (\gamma/\epsilon_0 x)^{1/2}$, these balances give

$$r_s \sim \frac{\epsilon_0^{1/8}\rho^{1/4}Q^{3/4}}{\gamma^{1/8}I^{1/4}x^{1/8}} \quad \text{and} \quad \sigma \sim \frac{\epsilon_0^{1/8}\rho^{1/4}I^{3/4}}{\gamma^{1/8}Q^{1/4}x^{1/8}}. \tag{2.8}$$

The electric field induced at the conical meniscus by the charge of the quasi-equipotential region of the jet where Equation 2.6 applies is of the order of the Taylor field, while the field induced by the far jet where Equation 2.7 applies is small compared to the Taylor field. The effect of the charge in the first of these regions may stabilize the conical meniscus, and the disappearance of this region when Q decreases to values of order ϵQ_0, together with the decrease of the field due to the charge of the spray drops, may unleash the instability of the conical meniscus as a hydrostatic configuration and account for the minimum experimental flow rate.

In this scenario, stabilization of the meniscus requires space charge, either in the jet or in the drops. The density of this space charge increases with the flow rate, and, if the charge extends to regions large compared to the cone-to-jet transition region or the current transfer region considered here, then its presence must be compensated by increasing the applied voltage, in order for the field around the meniscus to be able to form a cone. This may explain why naturally occurring conical tips, for example, in drops undergoing Coulomb explosions, often emit flow rates of the order of the experimental minimum, of $O(\epsilon Q_0)$ [32, 33]. Emission starts in this case when the electric field around the drop is enough to create a conical tip, but the charge of the emitted droplets tends to decrease this field and thus limits the emitted flow rate.

2.2.3 Very Viscous Liquids

Some of the order-of-magnitude estimations discussed above need be revised when viscous effects dominate the motion of the liquid. In the far jet, the balance of inertia and electric forces must be replaced by a balance of axial viscous and electric forces, which reads $\mu r_s^2 v/x^2 \sim \sigma E_x r_s$ in the momentum equation integrated across the jet. When this is done, the results Eq. 2.7 change to

$$r_s \sim \frac{\epsilon_0^{1/4}\mu Q}{\gamma^{1/4}I^{1/2}x^{3/4}} \quad \text{and} \quad \sigma \sim \frac{\epsilon_0^{1/4}\mu I^{1/2}}{\gamma^{1/4}x^{3/4}}. \tag{2.9}$$

The electric potential of the field induced by the line of charge equivalent to this jet is $\varphi_1 = O(\Gamma/\epsilon_0) = O(\mu^2 Q/\epsilon_0^{1/2}\gamma^{1/2}R^{3/2})$ up to logarithms (here, as before, $\Gamma = 2\pi\sigma r_s$), and the perturbation of the angle of the conical meniscus due to this potential is $\delta(R) = O(\mu^2 Q/\gamma R^2)$.

The pressure and viscous stresses of the viscosity-dominated flow in the cone-to-jet transition region, of characteristic size R_h' say, are of order $\mu v_h'/R_h'$ with $v_h' \sim Q/R_h'^2$. The order-of-magnitude balances $\mu v_h'/R_h' \sim \gamma/R_h' \sim \epsilon_0 E_h'^2$ give $R_h' = (\mu Q/\gamma)^{1/2}$ and $E_h' = \gamma^{3/4}/\epsilon_0^{1/2}\mu^{1/4}Q^{1/4}$. The condition $t_r \sim t_e$, with $t_r = R_h'/v_h'$, for charge relaxation effects to be important in the transition region, is satisfied for $Q \sim \epsilon^2 Q_0'$ with $Q_0' = \epsilon_0^2\gamma^3/\mu^3 K^2$. Charge relaxation effects are small in the high flow rate regime $Q \gg \epsilon^2 Q_0'$, in which the radius of the jet decreases exponentially with axial distance x until the outer field enters the liquid and conduction current becomes surface convection current in a region where $x = O(R_h')$ (up to logarithms) and $r_s \sim \epsilon_0^{1/4}\mu^{1/8}Q^{3/8}/\gamma^{1/8}K^{1/4}$. The current carried by the jet is still $I \sim (\gamma KQ)^{1/2}$. (See Ref. [31] for details.)

2.3 APPLICATIONS AND EXTENSIONS

2.3.1 Multiplexing

A shortcoming of the electrospray technique for practically any application but mass spectrometry [34] is due to the small value of the minimum flow rate, especially for liquids of high electrical conductivity. Since small drops with a narrow size distribution are generated only when the electrospray works near its minimum flow rate, the throughput of high-quality spray that can be obtained from a single source is very small. In principle, the limitation may be overcome by using many sources in parallel, and this multiplexing approach has been studied by several researchers [35, 36–38, 39]. However, difficulties arise when the number of sources and the packing density are high, as is required to achieve reasonable spray densities and fluxes. All the sources of a multiplexed system should work in the same conditions to ensure monodispersity, which requires that the injected flow rate be evenly distributed among the sources and that the electrical conditions be identical for all the sources. The first condition may be satisfied by letting the flow rate reaching each source be determined by a large head loss in the line connecting the source to a common main, and designing the manifold so as to have the same head loss in all the lines. The fabrication may be simplified by drilling the lines in a dielectric material, without metallic tubes, and letting the conducting liquid filling the lines act as an electrode [35]. A potential advantage of these liquid needles is that they intensify the electric field around their tips and may lead to the formation of emitting sources at lower voltages than when the lines are drilled in a metal. Systems of hundreds of sources with compactness higher than 250 sources per square centimeter have been microfabricated on silicon wafers [36]. The second condition, the electric independence of the elementary sources despite the close packing of their charged menisci and jets, can be satisfied using an extractor electrode, which is a perforated metallic plate mounted at a distance from the spray sources that is comparable to the distance between sources. The extractor serves the dual purpose of reducing the electric interference between neighboring sources and shielding them from the cloud of charged drops behind this electrode.

The space charge due to this cloud increases with the conductivity and flow rate of the liquid and with the compactness of the multiplexed source, and conditions may be reached when small satellite drops generated during the breakup of the jets reverse their path and impact on the extractor, causing flooding and bridging the electrodes. Path reversal can be avoided by applying an additional voltage between the extractor and a collector electrode farther downstream, which pushes the charged drops away from the extractor. This, however, increases the complexity of the system and entails some energy consumption, as the electric current carried by the whole cloud may be significant.

The effect of the space charge of a single source has been investigated by Fernández de la Mora [40]. Deng and Gomez [38] carried out numerical simulations of a single source plume behind the extractor, using a Lagrangian model that accounts for the electrostatic forces of all the drops and their images in the extractor and collector electrodes. The computed distributions of primary and satellite drops are in good agreement with experimental visualizations, but the numerical burden has precluded extension of these computations to multiplexed sources to date. Realizing this complexity, Deng and Gomez [38] proposed a simplified model in which each plume is idealized as a uniform line of charge and the trajectory of a single satellite drop is computed in the field induced by all the lines. With some additional simplifications, this model yields reasonable estimates of the minimum voltage difference between extractor and collector required to prevent path reversal for given values of the total current of the spray, the area it covers, the velocity with which the drops cross the extractor, and the distance between extractor and collector. A limitation of the model is that the assumption of uniform lines of charge amounts to assuming constant velocity of the drops. This may be reasonable for the main drops but not for the satellites whose paths are about to reverse. As a consequence, the model leaves out the contribution of the satellite drops to the electric field.

2.3.2 Coaxial Jet Electrosprays

Loscertales *et al.* [41] proposed a spinneret design in which two immiscible liquids, at least one of which is an electrical conductor, are supplied at constant flow rates through two coaxial metallic tubes charged to a high voltage relative to another electrode. The liquids form a compound meniscus at the end of the tubes, which is elongated by the electric stresses acting at its surfaces. In certain ranges of voltages, flow rates, and physical properties of the liquids, a coaxial jet issues from the tip of the meniscus and is continuously stretched by the electric shear due to the action of the electric field on the electric charge that the field itself accumulates at the surfaces of the jet. As for a simple electrospray, the diameter of the jet may become very small compared to the diameters of the tubes before the jet undergoes a varicose instability that breaks it into compound drops or a bending instability that further stretches it. The bending instability, whereby the jet begins to spiral violently at some distance from the spinneret, is an essential mechanism in electrospinning and has been much studied for single liquid jets, as it determines the final diameter of the fiber that forms when the outer liquid (often a polymer solution or melt), or both liquids, solidify; see Reneker and Yarin [42] and references therein.

The techniques of coaxial jet electrospray and electrospinning are ideally suited to produce nearly monodisperse capsules and fibers of different materials with sizes in the range from tens of nanometers to tens of micrometers and controlled coating thicknesses. Such micro-/nanocapsules and fibers are of interest for many applications, including targeted drug delivery, encapsulation of food additives, deposition of catalysts and fabrication of catalyst-bearing fibers, optical waveguides and nanocables, and fluidic devices, among others.

The operation of a coaxial jet electrospray depends on many parameters, and the domain of the parameter space where a coaxial cone-jet of a given couple of liquids can be realized is unknown in general. However, the notions of driving surface and driving liquid introduced by López-Herrera *et al.* [43] help classifying the functioning modes of these devices. The main idea is that the electric charge and the electric shear stress are often concentrated at one of the two interfaces, which is the outer interface if the outer liquid is a reasonably good conductor, and the inner interface if this liquid acts as a dielectric, in the sense that its electrical relaxation time is large compared to the residence time in the current transfer region. In the first case, the electric force acts on the outer interface and must be transmitted to the inner liquid by viscous stresses in the outer liquid in order to elongate the inner meniscus into a jet. In the second case, the electric force acts on the inner interface, but the outer liquid must still be dragged by viscous forces to form a continuous layer around the inner liquid.

In their experiments with different couples of liquids, Loscertales *et al.* [41] and López-Herrera *et al.* [43] found that the electric current increases as the square root of the flow rate of the driving liquid, which is the liquid through which conduction brings charge to the driving surface. The results for the size of the compound drops are more complex and depend on the properties and flow rates of both liquids; see Ref. [43] for details. Some numerical results showing qualitative agreement with the experiments are available for the case of two very viscous liquids in a simple geometrical configuration [44]. A recent review of the wide range of applications of the coaxial jet electrospray and electrospinning technique can be found in Barrero and Loscertales [45].

2.3.3 Electrodispersion in Dielectric Liquid Baths

Barrero *et al.* [13] showed that a cone-jet can be formed in a bath of a dielectric liquid immiscible with the conducting liquid to be atomized. This finding opens a way to generating emulsions with controllable distributions of drop sizes, which have potential applications to the synthesis of nanoparticles, the encapsulation of antibacterial agents, and the manufacturing of drug-laden particles for targeted delivery.

From their measurements of the electric-current/flow-rate characteristic for glycerine, ethylene glycol, and water electrosprayed in baths of hexane and heptane, Barrero *et al.* [13] concluded that the $I \propto Q^{1/2}$ law is also satisfied in a liquid bath. Marín [46] extended these results to a wide range of liquid conductivities, confirming the square root law for moderate flow rates, though with a proportionality constant different from that of Barrero et al, and showed that the electric current falls slightly below the square root law at high flow rates, for which the radius of the jet is not much smaller than the radius of the tube. Both authors find that the minimum flow rate at which a cone-jet can be established in a liquid bath is somewhat smaller than in air. The viscosity of the liquid of the bath is small compared to the viscosity of the conducting liquid in all these experiments, so that the presence of the bath has little effect on the current transfer region of the jet and on the current/flow-rate characteristic, though it can still affect the evolution of the far jet and the drift of the spray drops. Barrero *et al.* [13] clearly demonstrated the effect of the space charge due to the drops on the angle of a meniscus of water in heptane, which is smaller than the Taylor angle, in line with the theoretical results of Fernández de la Mora [40]. The effect of this space charge is dominant in the experiment of Larriba and Fernández de la Mora [47] with an ionic liquid (EMI-BF_4) in a bath of heptane, in which the electric current is almost independent of the flow rate and increases nearly quadratically with the voltage.

In a later work, Riboux *et al.* [48] determined the region of the voltage/flow-rate plane where a jet of glycerine in hexane undergoes asymmetric (whipping) oscillations. They found that the amplitude and frequency of the oscillations are reduced by the presence of the bath, which opposes a force to the lateral displacement of the jet, but that the bath increases the density of surface charge of the jet, and thus the whipping instability develops earlier than for a jet of glycerine in air. These authors also found that, for the relatively high flow rates used in their experiments, for which the radius of the jet is not much smaller than the radius of the tube, the current does not follow a square root law but increases more slowly, as the power 1/5 of the flow rate, and also increases as the power 4/3 of the voltage applied to the metallic tube through which the glycerine is injected.

Numerical computations for a jet issuing from an electrified tube [49] show that the electric current approaches a square root law for small values of the flow rate, but the current levels to a nearly uniform plateau for values of the flow rate similar or somewhat larger than in the experiments of Riboux et al, and tends to increase linearly with the flow rate for even larger values of this variable. Contrary to small flow rate solutions, which resemble the cone-jet regime discussed in Section 2.2, high-flow-rate solutions exist in a wide range of applied voltages. The computed current increases linearly with the voltage for small and moderate values of this variable and somewhat faster than linearly for very high voltages. These high flow rate results can be rationalized as follows.

The radius of the jet increases with the flow rate and decreases when the applied voltage increases, due to the increase in the electric shear stress that stretches the jet. The stretching is moderate in the conditions of the computations. If it is neglected altogether, so that $r_s \sim a$ where a is the radius of the tube, and if the electric field acting on the jet is assumed to be that of a needle, which is inversely proportional to the axial distance ($E_x \sim V/x$), then the density of surface charge beyond a certain charge relaxation region at the beginning of the jet is proportional to the applied voltage; $\sigma \approx \epsilon_0 E_n \sim \epsilon_0 V/a$. This result can be obtained from the relation $E_{x_{jet}} \approx [\ln(r_s/x) + O(1)]\, d(r_s E_r)/dx$ between the axial and radial components of the electric field induced at the surface of the jet by its electric charge (idealized as a line of charge) in the region where the induced axial field is of the order of the field of the needle and partially balances it. The conduction and convection contributions to the current can then be estimated as $I_b \sim Ka^2V/x$ and $I_s \sim \epsilon_0 VQ/a^2$. The condition $I_s \sim I_b$ gives $x \sim Ka^4/\epsilon_0 Q$ for axial extent of the current transfer region, and $I \sim \epsilon_0 VQ/a^2$ for the electric current carried by the jet.

The computed length of the current transfer region, defined as the distance to the tube at which $I_s = I_b$, first increases with the flow rate, extending into the region where the electric field decreases as the inverse of x and the estimations done in the preceding paragraph are applicable, and then decreases. The convection current rises rapidly at the beginning of the jet when the flow rate is large, and thus the surface charge needs not screen a long stretch of the jet from the applied field. However, the conduction current remaining in the jet decreases slowly with streamwise distance in the absence of noticeable stretching, and its contribution to the total current is sensitive to the conditions of the jet far downstream. The length of the current transfer region increases with the conductivity of the liquid and decreases when the applied voltage increases, at least for large values of this variable and moderate flow rates. A more detailed account of the high-flow-rate regime of these electrified jets will be given elsewhere [50].

In an attempt to account for the effect of a moving liquid bath, Gundabala *et al.* [51] set up an electrospray in a microfluidic device where a stream of very viscous conducting oil is surrounded by a coflow of dielectric oil. They found that the current depends weakly on the flow rate and increases linearly with the applied voltage. The fact that this current is not due to conduction in the liquid but to convection of the surface charge by the electric-shear-induced surface velocity was proved by an ingenious combination of experimental measurements and analysis. Qualitative analysis [52] suggests that, in the conditions of this experiment, in which the viscous force of the liquid bath on the current transfer region of the jet plays an important role, a straight stationary jet cannot exist at distances from the injection tube large compared to its diameter. The jet must either break up or undergo whipping, in agreement with the experimental results.

2.4 CONCLUSIONS

The literature on the cone-jet mode of an electrospray has been surveyed. Order-of-magnitude estimates have been presented, which suggest that the flow around the tip of the electrified meniscus of an electrospray has a two-region structure for values of the flow rate large compared to the experimental minimum. This structure comprises a cone-to-jet transition region where the hydrodynamic stresses on the liquid surface first becomes of the order of the electric and surface tension stresses, and a current transfer region spanning a stretch of the jet where convection of the surface charge takes over conduction and the total current of the spray is fixed. The two regions nearly merge, and charge relaxation effects come into play, when the flow rate is decreased to values of the order of the minimum estimated by Fernández de la Mora and Loscertales [10]. A few applications and extensions of the basic technique have been discussed. The reader is referred to the recent reviews by Fernández de la Mora [1] and Barrero and Loscertales [45] for comprehensive accounts of the current research and developments in this area.

REFERENCES

[1] Fernández de la Mora J. The fluid dynamics of Taylor cones. Annu Rev Fluid Mech 2007;39:217–243.

[2] Cloupeau M, Prunet-Foch B. Electrostatic spraying of liquid. Main functioning modes. J Electrostatics 1990;23:165–184.

[3] Cloupeau M, Prunet-Foch B. Electrohydrodynamic spraying functioning modes: a critical review. J Aerosol Sci 1994;25:1021–1036.

[4] Jaworek A, Krupa A. Classification of the modes of EHD spraying. J Aerosol Sci 1999;30:873–893.

[5] Cloupeau M, Prunet-Foch B. Electrostatic spraying of liquid in cone-jet mode. J Electrostatics 1989;22:135–159.

[6] Zeleny J. On the conditions of instability of electrified drops, with applications to the electrical discharge from liquid points. Proc Camb Philol Soc 1915;18:1–6.

[7] Zeleny J. The role of surface instability in electrical discharges from drops of alcohol and water in air at atmospheric pressure. J Franklin Inst 1935;219:659–675.

[8] Taylor GI. Disintegration of water drops in an electric field. Proc R Soc Lond A 1964;280:383–397.

[9] Smith DPH. The electrohydrodynamic atomization of liquids. IEEE Trans Ind Appl 1986;IA22:527–535.

[10] Fernández de la Mora J, Loscertales IG. The current emitted by highly conducting Taylor cones. J Fluid Mech 1994;260:155–184.

[11] Chen D-R, Pui DYH. Experimental investigation of scaling laws for electrospraying: dielectric constant effects. Aerosol Sci Technol 1997;27:367–380.

[12] Gañán-Calvo AM, Dávila J, Barrero A. Current and drop size in the electrospraying of liquids. Scaling laws. J Aerosol Sci 1997;28:249–275.

[13] Barrero A, López-Herrera JM, Boucard A, Loscertales IG, Márquez M. Steady cone-jet electrosprays in liquid insulator baths. J Colloid Interface Sci 2004;272:104–108.

[14] Higuera FJ. Breakup of a supported drop of a viscous conducting liquid in a uniform electric field. Phys Rev E 2008;77:016314.

[15] Rosell-Llompart J, Fernández de la Mora J. Generation of monodisperse droplets 0.3 to 4 μm in diameter from electrified cone-jets of highly conducting and viscous liquids. J Aerosol Sci 1994;25:1093–1119.

[16] Higuera FJ. Numerical computation of the domain of operation of an electrospray of a very viscous liquid. J Fluid Mech 2010;648:35–52.

[17] Cherney LT. Structure of Taylor cone-jets: limit of low flow rates. J Fluid Mech 1999;378:167–196.

[18] Guerrero I, Bocanegra R, Higuera FJ, Fernández de la Mora J. Ion evaporation from Taylor cones of propylene carbonate mixed with ionic liquids. J Fluid Mech 2007;591:437–459.

[19] Melcher JR, Taylor GI. Electrohydrodynamics: a review of the role of interfacial shear stress. Annu Rev Fluid Mech 1969;1:111–146.

[20] Saville DA. Electrohydrodynamics: the Taylor-Melcher leaky dielectric model. Annu Rev Fluid Mech 1997;29:27–64.

[21] Batchelor GK. *An Introduction to Fluid Dynamics*. Cambridge: Cambridge University Press; 1967. p 132.

[22] Landau LD, Lifshitz EM. *Electrodynamics of Continuous Media*. Oxford: Pergamon Press; 1960.

[23] Pantano C, Gañán-Calvo AM, Barrero A. Zeroth order, electrohydrostatic solution for electrospraying in cone-jet mode. J Aerosol Sci 1994;25:1065–1077.

[24] Miksis MJ. Shape of a drop in an electric field. Phys Fluids 1981;24:1967–1972.

[25] Wohlhuter FK, Basaran OA. Shapes and stability of pendant and sessile drops in an electric field. J Fluid Mech 1992;235:481–510.

[26] Ashley H, Landahl M. *Aerodynamics of Wings and Bodies*. Chapter 6. Reading (MA): Addison-Wesley; 1965.

[27] Hinch EJ. *Perturbation Methods*. Cambridge: Cambridge University Press; 1991.

[28] Barrero A, Gañán-Calvo AM, Dávila J, Palacios A, Gómez-González E. The role of the electrical conductivity and viscosity on the motions inside Taylor cones. J Electrostatics 1999;47:13–26.

[29] Fernández de la Mora J, Navascués J, Fernández F, Rosell-Llompart J. Generation of submicron monodisperse aerosols in electrosprays. J Aerosol Sci 1990;21 Suppl 1: S673–S676.

[30] Higuera FJ, Barrero A. Electrosprays of very polar liquids at low flow rates. Phys Fluids 2005;17:018104.

[31] Higuera FJ. Flow rate and electric current emitted by a Taylor cone. J Fluid Mech 2003;484:303–327.

[32] Fernández de la Mora J. On the outcome of the Coulombic fission of a charged isolated drop. J Colloid Interface Sci 1996;178:209–218.

[33] Grimm RL, Beauchamp JL. Dynamics of field-induced droplet ionization: time-resolved studies of distortion, jetting and progeny formation from charged and neutral methanol droplets exposed to strong electric fields. J Phys Chem B 2005;109:8244–8250.

[34] Fenn JB, Mann M, Meng CK, Wong SF, Whitehouse CM. Electrospray ionization for mass spectrometry of large biomolecules. Science 1989;246:64–71.

[35] Bocanegra R, Galán D, Márquez M, Loscertales IG, Barrero A. Multiple electrosprays emitted from an array of holes. J Aerosol Sci 2006;36:1387–1399.

[36] Deng W, Klemic JF, Li X, Reed M, Gomez A. Increase of electrospray throughput using multiplexed microfabricated sources for the scalable generation of monodisperse droplets. J Aerosol Sci 2006;37:696–714.

[37] Deng W, Klemic JF, Li X, Reed M, Gomez A. Liquid fuel combustor miniaturization via microfabrication. Proc Combust Inst 2007;31:2239–2246.

[38] Deng W, Gomez A. Influence of space charge on the scale-up of multiplexed electrosprays. J Aerosol Sci 2007;38:1062–1078.

[39] Velásquez-García LF, Akinwande AI, Martínez-Sánchez M. A planar array of microfabricated electrospray emitters for thruster applications. J Microelectromech Syst 2006;15:1272–1280.

[40] Fernández de la Mora J. The effect of charge emission from electrified liquid cones. J Fluid Mech 1992;243: 561–573.

[41] Loscertales IG, Barrero A, Guerrero I, Cortijo R, Márquez M, Gañán-Calvo AM. Micro/Nano encapsulation via electrified coaxial liquid jets. Science 2002;295:1695–1698.

[42] Reneker DH, Yarin AL. Electrospinning jets and polymer nanofibers. Polymer 2008;49:2387–2425.

[43] López-Herrera JM, Barrero A, López A, Loscertales IG, Márquez M. Coaxial jets generated from electrified Taylor cones. Scaling laws. J Aerosol Sci 2003;34:535–552.

[44] Higuera FJ. Stationary coaxial electrified jet of a dielectric liquid surrounded by a conductive liquid. Phys Fluids 2007;19:012102.

[45] Barrero A, Loscertales IG. Micro- and nanoparticles via capillary flow. Annu Rev Fluid Mech 2007;39:89–106.

[46] Marín AG. Generación y dinámica de chorros electrificados. Aplicación a la síntesis de emulsiones [PhD thesis]. Spain: University of Seville; 2008.

[47] Larriba C, Fernández de la Mora J. Electrospraying insulating liquids via charged nanodrops injection from the Taylor cone of an ionic liquid. Phys Fluids 2010;22:072002.

[48] Riboux G, Marín AG, Loscertales IG, Barrero A. Whipping instability characterization of an electrified visco-capillary jet. J Fluid Mech 2011;671:226–253.

[49] Higuera FJ. Electric current of an electrified jet issuing from a long metallic tube. J Fluid Mech 2011;675:596–606.

[50] Ibáñez SE, Doctor Thesis, Universidad Politécnica de Madrid 2015.

[51] Gundabala VR, Vilanova N, Fernández-Nieves A. Current-voltage characteristic of electrospray processes in microfluidics. Phys Rev Lett 2010;105:154503.

[52] Higuera FJ. Electrodispersion of a liquid of finite electrical conductivity in an immiscible dielectric liquid. Phys Fluids 2010;22:112107.

3

FLUID FLOWS FOR ENGINEERING COMPLEX MATERIALS

IGNACIO G. LOSCERTALES
Escuela Técnica Superior de Ingenieros Industriales, Universidad de Málaga, Málaga, Spain

3.1 INTRODUCTION

Production of droplets and particles of micrometer or even nanometer size with a narrow size distribution are of interest for engineering soft materials with many applications in science and technology. In particular, compound particles, such that each particle is made of a small amount of a certain substance surrounded by another one, are of particular importance for encapsulation of food additives, targeted drug delivery, and special material processing, among other technological fields. In pharmaceutics, for instance, synthesizing a new drug is only the beginning of the design process of a new medicine. Indeed, an efficient medical treatment requires the appropriate transport of the drug to the right tissues and organs, and the appropriate control of its delivery rate. Clearly, drug transport and delivery imposes rather severe constraints upon both the size and the structure of the capsule. For example, the size of the capsules cannot be larger than a couple of microns if they have to move across the intestine wall or if they have to be inhaled up to the alveoli, whereas only those with a given size below some 300 nm will be taken up by the liver, the spleen, the bone marrow, the heart, and so on [1]. Finally, the amount of drug in the capsule and its delivery profile determines the capsule structure. In biotechnology, encapsulation of both cells and genetic material are of vital importance. As in the previous case, the process requires a high degree of control of the capsule size and structure, but in this case the sizes sought are of tens of microns. Although the substance to be encapsulated may be either solid or liquid, the encapsulating agent is usually processed in liquid form (i.e., polymer solution, sol–gel). A key issue in synthesizing these complex materials is the way employed to form the micrometer-/nanometer-sized capsules from the bulk encapsulating liquid, with controllable and adjustable coating thicknesses. Also, other types of core–shell structures, such as nanofibers, coaxial nanofibers and nanotubes, are of great interest for applications in material science.

Fluids, Colloids and Soft Materials: An Introduction to Soft Matter Physics, First Edition. Edited by Alberto Fernandez Nieves and Antonio Manuel Puertas.

Top-down methods to produce micro- and nanoparticles require dividing a macroscopic (i.e., millimetric) piece of matter, generally a liquid, into tiny offsprings of micro- or nanometric size. Surface tension strongly opposes the huge increase in area inherent to this dividing process. Thus, to produce such small particles, energy must be supplied to the system to create interface. This energy raise is usually the result of mechanical work done by an external force field, that is, hydrodynamic forces, electric forces, etc. Two kinds of approaches can be distinguished depending on how the energy is supplied.

In one approach, very inhomogeneous and random extensional and shear flows are employed to generate the forces that inject the required energy to create additional interface between two immiscible liquids. Bulk mechanical emulsification belongs to this category. As a consequence of the broad range of frequencies into which the energy is put, the offspring droplets generally present a very broad size distribution. Nevertheless, a rather good degree of monodispersity might sometimes be achieved for a particular combination of the emulsification parameters (shear rate, rotation speeds, temperature, etc.) and a given combination of substances, as it is well known in emulsion polymerization and synthesis of monodisperse colloids (see, for instance, [2] and references there in). However, such a condition might not exist if one of the substances is changed, if a new one is added, or if a different droplet size is desired. The same occurs if capsules instead of droplets have to be formed. Furthermore, in many instances, the formation of the structure depends on chemical interactions, usually preventing the process from being applicable to a broad range of different substances.

In the other approach, which has the advantage of being based on purely physical mechanisms, the flow is designed such that the forces stretch, steadily and smoothly, the fluid interface without breaking it until at least one of its radii of curvature reaches a well-defined micro- or nanoscopic dimension d; at this point, the spontaneous breakup of the stretched interface by capillary instabilities yields rather monodisperse particles with size of the order of d. This type of flows is known as capillary flows due to the paramount role played by surface tension. For example, the formation and control of single and coaxial jets with diameters in the micrometer/nanometer range and its eventual varicose breakup leads to particles without structure (single jets) or compound droplets (coaxial jet) with the outer liquid encapsulating the inner one. In addition, if solidification of the liquid occurs before the breaking of the jet, one obtains fibers (single jet) or coaxial nanofibers or nanotubes (coaxial jet). The mean size of the particles obtained with these methods ranges from hundreds of micrometers to several nanometers, although the nanometric range is only reached when electric fields are employed. The particles obtained using this approach are, in general, quite monodisperse, and, in the case of capsules, this is true for both the capsule size and the shell thickness. The features provided by this approach make it particularly attractive for engineering unique materials with many technological applications.

Capillary flows capable of stretching out one or more interfaces up to the micro- or submicron dimension have been the subject of considerable research, both experimental and theoretical, in the last few years. Although the Reynolds number of these capillary flows, defined as $Re = (Vd)/\nu$, with ν the fluid kinematic viscosity, V a characteristic velocity, and d a characteristic channel diameter, is of some tens and smaller, the numerical simulation of some of them is quite complex due to (i) the disparity of length scales that can vary more than three orders of magnitude, (ii) the existence of a free surface that must be consistently determined from the solution of the problem, and (iii) the region where the interface breaks is time dependent in spite of the steady-state character of the flow upstream of the breaking zone.

In this chapter, we review the outcome of different capillary flows for stretching fluid interfaces in a smooth and steady manner down to micrometric dimensions and below. Section 3.2 contains a brief description of single-phase particles (without inner structure) that may be obtained through these types of capillary flows. Applications of these flows to produce micro- and nanoparticles with well-defined core–shell or even more complex structures are presented in Section 3.3. Finally, Section 3.4 summarizes the main advantages and drawbacks of these flows regarding the synthesis of nanometric materials.

3.2 SINGLE FLUID MICRO- OR NANOPARTICLES

Before describing the characteristics of the fluid particles that may be obtained with capillary flows, it seems convenient to very briefly define such flows. In general, there are two ways for stretching fluid interfaces down to micrometric or submicrometric dimensions. In the first one, forcing liquids through openings in solid walls with characteristic dimension d brings the curvature of the interface down to that size; for example, forcing a fluid through a pipe or through a membrane with characteristic diameter or pore size d. For practical purposes, however, such small apertures are prone to clogging for sizes below a few microns. The second way, however, uses suitable force fields instead of walls to bring the curvature of the interface down to d, where d is much smaller than any boundary dimension. These forces are, generally, surface tension and fluid dynamic forces such as pressure, inertial, and viscous forces, although electric or magnetic forces can also be used when the fluid is able to react to such fields.

It is worth noting that the fluid to be dispersed may be either mono- or multicomponent, that is, it could be a gas, a liquid, a suspension, an emulsion, or a solution. Therefore,

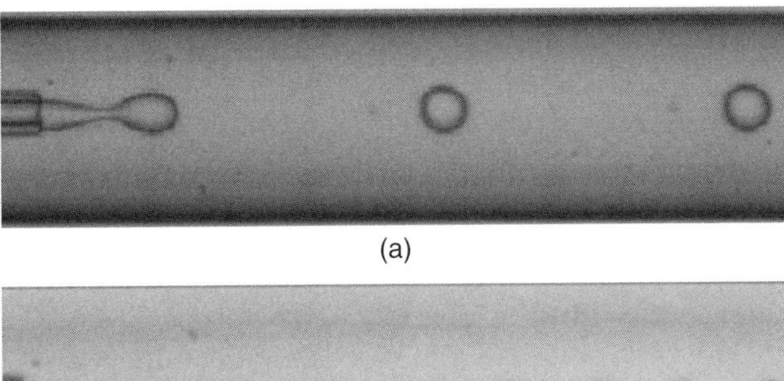

Figure 3.1 Co-flow. (a) Dripping mode. (b) Jetting mode $\alpha = 0$.

single- or multicomponent entities can be obtained using these flows [3].

3.2.1 Flows Through Micron-Sized Apertures

A simple example of these flows consists in the injection of a fluid of density ρ and viscosity μ at a velocity v through a hole of diameter d in a host fluid of density ρ_o and viscosity μ_o, immiscible with the injected one, which moves at a velocity v_o at an angle α with the other fluid velocity. Usually, the hole is just the tip of a micrometric needle immersed in the host fluid, as in the cases shown in Figure 3.1 and extensively discussed in Chapter 1, although other geometrical configurations, such as the pores of a porous membrane, might also be used.

This configuration, known as *co-flow*, exhibits a very rich diversity of flows that may be generated depending on the range of α values, the Weber, $We = \rho v^2 d/\gamma$, and capillary, $Ca = \mu v/\gamma$ numbers, and the Reynolds number of the host fluid flow, $Re_o = \rho_o v_o d/\mu_o$. Globally, the resulting flow is usually classified as *dripping* (Fig. 3.1a) or *jetting* (Fig. 3.1b). The formation of jets and drops (or bubbles) at the exit of the tube and the transition between these two modes have been the subject of numerous investigations [4–6].

Droplets generated in the dripping mode, as those shown in Figure 3.2, are in general much more monodisperse than those generated in the jetting mode. In particular, Umbanhowar *et al.* [8] have recently reported a method to produce highly monodisperse emulsions (standard deviation less than 3%), which consists in detaching droplets from a capillary tip by the action of a co-flowing stream ($\alpha = 0$). The host fluid drags the meniscus formed at the needle exit and causes its detachment to generate a drop with diameter of order d. The co-flowing method has been also exploited to generate highly monodisperse micron size droplets of nematic liquid crystal to form 2D and 3D arrays for electro-optical applications [7].

The fact that droplet pinch off occurs at distances from the needle exit of the order of d severely narrows the breakup wavelength range. Indeed, the needle diameter d acts as a wave filter, efficiently killing those wavelengths slightly away from a dominant one, which is of the order of d. This filtering effect is responsible for the extremely high degree of monodispersity of the detached droplets. In the jetting mode, however, the pinch off occurs at distances from the needle exit that are much larger than d allowing the broadening of the breakup wavelength range. Nonetheless, relatively monodisperse droplets are still obtained from its capillary breakup since the perturbation growth rate versus the perturbation wavelength usually exhibits a rather sharp maximum.

To decrease the possible clogging of the nozzle when producing micron-sized droplets, the *co-flow* is generally

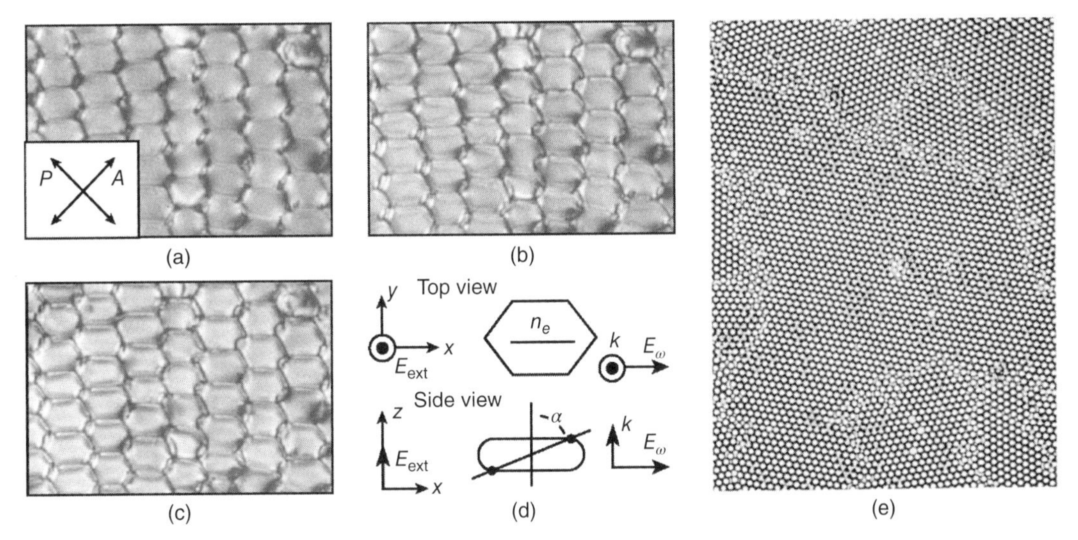

Figure 3.2 Monodisperse droplets by *co-flow* in dripping mode. Photomicrographs of a hexagonal-close-packed monolayer of liquid crystal droplets of roughly 10 μm under different electric fields. (a) 0 V/μm, (b) 0.03 V/μm, (c) 0.15 V/μm, (d) experimental sketch. Source: From Ref. [7]. A monolayer of 10 μm droplets of silicone oil in water is shown in (e). Source: From Ref. [8].

employed in the jetting regime, such that the jet emanating from the nozzle thins downstream due to the drag exerted by the outer fluid. De Castro *et al.* [9] and Marín *et al.* [10] recently generalized the drop diameter d scaling law for the jetting regime to cases where the jet either thinners or thickens; for a jet that thinners, one has

$$\frac{d}{D} = \frac{1}{D}\left(\frac{3\pi}{2k^*}\right)^{1/3}\left(\frac{4Q_i}{\pi v_o}\right)^{1/2}.$$

For a given external flow (i.e., given v_o), the minimum droplet diameter that may be attained would be given by the minimum inner liquid flow rate Q_i. This condition, together with the requirement that the outer fluid capillary number be of order one, $Ca_o \sim 1$, for jetting to be achieved, leads to $d_{\min} \sim \frac{\mu^2}{\gamma\rho}$, which may be submicronic for low viscosity liquids such as water. An example of a submicronic emulsion formed by the breakup of a jet that thins in *co-flow* is shown in Figure 3.3.

Other extensional flows within micron-sized channels have been also used to break single droplets into two daughter droplets whose size may be precisely controlled [11], as shown in Figure 3.4. In this implementation, an emulsion of droplets continuously flows across a *T*-junction; the pressure-driven extensional flow stretches each drop, which subsequently splits into two smaller droplets flowing through each branch of the T-junction.

Finally, it is worth mentioning the recent use of *co-flow* to create more complex geometries of liquids. Pairam and Fernandez-Nieves [12] reported the use of an injecting tip, static with respect to a rotating reservoir, which allowed the formation of toroidal drops, as shown in Figure 3.5. This novel type of drops may be stabilized by an appropriate choice of surfactants or by cleverly tuning the rheological properties of the continuous phase. Even though the smallest characteristic length is in the millimeter range, these unusual drops allow studying fundamental problems where topology plays a role.

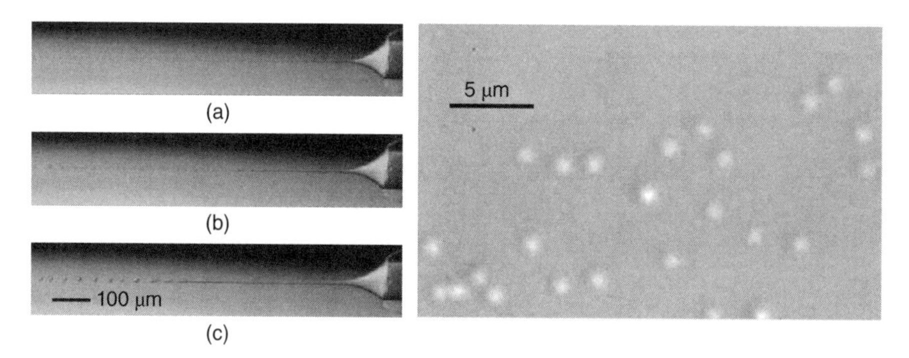

Figure 3.3 Monodisperse droplets by *co-flow* in jetting mode. Glycerin (outer fluid) and silicone oil (inner fluid) at different flow rate ratios. (a) $\frac{Q_i}{D^2U_o} = 3.42 \times 10^{-4}$, (b) $\frac{Q_i}{D^2U_o} = 9.12 \times 10^{-4}$, (c) $\frac{Q_i}{D^2U_o} = 2.62 \times 10^{-3}$. Source: From Ref. [10].

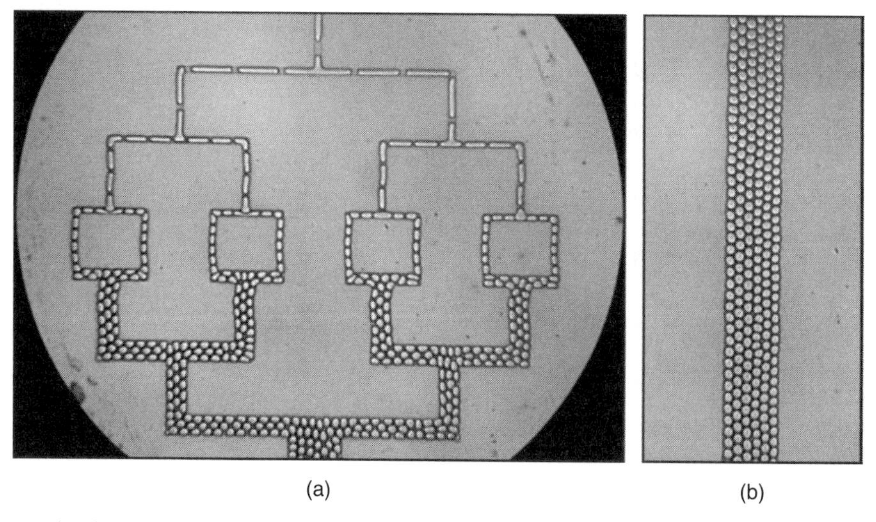

Figure 3.4 Monodisperse droplets by flow across a T-junction (flow from top to bottom). (a) Drops are formed by subsequent breakup at the T's. (b) The drops are flushed away from the device [11].

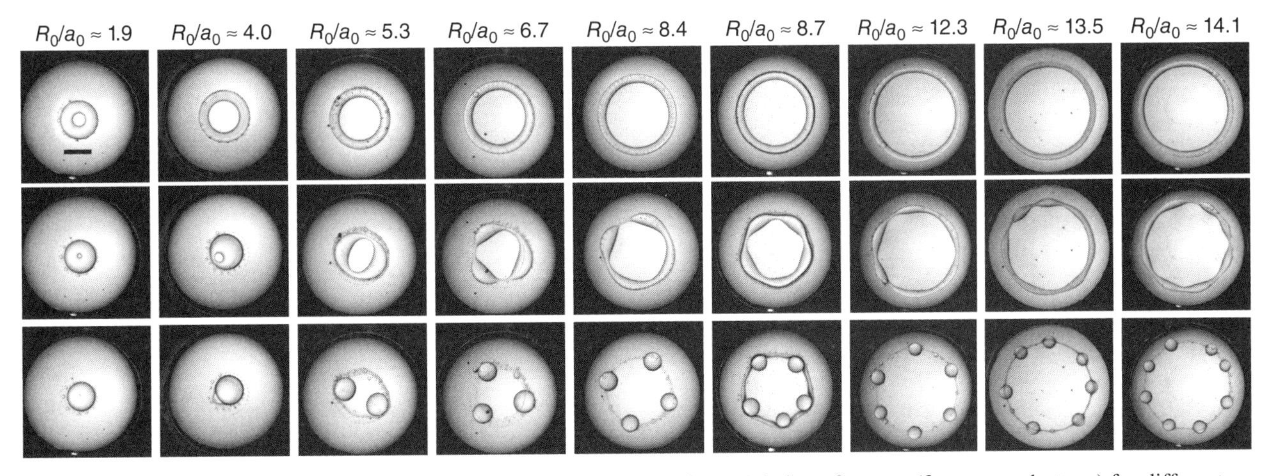

Figure 3.5 Toroidal drops by co-flow in jetting mode. Snapshots for the time evolution of a torus (from top to bottom) for different aspect ratios. The scale bar in the top left corresponds to 5 mm. In these experiments, the injected liquid is glycerin and the liquid in the continuous phase is silicone oil ($\mu_i/\mu_o = 1/30$). Source: From Ref. [12].

3.2.2 Microflows Driven by Hydrodynamic Focusing

In this case, the interface between two fluids is stretched out by a highly accelerated converging motion of one of them that sucks the other one toward the converging point.

One of the earliest implementation of this type of flows is the so-called *selective withdrawal*, which dates back to the end of the 1940s [13, 14]. The technique has been largely employed in the field of geophysical flows before being applied to coating microparticles [15]. In its simpler version, shown in Figure 3.6a, the tip of a tube of diameter D is located at a height H above an interface separating two immiscible liquids. By applying steady-state suction throughout the tube, the resulting converging flow of the lighter fluid (the focusing liquid in this case) sets the other liquid into motion. For sufficiently small values of the suction forces, only the lighter liquid is withdrawn throughout the tube: the hydrodynamic forces cannot overcome the capillary forces and the deformed interface eventually comes to rest. An increase in the suction forces leads to a transition where the heavier liquid is also withdrawn in the form of a steady-state thin jet of diameter d much smaller than D, co-flowing with the focusing liquid (the lighter one). The capillary breakup of this jet (not shown) gives rise to a stream of droplets with a mean diameter of the order of that of the jet diameter. Note that, for a given couple of liquids and a given tube diameter, there are two controlling parameters: the pressure drop along the tube, ΔP, that controls the flow rate Q through the tube and the distance between the tube exit and the interface, H. For a given value of H, increasing ΔP results in a thicker jet while for a given ΔP, increasing H results in a thinner one. In terms of dimensionless parameters we have that the dimensionless jet diameter d/D depends on the Reynolds number Re_o and H/D, where $Re_o = \rho_o Q/(\mu_o D)$ is based on the tube diameter D. We note that for a given H/D, no steady-state jet is formed unless the Reynolds number becomes larger than a critical value (i.e., critical flow rate Q).

Another implementation of this type of flows is the so-called *flow focusing* procedure [16], where a pressure drop ΔP across a thin plate orifice of diameter D provokes the converging motion of the focusing fluid. A second fluid is injected at a rate q through a tube of diameter D_t, whose end is located at a distance H in front of the orifice;

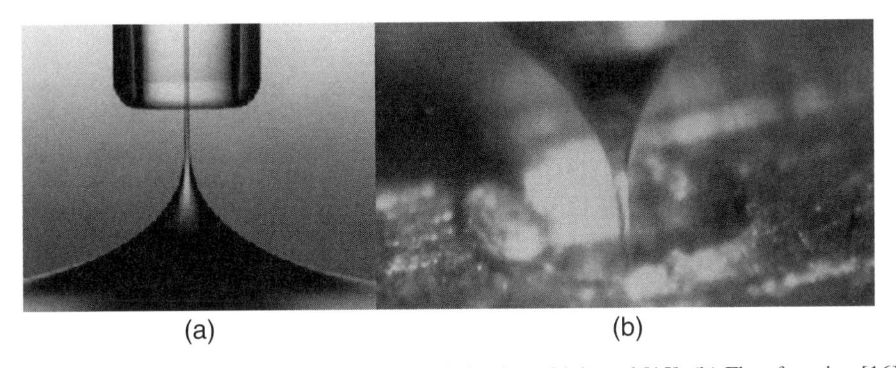

Figure 3.6 Hydrodynamic focusing flows. (a) Selective withdrawal [15]. (b) Flow focusing [16].

$D_t \sim H \sim D$. For a given value of H and for an appropriate range of values of both q and ΔP, the interface at the tube exit develops a cusp-like shape with a very thin steady-state jet of diameter d or smaller is issued; see Figure 3.6b. This jet eventually breaks up into a stream of droplets with a mean diameter that can be much smaller than the orifice diameter $d \ll D$. Knight *et al.* [17] reported a 2D flow focusing instead of the axisymmetric shown in Figure 3.6b.

As in *selective withdrawal*, for a given ΔP there is a minimum flow rate $q_{\min}$ below which no steady jet can be formed. For this $q_{\min}$ the jet diameter reaches its minimum value $d_{\min}$, which is approximately given by the balance between the pressure drop ΔP and the surface tension pressure $\gamma/d_{\min}$; this yields $d_{\min} \sim \gamma/\Delta P$. For the case in which the focusing fluid is a gas of density ρ_g, the maximum value of ΔP is of the order of $\rho_g c^2$, where c is the characteristic sound velocity of the gas; thus, for typical values of γ, one obtains $d_{\min} \sim 1$ μm.

Note that for the flows considered in this section, the diameters of the tubes and the orifice are usually much larger than the jet diameter; therefore, the solid walls do not filter out many breakup wavelengths, and, consequently, the droplets thus formed present a broader size distribution than those obtained in *co-flow*. Furthermore, there is a slight difference between the two implementations described in the present section that might influence the size distributions of the resulting droplets, which has to do with the stability of the flow. Since in flow focusing the discharge of the focusing flow into a quiescent fluid occurs just after crossing the orifice, there is a shear layer that forms, which is unstable, and develops into turbulence. This might affect the breakup of the thin jet in cases where it occurs at distances larger than D downstream of the orifice.

Related successful experiments relevant for the production of emulsions [18] and microfoams [19] are shown in Figure 3.7. These use a focusing flow geometry integrated into a planar microchannel device. Results in Anna *et al.* [18] show that the drop size as a function of flow rates and flow rate ratios of the two liquids (the focusing and the focused ones) includes a regime where the drop size is comparable to the orifice width (dripping) and a second one (jetting) where the drop size is dictated by the diameter of a thin *focused* thread. As a result, drops much smaller than the orifice diameter can result.

3.2.3 Micro- and Nanoflows Driven by Electric Forces

The interaction of an intense electric field with the interface between a conducting liquid and a dielectric medium is known to exist since Gilbert [20] reported the formation of a conical meniscus when an electrified piece of amber was brought close enough to a water drop. The deformation of the interface is caused by the force that the electric field exerts on the net surface charge induced by the field itself. Experiments show that the interface reach a motionless shape if the field strength is below a critical value, while for stronger fields the interface becomes conical, issuing mass and charge from the cone tip in the form of a thin jet of diameter d. In the latter case, the cone-jet structure becomes steady if the mass and charge it emits is supplied to the meniscus at the same rate. Taylor [21] explained the conical shape of the meniscus as a balance between electrostatic and surface tension stresses; thereafter, the conical meniscus is referred to as the Taylor cone. The thin jet eventually breaks up into a stream of highly charged droplets of diameter of the order of d. This electrohydrodynamic steady-state process is called steady cone-jet electrospray [22], or just electrospray; see Figure 3.8a [23]. The same electrohydrodynamic flow can be also used to obtain very thin fibers if the jet solidifies before breaking into charged droplets; see Figure 3.8b [24]. This process, known as electrospinning, occurs when the working fluid is a complex fluid, such as the mixture of polymers of high molecular weight dissolved in volatile solvent [25–27].

Electrospray The electrospray has been applied for bio-analysis [28], fine coatings [29], synthesis of powders

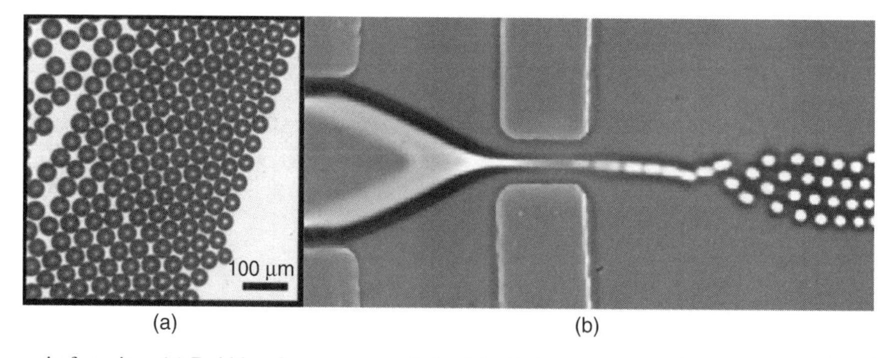

Figure 3.7 Hydrodynamic focusing. (a) Bubbles. Source: From Ref. [19]. (b) Emulsion in a 2D microchannel device. The width of the slit is 43 μm. Source: From Ref. [18].

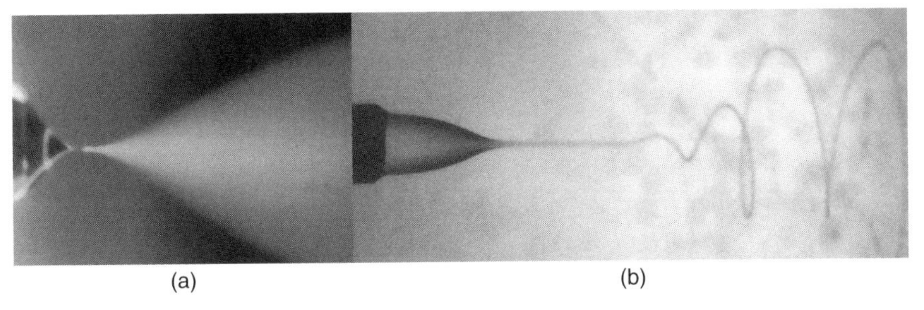

Figure 3.8 Microflows driven by electric forces. (a) Electrospray. Source: From Ref. [23]. (b) Electrospinning. Source: From Ref. [24].

[30], and electrical propulsion [31], among other technological applications. Recently, electrosprays in cone-jet mode have also been performed inside dielectric liquid baths for the production of fine emulsions [32]. Very recently, Gundabala *et al.* [33] have reported, for the first time, the implementation of liquid–liquid steady-state electrosprays in microfluidic devices. According to these authors, a crucial aspect of their embodiment is a flowing electrode capable of dragging away all the charge and mass emitted by the electrospray. Since it is a liquid–liquid system, aside of the electric effects also encountered in regular electrosprays, the microfluidic embodiment allows the simultaneous incorporation of hydrodynamics effects through the controlled motion of the continuous phase. This has become a very interesting approach to synthesize in one-step micro- and nanoemulsions.

In electrosprays, a flow rate q of a liquid with density ρ, viscosity μ, permittivity $\varepsilon\varepsilon_o$ (where ε_o is the vacuum permittivity), surface tension γ, and electrical conductivity K is fed through a capillary tube of diameter D_t connected to an electric potential V with respect to a grounded electrode. For a given liquid and a given tube-grounded electrode geometry, an electrospray forms at the exit of the tube for a certain range of values of both q and V.

From the point of view of materials production, the most outstanding feature of electrosprays compared to the previous microflows is the particle size it "naturally" yields. The minimum flow rate at which electrosprays can operate in steady-state conditions is approximately given by $q_{\min} \approx \gamma\varepsilon\varepsilon_o/(\rho K)$ (see Chapter 2); this minimum flow rate results in a minimum jet diameter

$$d_{\min} \approx \left(\frac{\gamma\varepsilon_o^2}{\rho K^2}\right)^{1/3} \varepsilon^n f(\varepsilon) \approx \varepsilon^n \left(\frac{\gamma\varepsilon_o^2}{\rho K^2}\right)^{1/3},$$

where $n \sim O(1)$. For liquids with electric conductivities of the order of 10^{-3} S/m, $d_{\min}$ is of the order of 1 μm, but if K takes values of the order of 1 S/m, then $d_{\min}$ becomes of the order of 10 nm.

On the other hand, since these highly charged droplets form upon the microjet breakup, the size distribution of these droplets is not as tight and monodisperse as droplets formed from an orifice (dripping), although there are regions within the (q, V) operating regime in which rather standard deviations from the droplet's mean size are smaller than 10%, as shown in the works by [34–36], among others.

Electrospinning Even though the flow is fundamentally the same as that of electrosprays, the rheological properties of these dissolutions, sometimes enhanced by the solvent evaporation from the jet, slow down and even prevent the growth of varicose instabilities; as it is well known, large values of liquid viscosity delay the jet breakup by reducing the growth rate of axisymmetric perturbations, so longer jets may be obtained. However, nonaxisymmetric perturbation modes can grow due to the net charge carried by the jet. Indeed, if a small portion of the charged jet moves slightly off axis, the charge distributed along the rest of the jet will push that portion further away from the axis, thus leading to a lateral instability known as whipping or bending instability (see [27] and references therein). A picture capturing the development of the whipping instability in a jet of glycerin in a heptane bath is shown in Figure 3.8b.

The chaotic, three-dimensional movement of the jet under this instability gives rise to a very large tensile stresses, which leads to a dramatic jet thinning. The solidification process, and thus the production of micro- and nanofibers, is enhanced by the spectacular increase in the rate of solvent evaporation due to the thinning process.

For the production of nanofibers, this technique is very competitive with other existing ones such as phase separation, self-assembly, and template synthesis, among others. As a result, electrospinning and whipping is the subject of an intense research activity nowadays [37].

Other Electric Regimes The parametrical space of electrohydrodynamics allows a large number of regimes aside of the steady cone-jet mode. Among those, it is worth mentioning the so-called electrodripping periodic regime [38] in which the meniscus subjected to an intense electric fields emits charge and mass in a periodic manner. This regime has attracted attention for applications in which the electrospray regime appears to deliver droplets much smaller than desired; this is the case of printing applications. In contrast to electrosprays, in which the droplet production frequency is of the order of the inverse of the electrical relaxation time,

$$f_{\mathrm{es}} \sim \frac{1}{t_e} = \frac{K}{\beta \varepsilon_o},$$

the electrodripping regime typically produces droplets at frequencies of the order of the inverse of the capillary time,

$$f_{\mathrm{ed}} \sim \frac{1}{t_c} = \left(\frac{\gamma}{\rho D^3}\right)^{1/2},$$

where the last equality holds if the viscosity is unimportant in the liquid motion inside the meniscus $\left(\frac{\mu}{\sqrt{\rho \gamma D}} \ll 1\right)$. Typically, $f_{\mathrm{ed}} \ll f_{\mathrm{es}}$ except in situations where D and/or K are sufficiently small. Also, the values of liquid flow rates for electrodripping are usually much larger than those used in electrosprays, which leads to the generation of much larger droplets. Other modes of electrodripping are actually under investigation [36, 39–41].

3.3 STEADY-STATE COMPLEX CAPILLARY FLOWS FOR PARTICLES WITH COMPLEX STRUCTURE

Micro- and nanoparticles with a well-defined core–shell structure (i.e., nanocapsules, hollow spheres, nanotubes, and coaxial nanofibers) may be also obtained from flows obeying the same basic principles than those reviewed in the previous sections; in this case, however, *two interfaces* separating *three fluid media* are required to produce the core–shell structure. The motion of the liquids must result in a coaxial stretching of the two interfaces. The breakup of the interfaces in this coaxial configuration may lead to core–shell particles. However, core–shell fibers can also be obtained from a coaxial jet if it solidifies before breaking.

These types of coaxial flows are governed by twice the number of parameters than those described in Section 3.2 and may exhibit even richer operating regimes. However, when seeking for steady-state situations, the possible regimes are not so large. The basic idea is to combine more complex capillary flows to obtain particles with more complex inner structure.

3.3.1 Hydrodynamic Focusing

Gañán-Calvo [16] reported the formation of a steady-state coaxial jet of water-based black ink (inner) and silicone oil (outer) by axisymmetric flow focusing; in his experiments, the two liquids are injected through coaxial capillary needles and they are focused by air. The inner jet (ink) is formed when the drag exerted on the water–silicone interface by the motion of the silicone overcomes the water–silicone interfacial tension. This author reported that the jet breakup resulted in the formation of compound spheres consisting of a silicone cover encapsulating a black ink core.

Much more recently, Utada *et al.* [42] introduced a fluidic device based on hydrodynamic focusing that generates double emulsions in a single step in the micrometric range. In their device, sketched in Figure 3.9, three immiscible fluids

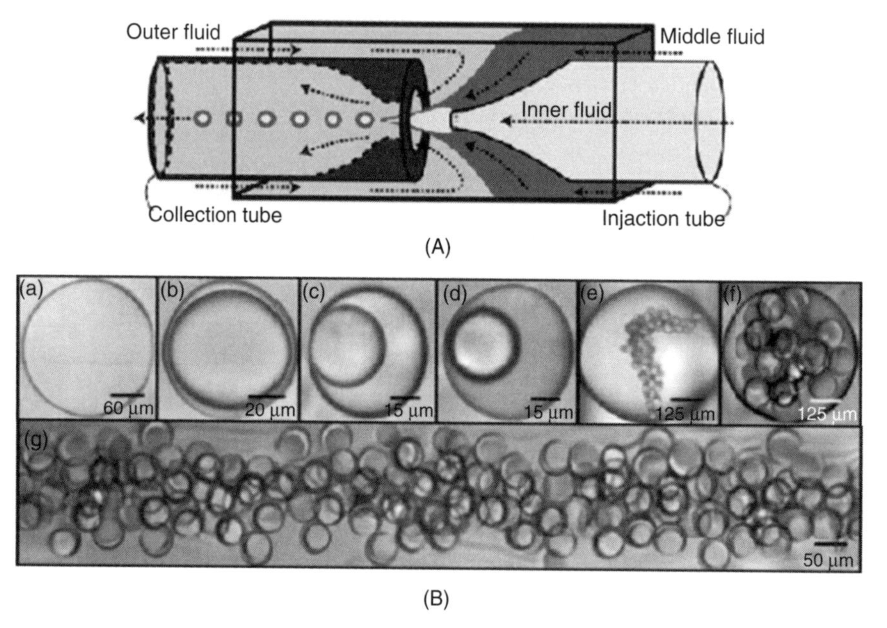

Figure 3.9 A: Microcapillary geometry for generation of double emulsions. B: (a–d) Double emulsions containing only one internal droplet. The thickness of the coating fluid on each drop can vary from extremely thin (less than 3 µm) as in (a) to significantly thicker. (e, f) Double emulsions containing many internal drops with different size and number distributions. (g) Double emulsion drops, each containing a single internal droplet, flowing in the collection tube (2) Electrospinning. Source: From Ref. [42].

are forced through a converging exit orifice. The converging flow of the outer fluid stretches out the two interfaces between the fluid media whose breakup by capillary instabilities forms core–shell drops. In steady-state conditions, two operative regimes, dripping and jetting were established. Dripping produces drops close to the entrance of the collection tube, within a single orifice diameter, analogous to a dripping faucet. In contrast, jetting produces a coaxial jet that extends three or more orifice diameters downstream into the collection tube, where it breaks into drops. For given dripping conditions, an increase in the flow rate of the focusing fluid (the outermost) beyond a threshold value causes the interface to abruptly lengthen, defining the transition to the jetting regime.

As explained in Section 3.2, droplets produced by dripping are typically highly monodisperse, while the jetting regime typically results in less monodisperse droplet size distributions. However, these authors find out a very narrow window of operational conditions in which jetting yields a monodispersity similar to that of dripping. The size distribution of the double emulsions is determined by the breakup mechanism, whereas the number of innermost droplets depends on the relative rates of drop formation of the inner and middle fluids. When the rates are comparable, the annulus and core of the coaxial jet break simultaneously, generating a core–shell drop.

This microfluidic system has recently being extended to include three basic building blocks with which much more complex microdroplets [43] may be formed. The building blocks, sketched in Figure 3.10, are a drop maker, a connector, and a liquid extractor; combinations of these enable the scale-up of the device to create higher order multicomponent multiple emulsions with very diverse structures. The number, ratio, and size of droplets, each with distinct contents being

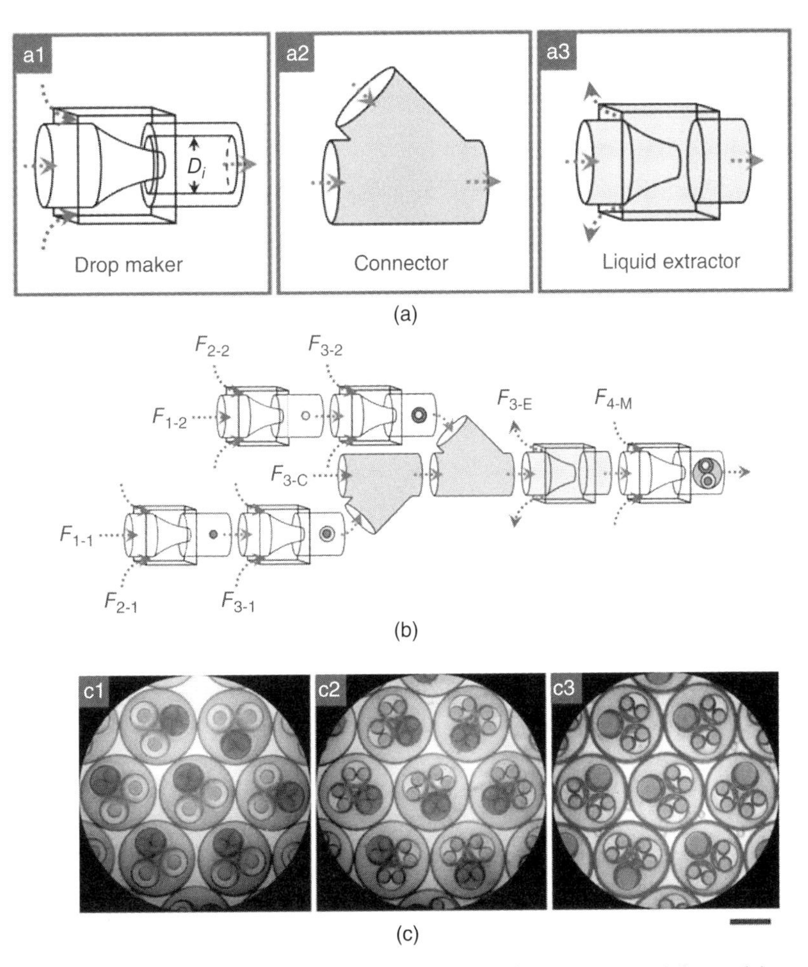

Figure 3.10 Extended microfluidic device for controlled production of sextuple-component triple emulsions. (a) Schematic diagram of functional building blocks: (a1) drop maker for generating droplets, (a2) connector for merging droplets from different drop makers, and (a3) liquid extractor for removing excess continuous phase. (b) Schematic diagram of the microfluidic device for generating sextuple-component triple emulsions. (c) Optical micrographs of monodisperse sextuple-component triple emulsions: (c1 and c2) O/W/O/W triple emulsions containing two different double emulsions and (c3) novel triple emulsions containing one W/O single emulsion and two O/W/O double emulsions. Scale bars are 400 μm. Source: From Ref. [43].

independently co-encapsulated in the same level, can be precisely controlled. More examples may be found in [44] and references therein.

Shum *et al.* [45] has somehow extended the use of microfluidics for controlled fabrication of monodisperse nonspherical particles from the microfabricated droplets. These droplets can be transformed to nonspherical particles through downstream simple, spontaneous processing steps, such as arrested coalescence, asymmetric polymer solidification, polymerization in microfluidic flow, and evaporation-driven clustering. Some examples are shown in Figure 3.11.

A very similar microfluidic device has been successfully used for the rapid fabrication of complex-shaped microfibers (e.g., single hollow, double hollow, and microbelt), with highly uniform structures [46]. Two laminar flows of a photocurable fluid and a liquid template (nonpolymerizing fluid) spontaneously form coaxial jet streams within a third fluid (outermost) that remain unbroken in microfluidic channels through the minimization of the interfacial energy between the fluids. Photopolymerization of the intermediate fluid leads to fiber formation. The formation and faith of the jet streams depends on the spreading coefficient, defined as $S_i = \gamma_{jk} - (\gamma_{ij} + \gamma_{ik})$, where $\gamma_{\xi\phi}$ is the interfacial tension between fluids ξ and ϕ. Thus, if S_L, S_P, and S_C are the spreading coefficients of the innermost liquid, the polymer (intermediate) and the continuous phase (the outermost), stable coaxial jets are obtained when $S_L < 0, S_P > 0$, and $S_c < 0$. Figure 3.12 shows the microfluidic system as well as microfibers of various morphologies.

In this device, the flow of the outermost (inert) fluid continuously drags the fiber out of the device through an exit port into a reservoir downstream.

3.3.2 Electrified Coaxial Jet

Particles with core–shell structure have also been obtained from electrified coaxial jets with diameter in the nanometer range [47]. In this technique, two immiscible liquids are injected at appropriate flow rates through two concentric capillary needles. At least one of the needles is connected to an electrical potential relative to a ground electrode. The needles are immersed in a dielectric host medium that may be a gas, a liquid, or vacuum. For a certain range of values of the electrical potential and flow rates, a compound Taylor cone is formed at the exit of the needles with an outer meniscus surrounding the inner one; see Figure 3.13. A liquid thread is issued from the vertex of each one of the two menisci, giving rise to a compound jet of two co-flowing liquids.

To obtain this compound Taylor cone, at least one of the two liquids must be sufficiently conducting. Similarly to simple electrosprays, the electric field pulls the induced net electric charge located at the interface between the conducting liquid and a dielectric medium, and sets this interface into motion; since this interface drags the bulk fluids, it may be called the *driving* interface. The driving interface may be either the *outermost* or the *innermost* one; the latter happens when the outer liquid is a dielectric. When the driving interface is the outermost, it induces a motion in the outer liquid that drags the liquid–liquid interface. When the drag overcomes the liquid–liquid interfacial tension, a steady-state coaxial jet may be formed. If the driving

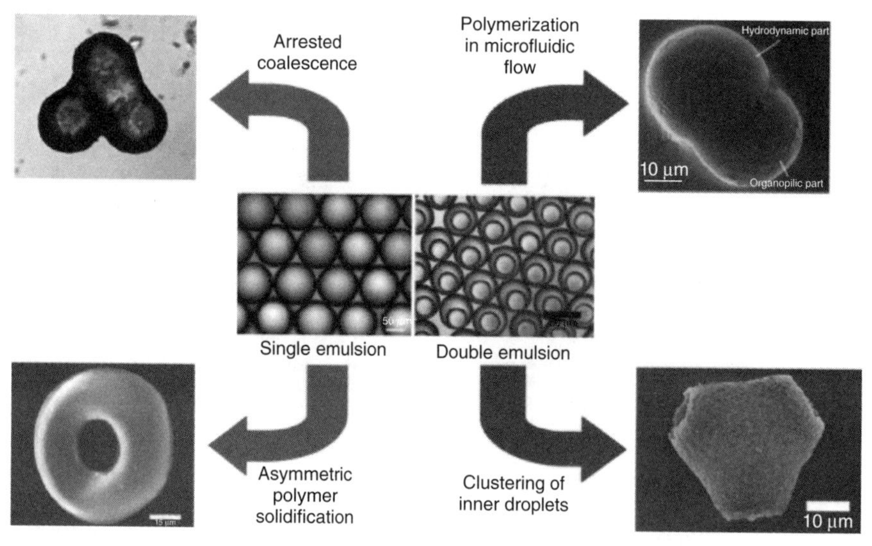

Figure 3.11 Schematic of the different strategies for converting emulsion precursors prepared with microfluidic devices into nonspherical particles (arrested coalescence, asymmetric polymer solidification, polymerization in microfluidic flow, clustering of inner droplets). Source: From Ref. [45].

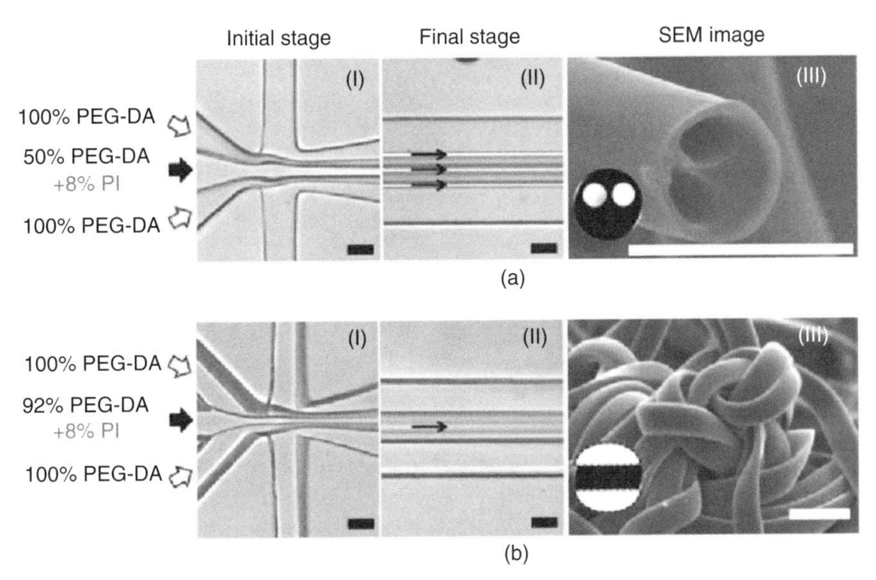

Figure 3.12 Fabrication of complex-shaped microfibers. (a) Hollow microfibers with two core regions. Polymerizing fluid (50% PEG-DA with 8% PI) and liquid template (100% PEG-DA) are introduced. The ratio of the three flow rates is QC:QP:QL ¼ 20:12:2 ml min#1. (b) The generation of thin microbelts. Polymerizing fluid (92% PEG-DA with 8% PI) and liquid template (100% PEG-DA) are introduced. Scale bars represent 50 μm. Source: From Ref. [46].

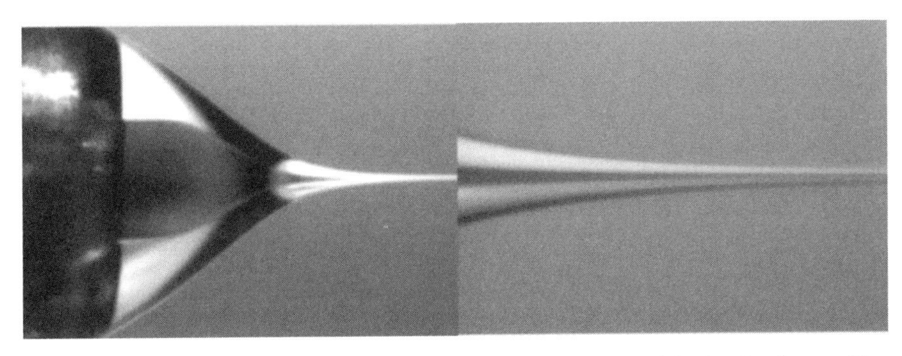

Figure 3.13 Compound Taylor cone and detail of the resulting electrified coaxial microjet. Source: From Ref. [47].

interface is the innermost, its motion is simultaneously diffused to both liquids by viscosity, setting both in motion to form the coaxial jet. Scaling laws showing the effect of the flow rates of both liquids on the current transported by these coaxial jets and on the size of the compound droplets has been recently investigated [48–51].

This technique has been used to generate, after coaxial jet breakup, core–shell micro- and nanocapsules [47, 52, 53] and microemulsions [54]. For example, Figure 3.14a shows 8-μm-size capsules of an aqueous solution coated by a photopolymer [47]; the insert shows details of the core–shell structure. Figure 3.14b shows hollow nanospheres of silicon oxide obtained after solvent extracting the inner liquid [52]; the inset shows the hollow structure. Note that the mean size of the capsules may be submicronic in contrast to the technique described in Section 3.3.1. However, the size distributions are broader than those obtained there; nonetheless, polydispersities of 10% can be obtained within certain ranges of the operating parameters.

Similarly to electrospinning, solidification of the outer liquid leads to hollow micro- and nanofibers [55, 56] while solidification of the two liquids leads to coaxial nanofibers [57–59]. This process has been termed co-electrospinning. Figure 3.15a shows hollow microfibers of silicon oxide, with a mean diameter of 500 nm and a wall thickness of 80 nm [55], while Figure 3.15b shows a coaxial nanofiber of polyethylene oxide (outer) and a mixture of polyethylene oxide and bromophenol (inner), with the outer and inner diameters of 100 and 13 nm, respectively [59].

3.4 SUMMARY

Some top-down methods to produce micro- and nanosized soft matter require dividing a macroscopic piece of liquid into tiny offspring of micro- or nanometric size. In Section 3.2, we reviewed two approaches, based on capillary flows, for stretching a fluid–fluid interface in a smooth and steady

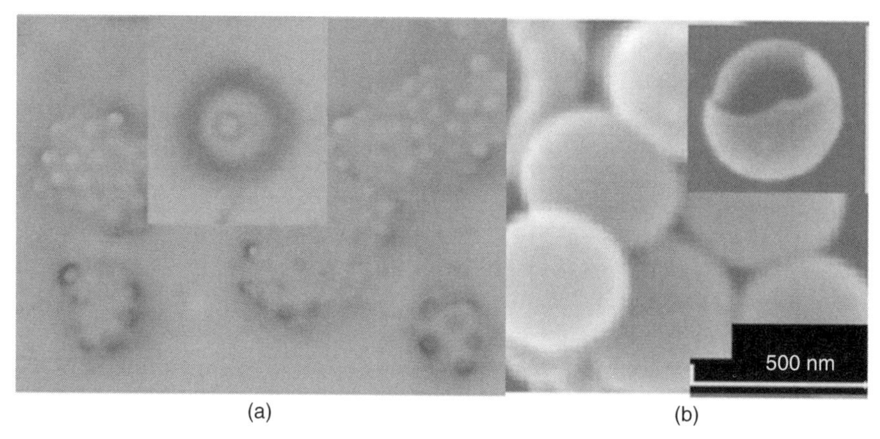

Figure 3.14 Microcapsules formed upon breakup of electrified coaxial jets. (a) 10 μm spheres containing a water core; insight: detail of a capsule. Source: From Ref. [47]. (b) Submicrometric hollow spheres; insight: detail of a broken capsule. Source: From Ref. [52].

Figure 3.15 (a) Hollow microfibers of silicon oxide precursor (sol–gel). Source: From Ref. [55]. (b) Coaxial nanofiber of two different polymers. Source: From Ref. [59].

manner down to a micro- or submicron size range. The breakup of the stretched interface gives rise to the micro- or nanoparticles. One of these approaches consists of forcing fluids through micrometric openings in solid walls, while in the other approach, the interface is stretched by the focusing effect of either hydrodynamic or electrical forces. The first approach may lead to emulsions and suspensions that are highly monodispersed in size (<3%), although their mean size is limited by the size of the opening, which is usually never smaller than a few microns due to possible clogging problems. The second approach avoids this problem, but droplet sizes well below 1 μm can only be achieved when the process is driven by electrical forces; the resultant polydispersity is usually larger in this second approach.

In Section 3.3, we described devices that generate capillary flows, which lead to micro- and nanoparticles with complex inner structure: nanocapsules, hollow spheres, nanotubes and coaxial nanofibers, multicore droplets, and so on. In this case, two interfaces separating three fluid media are required to produce the core–shell structure. The motion of the liquids must result in a coaxial stretching of the two interfaces whose breakup leads to core–shell particles. In particular, we reviewed a microcapillary device, based on hydrodynamic focusing, for producing highly monodisperse double emulsions in the micron size range. This device allows for a precise control of the capsule structure (single or multiple core) and can be expanded through more sophisticated microfluidic devices to form fibers. A different device uses electrical forces to generate coaxial jets with diameters in the micro- and nanometric size range. Micro- and nanocapsules are formed after jet breakup, whereas if the jet solidifies before that happens, coaxial nanofibers or nanotubes are obtained. The particle size dispersion in this case is not as good as that obtained with the use of microcapillary devices.

The reviewed approaches can produce micro- and nanoparticles, with or without inner structure. Generally, they allow for both a precise control of the particle size and a rather narrow size distribution. These characteristics make them quite attractive and competitive with other existing techniques. A noticeable feature of the approaches here described lies in the fact that the core–shell (or more complex) particles may be obtained in just one step; this is a clear advantage over multistep processes such as bulk emulsification techniques. However, the throughput of one of the devices based on the approaches discussed in this review is usually too small for many industrial purposes, restricting their use to analytical applications.

Increasing the production rate requires the operation of parallel devices. In the case of devices based on hydrodynamic focusing, the most promising way to build a multidevice is by using microfabrication techniques. However, some problems such as the control of the wetting properties of the channels should be solved to ensure the same operating condition for each device. In the case of devices based on electrocapillary flows, the main problems when trying to operate in parallel comes from the shielding effect of the space charge created by the highly charged aerosol and from the electric cross talk between neighboring devices. The design of efficient approaches for the operation of parallel devices will probably become a very active area of research in the near future.

ACKNOWLEDGMENTS

I would like to express my gratitude to Professor Barrero, who passed away in 2010, with whom I enjoyed working since 2001 and to whom I dedicate this chapter. I also acknowledge the support of the Spanish Ministry of Economy and Competitiveness under project DPI-2010-20450-C03-03, and of Yflow Inc. Fruitful discussions with Professor Higuera (at UPM, Madrid, Spain) and Fernandez-Nieves (at Georgia Tech, Atlanta, USA) are sincerely appreciated.

REFERENCES

[1] Gaumet M, Vargas A, Gurny R, Delie F. Eur J Pharm Biopharm 2008;69:1–9.
[2] Reese CE, Guerrero CD, Weissman JM, Lee K, Asher SA. J Colloid Interface Sci 2000;232:76–80.
[3] Sanders EH, Kloefkorn R, Bowlin GL, Simpson DG, Wnek GE. Macromolecules 2003;36:3803.
[4] Clanet C, Lasheras JC. J Fluid Mech 1999;383:307.
[5] Basaran OA. AIChE J 2002;48:1842.
[6] Milosevic IN, Longmire EK. Int J Multiphase Flow 2002;28:1853.
[7] Fernández-Nieves A, Link DR, Rudhardt D, Weitz DA. Phys Rev Lett 2004;92:105503.
[8] Umbanhowar PB, Prasad V, Weitz DA. Langmuir 2000;16:347.
[9] de Castro E, Gundabala V, Fernandez-Nieves A, Gordillo JM. New J Phys 2009;11:075021.
[10] Marín AG, Campo-Cortés F, Gordillo JM. Colloids Surf Physicochem Eng Aspects 2009;344:2.
[11] Link DR, Anna SL, Weitz DA, Stone HA. Phys Rev Lett 2004;92:054503.
[12] Pairam E, Fernandez-Nieves A. Phys Rev Lett 2009;102:234501.
[13] Craya A. L'Huille Blanche 1949;4:44.
[14] Debler WR. J Eng Mech Div Proc Am Soc Civil Eng 1959;85:51.
[15] Cohen I, Li H, Hougland JL, Mrksich M, Nagel SR. Science 2001;292:265.
[16] Gañán-Calvo AM. Phys Rev Lett 1998;80:285.
[17] Knight JB, Vishwanath A, Brody JP, Austin RH. Phys Rev Lett 1998;80:3863.
[18] Anna SL, Bontoux N, Stone HA. Appl Phys Lett 2003;82:364.
[19] Gordillo JM, Cheng Z, Gañán-Calvo AM, Márquez M, Weitz DA. Phys Fluids 2004;16:2828.
[20] Gilbert W. De Magnete, Magneticisque Corporibus, et de Magno Magnete Tellure. Peter Short (Ed.), London, 1600.
[21] Taylor GI. Proc R Soc London Ser A 1964;280:383.
[22] Cloupeau M, Prunet-Foch B. J Electrost 1989;22:135.
[23] Pantano C, Gañán-Calvo AM, Barrero A. J Aerosol Sci 1994;25:1065.
[24] Marín AG, Loscertales IG, Barrero A. Am Phys Soc Bull 57th Annual Meeting of the Division of Fluid Dynamics, Seattle, Washington, USA: 2004.
[25] Doshi J, Reneker DR. J Electrost 1995;35:151.
[26] Reneker DR, Yarin AL, Fong H, Koombhongse S. J Appl Phys 2000;87:4531.
[27] Fridikh SV, Yu JH, Brenner MP, Rutledge GC. Phys Rev Lett 2003;90:144502.
[28] Fenn JB, Mann M, Meng CK, Wong SK, Whitehouse C. Science 1989;246:64.
[29] Siefert W. Thin Solid Films 1984;120:267.
[30] Rulison AJ, Flagan RC. J Am Ceram Soc 1994;77:3244.
[31] Martínez-Sánchez, M, Fernández de la Mora, J, Hruby, V, Gamero-Castaño, M, Khayms, V. Proceedings of the 26th International Electric Propulsion Conference. Kitakyushu, Japan: Electric Rocket Propulsion Society; 1999. p 93–100.
[32] Barrero A, López-Herrera JM, Boucard A, Loscertales IG, Márquez M. J Colloids Interface Sci 2004;272:104.
[33] Gundabala VR, Vilanova N, Fernandez-Nieves A. Phys Rev Lett 2010;105:154503.
[34] Gomez, A, Tang, K. In: Semerjian HG, editor. Proceedings of the 5th International Conference on Liquid Atomization and Spray Systems, ICLASS-91, Gaithersburg, MD, USA: NIST Special Publication 813; 1991. p 805–812.
[35] Rosell-Llompart J, Fernández de la Mora J. J Aerosol Sci 1994;25(6):1093–1119.
[36] Vilanova N, Gundabala VR, Fernandez-Nieves A. Appl Phys Lett 2011;99:021910.
[37] Persano L, Camposeo A, Tekmen C, Pisignano D. Macromol Mater Eng 2013;298(5):504–520.
[38] Jaworek A, Krupa A. J Aerosol Sci 1999;30:873.
[39] Park J-U, Rogers JA. Nat Mater 2007;6:782.

[40] J. Choi, Y.-J. Kim, S.U. Son, K.C. An, S. Lee, NSTI-Nanotech 2010, Vol. 2 (2010)

[41] Lee S, Song J, Chung J. J Aerosol Sci 2012;52:89.

[42] Utada AS, Lorenceau E, Link DR, Kaplan PD, Stone HA, Weitz DA. Science 2005;308:537.

[43] Wang W, Xie R, Ju X-J, Luo T, Liu L, Weitz DA, Chu L-Y. Lab Chip 2011;11:1587, 10.1039/c1lc20065h.

[44] Kim S-H, Weitz DA. Angew Chem Int Ed 2011;50:1648.

[45] Shum HC, Abate AR, Lee D, Studart AR, Wang B, Chen C-H, Thiele J, Shah RK, Krummel A, Weitz DA. Macromol Rapid Commun 2010;31:108.

[46] Choi C-H, Yi H, Hwang S, Weitz DA, Lee C-S. Lab Chip 2011;11:1477, 10.1039/c0lc00711k.

[47] Loscertales IG, Barrero A, Guerrero I, Cortijo R, Márquez M, et al. Science 2002;295:1695.

[48] López-Herrera JM, Barrero A, López A, Loscertales IG, Márquez M. J Aerosol Sci 2003;34:535.

[49] Li F, Yin X-Y, Yin X-Z. Phys Fluids 2005;17:077104.

[50] Higuera FJ. Phys Fluids 2007;19:012102.

[51] Li F, Yin X-Y, Yin X-Z. J Fluid Mech 2009;632:199.

[52] Larsen G, Velarde-Ortiz R, Minchow K, Barrero A, Loscertales IG. J Am Chem Soc 2003;125:1154.

[53] Loscertales, IG, Márquez, M, Barrero, A. Materials Research Society Symposium Proceedings; 2004. Vol. 860.

[54] Gómez-Marín A, Loscertales IG, Barrero A. Phys Rev Lett 2007;98:014502.

[55] Loscertales IG, Barrero A, Márquez M, Spretz R, Velarde-Ortiz R, et al. J Am Chem Soc 2004;126:5376.

[56] Li D, Xia Y. Nano Lett 2004;4:933.

[57] Sun Z, Zussman E, Yarin AL, Wendorff JH, Greiner A. Adv Mater 2003;15:1929.

[58] Yu JH, Fridrikh SV, Rutledge GC. Adv Mater 2004; 16:1562.

[59] Díaz, J.E., Galán, D., Barrero, A., Márquez, M., Bedia, J., Rodríguez-Mirasol, J. and Loscertales, I. G. (2005) The European Aerosol Conference, Vol. 153.

SECTION II

COLLOIDS IN EXTERNAL FIELDS

4

FLUCTUATIONS IN PARTICLE SEDIMENTATION

P.N. SEGRÈ
Department of Physics, Oxford College of Emory University, Oxford, GA, USA

4.1 INTRODUCTION

In this chapter, we discuss some of the recent research and current issues concerning the dynamics of spheres sedimenting in Newtonian liquids. In Section 4.2, we briefly examine the theories and experiments for the mean sedimentation rate of suspensions. The functional dependencies of the settling rates on suspension concentration are quite different in Brownian and non-Brownian systems, and this difference can be related to the differing underlying particle structures during settling. In Section 4.3, we discuss in more detail recent developments concerning particle velocity fluctuations that occur during settling in non-Brownian systems. More than 20 years ago, theory and numerical simulation predicted that in dilute suspensions the velocity fluctuations about the mean would increase in an unbounded way with the size of the container, leading to the so-called divergence paradox. This spurred on considerable research in the field. The current understanding is that the predicted divergence does not occur because particles organize themselves into large correlated regions that screen out the long-range hydrodynamic interactions. The origin of the new hydrodynamic correlation length is still intensely debated. We focus in particular on recent efforts that connect it to the presence of small concentration gradients that emerge during settling.

4.2 MEAN SEDIMENTATION RATE

In Stokes flow at zero Reynolds number, an isolated sphere falls steadily through a suspended liquid at a terminal velocity determined by a balance between the weight of

Fluids, Colloids and Soft Materials: An Introduction to Soft Matter Physics, First Edition. Edited by Alberto Fernandez Nieves and Antonio Manuel Puertas.

the particle and the viscous drag on it [1]. For a sphere of radius a and density ρ_s, in a liquid of density ρ_0 and viscosity η, this balancing of the downward gravitational force, $f_g = (4\pi a^3/3)\Delta\rho g$, with an opposing viscous force,

$$f_d = -6\pi\eta a u_s, \qquad (4.1)$$

leads to an expression for the Stokes settling velocity of an isolated sphere [2],

$$u_s = \frac{(2/9)\Delta\rho g a^2}{\eta}, \qquad (4.2)$$

where $\Delta\rho \equiv \rho_s - \rho_0$.

The Stokes settling velocity result is strictly valid in the flow regime where viscous forces dominate over inertial ones. This is commonly referred to as the viscous, or Stokes flow, regime. The Reynolds number Re is the dimensionless ratio of the inertial to viscous force terms in the Navier–Stokes equation [1], $Re \sim \rho_0 u \cdot |\nabla u|/\eta|\nabla^2 u|$. For a particle moving in a liquid at speed u, by approximating $|\nabla u| \sim u/a$ and $|\nabla^2 u| \sim u/a^2$, one obtains a simple relation for the Reynolds number, $Re = \rho u a/\eta$. In this chapter, we will discuss exclusively the Stokes flow regime, where $Re \ll 1$. In regard to typical experiments involving the sedimentation of glass beads, this restricts the particle sizes to $a < 300$ nm in pure water, or $a < 3$ cm for pure glycerol.

One of the most easily accessible quantities in settling experiments is the mean sedimentation velocity u. In the laboratory, u can be measured by observing the falling of the upper sedimentation front, that is, the interface between batch suspension and clear liquid, as illustrated in Figure 4.1. The mean sedimentation rate can also be obtained by tracking individual beads in the falling particle column [3]. If the suspension is infinitely dilute, that is, very low volume fraction or $\phi \to 0$, the mean settling rate is given by the Stokes velocity, Eq. (4.2). At finite concentrations, however, even in very dilute systems, the mean settling rate is measurably slower. The general form of the sedimentation equation is given by Russel *et al.* [1]

$$u(\phi) = u_s H(\phi), \qquad (4.3)$$

where $H(\phi)$, called the hindered settling function, is a monotonically decreasing function of ϕ, with $H(0) = 1$.

The hindered settling function takes on different functional forms depending on the underlying particle structure during sedimentation. The particle structure can vary from random for Brownian systems, to partially ordered for non-Brownian spheres, to highly ordered for particles arranged in a sedimenting crystal lattice. Each of these cases will display a different ϕ dependence of the hindered settling function.

4.2.1 Brownian Sedimentation

A random dispersion occurs when the particles are small enough that thermal Brownian motion dominates over gravity-driven sedimentation. The Péclet number Pe helps to quantify the relative importance of these two terms. It is the ratio of the time it takes for a particle to diffuse its own radius, $\tau_B \propto a^2/D$, to the time it takes to fall its own radius $\tau_u = a/u_s$. The Péclet number then becomes $Pe = \tau_B/\tau_u \sim u_s a/D \sim \Delta\rho g a^4/k_B T$. Systems for which $Pe \ll 1$ are considered Brownian and those with $Pe \gg 1$ are considered non-Brownian. For the typical case of glass (silica) beads in water, the transition value $Pe = 1$ occurs for a bead size of only $a \sim 2$ μm.

A theory for the sedimentation velocity of dilute Brownian suspensions was developed by Batchelor [4]. He considered the case of equal-sized spheres sedimenting at low particle Reynolds numbers $Re \ll 1$. In addition, it was assumed that the particles are randomly distributed in space, the system size is infinite in the direction perpendicular to gravity, and only two-body hydrodynamic interactions are considered. The result is a calculation for the first-order correction to the Stokes velocity u_s:

$$u(\phi) = u_s[1 - (5 + \alpha)\phi], \qquad (4.4)$$

where

$$\alpha \equiv 3\int_2^\infty x[1 - g(x)]dx + \frac{15}{4}\int_2^\infty \frac{g(x)}{x^2}dx. \qquad (4.5)$$

Here $x \equiv r/a$ with r being the interparticle separation, and $g(x)$ the pair correlation function [1].

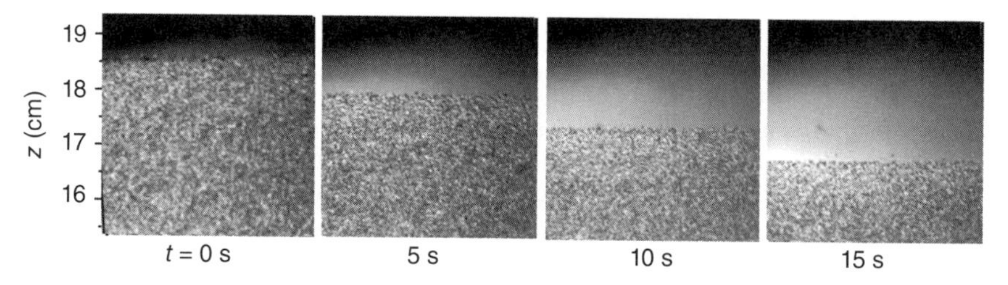

Figure 4.1 Photographs of the top portion of a sedimentation column comprised of $a = 154.5$ μm spheres, at a mean concentration of $\phi = 15\%$, falling in a water/glycerol mixture.

In a dilute Brownian suspension, the particles are randomly positioned so that $g(x)$ is constant for $x > 2$, and vanishes for $0 \leq x \leq 2$. Performing the integrations in Eq. (4.5), Batchelor [4] found

$$\frac{u(\phi)}{u_s} = 1 - 6.55\phi + O(\phi^2), \tag{4.6}$$

where $K = -6.55$ is known as the sedimentation coefficient. Batchelor's result indicates that sedimentation velocities decrease with increasing particle concentration. Numerous sedimentation experiments have been carried out in suspensions of Brownian hard spheres [5, 6]. In very dilute systems, $\phi \leq 10\%$, the departures from Stokes law are seen to be initially linear in ϕ, as predicted in Eq. (4.6), and the linear prefactor K is in good agreement with $K = -6.55$ [7]. The value of K gradually decreases to ≈ -4.7 for particle concentrations above $\phi \approx 20\%$. At higher concentrations, the linear Batchelor form underestimates the sedimentation velocity and second-order terms in ϕ become significant [8]. Hindered settling at moderate to high concentrations has been an active area of research by many authors [6, 8–11]. There are a number of empirical formulas for $H(\phi)$ that have been applied. Perhaps the best known is the Richardson–Zaki equation [12]

$$\frac{u(\phi)}{u_s} = (1 - \phi)^n, \tag{4.7}$$

where $n \approx 5$.

The Batchelor formalism for $u(\phi)$ can also be used for non-hard-sphere-like suspensions. For particles with attractive interactions, the increased close pairing of particles leads to an increased sedimentation rate with respect to that of hard spheres. Similarly, repulsive interactions decrease settling rates. Sangani and Acrivos [13] examined the settling of highly ordered arrangements of hard spheres. Unlike in random dispersions, in crystalline arrangements, the forces and velocities are the same for each particle. For regular periodic arrays, Sangani and Acrivos found solutions of the form

$$\frac{u(\phi)}{u_s} = 1 - \beta\phi^{1/3} + \cdots, \tag{4.8}$$

where $\beta = 1.76$ for a simple cubic lattice. The differences in the functional dependence on ϕ in Equations 4.6 and 4.8 is due to the differences in particle structure. Thies-Weesie *et al.* [7] found a similar form, $u(\phi)/u_s = 1 - 2.8\phi^{1/3}$, in dilute suspensions of highly charged spheres. The strong repulsive interactions force neighboring spheres to maintain as large a separation as possible. This results in a strong peak in the pair correlation function $g(r)$ at $r \sim a\phi^{-1/3}$, the mean interparticle spacing. The strong peak results in a change in the hindered settling function from a linear dependence on ϕ for the random case, to a $\phi^{1/3}$ dependence seen with highly ordered spheres.

4.2.2 Non-Brownian Sedimentation

The settling of collections of spheres large enough that the Péclet number $Pe \gg 1$ is referred to as non-Brownian sedimentation (typically for $a > 5$ μm). In this regime, Brownian motion is negligible. While in Brownian particle sedimentation it can be assumed that the particle configurations are random during settling, in the non-Brownian regime it is not known *a priori* what configurations the particles will exhibit during settling.

It has been known for a long time that the reduced settling functions for dilute Brownian and non-Brownian hard spheres are very different [14–16]. Benes *et al.* [17] conducted a recent exhaustive study of the settling of suspensions over a wide range of Péclet numbers for both Brownian and non-Brownian particles; see Figure 4.2(a). They verified the initial linear Batchelor form for $H(\phi)$ for Brownian particles but found a markedly different form in non-Brownian sedimentation. Overall, the normalized settling rates are significantly reduced as the Péclet number increases. For $Pe \gg 1$, they found the results in the dilute regime, $\phi < 10\%$, fit well to

$$\frac{u(\phi)}{u_s} = 1 - 1.2\phi^{1/3}\,. \tag{4.9}$$

The $\phi^{1/3}$ dependence is not consistent with the particles being randomly distributed in space.

In large Pe number sedimentation, without Brownian motion to keep the particle positions randomized during settling, these results show that some level of ordering does occur. In a second series of experiments, Benes *et al.* [17] used a single sample to explicitly demonstrate that the differences in hindered settling functions are due to differences in particle structure. They measured $u(\phi)/u_s$ as a function of applied shear rate Ω in a non-Brownian sample, shown in Figure 4.2b. The shearing action is perpendicular to the gravity direction and is used to effectively break up and randomize any particle correlations that may be present during settling. The velocities markedly increase when the randomizing shear is present, consistent with the particles becoming more randomly distributed.

An experimental determination of the particle microstructure during settling of large non-Brownian spheres is very difficult to achieve in large three-dimensional samples. Direct imaging has been limited to very dilute suspensions or those confined to quasi two-dimensional Hele-Shaw cells [18]. Numerical simulations have had more success at accurately tracking particle positions [19]. Nguyen and Ladd used Lattice-Boltzmann techniques and found [20, 21] that particle structure factors $S(k)$ show signs of microstructural rearrangements, and the suppression of long-wavelength density fluctuations, as sedimentation proceeds in a finite-sized container.

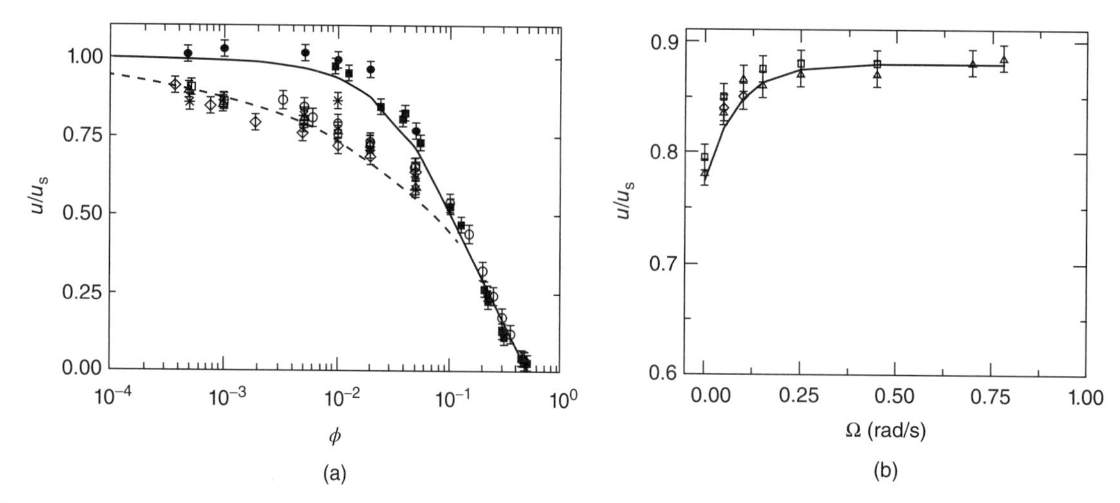

Figure 4.2 (a) Normalized sedimentation velocity as a function of volume fraction for hard spheres with the Péclet number $Pe = 0.048$ (solid squares), 0.68 (solid circles), 21 (stars), 1600 (open triangles), 5400 (open diamonds), 9.8×10^4 (open squares), and 8.5×10^7 (open circles). The solid line shows the Brownian form $u/u_s = (1 - \phi)^{6.55}$, and the dashed line shows the non-Brownian form $u/u_s = 1 - 1.2\phi^{1/3}$. Source: From Figure 1, Ref. [17]. (b) Normalized sedimentation velocity $u(\phi)/u_s$ as a function of cylinder rotation rate Ω. High rotation rates act to shear and randomize the particle positions. The volume fraction is $\phi = 5 \times 10^{-3}$, and $Pe = 5400$. Source: Adapted from Figure 2, Ref. [17].

4.3 VELOCITY FLUCTUATIONS

4.3.1 Introduction

One of the most active areas of research in particle sedimentation in recent years concerns the nature of the individual particle dynamics and fluctuations that occur during settling [22]. Even in the case of sedimentation of collections of monodisperse spheres, which individually possess the same Stokes settling velocity, there can exist large amplitude fluctuations in particle velocities about the mean settling rate. While moving through a liquid, particles interact with each other through direct and long-range hydrodynamic interactions. The constantly changing suspension microstructure gives rise to particle velocity fluctuations, which, in highly concentrated suspensions, can show variances that are larger than the mean settling rate [23].

Caflisch and Luke Divergence Paradox

Much of the recent activity in the field can be traced back to the remarkable prediction by Caflisch and Luke [24] for the *rms* velocity variance Δu $(= \sqrt{\langle (u - u_s)^2 \rangle}\ )$ in a collection of monodisperse spheres falling in a container of linear transverse dimension L. Considering only pairwise hydrodynamic interactions and assuming a random particle distribution in space, they calculated that the fluctuations scale as

$$\frac{\Delta u}{u_s} \propto \sqrt{\phi L/a}\,. \tag{4.10}$$

There are two remarkable features of this prediction. First, the fluctuation amplitudes are not universal for a given volume fraction; they explicitly depend upon the size of the container L. Second, the fluctuations grow with the container size and there is no upper bound to how large they can be. Since it does not seem physically reasonable that velocity fluctuations could, even in very dilute samples, diverge with the container size, the Caflisch and Luke (C+L) model yields what is often referred to as the "divergence paradox."

A physical description of how the divergence paradox arises was put forward by Hinch [25]. Under the assumption that the particles in a suspension are randomly distributed, Poisson statistics dictates that any subregion in the cell containing on average N particles will have a mean variance in particle number of $\Delta N = \sqrt{N}$. The variance in particle number leads to a variance in local mass density, which will drive fluctuations in local velocity. To show this, Hinch considered a typical region of linear size l, containing on average $\langle N\rangle \propto \phi l^3/a^3$ particles. The mean particle number variance is then $\Delta N = \sqrt{\langle N\rangle} \propto \sqrt{\phi l^3/a^3}$. The variance in buoyant weight of this region with respect to the average is then $\Delta F_g \propto \Delta N(\Delta\rho a^3)g \propto \Delta\rho g a^3\sqrt{\phi l^3/a^3}$, indicating that this region of size l either rises (for $\langle N\rangle - \Delta N$) or falls (for $\langle N\rangle\Delta N$) with respect to the mean sedimentation rate. The excess velocity is found by equating the excess buoyant weight ΔF_g with its Stokes drag, $\Delta F_\eta \propto \eta l \Delta u$, giving $\Delta u/u_s \propto \sqrt{\phi l/a}$. Assuming that the largest subregion in the cell obeying Poisson statistics is proportional to the cell size, $l \sim L$, this reasoning reproduces the Caflisch and Luke prediction Equation 4.10.

Figure 4.3 shows a schematic representation of the Hinch picture, sometimes referred to as the "blob' model.

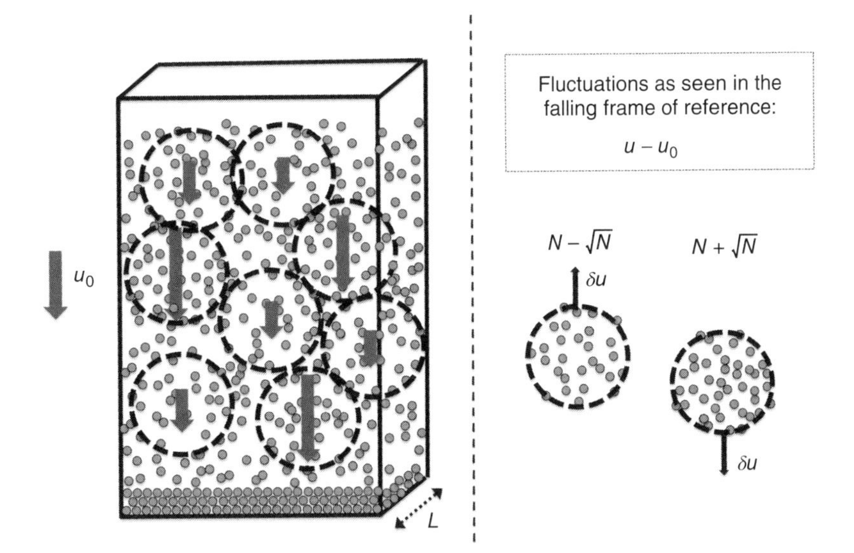

Figure 4.3 Cartoon representation of the Hinch blob model. LHS: Particles in subregions, or blobs, of volume L^3 in the sedimentation cell follow Poisson (random) statistics. RHS: In the blobs, the mean particle number is N and the mean variance $\sqrt{N}$. Relative to the mean fall speed, heavier blobs fall downward, lighter blobs rise upward.

The volume fraction is assumed to be very small so that the mean settling rate is equal to the Stokes velocity u_s. The cartoon portrays a typical snapshot in time and is not meant to describe how the configuration will evolve as time proceeds. The highlighted regions, or blobs, represent the largest regions that follow Poisson random statistics. Their size is related to the smallest cell dimension, in this case the cell thickness L. The velocity vectors shown indicate how regions with an *excess* of particles fall faster than u_s, while those with a *deficit* of particles fall slower than u_s. All the particles inside each blob move together owing to the long-range nature of the hydrodynamic interactions at low Reynolds numbers. In this cartoon, the normalized velocity variance would be of order $\Delta u/u_s \sim 0.5$.

4.3.2 Thin Cells and Quasi Steady-State Sedimentation

At the time of the Caflisch and Luke prediction in 1985, and for several years thereafter, there was an insufficient amount of experimental results against which one could readily test their model. Early evidence in support of the C+L predictions came from numerical simulations using the Lattice-Boltzmann technique and other methods [26–28]. The simulations used periodic boundary conditions in place of impenetrable top and bottom cell walls and started from initial configurations of particles obeying random Poisson statistics. They examined the particle velocities at early times soon after settling had begun. The results showed that the early time velocity variances did grow with the system size in the same way, $\Delta u \propto \sqrt{L}$, as C+L predicted.

Several initial experiments, on the other hand, gave seemingly conflicting results. In 1971, Koglin [29] found that the velocity variances in a dilute system increased with system size in agreement with the C+L divergence model. Later, Nicolai and Guazzelli [3, 30] used single particle tracking methods to extract particle velocities in samples at dilute to moderate concentrations. They examined a range of cell sizes that were somewhat larger (in terms of L/a) than that Koglin used and found that Δu did not display any clear dependence on cell size. Soon after, Segrè *et al.* [31] conducted experiments in cells whose sizes spanned both the Koglin and Guazzelli ranges. They observed a container-size dependence in the smallest cells, and a crossover to a size-independent region in the largest cells. Using a wide range of dilute volume fractions, it was found that the crossover cell size at which the fluctuations began to saturate occurred at $L \sim 60a\phi^{-1/3}$, corresponding to a cell dimension of $L \sim 40$ mean interparticle spacings. The saturation of the fluctuation values with cell size is indicative of a breakdown in the Caflisch and Luke model. The C+L model is based on the assumption that the particle configuration remains random at all times, and the observed deviations suggested that this is not the case for cells larger than $L/a \sim 60\phi^{-1/3}$.

To examine how the particle velocities and fluctuations are organized spatially during settling, Segrè *et al.* [31] used the technique of particle image velocimetry [32]; see Figure 4.4. PIV is capable of simultaneously measuring particle velocities over a wide section of the sample cells so as to produce velocity vector maps. Figure 4.5 shows typical results for two different volume fractions. At extreme dilution, $\phi = 0.01\%$, the velocities seen in Figure 4.5a appear nearly all uniform in amplitude and direction. By subtracting off the mean and enlarging the

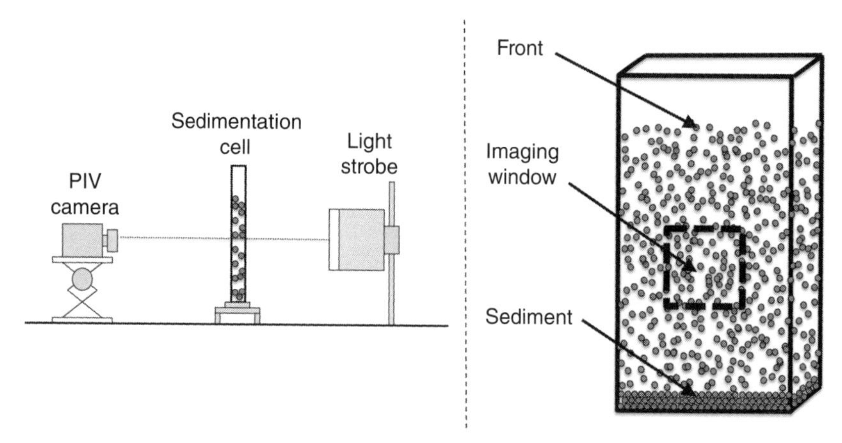

Figure 4.4 Schematic of a PIV setup used to record particle velocities during settling. The PIV camera records two sample images, closely spaced in time δt. A computer detects slight differences in the images, using cross-correlation techniques, to extract local particle displacements δr_i, and velocities $\delta u_i = \delta r_i/\delta t$, throughout the imaging window.

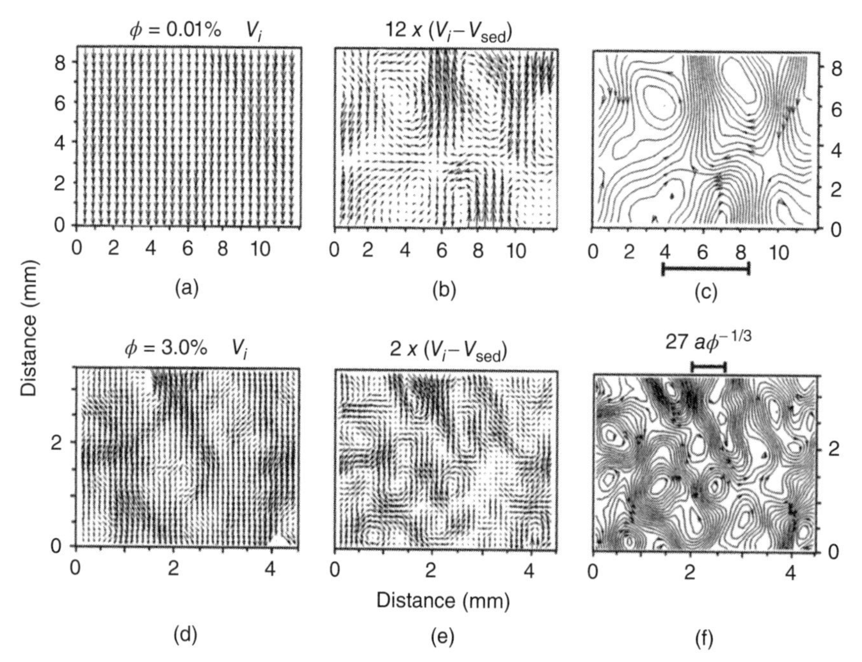

Figure 4.5 PIV results for two suspensions of sedimenting polystyrene spheres (a = 15.5 μm) at ϕ = 0.0001 (a)–(c) and ϕ = 0.03 (d)–(f). (a) and (d) Velocity vector maps measured near mid-height in sedimentation cells. Each vector V_i represents the velocity of a region of two to four particles. (b)–(e) Velocity fluctuations calculated by subtracting off the mean values, $V_{sed} = \langle V_i \rangle$ from (a) and (d). Note the different magnifications of the vector scales used for clarity. (c) and (f) Velocity fluctuation flow lines. The distance bars drawn are the corresponding lengths, $\xi = 27a\phi^{-1/3}$. Source: From Figure 1, Ref. [31].

scale, however, the fluctuations in velocity are clearly seen in Figure 4.5b. The dominant feature of the velocity maps are the swirling regions that extend over distances much larger than the individual particle sizes. At a higher concentration, $\phi = 3.0\%$, the general features are similar, but the fluctuations are larger in magnitude, while the swirling regions are smaller in size. The vector maps can be analyzed by calculating spatial correlation functions of the velocity fluctuations $\delta u_i = u_i - \langle u_i \rangle$. For correlations in the gravity, or z direction, one calculates the normalized value of $C_z(z) = \langle \delta u_z(0)\delta u_z(z)\rangle / \langle \delta u_z(0)^2 \rangle$, where $\langle \cdots \rangle$ represents a spatial average over a large region of the cell. The inset of Figure 4.6b shows results for correlation functions obtained by Snabre *et al.* [33] in both the vertical (z) and horizontal (x) directions. As is often found, the correlation functions decay nearly exponentially with distance, and simple fits yield characteristic correlation decay lengths ξ.

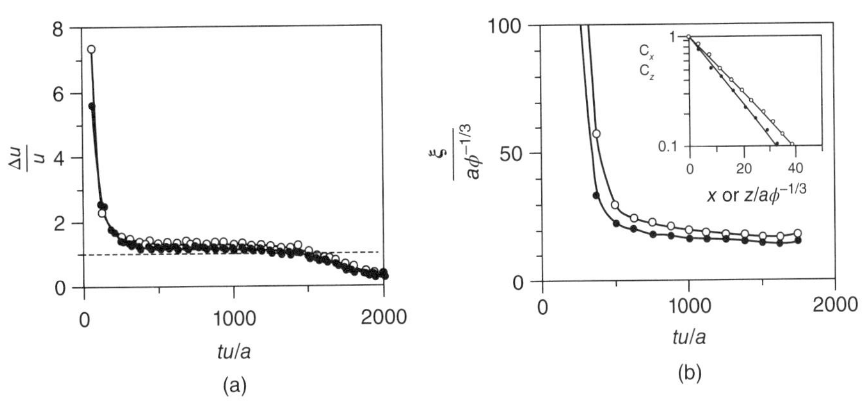

Figure 4.6 (a) Normalized amplitude $\Delta u/u$ of velocity fluctuations versus normalized time tu/a. (b) Normalized mean correlation length $\xi_x/a\phi^{-1/3}$ in x-direction (open circles) and z-direction (solid circles) as a function of the dimensionless time tu/a after the initial mixing of the $\phi = 40\%$ suspension. Inset: the normalized correlation functions $C_x(x)$ (open circles) and $C_z(z)$ (solid circles) as a function of the dimensionless distance $x/a\phi^{-1/3}$ or $z/a\phi^{-1/3}$ in the quasi-steady-state regime. Solid curves are the exponential ts of correlation functions giving the length scales $\xi_x/a\phi^{-1/3} \sim 14$ and $\xi_z/a\phi^{-1/3} \sim 17$ of velocity fluctuations in x- and z-directions. Source: Adapted from Figure 5, Ref. [33] .

The characteristic correlation lengths (in the vertical direction) found in the dilute experiments in Ref. [31] displayed a simple scaling with volume fraction,

$$\frac{\xi}{a} \approx 11\phi^{-1/3}, \tag{4.11}$$

corresponding to distances of $\xi \approx 6$ mean interparticle spacings (1 MIS $\approx n^{-1/3} \approx 1.6a\phi^{-1/3}$). The proportionality to the MIS yields a simple physical picture that explains why the swirls are visibly smaller at $\phi = 3.0\%$ than at $\phi = 0.01\%$ in Figure 4.5. In the more concentrated sample, the particles are simply much closer together on average ($300^{1/3} \approx 6.7$ times closer). Power law scalings were also found in the dilute regime [31] for the magnitude of the mean velocity variances (in the vertical direction) with

$$\frac{\Delta u}{u_s} \approx 2\phi^{1/3} . \tag{4.12}$$

The power law scalings for the spatial size and amplitude of the fluctuations can be connected together using the Hinch blob model by using ξ as the characteristic largest fluctuation size, rather than the larger cell size L. Inserting Equation 4.11 into Equation 4.10 yields $\Delta u/u_s \sim \sqrt{\phi\xi/a} \sim \phi^{1/3}$, in agreement with the observed scaling Equation 4.12.

All the results in this section refer to the thin cell regime (see Section 4.3.3 for more details on thin/thick cells). The key feature seen in thin cells is that the fluctuations show a quasi steady-state behavior soon after the initial sample preparation and the start of sedimentation. Figure 4.6 illustrates this behavior in the time-dependent values for Δu and ξ in the thin cell regime. Both quantities are measured near the middle of the cell and (initially) far from the falling top interface. At early times, Δu and ξ have very large magnitudes that rapidly decay. The system then reaches a steady state that persists over the duration of the settling until the top interface enters the observation window (at normalized times $\approx$ 1500–2000 in Fig. 4.6a). All the fluctuation values described in this section correspond to those measured in the steady-state regime in thin cells.

In the semidilute to concentrated samples [23], the velocity fluctuations display similar behavior to the dilute regime described above. The flow fields show swirling patterns such as those in Figure 4.5, but the fluctuation magnitudes are significantly enhanced as shown in Figure 4.7a. Remarkably, for $\phi > 0.10$, the fluctuations are seen to be as large as, or larger, than the mean settling rate itself. The correlation lengths ξ are also consistent with the power law scaling in Equation 4.11 up to $\phi = 0.50$. An extension of the Hinch model to concentrated systems $\phi \leq 0.50$ [23] successfully relates $\Delta u/\overline{u}$ to ξ.

Hydrodynamic Diffusion

Studies of the time evolution of individual particle motions during settling reveal that, relative to the mean settling velocity, individual particles execute random walks in time. The random motion is generated by the complex and constantly evolving hydrodynamic interactions that act on each sphere. Several experimental methods have been developed to determine the resulting hydrodynamic diffusion coefficients D. Nicolai *et al.* [3] used individual particle tracking methods to calculate diffusion coefficients from the time correlation function of the velocity fluctuations, that is, $D = \int_0^\infty \langle \delta u(0)\delta u(\tau)\rangle d\tau$, with results for D versus ϕ shown in Figure 4.7b. Martin *et al.* [34] determined D from an analysis of the steady-state concentration profile in fluidized beds. Segrè extracted D from the time evolution of the velocity fluctuation maps such as those seen in Figure 4.5b and e. The flow patterns are constantly evolving and show a

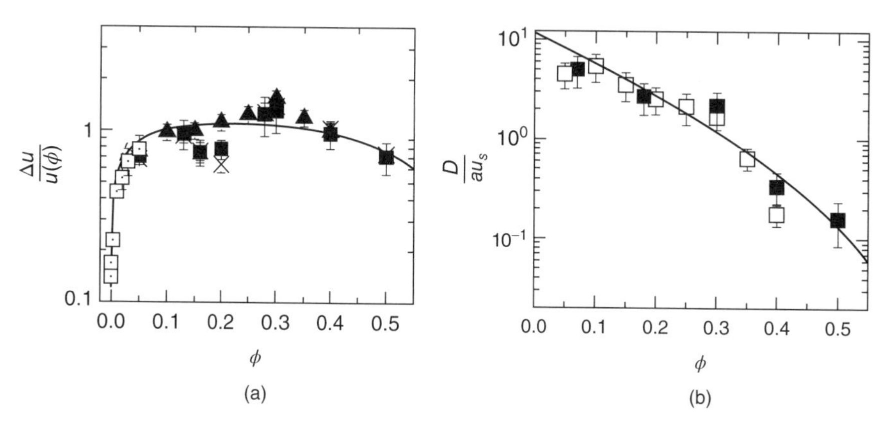

Figure 4.7 (a) Normalized velocity variances $\Delta u/u(\phi)$ as a function of ϕ in thin cells. These fluctuations are in the directions parallel (‖; filled squares, triangles, and open squares) and perpendicular (⊥ multiplied by 2, crosses) to gravity. (b) Normalized particle diffusion coefficients D/au_s versus ϕ during sedimentation. Solid squares are from Ref. [23], open squares are from Ref. [3]. Source: From Figures 2 and 4, Ref. [23].

characteristic decorrelation time $\tau_\xi \approx \xi/2.5\Delta u$. Combined with the velocity variances, one can estimate the diffusion coefficients as $D \sim \Delta u^2 \tau_\xi \sim \Delta u\xi/2.5$ [23]. In very dilute samples, this yields a limiting value of $D_0 \sim 10au_s$.

In a way reminiscent of thermal Brownian diffusion [1], Figure 4.7b shows that hydrodynamic diffusion coefficients trend to small values at high concentrations. In both cases, the increasing solution viscosity plays a major role in slowing down diffusion. Hydrodynamic diffusion is further reduced at high concentrations because of excluded volume effects that reduce particle number fluctuations [23]. An expression for D (the solid line in Fig. 4.7b), that has a similar form to the (Brownian) Stokes–Einstein equation $D = k_BT/6\pi\eta(\phi)a$, has also been found in experiments [23],

$$D = \frac{(0.4\Delta m_\xi g\xi)}{6\pi\eta(\phi)\xi}, \quad (4.13)$$

where $\Delta m_\xi g$ is the buoyant weight of the fluctuation region and $\eta(\phi)$ the concentration-dependent viscosity. From the Stokes–Einstein-like equation, one can identify the effective "Temperature," $k_BT = 0.4\Delta m_\xi g\xi$, as being related to the gravitational potential energy of the fluctuations.

4.3.3 Thick Cells, Time-Dependent Sedimentation, and Stratification

The key feature of the velocity fluctuations during sedimentation is the large-scale swirling patterns seen in Figure 4.5. The correlation length ξ captures the size of the swirls and is central to any description of sedimentation dynamics. It is important to remember that the Hinch model, while successful at describing the *relation between* ξ and the velocity variances, is not a theory that can *a priori* predict what the correlation lengths will be in the absence of knowledge of Δu. By the late 1990s, following the emergence of fluctuation data and empirical scaling relations, numerous theories were put forth [35–38] to account for the experimental results. While several theories predicted the power law scalings in Equations 4.11 and 4.12, there were nonetheless substantial differences in the various assumptions between them, and no clear consensus emerged as to the origin of the new hydrodynamic correlation length ξ.

In the search for the origin of ξ, around the year 2000, researchers began to reexamine the validity of two of the commonly used working assumptions in most models and experiments up to that date. (i) Steady-state behavior: the fluctuations, after the initial mixing transients die out, are assumed to show near constant values that persist for the duration of the settling. (ii) No stratification: the local average concentration ϕ is uniform with height during settling. The first assumption was derived from experimental observations of steady-state behavior, such as that seen in Figure 4.6, and had accurately described the existing experiments. Assumption (ii), on the other hand, had not been explicitly examined or tested in experiments. It implies that there is no mean concentration gradient, or stratification, in the vertical direction, that is, $d\phi/dz = 0$.

Time-Dependent Sedimentation

In 2001, experiments were conducted that demonstrated a violation of assumption (i), that sedimenting systems always reach steady-state behavior in time. Lei *et al.* [39] used direct particle imaging techniques to examine in detail whether particle positions are randomly distributed throughout the cell during sedimentation. They studied samples of monodisperse spheres settling at $\phi = 0.4\%$, recording the particle number fluctuations ΔN in different sized volumes throughout the cell. For a random sample, the mean number of particles is related to the linear sample

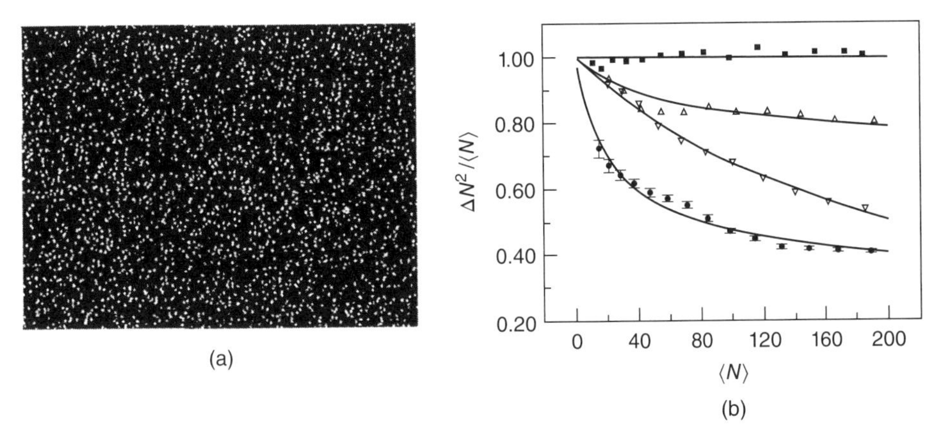

Figure 4.8 (a) Image taken during particle sedimentation in a thick cell of (normalized) dimension $d/a = 244$: a particle configuration taken 35 min after mixing. (b) Reduced number fluctuations, $\Delta N^2/\langle N \rangle$, as a function of the average number of particles in the test volume, $\langle N \rangle$, shown for different times after mixing: 2 min (solid squares), 4 h (up triangles), 6 h (down triangles), and 8 h (solid circles). The error bars represent 68% confidence estimate of the average value. Source: From Figures 1a and 3, Ref. [39].

size, $\langle N \rangle \sim \phi l^3/a^3$. Random Poisson statistics dictates that the relative number fluctuations $\Delta N^2/\langle N \rangle = 1$ for all volume sizes. Figure 4.8 shows their main result: At the earliest time, just after the initial sample preparation and before significant settling has taken place, the sample appears random at sample sizes tested. At later times, however, the fluctuations become suppressed in magnitude and show a clear trend that decreases with increasing sample volume (and size). The suppression of fluctuations over long distances observed here is consistent with the picture derived from the velocity fluctuation measurements—the particles organize themselves so that there is only a finite range, $\sim \xi$, over which Poisson statistics is valid. The most striking feature in Figure 4.8 is that the number fluctuations are continually decreasing in time—no steady-state profile is ever seen. The key difference between this experiment and the earlier ones [30, 31] that achieved steady states at similar concentrations is the relative size of the sample cell. Lei *et al.* [39] used a much larger cell than had been used before.

The question of how the cell dimensions affect the sedimentation dynamics was examined in great detail by Tee *et al.* [40] the following year, in 2002. They used two experimental techniques, dynamic light scattering (DLS) and particle imaging velocimetry (PIV), as well as numerical simulations, to measure the time dependence of the velocity fluctuations of dilute samples over a wide range of cell sizes. Figure 4.9 shows results from all three methods. It's clear that some samples reach steady states in time, and some do not. Tee *et al.* concluded that the key variable in determining whether or not steady state is reached is the relative size of the cell's smallest dimension, the normalized cell depth d/a. The velocity fluctuations in thin cells always settled to steady-state values. In thick cells, the fluctuations

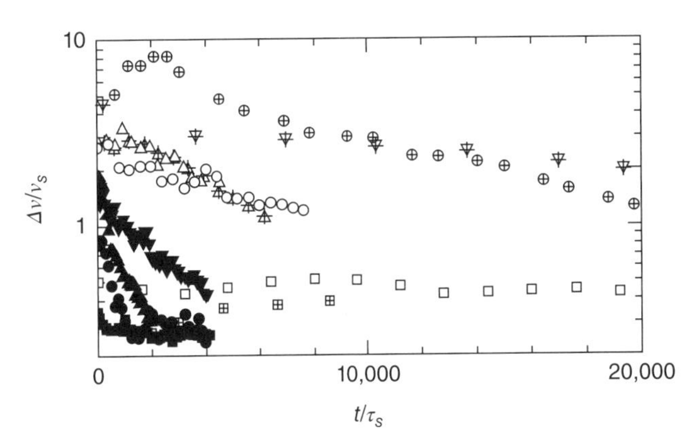

Figure 4.9 Velocity fluctuations as a function of time (in units of the Stokes time $\tau_s = a/v_s$). Solid, hatched, and open symbols correspond to PIV, dynamic light scattering (DLS), and simulations, respectively, for thick and thin cells of different heights. The velocity fluctuations continuously decay as a function of time for thick cells and reach constant values for thin cells. The DLS data continue to decay (not shown) for the full experimental runs. Data are labeled by cell dimensions ($d/a \times w/a \times h/a$) **PIV:** (square) $113 \times 1890 \times 11,300$, (circle) $113 \times 1890 \times 11,300$, (triangle) $226 \times 2260 \times 5280$, (inverted triangle) $226 \times 2260 \times 10,600$; **DLS:** (square) $103 \times 1920 \times 19,200$, (circle) $400 \times 7200 \times 28,400$, (triangle) $660 \times 7200 \times 12,000$, (inverted triangle) $1330 \times 16,700 \times 45,300$; **Simulations:** (square) thin cell, (circle) thick cell, (triangle) thick cell. Source: From Figure 1, Ref. [40].

continually decreased in time and never stabilized. The characteristic cell depth that demarcated the transition from thin to thick cell behavior occurred at $d/a \sim 140$.

In the thick cell regime, $d/a > 140$, the fluctuations decayed exponentially in time, $\Delta u(t) = \Delta u(0) \exp(-t/\tau)$ [40]. The decay times τ were not found to depend on the cell depth or width but were nearly linear in cell height h/a, with

$$\frac{\tau}{\tau_S} \sim \left(\frac{h}{a}\right)^{1.21\pm0.15}. \tag{4.14}$$

The values for the decay times illustrate why steady state is not achieved in thick cells—they are of similar order to the time it takes a bead to fall the entire height of the cell $t = \tau_s(h/a)$.

The significance of cell boundaries was also confirmed in 2005 by Nguyen and Ladd using Lattice-Boltzmann simulations [21]. Simulations done in the absence of cell walls, that is, with periodic boundary conditions, found that the velocity fluctuations were nearly constant in time with values that depended on the size of the container. When impenetrable cell walls were introduced, however, the fluctuations were seen to decay in time to a steady value that either (i) depended on the size of the container for $L < 15a\phi^{-1/3}$, or (ii) were independent of container size for $L > 15a\phi^{-1/3}$.

The potential significance of vertical concentration gradients on fluctuations was considered by Luke in 2000 [41]. He applied a linear analysis to a simple continuum model for the time evolution of sedimentation after initial randomization. He found that starting from an initial random (Poisson) state, "the suspension somehow evolves toward a stably stratified state." Moreover, he wrote that once in the stably stratified state for "a large container, the velocity fluctuations are at this point significantly reduced in comparison to the initial state." He determined that individual regions displaying a velocity fluctuation would decay in time in a stably stratified suspension as it moved up or down to its neutral buoyancy level. These arguments suggested that the presence of a concentration gradient can serve to cut off large-scale density fluctuations, that is, the correlation size ξ may be set by the value of the vertical stratification $d\phi/dz$. Luke's heuristic ideas stressed the importance of examining the time evolution of sedimentation but did not specify what exactly would create the stratification.

Stratification Scaling Model

Mucha and Brenner [42, 43] proposed a simple scaling argument (MB model) to show that even very small stratifications can markedly reduce velocity fluctuations. They considered a system with a linearly decreasing concentration with height, $\phi = \phi_0(1 - \beta z)$. The stratification $\beta \equiv -(1/\phi_0)\nabla\phi > 0$ so that the system is more concentrated at the bottom than the top (thus avoiding the Rayleigh–Taylor instability [44, 45]). Fluctuations occurring in a region of size ξ are assumed to follow Poisson statistics; $\Delta\phi \approx \sqrt{\phi a^3/\xi^3}$ and $\Delta u = u_s\sqrt{\phi\xi/a}$. In the *absence* of a stratification, concentration fluctuations are limited in how far they advect by the randomness in surrounding particle motions. A region of size ξ will only advect a distance $\sim \xi$ during its lifetime. With a stratification, however, this travel distance may be shortened if the fluctuation region reaches its neutral buoyancy point before traveling a distance ξ. A stratification can thus cut off fluctuations for scales bigger than those where the two changes in concentration are equal, that is, $(\xi\nabla\phi =)\ \xi\phi\beta \sim \Delta\phi$. The largest region for which motion occurs then is of order $\xi \sim \Delta\phi/\phi\beta$, or

$$\frac{\xi}{a} \sim \phi^{-1/5}[a\beta]^{-2/5}, \tag{4.15a}$$

$$\frac{\Delta u}{u_s} \sim \phi^{2/5}[a\beta]^{-1/5}, \tag{4.15b}$$

which relates the correlation size ξ, and velocity variance Δu, to the concentration stratification β.

These scaling results apply to the region of large stratification β, where $\xi < d$. For a homogeneous system, $\beta = 0$, the correlation length is set by the smallest cell dimension, the depth d. Setting $\xi = d$ in Equation 4.15a yields an estimate for the critical stratification,

$$a\beta_{\text{crit}} \sim \left(\frac{a}{d}\right)^{5/2}\phi^{-1/2}, \tag{4.16}$$

at which a crossover from cell depth to stratification-controlled fluctuations occurs. Equations 4.15a and 4.15b can also be written in the normalized forms

$$\frac{\xi}{d} \sim \left(\frac{\beta}{\beta_{\text{crit}}}\right)^{-2/5}, \tag{4.17a}$$

$$\frac{\Delta u^2}{\Delta u^2_{\text{Poisson}}} \sim \left(\frac{\beta}{\beta_{\text{crit}}}\right)^{-2/5}. \tag{4.17b}$$

The velocity fluctuations are normalized by the Poisson value $\Delta u^2_{\text{Poisson}} = u_s^2\phi d/a$ for a homogeneous system, where $\beta = 0$ and $\xi = d$.

Several experiments and simulations have tested the MB scaling model in dilute systems [43, 46–48]. Gomez *et al.* [48] conducted experiments and two types of numerical simulations; one starting from a uniform suspension and the other from an initially stratified suspension. All of the experiments and simulations found that the *initial* velocity variances were consistent with the Caflisch and Luke scaling Equation 4.10, with the correlation length being set by the cell dimension d. Figure 4.10 shows results from their simulations in initially stratified systems. The velocity fluctuations start out with the Caflisch and Luke values, then decay in time as settling proceeds. At long times, the fluctuation values reach a plateau, and the plateau values (shown in Fig. 4.11) decrease with increasing stratification. Tee *et al.* [43, 46] also performed experiments and numerical simulations with a uniform stratification imposed upon the initial suspension. As seen in Figure 4.11, their simulations show clear evidence of the proposed $(\beta/\beta_{\text{crit}})^{-2/5}$ scaling

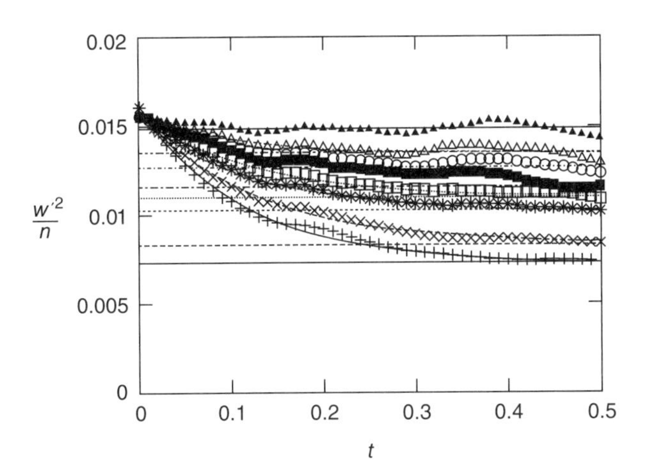

Figure 4.10 Numerical simulation results for the decay in time of velocity fluctuations with imposed concentration stratifications $(L/\phi)\nabla\phi = 0.04(+)$, $0.03(\Diamond)$, $0.02(*)$, 0.015(open square), 0.01(closed square), 0.075 (open circle), 0.005 (Δ) and 0 (closed triangle). Each simulation was made with 10^4 particles, and the cell dimensions correspond to the thin cell regime. Source: From Figure 6, Ref. [48].

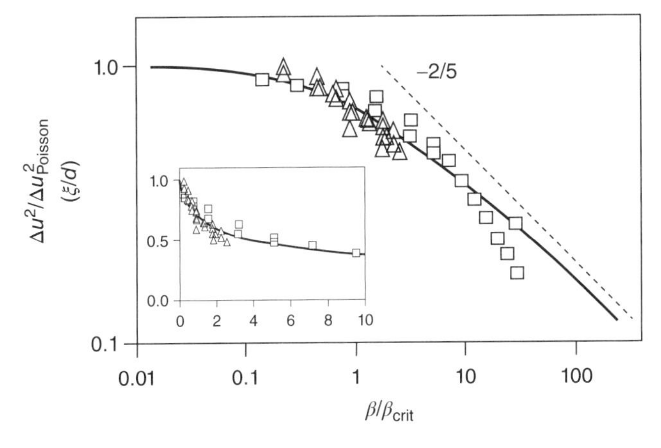

Figure 4.11 Averaged vertical velocity variances (rescaled by the Poisson value) versus normalized stratification. Simulation results from Figure 7, Ref. [48] (triangle), simulation results from Figure 4, Ref. [46] (square). The angled dashed line is the MB scaling model exponent $-2/5$. The solid line is Equations 4.20a and 4.20b with $c_3 = c_4 = 1.35$. Note the model line also represents the reduced correlation length since $\xi/d = \Delta u^2/\Delta u^2_{\text{Poisson}}$. Inset: Results plotted on a linear scale to highlight the initial decay for $\beta \sim \beta_{\text{crit}}$.

for the variance of the velocity fluctuations in the limit of large stratifications. Tee *et al.* also found evidence for the $(\beta/\beta_{\text{crit}})^{-2/5}$ scaling in experiments, although the fluctuation values were much reduced relative to the simulations. It is thought that the presence of undesired density gradients in the background liquid sucrose [22, 46] contributed to this discrepancy. Overall, these results confirm the concept introduced by Luke in 2000 [41], that in highly stratified systems the magnitudes of the velocity fluctuations depend upon the degree of concentration stratification.

The MB scaling results indicate that the onset of stratification-controlled fluctuations arises for very small concentration gradients. Assuming a uniform stratification at the critical value $\beta = \beta_{\text{crit}}$ throughout the cell, the fractional variation in concentration from the bottom of the cell to the top is $\sim -(h/\phi_0)\nabla\phi = h\beta_{\text{crit}} = (h/a)(a/d)^{5/2}\phi_0^{-1/2}$. For typical cell dimensions [40] $d/a : w/a : h/a = 150 : 750 : 1500$ at $\phi_0 = 1\%$, the total top to bottom fractional variation in ϕ is only 5%. Measuring such small gradients can be a challenge in experiments, particularly for somewhat polydisperse suspensions [21], where it can be difficult to separate the effects of particle size variations and concentration changes.

4.3.4 Stratification Model in a Fluidized Bed

The MB model Equations 4.15a and 4.15b, obtained on the basis of scaling arguments, can also be derived as the limiting cases of a general solution to the convection–diffusion equation. We consider the case of sedimentation in a fluidized bed [34, 49], where fluid is pumped upward through the sample cell at a velocity that matches the average sedimentation velocity of the falling particles. This produces a stable column of particles whose height is neither rising nor falling in the lab frame. The system is very dilute so that on average, relative to the fluid, particles move at speeds equal to the Stokes velocity u_s. We assume the system reaches steady-state sedimentation in which, at each height z in the lab frame, the (i) *time-averaged* local particle velocities vanish, $\langle v(z,t)\rangle_\tau = 0$, and (ii) the *time-averaged* local concentration is constant, $\langle\phi(z,t)\rangle_\tau = \phi(z)$ and $\langle\partial\phi(z,t)/\partial t\rangle_\tau = 0$.

We start from the general convection–diffusion equation,

$$-\frac{\partial\phi}{\partial t} = \nabla[\phi u - D\nabla\phi]\,, \tag{4.18}$$

and take the long time average $\langle\cdots\rangle_\tau$ of both sides. The LHS vanishes for reason (ii) above and integrating the RHS leads to $\langle\phi u\rangle_\tau - \langle D\nabla\phi\rangle_\tau = C$, where C is an integration constant. Expanding to first order in the fluctuations, $\phi(z,t) = \phi(z) + \delta\phi(z,t)$ and $v(z,t) = \delta v(z,t)$. In steady state, the *time-averaged* fluctuations vanish, $\langle\delta v(z,t)\rangle_\tau = \langle\delta\phi(z,t)\rangle_\tau = 0$. The advective term becomes $\langle\phi u\rangle_\tau = \langle[\phi + \delta\phi]\delta u\rangle_\tau = \langle\delta\phi\delta u\rangle_\tau$, which has a negative value because velocity and concentration fluctuations are anticorrelated (denser regions fall, lighter regions rise). To proceed further, we assume that there is a characteristic correlation length ξ for the velocity fluctuations and that Poisson statistics is valid up to this length. The Hinch model can then be used for the *rms* fluctuation magnitudes, $\Delta\phi = \sqrt{\phi a^3/\xi^3}$, and $\Delta u = u_s\sqrt{\phi\xi/a}$. The advective term then becomes $\langle\delta\phi\delta u\rangle_\tau \approx -\Delta\phi\Delta u \approx -u_s(\phi a/\xi)$. The hydrodynamic diffusivity is $D \approx \Delta u \xi$ [23], so that Equation 4.18 reduces to $-u_s(\phi a/\xi) - u_s(\sqrt{\phi \xi^3/a})\nabla\phi = C$. The constant C can now be found by considering the case where $\nabla\phi = 0$. As argued by Mucha and Brenner [42, 43], we assume that for a homogeneous system, the correlation length is the cell's thinnest dimension, the depth, so that for $\xi = d$ we find $C = -u_s\phi a/d$.

The solution to the convection–diffusion equation 4.18 then becomes, in the gradient notation of Mucha and Brenner $\beta \equiv -(1/\phi)\nabla\phi$,

$$\frac{\xi/a}{(1-\xi/d)^{2/5}} = c_1\phi^{-1/5}(a\beta)^{-2/5}, \qquad (4.19a)$$

$$\frac{\Delta u/u_s}{(1-\Delta u^2/\Delta u^2_{\text{Poisson}})^{1/5}} = c_2\phi^{2/5}(a\beta)^{-1/5}, \qquad (4.19b)$$

where c_1 and c_2 are fit constants expected to be of order unity. Equations 4.19a and 4.19b are the main results. They are general relations between the correlation lengths and velocity fluctuations ξ and Δu, and the concentration gradient β. In a homogeneous system, $\beta = 0$, the fluctuations reach their maximum values, $\xi = d$ and $\Delta u = \Delta u_{\text{Poisson}}$. Both quantities are reduced when a stratification is present. In the limit of large stratifications, as considered in the MB scaling model above, where $\xi/d \ll 1$ and $\Delta u^2/\Delta u^2_{\text{Poisson}} \ll 1$, Equation 4.19 reproduces the Mucha and Brenner scaling Equations 4.15a and 4.15b. The model Equation 4.19 can also be written in the normalized forms

$$\frac{\xi/d}{(1-\xi/d)^{2/5}} = c_3(\beta/\beta_{\text{crit}})^{-2/5}, \qquad (4.20a)$$

$$\frac{\Delta u^2/\Delta u^2_{\text{Poisson}}}{(1-\Delta u^2/\Delta u^2_{\text{Poisson}})^{2/5}} = c_4(\beta/\beta_{\text{crit}})^{-2/5}, \qquad (4.20b)$$

which also reproduce the MB model scaling Equation 4.17 in the large stratification limit.

Figure 4.11 plots the model Equations 4.20 for ξ/d and $\Delta u^2/\Delta u^2_{\text{Poisson}}$ versus the reduced stratification $\beta/\beta_{\text{crit}}$. (Note that there is only one theory line for both values, since $\Delta u^2/\Delta u^2_{\text{Poisson}} = \xi/d$.) Also plotted are the simulation results for moderate stratifications from Ref. [48], and large stratifications from Ref. [46]. The model captures the trends in both data sets fairly well. As described by Mucha and Brenner [42, 43], the onset of stratification-controlled fluctuations is seen to occur at $\beta/\beta_{\text{crit}} \sim 1$. The model also shows that to access the $(-2/5)$ MB power law scaling regime requires very large stratifications $\beta/\beta_{\text{crit}} > 10$. Smaller values result in a significantly weaker dependence on stratification. (The local power law slopes $\Delta u^2/\Delta u^2_{\text{Poisson}} \propto (\beta/\beta_{\text{crit}})^{-\alpha}$ are $\alpha \approx 0.09, 0.22, 0.33$, and 0.38 for $\beta/\beta_{\text{crit}} = 0.1, 1, 10$, and 100.)

To access the MB scaling regime predicted for $\beta/\beta_{\text{crit}} > 10$ requires large system sizes as well. Consider a typical cell of dimensions $d : w : h = 1 : 5 : 10$, containing a sample at average concentration $\phi_0 = 0.01$ and a large concentration gradient of order $-\nabla\phi \sim \phi_0/h$. The reduced stratification is $\beta/\beta_{\text{crit}} \sim (1/h)/a^{3/2}d^{-5/2}\phi_0^{-1/2} \sim 0.01(d/a)^{3/2}$. To reach $\beta/\beta_{\text{crit}} = 10$ then requires a cell depth $d/a \approx 100$, and $N = (\phi_0 dwh/(4/3)\pi a^3) \approx 1 \times 10^5$ particles in the cell. For $\beta/\beta_{\text{crit}} = 100$, $d/a \approx 465$, and $N \approx 10 \times 10^6$ particles, a very large number that can be realized in experiment, but that has yet to be achieved in simulations.

The physical picture behind the stratification model Equations 4.19 and 4.20 relate to the conditions necessary to obtain a steady-state concentration profile in a fluidized bed. Steady state is achieved when there is a balance between gradient diffusion that drives particles upward and velocity fluctuations that drive a net particle flux downward. To illustrate the latter flux, consider the two typical fluctuations on the RHS of Figure 4.3. Both fluctuations move at the same speed Δu, but the falling one is denser than the rising one, so there is a net advective flux downward.

4.4 SUMMARY

This chapter has focused on some of the current issues and understanding of the sedimentation dynamics of non-Brownian spheres at low Reynolds numbers. The complexities of such systems relate to the long-range nature of the many-body hydrodynamic interactions between settling particles. Significant progress has been made in the field since the original proposal by Caflisch and Luke more than 20 years ago that the velocity fluctuations would increase in an unbounded way with the size of the container. Our current understanding is that this predicted divergence does not occur because particles organize themselves into large correlated regions, of size $\xi \sim 10 - 20a\phi^{-1/3}$, that screen out the long-range hydrodynamic interactions. The size of the characteristic swirling regions is at the heart of all of the fluctuation phenomena. Its value has been shown to directly determine the magnitude of the velocity fluctuations Δu, the rate of hydrodynamic diffusion D, and even whether or not a sedimenting column will reach steady state before all the particles reach the cell bottom. Despite its central importance, there still remain many unanswered questions as to what the origin of ξ is.

Some of the more promising recent work, specifically the Mucha and Brenner scaling model, has shown that that ξ can be influenced by the presence of very small vertical gradients in particle concentration that arise during settling. The strongest evidence for this comes from numerical simulations on systems, where various stratifications are externally imposed upon the system. Prior to the realization that small concentration gradients may be of primary importance, the majority of experiments that measured ξ did not search for or measure $\nabla\phi$, and more work needs to be done in this area in the future to fully test these new ideas. To be fair, it should also be pointed out that the proposed MB scaling model, and the fluidized bed model presented in Section 4.3.4, do not by themselves constitute first principles theories for the correlation lengths ξ or the stratification $\nabla\phi$. That is because these models both express ξ *in terms of* a given stratification $\nabla\phi$, whose value itself has to be input from additional measurements. We are not presently aware of any theory that has been shown to predict a priori the level of concentration stratification during sedimentation, and it could be very fruitful for more research to be done in this area.

REFERENCES

[1] Russel WB, Saville DA, Schowalter WR. *Colloidal Dispersions*. Cambridge: Cambridge University Press; 1989.

[2] Batchelor GK. *An Introduction to Fluid Dynamics*. Cambridge: Cambridge University Press; 1967.

[3] Nicolai H, Herzhaft B, Hinch EJ, Oger L, Guazzelli E. Particle velocity fluctuations and hydrodynamic self-diffusion of sedimenting non-Brownian spheres. Phys Fluids 1995;7:12.

[4] Batchelor GK. Sedimentation in a dilute dispersion of spheres. J Fluid Mech 1972;52:245.

[5] Xu W, Nikolov AD, Wasan DT. The effect of many-body interactions on the sedimentation of monodisperse particle dispersions. J Colloid Interface Sci 1998;197:160.

[6] Di Felice R. The sedimentation velocity of dilute suspensions of nearly monosized spheres. Int J Multiphase Flow 1999;25:559.

[7] Thies-Weesie DME, Philipse AP, Nagele G, Mandl B, Klein R. Nonanalytical concentration dependence of sedimentation of charged silica spheres in an organic solvent: experiments and calculations. J Colloid Interface Sci 1995;87:43.

[8] Vesaratchanon S, Nikolov A, Wasan DT. Sedimentation in nano-colloidal dispersions: effects of collective interactions and particle charge. Adv Colloid Interface Sci 2007;134:268.

[9] Batchelor GK. Sedimentation in a dilute polydisperse system of interacting spheres. Part 1. General theory. J Fluid Mech 1982;119:379.

[10] Batchelor GK, Wen CS. Sedimentation in a dilute polydisperse system of interacting spheres. Part 2. Numerical results. J Fluid Mech 1982;124:495.

[11] Jansen JW, de Kruif CG, Vrij A. Attractions in sterically stabilized silica dispersions: IV. Sedimentation. J Colloid Interface Sci 1986;114:501.

[12] Richardson JF, Zaki WN. Sedimentation and fluidisation: Part 1:. Trans Inst Chem Eng 1954;32:35.

[13] Sangani AS, Acrivos A. Slow flow through a periodic array of spheres. Int J Multiphase Flow 1982;8:343.

[14] Davis RH, Acrivos A. Sedimentation of noncolloidal particles at low Reynolds numbers. Annu Rev Fluid Mech 1985;17:91.

[15] Lynch ED, Herbolzheimer E. Formation of microscale structure in sedimenting suspensions. Bull Am Phys Soc 1983;28:1365.

[16] Oliver DR. The sedimentation of suspensions of closely-sized spherical particles. Chem Eng Sci 1961;15:230.

[17] Benes K, Tong P, Ackerson BJ. Sedimentation, Peclet number, and hydrodynamic screening. Phys Rev E 2007;76:056302.

[18] Rouyer F, Lhuillier D, Martin J, Salin D. Structure, density, and velocity fluctuations in quasi-two-dimensional non-Brownian suspensions of spheres. Phys Fluids 2000;12:958.

[19] Yin X, Koch DL. Hindered settling velocity and microstructure in suspensions of solid spheres with moderate Reynolds numbers. Phys Fluids 2007;19:093302.

[20] Nguyen N-Q, Ladd AJC. Microstructure in a settling suspension of hard spheres. Phys Rev E 2004;69:050401.

[21] Nguyen N-Q, Ladd AJC. Sedimentation of hard-sphere suspensions at low Reynolds number. J Fluid Mech 2005;525:73.

[22] Guazzelli E, Hinch J. Fluctuations and instability in sedimentation. Annu Rev Fluid Mech 2011;43:97.

[23] Segrè PN, Liu F, Umbanhower P, Weitz DA. An effective gravitational temperature for sedimentation. Nature 2001;409:594.

[24] Caflisch RE, Luke JHC. Variance in the sedimentation speed of a suspension. Phys Fluids 1985;28:259.

[25] Hinch EJ. In: Guyon E, Nadal J.P., Pomeau Y., editors. *Disorder in Mixing*. Dordrecht: Kluwer Academic; 1988.

[26] Koch DL. Hydrodynamic diffusion in a suspension of sedimenting point particles with periodic boundary conditions. Phys Fluids 1994;6:2894.

[27] Ladd AJC. Hydrodynamic screening in sedimenting suspensions of non-Brownian spheres. Phys Rev Lett 1996;76:1392.

[28] Ladd AJC. Sedimentation of homogeneous suspensions of non-Brownian spheres. Phys Fluids 1997;9:491.

[29] Koglin JE. [PhD thesis]. University of Karlsruhe: Germany; 1971.

[30] Nicolai H, Guazzelli E. Effect of the vessel size on the hydrodynamic diffusion of sedimenting spheres. Phys Fluids 1995;7:3.

[31] Segrè PN, Herbolzheimer E, Chaikin PM. Long range correlations in sedimentation. Phys Rev Lett 1997;79:2574.

[32] Adrian RJ. Particle-imaging techniques for experimental fluid mechanics. Annu Rev Fluid Mech 1991;23:261.

[33] Snabre P, Pouligny B, Metayer C, Nadal F. Size segregation and particle velocity fluctuations in settling concentrated suspensions. Rheol Acta 2009;48:855.

[34] Martin J, Rakotomalala N, Salin D. Hydrodynamic dispersion of noncolloidal suspensions: measurement from Einstein's argument. Phys Rev Lett 1995;74:1347.

[35] Levine A, Ramaswamy S, Frey E, Bruinsma R. Screened and unscreened phases in sedimenting suspensions. Phys Rev Lett 1998;81:5944.

[36] Tong P, Ackerson BJ. Analogies between colloidal sedimentation and turbulent convection at high Prandtl numbers. Phys Rev E 1998;58:6931.

[37] Brenner MP. Screening mechanisms in sedimentation. Phys Fluids 1999;11:754.

[38] Ramaswamy S. Issues in the statistical mechanics of steady sedimentation. Adv Phys 2001;50:297.

[39] Lei X, Ackerson BJ, Tong P. Settling statistics of hard sphere particles. Phys Rev Lett 2001;86:3300.

[40] Tee SY, Mucha PJ, Cipelletti L, Manley S, Brenner MP, Segrè PN, Weitz DA. Nonuniversal velocity fluctuations of sedimenting particles. Phys Rev Lett 2002;89:054501.

[41] Luke JHC. Decay of velocity fluctuations in a stably stratified suspension. Phys Fluids 2000;12:1619.

[42] Mucha PJ, Brenner MP. Diffusivities and front propagation in sedimentation. Phys Fluids 2003;15:1305.

[43] Mucha PJ, Tee SY, Shraiman BI, Brenner MP. A model for velocity fluctuations in sedimentation. J Fluid Mech 2004;501:71.

[44] Voltz C, Pesch W, Rehberg I. Rayleigh-Taylor instability in sedimenting systems. Phys Rev E 2001;65:011404.

[45] Patrick Royall C, Dzubiella J, Schmidt M, van Blaaderen A. Nonequilibrium sedimentation of colloids on the particle scale. Phys Rev Lett 2007;98:188304.

[46] Tee S-Y, Mucha PJ, Brenner MP, Weitz DA. Velocity fluctuations of initially stratified sedimenting spheres. Phys Fluids 2007;19:113304.

[47] Gomez DC, Bergougnoux L, Hinch J, Guazzelli E. On stratification control of the velocity fluctuations in sedimentation. Phys Fluids 2007;19:098102.

[48] Gomez C, Bergougnoux L, Guazzelli E, Hinch EJ. Fluctuations and stratification in sedimentation of dilute suspensions of spheres. Phys Fluids 2009;21:093304.

[49] Xue J-Z, Herbolzheimer E, Rutgers MA, Russel WB, Chaikin PM. Diffusion, dispersion, and settling of hard spheres. Phys Rev Lett 1992;69:1715.

5

PARTICLES IN ELECTRIC FIELDS

Todd M. Squires
Department of Chemical Engineering, University of California, Santa Barbara, CA, USA

Electrostatic fields often play an extremely important role in soft matter: they are widely used to stabilize colloids against aggregation, give stiffness to polyelectrolytes, enable ion exchange, and impart unique properties to complex coacervates.

In other cases, kinetic phenomena arise due to electric fields—the electrophoretic motion of colloids and polymers is widely used to characterize surface charge and represented the core technology behind the original sequencing of the human genome. Electro-osmotic flows (EOFs) arise when applied electric fields force ions (and the fluids around them) into motion [1–4]; conversely, fluid flows sweep ions into motion, establishing streaming currents or potentials. Diffusiophoresis and diffusio-osmosis represent intimately related but distinct phenomena: particle or fluid motion due to gradients in ionic strength.

These are all linear electrokinetic effects—in which the motion or flow varies linearly with the applied field. In addition, nonlinear electrokinetic effects can play an important role in structure formation, particle motion, suspension behavior, and material rheology. Examples include dielectrophoresis (DEP), where an applied field induces an electric dipole in a particle, which is then forced by an electric field gradient, driving motion; AC and

Fluids, Colloids and Soft Materials: An Introduction to Soft Matter Physics, First Edition. Edited by Alberto Fernandez Nieves and Antonio Manuel Puertas.

induced-charge electrokinetics, wherein applied electric fields *induce* the formation of charge double layers around polarizable surfaces, then force those *induced* double layers into EOF, and magneto- and electrorheological fluids, in which applied fields change the microstructural morphology of a material, and thus its viscoelastic properties.

In this chapter, we give a basic treatment of these various effects—all of which involve the interaction between applied fields, fluid flows, particle motion, and structure formation. We start with electrostatics in electrolytic solutions, the classic description of which gives the Poisson–Boltzmann (PB) equation as well as its modifications. We then discuss linear electrokinetic effects and introduce the Poisson–Nernst–Planck–Stokes (PNPS) equations, from which follow the classic description of all electrokinetic effects. We then describe the thin double-layer limit of the classic electrokinetic effects—examples include electro-osmosis, electrophoresis, and diffusiophoresis. These classic descriptions neglect the influence of ion transport within the electric double layer (called surface conduction). Because significant qualitative changes in behavior occur when surface conduction becomes dominant, we describe some nontrivial effects of surface conductivity—maxima in electrophoretic mobility and the generation of large-scale salt gradients in solution, called concentration polarization. Finally, we describe two nonlinear electrokinetic effects—induced-charge electrokinetics (ICEK) (in which an applied field *induces* a charge cloud around a polarizable object, then drives that locally charged fluid into electrokinetic motion), and DEP, in which an applied field interacts with the electrostatic dipole it induces within a particle to drive motion.

In all cases, we start by presenting what are now fairly standard arguments and demonstrations but with particular emphasis on physically intuitive, paradigmatic pictures. We try to provide some additional insight whenever possible and generally follow with alternate descriptions or approaches that are less well known but that enable extensions to the standard approach. These "add-ons" are meant to be more advanced and so can be skipped if desired.

Finally, our presentation of many of these topics directly follows, and cites, our own work on these topics. This is by no means meant to imply that our work (or the work presented here) was the first to understand these topics. Rather, this reflects our own (hopefully understandable) bias on these topics—in many cases, what is covered here are those problems we posed in order to teach ourselves how to think about the fundamental physical processes at hand. An example includes induced-charge electro-osmosis, which followed from work on AC electrokinetics [5, 6], further back to colloidal aggregation and patterning near electrodes [7], and even further back to Soviet work on the flow around, and interactions between, metallic colloids [8–10]. The electrophoretic mobility maximum was understood by Dukhin and Derjaguin [11], O'Brien and White [12], and O'Brien and Hunter [13]; we here show our work involving slipping particles to emphasize the general understanding that follows from identification of the appropriate Dukhin number. Work on concentration polarization was long known from the membrane community, and more recently from micro- and nanofluidic electrokinetics; we present our work as (in our minds) the simplest systems in which to understand concentration polarization as driven by physics within thin electric double layers.

5.1 ELECTROSTATICS IN ELECTROLYTES

While water may at first glance to be fairly featureless, it generally contains a significant concentration of (charged) ions—examples include added salts (Na^+ and Cl^-) and free protons or hydroxyls in acidic or basic solutions. Even pure, pH 7 water contains 10^{-7} M H^+ and OH^- ions. The ubiquity of ions in aqueous solution follows from how effectively water dissolves ions: because the H_2O molecule is so polar, it naturally aligns and surrounds charged sites to form polarized hydration shells. This hydration shell has the effect of stabilizing the ions in solution against "aggregation" with oppositely charged ions, by forming a sort of armor that prevents anions and cations from recombining [14], much like a polymer brush can stabilize colloids against flocculation. Nonpolar solvents (e.g., oils) are far worse solvents for ions, so that dissolved "ions" take the form of inverse micelles, with ions in a hydrophilic domain [15].

For related reasons, most surfaces develop a charge when in contact with water [16–19]: surface groups may dissociate as water molecules hydrate the dissociated ion, and ions of one species may have a higher binding affinity for certain surfaces than those of the opposite sign. Electrically neutral surfaces are the exception, rather than the rule, when in contact with water.

When an electrolyte is in contact with a charged surface, oppositely charged *counterions* experience an electrostatic attraction to the surface, whereas like-charged *co-ions* are repelled (Fig. 5.1). As they become more and more concentrated, counterions experience an entropic (diffusive) repulsion from the wall. These competing forces balance at a particular length scale—the Debye screening length λ_D—over which the electrostatic effects of the surface charge are exponentially screened. Since this screening length—and thus the range over which the surface charge is "felt"—typically ranges from a few to tens of nanometers, these strong surface charges are easily missed in the macroscopic world. The surface charge and screening cloud is called the *electric double layer* (EDL) and is described next.

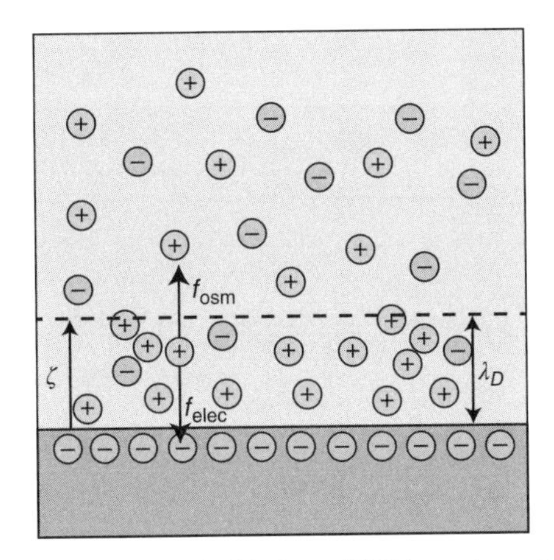

Figure 5.1 An electric double layer (EDL) consists of a charged surface in contact with an electrolyte, wherein dissolved ions rearrange to "screen" the surface charge. In particular, oppositely charged "counterions" are electrostatically attracted toward the wall, and similarly charged "co-ions" are repelled. Entropic (diffusive) forces act to drive the ions to smooth out the concentration gradients generated by the electrostatic attraction, and the two balance at the characteristic Debye "screening" length λ_D, which ranges from subnanometer to hundreds of nanometers in aqueous electrolytes.

5.1.1 The Poisson–Boltzmann Equation

This competition of forces within the electric double layer is described mathematically by the Poisson–Boltzmann equation. We start with the standard approach for its derivation, then discuss the assumptions that go into this approach, and the questions that arise in making these assumptions.

For simplicity, we consider a monovalent, binary electrolyte (two species of ions, cations and anions, with charge $+e$ and $-e$, respectively). It is straightforward to generalize this treatment for more general electrolytes, and we simply state the more general formulas.

Ions that are distributed with number density $n_\pm(\mathbf{r})$ establish an electrostatic field $\phi(\mathbf{r})$, as governed by Poisson's equation

$$\nabla^2\phi = -\frac{\rho_E}{\epsilon} = -\frac{e(n_+ - n_-)}{\epsilon}, \tag{5.1}$$

where ϵ is the dielectric permittivity of water and e is the elementary charge. Since ions are charged, they also respond to electrostatic fields. So, given an electrostatic potential $\phi(\mathbf{r})$, anions and cations in equilibrium obey the Boltzmann distribution

$$n_\pm(\mathbf{r}) = n^0 \exp\left(\mp\frac{e\phi(\mathbf{r})}{k_BT}\right), \tag{5.2}$$

where k_BT represents thermal energy and where n^0 refers to the concentration of both anions and cations in an electroneutral reservoir, where the electrostatic potential ϕ is zero. Now, assuming that the ions respond to the *mean field* that is established statistically by those same ions, we simply insert $n_\pm$ from Equation 5.2 into Poisson's equation 5.1, to obtain the Poisson–Boltzmann equation

$$\nabla^2\phi = -\frac{n^0 e}{\epsilon}\left(\exp\left(-\frac{e\phi}{k_BT}\right) - \exp\left(\frac{e\phi}{k_BT}\right)\right)$$
$$\equiv \frac{2n^0 e}{\epsilon}\sinh\left(\frac{e\phi}{k_BT}\right). \tag{5.3}$$

The (dimensionless) quantity $e\phi/k_BT$ gives the electrostatic energy relative to a thermal voltage

$$\phi_{\text{th}} = \frac{k_BT}{e}, \tag{5.4}$$

which is approximately 26 mV at room temperature. Its physical importance is as follows: say that the electrostatic potential $\phi(\mathbf{r}_0) > 0$ is positive at a position $\mathbf{r}_0$. If $e\phi(\mathbf{r}_0)/k_BT \gg 1$, then the electrostatic energy required to bring a (positively charged) cation from "infinity" to $\mathbf{r}_0$ is much greater than what thermal energy has available to drive it there; consequently, cations are only found at $\mathbf{r}_0$ in exceedingly rare cases. Conversely, the energy required to pull anions away from $\mathbf{r}_0$ exceeds what thermal energy has to offer, and so anions are strongly attracted to $\mathbf{r}_0$.

Scaling all potentials ϕ with ϕ_{th} gives

$$\nabla^2\left(\frac{e\phi}{k_BT}\right) = \kappa^2 \sinh\frac{e\phi}{k_BT}, \tag{5.5}$$

where

$$\kappa^{-1} = \lambda_D = \sqrt{\frac{\epsilon k_BT}{2n^0e^2}} \tag{5.6}$$

is the Debye "screening" length. In room temperature water, λ_D ranges from 1 μm (for pure, pH 7 water) to less than a nanometer ($\lambda_D = 0.3$ nm for 1 M aqueous salt solution).

When applied potentials are everywhere small $\phi \ll \phi_{\text{th}}$, the nonlinear Poisson–Boltzmann equation can be linearized (via $\sinh x \approx x$) to give

$$\nabla^2\phi = \kappa^2\phi. \tag{5.7}$$

The electrostatic potential surrounding a planar surface (at $z = 0$) with potential ζ relative to an adjacent electrolyte ($z > 0$), yielding

$$\phi(z) = \zeta e^{-\kappa z}, \tag{5.8}$$

which describes a diffuse double layer with charge density

$$\rho_E(z) = -\epsilon\epsilon_0\nabla^2\phi = -\epsilon\epsilon_0\kappa^2\zeta e^{-\kappa z}. \tag{5.9}$$

The charge in the double layer, when integrated, contains an area charge density

$$\sigma_{\text{DL}} = -\epsilon\epsilon_0\kappa\zeta, \tag{5.10}$$

TABLE 5.1 Solutions to the Poisson–Boltzmann Equation for the Electrostatic Potential ϕ, Ionic Charge Density ρ_E, and Surface Charge Density σ_w

Geometry	Electrostatic Potential $\phi(z)$ or $\phi(r)$	Ionic Charge Density $\rho_E(z)$ or $\rho_E(r)$	Surface Charge Density σ_w
Plane	$\zeta e^{-\kappa z}$	$-\frac{\epsilon_w}{\lambda_D^2}\zeta e^{-\kappa z}$	$\frac{\epsilon_w}{\lambda_D}\zeta$
Plane $\left(\text{nonlinear; } \Gamma = \tanh\frac{e\|\zeta\|}{4k_BT}\right)$	$\frac{2k_BT}{e}\text{sign}(\zeta)\ln\frac{1+\Gamma e^{-\kappa z}}{1-\Gamma e^{-\kappa z}}$	$-\frac{4k_BT}{e}\frac{\epsilon_w}{\lambda_D^2}\text{sign}(\zeta)\Gamma e^{-\kappa z}\frac{(1+\Gamma^2 e^{-2\kappa z})}{(1-\Gamma^2 e^{-2\kappa z})^2}$	$2\frac{\epsilon_w}{\lambda_D}\frac{k_BT}{e}\sinh\frac{e\zeta}{2k_BT}$
Inside cylinder, radius a	$\zeta\frac{I_0(\kappa r)}{I_0(\kappa a)}$	$-\frac{\epsilon_w}{\lambda_D^2}\zeta\frac{I_0(\kappa r)}{I_0(\kappa a)}$	$\frac{\epsilon_w}{\lambda_D}\frac{I_1(\kappa a)}{I_0(\kappa a)}\zeta$
Outside cylinder, radius a	$\zeta\frac{K_0(\kappa r)}{K_0(\kappa a)}$	$-\frac{\epsilon_w}{\lambda_D^2}\zeta\frac{K_0(\kappa r)}{K_0(\kappa a)}$	$\frac{\epsilon_w}{\lambda_D}\frac{K_1(\kappa a)}{K_0(\kappa a)}\zeta$
Outside sphere, radius a	$\zeta\frac{a}{r}e^{-\kappa(r-a)}$	$-\frac{\epsilon_w}{\lambda_D^2}\zeta\frac{a}{r}e^{-\kappa(r-a)}$	$\frac{\epsilon_w}{\lambda_D}\frac{1+\kappa a}{\kappa a}\zeta$

which is equal and opposite to the surface charge density of the charged surface that establishes it:

$$\sigma_w = -\sigma_{\text{DL}} = \frac{\epsilon\epsilon_0}{\lambda_D}\zeta. \tag{5.11}$$

At low potentials, thin electric double layers act like parallel-plate capacitors, with a potential drop ζ that occurs between plates separated by λ_D, across a dielectric medium with permittivity ϵ.

Analytical solutions to the linearized Poisson–Boltzmann equation exist for various special geometries—much like for Laplace's equation—with key examples given in Table 5.1. More detailed discussions can be found in textbooks focusing on microfluidics [20, 21], biological physics [22], or colloids [17].

The Poisson–Boltzmann equation can be generalized to allow for N different ion species, each with valence z_i and bulk concentration n_i^0, quite simply as

$$\nabla^2\tilde{\phi} = \frac{e^2}{\epsilon k_BT}\sum_{i=1}^{N} n_i^0 z_i \exp(z_i\tilde{\phi}), \tag{5.12}$$

where

$$\tilde{\phi} = \frac{e\phi}{k_BT} \tag{5.13}$$

is the potential made dimensionless by scaling with the thermal potential ϕ_{th}. Because the reservoir (where $\phi = 0$) must be electroneutral, $\Sigma_{i=1}^N n_i^0 z_i = 0$. In the limit of low potentials, where the exponentials can be Taylor-expanded, Equation 5.12 simply reduces to Equation 5.7, with

$$\kappa^{-1} = \lambda_D = \sqrt{\frac{\epsilon k_BT}{\Sigma n_i^0 z_i^2 e^2}}. \tag{5.14}$$

Note, too, that the condition for linearization becomes more stringent when the valence z gets higher: the thermal potential for an ion of valence z_i becomes k_BT/z_ie.

Lastly—the nonlinear PB equation for a binary electrolyte (5.5) can be solved exactly over a planar interface, by multiplying both sides by $d\tilde{\phi}/dz$ and integrating (defining the integration constant so that $\phi' = 0$ when $\phi = 0$), giving

$$\left(\frac{d\tilde{\phi}}{dz}\right)^2 = 2\kappa^2(\cosh\tilde{\phi} - 1) = 4\kappa^2\sinh^2\left(\frac{\tilde{\phi}}{2}\right). \tag{5.15}$$

Although this can be integrated directly (results in Table 5.1), the surface charge density (and thus capacitance) can be immediately derived using the field at the surface ($z = 0$), via

$$\sigma_w = -\epsilon\frac{d\phi}{dz}\bigg|_{z=0} = 2\kappa\epsilon\frac{k_BT}{e}\sinh\left(\frac{\tilde{\zeta}}{2}\right), \tag{5.16}$$

where

$$\tilde{\zeta} = \frac{e\zeta}{k_BT}, \tag{5.17}$$

is the surface potential scaled by ϕ_{th}. The differential capacitance,

$$C(\zeta) = \frac{d\sigma_w}{d\zeta} = \kappa\epsilon\cosh\left(\frac{\tilde{\zeta}}{2}\right), \tag{5.18}$$

has a constant low-$\tilde{\zeta}$ limit of ϵ/λ_D, as found in Equation 5.11. At high $\tilde{\zeta}$, however, the surface charge (and thus capacitance) increases exponentially with ζ.

5.1.2 Assumptions Underlying the Poisson–Boltzmann Equation

While the derivation may seem straightforward, several assumptions are hidden in the Poisson–Boltzmann equation.

First and foremost, we have assumed that a thermodynamically large number (say N cations and N anions) of ions, with $6N$ total degrees of freedom, can be accurately represented in terms of a continuum field $n_{\pm}(\mathbf{r})$ for each ion species, instead of in terms of the specific positions $\mathbf{r}_i^{\pm}(t)$ of all $2N$ ions at any given time t. Tracking $2N$ ion positions would, of course, be impossibly cumbersome. Given the importance of thermal fluctuations on the nanoscale, however, we adopt a statistical approach for the ion distributions, where ions have ensemble-averaged probability densities $n_{\pm}(\mathbf{r}, t)$. Furthermore, we assume that the electrostatic field experienced by any individual ion is equal to the *mean field* set up by the *ensemble-averaged* distribution of ions around them.

There are several circumstances under which the fundamental assumptions that underlie the Poisson–Boltzmann equation are violated.

1. *Noncontinuum effects.* The Poisson–Boltzmann equation assumes the material properties of the liquid electrolyte to retain their continuum (large-scale) values, even when, for example, interion distances within the EDL can approach molecular dimensions. For example, the water that forms hydration shells around ions or surfaces differs in its mechanical and dielectric response than bulk, liquid water.
2. *Steric effects between ions.* Poisson–Boltzmann assumes ions to be point-like and noninteracting, and that ions only interact with the mean electrostatic field, rather than with specific neighboring ions. Equation 5.2 predicts ion densities that increase exponentially with electrostatic potential, which becomes physically unreasonable for even moderate values of ϕ. For example, the concentrations predicted by PB give an ionic volume fraction (assuming ions as hard spheres, each with a volume V)

$$\nu = n_+ V = n^0 V \exp\left(\frac{e\phi}{k_B T}\right), \tag{5.19}$$

which becomes fully saturated with ions ($\nu = 1$) at a critical potential ϕ_c, given by

$$\phi_c = \frac{k_B T}{e} \ln\left(\frac{2}{\nu^B}\right), \tag{5.20}$$

where $\nu^B = 2n^0 V$ is the volume fraction of ions in the bulk of the solution. Clearly, the ideal, point-like ion assumption made in the standard Poisson–Boltzmann equation breaks down for ϕ anywhere near ϕ_c [23]. In what follows, we describe modifications to the Poisson–Boltzmann equation that account for steric effects between ions in a mean-field way.

3. *Non-mean field electrostatic interactions.* Another assumption is that the ions respond only to the ensemble-averaged field—*irrespective* of what ions are nearby. When ions interact strongly enough with their neighbors—so that a nontrivial fraction of their interaction energy comes from non-mean-field interactions, then the mean-field theory will break down. This can lead to a variety of qualitative changes to the basic predictions of the Poisson–Boltzmann equation.

In the bulk, highly valent ions can act in a correlated manner, even in the bulk electrolyte—meaning that the distribution of one ion species depends not purely on $\phi(\mathbf{r})$, but on the local ion density. As a somewhat absurd extreme, consider a colloidal sphere of radius a, with some surface charge density σ_w and thus total "valence" $4\pi a^2 \sigma_w$. Such colloids are surrounded by their own screening clouds—see Table 5.1 for the linearized solutions. Clearly, the ion distribution around such a "highly valent" colloid differs significantly from the distribution that would exist if the colloid were not present, and so one cannot treat a highly valent colloid as simply another ion species in the Poisson–Boltzmann equation.

Surfaces can induce significant qualitative changes as well. Rather than simply screening a surface charge in a monotonic way, multivalent ions can overcharge surfaces—effectively reversing their sign [24]. In addition, the strong repulsion between counterions attracted to a surface can cause them to arrange in a correlated way—even crystallizing due to mutual repulsion [25]. The intricate interplay between strongly charged surfaces and highly valent ions can lead to a variety of counterintuitive effects, including attractions between like-charged species. A well-known example is the condensation of DNA in the presence of multivalent ions [26].

5.1.3 Alternate Approach: The Electrochemical Potential

Before moving on to *electrokinetic* effects, we briefly discuss a complementary—and more easily generalizable—description of ion electrostatics.

Instead of assuming a Boltzmann distribution *a priori*, we describe the free energy of each ion in terms of its chemical potential—here generalized to an "electrochemical" potential to account for electric fields. The electrochemical potential for an ideal solution of ions (i.e., point-like and noninteracting) is given by

$$\mu_{\pm} = \pm e\phi + k_B T \ln n_{\pm}. \tag{5.21}$$

The first term represents the free energy of an ion of charge $\pm e$ in an electrostatic field ϕ, with a sign that depends on the ion charge, and the second term represents the free energy of mixing for an ideal solution.

In equilibrium, the electrochemical potential of each ion must be constant: any gradients in the potential energy of an ionic species would drive ion fluxes moving to restore equilibrium. Taking $\mu_\pm^0$ as a reference electrochemical potential in the reservoir, where the electrostatic potential ϕ is zero and there are no concentration gradients, gives

$$\mu_\pm^0 = k_B T \ln n_\pm^0. \tag{5.22}$$

Because the electrochemical potential of ions within the double layer is identical to that outside the double layer, we find

$$\mu_\pm^0 = k_B T \ln n_\pm^0 = \pm e\phi + k_B T \ln n_\pm, \tag{5.23}$$

which are solved by the Boltzmann distribution (5.2), as expected.

Although the electrochemical potential framework loses some of the "mechanical" feel of the standard Poisson–Boltzmann equation, it has the benefit of being more easily generalizable. For example, we discussed situations where the Poisson–Boltzmann equation fails due to steric interactions between ions (Eq. 5.20). A natural extension is to maintain the mean-field approximation, but to account for finite-ion-size effects by modifying the electrochemical potential. A host of so-called modified Poisson–Boltzmann equations have been derived, going back over half a century to Bikerman [27], with Bazant *et al.* [23] providing a comprehensive review and discussion of the intellectual history of the approach. In particular, Bikerman accounted for the nonideality of the (concentrated) ions within a highly charged double layer by adding a simple excess chemical potential,

$$\mu_{\text{ex}} = -k_B T \ln(1 - \nu) = -k_B T \ln(1 - n_+ V_+ - n_- V_-), \tag{5.24}$$

to the ideal solution electrochemical potential (5.23). This form corresponds to ions that are constrained to occupy a lattice. The resulting modified Poisson–Boltzmann equation can be solved exactly, giving

$$n_\pm = \frac{n_\pm^0 \exp(\mp\tilde{\phi})}{1 + 2\nu^B \sinh^2(\pm\tilde{\phi}/2)}, \tag{5.25}$$

where we have scaled potentials by the thermal potential, as in Equation (5.13). In reality, of course, ions are not constrained to reside on a lattice, and so other forms for the excess chemical potential can be adopted to account for the nonideal nature of ions in concentrated solutions. For example, the Carnahan–Starling equation of state for hard sphere suspensions has also been used [23].

Several features in the ion distribution (5.25) are notable: first, when the bulk volume fraction ν^B, or the electrostatic potential ϕ, is low enough that $\nu^B \sinh^2(\pm e\phi/2k_B T) \ll 1$ (which itself corresponds to a low volume fraction ν), Bikerman's distribution (5.25) simply reduces to the standard Boltzmann distribution (5.2). When the potential becomes large, however, the second term in the denominator of (5.25) dominates, giving

$$n_+ \approx \frac{2n_+^0}{\nu^B} \exp(-2\tilde{\phi}), \tag{5.26}$$

$$n_- \approx \frac{2n_-^0}{\nu^B} = \frac{1}{V}, \tag{5.27}$$

for co-ions and counterions, respectively, assuming the double layer to be positively charged. The density of co-ions becomes exponentially small, whereas the density of counterions saturates at close packing $n_- V \approx 1$. This stands in marked contrast with the ideal solution theory, which predicts a counterion density that increases exponentially with ϕ, without bound.

5.1.4 Electrokinetics

We now turn to *electrokinetic* effects, in which externally applied fields interact with electrical double layers to drive fluid flows or particles into motion. The reverse is also true: particle motions (e.g., sedimentation) or imposed flows (e.g., pressure-driven flows) interact with EDLs to establish electric fields or potentials. While the most commonly discussed field is undoubtedly the electric field, it is also possible to impose ion concentration gradients or thermal gradients. For a more in-depth discussion of the zoology of electrokinetic effects, see Ref. [16].

Here we mention several of the most commonly discussed *linear* electrokinetic effects (i.e., effects that arise in direct proportion to an applied field). An applied electric field forces the electrically charged fluid within the EDL to drive an *electro-osmotic flow* (EOF) over a charged surface, shown in Figure 5.2a. An applied electric field forces an EOF around a charged particle, causing that particle to migrate through the fluid via *electrophoresis* (Fig. 5.2b). A fluid flow over a charged surface advects the charge within the EDL to drive a *streaming current* (Fig. 5.2c). An ionic strength gradient in the bulk causes a gradient in the EDL thickness. Entropic forces drive ions within the EDL *down* the concentration gradient, exerting a net entropic force along the EDL that drives *diffusio-osmotic* or *chemiosmotic* flow (Fig. 5.2d). The EDL around a particle suspended in an electrolyte with a concentration gradient is driven into chemiosmotic flow, causing a slip velocity *down* the concentration gradient, which drives the particle *up* the concentration gradient via *diffusiophoresis*, or *chemiphoresis* (Fig. 5.2e). Dense particles sedimenting under gravity through an electrolyte establish a *sedimentation potential*: ions within the EDL are advected toward the rear of the

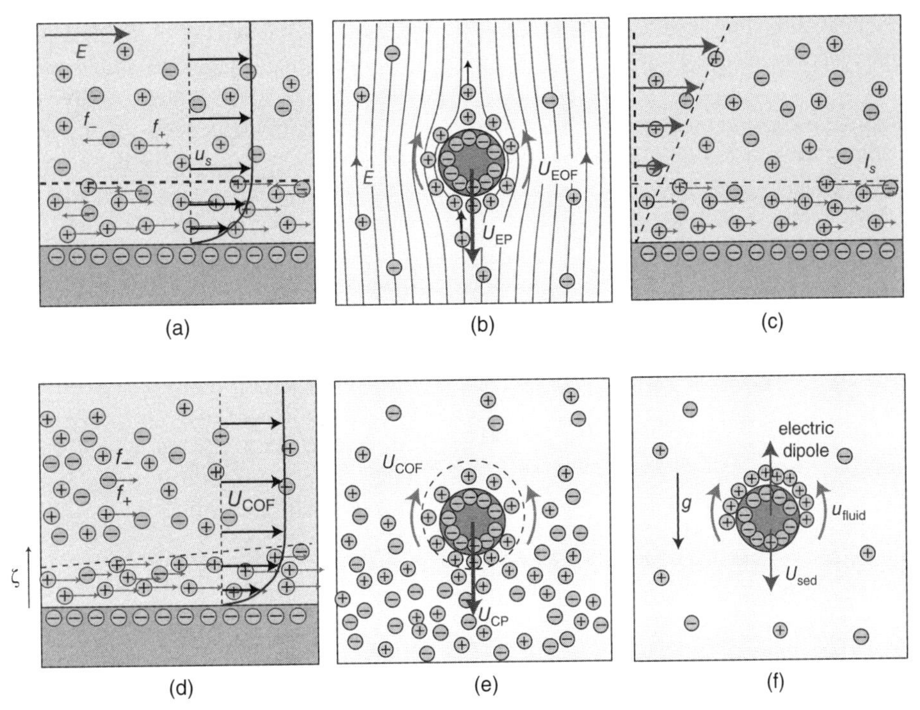

Figure 5.2 *Six paradigmatic electrokinetic effects*. Clockwise, from top: (a) *Electro-osmotic flows* arise when an applied electric field forces the charged EDL, driving a fluid flow. (b) *Electrophoresis* refers to the motion of a charged particle under an applied electric field. (c) *Streaming currents and potentials* arise when fluid flows (pressure-driven or shear) convectively transport ions within an EDL along a surface. (d) *Diffusio-osmotic* and *Chemi-osmotic flows* occur when gradients in solute (often electrolyte) concentration drive the EDL, and the fluid around it, into motion. (e) *Diffusiophoresis* and *Chemiphoresis* refers to the motion of a particle under the influence of a gradient in solute (often electrolyte) concentration. (f) *Sedimentation potentials* are established when a charged particle sediments under gravity. The fluid flow around the sedimenting particle drives a streaming current along the particle surface, which develops an electrostatic potential to drive a return current through the bulk.

moving particle due to fluid shear stresses, which imparts an electrostatic dipole to the overall electrolyte (Fig. 5.2f). *Electro-viscous* effects refer to the additional retardation or dissipation that occurs around a charged particle or surface due to the deformation (and response) of the electric double layer. The diffusivity of a colloid, for example, is reduced due to electroviscous effects.

5.2 THE POISSON–NERNST–PLANCK–STOKES EQUATIONS

Electrokinetic effects require the computation of the *dynamics* of ions and fluids, and thus necessitate the full "standard" set of equations for ion dynamics. These include the Poisson equation from electrostatics, the Nernst–Planck equations for ion transport, and the Stokes equations for viscous fluid flows.

The force on each ion is given by the negative gradient of its electrochemical potential, which for ideal point-like ions (Eq. 5.21) gives

$$f_{\pm}(\mathbf{r}) = -\nabla\mu_{\pm}(\mathbf{r}) = \mp e\nabla\phi(\mathbf{r}) - \frac{k_BT}{n_{\pm}}(\mathbf{r})\nabla n_{\pm}(\mathbf{r}). \quad (5.28)$$

The hydrodynamic mobility of each ion, $b_{\pm}$, then relates the force that is exerted upon each ion to its velocity $\mathbf{v}_{\pm}(\mathbf{r})$ *relative to* the local fluid, which itself may have a velocity $\mathbf{u}(\mathbf{r})$

$$\mathbf{v}_{\pm}(\mathbf{r}) - \mathbf{u}(\mathbf{r}) = \mp eb_{\pm}\nabla\phi(\mathbf{r}) - \frac{k_BTb_{\pm}}{n_{\pm}}(\mathbf{r})\nabla n_{\pm}(\mathbf{r}). \quad (5.29)$$

Finally, the local flux of each ion species is given by its velocity $\mathbf{v}_{\pm}$ times its number density:

$$\mathbf{j}_{\pm}(\mathbf{r}) = \mp eb_{\pm}n_{\pm}(\mathbf{r})\nabla\phi(\mathbf{r}) - k_BTb_{\pm}\nabla n_{\pm}(\mathbf{r}) + \mathbf{u}n_{\pm}(\mathbf{r}). \quad (5.30)$$

Ion conservation thus gives the Nernst–Planck equations:

$$\frac{\partial n_{\pm}(\mathbf{r})}{\partial t} = -\nabla\cdot\mathbf{j}_{\pm}(\mathbf{r}) = \pm eb_{\pm}\nabla\cdot(n_{\pm}(\mathbf{r})\nabla\phi(\mathbf{r})) + k_BTb_{\pm}\nabla^2 n_{\pm}(\mathbf{r}) - \mathbf{u}\cdot\nabla n_{\pm}(\mathbf{r}). \quad (5.31)$$

The first term corresponds to the *electromigrative* flux of ions that are driven *through* their local fluid environment by the electrical force exerted upon them. Using the Stokes–Einstein relation, $D_{\pm} = k_BTb_{\pm}$, the second term simply represents Fickian diffusion. The third term represents ion advection—*electroconvection*—with the flowing fluid.

The electrostatic field is given by Poisson's equation (5.1), and the fluid flow obeys the Stokes equations, with body forces on the fluid given by

$$\mathbf{f}_{\text{fluid}}(\mathbf{r}) = -\nabla\mu_+(\mathbf{r})n_+(\mathbf{r}) - \nabla\mu_-(\mathbf{r})n_-(\mathbf{r}) \tag{5.32}$$

to give

$$-\nabla P(\mathbf{r}) - \eta\nabla^2\mathbf{u}(\mathbf{r}) - \rho_E\nabla\phi(\mathbf{r}) - k_BT\nabla n(\mathbf{r}) = 0, \tag{5.33}$$

$$\nabla\cdot\mathbf{u} = 0, \tag{5.34}$$

where

$$\rho_E(\mathbf{r}) = e(n_+(\mathbf{r}) - n_-(\mathbf{r})) \tag{5.35}$$

is the local charge density, and

$$n(\mathbf{r}) = n_+(\mathbf{r}) + n_-(\mathbf{r}) \tag{5.36}$$

is the local ion concentration. Note that the electrochemical potential neatly accounts for both electrostatic forces and osmotic pressure gradients.

The Poisson–Nernst–Planck–Stokes equations represent a set of coupled, nonlinear partial differential equations: the ion fields (5.31) are driven by nonlinear interactions between ion, electric, and fluid fields; the fluid velocity (5.34) is forced by the interaction of the ion and electrostatic fields.

Hidden in these equations are the same assumptions that underlie the Poisson–Boltzmann equation: point-like ions that interact with the mean electrostatic field (and not with each other directly), within a fluid whose properties are homogeneous and given by their macroscopic, continuum values. If a modified Poisson–Boltzmann equation—which accounts for ion interactions with a modified chemical potential—more accurately describes the EDL, then the equivalent modifications can be made to the PNPS equations by simply modifying the ion forces (5.28) appropriately.

5.3 ELECTRO-OSMOTIC FLOWS

EOF arise when an externally applied electric field exerts a force on the charged liquid within the EDL surrounding a solid surface, driving the fluid itself into motion. Figure 5.2a shows a schematic: with all driving force exerted within a distance of order λ_D from the surface, the EOF velocity profile increases from a nonslip surface to an apparent "slip velocity" outside the EDL. Since the EDL is often (but not always) orders of magnitude thinner than the geometric length scales of many experimental systems (microfluidic or colloidal), details of the EOF distribution within the EDL are often neglected, and EOF instead treated by simply imposing an electrokinetic "slip" velocity at the surface, whose magnitude depends upon the local electric field and surface charge density. We therefore compute the EOF over a flat surface, which also represent the EOF over a surface whose radius of curvature is much greater than λ_D.

The standard approach to computing the EOF profile is remarkably general. An applied field $\mathbf{E}$ exerts a force $\rho_E\mathbf{E}_{\text{app}}$ on the charged fluid within the double layer, which thus enters the force balance inherent in the Stokes equations for viscous flows in the zero Reynolds number limit. Here, we simply consider a uniformly charged, planar surface oriented in the x–y plane, with electrolyte located at $z > 0$, under an applied field $\mathbf{E} = E_{\text{app}}\hat{x}$. In this case, the forced Stokes equations become

$$0 = \eta\frac{\partial^2 u_x}{\partial z^2} - \frac{\partial P}{\partial x} + \rho_E E_{\text{app}}. \tag{5.37}$$

Assuming there to be no pressure gradient at infinity—which is tantamount to specifying "open" ends through which fluid flows freely—and using Poisson's equation (5.1), (5.37) becomes

$$\frac{\partial^2 u_x}{\partial z^2} = \frac{\epsilon E_{\text{app}}}{\eta}\frac{\partial^2\phi}{\partial z^2}. \tag{5.38}$$

Even without specifying the model of the double layer (i.e., linear or nonlinear Poisson–Boltzmann, or one of the modified PB equations), this equation can be integrated twice to obtain

$$u_x(z) = \frac{\epsilon E_{\text{app}}}{\eta}(\phi(z) - \phi(0)), \tag{5.39}$$

where the no-slip boundary condition is enforced at $z = 0$. The electrokinetic "slip" velocity that results is then

$$u_x(z \to \infty) \equiv U_{\text{EOF}} = -\frac{\epsilon\zeta}{\eta}E_{\text{app}}, \tag{5.40}$$

known as the Helmholtz-Smoluchowski slip velocity, where

$$\zeta \equiv \phi(0) - \phi(z \to \infty) \tag{5.41}$$

is called the "zeta potential."

Several assumptions lay hidden within this approach. First, the physical properties of the electrolyte—namely, viscosity and permittivity—are assumed to remain constant and equal to their bulk values throughout the entire double layer. This assumption is likely to fail within concentrated or highly charged double layers—due to the strength of the hydrodynamic and interion interactions in a concentrated suspension of ions, the molecular structure of water near the surface, the inhomogeneity of the charge distribution, the roughness of the surface, and so on. Second, a tacit assumption is that the electrostatic potential $\phi = \zeta$ at the location where the no-slip boundary condition is applied (known as the "shear plane," here taken to be $z = 0$) is the same as the electrostatic potential at the wall. This would not be the case if, for example, a "Stern layer" of physically adsorbed ions (e.g., partially hydrated, but immobilized) were present, nor for rough surfaces. Finally, it assumes that the standard no-slip boundary condition applies, although that can be relaxed rather easily [28, 29].

5.3.1 Alternate Approach: The Electrochemical Potential

The electrochemical potential can be used to recast electro-osmosis in terms that are more easily generalizable, for example, for diffusio-osmosis or for systems with strong surface conduction. What makes this approach particularly simple is that—for thin double layers in near-equilibrium—the chemical potential of each ion species is approximately constant *across* the double layer. Consider an externally imposed chemical potential gradient $\nabla\mu_i^B$ in the bulk—examples include an applied electric field as in electro-osmosis or electrophoresis, or an imposed concentration gradient in diffusiophoresis. If this driving force is suitably weak, the electrochemical potential of each ion species remains approximately constant *across* the double layer, even in spite of the gradient *along* the double layer. Therefore, the force on each ion *within* the double layer,

$$F_i(z) = -\nabla\mu_i(z), \tag{5.42}$$

can then be simply evaluated using the chemical potential gradient *outside* the double layer via

$$F_i = -\nabla\mu_i^B. \tag{5.43}$$

While forces *do* exist on individual ionic species in the bulk fluid, they generally do not drive a flow: a simple Ohmic current is driven under an applied electric field, and a macroscopic osmotic pressure gradient is established under a concentration gradient. Electrokinetic motion is instead driven by the excess force on the surplus (or deficit) ions within the double layer, $\Delta n_i(x, z) = n_i(x, z) - n_i^B(x)$, giving a body force

$$f_i = \sum_{i=1}^{N} F_i\Delta n_i = -\sum_{i=1}^{N} \Delta n_i \frac{\partial\mu_i^B}{\partial x}. \tag{5.44}$$

Each ion species for which a chemical potential gradient $-\nabla\mu_i^B$ exists outside the double layer, then, drives an electro-osmotic flow according to

$$0 = \eta\frac{\partial^2 u_x}{\partial z^2} - \sum_{i=1}^{N} \Delta n_i(z)\frac{\partial\mu_i^B}{\partial x}, \tag{5.45}$$

giving

$$u_x(z) = -\sum_{i=1}^{N} \frac{1}{\eta}\frac{\partial\mu_i^B}{\partial x}\int_0^z \int_{z'}^{\infty} \Delta n_i(z'')dz''\, dz'. \tag{5.46}$$

The equivalent of the Helmholtz–Smoluchowski slip velocity is then found in the limit $z \to \infty$ to be

$$U_x = -\sum_{i=1}^{N} \frac{1}{\eta}\frac{\partial\mu_i^B}{\partial x}\int_0^{\infty} z'\Delta n_i(z')dz' \equiv \sum_{i=1}^{N} M_i\left(-\frac{\partial\mu_i^B}{\partial x}\right). \tag{5.47}$$

where

$$M_i = \frac{1}{\eta}\int_0^{\infty} z'\Delta n_i(z')dz'. \tag{5.48}$$

These integrals can be computed exactly for symmetric, monovalent ions distributed according to the nonlinear Poisson–Boltzmann equation, giving

$$M_{\pm}^{GC} = \frac{\epsilon}{e\eta}\left(\mp\frac{\zeta}{2} + 2\frac{k_BT}{e}\ln\cosh\frac{\tilde{\zeta}}{4}\right), \tag{5.49}$$

where $\tilde{\zeta}$ is the scaled ζ-potential, given by Equation (5.17). The Helmholtz–Smoluchowski equation (5.40) for the electro-osmotic slip velocity arises naturally from Equation 5.49 and Equation 5.47 when a constant electric field ($\mu_+ = e\phi = -eEx$ and $\mu_- = -e\phi = eEx$) is applied.

If, on the other hand, an ionic strength *gradient* is imposed,

$$\nabla n_B^+ = \nabla n_B^- \equiv \nabla n_B^0, \tag{5.50}$$

then the chemical potential gradients for the two ionic species

$$\nabla\mu_{\pm}^B = k_BT\frac{\nabla n_{\pm}}{n_{\pm}}, \tag{5.51}$$

are identical. The slip velocity that results from an external concentration gradient, computed using Equation 5.51 in Equation 5.47, gives a *diffusio-osmotic*, or *chemiosmotic* slip velocity [11, 30, 31]

$$U_{\rm DOF} = \sum_{i=1}^{2} M_i^{GC}\nabla\mu_i^B = -\left(4\frac{\epsilon}{\eta}\left(\frac{k_BT}{e}\right)^2 \ln\cosh\frac{\tilde{\zeta}}{4}\right)\nabla\ln n_0^B. \tag{5.52}$$

For $\tilde{\zeta} \ll 1$ (or $\zeta \ll \phi_{\rm th}$), one can expand $\ln\cosh x \sim \ln(1 + x^2/2) \sim x^2/2$ to give

$$U_{\rm DOF} \approx -\frac{\epsilon\zeta^2}{8\eta}\nabla\ln n_B^0. \tag{5.53}$$

Diffusio-osmotic flows are driven by osmotic pressure gradients $k_BT\nabla n_B^0$, rather than by electric forces. The osmotic pressure inside the EDL at high ionic strength is higher than in regions of the EDL at low ionic strength. This gradient leads to a net force on the EDL, down the osmotic pressure gradient, from high ionic strength to low. The *sign*

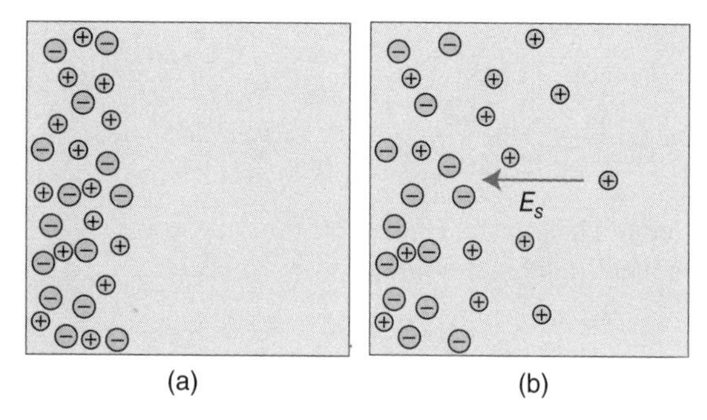

Figure 5.3 *Ambipolar diffusion.* The diffusive relaxation of an electrolyte whose ion diffusivities differ—here cations are fast and anions are slow—with an initial gradient of ionic strength (a). The faster species outruns the slower, spontaneously generating an electric field E_s that retards the more mobile species and speeds the less mobile ions (b). The two ion species then diffuse in a correlated manner, and the field E_s can drive an electro-osmotic (electrophoretic) component of diffusio-osmosis (diffusiophoresis).

of the charge in the EDL does not matter (note the DOF mobility is *even* in ζ), just the excess salt in the EDL.

Ionic strength gradients can also drive a variant of electro-osmosis, when the different ions have different diffusivities [30, 32]. Figure 5.3 shows an example where the cation is the more highly mobile ion, which diffusively relaxes more quickly. In doing so, it outruns the less mobile anion, driving a charge separation that gives rise to an electric field in the bulk, given by

$$E_{\nabla n} = \frac{k_B T}{e} \left(\frac{D_+ - D_-}{D_+ + D_-} \right) \nabla (\ln n_B^0). \tag{5.54}$$

This field speeds the relaxation of the slow species and retards the relaxation of the fast species. The two species thus diffuse in a correlated manner—called ambipolar diffusion [32, 33]—and an ionic strength gradient naturally establishes an electric field $E_{\nabla n}$ in the bulk that drives an EOF via Equation (5.40).

The value of the chemical potential picture is its simplicity and generality: a more detailed mechanical treatment of diffusio-osmosis [30, 34, 35] is more involved, requiring weak osmotic pressure gradients within the EDL to be computed due to the changing screening length along the surface. Here, on the other hand, one can simply compute the gradients *outside* the double layer, and use the general mobility M_i for each ionic species. Second, this picture holds for *any* mean-field description of the double layer, wherein the physical properties of the electrolyte are also assumed to remain constant. Gradients in viscosity, treated in a mean-field fashion—could be easily accommodated by moving $\eta(z)$ into the integral in Equation (5.47). The slip velocities for ions in modified Poisson–Boltzmann equations, for example, Bikerman's, can be evaluated using Equation (5.47), typically numerically, once the ion distributions $\Delta n_\pm$ are known. This approach is valuable when we consider effects of surface conduction (finite Dukhin number effects), as different ion species are affected differently in this regard.

5.4 ELECTROPHORESIS

We now turn to electrophoresis, in which charged particles are driven into motion by applied electric fields. Electrophoresis is by now a standard technique to characterize the surface (ζ) potential of colloids, by measuring the electrophoretic mobility,

$$M_{\text{EP}} = \frac{U_{\text{EP}}}{E}. \tag{5.55}$$

Electrophoresis is also widely used in separations—in particular, of nucleic acids and proteins—and represents a workhorse of molecular biology. The original sequencing of the human genome was accomplished with electrophoretic separations.

Physically, electrophoresis would seem rather simple: an electric field E forces a particle with charge Q through a fluid. A common—but incorrect—assumption is that the particle moves simply according to Stokes drag. In fact, the process is complicated by the presence of the EDL surrounding the particle: not only is there a QE force on the particle, but there is an equal and opposite $-QE$ force on the double layer; the particle/EDL ensemble, then, experiences identically zero force.

So why does the particle/EDL ensemble move, if there is no force on it? After all, a nonionized neon atom does not translate in an electric field; the electron distribution polarizes under the applied field, but the nucleus/electron cloud experiences no total force and thus does not translate. And, in fact, if the EDL itself were surrounded by a sort of membrane that did not pass ions (Fig. 5.4), the particle/EDL ensemble would also not translate. No such membrane exists, however. On the contrary, a reservoir of counterions and co-ions surrounds the particle/EDL, and ions can exchange between the bulk and EDL freely. It can happen, though, that there are so few ions in bulk that this exchange does not happen as quickly as "standard" electrophoresis would require—this is the realm of strong surface conduction (or high Dukhin number) and is discussed below.

A variety of subtleties arise in electrophoresis, owing to the "zero net force" exerted by the field on the particle/EDL: (i) the flow outside the EDL differs fundamentally and qualitatively from its Stokes-drag counterpart in force-driven motion, usually decaying with distance like r^{-3}, but at least as fast as r^{-2}, compared with the slower r^{-1} decay in sedimentation [36]; (ii) the electrophoretic mobility of a particle with thin double layer and uniform ζ potential is independent of its size, shape, or even concentration;

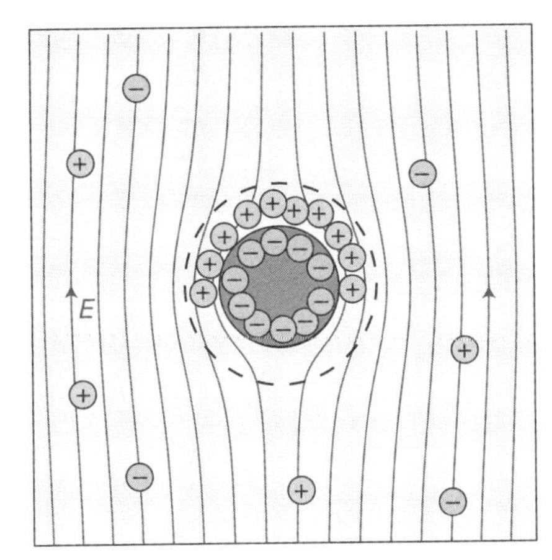

Figure 5.4 If ions were physically prevented from entering or leaving the EDL, for example, by a fictitious membrane (dashed line), electrophoresis would not occur. The EDL would be polarized by the electric field, but the (charge neutral) particle + EDL ensemble would experience no net force, and the ensemble would not move. This picture is relevant for very highly charged particles, for which surface conduction within the EDL is strong and the Dukhin number is high. In that case, there are so few counterions in the bulk, compared with the number required for "standard" electrophoresis, that the electrophoretic mobility of the particle is limited by bulk counterion transport processes.

and (iii) the electrophoretic mobility can be *nonmonotonic* with ζ.

At times, it will seem that perfectly reasonable calculations, well posed and accurately solved, give contradictory results. A large part of this difficulty is that there are *two* dimensionless parameters: κa which gives the sphere radius, relative to the double-layer thickness, and $\tilde{\zeta} = e\zeta/k_BT$, which gives the zeta potential, relative to the thermal voltage. The limit of $\kappa a \to \infty$, in particular, is mathematically singular, so that making the approximation $\kappa a \to \infty$ *before* calculating for a particular $\tilde{\zeta}$, can give a different result than imposing $\tilde{\zeta}$ first, then taking $\kappa a \to \infty$. We explore these effects in what follows, and see that in fact a dimensionless combination of these groups—called the Dukhin number—controls such systems.

5.4.1 Electrophoresis in the Thick Double-Layer Limit

One of the first rigorous theories for electrophoresis was given by Henry [37], who explicitly computed the hydrodynamic flows associated with a particle moving electrophoretically. Specifically, he assumed the double layer around a spherical particle of radius a to retain its equilibrium structure—obeying the linearized Poisson–Boltzmann equation (5.7)—even in the presence of an applied electric field. The electric field then interacts with the charge density ρ_E in the EDL to exert a body force on the fluid, which establishes a flow governed by the forced Stokes equations (5.34). Solving the Stokes equations with no-slip boundary conditions on the sphere surface ($r = a$), to approach a uniform (electrophoretic) velocity far away ($r \to \infty$), and requiring no net force on the system, gives an electrophoretic mobility

$$M_E = \frac{\kappa^2}{6}\frac{\epsilon\epsilon_0\zeta}{\eta}\int_a^\infty \frac{a^6 - a^4r^2 + 4ar^5 - 4r^6}{r^5} e^{-\kappa(r-a)}\, dr. \tag{5.56}$$

In the limit of small screening length $\kappa a \gg 1$, the mobility can be expressed as

$$M_E = \frac{\epsilon\epsilon_0\zeta}{\eta}\left(1 - \frac{3}{\kappa a} + \frac{25}{(\kappa a)^2} - \frac{220}{(\kappa a)^3} + \cdots\right), \tag{5.57}$$

and in the opposite limit of large screening length $\kappa a \ll 1$, the mobility is found to be

$$M_E = \frac{2}{3}\frac{\epsilon\epsilon_0\zeta}{\eta}\left(1 + \frac{(\kappa a)^2}{16} - \frac{5(\kappa a)^3}{48} - \frac{(\kappa a)^4}{96} + \frac{(\kappa a)^5}{96} - \frac{(\kappa a)^6 - 12(\kappa a)^4}{96} e^{\kappa a}\int_{\kappa a}^\infty \frac{e^{-t}}{t}dt\right), \tag{5.58}$$

The large screening length limit ($\kappa a \to 0$) is equivalent to no screening and represents Stokes drag $U = F/(6\pi\eta a)$ for a particle of surface potential $\zeta = Q/(4\pi\epsilon\epsilon_0 a)$.

5.4.2 Electrophoresis in the Thin Double-Layer Limit

In systems where the DL is thin compared with other macroscopic geometric dimensions (e.g., radius of curvature of a colloid, height or width of a channel, radius of a pore), a common approach is to simply solve the *unforced* Stokes equations everywhere in the flow and to impose the EOF slip velocity (Eq. 5.47), as appropriate for electro-osmosis (Eq. 5.40) or diffusio-osmosis (Eq. 5.52), as a boundary condition on the unforced Stokes flow outside the EDL.

Here, we calculate the electrophoretic mobility M_E of such a particle, in the limit where the local radius of curvature of the surface is everywhere much larger than the screening length, so that the EDL occupies an infinitesimally thin region along the particle surface. We follow the approach developed by Morrison [38], who demonstrated that M_E is given simply by $\epsilon\zeta/\eta$—irrespective of the size, shape, or even number of particles. In addition, the electrostatic and hydrodynamic fields are directly proportional to one another, with the same boundary conditions to within a multiplicative constant. Morrison's argument holds more generally for surface-force-driven motion [31]—for example, diffusiophoresis—as well as for electro-osmosis [39].

The governing equations and boundary conditions for the electrostatic potential ϕ and the fluid velocity $\mathbf{u}$ are shown in Table 5.2. In the bulk fluid outside of the charge cloud,

TABLE 5.2 Governing Equations and Boundary Conditions for the Electric Field and Potential Flow in Electrophoresis with Thin Double Layers and Uniform ζ [31, 38].

	Electrostatic Field $\mathbf{E} = -\nabla\phi$	Fluid Flow $\mathbf{u} = \nabla\psi$
Field equation	$\nabla^2\phi = 0$	$\nabla^2\psi = 0$
B.C. $r \to \infty$	$\nabla\phi \to -\mathbf{E}_\infty$	$\nabla\psi \to -\mathbf{U}_e$
No-flux B.C. $r = a$	$\hat{\mathbf{n}} \cdot \nabla\phi = 0$	$\hat{\mathbf{n}} \cdot \nabla\psi = 0$
Tangential B.C. $r = a$	$\hat{\mathbf{t}} \cdot \nabla\phi = -E_\parallel$	$\hat{\mathbf{t}} \cdot \nabla\psi = -\frac{\epsilon\epsilon_0\zeta}{\eta}E_\parallel$

Because the electrostatic and hydrodynamic potentials obey the same equation, with the same boundary conditions (up to a single, multiplicative constant), the solutions are simply proportional to each other, $\psi = (\epsilon\epsilon_0\zeta/\eta)\phi$. The electrophoretic mobility in this limit is thus independent of the particle size, shape, or even concentration.

ϕ obeys Laplace's equation, and the electric field approaches $\mathbf{E}_\infty$ far from the particle. In the absence of electrochemical reactions at the particle surface, the electric field normal to the surface must be zero; otherwise, a steady-state current $\mathbf{j}_\perp = \sigma\mathbf{E}_\perp$ would be driven into the EDL and particle. Finally, the electric field tangent to the surface is defined to be $\mathbf{E}_\parallel$.

We examine the fluid velocity in a frame moving with the particle, so that $\mathbf{u} \to -\mathbf{U}_e$ far from the particle. The fluid velocity obeys Stokes equations with an electrical body force $\rho_E\mathbf{E}$. In the thin EDL limit, the charge cloud looks locally planar, so that the fluid just outside of the EDL is given by the Helmholtz-Smoluchowski slip velocity (Eq. 5.40), and is proportional to the local tangential electric field. Finally, the fluid velocity normal to the surface must be zero.

Remarkably, the Stokes equations with the above boundary conditions admit *potential flow* (irrotational) solutions:

$$\mathbf{u} = \nabla\psi, \tag{5.59}$$

where ψ is the hydrodynamic velocity potential. One would never expect this to work—after all, potential flows are irrotational and used to describe inviscid flows. Potential flows do not allow both no-flux and no-slip conditions to be imposed, which is what viscous (Stokes) flows require. Only if one is exceedingly lucky, can a potential flow obey both boundary conditions. In the case of electrophoresis [38], or in surface-force driven "phoretic" motions [31], Morrison showed that one is indeed so lucky. The tangential electric field is not specified *a priori* in this problem but is defined to be whatever it works out to be, given Laplace's equation with a no-flux condition at the boundaries. The tangential slip velocity at the boundaries, on the other hand, *is* imposed explicitly; however, the imposed flow is simply proportional to the tangential electric field that solves Laplace's equation. The equations and boundary conditions for the fluid and electric potentials (ψ and ϕ) are therefore nearly identical, differing only by a multiplicative constant

$$\psi = \frac{\epsilon\epsilon_0\zeta}{\eta}\phi. \tag{5.60}$$

The electrophoretic velocity is thus given by

$$\mathbf{U}_e = \frac{\epsilon\epsilon_0\zeta}{\eta}\mathbf{E}_\infty, \tag{5.61}$$

and the electrophoretic mobility by

$$M_E = \frac{\epsilon\epsilon_0\zeta}{\eta}, \tag{5.62}$$

as found by Henry for a sphere in the $\kappa a \to \infty$ limit (Eq. 5.57).

Nowhere in this argument have we had to specify the size, shape, or even number of particles—indicating that the electrophoretic mobility in the thin EDL limit is *independent* of the size, shape, or even concentration of the particles! Rather, it depends only on a constant ζ and asymptotically thin double layer ($\kappa a \to \infty$). For this reason, electrophoretic separation of DNA by size is typically performed in a gel, so that friction increases with size [40, 41]. Furthermore, there is identically *no* interaction between particles moving under electrophoresis. The hydrodynamic and electrostatic interactions cancel each other identically, in striking contrast with sedimentation, where long-range hydrodynamic interactions lead to strong velocity fluctuations [36]. This remarkable result has been shown experimentally as well: Zukoski and Saville [42] measured the electrophoretic mobility of red blood cells in suspension to be *independent* of cell concentration up to 70% by volume!

Morrison's argument can be generalized to treat diffusiophoresis as well [30, 31], by replacing the electrostatic potential (ϕ) in Table 5.2 with the chemical potentials μ_i of all ion species. Notably, all obey $\nabla^2\mu_i = 0$ in bulk, with no normal gradient at the EDL surface, some (determined) tangential gradient, and a prescribed far-field gradient. The hydrodynamic potential established by the EOF due to each ion species is then proportional to the chemical potential of that ion, with proportionality constant M_i. For a binary electrolyte and the Gouy–Chapman equation, the diffusiophoretic mobility is

$$M_D = M_+^{GC} + M_-^{GC} = \left(4\frac{\epsilon}{\eta}\left(\frac{k_BT}{e}\right)^2 \ln\cosh\frac{\zeta}{4}\right), \tag{5.63}$$

as found previously [30].

Precision experiments for diffusiophoresis can be quite difficult, as it is more difficult to impose a large concentration gradient than it is to apply a large applied field. Furthermore, even weak convective flows can disrupt the

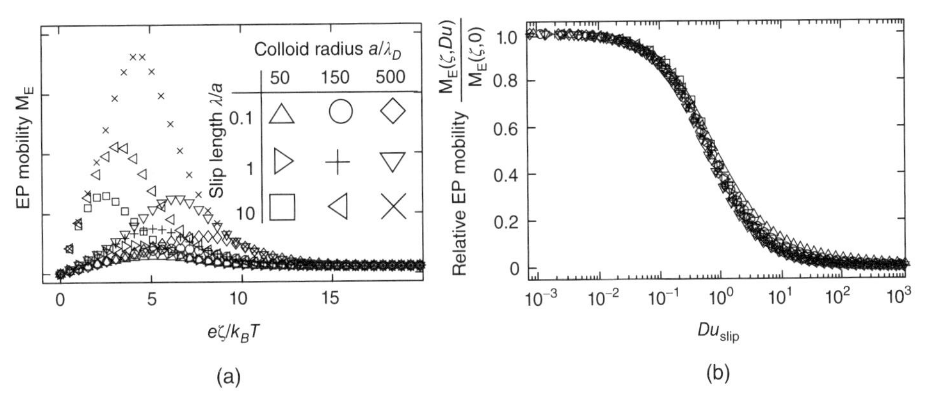

Figure 5.5 (a) The electrophoretic mobility of particles of radius a and solid/liquid "slip" length λ is nonmonotonic with ζ: for all slip lengths, the mobility M_E initially grows linearly with ζ, then reaches a maximum and subsequently decreases. Particles with longer slip lengths have greater electrophoretic mobilities initially—as expected—but attain a maximum mobility at lower ζ. (b) The same computed mobilities, replotted to emphasize the central role that surface conductivity plays in determining the electrophoretic mobility. Normalizing M_E by the low-ζ limit and plotting against the Dukhin number—suitably modified to account for the additional electroconvective surface current due to slip—reveals all computed mobilities to collapse onto a single master curve.

assumed diffusion. Recently, elegant microfluidic devices made of agarose enabled gradients to be imposed without convection, enabling diffusiophoretic manipulation of particles [43, 44] as well as direct measurements of M_D for electrolytes and solvents [45].

5.4.3 Electrophoresis for Arbitrary Charge and Screening Length

We have now seen two computations for electrophoresis—Henry's, involving a particle with an equilibrium, low-$\tilde{\zeta}$ double layer with arbitrary screening length, and Morrison's, involving a particle with an arbitrary ζ, but asymptotically thin double layer ($\kappa a \to \infty$). Moreover, the two approaches agree in their common limit. One might thus suspect that these two theories accurately cover the range of possibilities.

However, each of these approaches makes crucial simplifications *before* solving the problem: Henry assumes the EDL to retain its equilibrium structure and Morrison assumes the EDL to be infinitely thin. Both assumptions can be incorrect and cause qualitative changes in the mobility. This was made clear when Dukhin (summarized in Ref. [34]) and O'Brien and White [12] showed that assuming the EDL to remain unperturbed in the presence of an applied electric field violates ion conservation! The ionic current in regions with large ion density is larger than in regions of low ion density, so that a nonzero divergence $\nabla \cdot \mathbf{j}_\pm$ of ions occurs, and polarizes the double layer.

Full computations of the electrophoretic mobility of spherical colloids (e.g., [12, 34]) started with the full Poisson–Nernst–Planck–Stokes equations, which were then linearized assuming a weak applied field $\delta\mathbf{E}$ (i.e., linear response). This approach naturally leads to a "leading-order" (where $\delta\mathbf{E} = 0$) solution that gives the equilibrium EDL with no fluid velocity; and an $\mathcal{O}(\delta\mathbf{E})$ correction to the ion densities $\delta n_\pm$, electric field, and fluid velocity field $\delta\mathbf{u}$. Such computations give electrophoretic mobilities as shown in Figure 5.5: M_E generally increases linearly with ζ but reaches a *maximum*, beyond which it decays. The thinner the EDL, the larger the ζ, where M_E is maximal. Even with a screening length that is 500 times smaller than the colloid radius, Morrison's result (constant M_E, linearly dependent on ζ and independent of a) only holds up to a certain ζ, beyond which it fails.

In fact, Figure 5.5 shows electrophoretic mobilities calculated by Khair and Squires [46] for particles with varying degrees of surface solid/liquid slip, which had been postulated to significantly enhance electrokinetic effects [29]. Indeed, more "slippery" particles do indeed show greater electrophoretic mobility for low enough ζ but exhibit a maximum in M_E at lower ζ.

The resolution to these apparently contradictory results can be understood in terms of surface conductivity: the EDL—no matter how thin—carries an extra current above and beyond what would be carried by a "bulk" electrolyte in its absence. The surface conduction is defined as the *excess* conductivity in the EDL, above and beyond what would have been in the bulk at that point. Historically, this has been defined in terms of the excess electric current (here for symmetric, monovalent ions):

$$J_s = \int_0^\infty [(n_+(z) + n_-(z) - 2n^0)be^2 E_\| + \rho_E(z)u_\|(z)]dz \equiv \sigma_s E_\|, \tag{5.64}$$

where we note that J_s represents the *integral* of the excess current density j_s within the EDL. The first term represents surface conduction due to electromigration, and the second to electroconvection with the local fluid flow. This expression can be evaluated exactly for the Gouy–Chapman EDL (nonlinear Poisson–Boltzmann solution over a flat plate), using the EOF velocity, giving

$$\sigma_s = \frac{8(1+m)n^0 e^2 b}{\kappa}\sinh^2\left(\frac{\tilde{\zeta}}{4}\right). \tag{5.65}$$

Here the parameter

$$m = 2\frac{\epsilon_w}{\eta}\left(\frac{k_B T}{e}\right)^2 D \tag{5.66}$$

relates the relative importance of electroconvection and electromigration for ions. Henry's and Morrison's results—Equations 5.56 and 5.62, respectively—hold when this surface conduction is relatively weak. When surface conduction becomes significant, however, the mobility changes qualitatively.

One can understand the role of surface conductivity and ion conservation in Figure 5.6. If the standard (no normal field) assumption held, then ion conservation would be violated: the "excess" current into the box is zero, whereas the excess current out of the box is $\sigma_s E_{\|}$. Ions must therefore be injected into the box from the bulk in order to conserve ions, which therefore requires a nonzero $E_{\perp}$. Far enough "down-field" from the surface charge discontinuity, we expect the standard (no flux) condition to hold. The question, though, is how far down field? The field in the bulk scales like $\mathcal{O}(E)$; so the current density into the EDL from the bulk is $\mathcal{O}(\sigma_B E)$; we need this flux to persist over a control volume whose "length" L is sufficient for the bulk current entering to be equal to the surface current leaving. This length scale, called the electrokinetic healing length λ_h by Khair and Squires [47], or the "Dukhin length" by Bocquet and Charlaix [48],

$$\lambda_h = \frac{\sigma_s}{\sigma_B}, \tag{5.67}$$

gives the length "downcurrent" of any change in the surface conduction over which the fields and flows "heal." For the Gouy–Chapman double layer,

$$\lambda_h = \frac{4(1+m)}{\kappa}\sinh^2\left(\frac{|\tilde{\zeta}|}{4}\right). \tag{5.68}$$

In aqueous solutions, m is often an $\mathcal{O}(1)$ number, whereas in nonpolar media, the relatively large and slow ions would give a larger m.

5.4.4 Concentration Polarization

Surface conductivity has an additional effect, beyond changing the "no normal flux" nature of the external field. The excess current consists of predominantly one species of ion—the counterions. Say the surface current increases, as shown in Figure 5.6. A flux of counterions (here positive) into the EDL is required, as described above. However, the bulk co-ions driven *away* from the EDL are not replenished by a co-ion flux coming out of the EDL, since there are relatively few co-ions within the EDL. The charge separation that thus results—positive and negative ions are driven away from each other—gives rise to a net "sink" for salt. The ionic strength in solution, then, begins to decrease gradually. The bulk salt fields are transported by convection and diffusion until a steady-state flux with the EDL is reached.

This phenomenon—called *concentration polarization*—can be understood more easily in Figure 5.6b and c, in which a surface charge density switches from positive to negative. A field applied in the direction from positive to negative surface charge gives rise to surface currents directed *away from* the discontinuity in both directions. On the left, (negative) counterions are pulled in from solution, and on the right, (positive) counterions are pulled in from solution. Thus, both ion species are depleted from the region around the discontinuity, leading to a steady salt sink, of width $\mathcal{O}(\lambda_h)$, above the EDL. A region of low ionic strength develops—via diffusion and possibly convection—above

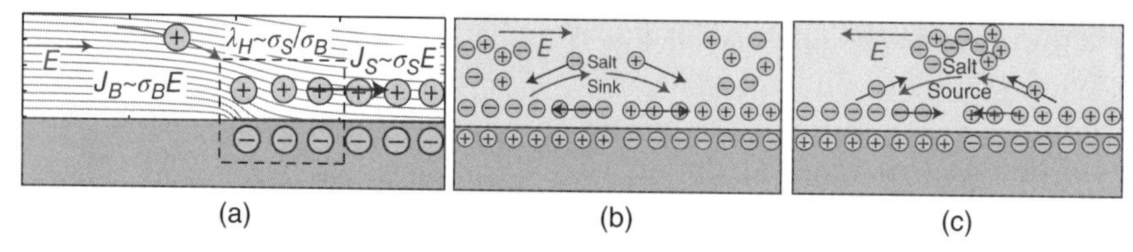

Figure 5.6 *Surface conduction: healing and concentration polarization.* The EDL has an excess conductivity σ_s, above that of the bulk, which can profoundly influence the electric and salt fields outside the EDL. (a) The EOF over a surface charge discontinuity, even if the EDL is thin, gives rise to a discontinuity in current $\sigma_s E$ due to the abrupt change in σ_s. Ion conservation within the (dashed) control volume requires ions to be driven into the EDL from the bulk. Since the bulk current scales like $\sigma_B E$, a downstream region of "healing length" $\lambda_h = \sigma_s/\sigma_B$ is required to permit enough normal current to conserve ions. Source: Adapted from Ref. [47]. (b) *Concentration polarization.* When the relative EOF on either side of a surface charge discontinuity is driven away from the discontinuity, counterions must be driven into the EDL on either side of the discontinuity, establishing a *sink* for salt and depleting the local ionic strength. (c) When a relative EOF occurs toward a surface charge discontinuity, counterions of both signs are ejected from the respective EDLs, establishing a salt *source* that enhances the local ionic strength.

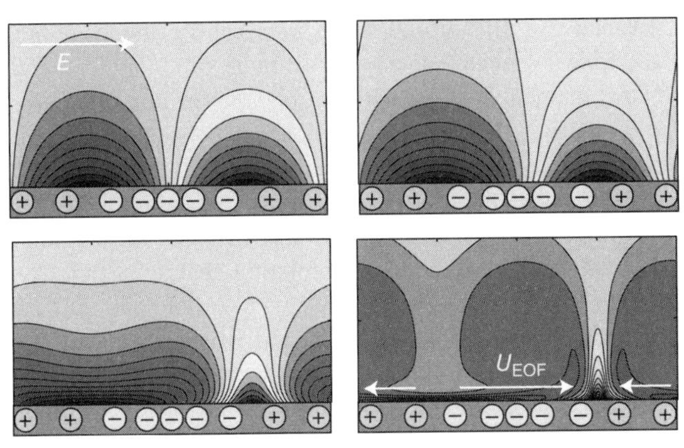

Figure 5.7 Concentration polarization over a planar surface with sinusoidally varying surface charge density, for increasingly strong electro-osmotic flow (characterized by a Péclet number $Pe = U_{EOF}L/D$, where L is the charge density wavelength). Gradients in surface current within the EDL establish sources and sinks of salt, effectively injecting and withdrawing salt into and out of the bulk. The salt field evolves as the salt diffuses and when it is convected by the EOF. (a) $Pe = 0.1$; (b) $Pe = 1.0$; (c) $Pe = 10$; (d) $Pe = 100$. Source: Adapted from Ref. [49].

this region. Conversely, a field applied in the opposite direction (Fig. 5.6c) gives rise to surface currents that *decrease* as they approach the discontinuity, necessitating the *ejection* of counterions on both sides of the discontinuity. The outflux of both ion species from the EDL gives rise to a salt *source*, of width λ_h, above the discontinuity, and a region of greater ionic strength gradually develops. Figure 5.7 shows the concentration polarization above a planar surface with sinusoidally varying surface charge [49], in which the salt sources and sinks established by the EDL give rise to a steady salt profile that reflects salt transport due to diffusion and convection with the EOF. High-salt "plumes" are a somewhat generic feature of such systems (e.g., [50]).

We now return to electrophoresis, where it should be clear how surface conduction can reduce the electrophoretic mobility. With colloidal electrophoresis, it is geometric curvature—rather than surface charge gradients—that gives rise to gradients in surface current. There is identically zero surface current at the "poles" of the colloid, and the surface current is maximal at the "equator" (Fig. 5.8a). Ions must thus be supplied from the bulk to conserve ions (Fig. 5.8b), necessitating a normal field. This normal field reduces the tangential field—and thus the EOF driving electrophoretic motion. This normal field also drives concentration polarization, which drives the EDL into chemiosmotic flow, opposite the original EOF (Fig. 5.8c).

So when does Morrison's argument work, and Smoluchowski's simple formula hold? In other words, when is surface conductivity weak enough (in a dimensionless sense) that it can be neglected? When $\lambda_h \ll a$, the field heals quickly, and this assumption is reasonable. If, on the other hand, the healing (Dukhin) length is of the same order as, or larger than, the colloid radius a, then the electric and concentration fields are nowhere like Morrison's standard no-flux condition would require. The relative importance of surface conduction, relative to conduction in the bulk, is given by the Dukhin number

$$Du = \frac{\lambda_h}{a}. \tag{5.69}$$

For the Gouy–Chapman model of the double layer, the Dukhin number is given by Equation (5.68) to be

$$Du_{GC} = \frac{4(1+m)}{\kappa a}\sinh^2\left(\frac{e|\zeta|}{4k_BT}\right). \tag{5.70}$$

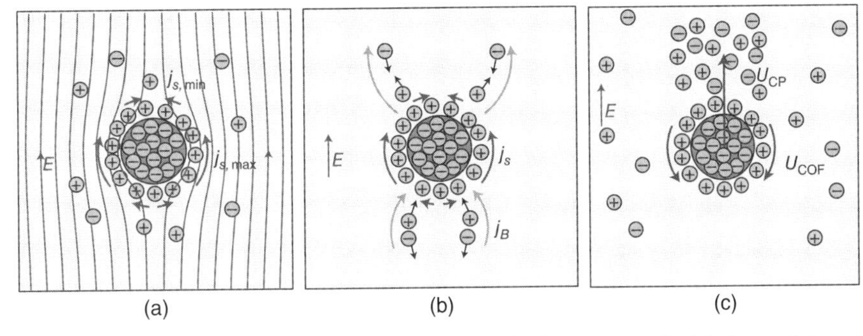

Figure 5.8 *Surface conduction retards electrophoresis.* (a) When the EDL is thin compared with the particle radius ($\kappa a \gg 1$), a frequent assumption is that the applied electric field has no normal component at the particle surface. However, the excess ion current j_s due to the double layer itself is inhomogeneous: zero at the field "poles," and maximal at the particle "equator." (b) Ions must stream into and out of the EDL from the bulk in order to conserve ions, which necessitates a nonzero *normal* field into the EDL. This normal component has two effects: (i) it reduces the tangential component, thereby reducing the EOF slip velocity and thus the EP velocity, and (ii) depletes the salt at one pole while increasing the salt concentration at the other pole (concentration polarization). (c) Concentration polarization drives a *Chemi-osmotic slip* velocity within the EDL, and thus a *Chemi-phoretic* velocity that is directed *opposite* to the original EP velocity. Both effects combine to retard the EP mobility, leading to the maximum observed in Figure 5.5, which occurs when the dimensionless surface conductivity (the Dukhin number) becomes $\mathcal{O}(1)$.

At a fixed EDL thickness (κa), increasing ζ increases the surface conductivity, until eventually Du exceeds one, and surface conduction effects become strong. Morrison's calculation, on the other hand, starts with the assumption that $\kappa a \to \infty$, so that Du is assumed to always be zero, and surface conduction never enters.

At high Du (well past the maxima in Fig. 5.5a), the EP mobility decays nearly to zero. This may seem counterintuitive: a particle with extremely high charge barely moves under an applied field. However, for the particle/EDL ensemble to migrate electrophoretically, counterions be exchanged between the EDL and bulk fast enough to conserve ions. When the EDL becomes charged highly enough, there simply are not enough counterions in the bulk to allow this exchange at the rate required by EP. In that case, the picture resembles Figure 5.4, in that the diffuse counterions are effectively bound to the particle, so that they polarize but do not drive steady electrophoretic motion.

Figure 5.5 demonstrates the central role played by the Dukhin number (rather than ζ) plays in determining the breakdown of Morrison's argument. In Figure 5.5, the electrophoretic mobility is scaled by the simple flat-plate result (which would hold when $Du = 0$),

$$M_R = \frac{M_E(\tilde{\zeta}, \lambda/a, \kappa a)}{M_e(\tilde{\zeta}, \lambda/a, \kappa a \to \infty)} \tag{5.71}$$

and is plotted not against $\tilde{\zeta}$, but against a modified Dukhin number that accounts for the finite solid/liquid slip λ/a:

$$Du_{\text{slip}} = Du_{\text{GC}} + 2m\frac{\lambda}{a}\sinh^2\frac{\tilde{\zeta}}{2}. \tag{5.72}$$

The second term represents the additional surface conductivity that arises when counterions are advected by the (faster) slipping flow. Notably, the wide variety of mobilities plotted in Figure 5.5a—which span orders of magnitude in slip length (λ/a), screening lengths (κa) and surface potential $\tilde{\zeta} = e\zeta/k_BT$, all collapse onto a single master curve when plotted in this way (Fig. 5.5b) [51].

More generally, one can understand many recent results in which somewhat surprising effects have arisen within nanopores and nanochannels using these pictures. One can picture a nanochannel as simply Figure 5.6 with a second charged surface some distance H above the one pictured. The electrical conductance of nanochannels and nanopores depends on ionic strength in a curious way: at high ionic strength, the conductance is simply due to conduction through the bulk of the pore. At low ionic strength, however, surface conduction takes over as the dominant contribution, and effectively "short circuits" the resistance of the nanopore. This occurs when $2\sigma_s \gg \sigma_B H$ for a nanochannel, or $2\pi R\sigma_s \gg \sigma_B \pi R^2$ for a nanopore: Dukhin numbers $\sigma_s/\sigma_B H$ and $\sigma_s/\sigma_B R$ control this behavior. In addition, Bocquet and coworkers showed that entrance effects associated with high-Du nanopores can be much stronger and longer ranged than a naive estimate would indicate [52]. Nanochannels or pores with anisotropic surface charge (one end positive, the other negative, as in Fig. 5.6b and c) give rise to anisotropically conducting nanochannel "diodes," akin to semiconductor p–n junctions. A field applied in the direction of Figure 5.6b drives a salt sink in the middle of the nanochannel, effectively depleting charge carriers from the pore and suppressing conductance. A field applied in the opposite direction, on the other hand, drives a concentration polarization that floods the center of the pore with charge carriers, increasing the conductance of the pore [53]. Bipolar membranes, consisting of an anionic membrane attached (in series) to a cationic membrane, give analogous conductivity. Finally, concentration polarization effects around nanochannels [54] and even wires [55] enables water desalination, in a manner related to electrodialysis.

On a more detailed level, the physical restriction is more severe than conservation of current—in fact, each ion species should be conserved individually. One can thus define more generally an excess surface flux of each ion species, via

$$\begin{aligned}\mathbf{j}_s^i &= \int_0^\infty \left[(n^i(z) - n^0)b_i\left(-\nabla_\parallel \mu_B\right) + \left(n^i(z) - n^0\right)u(z)\right]dz \\ &\equiv -\Sigma_s^i \nabla_\parallel \mu_B^i.\end{aligned} \tag{5.73}$$

At any thin double layer, then, the boundary conditions to impose on each chemical potential field μ_B^i would be

$$\nabla_\parallel \cdot \left(\Sigma_s^i \nabla_\parallel \mu_B^i\right) = n_0^i b_i \hat{\mathbf{n}} \cdot \nabla \mu_B^i\Big|_{z=0}. \tag{5.74}$$

The first term represents the divergence of the surface flux of each ion species, which could occur either because the driving force ($\nabla \mu_B^i$) changes—as occurs around curved surfaces—or because the surface conductivity Σ_s^i itself changes—as when the surface charge density changes. Nondimensionalizing (Eq. 5.74) by characteristic values of Σ_s^i and distances L over which either Σ_s^i or μ_B^i change gives rise to

$$Du^i \tilde{\nabla}_\parallel \cdot \left(\tilde{\Sigma}_s^i \tilde{\nabla}_\parallel \mu_B^i\right) = \hat{\mathbf{n}} \cdot \tilde{\nabla} \mu_B^i\Big|_{z=0}, \tag{5.75}$$

where the Dukhin number of ion species i is

$$Du^i = \frac{\Sigma_s^i}{\Sigma_B^i L}. \tag{5.76}$$

In fact, so-called "high-Du" systems involve Du_{counter} being large for the counterion(s) only. Co-ions have vanishingly small $Du_{\text{co-ion}}$.

When $Du^i \ll 1$, the left-hand side of Equation 5.75 is nearly zero, giving the no-flux boundary condition often assumed in the thin EDL limit. When $Du^i \gg 1$, however, the right-hand side is much smaller than the left, leaving instead a *divergence-free* requirement for the surface current of that particular ion. For electrophoresis, a divergence-free counterion surface current is necessarily zero—leading to the vanishing electrophoretic mobility in Figure 5.5.

5.5 NONLINEAR ELECTROKINETIC EFFECTS

Thus far, we have discussed only *linear* electrokinetic effects—electrokinetic responses that arise in direct proportion to the applied field. As with other linear response phenomena, linear electrokinetic effects represent minor changes from equilibrium.

We now turn to *nonlinear* electrokinetic effects. These effects arise when an applied field provokes some change in the system, then forces the changed system. We discuss two examples in particular:

- *Induced-charge electrokinetics* arise when an applied field *induces* an EDL around a polarizable object, then drives that induced EDL into electrokinetic motion. Induced-charge electro-osmotic (ICEO) flows arise around particles and surfaces, while particles can move via *induced-charge electrophoresis*. ICEK effects encompass a broad class of phenomena, including nonlinear interactions between suspended colloids [56], colloidal self-assembly at electrode interfaces [7], AC electro-osmosis effects [5], microfluidics [1, 2], and influencing the dielectrophoretic motion of colloids [57, 58].
- *Dielectrophoresis* a suspended particle in the presence of an electric field attains an electric dipole; any gradient in the applied field would then exert a force upon this induced dipole and drive the particle into motion. Dielectrophoretic interactions between particles give rise to *electrorheological fluids*, whose rheological properties (e.g., shear viscosity) can be tuned electrically.

A significant feature of nonlinear EK phenomena is that they persist even in AC fields. This stands in contrast with linear electrokinetic effects, which time-average to zero in AC fields.

5.5.1 Induced-Charge Electrokinetics

Thus far, we have operated under the assumption that there must be no ionic flux into the EDL; or, when the Dukhin number is large enough, that the ionic flux make up for the divergence of the surface current. Indeed, in steady state, this must be the case.

This assumption, however, might seem at odds with the standard dielectric boundary conditions familiar from introductory electromagnetism. In particular, the electric displacement field must be continuous at any dielectric boundary (say $z = 0$):

$$\epsilon_p \frac{\partial \phi}{\partial z}\bigg|_{z=0} = \epsilon_w \frac{\partial \phi}{\partial z}\bigg|_{z=0}. \tag{5.77}$$

Furthermore, the electrostatic potential everywhere within a conducting body must be constant, implying that the *only* field possible at the conductor/electrolyte interface is a normal field. How can these two assumptions be reconciled?

The field around an uncharged, conducting cylinder immersed in a uniform electric field $\mathbf{E}$ is shown in Figure 5.9a, with the equipotential condition establishing purely perpendicular fields at the cylinder surface. While this is the steady-state field in vacuum, ionic currents are driven along fields in electrolytes. Consequently, ions driven along field lines terminating at the conductor surface accumulate at the conductor/electrolyte surface in a dipolar fashion:

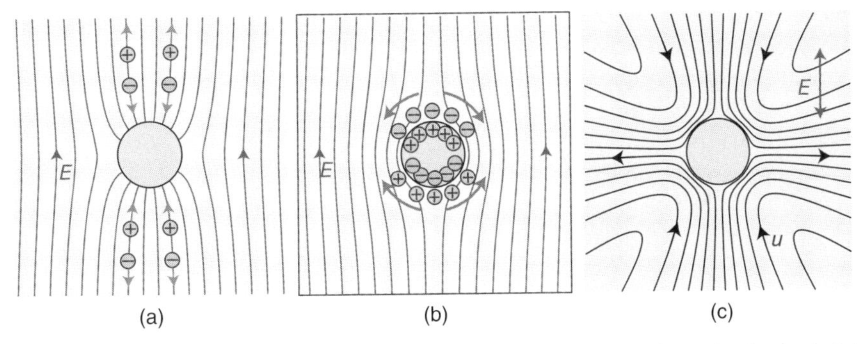

Figure 5.9 *Induced-charge electro-osmosis.* (a) An electric field applied around an uncharged, conducting body initially makes right angles with the particle surface in order to maintain the equipotential boundary condition. Ions in solution, however, travel along those field lines, reach the particle surface but are unable to cross it, in the absence of electrochemical reactions. (b) Ions accumulate in a dipolar *induced* double layer on either side, reaching steady state when all field lines are expelled from the induced EDL. (c) An induced-charge electro-osmotic flow arises when the applied field acts on the EDL that it has induced. Because the ICEO flow depends on the product of the field and its induced charge ($u \sim E^2$), a steady ICEO flow is driven even in an AC electric field.

cations are pushed along electric fields and accumulate on the side of the conductor nearest the field source (the "south pole" in Fig. 5.9a), whereas anions are pulled against electric fields and accumulate nearest the field sink (the "north pole" in Fig. 5.9a).

Under certain conditions (depending on ion and metal species, solution and electrode potentials), electrochemical reactions can occur that effectively "consume" these ions. Electrolysis (water splitting) is one example: hydronium ions (H^+) react to form H_2 molecules so long as electrons are supplied by the conductor and hydroxyl ions (OH^-) combine to form O_2 and water when the conductor can remove electrons. Furthermore, various ions can react with metal electrodes—for example, chlorine ions Cl^- dissolve from, and intercalate into, Ag/AgCl electrode pairs.

When conditions do not allow for electrochemical interactions, however (called ideal polarizability), ions can do nothing other than accumulate into an *induced* EDL. Image charges form within the conductor to maintain the equipotential condition, with an *induced* surface charge density. Steady state is reached when the induced EDL is strong enough that all normal field lines are expelled (Fig. 5.9b).

ICEK flows result when this *induced* EDL is forced into EOF by the field itself and are generally directed from the "poles" of the conductors toward the "equator" (Fig. 5.9c) [59]. Since the induced ζ-potential is given by the difference between conductor potential and the potential in the bulk, just outside the EDL, the induced ζ-potential difference between the two poles scales as $\zeta_i \sim Ea$. Standard EOF, then, gives an ICEO flow velocity of order

$$U_{\text{ICEO}} \sim \frac{\epsilon \zeta_i E}{\eta} \sim \frac{\epsilon E^2 a}{\eta}. \tag{5.78}$$

ICEK flows depend upon the square of the electric field: one power of E induces the EDL and the other power of E drives that induced EDL into EOF. Reversing the field direction also reverses the charge of the induced EDL, yet the flow is given by their product. Therefore, nonzero, time-averaged ICEK effects persist even under AC electric fields, so long as the induced EDL has time to form. The timescale for induced EDL formation can be understood by viewing the EDL as a capacitor (which, in the low-ζ limit, has constant capacitance ϵ_w/λ_D) which must be charged and discharged through a bulk resistor (the electrolyte, with conductivity σ_B and "length" $\mathcal{O}(a)$). An RC charging time,

$$\tau_c = \frac{\lambda_D a}{D}, \tag{5.79}$$

is thus required to induce the EDL around an ideally polarizable metal sphere [59, 60].

ICEK flows around metallic colloids were first described in the Soviet literature [8–10, 56], both theoretically and experimentally. In the microfluidic context, flows with the same physical basis were explored as they arose over coplanar electrodes driven by AC electric fields [5, 6].

Dielectric coatings—or any contaminant—can rather dramatically suppress ICEO flows [59, 61]. Consider a dielectric shell of thickness d and permittivity ϵ_d coating the conductor. After the induced EDL forms, the potential drop between the (equipotential) conductor and the bulk electrolyte occurs in two steps: an induced EDL, with potential drop ζ_i, and the (inert) dielectric, resulting in an induced ζ_i given by

$$\zeta_i = \frac{\zeta_i^0}{1+\delta}, \tag{5.80}$$

where

$$\delta = \frac{C_d}{C_{\text{DL}}} = \frac{\epsilon_w}{\epsilon_p}\frac{d}{\lambda_D}. \tag{5.81}$$

More generally, the ICEO flow around a dielectric colloid is given by Equation 5.80, in the limit where the "film" constitutes the entire particle ($d = a$). In the event where $\delta \gg 1$, the ICEO flow scales like

$$U_{\text{ICEO}} \sim \frac{\epsilon_w E^2 a}{\eta \delta} \sim \frac{\epsilon_p E^2 \lambda_D}{\eta}. \tag{5.82}$$

Finally, anisotropic, or anisotropically conducting, particles give rise to anisotropic ICEO flows that can drive a net directed flow (for fixed conductors) or drive induced-charge electrophoretic (ICEP) motion of suspended particles [57, 62].

5.5.2 Dielectrophoresis

Lastly, we turn to a discussion of DEP, which differs from the rest of the "phoretic" phenomena in that it is not driven by forces within the EDL but rather by forces on particles themselves.

DEP can be understood directly from classical electrostatics, without any real need for sophisticated theories (or even understanding) of ion transport, EDLs, and so on. An object with some shape, size, dielectric and electrical properties, immersed in an electric field $\mathbf{E}$ establishes an induced electrostatic dipole $\mathbf{p}$

$$\mathbf{p} = \mathbf{\Pi} \cdot \mathbf{E}, \tag{5.83}$$

where $\mathbf{\Pi}$ is a polarization tensor specific to the particle. For a solid, homogeneous, spherical particle of radius a, the polarization $\mathbf{\Pi}$ is isotropic:

$$\mathbf{p} = 4\pi \epsilon_f \left(\frac{\epsilon_p - \epsilon_f}{\epsilon_p + 2\epsilon_f} \right) a^3 \mathbf{E}, \tag{5.84}$$

where ϵ_p and ϵ_f are the dielectric permittivity of the particle and fluid, respectively, and the induced dipole is oriented parallel to the applied field. Nonspherical or electrically anisotropic particles, on the other hand, have principal axes of polarization given by eigenvectors of $\mathbf{\Pi}$. For simplicity, we focus on isotropic particles.

Since the potential energy of a dipole does not change when it translates in a uniform field, a uniform field $\mathbf{E}$ does not exert a force on induced dipole $\mathbf{p}$ and leads to zero particle translation. Instead, it is field *gradients* that cause a particle to migrate via DEP [63, 64]. The dielectrophoretic force is given by

$$\mathbf{F} = \mathbf{p} \cdot \nabla \mathbf{E} = 2\pi\epsilon_f a^3 \left(\frac{\epsilon_p - \epsilon_f}{\epsilon_p + 2\epsilon_f} \right) \frac{\nabla E^2}{2}. \tag{5.85}$$

Physically, a sphere that is more polarizable than the fluid ($\epsilon_p > \epsilon_f$) induces a dipole oriented with the applied field. Such particles move *up* field strength gradients, in what is called *positive DEP*. Spheres that are *less* polarizable than the fluid, on the other hand, induce dipoles oriented *against* the applied field, and experience DEP forces directed *down* field gradients (*negative DEP*).

More generally, the dielectric response of even spherical particles is more complicated than Equation 5.84. Both ϵ_p and ϵ_f are, in general, frequency dependent and complex, and may additionally contain a conductive component σ,

$$\hat{\epsilon}_p(\omega) = \epsilon_p^*(\omega) + i\frac{\sigma_p^*(\omega)}{\omega}, \tag{5.86}$$

which will generally depend on frequency ω. Furthermore, EDLs that surround charged particles polarize under applied fields, further contributing to the induced dipole moment in a frequency-dependent manner. Under a field of a given frequency, then, the dipole moment is given by

$$\mathbf{p}_0(\omega)e^{-i\omega t} = 4\pi\epsilon_f f_{\mathrm{CM}}(\omega)a^3\mathbf{E}_0(\omega)e^{-i\omega t}, \tag{5.87}$$

where

$$f_{\mathrm{CM}} = \frac{\hat{\epsilon}_p(\omega) - \hat{\epsilon}_f(\omega)}{\hat{\epsilon}_p(\omega) + 2\hat{\epsilon}_f(\omega)} \tag{5.88}$$

is the Clausius–Mossotti factor.

The time-averaged DEP force on such a sphere is then

$$\langle \mathbf{F} \rangle(\omega) = 2\pi\epsilon_f \Re[f_{\mathrm{CM}}]a^3 \frac{\nabla E^2}{2}, \tag{5.89}$$

giving a DEP velocity

$$\mathbf{U}_0(\omega) = \frac{\epsilon_f a^2}{3\eta}\Re[f_{\mathrm{CM}}]\frac{\nabla E^2}{2}, \tag{5.90}$$

where $\Re$ denotes the real part. Notably, a nonzero DEP velocity is driven even with an AC applied field. When $\Re[f_{\mathrm{CM}}] > 0$, the particle experiences positive DEP, and vice versa. The frequency dependence of the Clausius–Mossotti factor f_{CM} has been used, for example, to sort live cells from dead ones, by operating at frequencies where live cells experienced positive DEP, whereas dead ones experienced negative DEP [65].

5.5.3 Particle Interactions and Electrorheological Fluids

We noted that a single particle does not move dielectrophoretically under a uniform applied field. Multiple suspended particles, on the other hand, do interact dielectrophoretically and move relative to each other, even in a uniform electric field. An electric dipole is induced in each particle—as given by Equation 5.87—which, in turn, establishes field gradients that exert DEP forces on the dipoles induced in the surrounding particles. Because the field around a dipole decays like r^{-2}, the field gradient decays like r^{-3}, and the interaction force between two particles separated by a distance d scales like $p_1 p_2/d^3$. Notably, DEP interactions between two particles are attractive when the particles are lined up along with the field, whereas DEP forces are repulsive when particles are aligned perpendicular to the field. Particles tend to chain up along field lines under DEP; paramagnetic particles chain in analogous fashion under an applied magnetic field. Suspensions of electrically or magnetically polarizable particles—called electro- or magnetorheological fluids—exhibit a tunable viscous response: an applied field forms chains of particles that span the system, and that must be ruptured in order for the fluid to shear [66].

Lastly, we note that ICEK and DEP effects can combine to give a further rich variety of physics. For example, ICEO flows around a particle in a gradient field give rise to ICEP motion, with a magnitude that can be comparable to DEP motion. In fact, the ICEP and DEP contributions to the motion of an ideally polarizable metal sphere in the zero-frequency limit precisely cancel [57, 58]! More generally, fluid flows around particles moving via ICEP and DEP decay with different rates, so that interactions can be quite involved [67, 68]. Furthermore, AC fields have been used to pattern colloids on electrodes, essentially using a combination of (wall-induced) ICEP and DEP interactions [7]. Particles in concentrated suspension have been shown to undergo an order to disorder transition below a critical field [69]. The list goes on.

5.6 CONCLUSIONS

To summarize, we have tried here to give a sense for the rich variety of ways in which fields, flows, and ions can interact to give rise to various electrokinetic effects. While the number of these effects may seem impossibly vast, we hope to have conveyed the sense that understanding a few basic principles enables one to think intuitively and physically about almost all of them. Given the central presence of liquids (and therefore ions) in soft matter, it is no surprise that many of these effects can arise and play a significant role in the evolution of these systems.

We sincerely hope that this work will enable researchers in soft matter to recognize electrokinetic effects when they arise, exploit them when appropriate, and to be creative in their use.

REFERENCES

[1] Stone HA, Stroock AD, Ajdari A. Engineering flows in small devices: microfluidics toward a lab-on-a-chip. Ann Rev Fluid Mech 2004;36:381–411.

[2] Squires TM, Quake SR. Microfluidics: fluid physics at the nanoliter scale. Rev Mod Phys 2005;77(3):977–1026.

[3] Schoch RB, Han JY, Renaud P. Transport phenomena in nanofluidics. Rev Mod Phys 2008;80(3):839–883.

[4] Kirby BJ. *Micro- and Nanoscale Fluid Mechanics: Transport in Microfluidic Devices*. Cambridge: Cambridge University Press; 2010.

[5] Ramos A, Morgan H, Green NG, Castellanos A. AC electric-field-induced fluid flow in microelectrodes. J Colloid Interface Sci 1999;217(2):420–422.

[6] Ajdari A. Pumping liquids using asymmetric electrode arrays. Phys Rev E 2000;61(1):R45–R48.

[7] Prieve DC, Sides PJ, Wirth CL. 2-D assembly of colloidal particles on a planar interface. Curr Opin Colloid Interface Sci 2010;15(3):160–174.

[8] Dukhin AS, Murtsovkin VA. Pair interaction of particles in electric-field. 2. Influence of polarization of double-layer of dielectric particles on their hydrodynamic interaction in a stationary electric-field. Colloid J USSR 1986;48(2):203–209.

[9] Gamayunov NI, Murtsovkin VA, Dukhin AS. Pair interaction of particles in electric-field. 1. Features of hydrodynamic interaction of polarized particles. Colloid J USSR 1986;48(2):197–203.

[10] Levich VG. *Physicochemical Hydrodynamics*. Englewood Cliffs (NJ): Prentice-Hall, Inc.; 1962.

[11] Dukhin SS, Derjaguin BV. Electrokinetic phenomena. In: Matijevic E, editor *Surface and Colloid Science*. Volume 7. New York: John Wiley and Sons; 1974.

[12] O'Brien RW, White LR. Electrophoretic mobility of a spherical colloidal particle. J Chem Soc Faraday Trans 2 1978;74:1607–1626.

[13] O'Brien RW, Hunter RJ. The electrophoretic mobility of large colloidal particles. Can J Chem-Rev Can Chim 1981;59(13):1878–1887.

[14] Morrison ID. Electrical charges in nonaqueous media. Colloids Surf A 1993;71:1–37.

[15] Hsu MF, Dufresne ER, Weitz DA. Charge stabilization in nonpolar solvents. Langmuir 2005;21(11):4881–4887.

[16] Lyklema J. *Fundamentals of Interface and Colloid Science*. Volume 2. London: Academic Press; 1995.

[17] Russel WB, Saville DA, Schowalter WR. *Colloidal Dispersions*. Cambridge: Cambridge University Press; 1989.

[18] Probstein RF. *Physicochemical Hydrodynamics*. New York: John Wiley and Sons; 1994.

[19] Hunter RJ. *Foundations of Colloid Science*. 2nd ed.. Oxford: Oxford University Press; 2000.

[20] Kirby BJ. *Micro- and Nanoscale Fluid Mechanics: Transport in Microfluidic Devices*. Cambridge: Cambridge University Press; 2009.

[21] Bruus H. *Theoretical Microfluidics*. Oxford: Oxford University Press; 2008.

[22] Andelman D. Electrostatic properties of membranes: the Poisson–Boltzmann theory. In: Lipowsky R, Sackmann E, editors. *Handbook of Biological Physics*. North Holland: Elsevier Science; 1995. p 603–641.

[23] Bazant MZ, Kilic MS, Storey BD, Ajdari A. Towards an understanding of nonlinear electrokinetics at large applied voltages in concentrated solutions. Adv Colloid Interface Sci 2009;152:48–88.

[24] Grosberg AY, Nguyen TT, Shklovskii BI. Colloquium: the physics of charge inversion in chemical and biological systems. Rev Mod Phys 2002;74(2):329–345.

[25] Moreira AG, Netz RR. Strong-coupling theory for counter-ion distributions. Europhys Lett 2000;52(6):705–711.

[26] Gelbart WM, Bruinsma RF, Pincus PA, Parsegian VA. DNA-inspired electrostatics. Phys Today 2000;53(9): 38–44.

[27] Bikerman JJ. Structure and capacity of the electrical double layer. Philos Mag 1942;33:384.

[28] Muller VM, Sergeeva IP, Sobolev VD, Churaev NV. Boundary effects in the theory of electrokinetic phenomena. Colloid J USSR 1986;48(4):606–614.

[29] Joly L, Ybert C, Trizac E, Bocquet L. Hydrodynamics within the electric double layer on slipping surfaces. Phys Rev Lett 2004;93:257805.

[30] Prieve DC, Anderson JL, Ebel JP, Lowell ME. Motion of a particle generated by chemical gradients. Part 2. Electrolytes. J Fluid Mech 1984;148:247–269.

[31] Anderson JL. Colloid transport by interfacial forces. Annu Rev Fluid Mech 1989;21:61–99.

[32] Deen WM. *Analysis of Transport Phenomena*. New York: Oxford University Press; 2011.

[33] Cussler EL. *Diffusion: Mass Transfer in Fluid Systems*. Cambridge: Cambridge University Press; 2009.

[34] Dukhin SS. Nonequilibrium electric surface phenomena. Adv Colloid Interface Sci 1993;44:1–134.

[35] Deryaguin BV, Dukhin SS, Korotkova A.A.. Diffusiophoresis in electrolyte solutions and its role in the mechanism of film formation from rubber latexes by the method of ionic deposition. Kolloid Zh 1961;23:53.

[36] Guazzelli E, Hinch EJ. Fluctuations and instability in sedimentation. Ann Rev Fluid Mech 2011;43:97–116.

[37] Henry DC. The cataphoresis of suspended particles. Part I. The equation of cataphoresis. Trans Faraday Soc 1931;133:106–129.

[38] Morrison FA. Electrophoresis of a particle of arbitrary shape. J Colloid Interface Sci 1970;34(2):210–214.

[39] Cummings EB, Griffiths SK, Nilson RH, Paul PH. Conditions for similitude between the fluid velocity and electric field in electroosmotic flow. Anal Chem 2000;72(11):2526–2532.

[40] Viovy JL. Electrophoresis of DNA and other polyelectrolytes: physical mechanisms. Rev Mod Phys 2000;72(3):813–872.

[41] Dorfman KD. DNA electrophoresis in microfabricated devices. Rev Mod Phys 2010;82(4):2903–2947.

[42] Zukoski CF, Saville DA. Electrokinetic properties of particles in concentrated suspensions. J Colloid Interface Sci 1987;115(2):422–436.

[43] Palacci J, Cottin-Bizonne C, Ybert C, Bocquet L. Osmotic traps for colloids and macromolecules based on logarithmic sensing in salt taxis. Soft Matter 2012;8(4):980–994.

[44] Abecassis B, Cottin-Bizonne C, Ybert C, Ajdari A, Bocquet L. Boosting migration of large particles by solute contrasts. Nat Mater 2008;7(10):785–789.

[45] Paustian JS, Nery-Azevedo R, Lundin S-TB, Gilkey MJ, Squires TM. Microfluidic microdialysis: spatiotemporal control over solution microenvironments using integrated hydrogel membrane microwindows. Phys Rev X. 2013;3(4):041010.

[46] Khair AS, Squires TM. The influence of hydrodynamic slip on the electrophoretic mobility of a spherical colloidal particle. Phys Fluids 2009;21:042001.

[47] Khair AS, Squires TM. Surprising consequences of ion conservation in electro-osmosis over a surface charge discontinuity. J Fluid Mech 2008;615:323–334.

[48] Bocquet L, Charlaix E. Nanofluidics, from bulk to interfaces. Chem Soc Rev 2010;39(3):1073–1095.

[49] Khair AS, Squires TM. Fundamental aspects of concentration polarization arising from nonuniform electrokinetic transport. Phys Fluids 2008;20(8):20.

[50] Leinweber FC, Tallarek U. Concentration polarization-based nonlinear electrokinetics in porous media: induced-charge electroosmosis. J Phys Chem B 2005;109(46):21481–21485.

[51] Khair AS, Squires TM. Steric e?ects on electrophoresis of a colloidal particle. J Fluid Mech 2009;640:343–356.

[52] Lee C, Joly L, Siria A, Biance A-L, Fulcrand R, Bocquet L. Large apparent electric size of solid-state nanopores due to spatially extended surface conduction. Nano Lett 2012. DOI: ASAP:dx.doi.org/10.1021/nl301412b.

[53] Karnik R, Duan CH, Castelino K, Daiguji H, Majumdar A. Rectification of ionic current in a nanofluidic diode. Nano Lett 2007;7(3):547–551.

[54] Kim SJ, Ko SH, Kang KH, Han J. Direct seawater desalination by ion concentration polarization. Nat Nanotechnol 2010;5(4):297–301.

[55] Porada S, Sales BB, Hamelers HVM, Biesheuvel PM. Water desalination with wires. J Phys Chem Lett 2012;3:1613–1618.

[56] Murtsovkin VA. Nonlinear flows near polarized disperse particles. Colloid J 1996;58(3):341–349.

[57] Squires TM, Bazant MZ. Breaking symmetries in induced-charge electro-osmosis and electrophoresis. J Fluid Mech 2006;560:65–101.

[58] Shilov VN, Simonova TS. Polarization of electric double-layer of disperse particles and dipolophoresis in a steady (DC) field. Colloid J USSR 1981;43(1):90–96.

[59] Squires TM, Bazant MZ. Induced-charge electro-osmosis. J Fluid Mech 2004;509:217–252.

[60] Simonov IN, Shilov VN. Theory of the polarization of the diffuse part of a thin double layer at conducting, spherical particles in an alternating electric field. Colloid J USSR 1973;35(2):350–353.

[61] Pascall AJ, Squires TM. Induced charge electroosmosis over controllably-contaminated electrodes. Phys Rev Lett 2009;104:088301.

[62] Gangwal S, Cayre OJ, Bazant MZ, Velev OD. Induced-charge electrophoresis of metallodielectric particles. Phys Rev Lett 2008;100(5):058302.

[63] Pohl HA. *Dielectrophoresis: The Behavior of Neutral Matter in Nonuniform Electric Fields*. Cambridge: Cambridge University Press; 1978.

[64] Morgan H, Green NG. *AC Electrokinetic: Colloids and Nanoparticles*. Philadelphia (PA): Research Studies Press; 2003.

[65] Markx GH, Talary MS, Pethig R. Separation of viable and non-viable yeast using dielectrophoresis. J Biotechnol 1994;32:29.

[66] Gast AP, Zukoski CF. Electrorheological fluids as colloidal suspensions. Adv Colloid Interface Sci 1989;30(3-4):153–202.

[67] Saintillan D, Darve E, Shaqfeh ESG. Hydrodynamic interactions in the induced-charge electrophoresis of colloidal rod dispersions. J Fluid Mech 2006;563:223–259.

[68] Rose KA, Hoffman B, Saintillan D, Shaqfeh ESG, Santiago JG. Hydrodynamic interactions in metal rodlike-particle suspensions due to induced charge electroosmosis. Phys Rev E 2009;79(1):011402.

[69] Mittal M, Lele PP, Kaler EW, Furst EM. Polarization and interactions of colloidal particles in ac electric fields. J Chem Phys 2008;129(6):7.

6

COLLOIDAL DISPERSIONS IN SHEAR FLOW

Minne P. Lettinga[1,2]

[1] *Forschungszentrum Jülich, Institute of Complex Systems (ICS-3), 52425 Jülich, Germany*

[2] *Laboratory for Soft Matter and Biophysics, KU Leuven, Celestijnenlaan 200D, B-3001 Leuven, Belgium*

6.1 INTRODUCTION

One of the key tasks in physical chemistry is to predict whether dispersions are stable, metastable, or unstable so, in other words, to predict the phase diagram. The factors that determine the stability are the interaction between the particles that constitute the dispersion and their shape. The beauty of colloid science is that colloidal chemistry can be used to exactly program the interaction between particles and their shape: particles with all kinds of shapes can be made attractive, repulsive, or hard-core-like. In addition, colloids are experimentally easily accessible since their size is typical of the order of visible light, so that the light scattering and microscopic techniques can be used to probe the systems. Thus, thermodynamic theories based on particle–particle interactions can be used to calculate phase diagrams, which can be readily tested experimentally. While this explains why colloidal dispersions are popular

Fluids, Colloids and Soft Materials: An Introduction to Soft Matter Physics, First Edition. Edited by Alberto Fernandez Nieves and Antonio Manuel Puertas.

in fundamental research, they are also of huge practical interest, since many examples can be found in daily life. Standard examples are blood, paints, and dairy products, but the list is inexhaustible. Colloidal dispersions are thus a core class of soft matter materials.

The main characteristic of soft matter systems is the ease with which they can be perturbed, since the interactions are of the order of 1 kT. As a consequence, when studying phase behavior, extreme care needs to be taken that the systems are not perturbed by, for example, tumbling, vibrations, temperature gradients, EM fields, and gravity. In the real world of industrial processing, cosmetic and food industry, biology, and so on, soft matter systems will often be exposed to external fields. The question is whether the stability and structure of the colloidal dispersions change, and therefore phase diagrams, when systems are subjected to an external field. The most prominent external field is shear flow, because any motion of the sample will most likely induce shear flow. Flow can be desirable when material is transported from A to B. A good example here is blood circulation, where highly concentrated dispersions of more or less deformable cells are pushed through narrow channels [1]. Flow can be partly wanted and partly unwanted as is the case for paints for which the flow behavior sets the final product as well as the pathway to get there [2]. Sometimes, there is no control as can be the case when putting tomato ketchup on a hamburger. Sheared colloidal dispersions are also of fundamental interest. First, the interplay between thermodynamic instabilities and hydrodynamic instabilities guaranties a plethora of novel effects; see Ref. [3] for an extensive introduction. Second, shear flow perturbs the interaction between the particles and with that also the phase behavior. Understanding both aspects on a fundamental level will in return give insights in practical questions regarding the force that is needed to shear a sample and the stability of the resulting flow.

The goal of this chapter is to give the reader a flavor of the complexity of sheared colloidal dispersions. We do this on the hand of a few examples for which it is at least to some extent possible to do theory so that theory and experiments can be compared. For each example, we first describe the equilibrium phase behavior and then show how this phase behavior is affected by shear flow. Thus, nonequilibrium phase diagrams will be constructed indicating where the different phases are located as a function of the relevant parameter such as the colloid concentration or interaction and the shear rate. We also discuss how this nonequilibrium phase behavior perturbs the flow behavior of the system including flow instabilities.

We discuss systems with increasing degree of complexity. We start with dispersions of colloidal hard spheres, which form the benchmark of colloidal science. This system undergoes a phase transition from the liquid phase, where the Brownian particles are disordered, to a crystalline state, which takes place at a volume fraction of 49%. At even higher concentrations, for volume fractions larger than 0.58, the system will be structurally arrested and form a glass, which are discussed in Chapters 12 and 13. When the liquid phase is subjected to shear flow, already many interesting features are observed such as the divergent viscosity with increasing concentration [4], which is relevant for blood rheology [5], and extreme shear thickening [6], as used for body armor [7]. We focus, however, on the *effect of shear flow on the crystallization of colloidal spheres*, that will be discussed in Section 6.3. We exploit here the fact that colloidal spheres can be charged and therefore repulsive, so that the crystallization takes place at lower concentrations.

Colloidal spheres can also be made attractive, either by grafting them with a small polymer layer for which the solvent conditions can be tuned by temperature or pressure, or by adding small polymers in the solution that push the colloids together due to an osmotic pressure imbalance, the depletion force. Attractive colloids can undergo a gas–liquid phase separation, just as a van der Waals gas, resulting in a colloid-poor top phase and a colloid-rich bottom phase. The critical fluctuations just before phase separation sets in are slow and very long ranged and, therefore, highly susceptible to shear flow. In Section 6.4, we discuss how this interaction affects the gas–liquid phase boundaries.

The phase behavior of nonspherical particles, in particular rod-like colloids, in shear flow is more complex. The phase behavior of colloidal rods in equilibrium is already very rich because two degrees of freedom play a role, namely, positional and orientational ordering, while the anisotropy in the shape results in a cascade of positionally ordered states. The effect of shear flow is most apparent in the orientational and positional disordered isotropic phase. Here, shear flow will perturb the orientation ordering of the system and will induce the formation of the orientational ordered but positional disordered nematic phase. The interaction between shear flow and the nematic phase, on the other hand, is very complex. In Section 6.5, we discuss how the different dynamic responses of the isotropic and nematic phase to shear flow affect the nonequilibrium phase behavior. We start, however, by introducing the basic concepts of rheology, which are needed to understand the nonequilibrium phase behavior.

6.2 BASIC CONCEPTS OF RHEOLOGY

6.2.1 Definition of Shear Flow

The science of flowing materials is called rheology. In classical rheological experiments, the force is measured that is required to strain a sample or, vice versa, the strain is measured that results from applying a force. The simplest and most applied deformation is the shear strain $\gamma = L/h$ when sample between two parallel walls is deformed by

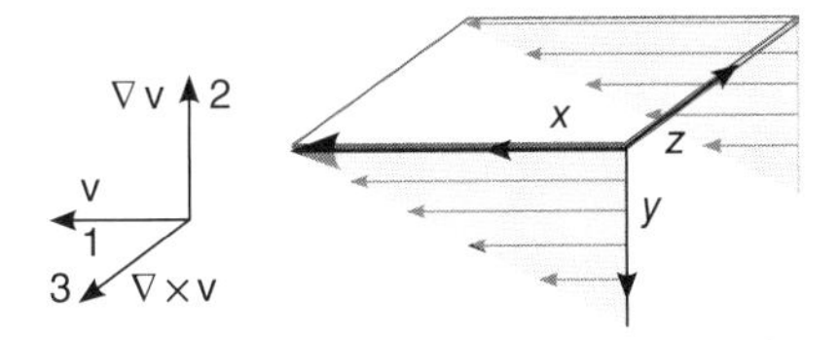

Figure 6.1 Definition of the shear directions: 1, flow direction; 2, gradient direction; 3, vorticity or neutral direction.

moving one wall relative to the other over a distance L in the x-direction, while both walls are a distance h apart in the y-direction. For a small volume element, this can be written as the gradient in the deformation field u_x, so that $\gamma = \frac{\partial u_x}{\partial y}$. A flow $\mathbf{v} = v\hat{x}$ will be induced when the sample is continuously strained. The flow rate depends on the distance from the moving wall y so that a shear rate can be defined as $\dot{\gamma} = \frac{\partial v_x}{\partial y}$. Thus, the y-direction is called the gradient direction. The third direction is unperturbed and is called the neutral direction, or also the vorticity direction since it is set by $\nabla \times \mathbf{v}$. The different directions are schematically shown in Figure 6.1.

The deformation rate tensor $\mathbf{\Gamma}$ describes the rate of deformation in three dimensions. For simple shear flow, it reads

$$\mathbf{\Gamma} = \dot{\gamma}\hat{\mathbf{\Gamma}} == \dot{\gamma}\begin{pmatrix} 0 & 1 & 0 \\ 0 & 0 & 0 \\ 0 & 0 & 0 \end{pmatrix}. \tag{6.1}$$

The most easily measurable force in rheology is generally the force that is exerted by the sample on the plane of the moving wall that is moving in the 1 (= x) direction, which has the normal parallel to the gradient in the flow which is the 2 (= y) direction. This stress is called the shear stress and is denoted by σ_{12}, where the indices refer to the plane on which the force is exerted (1) and the direction in which the force acts (2). σ_{12} is mostly referred to as the stress σ (or also often τ). Stresses in other directions are also important, especially for complex systems, but are often difficult to measure.

When the shear stress is proportional to the strain γ, then the sample behaves like a solid and the proportionality constant G is the modulus, as introduced by Hooke:

$$\sigma_{12} = G\gamma. \tag{6.2}$$

When the shear stress is proportional to the rate of deformation $\dot{\gamma}$, then the sample behaves like a fluid and the proportionality constant η is the viscosity, as introduced by Newton:

$$\sigma_{12} = \eta\dot{\gamma}. \tag{6.3}$$

Often a material is not a solid and not a liquid but something in between. These so-called viscoelastic materials are best characterized by subjecting them to an oscillatory shear flow given by a time-dependent strain of

$$\gamma(t) = \gamma_0 \sin(\omega t). \tag{6.4}$$

This is the so-called dynamic test. When the strain amplitude γ_0 is not too high so the sample does not change its character, then the response can be written as

$$\sigma_{12}(t) = G^*\gamma_0 \sin(\omega t + \delta) = [G' \sin(\omega t) + G'' \cos(\omega t)]\gamma_0. \tag{6.5}$$

The sample can thus be characterized by the modulus G^* and the phase angle, or by the storage modulus G' and loss modulus G'' (see Chapter 9). The system behaves like a solid when $G' \gg G''$ and $\delta = 0°$, and like a fluid when $G' < G''$ and $\delta = 90°$. The system is viscoelastic when $G' \approx G''$ or $0 < \delta < 90°$. Thus, rheology is often used to probe the state of a material.

Standard geometries that are used to impose shear flow are depicted in Figure 6.2. The geometries can be categorized in pressure flow, where shear is generated by pushing sample through a channel, and drag flows, where shear is generated between a moving and a fixed wall. The choice of the geometry depends on many different aspects. For details on the advantages and disadvantages of the different shear geometries from a pure rheological point of few, refer to Chapter 5 of Ref. [8].

Pressure-drop flow is often used for imaging, basically because it is very easy to set up. As can be appreciated from Figure 6.2a and b, shear flow is parabolic throughout the cell but will be linear close to the wall. In general, however, drag geometries allow for a better control over the shear and the shear rate is more (or less, see Ref. [8]) homogeneous. In the coming paragraphs, we refer to these geometries when discussing experiments.

While for a decent dynamic test care needs to be taken that the material is not influenced by shear flow, in this chapter we are interested in exactly that: how is phase behavior influenced by shear flow? Although there are rheological signatures for shear-induced phase transitions, the interpretation is often ambiguous. Therefore, it is important that the structure of the suspensions can be probed while shearing. This is an important factor that also determines the choice of the shear geometry, since in practice it is difficult to obtain reliable rheological and structural data simultaneously.

6.2.2 Scaling the Shear Rate

The implicit assumption in Newton's law for fluids is that the fluid behaves like a continuum that can be described by a single friction coefficient, namely, the viscosity η. The continuum does not change its properties when subjected to shear flow because shear flow does not affect the particles constituting the fluid. To satisfy this condition, the memory of the

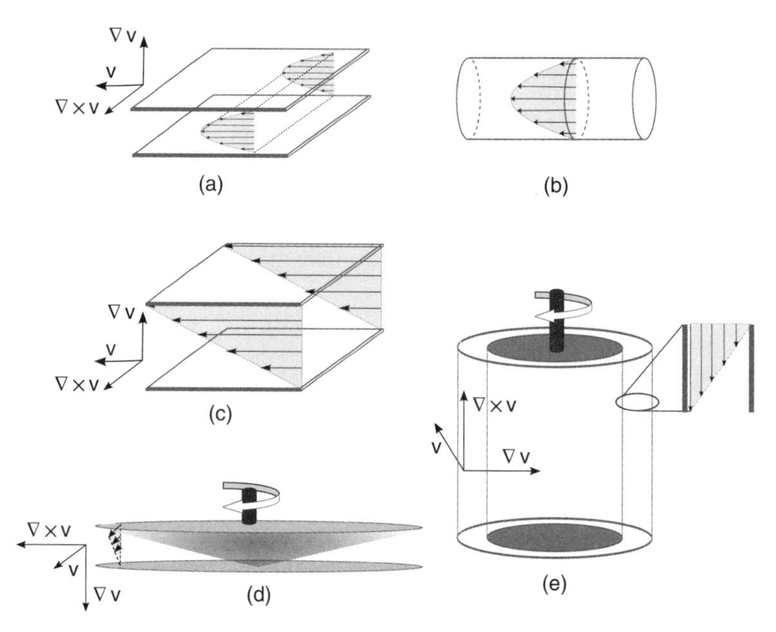

Figure 6.2 Shear geometries. (a) Slit flow; (b) Poiseuille flow; (c) flat Couette or plate–plate cell; (d) cone–plate cell; and (e) Couette cell

fluid should be very short compared to the applied deformation rate. Thus, the relevant relaxation time of the system τ needs to be defined in order to estimate the effect of shear flow. This is expressed by the Péclet number, which measures the shear rate relative to the respective relaxation process. This resulting scaled shear rate is called the Péclet number and is defined as $Pe = \dot{\gamma}\tau$. Similarly, the frequency in case of a dynamic test can be scaled, which is then called the Deborah number $De = \omega\tau$. Changes in the intermolecular structure of the solvent molecules are very quickly lost for simple fluids due to the diffusion of the molecules so that the relaxation time is very fast and $Pe \ll 1$. The diffusion rate D_0 for colloidal spheres is

$$D_0 = k_B T / 6\pi\eta R\,, \tag{6.6}$$

where k_B is Boltzmann's constant, R is the radius of the particle, and η is the viscosity. Thus, the time for a particle to diffuse its own diameter is $\tau = 1/R^2 D_0 = 6\pi\eta R^3/k_B T$. In order for the shear rate to take effect, it has to be of the same timescale so $Pe = \tau\dot{\gamma} = (6\eta\dot{\gamma})/\langle k_B T/R^3\rangle \approx 1$. For a colloid of $R = 500$ nm in water, this means that a shear rate of 0.6 s^{-1} is sufficient to compete with Brownian motion. When considering colloidal rods, shear flow will compete with the rotational motion of the rods. For slender rods of 1 μm, the rotational diffusion in water at room temperature is $D_R = 0.062\ s^{-1}$. So in order to have an appreciable alignment, a shear rate of $\dot{\gamma} > \mathcal{O}(10\ s^{-1})$ needs to be applied.

So far, we considered the competition between shear flow and Brownian motion. Often, however, particles interact with each other and form structures. Interactions generally slow down the dynamics, so that shear rate now needs to be scaled by the dynamics of the structure in order to appreciate the effect of shear on the structure and as a result on the phase behavior. In the remaining of this chapter, we give a few insightful examples of how shear flow and structure are coupled.

6.2.3 Flow Instabilities

As we mentioned earlier, for simple fluids the stress is proportional to the applied shear rate. But the colloidal dispersions that are discussed in this chapter, as well as many other materials, are not simple fluids and behave "non-Newtonian." The reason for this was just discussed above: if the shear rate is faster than the dynamics in the system, then the system will be perturbed by the shear rate and change its properties. The direct consequence will be that the viscosity of the sample will change with the shear rate that is applied. This will be apparent when taking a flow curve where the stress, and hence the viscosity, is measured as a function of shear rate.

The simplest flow curve is a straight line, where the viscosity does not change with shear rate or the stress increases linearly. The main other examples are a shear thinning flow behavior, where the viscosity decreases with increasing shear rate, and a shear thickening flow behavior, where the viscosity increases with increasing shear rate, as shown in Figure 6.3. Strong shear thinning is a very desirable feature for industrial applications. The flow behavior of shampoo, for example, is mainly engineered by using surfactant wormlike micelles: at low shear it behaves like a gel, at high shear it flows very easily. The same features are also very useful in, for example, the oil industry [9]. Shear thickening

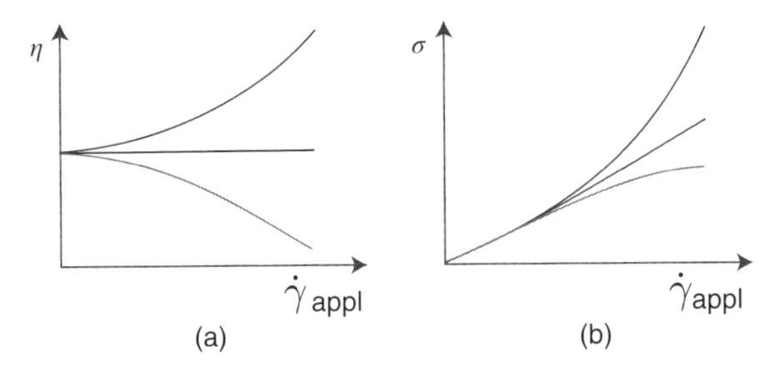

Figure 6.3 Cartoon of three different flow curves in viscosity versus shear rate representation and stress versus shear rate representation: Newtonian (black); shear thinning (red); shear thickening (blue). *(See color plate section for the color representation of this figure.)*

is used for dampening materials such as body armor. When a bullet enters a jacket with shear thickening material at high speed, then this material stiffens and the bullet is slowed down [10].

When the flow behavior is extremely non-Newtonian, then the stress behavior can cause mechanical instabilities in the shear cell. This means that instead of a smooth linear flow profile between the moving wall and the standing wall, as is required for Newtonian flow, the flow will be structured: regular or irregular (turbulent) flow profiles will develop in either the vorticity direction or the gradient direction or in both.

Flow instabilities can be predicted on the base of constitutive models that describe shear thickening and thinning behavior. Constitutive models that describe extreme shear thickening assume a functional form as depicted in Figure 6.4a, where the dashed line represents meta- or unstable states. A stable state, assuming laminar flow, would be achieved when alternating bands are formed with different viscosity, stacked along the vorticity direction [11]. The bands have a different stress since they are subjected to the same shear rate, as indicated by the dotted lines in the right panel of Figure 6.4a. This "classical" picture of vorticity banding is shown schematically on the right of Figure 6.4a. The question is, however, what mechanism drives the formation of these bands and whether the flow within the bands is laminar with equal shear rates.

Constitutive models that describe extreme shear thinning assume a functional form as given in Figure 6.4b, where again the dashed line represents meta- or unstable states: the part where the stress decreases with increasing shear rate renders the system mechanically unstable. "Unstable" means in this case that small deviations from the linear velocity profile will grow. From a theoretical point of view, this is very similar to an equilibrium gas–liquid phase separation [12, 13], as discussed in Section 6.4. In the "classical" picture of gradient band formation, at the end of the shear-driven phase separation, the system splits up into a region of a low viscosity and high shear rate close to the moving wall and a region with a high viscosity and low shear rate close to the static wall such that the stress in both bands is constant, see the left side of Figure 6.4b.

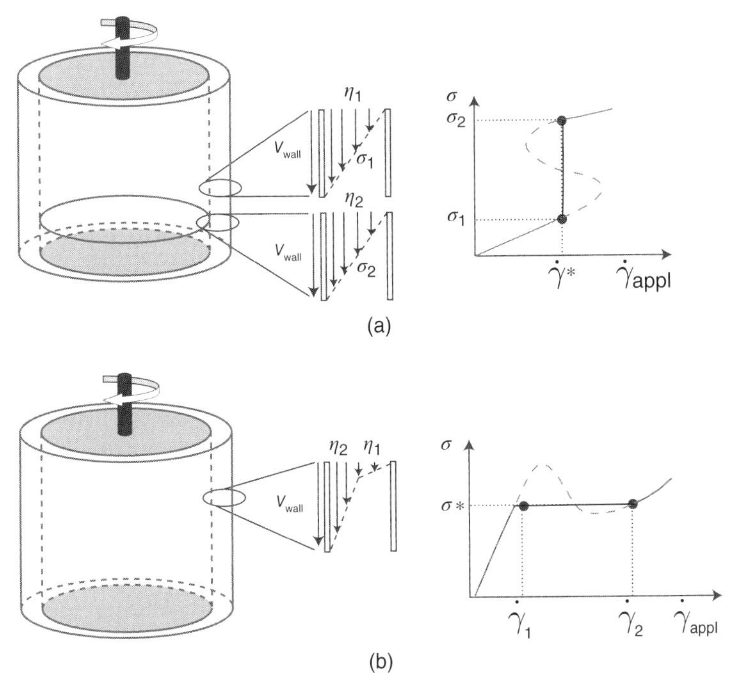

Figure 6.4 Two types of shear banding: in the vorticity direction (a) and in the gradient direction (b). The flow profiles are indicated in the middle by the different lengths of the arrows. To the right flow curves underlying the shear band formation at the right.

Gradient banding in shear flow is far better documented [14] and understood [15, 16] than vorticity banding. In principle, its occurrence is independent of the structural changes in the fluid underlying the shear thinning behavior, as long as the shear thinning is substantial. Thus, gradient banding has been observed for systems ranging from hard sphere colloidal crystals [17] and soft colloidal glasses [18] to associative polymer networks [19], entangled polymer solutions [20, 21], and DNA solutions [22, 23]. Shear thinning behavior is far more pronounced for dispersions of living polymers, which consist of monomers that continuously break and recombine. This class of polymers has two mechanisms to release stress: reptation and the break up and recombination of the polymers. For the case where the average breaking time is much faster than the reptation time, it can be shown that this results in a unique relaxation mechanism and strong shear thinning [24, 25]. Surfactant wormlike micelles are self-assembled particles, and display strong shear thinning, which is easily tunable [26, 27]. Therefore, wormlike micelles are the "working horse" for experimentalists in the field of flow instabilities. The first experimental proof for the occurrence of structure formation in the fluid were birefringence measurements showing the split up of fluid into regions with low and high birefringence, that is, probably into regions with high and low viscosity, respectively [28]. This strongly suggests that the shear rate in the two regions is different. The existence of shear banding in this class of systems has first been shown by Callaghan *et al.* in a pipe-flow geometry using Nuclear Magnetic Resonance Microscopy [29]. Using Heterodyne Dynamic Light Scattering, Salmon *et al.* [30] demonstrated for the first time that indeed a scenario is at hand where the systems splits up in two bands with shear rates of $\dot{\gamma}_1$ and $\dot{\gamma}_2$ using the Couette geometry. The contribution of both bands is set by the lever rule, similar to the lever rule for gas–liquid phase separation. The mechanisms of shear band formation and the relation with the birefringent structures was nicely demonstrated both by Hu and Lips [31] and Miller and Rothstein [32]. Both groups used Particle Imaging Velocimetry (PIV) in combination with small-angle light scattering and birefringence imaging. Even more structural insight was supplied by using a combination of small-angle neutron scattering and shear flow [33].

Vorticity banding is less common than gradient banding. A detailed recent discussion of elastic instabilities that drive the formation of vorticity bands can be found in Ref. [34]. The classical picture of vorticity banding due to shear thickening is more the exception than the rule, although again for wormlike micelles the predicted behavior has been found [35]. Another, more common, mechanism is that normal stresses are generated through the nonlinear elastic deformation of the dispersed mesoscopic entities, similar to the Weissenberg effect in polymeric systems [36]. The role of the polymer chains in the classic Weissenberg effect is now played by large-scale networks, such as carbon nanotube dispersion [37] and clay gels [38], or inhomogeneities formed during the initial stages of phase separation as found for polymer–protein mixtures [39] and coexisting isotropic-nematic phase of colloidal rods [40]. These structures will be nonuniformly stretched due to the curved streamlines that are present in most rheological devices, thus generating hoop-stresses that give rise to elastic normal forces [41]. Another mechanism relates the formation of vorticity bands shear bands to gradient shear banding. There is experimental evidence [42, 43] and theoretical justification [44] that vorticity bands are formed due to instabilities in the interface between two gradient bands. All of these scenarios lead to a rolling motion within the bands, superimposed on the laminar flow. This rolling motion is absent in the classic picture, depicted in Figure 6.4a. Note also that these are pure elastic instabilities and that inertia does not play a role as is the case for the well-known Taylor instability [45].

6.3 EFFECT OF SHEAR FLOW ON CRYSTALLIZATION OF COLLOIDAL SPHERES

Colloidal particles with the simplest interaction potential and the simplest shape are colloidal hard spheres. Despite this conceptual simplicity, it is impossible to calculate the phase diagram with analytic theories. It was only in 1957 that one of the first computer simulations were used to show that hard spheres undergo a transition from a liquid phase, where particles can be at any position in space, to a solid phase, where the particles are bound to their crystal lattice positions [46]. Since hard spheres have no potential energy, their phase behavior is determined entirely by entropy and depends only on volume fraction φ [47]. These systems can still order, and thus reduce the entropy, because locally the particles will have more free volume, thus increasing the entropy. Fluid and crystal coexist between the freezing point $\varphi_f = 0.494$ and the melting point $\varphi_m = 0.545$, while the pure crystal is stable for $\varphi > \varphi_m$, as shown again by simulations [48]. The volume fractions that border the phase coexisting region are φ_f and φ_m are called the binodal points. Colloidal systems proved to be ideal to study these most fundamental transitions, once chemists were able to synthesize particles that behave almost like billiard balls. The complete phase diagram was experimentally determined using dispersions of polymer-coated polymethylmethacrylate particles, including a transition to a glassy state of random close-packed spheres at even higher concentrations [49]. Nonetheless, colloids will always have some softness, albeit due to the repulsive polymer layer that is grafted on their surface or due to the surface charge. For extremely charged colloids, it was even possible to make crystals with 1% of the volume occupied by particles [50]. Scaling of the volume fraction is thus performed by using an effective diameter. The other appeal of colloidal spheres is that they can be directly visualized of one of the most fundamental processes of structure formation [51], which allows comparison with simulations [52]. A nice overview on crystallizing colloids is given in Ref. [53].

We discuss here the effect that shear flow has on the liquid–crystal phase transition. The flow behavior of colloid hard sphere dispersions in the liquid phase is a very extended field of research, both for practical and fundamental reasons. While at low concentrations the viscosity is merely increasing [54], colloidal dispersions attain a highly nonlinear viscoelastic behavior at high concentrations, approaching the liquid–crystal transition. This is due to the fact that the shear-induced distortion on the particle level leads to extreme shear thickening behavior [55]. Shear flow per definition affects the crystal structure because particles are displaced from their fixed lattice positions. Thus, the rate of deformation competes with the rate that crystals are formed. In this way, the effect of shear flow is connected to the kinetics of the phase transition, so to the rate at which crystals form. As we will see, there is a cascade of different states into which the quiescent crystal can be transformed, while crystallization of a quiescent amorphous dispersion can also be induced by shear flow. We discuss two types of nonequilibrium phase diagrams. First, the effect of oscillatory shear flow on the structure and phase behavior is discussed and then the effect of steady shear flow. It will be shown that the transition between different states has a pronounced influence on the flow behavior of these highly structured dispersions. The sequence of transitions and the phase, which will be described below, are neither complete nor universal. Subtle changes in the particle interaction can already alter the shear response. The aim is, however, to convey an idea about the physical processes at hand. For an overview on the structure of colloidal spheres in shear flow, we refer to Ref. [56].

6.3.1 Equilibrium Phase Behavior

According to *Classical Nucleation Theory* (CNT, see [57] for a review), the kinetics of crystallization is determined by two processes. First, a crystal nucleus needs to be formed out of the isotropic background. This is a process with an activation energy barrier and therefore it has an induction time τ_{ind}. The activation barrier is due to the balance between the energy penalty for the formation of a sharp interface and the energy gain due to the formation of the new phase with lower free energy. The decrease in the bulk energy depends on the difference between the solid and liquid chemical potentials $\Delta\mu$ times the volume of the newly formed nucleus, while the increase in surface energy depends on the solid–liquid interfacial tension γ times surface of the nucleus. Both contributions result in a free energy barrier ΔG_{crit} for the formation of a nucleus

$$\Delta G_{\text{crit}} = \frac{16\pi}{3}\gamma^3/(\rho_s\Delta\mu)^2, \qquad (6.7)$$

where ρ_s is the number density of the solid.

The rate I at which nuclei are formed depends exponentially on ΔG_{crit} and on the second process of crystallization, namely, the rate at which particles diffuse to the formed nucleus, given by the kinetic factor κ, which is connected to the short-time self-diffusion constant of the colloids. I can now be written as

$$I_{\text{growth}} = \kappa \exp -\Delta G_{\text{crit}}/k_B T\,, \qquad (6.8)$$

where T is the absolute temperature, k_B is Boltzmann's constant. Equation 6.8 contains the functional dependence of the nucleation rate on the relevant physical quantities $\Delta\mu$, κ, and γ. *A priori* knowledge of these quantities is needed to make quantitative predictions of the nucleation rate. Since to date there is no theory that provides "first principles" predictions of these parameters, estimates need to be made depending on the exact system. For hard spheres, it is predicted that the nucleation rate displays a maximum as a function of the volume fraction: while the driving force $\Delta\mu$ increases with increasing concentration, the diffusion slows down until it is arrested at the glass transition [58]. The slowing down is less dramatic for charged spheres and therefore no maximum is expected. The general trend was confirmed by a vast amount of light scattering experiments [59–61]. Experiments and simulations can be compared after performing a careful scaling of the experimental volume fraction, as described above. A reasonable agreement is found at high concentrations, but the nucleation rate found in simulations at low volume fractions approaching the freezing point φ_f are several orders of magnitude slower. The exact reason for this discrepancy is yet unknown.

6.3.2 Nonequilibrium Phase Behavior

There are many ways how shear flow can perturb the crystallization process, as depicted in Figure 6.5. Shear flow can *enhance* the growth rate by convecting particles to the crystal nucleus, as shown in Figure 6.5a. Another way how shear flow enhances mobility is by breaking cages that form at high concentrations, as shown in Figure 6.5b. This is especially of importance for hard spheres, which have the tendency to jam into a glassy state. In both cases, the diffusion of the constituents is enhanced so that the diffusive term κ in Equation 5.8 is replaced by a convective shear-dependent term. Finally, shear flow can also induce ordering by aligning spheres into strings. This potentially plays a role in decreasing the energy barrier ΔG_{crit} since it creates already an interface, as shown in Figure 6.5c. On the other hand, shear flow can also *frustrate* crystallization

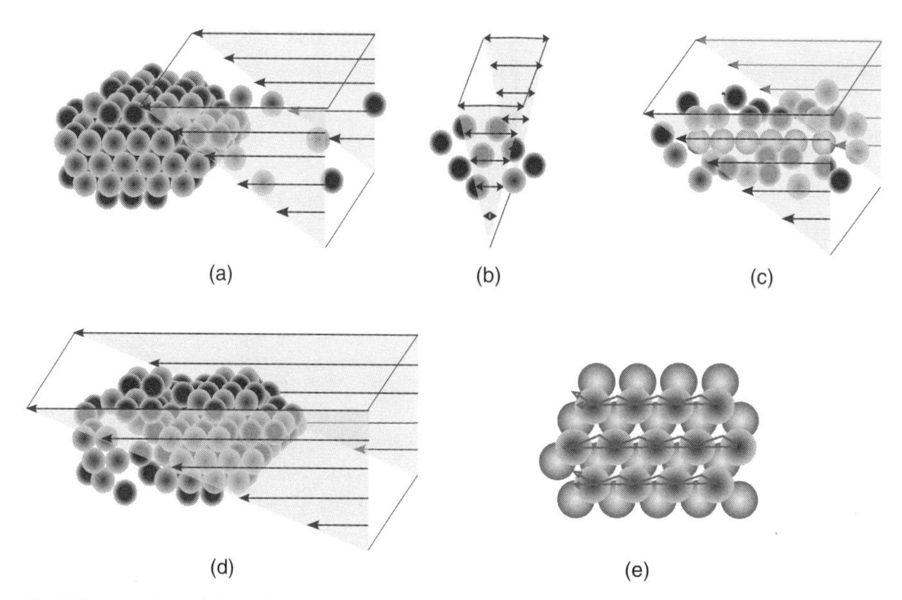

Figure 6.5 Sheared colloidal crystal: positive effects due to (a) convection of particles to nucleus, (b) cage breaking, and (c) alignment of spheres; negative effects due to (d) erosion and (e) registered motion.

and/or destroy crystals. First, when a sample is crystallizing but not yet fully crystalline, then the surface of the crystal nuclei can be eroded by the shear flow, as shown in Figure 6.5d. Second, when fully crystalline samples are subjected to shear flow, then the crystal will have to deform by the shear and finally yield. This effect is dramatic when the crystallite has the wrong orientation with respect to the shear flow, which will readily cause the destruction of the crystallite by shear. Sheared crystals will therefore only grow when they have the right orientation to accommodate deformations. In Figure 6.5e, two mechanisms are depicted that might take place. At low shear rates, when the system is given sufficient time, layers of crystals may slide over each other by a registered motion. Here, the spheres jump between the points in the crystal lattice in a zigzag motion. At high shear rates, layers will slip over each other. This motion can only be accommodated when the crystal is somewhat deformable, so when the interaction potential between the particles is soft, or when the concentration is not too high.

The experimental signature of crystals is very clear when using scattering techniques, since Bragg diffraction from crystals causes strong scattering peaks. These peaks emerge out of a background diffuse "Debye–Scherrer" ring, characterizing the liquid structure factor. A hexagonal pattern will be observed when a sample consists of a single crystal. Samples will often be polycrystalline so that scattering will take place from crystallites with many different orientations, resulting in a diffuse ring. Since the emergence of fluorescently labeled colloids in combination with confocal microscopy, real space measurements reveal details that are lost in ensemble averaged techniques like scattering. Macroscopic measurements are also very important, in particular, rheological measurements, where, for example, the viscosity is measured as a function of shear rate. The above-mentioned processes can thus be studied with this set of experimental tools.

Pioneering work on crystallization of colloidal hard spheres in shear flow is due to Bruce Ackerson and coworkers, employing the same grafted PMMA spheres as were used to determine the equilibrium phase diagram [49]. The main experimental technique that was applied was light and neutron scattering in combination with steady and oscillatory shear flow, as schematically shown in Figure 6.6. Here, the incident radiation propagates along the centerline of the shear cell or along the tangent, probing the flow-vorticity and flow-gradient plane, respectively. In principle, all possible effects that shear flow has on crystallization have been found by him and his coworkers. Their findings are summarized in a seminal paper in 1990 [62] and they will be discussed below.

The most subtle method of probing the effect of shear flow on crystallization is by applying oscillatory shear; see Equation 6.4. In this way, the effect of the strain deformation that is needed to aid (Fig. 6.5a–c) or frustrate crystallization (Fig. 6.5d and e) and time needed to take effect can be varied independently, although the frequency was fixed in the experiments discussed below. Oscillatory shear flow will induce crystallization at volume fractions as low as $\varphi = 0.47$, for which the equilibrium sample is liquid-like ($\varphi < \varphi_f$), but only at sufficiently high strain amplitudes of $\gamma_0 > 1$. This means that both the Péclet number and the deformation need to be high enough. At strain amplitudes of $\gamma_0 > 2$, layer structure is observed and at even higher strain amplitudes the structure melts. For a volume fraction of $\varphi = 0.50$ phase coexistence is observed in equilibrium ($\varphi_f < \varphi < \varphi_m$) and the same transitions as described before,

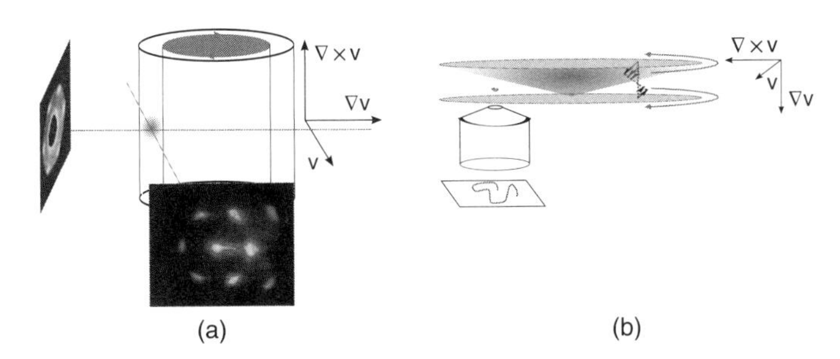

Figure 6.6 Geometries where sheared structures can be probed *in situ* using scattering (a) and confocal microscopy (b).

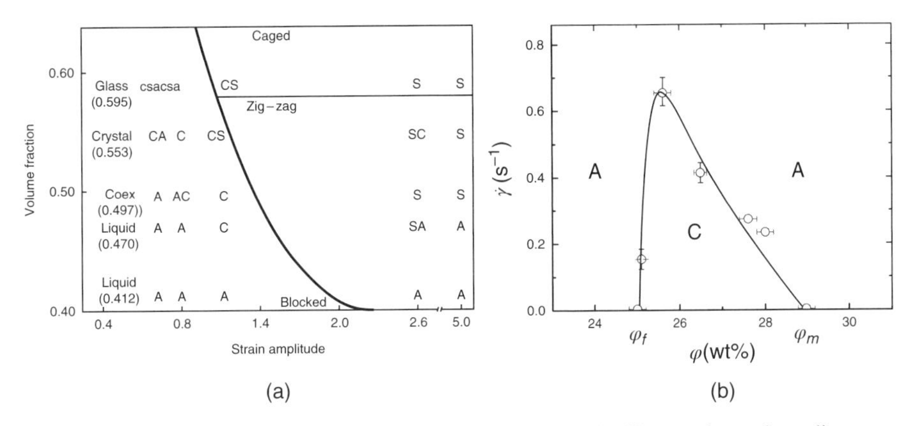

Figure 6.7 Nonequilibrium phase diagrams for crystallizing colloidal spheres. (a) Oscillatory shear phase diagram as a function of volume fraction of hard spheres and strain amplitude at 3.5 Hz. Source: Reprinted with permission from Ref. [62]. Copyright 1990 The Society of Rheology. The symbols correspond to crystalline order (C), layer ordering (S), and amorphous order (A). The first symbol in a grouping indicates the predominant ordering observed. (b) Steady shear phase diagram as a function of volume fraction of charged spheres and shear rate. The symbols are the same as for (a). Source: Adapted with permission from Ref. [63]. Copyright 2005 American Chemical Society.

just at lower strain amplitudes. For $\varphi = 0.553$, the sample is fully crystalline at equilibrium ($\varphi > \varphi_m$), while already small strain amplitudes force the system into an oriented face-centered cubic (fcc) crystal structure. At higher strain amplitudes, the sample yields by forming layers that slide over each other. For a volume fraction of $\varphi = 0.595$, the equilibrium spheres are randomly close-packed in a glassy state. The glassy state and the liquid state as found for $\varphi = 0.47$ are structurally the same. Also for $\varphi = 0.595$, crystal formation is induced by shear. In this case, shear-induced crystallization is due to cage breaking as schematically shown in Figure 6.5b, and not due to alignment of the spheres as shown in Figure 6.5c, which is the mechanism expected for the concentrated liquid phase. At high strain amplitudes, the system yields by forming layers that zigzag over each other, since there is not enough space for sliding. A nonequilibrium phase diagram was thus obtained by varying the volume fraction of the colloids and the amplitude of the applied shear γ_0, as depicted in Figure 6.7a.

Ackerson [62] realized that different transitions between dynamical states are accompanied by different crystal structures. The transitions were interpreted by using models that describe what strain can be accommodated before particles bump into each other, which depends on the crystal structure. These models can be best validated by observing the structures in real space. First attempts showed an intriguing shear-induced buckled state for an equilibrium fcc structure in confinement [64]. Later studies use a counter-rotating shear cell in combination with confocal microscopy, as depicted in Figure 6.6b, in which a stationary layer can be positioned away from the glass wall so that the particles stay in the field of view [65, 66]. With this technique, the existence of the oscillating twinned fcc phase, the sliding layer phase, and the shear-induced strings have been confirmed. In addition, a state where the crystal makes an angle with the flow direction was found, while nonequilibrium Brownian dynamics simulations complied with the experimental findings [66].

Ackerson *et al.* also studied the effect of steady shear flow on crystallization of colloidal spheres. They observed that layering is present at high concentrations and shear flow, but there is always evidence for the existence of an amorphous distorted structure. A transition from a random close packed crystal to an oriented fcc is observed only at low shear rates. The same sequence of transitions was observed for crystals consisting of micelles of block-copolymers using small-angle X-ray and neutron scattering [67]. These

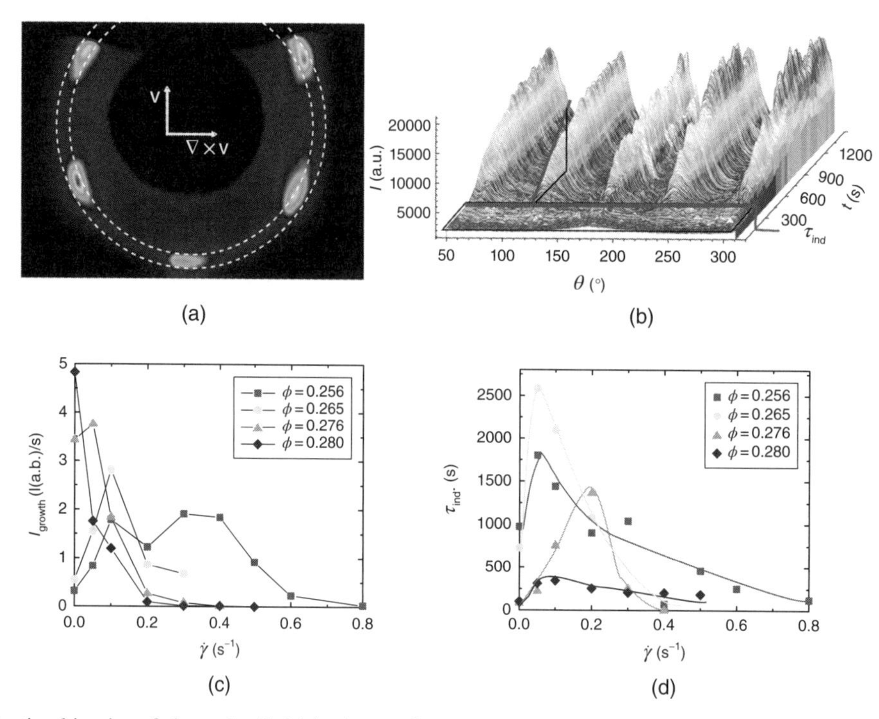

Figure 6.8 Crystallization kinetics of charged colloidal spheres after a quench from a high shear rate where the crystal structure is melted to a final shear rate. (a) Light scattering pattern using a geometry as shown in Figure 6.6a, probing the flow-vorticity plane. The pattern is taken 1500 s after a quench to a shear rate of 0.05 s^{-1}. (b) Azimuthal profiles taken at the scatter angle of the first-order Bragg peaks, dashed lines in (a), as a function of time after the quench to $\dot{\gamma} = 0.05$ s^{-1}. The growth rate I_{growth} is obtained from the slope of the red line and induction time τ_{ind} is obtained from the start of growth. τ_{ind} and I_{growth} are plotted in (c) and (d), respectively, as a function shear rate φ for various volume fractions. Source: Adapted with permission from Ref. [63]. Copyright 2005 American Chemical Society. (*See color plate section for the color representation of this figure.*)

systems are basically small colloidal spheres (radius of 12 nm) with soft interactions. The transitions were located at much higher shear rates, which is related to the small size and fast relaxation of the particles; see Equation 6.6. The zigzag motion was observed in real space using the counter-rotating shear cell in combination with a fast confocal microscope. This technique has the advantage that the local motion of the particles can be followed. As a consequence, it could be shown how the crystal shear melts when the particles have random displacements of more than 12% of the particle separation, reminiscent of the Lindemann criterion for melting in equilibrium systems [65].

These dynamic transitions underlie the nonequilibrium phase behavior of colloidal crystals in shear flow. To obtain the full phase diagram, one needs to determine the shear rate dependence of the nonequilibrium binodal points $\varphi_f(\dot{\gamma})$ and $\varphi_m(\dot{\gamma})$. The best way to obtain these points is by monitoring the kinetics of the crystallization. In Ref. [63], the crystallization kinetics of charged silica spheres in shear flow was studied. This system freezes at a volume fraction as low as $\varphi_f = 0.25$ while at $\varphi = 0.29$ it forms an amorphous glass [68]. The advantage of charged systems is that the crystals are much softer and form faster. Shear flow can be conveniently used to create a melted initial state using a high shear rate, followed by a quench at $t = 0$ to the shear rate for which the crystallization will be studied. Figure 6.8a displays a typical scatter pattern as formed 1500 s after a quench to a shear rate of 0.05 s^{-1}, showing clear Bragg peaks of an oriented crystal. The kinetics of the formation of this structure can be followed when plotting the azimuthal profile taken at the peak position; see Figure 6.8b. The nucleation time τ_{ind} and growth rate I_{growth} can be obtained from the evolution of the azimuthal profile: the growth rate relates to the slope of the total intensity as a function of time and the induction time is obtained from the crossing of this slope with the background; see Figure 6.8b. The thus obtained induction times τ_{ind} and growth rates are plotted in Figure 6.8c and d, respectively, as a function of shear rate for various concentrations. The phase boundary at low concentration is determined from these plots by extrapolation to the concentration where the inverse of induction time τ_{ind} goes to zero ($1/\tau_{ind} \rightarrow 0$), while the boundary at high concentration and shear rate is found by extrapolation to the point where the growth rate goes to zero ($I_{growth} \rightarrow 0$).

The resulting nonequilibrium phase diagram of sheared charged colloids, as plotted in Figure 6.7b, is strongly peaked due to the competing effects of shear flow. The phase boundary at low concentrations is almost vertical and therefore independent of the shear rate. This implies that at

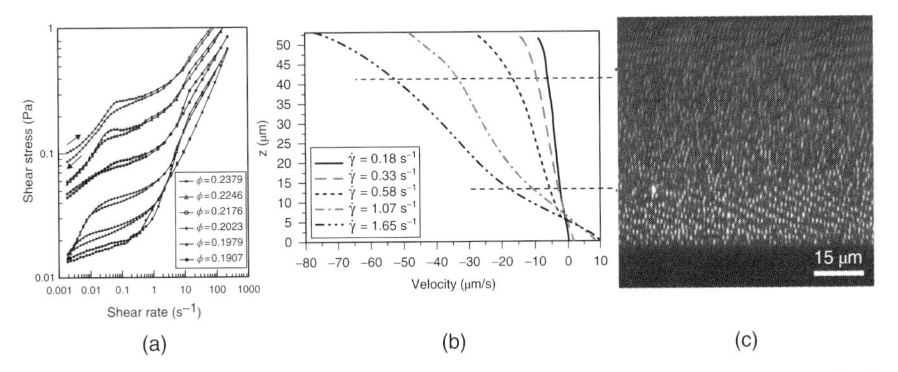

Figure 6.9 (a) Flow curves for charged silica spheres at different volume fractions showing strong shear thinning when increasing the shear rate. Source: Reprinted (adapted) with permission from Ref. [17]. Copyright 1994 American Chemical Society. (b, c) Shear banding in colloidal crystals observed with *in situ* confocal microscopy measurements, using a cone-plate shear cell as depicted in Figure 6.6b. Source: Reprinted from Ref. [74]. (b) Flow profiles of a $\varphi = 0.50$ dispersion of silica particles measured for various applied shear rates. The local shear rate was significantly higher at both the top and bottom plates. (c) Confocal microscopy image of the velocity-gradient plane taken immediately after cessation of the flow, using a cone-plate geometry as shown in Figure 6.6b. At both walls, the particles arranged in crystalline layers, but in the middle, the structure was liquid-like. Horizontal dashed lines indicate these boundaries and show the correspondence with the flow profiles.

low concentrations, the chemical potential difference $\Delta\mu$ and interfacial tension γ between the liquid and crystal, see Equation 6.7, are almost independent of the shear rate. Indeed, at low concentrations the induction time τ_{ind} is not very sensitive to changes in the shear rate (see red curve in Figure 6.8). For higher concentrations, the induction time exhibits a maximum as a function of the shear rate. The increase in τ_{ind} with increasing shear rate for the lower rates is probably connected to the suppression of nuclei formation, as found in simulations [69]. The decrease in τ_{ind} at higher shear rates is probably related to shear-induced alignment of the spheres, lowering the surface tension γ. Without shear flow τ_{ind} decreases with increasing concentration, which complies with a decrease in the energy barrier. The growth rate without shear flow increases monotonically as a function of concentration, which complies with an increased concentration of spheres and therefore of $\Delta\mu$; see Equation 6.8. The fact that the growth rate displays a maximum at low concentrations as a function of shear rate nicely illustrates the competing effects of shear flow. Low shear rates enhance transport of particles toward the nucleus, as illustrated in Figure 6.5a, and therefore the kinetic factor κ, while at high shear rates erosion takes place, as shown in Figure 6.5d. For higher concentrations, there are more particles available for the crystallite so that shear flow only erodes the crystallites, as illustrated by the monotonous decay of the growth rate as a function of shear rate.

6.3.3 The Effect on Flow Behavior

We have so far described a number of structural changes that can be induced by shear flow, either creating or destroying crystals. These changes are accompanied by changes in the mechanical properties of the systems. Indeed, dynamic rheological tests show that the modulus G^* around the crystallization point is lower in the crystal phase than in the liquid phase, while in both phases storage modulus G' is higher than the loss modulus G'' [70], indicating a transition between two solid states.

Steady shear experiments show that the softening of the material is also apparent from the strong shear thinning, which has been seen for a variety of systems [17, 71, 72]. An example of a flow curve that exhibits shear thinning is shown in Figure 6.9a for a system of charged spheres at various concentrations. For all of these systems, the shear thinning is accompanied by the formation of regions where the sample is crystalline and regions where the sample is amorphous. As explained in Section 6.2.3, it is expected that a strong shear thinning system will form shear bands with different shear rates. With the development of *in situ* confocal microscopy, it is possible to determine the structure at the same time as the flow profile. For both oscillatory [73] and steady shear flow, it was observed that the high shear rate and low viscosity region is the crystalline region; the low shear rate and high viscosity region has an amorphous structure [74]; see Figure 6.9b and c.

The occurrence of such a pronounced flow behavior depends on many factors. If the spheres are very charged, as the system used in Section 6.3.2, then the interaction between the spheres is so soft that the stress barrier for the formation of structurally different sample is low. The history of the sample is important as can be deduced from the fact that the flow curve obtained by increasing the shear rate is different from the flow curve obtained by decreasing the shear rate; see Figure 6.9a. This also shows that the flow behavior depends on the initial condition of the sample, which could be a polydomain crystal, amorphous or pure crystal, and so on. Although all these subtleties are very important, for now it is sufficient to realize that indeed the macroscopic flow behavior is set by the formation and destruction of crystalline microstructure.

6.4 EFFECT OF SHEAR FLOW ON GAS–LIQUID PHASE SEPARATING COLLOIDAL SPHERES

The phase behavior of particles with an attractive interaction potential is very different from repulsive or hard spheres and therefore also the effect of shear flow is expected to be very different. Chapter 14 describes the different effects that attraction can have, focusing on gelation, which typically occurs when particles have a short-ranged and deep-interaction potential. Moderate attractions cause the system to separate in a phase with low density, the gas phase, and a phase with high density, the liquid phase. The gas–liquid phase separation of colloids is very similar to the molecular gas–liquid phase separation but also to binary liquids and polymer solutions. All of these classes of systems have a critical point where the binodal and spinodal lines meet.

The spinodal line separates the metastable region from the unstable region in the phase diagram. Within the unstable region, an initially homogeneous system evolves continuously toward the demixed state. In contrast, the binodal separates the stable from the metastable region of the phase diagram. Demixing from the metastable region evolves through nucleation-and-growth, where on formation of the nucleus a discontinuous jump to a locally relatively high concentration occurs, as we saw for crystallization. At the spinodal, the (effective) diffusion coefficient D_{eff} vanishes, which is commonly referred to as "critical slowing down." Slowing down of dynamics at the spinodal is due to cancellation of the Brownian forces (which tend to homogenize the system) and the attractive interactions between the colloidal particles (which favor demixing). The driving net force for colloidal-particle displacements is thus very small, leading to very slow dynamics. In addition, the range of interactions between two colloidal particles as mediated through many intermediate colloids diverges at the spinodal. The divergence of this interaction range, the so-called correlation length ξ, was first described by Ornstein and Zernike [75]. Both of these effects are quantified by a vanishing concentration derivative of the osmotic pressure, as will be discussed later. While the effective diffusion coefficient is zero at the spinodal, it becomes negative on crossing the spinodal into the unstable region, which mathematically describes the continuous growth of inhomogeneities. The initial stage of spinodal decomposition is thermodynamically described for the first time in Ref. [76], which is now known as the Cahn–Hilliard theory.

Due to the very slow dynamics near the spinodal, unusually small shear rates are sufficient to significantly distort microstructural order. This is mathematically quantified by an effective Péclet number that contains both the effective diffusion coefficient and the correlation length, as will be seen later. Onuki and Kawasaki discussed the effect of shear on critical systems, by using a renormalization group approach. It was shown that phase separation is suppressed by the effect that shear flow has on the life time of concentration fluctuations [77]. Experiments on semidilute polymer solutions revealed that shear flow can also enhance concentration fluctuations, which induces phase separation [78]. This is in accordance with dynamic theories where shear-induced stresses generated by internal degrees of freedom and entanglements of the polymers were implicitly included. It was shown that the spinodal is suppressed in the vorticity direction and enhanced in the gradient direction [79, 80]. These theories are to some extent applicable to colloidal dispersions [81]. A nice review of the critical behavior in shear flow is given by Onuki [82].

We focus in this chapter on the phase behavior of colloids in shear flow, so that the interest here is in the development of a nonequilibrium theory for concentrated colloidal systems with attractive intercolloidal interactions. The way to progress is by using microscopic theories that describe the distribution of positions for individual particles and how this distribution evolves in time. In other words, an equation of motion for the probability density function of the particle positions needs to be developed. This type of theory has the advantage that it is straightforward to incorporate shear flow, which is (at least in principle) not possible through a quasithermodynamic approach as shear flow is a nonconservative external field. The equation of motion for the position coordinates of an assembly of many colloidal particles is the so-called Smoluchowski equation. In the following, we first describe how critical slowing down (quantified through the effective diffusion coefficient D_{eff}) and the divergent correlation length ξ can be obtained from the Smoluchowski equation for attractive colloidal spheres without shear flow and then how these parameters are affected by shear flow. Details of this theory can be found in Refs [54, 83–85]. This theory will be contrasted to light scattering experiments. We finish this section, describing the effect of shear flow on demixing systems and see how this affects the stability of the system.

6.4.1 Equilibrium Phase Behavior

The Smoluchowski equation of motion for the probability density function $P \equiv P(\mathbf{r}_1, \mathbf{r}_2, \cdots, \mathbf{r}_N, t)$ of the position coordinates $\mathbf{r}_j$, $j = 1, 2, \cdots, N$, of all N colloidal particles in the system reads, with the neglect of hydrodynamic interaction, is given by

$$\frac{\partial}{\partial t} P = D_0 \sum_{j=1}^{N} \nabla_{r_j} \cdot [\beta[\nabla_{r_j} \Phi]\, P + \nabla_{r_j} P]. \tag{6.9}$$

Equation 6.9 relates the probability of finding a particle in a certain volume to the forces acting on it, either mediated by the other particles, first term on the right, or by the randomizing Brownian diffusive forces, second term on the right. Here, D_0 is the Stokes–Einstein diffusion coefficient, $\beta = 1/k_B T$ (with k_B Boltzmann's constant and T the temperature). $\Phi \equiv \Phi(\mathbf{r}_1, \mathbf{r}_2, \cdots, \mathbf{r}_N)$ the potential energy of the assembly of colloidal particles.

In order to predict if a system is stable, one needs to check if the average distance between the particles is stable. The correlation between the relative position of two particles is described by the pair correlation function $g(t)$, which is defined as

$$P_2(\mathbf{r}_1, \mathbf{r}_2, t) \equiv \int d\mathbf{r}_3 \cdots \int d\mathbf{r}_N\, P(\mathbf{r}_1, \mathbf{r}_2, \mathbf{r}_3, \cdots, \mathbf{r}_N, t) \equiv P_1(\mathbf{r}_1, t)\, P_1(\mathbf{r}_2, t)\, g(\mathbf{r}_1, \mathbf{r}_2, t)\,. \tag{6.10}$$

Note that the reduced pdf for a single particle P_1 is per definition $P_1 = 1/V = \rho/N$ (where V is the volume of the system, N the total number of spherical colloids, and $\rho = N/V$ the number density). The potential is taken to be the sum of so-called pair-interaction potentials $V(r_{ij})$, that is (with $r_{ij} = \mid \mathbf{r}_i - \mathbf{r}_j \mid$)),

$$\Phi(\mathbf{r}_1, \mathbf{r}_2, \ldots, \mathbf{r}_N) = \sum_{i,j=1\ ,i<j}^{N} V(r_{ij})\,. \tag{6.11}$$

This is the point where the attraction between particles comes into play. Integration of the Smoluchowski equation (6.9) gives (with $\mathbf{R} = \mathbf{r}_1 - \mathbf{r}_2$ the distance between two spherical colloids and ∇ the gradient operator with respect to $\mathbf{R}$),

$$\frac{\partial}{\partial t} g(\mathbf{R}) = 2\, D_0 \nabla \cdot \{\nabla g(\mathbf{R}) + \beta\, g(\mathbf{R}) [\nabla V(R) - \mathbf{F}^{\text{ind}}(\mathbf{R})]\}. \tag{6.12}$$

Here, $\mathbf{F}^{\text{ind}}$ is the "*indirect force*" between two particles, which is mediated via a third particle, and is equal to

$$\mathbf{F}^{\text{ind}}(\mathbf{R}) = -\rho \int d\mathbf{r}\, [\nabla V(r)]\, \frac{g_3(\mathbf{r}, \mathbf{R})}{g(\mathbf{R})}\,, \tag{6.13}$$

where $\rho = N/V$ is the particle number density and $g_3(\mathbf{r}, \mathbf{R})$ is the three-particle correlation function; $\mathbf{r}$ is the distance between particle 1 and a third particle, $\mathbf{r} = \mathbf{r}_1 - \mathbf{r}_3$.

For critical phenomena, the indirect force is essential, being responsible for the long-ranged interactions between two spheres as mediated through the remaining spheres. Thus, we want to find a solution for Equation 6.12 for distances $R \gg R_V$, where R_V is the range of the pair interaction. Equation 6.12 can only be solved; however, when g_3 can be expressed in terms of $g(\mathbf{R})$.

The simplest closure is the well-known superposition approximation

$$g_3(\mathbf{r}, \mathbf{R}) = g(\mathbf{r}) g(\mathbf{R}) g(\mathbf{r} - \mathbf{R}). \tag{6.14}$$

The problem with this closure is that it does not lead to the Ornstein–Zernike form for the pair correlation function with a correlation length that diverges at the critical point, because the influence of a third particle on the correlation between the other two particles is neglected. To fix this, we note that the pair correlation function in the integral in Equation 5.13 is multiplied by the pair-force $\nabla V(r)$. Since the range of $\nabla V(r)$ is given by R_V, we need a closure relation only for this small distances $r \le R_V$. The effect of the presence of the third particle at position $\mathbf{r}_2$ is that it enhances the local density around the two neighboring particles at $\mathbf{r}_1$ and $\mathbf{r}_3$. This density enhancement is equal to

$$\Delta\rho \approx \rho \left[g\left(\mathbf{R} - \frac{1}{2}\mathbf{r}\right) - 1 \right] = \rho\, h\left(\mathbf{R} - \frac{1}{2}\mathbf{r}\right), \tag{6.15}$$

where $h = g - 1$ is the total correlation function. $\mathbf{R} - \frac{1}{2}\mathbf{r}$ is the point in between the positions of particles 1 and 3, so of order $\approx R_V$, relative to the position of particle 2 at a distance $R \gg R_V$. The total correlation function at these large separations is a very smooth function around zero. Close to the critical point, the total correlation function becomes very long ranged and decays monotonically, for distances much larger than R_V, to zero. This will be confirmed later. As a consequence, we evaluate the $g(\mathbf{r})$ in Equation 6.14 at a density $\rho + \Delta\rho \approx \rho(\mathbf{R} - \frac{1}{2}\mathbf{r})$ at the position in between the two neighboring particles 1 and 3. Since for $\mathbf{R} \gg R_V$ and $\Delta\rho/\rho \ll 1$, we can expand with respect to $\Delta\rho/\rho \ll 1$ to first order, yielding

$$g_3(\mathbf{r}, \mathbf{R}) = \left\{ g(\mathbf{r}) + \frac{dg(\mathbf{r})}{d\rho}\, \rho\, h\left(\mathbf{R} - \frac{1}{2}\mathbf{r}\right) \right\} g(\mathbf{R} - \mathbf{r})\, g(\mathbf{R})\,. \tag{6.16}$$

This is exactly the relation that was first proposed by Fixman [85].

The final Smoluchowski equation can now be obtained (i) by substituting the closure relation Eq. (6.16) into Eq. (6.13); (ii) applying a Taylor expansion of $h(\mathbf{R})$ to within linear order, since, as we mentioned before, $h(\mathbf{R})$ is very small for the large separations of interest. This leads to

$$\frac{\partial}{\partial t} h(\mathbf{R}) = 2\, \beta\, D_0\, \nabla \cdot \left(\nabla \left[\frac{d\Pi}{d\rho}\, h(\mathbf{R}) - \Sigma\, \nabla^2 h(\mathbf{R}) \right] \right). \tag{6.17}$$

We introduced here two new quantities. From the above derivation, we find that Π is given by

$$\Pi = \bar{\rho}\, k_B T - \frac{2\pi}{3}\rho^2 \int_0^{\infty} dr\, r^3\, \frac{dV(r)}{dr} g^{\text{eq}}(r). \tag{6.18}$$

This is precisely the equilibrium statistical mechanical well-known expression for the osmotic pressure. The second quantity, Σ, is given by

$$\Sigma = \frac{2\pi}{15}\rho \int_0^\infty dr\, r^5 \frac{dV(r)}{dr} \left\{ g^{eq}(r) + \frac{1}{8}\rho \frac{dg^{eq}(r)}{d\rho} \right\}. \quad (6.19)$$

This is a microscopic expression for the famous Cahn–Hilliard theory free energy square-gradient coefficient.

Equation 6.17 is the fundamental equation of motion that we use to analyze the effect of shear flow on long-ranged critical structure. In equilibrium $\partial h/\partial t = 0$, this differential equation has a solution (C is a constant):

$$h(R) = C\, \frac{\exp\{-R/\xi\}}{R}\,, \quad (6.20)$$

where *the correlation length* ξ is equal to

$$\xi = \sqrt{\frac{\Sigma}{d\Pi/d\rho}}\,. \quad (6.21)$$

According to Equation 6.20 for the total correlation function, the correlation length is a measure for the range over which two spheres interact with each other, via intermediate spheres. Hence, it is one of the parameters we set out to determine. Since $d\Pi/d\rho$ vanishes at the critical point, the correlation length diverges as expected. Equation 6.21 thus confirms our earlier assumption that at large distances $h(R)$ monotonically decreases smoothly to zero.

In principle, $h(R)$ can be measured microscopically. In the case where collective behavior is of interest, however, it is better to choose a scattering approach for experiments, where a full ensemble of particles is probed in reciprocal space. In scattering experiments, the structure factor is measured, which is defined as

$$S^{eq}(\mathbf{k}) = \frac{1}{N}\sum_{i,j=1}^{N} \langle \exp\{\mathbf{k}\cdot(\mathbf{r_i} - \mathbf{r_j})\}\rangle = 1 + \rho h(\mathbf{k})\,, \quad (6.22)$$

with $h(\mathbf{k})$ the Fourier transform of $h(\mathbf{r})$. Performing this Fourier transformation of $h(R)$ in Equation 6.20 we find,

$$S^{eq}(k) = C\, \frac{4\,\pi}{\xi^{-2} + k^2}\,. \quad (6.23)$$

Since $S(k = 0) = 1/[\beta\; d\Pi/d\rho]$, the constant C is found to be equal to $1/[4\pi\;\beta\;\xi\; d\Pi/d\rho] = 1/[4\pi\;\beta\;\Sigma]$, and hence,

$$S^{eq}(k) = \frac{1}{\beta\;\Sigma}\,\frac{1}{\xi^{-2} + k^2}\,. \quad (6.24)$$

This well-known form for the small-wave vector dependence of the structure factor near the critical point is commonly referred to as *the Ornstein–Zernike structure factor*. The Ornstein–Zernike structure factor implies that a plot of $1/S(k)$ versus k^2 is linear, with a slope equal to $\beta\;\Sigma$ and an intercept at $k = 0$ equal to $\beta\;\Sigma\;\xi^{-2}$. The above analysis is only valid when the correlation length ξ is much larger than the range of the pair-interaction potential R_V. This imposes, through Equation 6.21, in which part of the phase diagram the above theory can be applied, namely, where $d\Pi/d\rho$ is very small.

Having derived a microscopic expression for the diverging density fluctuations close to the critical point, we are left with the question about the relaxation dynamics of microstructural order close to the critical point. This is most conveniently described in terms of the Fourier transformed equation of motion (Eq. 6.17), which reads in terms of the structure factor,

$$\frac{\partial S(\mathbf{k})}{\partial t} = -2\; D^{eff}(k)\; k^2\; [S(\mathbf{k}) - S^{eq}(k)]\,, \quad (6.25)$$

where S^{eq} is the equilibrium structure factor in Equation 6.24, and the "*effective diffusion coefficient*" D^{eff} is given by,

$$D^{eff}(k) = D_0\;\beta \left[\frac{d\Pi}{d\rho} + \Sigma k^2\right] = D_0\;\beta\;\Sigma\left[\xi^{-2} + k^2\right]. \quad (6.26)$$

The solution of Equation 6.25 is $\sim \exp\{-2\; D^{eff}(k)\; k^2\; t\}$. The effective diffusion coefficient characterizes the relaxation rate of microstructural order with a wavelength equal to $2\pi/k$. The effective diffusion coefficient becomes very small as compared to the single-particle diffusion coefficient D_0 on approach of the critical point, since the correlation length diverges and Σ is well behaved near the critical point. The dynamics of long wavelength structures (density fluctuations) therefore slows down, which is commonly referred to as *critical slowing down*.

Dispersions of colloidal spheres are ideal to experimentally test the relations (Eqs. 6.25 and 6.26), because attractions between the colloids can be easily induced. In general, there are two ways to achieve attractions. The first way is by grafting a small polymer layer on the surface of the colloids so that the particles become attractive when the solvent quality of the polymers is reduced, generally by reducing the temperature. The second way is by adding small polymers to the dispersion that act as depletion agents [86]. Changes in the amount and size of the polymers lead to changes in the range and depth of the interaction potential. It thus has the same effect as that of changing the temperature for most critical systems, such as critical fluid mixtures [87] and polymer dispersions [88]. The phenomena that are described in this section for colloids can also be observed for these systems, which have however a different molecular origin of the critical behavior.

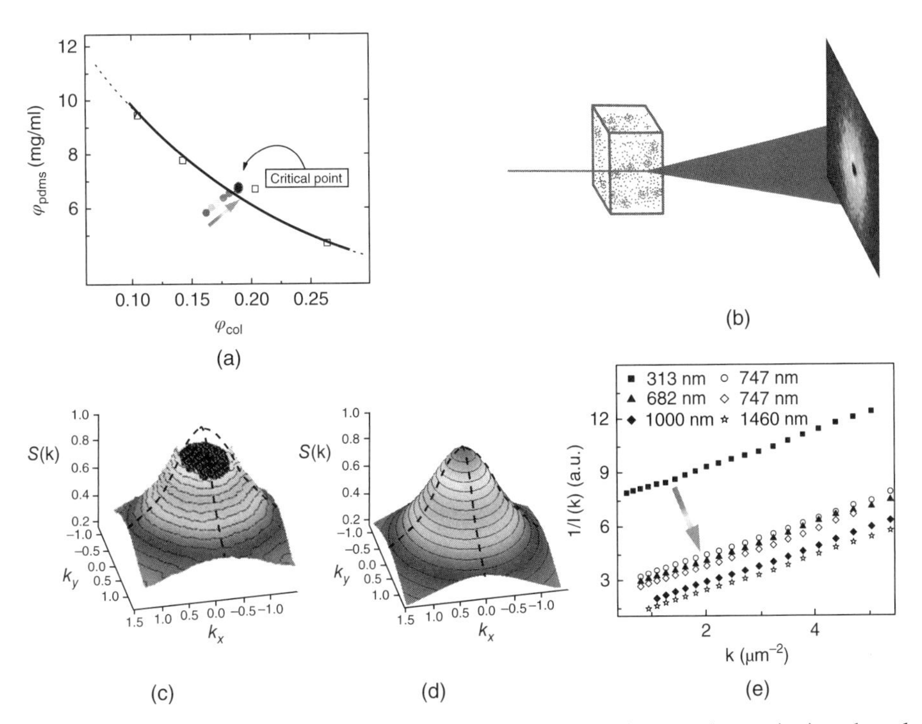

Figure 6.10 (a) The equilibrium phase diagram of mixtures of silica spheres (102 nm diameter, the x-axis gives the volume fraction φ_{col}) and polydimethylsiloxane (molecular weight of 206 kg/mol, concentration is given on the y-axis in mg/ml). The range of the interaction potential $R_V \approx 25$nm, since the radius of gyration is 23 nm. Source: The figure is adapted with permission from Ref. [89]. (b) Schematic small-angle light scattering setup to determine the critical structure (c), which can also be calculated using Equation 6.24 (d). The correlation lengths are calculated from the angular averaged structure factors (d), which are schematically shown in (e). Source: The figures (c,d,e) are adapted from Ref. [90] ©IOP Publishing. Reproduced by permission of IOP Publishing. All rights reserved. (*See color plate section for the color representation of this figure.*)

Experiments on mixtures of silica spheres grafted with stearyl alcohol and polydimethylsiloxane dispersed in cyclohexane are shown in Figure 6.10. These spheres behave as almost perfect hard spheres, while depletion attraction is induced by adding the polymer. The critical point of such systems can be found in two steps. In the first step, polymer and colloid concentration are tuned such that the mixture phase separates exactly in an equal amount of the gas phase and the liquid phase. In the second step, the system is diluted up to the point that the system does not phase separate any more, which is the critical point. The equilibrium phase diagram is plotted in Figure 6.10a. The distance to the critical point can be tuned by first preparing a critical composition ratio of silica to polymer, that phase separates in equal parts of the liquid and fluid phase as indicated by the arrow in Figure 6.10. Sequentially, this dispersion can be diluted until the critical point is crossed and concentrated again, approaching the critical point. A schematic small-angle light scattering (SALS) setup with which the distance to the critical point can be probed is shown in Figure 6.10b. The intensity $I(\mathbf{k})$ measured with SALS is directly proportional to the structure factor $S^{eq}(\mathbf{k})$, as plotted in Figure 6.10c, together with a calculated structure factor using Equation 6.24 in Figure 6.10d. The correlation length ξ, which characterizes the distance to the critical point, can be exactly determined by plotting the reciprocal of the azimuthal averaged intensity $1/I(k)$ versus k^2, see Equation 6.24, as shown in Figure 6.10e.

6.4.2 Nonequilibrium Phase Behavior

Let us recall that the Péclet number describes the rate of deformation relative to the relaxation rate of the relevant length scale. For individual spherical particles, these are the diffusion time D_0 and the radius of the spheres a. For interacting particles, the relevant length scale is the range of interaction R_V. Thus, to calculate the relevant bare Péclet number the shear rate needs to be scaled with the time it takes to diffuse over this length scale:

$$Pe = \frac{\dot{\gamma} R_V^2}{2D_0}. \tag{6.27}$$

For critical systems, the situation changes dramatically. The relevant length scale is now given by the correlation length ξ, while the relaxation rate is given by D^{eff}. When approaching the critical point, these parameters diverge and approach zero, respectively. Hence, a new definition of the

Péclet number is required that is given by an effective (or, dressed) Péclet number λ that is equal to

$$\lambda = \frac{\dot{\gamma}\,\xi^2}{2\,D^{\mathrm{eff}}(k=0)}\,. \tag{6.28}$$

The question now arises as to how large the actual shear rate $\dot{\gamma}$ should be to significantly affect the location phase transition lines. First, we note that close to the critical point the dressed Péclet number is very large for shear rates where the bare Péclet number is still small, since $D^{\mathrm{eff}} \ll D_0$ and $\xi \gg R_V$. We, therefore, assume that $g(r)$ for $r \le R_V$ is not affected by the shear flow:

$$g(\mathbf{r}) = g^{\mathrm{eq}}(\mathbf{r})\,, \tag{6.29}$$

where g^{eq} is the equilibrium pair correlation function, that is, the correlation function of the quiescent, unsheared system. Next, a term needs to be added in the original Smoluchowski equation for the probability density function of the ensemble of particles, Equation 6.9, which describes the convection of particles in and out of a volume by shear flow. This term is given by $\dot{\gamma}\sum_{j=1}^{N}\nabla_j\cdot[P\hat{\boldsymbol{\Gamma}}\cdot\mathbf{r}_j]$, where $\hat{\boldsymbol{\Gamma}}$ is the deformation tensor as introduced earlier (Eq. 6.1). This term translates to $\dot{\gamma}\,\nabla\cdot\{h(\mathbf{r})\hat{\boldsymbol{\Gamma}}\cdot\mathbf{r}\}$, following the procedure described in the previous section, so that the Smoluchowski equation for $h(\mathbf{R})$ now reads

$$\frac{\partial}{\partial t}h(\mathbf{R}) = 2\,\beta\,D_0\,\nabla\cdot\left(\nabla\left[\frac{d\Pi}{d\rho}h(\mathbf{R}) - \Sigma\,\nabla^2 h(\mathbf{R})\right]\right) - \dot{\gamma}\,\nabla\cdot\left\{h(\mathbf{R})\hat{\boldsymbol{\Gamma}}\cdot\mathbf{R}\right\}\,, \tag{6.30}$$

which equation of motion is quite similar to that proposed by Onuki and Kawasaki [77]. With this equation, the coupling between thermodynamics and shear flow is made within the limits that shear flow is not allowed to be so high that it affects the short-ranged correlations.

Again, Fourier transformation of Equation 6.30 leads to an expression that can be easily experimentally tested. In steady flow conditions, this gives

$$0 = \lambda K_1\frac{\partial S(\mathbf{K})}{\partial K_2} - K^2(1+K^2)\,[S(\mathbf{K}) - S^{\mathrm{eq}}(K)]\,, \tag{6.31}$$

where $\mathbf{K} = \mathbf{k}\,\xi$ is the dimensionless wave vector and the index 1 indicates the flow direction, using the convention given in Figure 6.1. The solution of this equation is

$$S(\mathbf{K}) = \frac{\xi^2}{\beta\Sigma}\frac{1}{\lambda K_1}\int_{K_2}^{\pm\infty} dX\,(K^2 - K_2^2 + X^2) \times \exp\{-F(\mathbf{K}\mid X)/\lambda\,K_1\}\,, \tag{6.32}$$

with

$$F(\mathbf{K}\mid X) = (X-K_2)(K^2-K_2^2)(1+K^2-K_2^2) + \frac{1}{3}(X^3-K_2^3)(1+2K^2-2K_2^2) + \frac{1}{5}(X^5-K_2^5)(X^2-K_2^2)\,, \tag{6.33}$$

where Equation 6.24 for the equilibrium structure factor has been used. The integration limit in Equation 6.32 is $+\infty$ when $\lambda K_1 > 0$, and $-\infty$ when $\lambda K_1 < 0$. In order to appreciate the effect of shear flow, we plot in Figure 6.11 the distortion of the structure factor in terms of

$$\Delta S(\mathbf{K}) = S_{\mathrm{eq}}(\mathbf{K}) - S(\mathbf{K}) \tag{6.34}$$

for experiment, where the system was used as described in Figure 6.10, as well as theory.

The distortion of the critical structure factor can be interpreted as suppression of critical fluctuations by convection due to shear. This distortion only takes place in flow and gradient direction, while the structures in the vorticity direction, where $k_{1,2} = 0$, remain unaltered. The distortion of the critical fluctuations in the flow direction is directly experimentally accessible from the distorted structure factor. This can be seen in Figure 6.11b and c, where the shaded areas indicate the difference between the equilibrium and sheared structure factor. The distorted structure factor leads to a diminished correlation length, which, according to the interpretation in the equilibrium phase, reflects an increase in the distance to the critical point. The question is if a system can be further away from the critical point only in specified directions.

Careful inspection of Figure 6.11b shows, however, that $\Delta S \neq 0$ along the cut where $k_1 = 0$. Figure 6.11d clearly shows that ΔS increases with increasing shear rate. This means that there is an overall shear-induced shift of the critical point, as schematically shown in Figure 6.11e, contrary to the theoretical prediction. Thus, with increasing shear rate, the system moves away from the critical point. We now give an argument about the origin of this shear-induced shift of the phase boundary.

The core of the argument lies in the assumption we made earlier that the pair correlation function is not affected by shear flow. This assumption is false when the bare Péclet number given by Equation 6.27 is around one. To incorporate short-ranged distortions, instead of Equation 6.29, we expand the correlation function to first order in the bare Péclet number as

$$g(\mathbf{r}') = g^{\mathrm{eq}}(r')\left\{1 + |\,Pe\,|\,f_0(r') - Pe\,\frac{\mathbf{r}'\cdot\hat{\mathbf{E}}\cdot\mathbf{r}'}{r'^2}f_1(r')\right\}, \quad \text{for} \quad r' \le R_V, \tag{6.35}$$

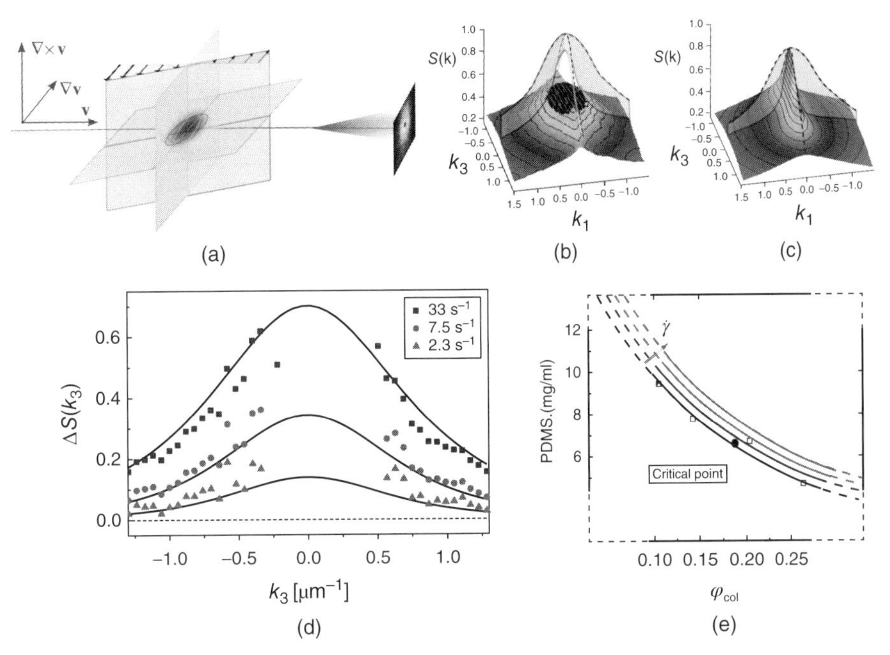

Figure 6.11 (a) Geometrical setup of a sheared critical system. The same experimental system is used as in Figure 6.6. Experimental (b) and theoretically calculated (c) sheared critical structure factor detected in the flow (1,**v**) and vorticity (3, $\nabla\times$**v**) scattering plane. The equilibrium correlation length is 750 nm and the dressed Péclet number is $\lambda = 4$. The shaded surfaces show the difference with the equilibrium structure factor. (d) The difference between the equilibrium and sheared structure factor for different shear rates, taken at the axis where no distortion is expected ($k_1 = 0$). The solid lines are fits to the theory, using Equation 6.38. The dashed line indicates the theoretically predicted absence of distortion in this direction. Source: Figure adapted from Ref. [90] ©IOP Publishing. Reproduced by permission from IOP Publishing. All rights reserved. (e) Sketch of the effect of shear flow on the phase boundary. Source: Adapted with permission from Ref. [89]. (*See color plate section for the color representation of this figure.*)

where (the superscript T stands for transposition),

$$\hat{\mathbf{E}} = \frac{1}{2}\left[\hat{\mathbf{\Gamma}} + \hat{\mathbf{\Gamma}}^T\right] \tag{6.36}$$

is the symmetric part of the dimensionless velocity gradient tensor. The functions f_0 and f_1 describe the isotropic and anisotropic response of the short-ranged part of the pair correlation function to shear flow, respectively. The functional form of the anisotropic response can be understood as follows. The velocity gradient tensor can be decomposed as $\hat{\mathbf{\Gamma}} = \frac{1}{2}[\hat{\mathbf{\Gamma}} + \hat{\mathbf{\Gamma}}^T] + \frac{1}{2}[\hat{\mathbf{\Gamma}} - \hat{\mathbf{\Gamma}}^T]$. The second term describes a rotational flow, which has no effect on the anisotropy of the structure factor. The first term describes an extensional flow, which gives rise to an increased concentration along the $x = y$ direction and a decrease along the $x = -y$ direction. This is precisely what is described by the anisotropic contribution in Equation 6.35.

For symmetry reasons, such a linear isotropic distortion must be proportional to the absolute value of the shear rate. This in turn implies that the pair correlation function for short distances is a nonanalytic function of the shear rate. The probable reason for this is that in the equation of motion for g for short distances, integrals containing triplet and four-point correlation functions appear, and these integrals extend over all space. Such integrals also probe the long-ranged behavior of g, which is nonanalytic in the shear rate. Since close to the critical point such long-range correlations are significant, it might be that these terms give rise to the nonanalytic behavior of g for small distances. The nonanalytic form in Equation 6.35 is thus probably due to coupling of singularly distorted long- to short-ranged correlations.

Repeating the analysis leading to Equation 6.32, but now using the Ansatz in Equation 6.35 instead of assumption (iii) (see Eq. 6.29), two additional contributions to Equation 6.31 for the structure factor are found:

$$\begin{aligned} 0 = \lambda K_1 \frac{\partial S(\mathbf{K})}{\partial K_2} &- K^2\left(1 + \alpha_{\text{iso}} + K^2\right) S(\mathbf{K}) \\ &+ K^2\left(1 + K^2\right) S^{\text{eq}}(K) + \alpha_{\text{aniso}}\, K_1 K_2 [S(\mathbf{K} - 1]\,, \end{aligned} \tag{6.37}$$

where, as before, $\mathbf{K} = \mathbf{k}\,\xi$ is the dimensionless wave number and λ is the dressed Péclet number as defined in Equation 6.28. There are two new parameters, α_{iso} and α_{aniso}, which can be expressed in terms of integrals of the equilibrium pair correlation function [90]. α_{iso} measures the susceptibility of isotropic short-ranged distortions and α_{aniso} of the anisotropic distortions.

The solution of Equation 6.37 is (we use for convenience the same notation for S and F as in Section 6.4.1, where distortion of short-ranged correlations were neglected)

$$S(\mathbf{K}) = \frac{1}{\lambda K_1}\int_{K_2}^{\pm\infty} dX \Big\{ \left(K^2 - K_2^2 + X^2\right)\left(1 + K^2 - K_2^2 + X^2\right) \times S^{eq}\left(\sqrt{K_1^2 + X^2 + K_3^2}\right) - \alpha_{aniso}\, K_1 X \Big\} \times \exp\{-F(\mathbf{K} \mid X)/\lambda\, K_1\}\,, \qquad (6.38)$$

with,

$$F(\mathbf{K} \mid X) = (X - K_2)\left(K^2 - K_2^2\right)\left(1 + \alpha_{iso} + K^2 - K_2^2\right) + \frac{1}{3}\left(X^3 - K_2^3\right)\left(1 + \alpha_{iso} + 2K^2 - 2K_2^2\right) + \frac{1}{5}\left(X^5 - K_2^5\right) - \alpha_{aniso}\, K_1\left(X^2 - K_2^2\right). \qquad (6.39)$$

The integration limits are the same as in Equation 6.32. Equation 6.38 can be used to fit ΔS as is shown by the solid lines Figure 6.11d. The resulting shear rate dependence of α_{iso} is indeed linear. The original equation (Eq. 6.37) is recovered when the short-ranged distortions are neglected, so $\alpha_{iso} = 0 = \alpha_{aniso}$. The finite distortions of long-ranged microstructural order for $K_1 = 0$ is solely due to the coupling to short-ranged distortions. The structure factor for wave vectors with $K_1 = 0$ attains again the Ornstein–Zernike form Equation 6.24,

$$S(\mathbf{K}) = \frac{1}{\beta\,\Sigma}\,\frac{1}{(\xi^{eff})^{-2} + k^2} \quad (K_1 = 0)\,, \qquad (6.40)$$

but with a shear-rate-dependent "effective correlation length,"

$$\xi^{eff} = \xi/\sqrt{1 + \alpha_{iso}}\,. \qquad (6.41)$$

This result confirms the experimental result presented earlier that the overall structure factor is isotropically distorted. We now learn that this is due to local distortions of the local structure, with the effect that the location of the critical point is shifted on applying shear flow, as illustrated in Figure 6.11e.

6.4.3 The Effect on Flow Behavior

When discussing the effect that dispersions of sticky spheres have on the stability of the flow, one should distinguish between the regime above and below the spinodal line. In other words, one needs to know if the system is still mixed but in a state where it is critically slowing down or if the system undergoes spinodal decomposition. For the flow behavior of critical mixtures, the main results are based on the work discussed above. Theoretically, the viscosity of such dispersions can be calculated as a function of the distance to the critical point. This calculation is involved [84] and will not be presented here. The prediction is, however, that the viscosity more strongly diverges as compared to molecular systems due to long-ranged hydrodynamic interactions between the colloidal particles. Such a strong critical enhancement of the shear viscosity of critical colloidal suspensions has been confirmed in experiments [91]. Flow curves at different points toward the critical point have however not been obtained, although it is very conceivable that such samples are extremely shear thinning, which might result in flow instabilities.

In contrast, there is a vast amount of work done on systems that phase separate in shear flow. The morphology of the structures that are formed during phase separation and the way this morphology is affected by shear flow strongly depends on the composition of the sample and the depth of the quench, which determines the final composition of the two phases [92]. We consider here the simplest case, namely, gas–liquid phase separating systems in shear flow, where both phases dynamically behave the same. When a system is quenched just beyond the critical point, then initially a bicontinuous structure will be formed: any fluctuation of arbitrary small amplitude and sufficiently long wavelength will continuously grow resulting in spinodal decomposition. The length scale of the fast-growing fluctuation is set by the competition between the diffusion that favors small length scales and minimization of density gradient that favors large length scales. Capillary forces will drive further coarsening at the moment the interfaces are so sharp that they reach their thermodynamic interfacial tension, lowering the free energy by reduction of the total interfacial area [93].

Shear flow will compete with this coarsening process. While the growth rate in the neutral vorticity direction is unhindered, the growth rate in the flow direction coarsening is enhanced by the shear flow. At later times and sufficiently high shear rates, the spinodal structure loses its continuous character and no interconnected structures are observed anymore in the gradient and vorticity direction, leading to an anisotropic domain structure. Such string-like structures have been observed for various systems such as binary fluids [94], polymer solutions [95], mixtures of polysaccharides (dextran) and proteins (gelatin) [39], and colloidal sphere–polymer mixture [96] for which we described in Section 6.4.3 the flow behavior at the stable side of the critical point. For the last two systems, the structures are shown in Figure 6.12. The string-like structures could be considered as small local shear bands, but they do not merge in the gradient nor the vorticity direction. The bands represent a true stable state that is uniquely defined because the same state is reached when quenching from different initial shear rates, both above and below the final shear rate.

The origin of the stability of these structures as well as the shear rate dependence of the diameter of the strings L, given

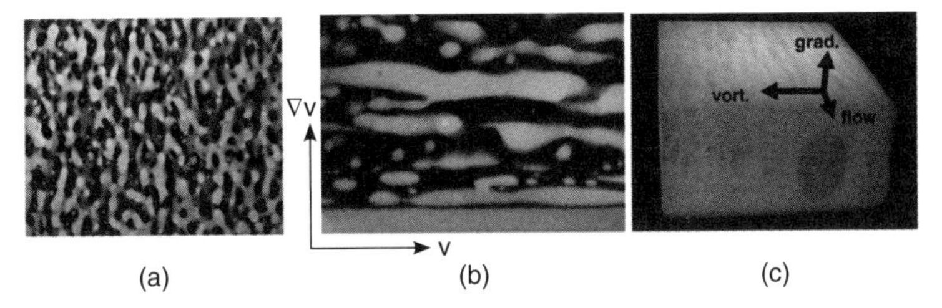

Figure 6.12 Confocal images of fluorescently labeled PMMA spheres with polystyrene polymers of (a) an equilibrium phase separating system and (b) the same system under a shear flow of 1.4 s^{-1}. Source: The figures are taken from Ref. [96] ©IOP Publishing. Reproduced by permission from IOP Publishing. All rights reserved. (c) 3D image of a demixing system of a dextran–gelatin dispersion under a shear flow of 9 s^{-1}. Source: Images taken from Ref. [39]. All images were taken using a cone-plate setup as shown in Figure 6.6b.

by $L \propto (\dot{\gamma} \cdot \tau)^{-\alpha}$, is still under debate, where τ is related to the growth rate. It is expected that $\alpha = 1$ [96], by balancing the capillary force-driven coarsening rate with the shear deformation rate, but for the polymer solution [95] as well as the colloid mixture [96], a much weaker dependence was found of $\alpha \approx 1/3$. In the absence of shear, a long cylinder is unstable against long wavelength undulations and would immediately break up into small droplets [97]. This suggests that the strings can be stable only when undulations of the surfaces of the strings, which are prominent due to the low interfacial tension of the systems close the critical point, are strongly prohibited by shear. The strings are therefore the result of the interplay between hydrodynamic and thermodynamic forces.

At very high shear rates, the strings can become as thin as the interface so that the two phases will finally merge. This has been observed for the polymer solutions [95] as well as liquid mixtures [87, 98]. Thus, a homogeneous system is obtained where in quiescent conditions the system would phase separate, so that one can say that the critical point has shifted, as predicted by Onuki and Kawasaki [77]. For colloidal systems, we discussed this shear-induced shift in the previous section, approaching the critical point from the one-phase region and not the two-phase region as discussed here. Disappearance of the string structure for colloidal systems at high shear rates has not been observed probably because the required shear rates were so high that no stable flow could be guaranteed anymore.

6.5 EFFECT OF SHEAR FLOW ON THE ISOTROPIC–NEMATIC PHASE TRANSITION OF COLLOIDAL RODS

Almost all molecules, proteins, and other types of particles that are present in nature are not spherical. It is, therefore, appropriate to increase the level of complexity and move from spherical colloids, as discussed in the last two sections, to rods, which are the simplest nonspherical particles. Compared to spheres, rods have a far more rich phase behavior since rods have both positional and orientational degrees of freedom. The simplest transition of rod-like particles is the transition from the *isotropic* phase (I), displaying orientational and positional disorder, to the *nematic* phase (N), displaying a mean averaged orientation, but still no long-ranged positional order. Orientational order manifests itself when the dispersion is birefringent due to the difference in the refractive index parallel and perpendicular to the average orientation of the rods. Observations of birefringent structures have been first described by Zocher [99] who studied dispersions of V_2O_5 and FeOOH. These systems consist of inorganic needle-shaped particles, as was found by ultramicroscopy, introduced by Zsigmondy in the early 20th century [100]. Not much later, in 1936, similar observations were made on dispersions of tobacco mosaic virus by Bawden *et al.* [101]. At that time, it was surprising that dilute low viscous dispersions of colloidal particles could show birefringence, since birefringence was associated with mineral crystals or thermotropic liquid crystals crystals, as discussed already in Chapter 16.

It was Onsager who reasoned on the basis of purely geometrical considerations that dispersions of hard rods should also undergo a phase transition from the isotropic phase to the nematic phase [102]. At this phase transition, the system will gain positional entropy, since the accessible volume is higher when rods are aligned in the nematic phase, at the cost of orientational entropy. He predicted that this is a first-order phase transition with a discontinuity in orientation as well as concentration. As a result there is a concentration range where the isotropic and nematic phases coexist. The concentration and orientational order parameter of the coexisting phases characterize the *binodal* points. These are also the points after which the dispersion becomes metastable to fluctuations in the orientation when increasing the concentration from an initially isotropic phase or decreasing the concentration from an initially nematic phase. Similarly, spinodal points can be defined and calculated, which mark the concentration where the dispersion becomes unstable and each fluctuation in orientation will result in phase separation.

For dispersions of rods, one expects that external fields will affect the behavior also on a single particle level. Where so far we scaled the shear rate with translational diffusion, now the applied field, and therefore the applied torque on the rod, is competing with the rotational diffusion of the rod, orientating the rod when the field is sufficiently strong. This has already been shown in 1902 by Majorana who observed that dispersions of colloidal ferroxides become birefringent when subjected to a magnetic field [103]. Not much later, in 1912, Zocher found that flow could also induce birefringence [104], while Bawden *et al.* demonstrated the peculiarity of the fact that fluids can show birefringence, by publishing a photo of a goldfish in a bowl with tobacco mosaic viruses between crossed polarizers [101]. The obvious question is how alignment induced by an external field influences the location of the I–N transition, since random fluctuations in the orientation are not needed anymore to form a nematic phase from the isotropic phase. In other words, the phase diagram of colloidal rods might change when rods are subjected to an external field. Indeed, studies on rod-like viruses, in this case *fd* virus which we will further introduce later, in a magnetic field showed that the I–N spinodal point shifts to lower concentrations [105]. This means that the external field stabilizes the nematic phase.

In this section, we discuss how the isotropic–nematic phase transition is influenced by shear flow. As in the previous sections, we start off with a theoretical treatment of the isotropic phase behavior. Again, we use a dynamical approach based on the Smoluchowski equation, which now describes the equation of motion of the orientational probability density distribution. This equation will first be used to determine the I–N and the N–I spinodal points. Next, we show how the nonequilibrium spinodal lines can be calculated using the same approach, but now including the effect of shear flow on the orientation of the rods. This part is based on Refs [106, 107] and follows an approach similar to Ref. [108], while similar results were found for thermotropic liquid crystals in shear flow [109].

In order to calculate the binodal lines, one has to take the gradients of concentration and orientation at the interfaces into account, as well as the fact that both phases have a different dynamical behavior. While rods in the isotropic phase simply display alignment with increasing shear rate, the flow behavior of the nematic phase is very rich and involves several dynamical transitions [110, 111]. It is predicted that the coexistence of these phases leads to flow instabilities, where shear bands can form either in the flow or in the vorticity direction, while the bands consist of the flow-aligning isotropic phase or the nematic phase [11]. As mentioned in Section 6.2.3, flow instabilities have been observed for many different systems. We choose to restrict our discussion to experiments on almost ideal colloidal rods, the above-mentioned *fd* virus, which are extremely slender and relatively stiff. Other systems, especially wormlike micelles, do have features similar to colloidal rods, but the origin of flow instabilities can be manyfold and are not necessarily related to the proximity of the I–N transition. We show how experiments and simulations on colloidal rods can be used to determine the nematic–isotropic nonequilibrium spinodal as well as the binodal line that separates the phase coexistence region from the homogeneous mixed phase. It will be shown that the phase coexistence persists as long as the nematic phase is in a tumbling state. Moreover, a flow instability does exist related to the I–N coexistence, but it is different from the theoretically predicted gradient shear bands due to the strong shear thinning close to the nonequilibrium point [11, 107].

6.5.1 Equilibrium Phase Behavior: Isotropic–Nematic Phase Transition from a Dynamical Viewpoint

To predict the equilibrium phase behavior, we first derive an equation of motion of the relevant order parameter, which we then extend to incorporate the effect of shear flow, as we did for critical colloid–polymer mixtures. In this case, we derive an equation of motion that describes the fluctuations in the orientational distribution function (ODF) of the rods, given by $P(\hat{\mathbf{u}}, t)$, since the orientational order parameter changes at the I–N transition. $\hat{\mathbf{u}}$ is the vector that describes the orientation of the long axis of the rod; see Figure 6.13a.

Similar to the equation of motion for the probability density function that was used to derive the critical slowing down of attractive colloids, see Equation 6.9, the Smoluchowski approach will be used to derive the equation of motion of the orientation of the rods:

$$\frac{\partial}{\partial t} P(\hat{\mathbf{u}}, t) = D_r \hat{\mathcal{R}} \cdot \left\{ \hat{\mathcal{R}} P(\hat{\mathbf{u}}, t) - \beta\, P(\hat{\mathbf{u}}, t)\, \overline{\mathbf{T}}(\hat{\mathbf{u}}, t) \right\}. \quad (6.42)$$

This equation describes the change in the orientation of the rods along the surface of the unit sphere, which depends on the rotational diffusion coefficient of the rods, first term on the right, and the torque exerted on the rods, second term on the right; see Figure 6.13b. $\hat{\mathcal{R}}_j$ is the rotation operator given by $\hat{\mathcal{R}}_j\,(\cdots) = \hat{\mathbf{u}}_j \times \nabla_{\hat{\mathbf{u}}_j}\,(\cdots)$, D_r is the rotational diffusion coefficient of the rods at infinite dilution. Similar to

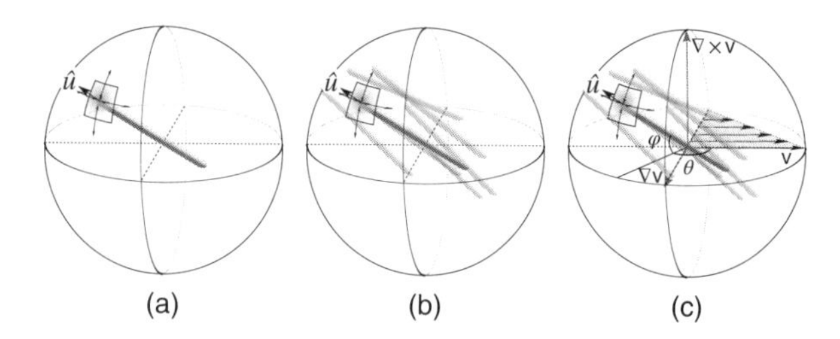

Figure 6.13 (a) Definition of the unit vector $\hat{\mathbf{u}}$. The arrows out of the little surface area indicate the rotational Brownian diffusion given by the first term in Equations 6.42 and 6.50. In concentrated solutions, (b) neighboring rods exert a torque on the rod influencing the rotational diffusion given by the second term in Equations 6.42 and 6.50. Shear flow (c) adds another torque term given by the third term in Equation 6.50. (*See color plate section for the color representation of this figure.*)

the attractive spheres discussed earlier, the torque in equilibrium is due to the interaction with the other rods by (with $\mathbf{R} = \mathbf{r}_1 - \mathbf{r}_2$ and $\hat{\mathbf{u}}' = \hat{\mathbf{u}}_2$)

$$\overline{\mathbf{T}}(\hat{\mathbf{u}}, t) = -\overline{\rho} \int d\mathbf{R} \oint d\hat{\mathbf{u}}' P(\hat{\mathbf{u}}', t) g(\mathbf{R}, \hat{\mathbf{u}}, \hat{\mathbf{u}}', t) \hat{\mathcal{R}} V(\mathbf{R}, \hat{\mathbf{u}}, \hat{\mathbf{u}}'), \tag{6.43}$$

where $\overline{\rho} = N/V$ is the number density of rods, $g(\mathbf{R}, \hat{\mathbf{u}}, \hat{\mathbf{u}}', t)$ is the pair correlation function, and $V(\mathbf{R}, \hat{\mathbf{u}}, \hat{\mathbf{u}}')$ is the interaction potential between two rods at distance $\mathbf{R}$ and orientations $\hat{\mathbf{u}}$ and $\hat{\mathbf{u}}'$.

The crucial difference with the attractive spheres treated earlier is that the rods have pure hard-core interactions. This means that the interaction energy is infinite when the rods overlap and zero when they do not overlap. If the rods are in addition also very long and thin, then it is sufficient to consider only pair interactions. The condition is that the aspect ratio $\frac{L}{d} > 100$, where L is the length of the rod and d its thickness. This consideration is the same one that leads to Onsager's prediction for the I–N phase transition. So at this point no closure relation is needed, as it was for the attractive spheres. The integral in Equation 6.43 now reduces to

$$\begin{aligned}\overline{\mathbf{T}}(\hat{\mathbf{u}}, t) &= -\hat{\mathcal{R}}\ V^{\mathrm{eff}}(\hat{\mathbf{u}}, t) \\ &= -\hat{\mathcal{R}}\ 2dL^2\beta^{-1}\overline{\rho} \oint d\hat{\mathbf{u}}'\ P(\hat{\mathbf{u}}', t) \mid \hat{\mathbf{u}} \times \hat{\mathbf{u}}' \mid, \end{aligned} \tag{6.44}$$

where the term $\mid \hat{\mathbf{u}} \times \hat{\mathbf{u}}' \mid$ is a general expression for the volume that is excluded by two rods; see Figure 6.14. *The effective potential* V^{eff} is commonly referred to as *the Doi–Edwards potential*, but goes back to Onsager.

The ODF can generally not be measured directly. Therefore, we need to relate it to a measurable quantity. Moreover, later on we want to relate the degree of ordering with the

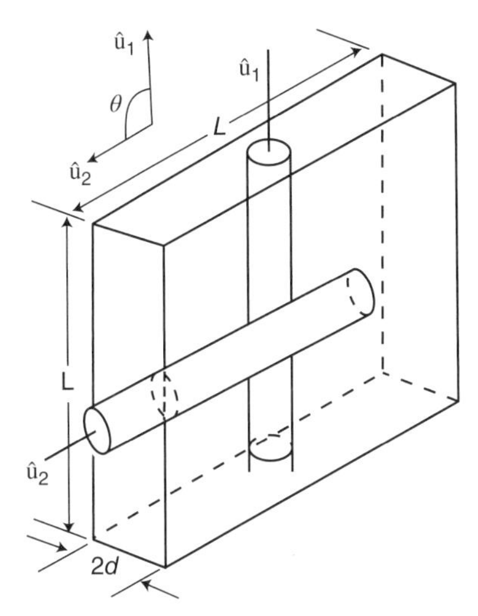

Figure 6.14 Cartoon of the excluded volume effect in the isotropic phase, which is given by $V_{\mathrm{exc}} = \sin(\theta)\ dL^2$.

stress in the system when it is sheared. Both the ordering and the stress need to be described by tensors. The ordering tensor $\mathbf{S}$ is defined by an integral over the dyadic product $\hat{\mathbf{u}}\hat{\mathbf{u}}$,

$$\mathbf{S} = \oint d\hat{\mathbf{u}} \quad \hat{\mathbf{u}}\hat{\mathbf{u}} P(\hat{\mathbf{u}}, t). \tag{6.45}$$

In experiments often a scalar is measured, for example, the birefringence of a sample, which gives the scalar S as defined in Equation 16.1. Here we use λ which is the highest eigenvalue of $\mathbf{S}$, $\lambda = 2\pi \int_0^\pi \cos(\theta)^2 P(\theta) \sin(\theta) d\theta$. For perfectly parallel rods $\lambda = 1$, while in the isotropic phase $\lambda = \frac{1}{3}$. Instead of λ, the scalar order parameter $S = \frac{3}{2}(\lambda - \frac{1}{3})$ is often used, which is 0 in the isotropic state and 1 for perfect alignment. Note that S is the same as the average over the second Legendre polynomial P_2.

An equation of motion for the ordering tensor S can now be found using Equation 6.44 in Equation 6.42 with a proper expansion for $\mid \hat{\mathbf{u}} \times \hat{\mathbf{u}}' \mid$, and operating on both sides of Equation 6.42 with $\oint d\hat{\mathbf{u}}\ (\hat{\mathbf{u}}\hat{\mathbf{u}})$ following Doi and Edwards; see Ref. [112]:

$$\frac{d}{dt}\mathbf{S} = -6D_r \left\{ \mathbf{S} - \frac{1}{3}\hat{\mathbf{I}} + \frac{L}{D}\varphi \left(\mathbf{S}^{(4)} : \mathbf{S} - \mathbf{S} \cdot \mathbf{S} \right) \right\}. \tag{6.46}$$

As in the equation of motion of the pair correlation function of attractive spheres, Equation 6.12, there is an unknown parameter on the right side, in this case $\mathbf{S}^{(4)} = \oint d\hat{\mathbf{u}}\ \hat{\mathbf{u}}\hat{\mathbf{u}}\hat{\mathbf{u}}\hat{\mathbf{u}} P(\hat{\mathbf{u}}, t)$. Hence, at this point a closure relation is needed. The simplest closure relation is $\mathbf{S}^{(4)} = \mathbf{SS}$. As we will discuss later, this closure can lead to wrong results when applying shear. In equilibrium, however, it works fine. To find the I–N transition, we seek for a small fluctuation $\delta\lambda$ in the ordering around homogeneous isotropic state, where $\lambda = \frac{1}{3}$. Only considering first-order terms in $\delta\lambda$ results in an equation of motion for the perturbation of the stationary state

$$\frac{d\delta\lambda}{dt} == -6D_r \left\{ 1 - \frac{1}{4}\frac{L}{d}\phi \right\} \delta\lambda = -6D_r^{\mathrm{eff}}\delta\lambda. \tag{6.47}$$

Here, D_r^{eff} is the effective rotational diffusion coefficient:

$$D_r^{\mathrm{eff}} = D_r \left\{ 1 - \frac{1}{4}\frac{L}{d}\varphi \right\}. \tag{6.48}$$

The solution of this equation of motion for the orientational order parameter is now simply given by

$$\delta\lambda(t) = \delta\lambda(t=0)\ \exp\left\{ -6D_r^{\mathrm{eff}} t \right\}. \tag{6.49}$$

This equation shows how fast a fluctuation in the local ordering will be randomized by the rotational diffusion of the rods. Equation 6.49 contains important information about the phase behavior of the rods. It shows namely that $D_r^{\mathrm{eff}} \to 0$ when $\varphi\frac{L}{d} \to 4$. Hence, this system shows critical slowing down of the collective rotational diffusion.

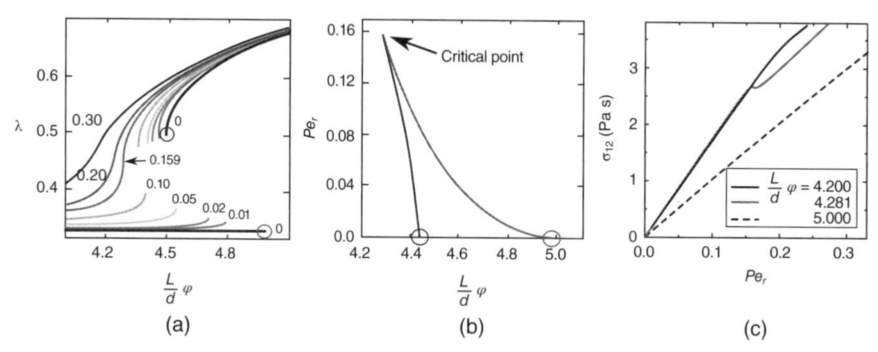

Figure 6.15 (a) Bifurcation diagram of the nematic order parameter λ, the highest eigen value of stationary solutions of **S**, versus the scaled volume fraction $\frac{L}{d}\varphi$, as obtained from the equation of motion (Eq. 6.51) hard rods in shear flow. The numbers refer to values of the bare rotational Péclet number. (b) The nonequilibrium spinodal lines as obtained from (a) by taking the nonequilibrium spinodal points. The critical concentration is defined by the point where both spinodal lines meet and given by a critical concentration of $\frac{L}{d}_{\text{crit}}\varphi = 4.281$ and critical Péclet number of $Pe_r^{\text{crit}} = 0.159$. The circles indicate the equilibrium I–N (red) and N–I (blue) spinodal points and are per definition the end of the isotropic and nematic branch as calculated by Kayser and Raveché [113]. (c) Theoretical flow curves plotting the shear stress σ_{12} versus the bare Péclet number Pe_r. At the critical concentration of $\frac{L}{d}\varphi_r$ the flow curve contains an unstable part where the slope in σ_{12}. Source: Reprinted and adapted from Ref. [107]. Copyright 2003 with permission from Elsevier. (*See color plate section for the color representation of this figure.*)

Fluctuations will grow and the aligned nematic phase is formed into the isotropic phase when $D_r^{\text{eff}} < 0$. Note the similarities between the attractive spheres and the hard rods. For both systems, an effective diffusion coefficient can be defined (compare Eq. 6.48 with Eq. 6.26), which approaches zero at the phase transition.

The isotropic–nematic spinodal point where $D_r^{\text{eff}} = 0$ is found at the scaled concentration $\varphi\frac{L}{d} = 4$. The results agree with the original thermodynamic approach of Onsager [102]. This is not surprising given that the same assumptions were made for the interaction potential between the particles. The equilibrium bifurcation diagram is calculated and plotted as thick lines in Figure 6.15a. It is given by the highest eigenvalue λ of stable stationary solutions S_0 versus the concentration, similar to Ref. [113]. Interestingly, the isotropic–nematic spinodal point is located at a higher volume fraction than the nematic–isotropic spinodal point, which is the volume fraction where the homogeneous nematic phase becomes unstable upon diluting the dispersion. The binodal points could, in principle, be calculated from a similar Smoluchowski approach for inhomogeneous systems, but this has so far not been done. The reason is that this approach considers fluctuations in a homogeneous background phase, whereas the binodal points describe phase coexistence between phases with different concentration and orientation.

6.5.2 Nonequilibrium Phase Behavior of Sheared Rods: Theory

In order to calculate the nonequilibrium phase behavior of dispersions of colloidal rods, different aspects need to be taken into consideration. First, shear-induced alignment is expected to enhance the I–N phase transition. Second, the different dynamical behavior of the isotropic and nematic phase needs to be considered. To start off, the shear term needs to be introduced in the equation of motion of the ODF, giving

$$\frac{\partial P(\hat{\mathbf{u}},t)}{\partial t} = D_r\hat{\mathcal{R}}\cdot\left\{\hat{\mathcal{R}}P(\hat{\mathbf{u}},t) + DL^2\bar{\rho}P(\hat{\mathbf{u}},t)\hat{\mathcal{R}}\oint d\hat{\mathbf{u}}'\ P(\hat{\mathbf{u}}',t)|\hat{\mathbf{u}}'\times\hat{\mathbf{u}}|\right\} - \dot{\gamma}\hat{\mathcal{R}}\cdot P(\hat{\mathbf{u}},t)\,\hat{\mathbf{u}}\times\left(\hat{\mathbf{\Gamma}}\cdot\hat{\mathbf{u}}\right). \tag{6.50}$$

The last term in Equation 6.50, the shear term, describes the rotation of the rods out of the small surface area due to the applied torque, as indicated in Figure 6.13c. This equation again forms the basis for the prediction of the nonequilibrium phase diagram.

The *spinodal* lines can now be calculated in a similar way to the derivation of the spinodal points in equilibrium, using a range of fixed shear rates $\dot{\gamma}$. Again both sides of Equation 6.50 are multiplied with $\hat{\mathbf{u}}\hat{\mathbf{u}}$ and the integral is taken to yield the equation of motion for the orientational ordering tensor:

$$\frac{d}{dt}\mathbf{S} = -6D_r\left\{\mathbf{S} - \frac{1}{3}\hat{\mathbf{I}} + \frac{L}{D}\varphi\left(\mathbf{S}^{(4)} : \mathbf{S} - \mathbf{S}\cdot\mathbf{S}\right)\right\} + \dot{\gamma}\left\{\hat{\mathbf{\Gamma}}\cdot\mathbf{S} + \mathbf{S}\cdot\hat{\mathbf{\Gamma}}^T - 2\mathbf{S}^{(4)} : \hat{\mathbf{E}}\right\}. \tag{6.51}$$

This equation was first put forward by Hess [114]. To proceed from this equation, again a closure relation is needed, expressing $\mathbf{S}^{(4)}$ in terms of $\mathbf{S}$. Compared to the equilibrium case, the final result depends much more on the chosen closure.

The following closure proved to be quite accurate [107]:

$$\mathbf{S}^{(4)} : \mathbf{M} = \langle \hat{u}\hat{u}\hat{u}\hat{u} \rangle : \mathbf{M}$$
$$= \frac{1}{5}\left\{ \mathbf{S}\cdot\overline{\mathbf{M}} + \overline{\mathbf{M}}\cdot\mathbf{S} - \mathbf{S}\cdot\mathbf{S}\cdot\overline{\mathbf{M}} - \overline{\mathbf{M}}\cdot\mathbf{S}\cdot\mathbf{S} \right.$$
$$\left. + 2\mathbf{S}\cdot\overline{\mathbf{M}}\cdot\mathbf{S} + 3\mathbf{SS} : \overline{\mathbf{M}} \right\} . \qquad (6.52)$$

Here, it is used that $\mathbf{S}^{(4)}$ occurs in Equation 6.51 in the doubly contracted form

$$\mathbf{A} \equiv \mathbf{S}^{(4)} : \mathbf{M} , \qquad (6.53)$$

where the second-order tensor $\mathbf{M}$ is either equal to $\mathbf{S}$ or $\hat{\mathbf{E}}$ and $\overline{\mathbf{M}}$ is the symmetric part of $\mathbf{M}$,

$$\overline{\mathbf{M}} \equiv \frac{1}{2}(\mathbf{M} + \mathbf{M}^T) . \qquad (6.54)$$

The stress tensor Σ_D, describing the macroscopic response to shear flow, can be calculated once the orientational ordering tensor $\mathbf{S}$, describing the microscopic state of the rods, is known. This relation is given by the Doi–Edwards relation

$$\Sigma_D = 2\,\eta_0\,\dot{\gamma}\,\hat{\mathbf{E}} + 3\bar{\rho}k_BT\left[\mathbf{S} - \frac{1}{3}\hat{\mathbf{I}} + \frac{L}{D}\varphi\left\{\mathbf{S}^{(4)} : \mathbf{S} - \mathbf{S}\cdot\mathbf{S}\right\}\right.$$
$$\left. + \frac{1}{6}Pe_r\left\{\mathbf{S}^{(4)} : \hat{\mathbf{E}} - \frac{1}{3}\hat{\mathbf{I}}\mathbf{S} : \hat{\mathbf{E}}\right\}\right] . \qquad (6.55)$$

The first term gives the solvent contribution to the stress, the second term the contribution of the individual rods, the third term the contribution of the interaction between the rods, while the last term is the extra stress as caused by the deformation of the flow field. Since a tensor cannot be measured, measurable scalar numbers need to be extracted, as was the case with the orientational ordering tensor. The most prominent scalar is the shear stress σ_{12}, as was introduced in Section 6.2.1, because it is the easiest to measure.

To obtain *the spinodal points in shear flow*, first the stationary solution $\mathbf{S}_0$ of Equations 6.51 and 6.52 needs to be calculated for a finite value of $\dot{\gamma}$, which is done numerically. The reason is that the highest eigenvalue of $\mathbf{S}$, λ, is higher than its isotropic value because shear flow will induce some alignment, so $\lambda > \frac{1}{3}$. Therefore, the sheared isotropic phase is called the para-nematic phase. Of course, the ordering of the nematic phase will also be influenced by shear flow. Next, the stability of the solution is tested by analyzing if fluctuations around S_0 grow or vanish, as in Equation 6.47, while approaching a spinodal line $D_r^{\text{eff}} \to 0$. Nonequilibrium bifurcation diagrams can be calculated following this procedure, as shown in Figure 6.15a, for various values of the bare Péclet number Pe_r, which is the shear rate scaled by the rotational diffusion of the rods D_r:

$$Pe_r = \dot{\gamma}/D_r . \qquad (6.56)$$

The nonequilibrium spinodal lines, depicted in Figure 6.15b, are obtained by plotting the end points of the nematic and para-nematic branches. The para-nematic - nematic and the nematic - para-nematic spinodal lines meet at a shear rate of $Pe_r^{\text{crit}} = 0.159$ and a scaled volume fraction of $\frac{L}{d}\varphi_{\text{crit}} = 4.281$. At this point, the para-nematic phase can, in principle, be continuously transformed into the nematic phase, and vice versa, by simultaneously changing the concentration and shear rate. Therefore, this point defines the *nonequilibrium critical point*. This is a very important concept for sheared dispersions. At this point, the shear forces are so large that rod–rod interactions are not anymore able to induce a discontinuous transition. This effect can be quantified by defining an effective rotational Péclet number as

$$Pe_{\text{eff}} = \dot{\gamma}_0/D_r^{\text{eff}} . \qquad (6.57)$$

The fact that $D_r^{\text{eff}} \to 0$ close to nonequilibrium critical point indeed suggests dramatic changes in the sample when subjected to shear flow. This is supported by the flow curve that is calculated using Equation 6.55, which displays a region where the slope of the stress versus the shear rate is negative ($\frac{\partial\sigma_{12}}{\partial\dot{\gamma}} < 0$); see Figure 6.15c suggesting that the flow is unstable. Another important conclusion from Figure 6.15b is that the homogeneous nematic phase remains stable up to lower volume fractions when applying a shear flow, while the homogeneous isotropic phase turns unstable at lower volume fractions. Thus, we reach the important conclusion that *shear flow acts to stabilize the homogeneous nematic phase, shifting the I–N spinodal to lower concentrations*.

Having defined the nonequilibrium spinodal lines, we have the question of how to determine the *nonequilibrium binodal lines*. The Smoluchowski approach is not applicable because two phases with different orientation and concentration coexist and therefore gradients in these parameters are very high. A free energy approach has been used to calculate sheared phase coexisting states [11], similar to Onsager's approach to calculate equilibrium binodal points [102]. It is questionable if this approach is appropriate because energy is continuously dissipated into the system. An additional problem is that two phases coexist with very different dynamic behavior [11]. While we treated the flow response of the isotropic phase above, the flow behavior of the nematic phase is very complex and deserves extra attention.

Hess was the first to theoretically show that the nematic director, which defines the average orientation of the rods, undergoes a tumbling motion when a nematic sample is subjected to shear flow [110]. The full 3D numerical solution of Equation 6.50 was first given by Larson [111], where the sequence of dynamic transitions was fully described. The reason for this complex behavior is that the angular distribution of rods in the nematic phase is actually quite wide. Even when the director of the nematic aligns with the

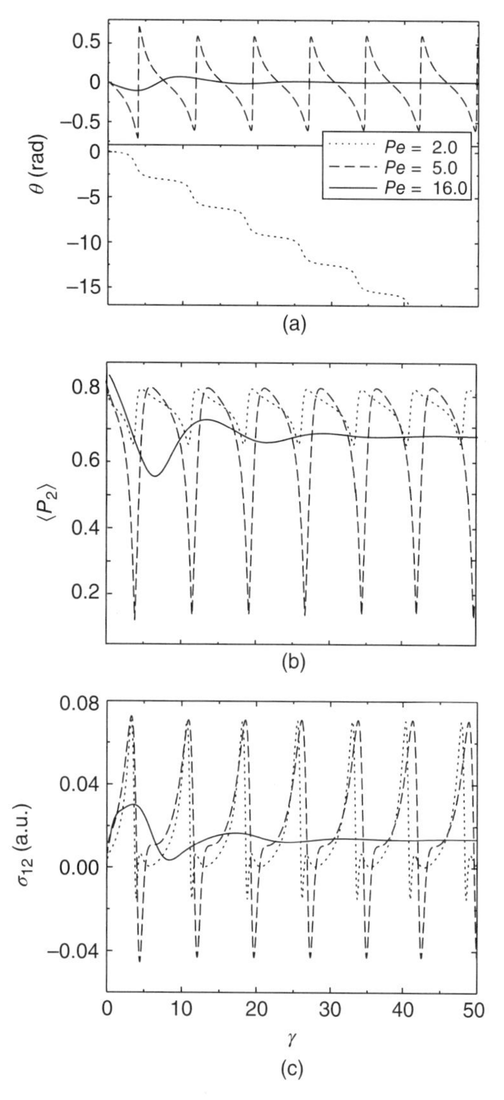

Figure 6.16 The response of the angle of the nematic director (a), the magnitude of the nematic order parameter (b), and the average stress (c) as a function of strain after a start-up of the flow. The dimensionless rod concentration is $\frac{L}{d}\varphi = 4.5$. Data are obtained by numerically solving Equation 6.50. The rods are initially placed in the flow-gradient plane. For the stress calculation, the solvent contribution was omitted. Source: Reprinted with permission from Ref. [115]. Copyright 2005 American Chemical Society.

flow, a torque will be exerted on the rods in the wings of the angular distribution, which will cause a rotation. While these rods rotate, they drag along the other rods due to the steric interactions, so that a full tumble of the director in the plane defined by the velocity and the velocity gradient vectors will occur. This can be seen in the lower panel of Figure 6.16a, which depicts the angle of the director with the flow direction. During the tumble, the distribution will widen when $\theta = 45°$, where the flow is extensional. This can be seen in Figure 6.16b, where the orientational order parameter $\langle P_2 \rangle = 0.5(3\lambda - 1)$ is plotted: for $Pe_r = 5.0$, $\langle P_2 \rangle$ approaches zero at regular points during the oscillation. At high shear rates, the widening is so drastic that rods will return back to the flow axis, as isotropic rods do, so that the director will wag instead of tumble; see dashed line in the upper panel of Figure 6.16b. All rods will flow align when the shear rate is so high that it overwhelms the Brownian diffusion, so $Pe_r \gg 1$. The sequence of dynamic transitions depends on the degree of ordering of the system and therefore, in the case of lyotropic liquid crystals, on the concentration. The corresponding stress responses, as calculated using the calculated response of **S** in Equation 6.55, are depicted in Figure 6.16c. The very clear oscillations in the stress suggest that rheology is a way to determine if a nematic sample is tumbling, wagging, or flow aligning. The problem is, however, that in a sample there are many nematic domains that will undergo oscillatory motions out of phase. In the experimental section we come back to this point, explaining how rheology can nevertheless be used to find the transition between these dynamic states.

We have now described the flow response of both the isotropic and the nematic phase. In order to find binodal lines, another degree of complexity needs to be added to the system, namely, the interface between these very differently behaving phases. Due to the complexity of the problem, no satisfactory calculations have been performed so far. To gain a better understanding, experiments and simulations are needed, as described in the next paragraph.

6.5.3 Nonequilibrium Phase Behavior of Sheared Rods: Experiment

In principle, there are three parameters that are relatively easy to access experimentally: the average orientation using birefringence, concentration differences using scattering, and the stress using rheology. The protocol to find spinodal and binodal lines is, in principle, very easy. Prepare a homogeneous dispersion; change the relevant parameters, that is, concentration or shear rate; determine if the sample phase separates using one of the mentioned techniques. If the system remains homogeneous, then the system is outside the two-phase region. If the system immediately phase separates, then it is in the unstable region. If the system phase separates, but after an induction time τ_{ind}, then it is metastable. The latter situation was encountered for the crystallization of repulsive spheres; see Section 6.3.2. Experimentally, there are three problems to solve. First, a very reproducible homogeneous state needs to be created. Second, either the concentration or the shear rate needs to be changed instantaneously. Third, gradients in orientation or concentration need to be measured. The great advantage of rods is that a homogeneous shear stabilized (para)nematic phase can be easily created by applying shear rates that are so high that all rods align, so $Pe_r \gg 1$, while

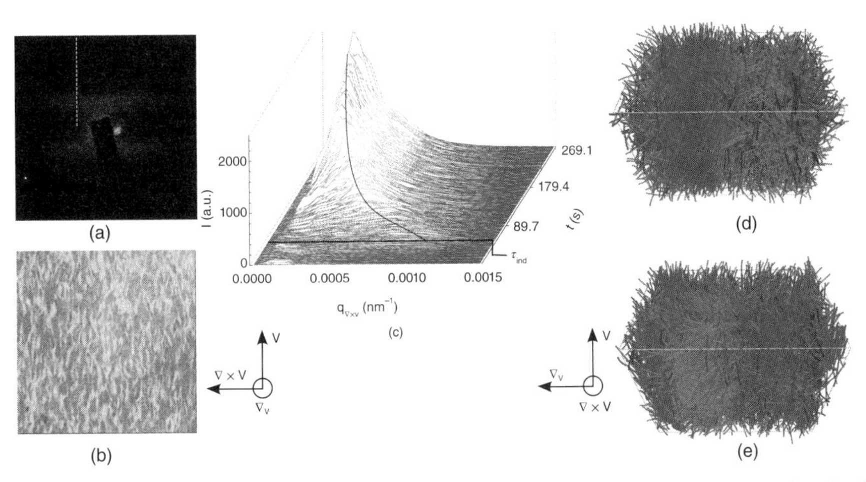

Figure 6.17 I–N phase coexisting *fd* suspension in shear flow probed by SALS (a) and polarization confocal microscopy (b). (c) Cross section of the scatter pattern (taken at the dashed line in (a)) as a function of time. The size of the form structure can be deduced from this plot as well as the induction time τ_{ind}. (d) MPCD simulation of a coexisting isotropic–nematic system of rods. The color indicates the orientation. (e) As (d) but now under shear flow. The original nematic phase is undergoing a tumbling motion while the isotropic flow aligns into a para-nematic phase. (*See color plate section for the color representation of this figure.*)

the shear rate can be instantaneously set to another value. Thus, the experimental protocol is to prepare a system in the two-phase region where isotropic and nematic phase coexist; subject the sample to a very high shear rate for long enough time, so that all structures disappear; quench the system to a low shear rate; observe the sample either by microscopy or light scattering, while measuring the stress if possible.

The only issue now is to find an appropriate system. This is not an easy task since there are many requirements to fulfill for a system to be called a model system. The first requirement stems from Onsager's assumption that only two rods interact at the same time. This is only valid for very slender rods with an aspect ratio of $\frac{L}{d} \approx 100$. We implicitly used this limit also in Equation 6.44. Such slender rods do exist, but they will always be somewhat flexible, which means that the persistence length is of the order of the contour length. The other issue is that it is very difficult to have monodisperse rods with equal length as well as thickness. There is actually only one particle that fulfills most of the requirements, namely, bacteriophage *fd* virus. It has a length of $L = 880$ nm and thickness of 6.6 nm and it is monodisperse by nature. At a neutral pH, it is charged and somewhat flexible with a persistence length of around 3 μm. The charge adds an electrostatic repulsion to the interaction, increasing the effective thickness as was already shown by Onsager. The flexibility needs to be accounted for explicitly to have an exact fit with theory for the equilibrium binodal points. A new model particle has been introduced in 2009. It is a mutant of the *fd*, where one amino acid of the more than 2000 coating proteins is changed. This particle is so stiff that dispersions of it show a complete match with Onsager theory without any fitting [116]. The results presented below, however, use *fd* as a model system, since the experiments were done before 2009 [115, 117]. To facilitate the experiments dextran was added. This is a small polymer that has a radius of gyration similar to the thickness of the rods. The addition of the polymer induces attraction between the rods due to depletion forces, very similar to what was used to obtain the critical systems discussed in Section 6.4.3. The attraction causes the biphasic region to widen, so that samples between a complete isotropic phase ($\varphi_{\text{nem}} = 0$) and complete nematic phase ($\varphi_{\text{nem}} = 1$) can easily be prepared.

Figure 6.17 shows snapshots of SALS and polarization confocal microscopy experiments on a dispersion of *fd* virus in the I–N phase coexisting region after a quench, that is, reduction of the shear rate. The bright regions in the microscopy image are elongated nematic droplets in a background of the para-nematic phase. The images were taken using a counter-rotating cone-plate shear cell as depicted in Figure 6.6b. The Fourier transfer of this image is similar to the SALS pattern of the same sample, which was taken using a Couette geometry as shown in Figure 6.6a.

Normally, the morphology of the formed structures gives insights into the kind of phase separation that takes place: a bicontinuous structure is formed when the sample phase separates by spinodal decomposition; droplets of the isotropic phase will form when the sample phase separates by nucleation and growth. When phase separation takes place while the sample is sheared, however, very stable, small and elongated structures are formed with interfaces both in the gradient and in the vorticity direction and ill-defined edges in the flow direction, very similar to the shear stabilized biphasic gas–liquid structures discussed in Section 6.4.2.

Since the sheared system hardly phase separates, it is difficult to measure the concentrations of the phase-separated sample, so no tie lines can be constructed to determine the

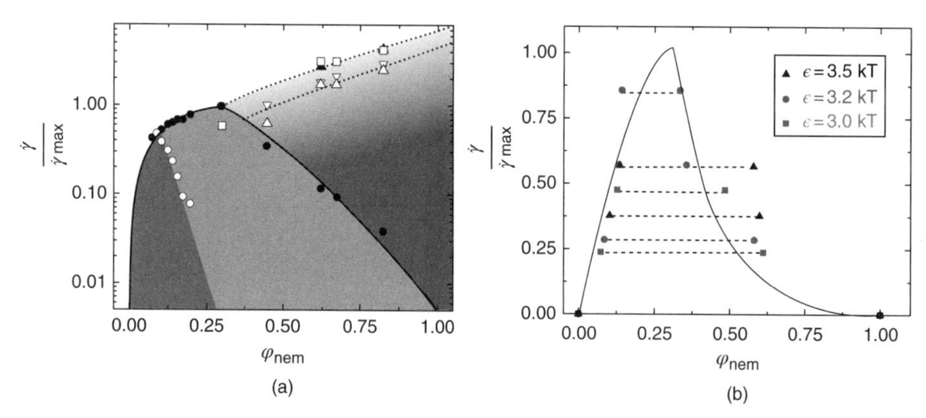

Figure 6.18 (a) Nonequilibrium experimental phase diagram for *fd* virus with 18 mg/ml dextran: binodal (•, solid line is a fit) and N–I spinodal (○) points obtained from SALS; tumbling-to-aligning transition, given by the points where all structures disappear (■) and where the response of the stress after flow reversal is overdamped (□); a wagging motion causes a local maximum in the viscosity (▽) and a minimum damping constant (△). The dotted lines are guides to the eye. The quenched nematic phase is unstable in the blue region and metastable in the green region. In the red region, the full nematic phase is in a tumbling state. (b) Nonequilibrium binodal as obtained from MPCD simulations. The symbols refer to simulations with different depth of the Lennard-Jones potential ϵ causing attractions between the rods. The full line is a fit to the experimental data, as plotted in (a). The dashed lines are the tie lines, connecting binodal points. The shear rate is scaled by the shear rate at the maximum of the binodal. The concentrations in experiment and simulations are given in terms of the fraction of the nematic phase in the quiescent sample. (*See color plate section for the color representation of this figure.*)

binodal. To determine the spinodal from the morphology of the sheared structures in the nonequilibrium two-phase region is also difficult since there are no clear differences. The best way to determine the nematic–isotropic binodal and spinodal is to determine the points, that is concentrations and shear rates, where $1/\tau_{\text{ind}} \to 0$ (binodal) and $\tau_{\text{ind}} \to 0$ (spinodal). In the example that is shown in Figure 6.17a the cross section of the scatter pattern shows a peak in the intensity that grows and shifts to small q values, indicating that a structure, which is initially about the size of a rod length, nucleates at $\tau_{\text{ind}} \approx 30$ s and grows in time; see Figure 6.17c. Binodal and spinodal points are plotted in Figure 6.18a as solid and open points, respectively, and mark the borders of the unstable region (blue) and the metastable region (green). We learn from this diagram that shear flow stabilizes the homogeneous nematic phase, since the spinodal line has a negative slope, but *inhibits* the formation of the nematic phase into the para-nematic phase, since the low concentration branch of binodal has a positive slope, contrary to earlier theoretical predictions [11]. The question is now what the origin of the inhibition of the formation of a uniform para-nematic phase is. The answer lies in the behavior of the nematic phase in shear flow.

Information on the tumbling behavior in the nematic phase is obtained by performing flow reversals, where the sign of the shear flow is changed at $t = 0$. This procedure supplies a well-defined time at which tumbling of domains is synchronized. At low shear rates, only a few oscillations are observed because there are still many nematic domains causing the tumbling to go out of phase; see the green and blue line in Figure 6.19, where the response of the viscosity to a flow reversal is plotted versus the acquired strain after reversal. At a higher absolute shear rate, the damping constant is much less, since more oscillations are observed, while the viscosity is higher, see the olive line in Figure 6.19a. These observations comply with a state where structure is almost lost, hence the low damping, while the increased viscosity is a signature of the widening of the orientational distribution as is theoretically predicted by the dashed line for $Pe = 5$ in Fig. 6.16. The oscillations are again more damped for even higher shear rates accompanied by a drop in the viscosity, confirming the theoretically predicted flow alignment. The bare Péclet numbers where these *dynamic* transitions take place can be plotted as a function of concentration or as a function of the equilibrium orientational order parameter $\langle P_2 \rangle_{\text{eq}}$, which is known for *fd* virus [118]. The latter way of plotting allows for a direct comparison of theory and simulations [119] because it is a way to account for the effect of flexibility. The result, plotted in Figure 6.19b, shows a very nice agreement between theory and experiment, where the shear rate needed to assure flow alignment increases with increasing orientational ordering.

Dynamic transitions as shown in Figure 6.19 can also be found for mixtures that are in phase coexistence in equilibrium as long as the nematic phase dominates the rheological response. These points complete the experimental phase diagram Figure 6.18a, bordering a region where the nematic phase is stable but tumbling or wagging (red), and a region where there is a homogeneous flow-aligned state. The fact that the tumbling-to-flow alignment transition line hits the maximum of the binodal line leads to an important conclusion. It shows that at phase coexistence the nematic phase

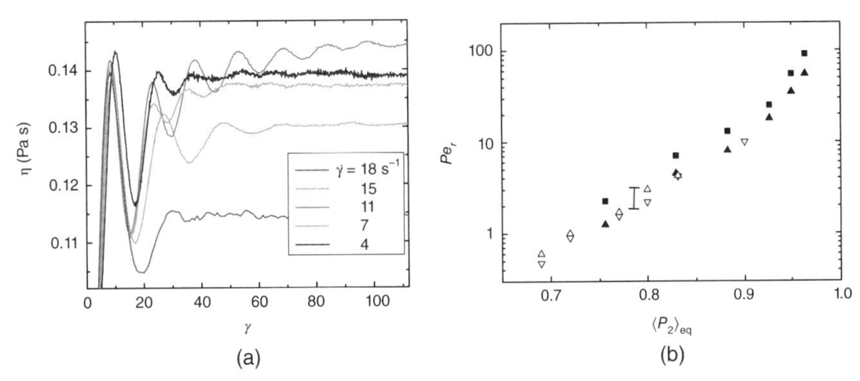

Figure 6.19 (a) Response of the viscosity after a flow reversal for various shear rates at a concentration nem = 0.67. The damping constant first decreases and then increases going toward flow alignment. The time is scaled with the applied shear rate, so the x-axis is given in strain units. (b) Dynamic phase diagram of the tumbling nematic phase as a function of the orientational order parameter at equilibrium. The experimental points indicate the Péclet numbers, where the viscosity shows a local maximum (▽) and a minimum damping constant (△). The theoretical points indicate tumbling to wagging (▲) and wagging to flow aligning (■) transitions. The region of the transition as inferred from simulations [119] is indicated by the vertical line. Source: Reprinted (adapted) with permission from Ref. [115]. Copyright 2005 American Chemical Society. (*See color plate section for the color representation of this figure.*)

is always tumbling, since all the binodal points at the high concentration side of the maximum in the binodal lie in the tumbling region. In contrast, the flow-aligned para-nematic phase is not tumbling, so that there is on average a huge interfacial tension between the two phases. This interfacial tension disappears only when both phases flow align and display the same dynamic behavior.

This scenario is confirmed by multiple particle collision dynamics (MPCD) simulations, which take the full hydrodynamic interactions into account. Figure 6.17d and e show snapshots of an I–N phase coexisting at equilibrium and after applying a shear flow. Contrary to the experiments, the simulation allows for the direct determination of the binodal points at a fixed shear rate, because the local concentration in the two phases can be determined. Before shear flow is switched on, a biphasic structure is formed with a nematic layer sandwiched between two isotropic layers. The interface between the two phases does not change position and has the normal in the gradient direction. Due to the size of the simulation box, which is considerably smaller than the length of the biphasic structures observed in experiments, one cannot discern if this structure can be categorized as a shear band. It can be observed that the concentration in the para-nematic phase increases, while the concentration in the nematic phase decreases. The final binodal points are plotted in Figure 6.18b for different attraction strengths, scaling the shear rate with the shear rate at the maximum of the binodal and the concentration by the volume fraction of the nematic phase in equilibrium. The solid line is the conjectured master curve that also fits the experimental data as plotted in Figure 6.18a and binodal points at other dextran concentrations when performing the same scaling. Moreover, it is observed that the nematic phase performs tumbling motion, one of which can be seen in Figure 6.17e, up to the point where the phase coexistence disappears, confirming the above sketched scenario.

When comparing the experimental phase diagram Fig. 6.18a with the theoretical phase diagram Fig. 6.15 for sheared colloidal rods, we notice that the story is not complete. We do not know where to locate the critical point, since the isotropic–nematic spinodal line could not experimentally be determined.

6.5.4 The Effect of the Isotropic–Nematic Transition on the Flow Behavior

In deriving the nonequilibrium phase diagram, we made the hidden assumption that the flow in the shear cell was linear. This assumption might not be valid since we argued in Section 6.2.3 that flow can become unstable when the sample is extremely shear thinning so that $\frac{\partial \sigma_{12}}{\partial \dot{\gamma}} < 0$. The flow curve that was calculated in Section 6.5.2 for a dispersion of rods at a critical concentration of $\frac{L}{d}\varphi_{\text{crit}} = 4.281$ indeed displays such a region; see Figure 6.15c. The region where $\frac{\partial \sigma_{12}}{\partial \dot{\gamma}} < 0$ is, however, very small in shear rate as well as volume fraction and located very close to the critical point. Therefore, it is not surprising that in experiments on *fd* virus dispersions as described above no gradient shear banding has been found, not even for a sample in the vicinity of the I–N binodal. It is not possible to produce a more concentrated homogeneous sample, and according to theory, this should be the right concentration to obtain the small decline in stress.

On the other hand, bands have been observed that were stacked in the vorticity or neutral direction [40]; see Figure 6.20b. This vorticity band formation is observed in a limited region *within* the biphasic region; see Figure 6.20a, where the sample is microscopically phase separated as shown in

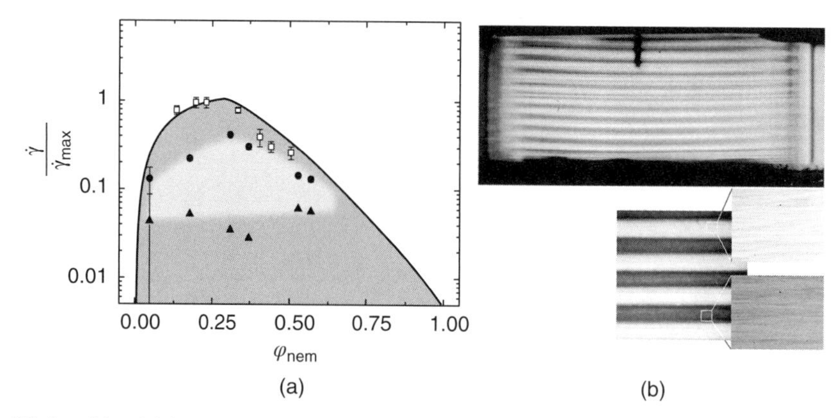

Figure 6.20 (a) Nonequilibrium binodal for the I–N transition of *fd* virus in shear flow, as Figure 6.18a, including the region where vorticity banding is observed. The dextran concentration was 10.6 mg/ml. The binodal line is the same as in Figure 6.18a. (b) Macro- and micrographs of vorticity bands all taken with almost crossed polarizers. The top picture shows that bands are formed throughout the Couette cell. The height of the cell being 3 cm. The lower figure shows a banded state for the same, where the bandwidth is about 2 mm. The two enlargements on the right show the inhomogeneities that are present within the bands. (*See color plate section for the color representation of this figure.*)

Figure 6.17. This shear band formation is therefore coupled to the presence of biphasic structures in the dispersion, as we have seen in Section 6.4.3 for demixing gas–liquid colloid–polymer mixtures. It therefore belongs to the class of systems where vorticity bands formation is connected to deformation of inhomogeneities, as introduced in Section 6.2.3. There are, however, important differences compared to the band formation for gas–liquid colloid sphere–polymer mixtures. The bands that are observed here contain inhomogeneities, whereas in most cases, like for the mixtures discussed in Section 6.4.3, the bands consist of a pure phase. Although the exact reason for this shear band formation is unknown, one can argue that the inhomogeneous deformation of the biphasic structures leads to hoop stresses similar to the Weissenberg effect for polymers, due to the fact that in Couette geometry there is a gradient in shear rate throughout the gap.

6.6 CONCLUDING REMARKS

This chapter forms an introduction in the field of nonequilibrium phase behavior of colloidal dispersions and the connection with flow instabilities. We first treated the most simple class of dispersions, namely, dispersions of hard or repulsive colloidal spheres. These systems crystallize at a critical concentration, and we showed that shear flow can enhance as well as frustrate this crystallization, based on early scattering experiments that were recently complemented with confocal microscopy in shear flow. Though concepts of CNT are helpful to understand crystallization, they cannot be used to predict this nonequilibrium phase behavior and one has to rely on simulations to support the interpretation of the results.

This situation is different for the second class of dispersions that were discussed, namely, dispersions of attractive spheres. For this system, we used a microscopic theory based on the Smoluchowski equation of motion on the level of the pair correlation function, which is directly related to the structure factor. The nice feature of the theory is that shear flow can readily be included, while the structure factor can be directly accessed with light scattering. Thus, we could show how the critical point shifts to higher concentrations when applying shear flow by directly comparing theory and experiment.

The most complex system we treated were dispersions of colloidal rods that display an isotropic–nematic phase transition. Again, the Smoluchowski approach was used, but now by deriving an equation of motion for the orientation of the particles. The theory predicts that the shear flow stabilizes the homogeneous nematic state, since the spinodal has a negative slope. This was confirmed by confocal and small-angle light scattering experiments. These experiments, aided by simulations, also showed that the nematic director tends to tumble while rods in isotropic phase flow align. As a consequence, the sheared nematic and isotropic phases can only merge when rods in both phases are flow aligning. Thus, a nonequilibrium phase diagram is obtained that is very similar to that of crystallizing sphere, with a low concentration branch of binodal that has a positive slope.

Although the selection of subjects was somewhat biased by research in our group, we do think it nicely illustrates how understanding can be gained for increasingly complex systems. We have shown that knowledge of the microstructure on the theoretical and experimental level is a prerequisite for the understanding of the nonequilibrium phase behavior and flow of these complex fluids. In real life, that is in industrial applications and in biology, the situation is often much more complex, given the constitution of the substances and the involved flow geometries. We believe, however, that the foundations put out here are crucial to start understanding more complex situations.

REFERENCES

[1] Merrill EW. Physiol Rev 1969;49:863.
[2] de Viguerie L, Ducouret G, Lequeux F, Moutard-Martin T, Walter P. C R Physique 2009;10:612.
[3] Larson RG. *The Structure and Rheology of Complex Fluids*, Topics in Chemical Engineering. Oxford: Oxford University Press; 1998.
[4] Cheng Z, Zhu J, Chaikin PM. Phys Rev 2002;65:041405.
[5] Pal R. J Biomech 2003;36:981.
[6] Frith WJ, d'Haene P, Buscall R, Mewis J. J Rheol 1996;40:531.
[7] Wagner NJ, Brady JF. Phys Today 2009;62:27–32.
[8] Macosko CW. *Rheology-Principles, Measurements, and Applications*. New York: Wiley-VCH; 1994.
[9] Maitland GC. Curr Opin Colloid Interface Sci 2000;5:301.
[10] Decker MJ, Halbach CJ, Nam CH, Wagner NJ, Wetzel ED. Compos Sci Technol 2007;67:565.
[11] Olmsted PD, Lu CYD. Phys Rev E 1999;60:4397.
[12] Olmsted PD. Europhys Lett 1999;48:399.
[13] Dhont JKG. Phys Rev E 1999;60:4534.
[14] Manneville S. Rheol Acta 2008;47:301.
[15] Dhont JKG, Briels WJ. Rheol Acta 2008;47:257.
[16] Olmsted PD. Rheol Acta 2008;47:283.
[17] Imhof A, van Blaaderen A, Dhont JKG. Langmuir 1994;10:3477.
[18] Holmes WM, Callaghan PT, Vlassopoulos D, Roovers J. J Rheol 2004;48:1085.
[19] Sprakel J, Spruijt E, Stuart MAC, Besseling NAM, Lettinga MP, van der Gucht J. Soft Matter 2008;4:1696.
[20] Tapadia P, Wang SQ. Phys Rev Lett 2003;91:198301.
[21] Tapadia P, Ravindranath S, Wang SQ. Phys Rev Lett 2006;96:196001.
[22] Boukany PE, Hu YT, Wang SQ. Macromolecules 2008;41:2644.
[23] Hu YT, Palla C, Lips A. Macromolecules 2008;41:6618.
[24] Spenley NA, Cates ME, McLeish TCB. Phys Rev Lett 1993;71:939.
[25] Cates ME. Macromolecules 1987;20:2289.
[26] Rehage H, Hoffmann H. J Phys Chem 1988;92:4712.
[27] Rehage H, Hoffmann H. Mol Phys 1991;74:933.
[28] Decruppe JP, Cressely R, Makhloufi R, Cappelaere E. Colloid Polym Sci 1995;273:346.
[29] Callaghan PT, Cates ME, Rofe CJ, Smeulders JBAF. J Phys II France 1996;6:375.
[30] Salmon JB, Colin A, Manneville S. Phys Rev Lett 2003;90:228303.
[31] Hu YT, Lips A. J Rheol 2005;49:1001.
[32] Miller E, Rothstein JP. J Non-Newtonian Fluid Mech 2007;143:22.
[33] Helgeson ME, Reichert MD, Hu YT, Wagner NJ. Soft Matter 2009;5:3858.
[34] Fardin MA, Ober TJ, Grenard V, Divoux T, Manneville S, McKinley GH, Lerouge S. Soft Matter 2012;8:10072.
[35] Boltenhagen P, Hu Y, Matthys EF, Pine DJ. Phys Rev Lett 1997;79:2359.
[36] Larson RG, Shaqfeh ESG, Muller SJ. J Fluid Mech 1990;218:573.
[37] Lin-Gibson S, Pathak JA, Grulke EA, Wang H, Hobbie EK. Phys Rev Lett 2004;92:048302.
[38] Pignon F, Magnin A, Piau JM. Phys Rev Lett 1997;79:4689.
[39] Tromp RH, Hoog EHAD. Phys Rev E 2008;77:031503.
[40] Kang K, Lettinga MP, Dogic Z, Dhont JKG. Phys Rev E 2006;74:026307.
[41] Pakdel P, McKinley GH. Phys Rev Lett 1996;77:2459.
[42] Fardin MA, Lopez D, Croso J, Grégoire G, Cardoso O, McKinley GH, Lerouge S. Phys Rev Lett 2010;104:178303.
[43] Fardin MA, Ober TJ, Gay C, Gregoire G, McKinley GH, Lerouge S. Soft Matter 2012;8:910.
[44] Fielding SM. Phys Rev E 2007;76:016311.
[45] Larson RG. Rheol Acta 1992;31:213.
[46] Alder BJ, Wainwright TE. J Chem Phys 1957;27:1208.
[47] Frenkel D. Physica A 1999;263:26.
[48] Hoover WG, Ree FH. J Chem Phys 1968;49:3609.
[49] Pusey PN, van Megen W. Nature 1986;320:340.
[50] Palberg T, Haertl W, Wittig U, Versmold H, Wuerth M, Simnacher E. J Phys Chem 1992;96:8180.
[51] Gasser U, Weeks ER, Schofield A, Pusey PN, Weitz DA. Science 2001;292:258.
[52] Auer S, Frenkel D. Nature 2001;409:1020.
[53] Gasser U. J Phys Condens Matter 2009;21:203101.
[54] Dhont JKG. *An Introduction to Dynamics of Colloids*. Amsterdam: Elsevier; 1996.
[55] Russel WB, Gast AP. J Chem Phys 1986;84:1815.
[56] Vermant J, Solomon MJ. J Phys Condens Matter 2005;17:R187.
[57] Oxtoby DW. J Phys Condens Matter 1992;4:7627.
[58] Russel WB. Phase Transit 1990;21:127.
[59] Dhont JKG, Smits C, Lekkerkerker HNW. J Colloid Interface Sci 1992;152:386.
[60] Schatzel K, Ackerson BJ. Phys Rev E 1993;48:3766.
[61] Harland JL, vanMegen W. Phys Rev E 1997;55:3054.
[62] Ackerson BJ. J Rheol 1990;34:553.
[63] Holmqvist P, Lettinga MP, Buitenhuis J, Dhont JKG. Langmuir 2005;21:10976.
[64] Cohen I, Mason TG, Weitz DA. Phys Rev Lett 2004;93:046001/1.
[65] Derks D, Wu YL, van Blaaderen A, Imhof A. Soft Matter 2009;5:1060.
[66] Besseling TH, Hermes M, Fortini A, Dijkstra M, Imhof A, van Blaaderen A. Soft Matter 2012;8:6931.
[67] Molino FR, Berret JF, Porte G, Diat O, Lindner P. Eur Phys J B 1998;3:59.

[68] Holmqvist P. Langmuir 2014;30:6678.
[69] Blaak R, Auer S, Frenkel D, Löwen H. Phys Rev Lett 2004;93:068303.
[70] Koumakis N, Schefield AB, Petekidis G. Soft Matter 2008;4:2008.
[71] Chen LB, Chow MK, Ackerson BJ, Zukoski CF. Langmuir 1994;10:2817.
[72] Eiser E, Molino F, Porte G, Diat O. Phys Rev E 2000;61:6759.
[73] Cohen I, Davidovitch B, Schofield AB, Brenner MP, Weitz DA. Phys Rev Lett 2006;97:215502.
[74] Wu YL, Derks D, van Blaaderen A, Imhof A. Proc Natl Acad Sci U S A 2009;106:10564.
[75] Ornstein LS, Zernike F. Proc K Akad Wet Amsterdam 1914;17:793.
[76] Cahn JW. J Chem Phys 1965;42:93.
[77] Onuki A, Kawasaki K. Ann Phys 1979;121:456.
[78] Wu XL, Pine DJ, Dixon PK. Phys Rev Lett 1991;66:2408.
[79] Onuki A. Phys Rev Lett 1989;62:2472.
[80] Milner ST. Phys Rev E 1993;48:3674.
[81] Tanaka H. Phys Rev E 1999;59:6842.
[82] Onuki A. J Phys Condens Matter 1997;9:6119.
[83] Dhont JKG, Verduin H. J Chem Phys 1994;101:6193.
[84] Dhont JKG. J Chem Phys 1995;103:7072.
[85] Fixman M. J Chem Phys 1962;36:1965.
[86] Israelachvili JN. *Intermolecular and Surface Forces: With Applications to Colloidal and Biological Systems*. London, Orlando (FL): Academic Press; 1985, ISBN: 0123751802 (U.S. alk. paper), note 84021556 Jacob N. Israelachvili. ill.; 24 cm. Bibliography: p 276–286. Includes index.
[87] Beysens D, Gbadamassi M, Boyer L. Phys Rev Lett 1979;43:1253.
[88] Helfand E, Fredrickson GH. Phys Rev Lett 1989;62:2468.
[89] Lenstra TAJ, Dhont JKG. Phys Rev E 2001;63:061401.
[90] Wang H, Lettinga MP, Dhont JKG. J Phys Condens Matter 2002;14:7599.
[91] Bodnar I, Verduin H, Dhont JKG. Phys Rev Lett 1996;77:5304.
[92] Tanaka H. Faraday Discuss 2012;158:371.
[93] Hohenberg PC, Halperin BI. Rev Mod Phys 1977;49:435.
[94] Chan CK, Perrot F, Beysens D. Phys Rev A 1991;43:1826.
[95] Hashimoto T, Matsuzaka K, Moses E, Onuki A. Phys Rev Lett 1995;74:126.
[96] Derks D, Aarts D, Bonn D, Imhof A. J Phys Condens Matter 2008;20:494222.
[97] Tomotika S. Proc R Soc Lond Ser A 1936;153:0302.
[98] Beysens D, Gbadamassi M, Moncefbouanz B. Phys Rev A 1983;28:2491.
[99] Zocher H, Anorg Z. Z Anorg Allg Chem 1925;147:91.
[100] Zsigmondy R. *Zur Erkenntnis der Kolloide : Ueber irreversible Hydrosole und Ultramikroskopie*. Jena: Fischer; 1919.
[101] Bawden FC, Pirie NW, Bernal JD, Fankuchen I. Nature 1936;138:1051.
[102] Onsager L. Ann NY Acad Sci 1949;51:627.
[103] Majorana Q. Il Nuovo Cimento 1902;4:374.
[104] Zocher H. Z Phys Chem 1912;98:293.
[105] Tang J, Fraden S. Phys Rev Lett 1993;71:3509.
[106] Lenstra TAJ, Dogic Z, Dhont JKG. J Chem Phys 2001;114:10151.
[107] Dhont JKG, Briels WJ. Colloids Surf, A 2003;213:131.
[108] See H, Doi M, Larson R. J Chem Phys 1990;92:792.
[109] Olmsted PD, Goldbart P. Phys Rev A 1990;41:4578.
[110] Hess S. Z Naturforsch A: J Phys Sci 1976;31:1034.
[111] Larson RG. Macromolecules 1990;23:3983.
[112] Doi M, Edwards SF. *The Theory of Polymer Dynamics*. Oxford: Oxford University Press; 1986.
[113] Kayser RF, Raveché HJ. Phys Rev A 1978;17:2067.
[114] Hess S. Z Naturforsch A: J Phys Sci 1976;31:1507.
[115] Lettinga MP, Dogic Z, Wang H, Vermant J. Langmuir 2005;21:8048.
[116] Barry E, Beller D, Dogic Z. Soft Matter 2009;5:2563.
[117] Ripoll M, Holmqvist P, Winkler RG, Gompper G, Dhont JKG, Lettinga MP. Phys Rev Lett 2008;101:168302.
[118] Purdy KR, Dogic Z, Fraden S, Rü HA, Lurio L, Mochrie SGJ. Phys Rev E 2003;67:031708.
[119] Tao Y, den Otter WK, Briels WJ. Phys Rev Lett 2005;95:237802.

7

COLLOIDAL INTERACTIONS WITH OPTICAL FIELDS: OPTICAL TWEEZERS

DAVID MCGLOIN, CRAIG MCDONALD, & YURI BELOTTI
SUPA, School of Science and Engineering, University of Dundee, Dundee, UK

7.1 INTRODUCTION

With so much going on at the microscale, it has been necessary to develop a suite of tools to probe particles that exist well beyond the range of human vision. One significant problem that is encountered in the move from the macro to the micro is that Brownian motion kicks in, and, therefore, particle localization becomes an issue. Localization allows ease of interrogation, the ability to perturb in a controlled manner, and enables repeatability to become a trivial part of the experimental design. It is compromised, however, due to the relatively small depth of field of high-magnification microscope systems. Hence, particles that have a natural tendency to move, or to sink or float within the sample you are studying, can be challenging to interrogate even briefly, never mind over significant timescales. If one moves to the nanoscale, then things become even worse.

There are a range of tools that allow this form of localization. These include micromanipulators [1], electrostatic traps [2], dielectrophoretic traps [3], magnetic tweezers [4], and microfluidic confinement [5] along with the subject of this chapter: optical tweezers.

Optical tweezers, as the name suggests, make use of light to trap and confine microparticles. In general, this means using laser light, with the consequence that, due to the diffraction limit, particles in the range of 0.5–20 μm are fairly trivial to trap. As we will see though, objects outside this range can also be manipulated with a little more effort. Also, optical tweezers generate forces in the picoNewton range over nanometers of displacement. This is a very convenient

Fluids, Colloids and Soft Materials: An Introduction to Soft Matter Physics, First Edition. Edited by Alberto Fernandez Nieves and Antonio Manuel Puertas.

force and positioning regime with which to study biological systems, especially molecular motors, and it is in this area that very significant studies have taken place, especially into motors such as kinesin [6], myosin [7], and dynein [8].

Optical tweezers have a rich history, with the development of the first optical traps in the 1970s [9]. These systems made use of *radiation pressure* and were able to hold particles against gravity or in dual-beam configurations. These systems enabled a range of initial studies to be carried out, especially in optical levitation of particles in air, but lacked the later precision and control of optical tweezers. These initial studies were also a precursor for another use of optical radiation pressure – the cooling and trapping of gaseous atomic samples [10].

The pioneer of optical tweezers was Arthur Ashkin, who in 1986 published the seminal paper in the field, "Observation of a Single-Beam Gradient Force Optical Trap for Dielectric Particles" [11]. This introduced the world to this new tool, which was able to confine particles in three dimensions, as opposed to the two dimensions of the radiation pressure force traps that preceded it. In doing so, a new precision technique was born that would have impact in a range of sciences, spanning biological, chemical, and physical, that would lead to new commercial instrumentation and spur on developments in related areas such as optical beam shaping. Indeed, Ashkin was particularly prescient in his predictions of the applications of his work in the original paper:

> "They also open up a new size regime to optical trapping encompassing macromolecules, colloids, small aerosols, and possibly biological particles."

In each of these he was right, and in fact his hesitancy over the biological work was quite unfounded.

In this chapter, we will review the basic physics and design of an optical tweezers system, and how they can be used in a range of studies, with emphasis on colloidal systems. In addition, we will examine how modern optical tweezers are able to take advantage of the ability to shape optical beams into near arbitrary shapes through a range of technological innovations from the acousto-optical modulator to the spatial light modulator. The exquisite control that such tools offer over the laser beam parameters is one of the reasons that optical tweezers, as opposed to other manipulation techniques, remain so popular. The flexibility that using light as a noncontact manipulation methodology offers is far in excess of that offered by using other types of fields and typically is rather more straightforward to implement.

In addition, we will also explore other optical techniques that make use of radiation pressure to manipulate particles – these are typically dual-beam traps and can offer some flexibility over optical tweezers for certain applications.

7.2 THEORY

Optical tweezers can be explained fairly simply using some high school physics – if we consider a microscopic glass bead that is interacting with a focused laser beam (typically, these are focused using a high (>1) numerical aperture microscope objective), then two things happen – first, the scattering light on the bead will tend to push the particle in the direction of the beam propagation. This is called radiation pressure and is the type of force first used by Ashkin in his original work to trap particles against gravity. It is also radiation pressure that was used in his dual-beam trapping experiments. Second, the light that impinges on the glass bead is also refracted through the sphere. The change in momentum that results from the change in direction of the light generates a force. As a typical laser beam has a Gaussian profile, the momentum change is uneven from light passing through the bead close to the center of the beam compared to light from the edge of the beam. This intensity gradient across the beam gives rise the term "gradient force" for the force that is produced. As can be seen from Figure 7.1, this leads the bead to be pulled toward the center, the most intense part, of the beam. Note that this is only true if the polarizability of the bead is positive (which will be true if the ratio of the refractive index of the bead to the refractive index of the surrounding medium, typically water, is >1). If it is negative (as in the case where the refractive indices ratio is <1, e.g., a bubble), then the forces act so as to repel the particle from the trap. One can also consider the trap to occur due to the interaction of the dipoles induced in the polarizable material with the optical field that leads to the trapping effect – the inhomogeneous field leads to the gradient force.

Optical tweezers arise from the gradient force, provided the force generated toward the beam focus is sufficient to overcome the radiation pressure force trying to push the particle out of the trap. Often, optical tweezers are designed to point in the upward direction to make use of gravity to help with this situation but, counterintuitively, the beam can also point downward and a particle will still be trapped. A particle in which the z-component (the beam propagation direction component) of the force overcomes the scattering radiation pressure force is said to be three-dimensionally trapped. Particles can be two-dimensionally trapped in the x–y plane but not in the z-direction, either by balancing the radiation pressure against gravity or by pushing a particle against the bottom of the chamber in which it is confined. The requirement on the z-direction gradient force to overcome the scattering force is why high numerical aperture focusing optics are typically used in optical tweezers, to enable a very high gradient force to be generated.

Once a particle is trapped it then obeys Hooke's Law, with displacements from the trap center experiencing a restoring force. Hence, for relatively small displacements of the order

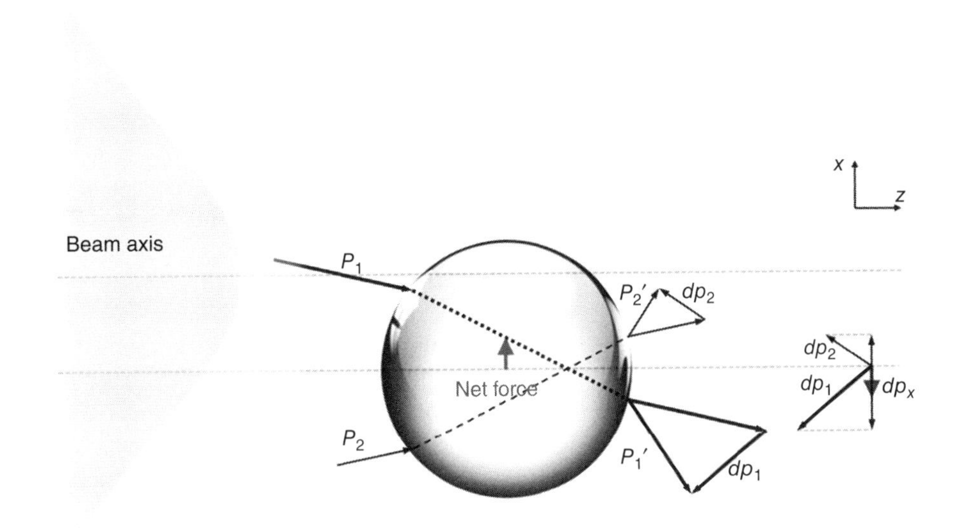

Figure 7.1 Optical trapping forces diagram. The figure outlines the momentum changes associated with the gradient force pulling the particle toward the axis of the beam, which is most intense at the centre. This assumes the particle has a higher refractive index that the surrounding. The force on the bead acts in the *x*-direction toward the axis. A similar argument can be used when considering the trapping forces in the *z*-direction.

of 150 nm, where the force is linearly proportional to the displacement, the force on a bead is given by

$$F = -kx. \tag{7.1}$$

This relationship, that of a linear spring, is one of the key factors in making optical tweezers such a powerful tool – as through high-precision position detection techniques it is then possible to measure forces very accurately. Typically, force measurements are in the picoNewton regime but exceptionally sensitive traps in the femtoNewton regime have also been demonstrated [12]. We will consider the practicalities of how forces are measured in Section 7.3.

If we have a need to calculate forces, then for particles that are considered "large" meaning they are much larger than the trapping laser light wavelength, in the so-called Mie regime, then we can make use of a simple ray optics approach [13]. Within the Rayleigh regime, where the particles are much smaller than the wavelength of the trapping light, we can make the assumption that the particle behaves as a point dipole, and hence the scattering force the particle experiences can be described by [14]

$$F_{\text{scat}} = \frac{n_m}{c}\sigma I = \frac{128\pi^5 r^6}{3c\lambda_0^4}\left(\frac{m^2-1}{m^2+2}\right)^2 n_m I, \tag{7.2}$$

where I is the incident intensity, σ is the scattering cross section of the sphere, with radius r. The refractive index of the medium is given by n_m, c is the speed of light in vacuum, m is the ratio of the refractive indices of the particle (n_p) and the medium, and the wavelength of the trapping light is λ_0. The gradient force is given by

$$F_{\text{grad}} = \frac{2\pi n_m r^3}{c}\left(\frac{m^2-1}{m^2+2}\right)\nabla I. \tag{7.3}$$

Of interest is the scaling of the two different forces with radius. The much more rapidly diminishing scattering force with radius means that materials that would be conventionally considered to be highly scattering at larger length scales, for example, gold, but which are highly polarizable, can be trapped fairly straightforwardly, if small enough, using an optical tweezers system. Hence, tweezers are routinely used to trap metallic nanoparticles [15, 16].

The intermediate size regime where the particle is neither much bigger nor much smaller than the wavelength of light is more complex to deal with and more complex electromagnetic treatments are required [17]. This is unfortunate as a significant number of experimental studies must work in this regime. Increasingly, though, there are open-source tools that can deal with such problems, with the Optical Tweezers Computational Toolbox [18], developed by Timo Nieminen from the University of Queensland, providing a fairly sophisticated *T*-matrix and Lorenz–Mie approach for a range of optical tweezers problems. For many situations, even within the intermediate regime, however, a ray optics approach will offer acceptable results, and toolboxes for this type of situation are also available [19].

7.3 EXPERIMENTAL SYSTEMS

7.3.1 Optical Tweezers

Optical tweezers are a relatively straightforward device to design and build, and it is this simplicity that makes them such a powerful tool. It does not take great optical engineering skills, nor in-depth knowledge of optical physics to set up a system and get it taking fairly sophisticated results. An example is a teaching system setup developed at Berkeley, which uses low-cost components and can be assembled for around \$15k not including an optical table [20]. Higher end systems would perhaps have more stable laser sources, higher precision nanopositioning stages and may opt for more stable optical mechanical pieces, but still this basic system is capable of measurements on molecular motors. Increasingly, control software for such systems can be found open source [21] and even more sophisticated instrument control software, such as that designed for holographic optical tweezers can now be found freely available online [22]. The implication is that with a modest investment into a system, a nonexpert user (with a friendly physicist to help with the alignment) could have a robust system up and running very quickly, one capable of research-grade measurements.

Design Considerations When constructing a set of optical tweezers, there are a number of different factors that must first be considered. The most important of which is the nature of the experiment, as this will dictate how simple or complex a system must be built, possibly requiring the addition of specialized, application-specific optics. How trap location will be controlled; whether or not force-measuring techniques are required; are there wavelength constraints; and how stable the overall system must be will all be determined by the nature of the experiment. However, regardless of the complexity of the system, the core of the system will, more often than not, be designed around a basic optical tweezers system, shown in Figure 7.2.

There are eight main components in a basic optical tweezers: a trapping laser, a high numerical aperture (NA) microscope objective, an illumination source, a camera, a dichroic mirror, steering mirrors, a beam steering lens relay system, and a sample of interest. The exact specifications of each of these components can vary greatly with application, and even with available lab space and budget. However, it is important to understand the overall need for each of these eight components and how, through the correct combination, a stable optical tweezers can be assembled.

A good trapping laser will produce a collimated, monochromatic, coherent Gaussian beam. This will allow the light to be focused to a diffraction-limited spot by the microscope objective. Possibly, the most obvious parameter to consider when selecting a trapping laser is the laser wavelength, as this will influence everything from the selection of optical components, to the cost of the system, to heating of the sample. Shorter wavelengths are best avoided if the tweezers are to be used for investigations into biological samples as they can prove harmful to the sample, particularly so when trapping cells directly. However, longer (typically in the near IR and above), biosafe, wavelengths tend to have a higher extinction coefficient of water, thus leading to

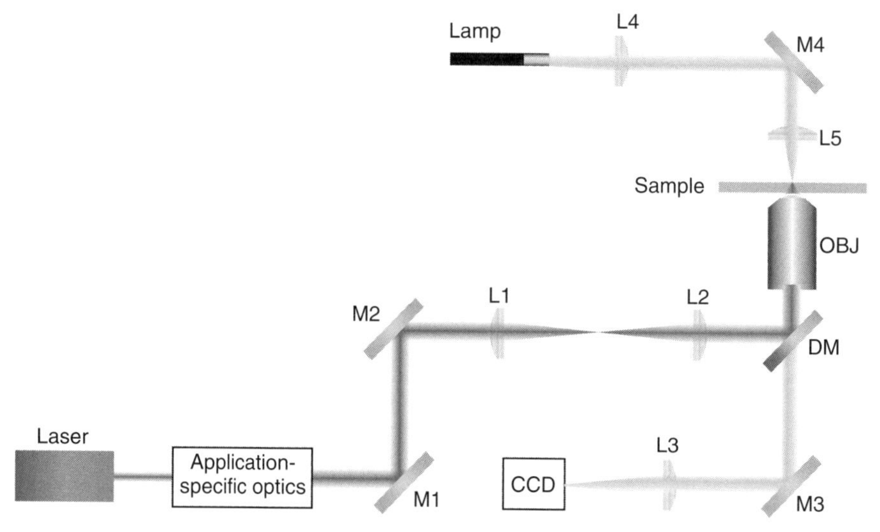

Figure 7.2 A basic optical tweezers system. After passing through any application specific optics, the laser is steered by mirrors M1 and M2, which act as the steering mirrors, through the beam steering lens relay system, formed by lenses L1 and L2, and is reflected by the dichroic mirror (DM) into the microscope objective (OBJ), which forms an optical trap in the sample. Lenses L3, L4 and L5 combine with the microscope objective and mirrors M3 and M4 to form the illumination path. This will image the sample on to the CCD camera.

greater heating effects felt by the sample [23]. The stability of the laser must also be considered, as pointing instabilities can lead to unwanted displacements of the optical trap in the trapping plane and power fluctuations can lead to temporal variations in the trap stiffness [24]. Output power stability is also important, as the maximum trap stiffness and, therefore, force that can be attained will be determined by the laser power and the throughput of the optical system.

A high NA microscope objective (NA > 1) is the most important element of an optical tweezers system as it is this that forms the trap. The NA and transmission of the objective will determine the efficiency with which input power is converted to trap stiffness. The maximum depth at which objects can be trapped will also be set by objective choice as it will depend on the working distance and the immersion medium of the objective. Trapping performance is hampered by spherical aberrations that arise due to the refractive index mismatch between objective immersion medium and the trapping medium, with the magnitude of these detrimental effects increasing with focal depth [24]. Trapping at depths of over 20 μm can be achieved by using water immersion objectives, rather than the more traditional oil immersion objectives, as they exhibit much lower spherical aberration [25]. By expanding the trapping laser to slightly overfill the back aperture of the microscope objective, a diffraction-limited focal spot is produced, thus generating a high-quality, three-dimensional optical trap.

The microscope objective, when combined with an illumination source and camera, also serves to image the sample in exactly the same manner as when used in a commercial microscope. However, this does pose the problem of having to combine the laser and imaging paths as they travel through the objective. To allow this, a dichroic mirror is used before the microscope objective so that only the laser wavelength is reflected and the remaining illumination light is transmitted to the camera. The illumination source, and the way in which it is used to illuminate the sample, is very important but also very flexible. Standard, white light illumination is suitable for the majority of applications, but other techniques, such as fluorescence imaging, can be combined with the tweezers to give a powerful investigative tool.

Essentially, that is all that is needed to create an optical trap. However, in terms of maneuverability, this will create a somewhat limited tweezing system. The trap will be stationary and any movement through the sample must be achieved by physically moving the sample. Clearly, this is not always practical and some form of control of the optical trap would be advantageous. By including steering mirrors, the trap can be moved laterally in the focal plane of the objective, allowing for control of a trapped object in two dimensions. However, if the steering mirrors are placed directly behind the objective, it can be clearly shown (Fig. 7.3) by ray optics that any tilt in these mirrors will cause the beam to shift from the back aperture of the objective. Any shift away

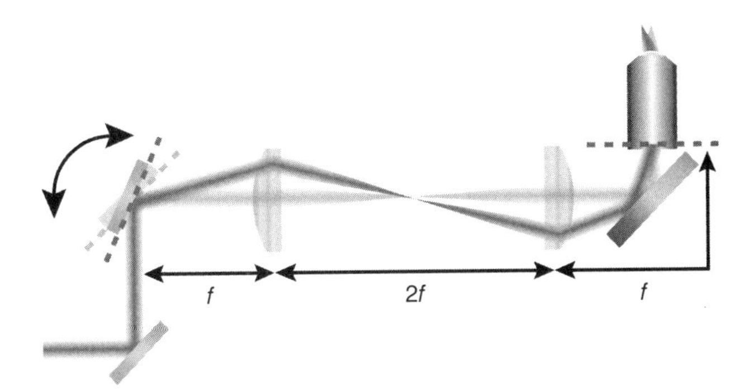

Figure 7.3 Beam steering lens relay system formed by use of a 4f optical system. Dotted lines denote conjugate planes. Notice the angular deviation in the mirror causes an angular displacement at the back aperture of the microscope objective, which translates to a lateral movement of the beam focus, without misalignment. Original mirror and beam position are included for reference but are less opaque for clarity.

from the center of the microscope objective will deform the focus, thus producing an asymmetric focal spot and, therefore, an asymmetric trap. By placing a $4f$ imaging system between the steering mirrors and the microscope objective, this problem can be avoided. Such a system is called $4f$ due to there being four focal lengths between the start and end of the system: one focal length between the mirror and the first lens, two focal lengths between the lenses, and then one focal length to the back aperture of the microscope objective. Through the inclusion of this beam steering lens relay system, the steering mirror becomes a conjugate plane to the back aperture of the microscope objective. A conjugate plane is defined as a plane where the intensity distribution is an image of the intensity distribution across a corresponding plane [26]. As the steering mirror and objective are now conjugate, any angular displacement at the mirror becomes an angular displacement at the back aperture, without any lateral movement. This angular displacement is then translated to the objective focus as a lateral movement in the focal spot. Hence, steering of the trap can be achieved without any degradation in trap performance.

It is worth noting that while each of these components is of great importance in a tweezers system, even the highest quality optics will be of little use if not properly mounted. As optical tweezers operate at such high magnifications, drift and vibration can present major issues when trying to perform precise measurements. While some work has been carried out into the suitability of constructing optomechanics from LEGO [27], the majority of optical tweezers are either integrated into commercial microscopes or custom-designed breadboard-based systems. Commercial microscopes offer many advantages, such as familiarity of design and high stability, especially of the sample and microscope objective. However, they are often not designed to be adapted to become an optical tweezers system. This could lead to convoluted designs required in order to integrate the laser, possibly negating the increased stability originally offered. The flexibility of custom-built systems can lead to a much simpler optical design and ease of alignment, but care must be taken during construction to avoid mechanical resonances.

7.3.2 Force Measuring Techniques

It is easy to see how optical tweezers could have become a novel toy for physicists to play within their labs, picking up, moving, and looking at different micron-scale objects, had it not been for their impressive ability to be used as a quantitative tool. In fact, due to the ability of optical tweezers to exert picoNewton forces, and then to accurately measure these forces, optical tweezers have become an indispensable quantitative tool in many labs.

At the core of all quantitative optical trapping is sensitive position detection, as most force measurements are made by indirectly measuring the position of the trapped object. As outlined in Section 7.2, the force (F) acting on a trapped bead is given by Equation 7.1, where k is the trap stiffness, essentially a measure of how well the particle is trapped, and x is the position of the particle relative to the center of the trap. This equation will hold so long as the particle is not displaced more than a few hundred nanometers from the trap center. Within this region, the potential energy of the sample follows a simple parabolic equation and the trap behaves harmonically. Therefore, the restoring force, which pulls the particle back to the trap center, is linear with respect to displacement. As the particle is acting in a similar way to an object on a spring, the restoring force can be described by an optical equivalent to Hooke's law (3.1), with the trap stiffness being thought of an analogous to the spring constant, Figure 7.4. Consequently, measuring the force on a particle is often split into two distinct steps: calibration of the optical trap, in order to calculate the trap stiffness, k, and measurement of the displacement of the trapped particle from the trap center, x.

Owing to the importance of accurately measuring both of these components, a smorgasbord of different techniques have been developed, improved, and refined over the years. The two most prominent techniques are video-based particle tracking and interferometric particle tracking.

Displacement Measurement Measuring the displacement of the particle using video-based techniques is not only relatively straightforward but also very convenient as most optical tweezers already have a camera to observe the sample. Once the magnification is known, that is, by calibrating the video picture against a known distance, such as a microscope graticule, then video particle tracking measurements are given in absolute distance units. This calibration is linear over the field of view of the camera, typically tens of microns, and is independent of sample depth. There is also a wide variety of commercial and open-source particle tracking software, which, through algorithms such as center of mass tracking, can give subpixel resolution. Due to the high magnification used in optical tweezers, this can correspond to a few nanometers and, most importantly, can easily be implemented to track multiple particles at the same time [23, 28].

Video-based techniques, however, are not without their disadvantages. Although effective particle tracking can be performed in two dimensions, extending the technique into 3D can prove challenging. Various postprocessing software packages have been developed to allow for an estimation of the z-position of the trapped particle, with techniques ranging from the approximate to the empirical approach. Approximation of the z-position can be performed by monitoring the brightness of the center of the particle [29], while more empirical approaches can involve the comparison between images of the trapped particle with previously taken images of the particle at known depths [30]. Changes to the microscope imaging system can help improve 3D tracking, with recent work using stereoscopic imaging proving to be very promising. Here, two images of the sample are produced and, by tracking the particle in each image in 2D, 3D positions for the particle can be determined. Such systems are able to track particles with nanometer accuracy over a 10 μm axial range [31].

High frame rate cameras are also required for video-based particle tracking, with higher frame rates giving higher temporal resolution. Burst frame rates of tens of kilohertz are achievable but are limited by computer memory (although this problem fades over time with improved on-board camera data handling and faster and more powerful computers using high bandwidth connections), leading to small sample times and the added issue of data management (data management is one of the big challenges in modern high-resolution microscopy). The main limitation to video-based tracking, however, is that it is restricted by the number of photons that can be detected by the camera. Higher frame rates mean shorter exposures, which implies higher illumination levels are needed. Therefore, as frame rate increases, spatial resolution decreases [24].

Interferometric particle tracking techniques, notably back-focal-plane interferometry (BFPI), tend to be more popular than video-based techniques due to their straightforward implementation and high spatial (<1 nm) and temporal (kHz) resolution [32]. Here, a position sensing photodetector, typically a quadrant photodiode (QPD), is placed at the back focal plane of the microscope's condenser lens. Light, having passed through the sample, is collected by the condenser and imaged on to the QPD, which produces voltage signals that are dependent on the intensity of the illuminating light. A QPD consists of four separate photodetectors in a 2×2 array which each produce a voltage dependent on their illumination. Each of these four voltages can be combined in different ways in order to give an x, y, and z position of the trapped object, Figure 7.5. As the trapped object is

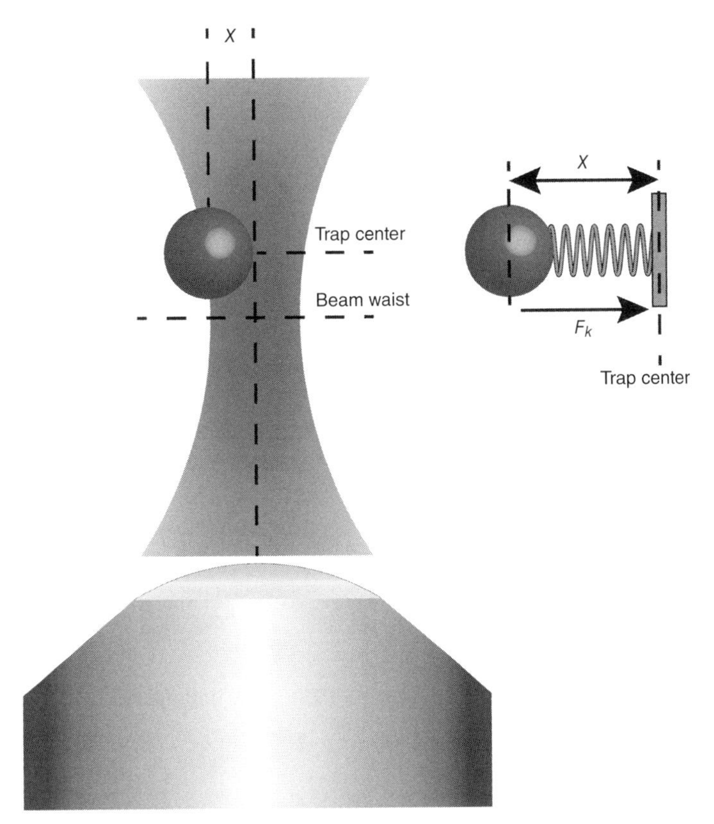

Figure 7.4 Close up view of a displaced bead in an optical trap. The laser in this case is traveling from the bottom to the top of the page, with the optical trap forming just downstream of the beam waist. The dielectric particle, displaced by a distance x from the trap centre, will feel a restoring force pulling it back to the trap centre. This restoring force is linear with respect to distance. Therefore, the trapped particle can be thought of as being attached to a spring with spring constant k.

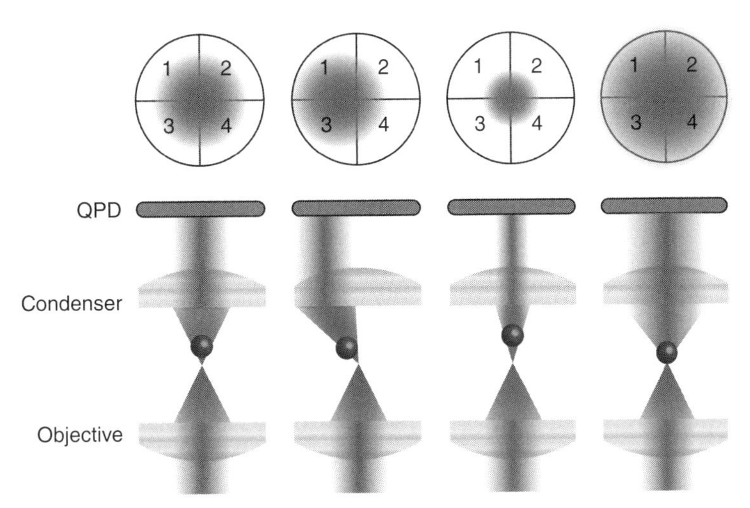

Figure 7.5 QPD placed in the back focal plane of the microscope condenser. In this case, voltage signals are given as $V_x=(V_1+V_3)-(V_2+V_4)$, $V_y=(V_1+V_2)-(V_3+V_4)$, $V_z=V_1+V_2+V_3+V_4$. From left to right: a particle sitting in the centre of the trap gives equal illumination on all four quadrants; a particle displaced from the trap centre will cause one side of the detector to collect more light than the other side; an increase in particle height will cause a focussing of the light on the detector; a decrease in the height causes an increase in the divergence of the light, resulting in a decreased signal.

displaced laterally from the center of the trap, the beam on the QPD is also displaced laterally by a proportional amount. As one side of the QPD is now collecting more light than the other, the x and y voltage signals, therefore, must change, Figure 7.5. Any axial change in the trapped object causes a change in the divergence of the scattered light. Hence, there is a corresponding change in the size of the beam on the detector. The z voltage signal is proportional to overall intensity on the detector. Consequently, when the particle focuses the beam on the detector, the voltage signal increases, and when the divergence of the beam is increased by the particle, the voltage signal decreases, Figure 7.5. By monitoring the change in the voltage signal, the change in the collected light can be measured and, therefore, the change in particle position can be inferred.

This sensitivity, particularly for axial position detection, can be attributed to the fact that the light collected by the detector is not a simple Gaussian distribution. What the detector sees is an interference pattern produced from light that travels through the sample unaffected and light that is scattered by the trapped object. The field that is produced at the back-focal-plane of the condenser is, essentially, the Fourier transform of the field in the sample plane and offers some advantages in position detection. Notably, any change in the object's position within the trap produces a phase shift in the field at the back-focal-plane of the condenser. While this does not provide any information regarding the position of the trap within the sample, it does give information about the relative displacement of the trapped particle from the laser beam, which is precisely the information required for force measurements [32].

Although being straightforward to implement, BFPI systems do present their own challenges: highly sensitive displacement measurements mean that the detector must be accurately aligned, and careful calibration is required to find the relationship between detector voltage and particle displacement. Calibration curves are often calculated by, first, fixing a latex bead to a glass microscope slide. The bead is then scanned, using a piezoelectric stage, back and forth, in a triangular wave fashion, through the beam, producing similar voltage traces to that shown in Figure 7.6. Calibration curves that are produced in this way have an approximately linear regime between the voltage maximum and minimum. By calculating the slope of this linear region (in volts/second) and combining with the known stage velocity, a calibration constant β can be calculated in volts/nm [33]. Calculation of the calibration curve is non-trivial as care must be taken to ensure that, while calibrating one axis, the signal from the other axes is minimized. The calibration constant can also change depending on where in the beam the stuck bead was scanned. Scanning the bead slightly higher or lower than where the trap center is will result in a different calibration constant. In addition, detector response is a function of bead size, so

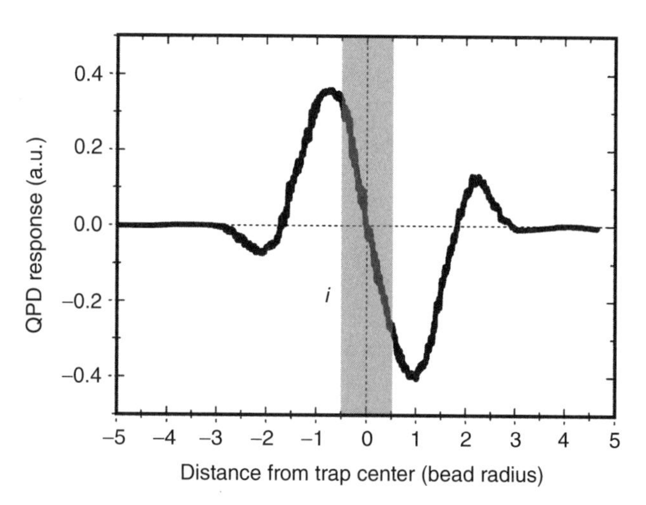

Figure 7.6 QPD response, in arbitrary units, as a function of particle distance from the trap centre, in bead radii. The graph shows the typical shape for a calibration curve for a QPD. The gray shaded region highlights the linear region of the graph. From the slope of the linear region, the calibration constant can be calculated.

must be calculated separately for objects of a different size.

A convenient, but indirect, method of calculating the calibration constant of the detector has been proposed by Schmidt *et al.* [33], which negates the need to scan a bead through the beam focus. By studying the positional fluctuations of an optically trapped particle, they developed a technique that calculates the detector's calibration constant to within 20% of the true value. This allows for a quick check of the system and can serve as a calibration method where precision is not a concern or where labs cannot get access to the expensive piezoelectric stages required for the active calibration method.

Trap Stiffness Measurement The second part of the puzzle when trying to measure a force using optical tweezers is determination of the trap stiffness, k. Several methods can be used to do so, all of which have their relative pros and cons, but all, in essence, follow the same underlying principle: apply known forces to a trapped object and then determine their effect. Trap stiffness can then be determined when both the forces and displacement produced are known. The many available methods can, in general, be classified into two main groups, which are defined by the origin of the applied external force: hydrodynamic or thermal [32].

Hydrodynamic methods, such as the Stokes Drag Force method, measure the displacement of the particle from the center of the trap as a function of drag force on the particle. By moving a trapped particle of radius R, at a controlled velocity v, through a sample liquid with dynamic viscosity η, the force acting on the sphere is given as

$$F_{\text{stokes}} = -6\pi\eta Rv. \tag{7.4}$$

This can be achieved either by movement of a motorized stage or flowing liquid through microfluidic devices. If the latter technique is used, care must be taken to compensate for the parabolic flow profile of the liquid through the channel, as flow velocity will be strongly dependent on channel depth. By measuring the displacement from the trap center of the particle and combining Equations 7.1 and 7.4, the trap stiffness can be determined. Although this may be the most direct method of trap stiffness measurement, the drag force technique is less reliable than techniques based on thermal motion of the particle.

Thermal motion–based techniques rely on the study of the Brownian motion of the trapped object and, therefore, do not require the application of an external force. There are two main methods based on thermal motion analysis: the equipartition method [34], which will give a fast estimation of the trap stiffness, and the power spectrum analysis method [35], which will give the most accurate and reliable trap stiffness measurement [32]. In order to record the Brownian motion of the trapped particle, a detector with a high bandwidth must be used. While possible with video-based particle tracking, it is here that the previously mentioned BFPI technique becomes invaluable. QPD-based systems have been reported, which have bandwidths of up to 1 MHz [36], making them ideal for the detailed study of Brownian dynamics.

For a particle bound in a harmonic potential, such as a bead trapped in by an optical tweezers, the equipartition theorem of energy is given as

$$\frac{1}{2}k_BT = \frac{1}{2}k\langle x^2\rangle, \tag{7.5}$$

where k_B is Boltzmann's constant and T is the temperature. By measuring the particle position fluctuations using a suitable position measuring technique, trap stiffness can be measured with no prior knowledge of the medium viscosity. However, the position detection system must be well calibrated in order to provide accurate positions. This technique is also highly susceptible to noise, which will increase the average squared displacement.

The power spectrum method involves recording the relative displacement of the particle from the trap center as a function of time. In the time domain, this data will have a Gaussian probability distribution, which is characteristic of the Brownian motion of a particle in a harmonic potential [35]. However, by computing the single-sided power spectrum of the position fluctuations – the Fourier transform of the position data – the characteristic shape of a Lorentzian function is obtained, Figure 7.7, and can be described by Equation 7.6. For a particle with diameter d, trapped with stiffness k, in a medium with viscosity η, the power spectrum, in units of distance^2/frequency, is

$$S(f) = \frac{S_0 f_c^2}{f_c^2 + f^2}, \tag{7.6}$$

where f is the frequency, $S_0 = 4\gamma k_B T/k^2$ is the zero-frequency intercept of $S(f)$, $f_c = k/2\pi\gamma$ is the corner frequency of the spectrum and $\gamma = 3\pi\eta d$ is the Stokes drag coefficient of the bead [11]. If the detector is calibrated, power spectrum units are given as nm^2/Hz, while an uncalibrated power spectrum, often the case where the data has been collected using BFPI, has units of V^2/Hz.

By performing a Lorentzian fit to the function, the corner frequency of the spectrum can be obtained and, therefore, the trap stiffness can be calculated. Moreover, this can be obtained independent of detector calibration. If the detector

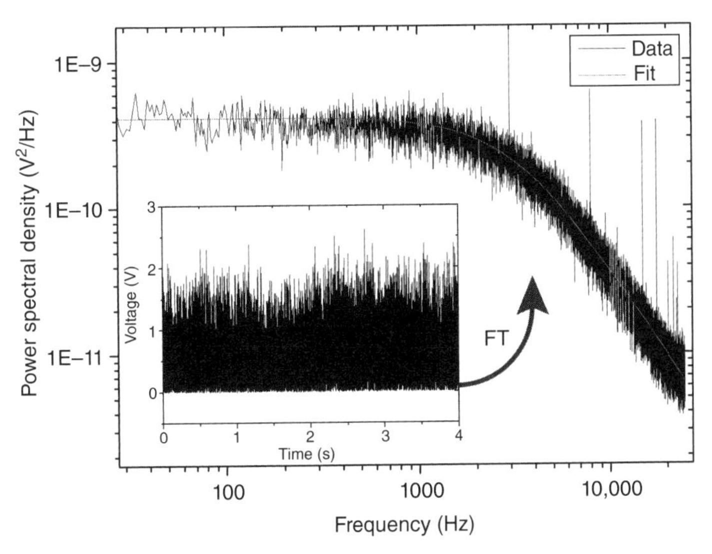

Figure 7.7 Power spectrum of a 1 μm bead trapped in water. Power spectrum taken from an uncalibrated QPD, hence units are in V^2/Hz rather than distance^2/Hz. Data, shown in black, shows that there is some high frequency noise present in the system. The Lorentzian fit, solid line, has been applied to the data and gives a corner frequency of ∼3200 Hz, which corresponds to a trap stiffness of ∼200 pN/μm.

has been calibrated, then viscous damping can be extracted from the power spectrum and, given a known particle radius, the viscosity of the surrounding medium can be determined. The corner frequency also serves to divide the Brownian motion into two distinct regimes. For $f \ll f_c$, the power spectrum exhibits a constant plateau, S_0, which reflects the confinement of the trapped particle. At $f \ll f_c$, the power spectrum falls off as S_0, which is representative of free diffusion and can be attributed to the sample moving as if the trap does not exist at short observation times [32].

Although the power spectrum method can prove to be the most accurate method for determining trap stiffness, in order to achieve high accuracy a wide variety of different effects must be considered and accounted for. For example, the force experienced by the trapped particle is dependent on trapping depth due to hydrodynamic interactions between the particle and the microscope coverslip, and so Faxén's correction must be applied. In addition, the position detection system can act as a low-pass filter and can, therefore, lead to an underestimate of the corner frequency of the power spectrum. This results in a measured trap stiffness that is lower than expected. Fortunately, Berg-Sørensen and Flyvbjerg have extensively studied the acquisition and accurate fitting of the Lorentzian for power spectrum analysis of optical tweezers [35]. In this paper they thoroughly analyze, far beyond the scope of this chapter, the theory behind power spectrum analysis, design constraints, and undesirable effects that can arise. Fortunately, they have also developed a freely available MATLAB program that can perform much of the data analysis required for power spectrum analysis [37].

7.3.3 Radiation Pressure Traps

Radiation pressure trap is a bit of a misnomer in most instances, as the ability to trap particles using the scattering force usually makes use of the gradient force as well. One of the clearest examples of this is in the dual-beam trap [38], which consists of a couple of identical, counter-propagating, divergent laser beams with Gaussian intensity profiles. Such traps can consist of more than two beams, resulting in a quad trap, for example [39]. In this configuration, the scattering forces balance each other in the middle of the trap because the two-beam trap geometry is symmetric, whereas the gradient forces pull particles toward the beam axis in order to provide a stable trap on the beam axis.

One of the major advantages of such a trap is that it can be formed making use of optical fibers as the beam delivery mechanism [38] and, as such, this makes it ideal for integration into microfluidic devices [40]. Clearly, it lacks the sophistication in most instances of optical tweezers, as it is difficult to control the beam profile through a multimode fiber, although recent advances have made progress in this area [41]. This means that such traps are typically limited to single particles. New techniques have demonstrated sophisticated control over trapped particles, such as rotation [42], which offers the option of using these traps for tomographic studies; integration with other imaging modalities, such as holographic microscopy [43] is also possible.

Although the total force acting on the center of mass is zero, if the trapped colloid is sufficiently elastic, the surface forces stretch the object along the beam axis, proportional to the intensity of the laser beam [44]. This gives rise to the name of this technique: the *optical stretcher*. A schematic is shown in Figure 7.8.

The stretching component of an optical stretcher is evident when we place a soft particle into the trap. One might imagine that such a particle would be squashed, in the same way as a grape would be squashed by fingers applying a force from each side, but in fact due to the way in which the light refracts through the particle it is stretched. This elongation is able to give us information about the mechanical state of, for example, a trapped cell,

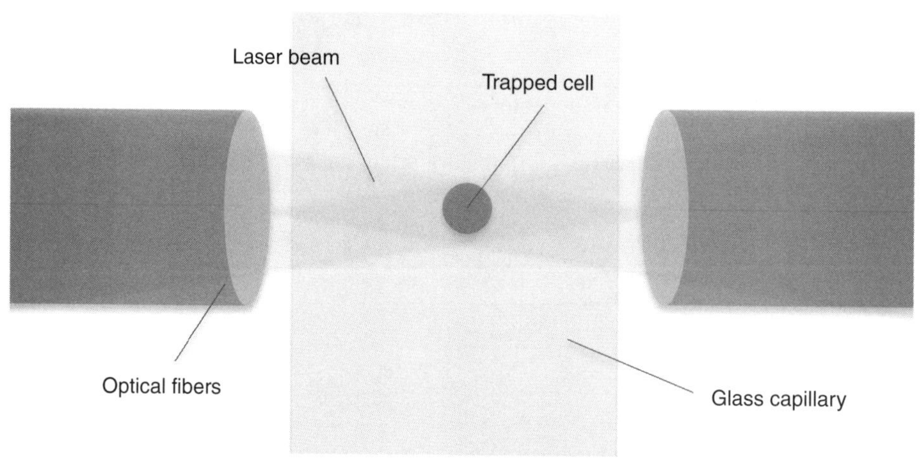

Figure 7.8 An optical stretcher. Two counter propagating laser beams trap a deformable colloidal particle. If enough power is applied then the particle will stretch.

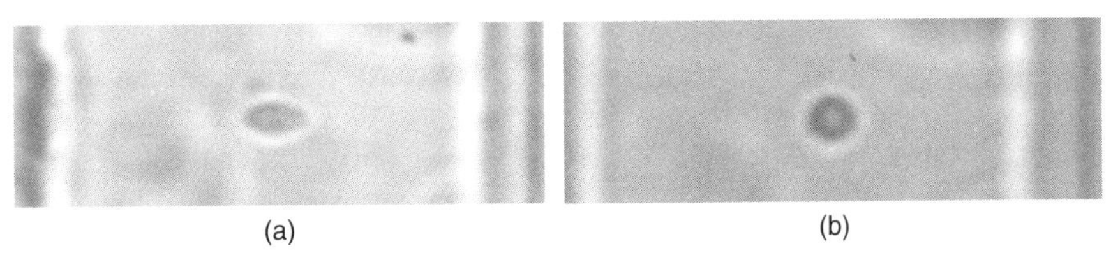

Figure 7.9 A red blood cell being stretched within an optical stretcher. (a) Cell in the stretched state; (b) shows the cell trapped at lower power in it's normal state.

and more specifically, about the cytoskeleton. An example of a stretched cell is shown in Figure 7.9.

When considering cells, we find that structural changes of the cytoskeleton cause changes in the viscoelastic response of the cell. The optical stretcher has been used to perform biomechanical measurements allowing, for instance, to classify normal, cancerous, and metastatic breast epithelial cells as a consequence of their optical deformability [45]. Due to the mechanical changes in cancer cells at different stages of their development, it offers an interesting "mechanical phenotyping" approach to cancer diagnosis in general [46].

Other interesting studies making use of this technology have included the ability to create a tunable laser from an optically stretched emulsion droplet [47], and their use in studying giant unilamellar vesicles demonstrating wider applicability than simply biological colloidal samples.

7.3.4 Beam Shaping Techniques

In addition to the flexibility offered by optical tweezers with regard to their integration with a myriad of microscopic techniques, there is an additional advantage over other micromanipulation technologies: the ability to control the shapes of the beams being used [48]. The power that derives from the use of devices such as spatial light modulators makes beam shaping facile.

Beam shaping typically relies on the ability to define the optical phase of a light beam, the phase being the part of the optical field description that determines how the beam evolves in space. It can also refer to the ability to move the beam in some controllable way as well, making use of fast beam steering mirrors [49] or acousto-optical deflectors (AODs) [50]. These latter mechanisms are powerful provided the beam steering is fast enough to allow the particle diffusion within the sample medium to be ignored. For example, to trap two particles, one simply needs to move the beam from one to the other before the particle has moved.

To give full three-dimensional control over the beam shape (if that is needed), then holographic beam shaping is usually employed. This makes use of a computer-controlled hologram (although static holograms can be used too) to interactively control the imposed phase of a beam.

The computer-controlled hologram is typically a nematic liquid crystal phase modulator, or spatial light modulator. The device is made up of a number of pixels, and each is able to alter the phase of an appropriately polarized incident optical field. This modulation is controlled via the application of a voltage to each pixel, which alters the orientation of the liquid crystals in each pixel. As the nematic crystals are birefringent, this leads to an alteration of the birefringence of the pixel, and hence a modulation of the phase of any light incident on the pixel. Full phase modulation between 0 and 2π is therefore possible with an appropriately designed device. Devices typically have pixel pitches of the order of $10\,\mu m^2$. Array resolutions range from 256×256 pixels up to high-resolution devices of 1920×1280. Typical frame rates are of the order of 60 Hz, but high-speed options are available in the 300 Hz+ range. This compares poorly with the AOD beam steering options, which easily access switching speeds above 1 kHz.

Modulators can also be used to control the amplitude of an incident beam, usually making use of ferroelectric liquid crystals. These can run at rates at the kilohertz level but are usually much less efficient than nematic devices. In situations where complex beams or large numbers of trap sites are needed, then nematic phase modulators are probably the best option. Where low power trapping is possible, ferroelectric devices offer a flexible option [51], and have been successfully employed in droplet microfluidics but are perhaps more suited to very low power applications such as atomic trapping [52].

The beam can be shaped in one of two ways. In the first, the phase of the beam shape that is desired is well known and has an exact solution. An example would be a hologram where the idea is to shift the beam off to one side. Doing so simply requires the application of the phase function for a prism to the SLM, and any incident beam will feel the appropriate spatial shift. Other solutions to the Helmholtz equation such as Laguerre–Gaussian beams [53] or Bessel beam [54] are trivially encoded.

More complex beam patterns have nonexact phase solutions, and so iterative techniques must be employed [55]. There are a myriad of options, and these were traditionally optimized on speed and image reconstruction quality, with speed being a key parameter for interactive solutions (note speed is not a huge issue for exact phase solution implementations in the main). However, with the advent of GPU-driven graphics acceleration, the majority of algorithms can be significantly sped up.

A typical iterative algorithm is the Gerchberg–Saxton [56], which has a range of variants. It makes use of the fact that a lens can carry out a Fourier transform of an input beam using an appropriate $2f$ geometry. It assumes that you know the target intensity that you desire to observe in the trapping plane – the focal plane of the microscope objective. It also assumes that you do not know the phase required to generate the desired intensity. As such one can start with a randomized phase. The algorithm then calculates what happens as the beam hits the hologram and is then propagated through a lens to form a modified intensity image. This is then compared to the desired image and assuming that they do not converge, the field amplitude is modified by removing the produced intensity with the desired one. This new field is then back propagated through the lens (computationally) and the process repeats, now using the newly generated phase and the initial input intensity onto the hologram. This process iterates until an appropriate convergence between the generated and desired hologram is achieved. Modifications of the algorithm include the weighting or mixing of the desired and transformed amplitudes, commonly known as the adaptive-additive algorithm [57].

Trap Control Methods for controlling and steering the laser with devices such as AODs or SLMs naturally lend themselves to computer control. This in turn allows for the development of flexible human–computer interfaces. The idea that optical tweezers are a useful quantitative tool for research, but one that is primarily developed and implemented by physicists, has led to a rise in user-friendly control techniques. These allow "nonexpert users," which is usually a misnomer, to control sophisticated systems in a straightforward way. The simplest form would be a point-and-click design, where the user sets trap location by clicking the location on a computer monitor. However, by turning to the computer and videogames industry, much more exciting forms of control can be developed.

The development and refinement of interface systems for intuitive control of virtual worlds has been a large area of development within the computing industry. Consequently, modern controllers are packed with high-end components that are available at a low cost due to the massive demand. By building on this hard work and exploiting the high-end devices, intuitive control systems for optical traps can be developed. Interfaces that provide force feedback allow users to, essentially, "feel" their microscopic sample [58]. The multitouch ability of Apple's iPad has allowed interfaces to be developed that can control multiple traps in two dimensions, with the "pinch" gesture being used to move particles in the third dimension [59]. Recent years have seen a shift in gaming industry to controllers where the user is the "controller" and optical tweezers have followed suit, with the development of an optical tweezers system, which is controlled with a Microsoft Kinect [60].

While such interfaces are not crucial to optical tweezers, an effective interface is vital to the wider acceptance of a device. This is especially true if optical tweezers are to be further adopted outside of the optics lab. With the many advantages they offer to, for example, colloidal, aerosol, and biological sciences, there is no reason why they should not be.

7.4 APPLICATIONS

Some interesting statistics, a citation search for "optical tweezers" gives around 5500 papers with a total number of citations of around 110,000. The number of papers published in each of the last 3 years is just over 550. If we extend this search to "optical tweezers + colloid" we find only 182 papers with around 5000 citations and about 20 papers per year in the last few years. So the use of the technique within the specific colloid community is clearly a niche one, even though nearly all optical tweezers experiments (unless you are trapping atoms [10]) use colloids of one form or another. As such here we will give a short overview of some of the literature and how one can carry out, for example, microrheology using these techniques. We will also explore the developing fields of nanoparticle manipulation, and aerosol trapping and analysis.

7.4.1 Colloidal Science

The use of optical tweezers in colloidal science is particularly useful for the measurement of interaction forces, as opposed to adhesion, binding, or structural forces involved in biological processes. A classical example of such work is that of Crocker *et al.* [61], in which long-range entropic forces are measured using, in this case, a line scanned optical tweezers. This was achieved with very high spatial resolution and enabled the form of the potential to be mapped out. This type of instrument is now well developed for such pair interaction studies [62]. Similar techniques making use of dual-beam tweezers have also been employed, studying, for example, forces between particles in liquid crystals [63]; techniques employing a single-beam tweezers and a micropipette have examined interaction forces in aqueous salt solutions in situations where the colloids have a specific charge [64], and demonstrate the flexibility of the technique, at least when looking at small numbers of colloids and their interactions. It is also possible to measure colloidal friction coefficients [65] and to explore interaction dynamics, not just in bulk solution but between liquids at an interface [66].

Other recent work has included the use of holographic tweezers [67] to trap nonspherical objects, acting as a deformable wall. This can be used to make measurements of osmotic pressure in highly confined geometries, which has significant application in biological systems. The ability of holographic optical tweezers to be tuned to complex particle shapes and numbers offers great flexibility for studying a much wider range of colloid types [68]. They also offer opportunities for colloidal sorting – particles flow through complex optical pattern and can be sorted via size or refractive index as the interaction with the lattice depends on these properties [69, 70].

Tweezers can also be used to probe the viscoelastic properties of the medium surrounding a trapped colloid [71]. This can be achieved using rotating particles [72], or simply by monitoring the motion of a trapped particle [73]. There is now freely available software that enables anyone to easily analyze such data and extract the viscoelastic properties desired [74, 75]. Much of the work in this area is devoted to the exploration of biological fluids and biopolymers, which are difficult to measure with more conventional techniques and which often have dynamic properties with their surroundings [76].

Another niche area worth mentioning in the context of optical–colloidal interactions is that of optical binding [23]. An optically bound array of particles involves, for example, a linear array of particles (much like an ion chain array) that are spatially separated. The separation occurs through the scattered light from one particle having some influence on the forces applied to a second in close proximity. It has been observed in simple chains [77, 78] using counter-propagating traps giving rise to interesting dynamical effects [79–81], as well as in single beams where the effects occur transverse to the beam [82, 83]. Such optical binding could lead to interesting ways in which to arrange and hold particles in complex forms, possibly in extremely large arrays over extensive distances [84].

7.4.2 Nanoparticles

The challenge with very small particles is their localization; while confined within a trap, they are free to move under Brownian motion, within the relatively large (compared to the particle size) focal region of the laser beam. However, the trapping of nanoparticles was one of the first experiments carried out using tweezers, being demonstrated in Ashkin's original optical tweezers paper [11], where small polystyrene beads with a diameter of 38 nm were studied, albeit briefly. It is always interesting to note how times change – such a demonstration would surely be hived off into a separate paper today. The ability to trap metallic particles provided a big impetus in the field, first demonstrated by Svoboda and Block in 1994 [14]. The use of metallic particles, whose small size gives rise to reduced scattering but with an enhanced polarizability, leads to a sufficient gradient force to enable a three-dimensional trap to occur. Much work has been done on looking at trapping different metallic types and other kinds of nanoparticles such as nanowires (not necessarily metallic) and quantum dots [85]. The bulk of work in this area is focused on metallic particles, primarily due to their other interesting properties related to the ability to excite surface plasmon resonances. This enables some modicum of control over properties such as optical force (and the sign of the force) along with the ability to control heating effects [86]. Heating always remains a significant issue in using metallic particles as sample heating can alter local viscosity [87]. In addition, there are applications in areas such as cell poration, whereby an optical trap can be used to position metallic particles at the cell membrane. This helps puncture the membrane to allow cellular uptake from the surrounding medium [88].

The bigger push at present in trapping nanoparticles comes in engineering the optical traps to deliver smaller trapping volumes, to enable better localization of particles. These typically take the form of metallic nanostructures, which produce high-intensity near-field optical patterns that are able to trap particles, often very close to a surface. These can take forms such as closely spaced nanodots [89], simple flat structures [90], and more complex geometries in the form of nanoantenna [91, 92]. Such traps can produce higher forces than in conventional tweezers, often with far lower power levels, due to the enhanced optical field effects that the plasmons generate. One of the main drawbacks of these techniques is their slight inflexibility – the substrates dictate the location of the trap and there is little scope for three-dimensional movement once the particle is trapped (there may also be issues with ease of integration with other imaging and spectroscopic modalities). This concern has been addressed recently by creating the nanoantenna patterns directly on the end of an optical fiber [93]. This enables coupling to the antenna in a facile way, which can then in turn be translated straightforwardly, allowing, for example, motion of a 50 nm particle to be moved over distances of 10 s of micrometers.

These more sophisticated techniques make use of an idea called self-induced back action [94], whereby the influence of the particle in the presence of the trapping structure alters the field to enhance the trapping strength. These techniques work very well on small dielectric particles, usually demonstrated on polystyrene and silica beads (pushing down toward 10-nm-diameter particles) [95], and are ideal candidates for single molecule studies [96] and possibly for studies on viruses [97].

Optical manipulation can also be used for "printing" nanoparticles, offering an interesting route to patterning metallic nanoparticles on surfaces [98], which may offer new ways to create novel sensors or optoelectronic devices.

7.4.3 Colloidal Aerosols

One of the more interesting developments in optical manipulation studies within the colloidal sphere has been the development of tools that can trap and probe airborne particles [99]. These aerosols are typically liquids but also include solid particles [100], and some techniques can even involve highly absorbing particles [101]. Droplets are typically generated using nebulizers [102] but can also be introduced in more integrated ways using, for example, surface acoustic wave nebulization [103]. The idea that one can trap in air is not new, and some of Arthur Ashkin's original work on radiation pressure trapping looked at airborne particles. Optical levitation was subsequently used to probe one of the very useful properties of such liquid droplets: their whispering gallery mode structure [104] that arises if light is injected into the droplet, which then can act as a time-evolving optical cavity. Levitation was traditionally used to explore areas such as Mie scattering from trapped objects [105] before finding a niche area exploring the dynamics of the droplets that were trapping [106]. Levitation has some advantages in terms of the particles sizes it can trap over optical tweezers, but the systems are larger and the trapping can be less stable. In 1997, Omori *et al.* [107] showed that it was also possible to trap airborne particles in a three-dimensional gradient trap, work which was not picked up in any seriousness until Jonathan Reid's group, at the University of Bristol showed that one could combine the trapping of liquid aerosols with cavity enhanced Raman spectroscopy, and make exceptionally precise measurements of the droplet size and composition. The cavity acts as an enhancement cavity for the Raman signal, which in turn overlays whispering gallery modes on the Raman signal. This allows the typical Raman chemical fingerprinting to take place, and by watching how the whispering gallery modes evolve in time, the droplet evaporation or growth can be followed. This work has now picked up apace, and there are a [104] range of innovations. The use of holographic traps was quickly developed to trap multiple particles [102], which enable droplet coagulation studies to be carried out [108]. The studies of droplet dynamics have become increasingly advanced with the role of relative humidity [109], formation of multiphase droplets [110], surface properties [111], laser heating [112], inorganic salt aerosols [113], glassy aerosol [114], and ionic liquid aerosols [115] all being explored. Studies have also swung back to make use of radiation pressure, especially Bessel beam traps [54], and have demonstrated precise refractive index retrieval [116] and enabled studies of glassy aerosol while new high stability traps for small submicron particles have also been proposed [39]. These tools offer a significant opportunity to study atmospherically relevant aerosols through confinement and the opportunity to study what would otherwise be fast-moving transient particles with relative ease.

There is also a move toward integrated devices for aerosol analysis. These make use of quasi-lab-on-a-chip devices [117] or can also make use of hollow core photonic crystal fiber [118], which offers a convenient mechanism for confining aerosols within a miniaturized sample chamber. The challenge for such devices is, usually, to keep the particles away from the chamber walls, but here the light forces aid the experimenter by helping to pull the particles toward the high intensity light that travels through the center of the devices. These types of devices should offer similar capabilities to microfluidic chips, except operating in air, as well as integration with optofluidic devices.

Interesting physics is also observable in airborne and vacuum traps. In conventional optical tweezers, where there is a viscous damping medium surrounding the trapped particle (typically water) then the particle motion is highly overdamped. However, in air, we can access underdamped dynamics, which allows observation of effects such as parametric resonance [119] and coupled hydrodynamic modes in multiparticle systems [120]. Theoretical considerations of particle dynamics in such traps have also been carried out, with the conclusion that airborne particles exist at the cusp of the region, where gradient forces and radiation pressure forces balance out, in many cases. This is primarily due to enhanced scattering forces related to the higher relative refractive index of particles in air [121].

Moreover, the use of low-pressure system allows other effects to come into play. An intriguing recent example is the use of a birefringent particle made from vaterite that is trapped in a circularly polarized light beam. This then acts to rotate the particle through the transfer of spin angular momentum. The lack of damping forces enable the trapped beads to get to incredibly high rotation rates of around 5 MHz (a claim that is made that this is the fastest man-made rotating object) [122] and ultimately the spheres are cooled down to 40 K through coupling between the particle motional degrees of freedom. This same experimental system can also be used to measure the viscosity of the surrounding medium [123]. Other very basic physics can be explored, with the advantage for many of these applications being that the trapped sphere is completely isolated from its surroundings, so there are no couplings to external decoherence mechanisms. These include cooling mechanisms [124] – with a view to cooling to the quantum ground state [125], cavity optomechanics [126, 127], measurement of the instantaneous velocity of Brownian particles [128] and the opportunity to use such trapped particles as ultraweak force sensors [129]. Indeed, such schemes have been proposed to detect gravitational waves [130] and may offer opportunities for exceptionally high-sensitivity measurements in the future. Such traps may also find themselves launched into space, where proposals and experimental sketches have suggested they could be used to create a new generation of space-borne telescopes [84] capable of resolving extra-solar planets directly.

A final application to consider also makes use of the fact transparent aerosols droplets form a high-Q cavity. This raises the possibility of creating a laser using an appropriately doped droplet [131]. Such optically trapped aerosol lasers offer new opportunities for using the cavities as sensors, including biosensors [132].

It is clear that colloidal interactions in air and in low-pressure systems have a significant amount of mileage in the coming years and offer one of the most exciting avenues for research, if one wishes to push at the technology and visionary science away from important applications in biology.

Photophoresis Another interaction that colloidal systems can have with light is through the process of *photophoresis*. This is a thermally driven mechanism induced by light, although it is also often used as another name for radiation pressure and is, for example, one mechanism that moves interstellar dust though the Universe [133]. It has also, in recent times, been used as an optical manipulation technique, specifically involving absorbing particles, and as such offers a mechanism to manipulate a class of particles that is not routinely open to optical tweezers.

Photophoresis drives particles through the generation of a thermal gradient across the particle due to optical heating of one side of the particle. The impact of molecules from the air surrounding the warmer side imparts a greater momentum than those on the cooler side and hence the particle will move in opposition to the thermal gradient. A group from the Australian National University has pioneered this technique, initially using shaped laser beams in the form of vortex beams [134], and has demonstrated that absorbing particles in the range 0.1–10 μm could be guided at velocities of 1 cm/s. By making use of a counter-propagating geometry, trapping was possible. The types of particle considered here included carbon nanoclusters, and as such these types of traps might be of interest to those looking at atmospheric nucleation particles. Related applications might also lead to new ways to load coherent X-ray analysis devices [135]. The work has been extended to trapping multiple particles in speckle fields [136] and also to very large particles, such as 100 μm carbon clusters [137]. These, in turn, can be precisely delivered to a specified location. Other work by the same group has explored the role of polarization in such optical traps, showing that radial polarization offers better trapping efficiencies, the use of single bottle beams for trapping [138].

Extensions to this work now include the ability to trap absorbing particles with a single focused laser beam [139] and the use of shaped light fields to create objects such as an "optical trampoline" [140] in which ink droplets are able to bounce away from a light sheet as they impinge on it – one could easily envisage using such a methodology to create lenses for absorbing microparticles. More recently, shaped beams formed through the process of conical refraction [141] have been used to store particles in conventional optical tweezers. Such beams form an optical bottle that can be "opened up" in the side by controlling the input polarization. By combining this idea with the above concepts on using bottle beams, it seems clear that complex and flexible modalities are possible for the movement and storage of such particles. Add this to the fact that such absorbing particle traps can then be coupled to the normal range of analytical techniques of an optical tweezers, such as Raman [142] and all the power of optical trapping localization functionality can be brought to bear on a wholly new class of particles.

7.5 CONCLUSIONS

It is interesting to note how optical manipulation techniques have developed over the past 40 years – there is little slowdown in the development of new techniques and improvements, and the ease of integration with other techniques offers a huge amount of flexibility to the researcher. There is little doubt that if a new imagining modality comes out, it will not be too long before a paper appears combining it with optical tweezers – such is the flexibility and opportunity for doing new science through localization. In the field of biophysics (not covered in this brief review), the tools are fairly mature, and a range of commercial systems exists for the nonexpert optical physicist. These systems are also suitable for more general colloidal research, although are not typically designed specifically with this field in mind – but there are increasingly sophisticated techniques being developed and open-source material becoming available. The entry level for a high-end research tool is becoming ever lower. A new system, specifically made for aerosol work, has recently been released by Biral. It is clear that wherever there is the need for the localization of microscopic particles, then optical manipulation has a significant role to play. There are exciting challenges ahead for work in airborne particles, whether one wishes to explore applied or fundamental aspects, while there are great technological challenges to solve to allow us to go smaller and control nanoparticles. Integration of optical manipulation devices with other manipulation technologies also looks to be an interesting avenue to pursue.

REFERENCES

[1] Wong PK, Ulmanella U, Ho CM. Fabrication process of microsurgical tools for single-cell trapping and intracytoplasmic injection. J Microelectromech Syst 2004;13(6):940–946.

[2] Swanson BD, Bacon NJ, Davis EJ, Baker MB. Electrodynamic trapping and manipulation of ice crystals. Q J Roy Meteorol Soc 1999;125:1039–1058.

[3] Zhang C, Khoshmanesh K, Mitchell A, Kalantar-zadeh K. Dielectrophoresis for manipulation of micro/nano particles in microfluidic systems. Anal Bioanal Chem 2009;396(1):401–420.

[4] Lin J, Valentine MT. High-force NdFeB-based magnetic tweezers device optimized for microrheology experiments. Rev Sci Instrum 2012;83(5):053905.

[5] Ainla A, Jeffries G, Jesorka A. Hydrodynamic flow confinement technology in microfluidic perfusion devices. Micromachines 2012;3(4):442–461.

[6] Blehm BH, Schroer TA, Trybus KM, Chemla YR, Selvin PR. In vivo optical trapping indicates kinesin's stall force is reduced by dynein during intracellular transport. Proc Natl Acad Sci U S A 2013;110(9):3381–3386.

[7] Takagi Y, Farrow RE, Billington N, Nagy A, Batters C, Yang Y, Sellers JR, Molloy JE. Myosin-10 produces its power-stroke in two phases and moves processively along a single actin filament under low load. Proc Natl Acad Sci U S A 2014;111(18):E1833–E1842.

[8] Lorch DP, Lesich KA, Lindemann CB, Hunt AJ. Non-processive force generation by mammalian axonemal dynein in situ on doublet microtubules. Cel Mol Bioeng 2013;6(4):431–440.

[9] Ashkin A. Acceleration and trapping of particles by radiation pressure. Phys Rev Lett 1970;24(4):156–159.

[10] Chu S. Nobel lecture: the manipulation of neutral particles. Rev Mod Phys 1998;70(3):685–706.

[11] Ashkin AA, Dziedzic JMJ, Bjorkholm JEJ, Chu SS. Observation of a single-beam gradient force optical trap for dielectric particles. Opt Lett 1986;11(5):288.

[12] Rohrbach A. Switching and measuring a force of 25 femtoNewtons with an optical trap. Opt Express 2005;13(2):9695–9701.

[13] Ashkin A. Forces of a single-beam gradient laser trap on a dielectric sphere in the ray optics regime. Biophys J 1992;61(2):569–582.

[14] Svoboda K, Block SM. Optical trapping of metallic Rayleigh particles. Opt Lett 1994;19(13):930–932.

[15] Huang LL, Guo HH, Li JJ, Ling LL, Feng BB, Li Z-YZ. Optical trapping of gold nanoparticles by cylindrical vector beam. Opt Lett 2012;37(10):1694–1696.

[16] Ohlinger A, Deak A, Lutich A, Feldmann J. Optically trapped gold nanoparticle enables listening at the microscale. Phys Rev Lett 2012;108(1):018101.

[17] Nieminen TA, du Preez-Wilkinson N, Stilgoe AB, Loke VLY, Bui AAM, Rubinsztein-Dunlop H. Optical tweezers: theory and modelling. J Quant Spectrosc Radiat Transfer 2014;146:59–80.

[18] Nieminen TA, Loke VLY, Stilgoe AB, Knöner G, Brańczyk AM, Heckenberg NR, Rubinsztein-Dunlop H. Optical tweezers computational toolbox. J Opt A Pure Appl Opt 2007;9(8):S196–S203.

[19] A. Callegari, M. Mijalkov, A. B. Gököz, and G. Volpe, "Computational toolbox for optical tweezers in geometrical optics," *arXiv.org*, vol. physics.optics. p. 5439, 2014 Feb 21.

[20] "Optical Trapping Lab: Design and Documentation." [Online]. Available: http://www.advancedlab.org/mediawiki/index.php/Design_and_Documentation_(OTZ). Accessed 2014 Jul 13.

[21] "Optical tweezers Control Software." [Online]. Available: http://openwetware.org/wiki/Koch_Lab:Publications/Drafts/Versatile_Feedback. Accessed 2014 Jul 13.

[22] "Optical tweezers software, University of Glasgow." [Online]. Available at http://www.gla.ac.uk/schools/physics/research/groups/optics/research/opticaltweezers/software/. Accessed 2014 Jul 13.

[23] Bowman RW, Padgett MJ. Optical trapping and binding. Rep Prog Phys 2013;76(2):026401.

[24] Neuman KC, Block SM. Optical trapping. Rev Sci Instrum 2004;75(9):2787.

[25] Lee WM, Reece PJ, Marchington RF, Metzger NK, Dholakia K. Construction and calibration of an optical trap on a fluorescence optical microscope. Nat Protoc 2007;2(12):3226–3238.

[26] Goodman JW. *Introduction to Fourier Optics*. Roberts & Company Publishers; 2005.

[27] Quercioli F, Tiribilli B, Mannoni A, Acciai S. Optomechanics with LEGO. Appl Opt 1998;37(16):3408–3416.

[28] Otto O, Gutsche C, Kremer F, Keyser UF. Optical tweezers with 2.5 kHz bandwidth video detection for single-colloid electrophoresis. Rev Sci Instrum 2008;79(2):023710.

[29] Crocker JC, Grier DG. Methods of digital video microscopy for colloidal studies. J Colloid Interface Sci 1996;179(1):298–310.

[30] Zhang Z, Menq C-H. Three-dimensional particle tracking with subnanometer resolution using off-focus images. Opt Pur Appl 2008;47(1):2361–2370.

[31] Bowman R, Preece D, Gibson G, Padgett M. Stereoscopic particle tracking for 3D touch, vision and closed-loop control in optical tweezers. J Opt 2011;13(4):044003.

[32] Verdeny I, Farré A, Soler JM, Lopez-Quesada C, Martin-Badosa E, Montes-Usategui M. Optical trapping: a review of essential concepts. Opt Pur Appl 2011;44(3):527–551.

[33] Allersma M, Gittes F, deCastro M, Stewart R, Schmidt C. Two-dimensional tracking of ncd motility by back focal plane interferometry. Biophys J 1998;74(2):1074–1085.

[34] Florin EL, Pralle A, Stelzer E, Hörber J. Photonic force microscope calibration by thermal noise analysis. Appl Phys Mater Sci Process 1998;66(7):S75–S78.

[35] Berg-Sørensen K, Flyvbjerg H. Power spectrum analysis for optical tweezers. Rev Sci Instrum 2004;75(3):594.

[36] Rohrbach A, Tischer C, Neumayer D, Florin E-L, Stelzer EHK. Trapping and tracking a local probe with a photonic force microscope. Rev Sci Instrum 2004;75(6):2197.

[37] Hansen PM, Tolić-Nørrelykke IM, Flyvbjerg H, Berg-Sørensen K. tweezercalib 2.1: faster version of MatLab package for precise calibration of optical tweezers. Comput Phys Commun 2006;174(6):518–520.

[38] Constable A, Kim J, Mervis J, Zarinetchi F, Prentiss M. Demonstration of a fiber-optical light-force trap. Opt Lett 1993;18(21):1867–1869.

[39] Thanopulos I, Luckhaus D, Preston TC, Signorell R. Dynamics of submicron aerosol droplets in a robust optical

trap formed by multiple Bessel beams. J Appl Phys 2014;115(15):154304.

[40] Lincoln B, Schinkinger S, Travis K, Wottawah F, Ebert S, Sauer F, Guck J. Reconfigurable microfluidic integration of a dual-beam laser trap with biomedical applications. Biomed Microdev 2007;9(5):703–710.

[41] Plöschner M, Straka B, Dholakia K, Čižmár T. GPU accelerated toolbox for real-time beam-shaping in multimode fibres. Opt Exp 2014;22(3):2933–2947.

[42] Black BJ, Mohanty SK. Fiber-optic spanner. Opt Lett 2012;37(24):5030–5032.

[43] Lee S-H, Grier DG. Holographic microscopy of holographically trapped three-dimensional structures. Opt Exp 2007;15(4):1505–1512.

[44] Guck J, Ananthakrishnan R, Mahmood H, Moon T, Cunningham C, Kas J. The optical stretcher: a novel laser tool to micromanipulate cells. Biophys J 2001;81(2):767–784.

[45] Guck J, Schinkinger S, Lincoln B, Wottawah F, Ebert S, Romeyke M, Lenz D, Erickson HM, Ananthakrishnan R, Mitchell D. Optical deformability as an inherent cell marker for testing malignant transformation and metastatic competence. Biophys J 2005;88(5):3689–3698.

[46] Runge J, Reichert TE, Fritsch A, Kas J, Bertolini J, Remmerbach TW. Evaluation of single-cell biomechanics as potential marker for oral squamous cell carcinomas: a pilot study. Oral Dis 2013;20(3):e120–e127.

[47] Aas M, Jonáš A, Kiraz A, Brzobohatý O, Ježek J, Pilát Z, Zemánek P. Spectral tuning of lasing emission from optofluidic droplet microlasers using optical stretching. Opt Exp 2013;21(18):21380–21394.

[48] Woerdemann M, Alpmann C, Esseling M, Denz C. Advanced optical trapping by complex beam shaping. Laser Photon Rev 2013;7(6):839–854.

[49] Speidel M, Friedrich L, Rohrbach A. Interferometric 3D tracking of several particles in a scanning laser focus. Opt Exp 2009;17(2):1003–1015.

[50] Brouhard GJ, Schek HTI, Hunt AJ. Advanced optical tweezers for the study of cellular and molecular biomechanics. IEEE Trans Biomed Eng 2003;50(1):121–125.

[51] Lanigan PMP, Munro I, Grace EJ, Casey DR, Phillips J, Klug DR, Ces O, Neil MAA. Dynamical hologram generation for high speed optical trapping of smart droplet microtools. Biomed Opt Exp 2012;3(7):1609.

[52] Boyer V, Chandrashekar CM, Foot CJ, Laczik ZJ. Dynamic optical trap generation using FLC SLMs for the manipulation of cold atoms. J Modern Opt 2004;51(14):2235–2240.

[53] Simpson NB, McGloin D, Dholakia K, Allen L, Padgett MJ. Optical tweezers with increased axial trapping efficiency. J Modern Opt 1998;45(9):1943–1949.

[54] McGloin D, Dholakia K. Bessel beams: diffraction in a new light. Contemp Phys 2005;46(1):15–28.

[55] Di Leonardo R, Ianni F, Ruocco G. Computer generation of optimal holograms for optical trap arrays. Opt Exp 2007;15(4):1913–1922.

[56] Sinclair G, Leach J, Jordan P, Gibson G, Yao E, Laczik Z, Padgett M, Courtial J. Interactive application in holographic optical tweezers of a multi-plane Gerchberg–Saxton algorithm for three-dimensional light shaping. Opt Exp 2004;12(8):1665–1670.

[57] Dufresne ER, Spalding GC, Dearing MT, Sheets SA, Grier DG. Computer-generated holographic optical tweezer arrays. Rev Sci Instrum 2001;72(3):1810.

[58] Pacoret C, Bowman R, Gibson G, Haliyo S, Carberry DM, Bergander A, Regnier S, Padgett M. Touching the microworld with force-feedback optical tweezers. Optics 2009;17(12):10259–10264.

[59] Bowman RW, Gibson G, Carberry D, Picco L, Miles M, Padgett MJ. iTweezers: optical micromanipulation controlled by an Apple iPad. J Opt 2011;13(4):044002.

[60] McDonald C, McPherson M, McDougall C, McGloin D. HoloHands: games console interface for controlling holographic optical manipulation. J Opt 2013;15(3):035708.

[61] Crocker JC, Matteo JA, Dinsmore AD, Yodh AG. Entropic attraction and repulsion in binary colloids probed with a line optical tweezer. Phys Rev Lett 1999;82(2):4352–4355.

[62] Rogers WB, Crocker JC. A tunable line optical tweezers instrument with nanometer spatial resolution. Rev Sci Instrum 2014;85(4):043704.

[63] Yada M, Yamamoto J, Yokoyama H. Direct observation of anisotropic interparticle forces in nematic colloids with optical tweezers. Phys Rev Lett 2004;92(18):185501.

[64] Gutsche C, Keyser U, Kegler K, Kremer F, Linse P. Forces between single pairs of charged colloids in aqueous salt solutions. Phys Rev E 2007;76(3):031403.

[65] Henderson S, Mitchell S, Bartlett P. Direct measurements of colloidal friction coefficients. Phys Rev E 2001;64(6):061403.

[66] Park BJ, Furst EM. Attractive interactions between colloids at the oil–water interface. Soft Matter 2011;7(17):7676–7682.

[67] Williams I, Oğuz EC, Bartlett P, Löwen H, Royall CP. Direct measurement of osmotic pressure via adaptive confinement of quasi hard disc colloids. Nat Commun 2013;4:2555.

[68] Lapointe CP, Mason TG, Smalyukh II. Towards total photonic control of complex-shaped colloids by vortex beams. Opt Exp 2011;19(19):18182–18189.

[69] MacDonald MP, Spalding GC, Dholakia K. Microfluidic sorting in an optical lattice. Nature 2003;426(6):421–424.

[70] Ladavac K, Kasza K, Grier DG. Sorting mesoscopic objects with periodic potential landscapes: optical fractionation. Phys Rev E 2004;70(1):10901.

[71] Puertas AM, Voigtmann T. Microrheology of colloidal systems. J Phys Condens Matter 2014;26(2):3101.

[72] Bennett JS, Gibson LJ, Kelly RM, Brousse E, Baudisch B, Preece D, Nieminen TA, Nicholson T, Heckenberg NR, Rubinsztein-Dunlop H. Spatially-resolved rotational microrheology with an optically-trapped sphere. Sci Rep 2013;3:1759.

[73] Watts F, Tan LE, Wilson CG, Girkin JM, Tassieri M, Wright AJ. Investigating the micro-rheology of the vitreous humor using an optically trapped local probe. J Opt 2013;16(1):015301.

[74] Tassieri M, Evans RML, Warren RL, Bailey NJ, Cooper JM. Microrheology with optical tweezers: data analysis. New J Phys 2012;14(11):115032.

[75] Tassieri M, Evans RML, Warren RL, Bailey NJ, Cooper JM. Microrheology with optical tweezers: data analysis. New J Phys 2012;14(11):115032.

[76] Shayegan M, Forde NR. Microrheological characterization of collagen systems: from molecular solutions to fibrillar gels. PLoS One 2013;8(8):e70590.

[77] Tatarkova SA, Carruthers AE, Dholakia K. One-dimensional optically bound arrays of microscopic particles. Phys Rev Lett 2002;89(28):283901.

[78] McGloin D, Carruthers AE, Dholakia K, Wright EM. Optically bound microscopic particles in one dimension. Phys Rev E 2004;69(2):021403.

[79] Metzger NK, Dholakia K, Wright EM. Observation of bistability and hysteresis in optical binding of two dielectric spheres. Phys Rev Lett 2006;96(6):068102.

[80] Metzger NK, Marchington RF, Mazilu M, Smith RL, Dholakia K, Wright EM. Measurement of the restoring forces acting on two optically bound particles from normal mode correlations. Phys Rev Lett 2007;98(6):068102.

[81] Taylor JM, Wong LY, Bain CD, Love GD. Emergent properties in optically bound matter. Opt Exp 2008;16(10):6921.

[82] Burns MM, Fournier J-M, Golovchenko JA. Optical matter: crystallization and binding in intense optical fields. Science 1990;249(4):749–754.

[83] Yan Z, Shah RA, Chado G, Gray SK, Pelton M, Scherer NF. Guiding spatial arrangements of silver nanoparticles by optical binding interactions in shaped light fields. ACS Nano 2013;7(2):1790–1802.

[84] Grzegorczyk TM, Rohner J, Fournier J-M. Optical mirror from laser-trapped mesoscopic particles. Phys Rev Lett 2014;112(2):023902.

[85] Bendix PM, Jauffred L, Norregaard K, Oddershede LB. Optical trapping of nanoparticles and quantum dots. IEEE J Select Topics Quantum Electron 2014;20(3):15–26.

[86] Seol Y, Carpenter AE, Perkins TT. Gold nanoparticles: enhanced optical trapping and sensitivity coupled with significant heating. Opt Lett 2006;31(16):2429.

[87] Lehmuskero A, Ogier R, Gschneidtner T, Johansson P, Käll M. Ultrafast spinning of gold nanoparticles in water using circularly polarized light. Nano Lett 2013;13(7):3129–3134.

[88] Arita Y, Ploschner M, Antkowiak M, Gunn-Moore F, Dholakia K. Laser-induced breakdown of an optically trapped gold nanoparticle for single cell transfection. Opt Lett 2013;38(17):3402.

[89] Grigorenko AN, Roberts NW, Dickinson MR, Zhang Y. Nanometric optical tweezers based on nanostructured substrates. Nat Photon 2008;2(6):365–370.

[90] Huang L, Maerkl SJ, Martin OJ. Integration of plasmonic trapping in a microfluidic environment. Opt Exp 2009;17(8):6018–6024.

[91] Roxworthy BJ, Ko KD, Kumar A, Fung KH, Chow EKC, Liu GL, Fang NX, Toussaint KC Jr. Application of plasmonic bowtie nanoantenna arrays for optical trapping, stacking, and sorting. Nano Lett 2012;12(2):796–801.

[92] Kang J-H, Kim K, Ee H-S, Lee Y-H, Yoon T-Y, Seo M-K, Park H-G. Low-power nano-optical vortex trapping via plasmonic diabolo nanoantennas. Nat Commun 2011;2:582.

[93] Berthelot J, Aćimović SS, Juan ML, Kreuzer MP, Renger J, Quidant R. Three-dimensional manipulation with scanning near-field optical nanotweezers. Nat Nanotech 2014;9(4):295–299.

[94] Juan ML, Gordon R, Pang Y, Eftekhari F, Quidant R. Self-induced back-action optical trapping of dielectric nanoparticles. Nat Phys 2009;5(12):915–919.

[95] Kotnala A, Gordon R. Quantification of high-efficiency trapping of nanoparticles in a double nanohole optical tweezer. Nano Lett 2014;14(2):853–856.

[96] Pang Y, Gordon R. Optical trapping of a single protein. Nano Lett 2012;12(1):402–406.

[97] Kotnala A, DePaoli D, Gordon R. Sensing nanoparticles using a double nanohole optical trap. Lab Chip 2013;13(20):4142.

[98] Do J, Fedoruk M, Jäckel F, Feldmann J. Two-color laser printing of individual gold nanorods. Nano Lett 2013;13(9):4164–4168.

[99] Mcgloin D, Burnham DR, Summers MD, Rudd D, Dewar N, Anand S. Optical manipulation of airborne particles: techniques and applications. Faraday Discuss 2007;137:335.

[100] Summers MD, Burnham DR, McGloin D. Trapping solid aerosols with optical tweezers: a comparison between gas and liquid phase optical traps. Opt Exp 2008;16(11):7739–7747.

[101] Pan Y-L, Hill SC, Coleman M. Photophoretic trapping of absorbing particles in air and measurement of their single-particle Raman spectra. Opt Exp 2012;20(5):5325–5334.

[102] Burnham DR, McGloin D. Holographic optical trapping of aerosol droplets. Opt Exp 2006;14(9):4175–4181.

[103] Anand S, Nylk J, Neale SL, Dodds C, Grant S, Ismail MH, Reboud J, Cooper JM, McGloin D. Aerosol droplet optical trap loading using surface acoustic wave nebulization. Opt Exp 2013;21(25):30148.

[104] Ashkin A, Dziedzic JM. Observation of resonances in the radiation pressure on dielectric spheres. Phys Rev Lett 1977;38(2):1351–1354.

[105] Trunk M, Lübben JF, Popp J, Schrader B. Investigation of a phase transition in a single optically levitated microdroplet by Raman–Mie scattering. Appl Opt 1997;36(15): 3305.

[106] Musick J, Popp J, Trunk M, Kiefer W. Investigations of radical polymerization and copolymerization reactions in optically levitated microdroplets by simultaneous Raman spectroscopy, Mie scattering, and radiation pressure measurements. Appl Spectrosc 1998;52(5):692–701.

[107] Omori R, Kobayashi T, Suzuki A. Observation of a single-beam gradient-force optical trap for dielectric particles in air. Opt Lett 1997;22(11):816.

[108] Butler JR, Wills JB, Mitchem L, Burnham DR, Mcgloin D, Reid JP. Spectroscopic characterisation and manipulation of arrays of sub-picolitre aerosol droplets. Lab Chip 2009;9(4):521.

[109] Walker JS, Wills JB, Reid JP, Wang L, Topping DO, Butler JR, Zhang Y-H. Direct comparison of the hygroscopic properties of ammonium sulfate and sodium chloride aerosol at relative humidities approaching saturation. J Phys Chem A 2010;114(48):12682–12691.

[110] Kwamena NOA, Buajarern J, Reid JP. Equilibrium morphology of mixed organic/inorganic/aqueous aerosol droplets: investigating the effect of relative humidity and surfactants. J Phys Chem A 2010;114(18):5787–5795.

[111] Davies JF, Miles REH, Haddrell AE, Reid JP. Influence of organic films on the evaporation and condensation of water in aerosol. Proc Nat Acad Sci U S A 2013;110(22): 8807–8812.

[112] Hunt OR, Ward AD, King MD. Laser heating of sulfuric acid droplets held in air by laser Raman tweezers. RSC Adv 2013;3(42):19448.

[113] Meresman H, Hudson AJ, Reid JP. Spectroscopic characterization of aqueous microdroplets containing inorganic salts. Analyst 2011;136(17):3487.

[114] Tong HJ, Reid JP, Bones DL, Luo BP, Krieger UK. Measurements of the timescales for the mass transfer of water in glassy aerosol at low relative humidity and ambient temperature. Atmos Chem Phys 2011;11(10):4739–4754.

[115] Moore LJ, Summers MD, Ritchie GAD. Optical trapping and spectroscopic characterisation of ionic liquid solutions. Phys Chem Chem Phys 2013;15(32):13489.

[116] Mason BJ, Walker JS, Reid JP, Orr-Ewing AJ. Deviations from plane-wave Mie scattering and precise retrieval of refractive index for a single spherical particle in an optical cavity. J Phys Chem A 2014;118(11):2083–2088.

[117] Horstmann M, Probst K, Fallnich C. An integrated fiber-based optical trap for single airborne particles. Appl Phys B 2011;103(1):35–39.

[118] Schmidt O, Garbos M, Euser T, Russell P. Reconfigurable optothermal microparticle trap in air-filled hollow-core photonic crystal fiber. Phys Rev Lett 2012;109(2): 024502.

[119] Di Leonardo R, Ruocco G, Leach J, Padgett M, Wright A, Girkin J, Burnham D, McGloin D. Parametric resonance of optically trapped aerosols. Phys Rev Lett 2007;99(1):010601.

[120] Yao AM, Keen SAJ, Burnham DR, Leach J, Leonardo RD, McGloin D, Padgett MJ. Underdamped modes in a hydrodynamically coupled microparticle system. New J Phys 2009;11(5):053007.

[121] Burnham DR, Mcgloin D. Modeling of optical traps for aerosols. Journal of the Optical Society of America B 2011;29(12):2856–2864.

[122] Arita Y, Mazilu M, Dholakia K. Laser-induced rotation and cooling of a trapped microgyroscope in vacuum. Nat Commun 2013;4:2374.

[123] Arita Y, McKinley AW, Mazilu M, Rubinsztein-Dunlop H, Dholakia K. Picoliter rheology of gaseous media using a rotating optically trapped birefringent microparticle. Anal Chem 2011;83(23):8855–8858.

[124] Barker PF. Doppler cooling a microsphere. Phys Rev Lett 2010;105(7):073002.

[125] Chang DE, Regal CA, Papp SB, Wilson DJ, Ye J, Painter O, Kimble HJ, Zoller P. Cavity opto-mechanics using an optically levitated nanosphere. Proc Natl Acad Sci U S A 2010;107(3):1005–1010.

[126] Lechner W, Habraken SJM, Kiesel N, Aspelmeyer M, Zoller P. Cavity optomechanics of levitated nanodumbbells: nonequilibrium phases and self-assembly. Phys Rev Lett 2013;110(14):143604.

[127] Guccione G, Hosseini M, Adlong S, Johnsson MT, Hope J, Buchler BC, Lam PK. Scattering-free optical levitation of a cavity mirror. Phys Rev Lett 2013;111(18):183001.

[128] Li T, Raizen MG. Brownian motion at short time scales. Annalen Der Physik 2013;525(4):281–295.

[129] J. Gieseler, M. Spasenovic, L. Novotny, and R. Quidant, Nonlinear mode-coupling and synchronization of a vacuum-trapped nanoparticle. *arXiv.org*, vol. cond-mat.mes-hall. 2014 Jan 23.

[130] Yin Z-Q, Geraci AA, Li T. Optomechanics of levitated dielectric particles. Int J Modern Phys B 2013;27(26):1330018.

[131] Karadag Y, Aas M, Jonáš A, Anand S, McGloin D, Kiraz A. Dye lasing in optically manipulated liquid aerosols. Opt Lett 2013;38(10):1669.

[132] Jonáš A, Aas M, Karadag Y, Manioğlu S, Anand S, Mcgloin D, Bayraktar H, Kiraz A. In vitro and in vivo biolasing of fluorescent proteins suspended in liquid microdroplet cavities. Lab Chip 2014;14(16):3093–3100.

[133] Teiser J, Dodson-Robinson SE. Photophoresis boosts giant planet formation. Astron Astrophys 2013;555:A98.

[134] Shvedov VG, Desyatnikov AS, Rode AV, Krolikowski W, Kivshar YS. Optical guiding of absorbing nanoclusters in air. Opt Exp 2009;17(7):5743.

[135] Eckerskorn N, Li L, Kirian RA, Küpper J, DePonte DP, Krolikowski W, Lee WM, Chapman HN, Rode AV. Hollow Bessel-like beam as an optical guide for a stream of microscopic particles. Opt Exp 2013;21(25):30492.

[136] Shvedov VG, Rode AV, Izdebskaya YV, Desyatnikov AS, Krolikowski W, Kivshar YS. Selective trapping of multiple particles by volume speckle field. Opt Exp 2010;18(3): 3137–3142.

[137] Shvedov VG, Rode AV, Izdebskaya YV, Desyatnikov AS, Krolikowski W, Kivshar YS. Giant optical manipulation. Phys Rev Lett 2010;105(11):118103.

[138] Shvedov VGV, Hnatovsky CC, Rode AVA, Krolikowski WW. Robust trapping and manipulation of airborne particles with a bottle beam. Opt Exp 2011;19(18):17350–17356.

[139] Lin J, Li Y-Q. Optical trapping and rotation of airborne absorbing particles with a single focused laser beam. Appl Phys Lett 2014;104(10):101909.

[140] Esseling M, Rose P, Alpmann C, Denz C. Photophoretic trampoline—interaction of single airborne absorbing droplets with light. Appl Phys Lett 2012;101(13):131115.

[141] O'Dwyer DP, Phelan CF, Ballantine KE, Rakovich YP, Lunney JG, Donegan JF. Conical diffraction of linearly polarised light controls the angular position of a microscopic object. Opt Exp 2010;18(26):1–8.

[142] Xie C, Dinno MA, Li Y-Q. Near-infrared Raman spectroscopy of single optically trapped biological cells. Opt Lett 2002;27(4):249–251.

SECTION III

EXPERIMENTAL TECHNIQUES

8

SCATTERING TECHNIQUES

LUCA CIPELLETTI[1], VÉRONIQUE TRAPPE[2], & DAVID J. PINE[3]

[1]*Laboratoire Charles Coulomb (L2C), UMR 5221 CNRS-Université de Montpellier, Montpellier, France*
[2]*Département de Physique, Université de Fribourg, Fribourg, Switzerland*
[3]*Center for Soft Matter Research, New York University, New York, NY, USA*

8.1 INTRODUCTION

Broadly speaking, probes of the structure of matter can be divided into two categories: direct methods, such as optical, electron, or atomic force microscopies that form real-space images of the structures being probed, and scattering methods, such as neutron, X-ray, or light scattering that measure the Fourier transform of the structures being probed. Superficially, it would seem that methods measuring the real-space structure would have the upper hand as there is no substitute for knowing the precise position of each and every particle as a function of time. However, scattering techniques retain their power and continue to be widely used. There are several reasons. First, scattering techniques can average over many more particles than can direct methods and thus often provide much better quantitative measurements of the average structural and dynamical properties of materials. Second, scattering techniques are more readily adapted to measure fully three-dimensional information. Direct methods, such as confocal microscopy, have made steady progress in this domain, but still can only probe relatively small volumes and slow dynamical processes, especially in 3D where resolving even millisecond dynamical processes remains a challenge. By contrast, scattering methods provide quantitative information from nanosecond to hour-long timescales, depending on the

Fluids, Colloids and Soft Materials: An Introduction to Soft Matter Physics, First Edition. Edited by Alberto Fernandez Nieves and Antonio Manuel Puertas.

scattering technique. This reveals a third and sometimes ill-appreciated limitation of direct microscopies, namely, the trade-off between probe volume and dynamical range. To obtain greater speeds using direct microscopy, smaller volumes must be probed, which necessarily limits how well average dynamical and structural properties can be measured. Such trade-offs are rarely a problem in scattering techniques, except at the outer extremes of the already broad dynamical range they cover. For these and other reasons, scattering techniques continue to provide some of the most useful and powerful probes of the structural and dynamical properties of materials.

Before going into the details of scattering techniques and to set the scene for the remainder of this chapter, we show in Figure 8.1 typical light scattering setups for wide-angle (a) and small-angle (b) measurements. Note that in both cases the detector collects light scattered in the far field, that is, in the focal plane of lens L1. Figure 8.2a shows a typical image of the instantaneous distribution of scattered intensity as measured by a small-angle apparatus equipped with a multielement detector, such as a charge-coupled device (CCD) camera. The intensity distribution fluctuates widely over very small changes in the scattering angle. These intensity fluctuations appear as small random flecks of light, known as *laser speckles*. Speckles arise from the interference between the electric field scattered by each particle, as it is discussed in Section 8.3. Figure 8.2b shows the dependence of the scattered intensity I on the scattering angle θ, measured along the diagonal line shown in a. $I(\theta)$ exhibits large fluctuations due to the speckles; however, there is clearly an overall decreasing trend with increasing θ. This can be better quantified by averaging $I(\theta)$ over many speckles, for example, over rings of pixels corresponding to nearly the same θ but a different azimuthal angle φ, as the portion of ring shown in Figure 8.2a. The speckle average yields the smoother $I(\theta)$ curve shown in Figure 8.2b as a thick line; in Section 8.3, we will see that valuable information on the shape and size of the scatterers and on their mutual interactions can be obtained from the average $I(\theta)$. This is the essence of static light scattering (SLS) [1]. The speckle image shown in Figure 8.2a is just a snapshot of the intensity distribution; over time, the speckle pattern is continuously renewed, as the scatterers move, for example, due to Brownian motion, and change their relative position, thereby modifying the interference pattern they generate. Figure 8.2c shows the temporal fluctuations of the intensity measured by a given CCD pixel. By averaging over time, one can usually recover the same smooth $I(\theta)$ curve as that obtained from a speckle average and shown in Figure 8.2b. This is particularly useful when only a point-like detector is available, as in the wide-angle setup of Figure 8.2a. Even more importantly, one can gain insight into the scatterers' dynamics by analyzing the fluctuations of I; this is the essence of the dynamic light scattering (DLS) method [2] that is discussed in Section 8.4.

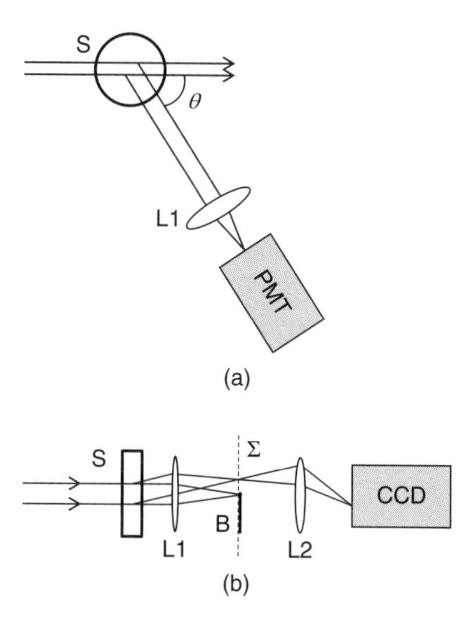

Figure 8.1 Schematic top views of (a) a wide-angle and (b) a small-angle light scattering setup. S, sample; θ, scattering angle; L1 Fourier lens, PMT, point-like detector (photomultiplier tube or avalanche photodiode); B, beam block; Σ, focal plane of L1; L2 lens that images Σ on the detector; CCD, multielement detector (CCD or CMOS).

8.2 LIGHT AND OTHER SCATTERING TECHNIQUES

The choice of the scattering technique is generally set by the characteristic length scales of the structures being probed. Scattering techniques generally work best when the wavelength of the radiation is about the same as the size of the structures that scatter the radiation. For X-ray and neutron scattering, the wavelength of the radiation is approximately 0.1 nm, making X-rays and neutrons ideal for probing the structure of materials on atomic and molecular length scales. For light scattering, the wavelength is about 500 nm, making light scattering ideal for probing the structure of soft materials such as colloids, emulsions, polymer solutions, and surfactant solutions, all of which have structures on length scales from a few hundred to thousands of nanometers. The above guidelines are, however, a little too restrictive. Advances in small-angle scattering methods over the past two or three decades have extended X-ray, neutron, and light scattering techniques so that they are capable of probing structures with length scales up to 10^2–10^4 times greater than wavelength of the scattered radiation. Thus, SLS can probe structures from a few hundred nanometers up to 10 μm or even 100 μm in some cases.

For those familiar with X-ray or neutron scattering, it is important to point out that light at optical wavelengths generally interacts much more strongly with matter than do X-rays or neutrons. Thus, while multiple scattering is seldom a concern in X-ray or neutron scattering, it is often

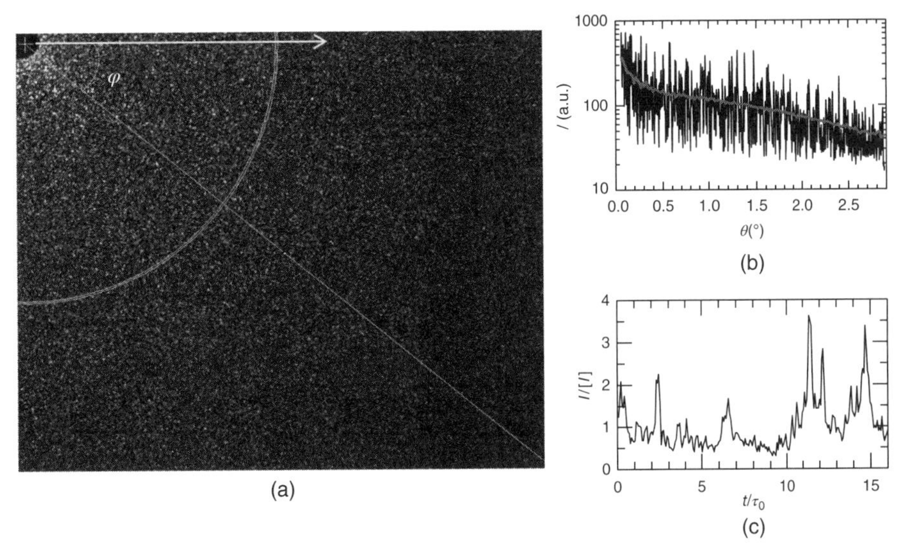

Figure 8.2 (a) Typical image of the instantaneous distribution of the light scattered at small angle as recorded by the setup shown in Figure 8.1b. The sample is a diluted suspension of particles with a diameter of 9.7 μm. The beam block is visible in the top-left corner of the image, and the cross indicates the position of the transmitted beam. Note the grainy appearance of the scattered intensity due to the speckles. (b) Thin line: intensity profile along the diagonal line shown in (a). The large fluctuations are due to the speckles. Thick line: profile obtained by averaging the intensity over rings of pixels such that depicted in (a), where only one ring is shown for clarity. (c) Temporal fluctuations of the scattered intensity measured in one single pixel. Intensity is normalized by its time average, t is normalized by the characteristic time of the fluctuations, obtained as discussed in Section 8.4. Source: Speckle image courtesy of E. Tamborini.

a problem for light scattering. Indeed, the scattering of light is typically 4–5 orders of magnitude stronger than the scattering of X-rays. This means that light scattering can be sensitive to very small concentrations of scatterers – very dilute solutions, for example – but it also means that multiple light scattering can be a concern even when the concentration of scatterers is not particularly high.

8.3 STATIC LIGHT SCATTERING

A schematic of the basic scattering geometry is shown in Figure 8.1a. Coherent light from a laser is directed toward the sample containing particles or other structures that scatter light. The incoming and scattered light are characterized by the wave vectors $\boldsymbol{k}_i$ and $\boldsymbol{k}_o$, respectively, pointing in the directions of propagation of the incoming and outgoing radiation, as shown in Figure 8.3. Their magnitudes are given by $k_i = k_o = 2\pi n/\lambda$, where n is the refractive index of the solvent and λ is the wavelength of the light in vacuum. The intensity of the scattered light is measured by a detector far from the sample cell or, equivalently, placed in the focal plane of a collecting lens as shown in Figure 8.1a. Such a measurement is assumed to be performed in the *single-scattering limit*, meaning that light is scattered no more than once before exiting the sample. This implies that most of the light exits the sample without having been scattered.

The basic principle underlying light scattering can be grasped by considering the intensity of the light scattered by

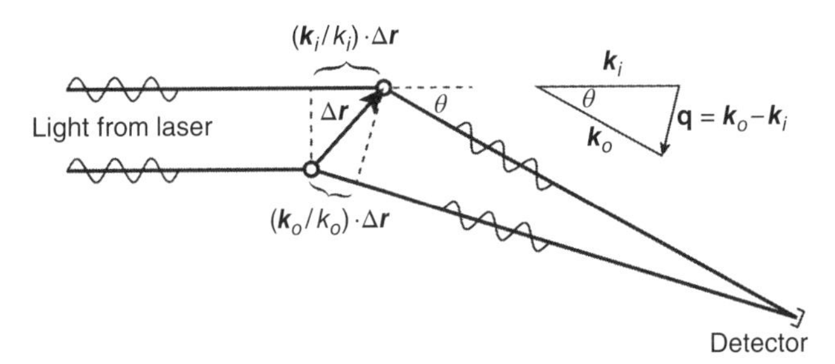

Figure 8.3 Scattering diagram for two particles within the sample cell.

two particles within the scattering volume. The key point is that light scattered from the top particle in Figure 8.3 travels a different distance from the laser to the detector than does light scattered from the bottom particle. From the scattering diagram in Figure 8.3, the difference in path lengths is given by $\Delta s = \boldsymbol{k}_o/k_o \cdot \Delta\boldsymbol{r} - \boldsymbol{k}_i/k_i \cdot \Delta\boldsymbol{r}$, where we have made the approximation that the light scattered from each of the two particles is parallel, which is valid when the size of the scattering volume is much smaller than the distance from the scattering volume to the detector. Thus, the phase difference between the two scattered waves is

$$\begin{aligned}\Delta\phi = k\Delta s &= (\boldsymbol{k}_o - \boldsymbol{k}_i)\cdot\Delta\boldsymbol{r}\\ &= \boldsymbol{q}\cdot\Delta\boldsymbol{r}, \qquad (8.1)\end{aligned}$$

where $k \equiv k_i = k_o$ and $\boldsymbol{q} \equiv \boldsymbol{k}_o - \boldsymbol{k}_i$. When $\Delta\phi = \pi$, the two waves interfere destructively and when $\Delta\phi = 0$, they interfere constructively. Therefore, when the distance between particles changes by $\Delta r \sim \pi/q$, the scattered light intensity changes from its most intense to zero. To get an idea of the range of distances Δr that can cause a significant change in the scattered intensity, we need to know the magnitude of q. From the scattering geometry shown in Figure 8.3, we see that this is given by

$$q = 2k \sin\frac{\theta}{2}. \tag{8.2}$$

Thus, the magnitude of q varies from a maximum value $q_{\max} = 2k = 2\pi n/\lambda$ for $\theta = 180°$ to 0 for $\theta = 0°$. As a practical matter, the scattering angle can never be set to zero in a real scattering apparatus because of interference from the unscattered beam. Depending on the instrument, the minimum value of θ can range from $\sim 15°$ for a wide-angle scattering instrument to as little as $0.1°$ for a well-designed small-angle scattering instrument. Thus, the magnitude of q can be as small as $q_{\min} = 2k\sin(\theta_{\min}/2) \approx k\theta_{\min}$. From these limits on the range of q, we see that the length scales that can be probed range from $\Delta r_{\min} \approx \lambda/(2n)$ to $\Delta r_{\max} \approx \pi/k\theta_{\min} \sim 10^2(\lambda/n)$. Note that there is an inverse relationship between the wave vector q and distance Δr over which light scattering is sensitive to particle positions.

We can obtain a quantitative expression for the electric field scattered through an angle θ into the detector by summing the scattered field from all N particles in the scattering volume

$$\boldsymbol{E}_d(\theta) = \sum_{i=1}^{N} \boldsymbol{E}_i(\theta) e^{i\boldsymbol{q}\cdot\boldsymbol{r}_i}. \tag{8.3}$$

If all N particles are identical spheres, then the scattered field $\boldsymbol{E}_i(\theta) = \boldsymbol{E}_s(\theta)$ is the same for all the particles and can be taken outside the sum

$$\boldsymbol{E}_d(\theta) = \boldsymbol{E}_s(\theta) \sum_{i=1}^{N} e^{i\boldsymbol{q}\cdot\boldsymbol{r}_i}. \tag{8.4}$$

The positions $\boldsymbol{r}_i$ in the phase factor $\boldsymbol{q}\cdot\boldsymbol{r}_i$ are measured with respect to an arbitrary fixed point, generally taken to be at the center of the scattering volume, but the scattered intensity is independent of this choice.

The quantity actually measured in a scattering experiment is proportional to the scattered intensity, which in turn is proportional to the square modulus of the electric field:

$$I_d(\theta) \propto |\boldsymbol{E}_d(\theta)|^2 = |\boldsymbol{E}_s(\theta)|^2 \sum_{i=1}^{N} e^{i\boldsymbol{q}\cdot\boldsymbol{r}_i} \sum_{j=1}^{N} e^{-i\boldsymbol{q}\cdot\boldsymbol{r}_j}, \tag{8.5}$$

$$\propto |\boldsymbol{E}_s(\theta)|^2 \sum_{i,j}^{N} e^{i\boldsymbol{q}\cdot(\boldsymbol{r}_i - \boldsymbol{r}_j)}. \tag{8.6}$$

The scattered intensity is thus seen to be sensitive to the relative positions of pairs of particles.

8.3.1 Static Structure Factor

Equation 8.6 gives the scattered intensity at an instant in time. An SLS experiment, however, measures the intensity *averaged* over some period of time. For the case of identical spheres, $\boldsymbol{E}_s(\theta)$ is constant so the time dependence of $I_d(\theta)$ comes solely from the sum over the time-dependent particle positions $\{\boldsymbol{r}_i(t)\}$ in the phase factor in Equation 8.6. More importantly, all the information about particle positions are contained in the sum over phase factors. For this reason, it is convenient to define a quantity called the *structure factor* $S(q)$, which is just the normalized ensemble average over the phase factors above:

$$S(q) = \frac{1}{N} \sum_{i,j}^{N} \langle e^{i\boldsymbol{q}\cdot(\boldsymbol{r}_i - \boldsymbol{r}_j)} \rangle. \tag{8.7}$$

In general, the structure factor is a function of the vector $\boldsymbol{q}$, but here we write it as a function of a scalar q as we are implicitly assuming that the particle positions are distributed isotropically, which is what one would expect for most liquid or glassy systems. For crystalline or nonisotropic systems, the vector $\boldsymbol{q}$ is retained.

The structure factor is particularly interesting because it depends only on the relative particle positions $\{\boldsymbol{r}_i - \boldsymbol{r}_j\}$ and not on scattering properties of the particles. The prefactor $\boldsymbol{E}_s(\theta)$ in Equation 8.6 does depend on the scattering properties but only those of a single particle. The scattering properties can be determined either from theory or from measurements on very dilute systems, where the scattering properties are independent of particle positions. Thus, if $\boldsymbol{E}_s(\theta)$ is known *a priori*, either by theory or measurement, static scattering measurements can be used to extract the structure factor $S(q)$.

Figure 8.4 shows the structure factor $S(q)$ as a function of qd for a suspension of hard spheres of diameter d. The structure factor is calculated from the Percus–Yevick model for three different volume fractions [3]. When interpreting scattering measurements, it is important to understand that scattering is caused by *fluctuations* in the dielectric constant, which in turn means fluctuations in the particle density. Thus, the lowest-q peak near $qd \approx 2\pi$ reflects the fact that the density fluctuates strongly on the length scale of the particle diameter d. The increase of the peak with increasing volume fraction ϕ near $qd \approx 2\pi$ reflects the increasing correlation between particles separated by one particle diameter as packing constraints become more important. By contrast, the suppression of $S(q)$ for small q reflects the fact that at long length scales, that is, on length scales averaged over many particle diameters, the density is very nearly uniform. In fact, in the limit that $q \to 0$, the static structure

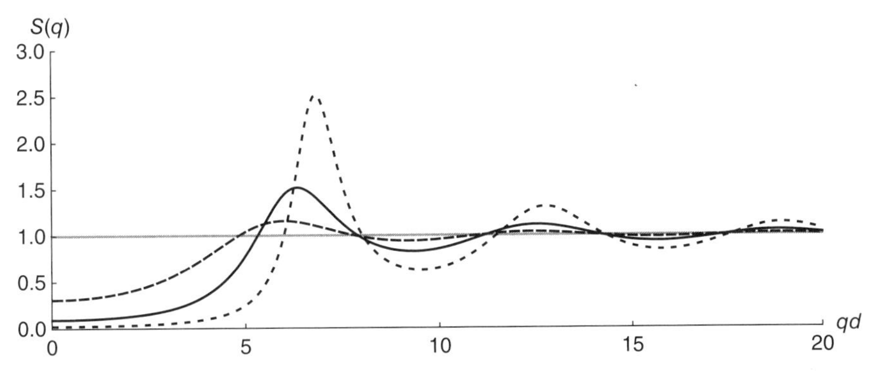

Figure 8.4 Structure factor $S(q)$ for hard spheres at three different volume fractions: $\phi = 0.15$, long dashes; $\phi = 0.30$, solid; $\phi = 0.45$ short dashes. From Percus–Yevick model. The structure factor is plotted as a function of qd, where d is the particle diameter.

factor becomes proportional to the osmotic compressibility χ_T of the scatterers, provided that the scatterers are identical.

8.3.2 Form Factor

According to Equation 8.6, the intensity of light scattered from a set of identical spheres is the product of the scattering from an individual sphere, $|\boldsymbol{E}_s(\theta)|^2$, and the static structure factor $S(q)$. The calculation of the scattered electric field $\boldsymbol{E}_s(\theta)$ can be quite complicated, even for a simple sphere. In fact, $\boldsymbol{E}_s(\theta)$ quite generally depends not only on the scattering angle θ shown in Figure 8.3 but also on the polarization of the light and a second scattering angle φ out of the scattering plane, that is, out of the plane defined by Figure 8.3 [1, 2]. In most conventional light scattering experiments, measurements are made only in the scattering plane ($\varphi = 0$) and the polarization of the incident light is fixed to be perpendicular to the scattering plane. For isotropic systems, light scattered within the scattering plane remains perpendicularly polarized. That is the case we consider here.

The usual practice is to define a *form factor* $F(\theta) \propto |\boldsymbol{E}_s(\theta)|^2$ that is proportional to the scattered intensity from an individual particle, although sometimes a form factor $f(\theta)$ proportional to the field is used, in which case $F(\theta) = |f(\theta)|^2$. Various normalization conventions are used but in any case the scattered intensity is given by

$$I(\theta) = A\,F(\theta)S(\theta), \tag{8.8}$$

where Equation 8.2 is used to write the structure factor as a function of θ rather than q. When performing scattering measurements, there are always a number of factors, such as optical, geometrical, and electronic, that make measuring the absolute scattering intensity impractical. The constant A is included to take these factors into account, which allows for some latitude in how $F(\theta)$ is normalized. A common convention is to choose the form factor such that when it is integrated over all scattering angles, it yields the *scattering cross section* of the scatterers, which is the geometrical equivalent area of a particle that blocks the same total amount of radiation. The scattering cross section for dielectric spheres of radius a generally goes like q^4a^6 for $qa \ll 1$ ($a \ll \lambda$) and asymptotes to $2\pi a^2$, twice the geometrical cross section, for $qa \gg 1$ ($a \gg \lambda$).

For uniform spheres, $F(\theta)$ is generally calculated using *Mie theory* [1]. Figure 8.5 shows $F(\theta)$ for scattering from colloidal spheres of different diameters. The dips in the scattered intensity are known as *Mie resonances* and are characteristic of particles with sharp boundaries and whose size is comparable to or larger than the wavelength of light. As the size of the particle becomes small compared to the wavelength of light, there are fewer and fewer Mie resonances until, for particles much smaller than the wavelength of light, the scattering becomes completely isotropic.

The relevance of Equation 8.8 to practical light scattering measurements is more limited than might appear. For scatterers whose size is comparable to the wavelength of light, almost any sample with a volume fraction ϕ exceeding even 0.01 will exhibit fairly strong multiple scattering even when the scatterers are index-matched to the solvent to 1%. For

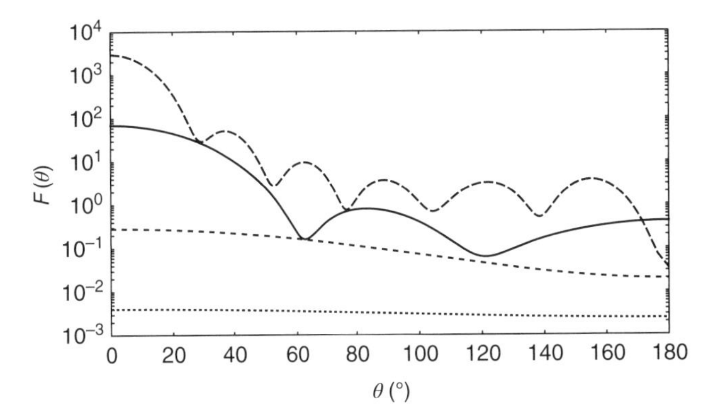

Figure 8.5 Form factors for scattering of light with a wavelength of 532 nm from a dilute suspension of polystyrene spheres of different diameters suspended in water: long dashes, 1 μm; solid, 0.5 μm; short dashes, 0.2 μm; dotted line, 0.1 μm.

such low volume fractions, $S(\theta)$ is very nearly unity over all scattering angles unless there are strong long-range interactions. At these very low volume fractions, $I(\theta) \approx A\,F(\theta)$ to a very good approximation. For particles whose size is much less than the wavelength of light, the scattering is much weaker and multiple scattering becomes less of an issue. In this case, $F(\theta)$ is very nearly independent of θ and $I(\theta) \approx A\,S(\theta)$. Therefore, in most cases, the scattered light intensity is either proportional to $F(\theta)$ or $S(\theta)$ when scattering measurements are performed in the single scattering limit.

8.4 DYNAMIC LIGHT SCATTERING

SLS from disordered liquid-like structures gives rise to average scattering patterns that vary smoothly with scattering angle θ or wave vector q, as illustrated in Figures 8.4 and 8.5. To the uninitiated, therefore, it often comes as some surprise that the unaveraged instantaneous scattering pattern from a collection of scatters has the speckled appearance shown in Figure 8.2a. As we have mentioned in Section 8.1, if the scatterers move relative to one another, these speckles of scattered light flicker on and off in time. DLS takes note of this fact and uses the time dependence of the flickering speckles to quantitatively characterize the underlying motion of the scatterers.

8.4.1 Conventional Dynamic Light Scattering

Although a typical light scattering experiment might collect light scattered from $\sim 10^4$ particles, the basic results can be understood by considering scattering from only two particles, as shown in Figure 8.6. The phase difference between two scattered waves at time t is given by Equation 8.1, namely $\Delta\phi = \boldsymbol{q} \cdot \Delta\boldsymbol{r}(t)$. As the two particles move, $\Delta\boldsymbol{r}(t)$ and the phase difference changes, and thus the intensity of the light scattered from the two particles into the detector changes. In a time τ, the change in the phase is $\Delta\phi(\tau) = \boldsymbol{q} \cdot [\Delta\boldsymbol{r}(t+\tau) - \Delta\boldsymbol{r}(t)]$. A significant change in the scattered intensity occurs when $\Delta\phi(\tau) \sim 1$, which means that $\Delta\boldsymbol{r}(t+\tau) - \Delta\boldsymbol{r}(t) \sim \lambda/\sin(\theta/2)$. Therefore, DLS measures particle motion on the length scale of the wavelength of light, or a little longer depending on the scattering angle.

We can make these ideas more quantitative, and in the process generalize the analysis to an arbitrary number of particles N, by writing the sum of the electric field over all particles that scatter light into the detector

$$\boldsymbol{E}_d(t) = \boldsymbol{E}_s \sum_{i=1}^{N} e^{i\boldsymbol{q}\cdot\boldsymbol{r}_i(t)}, \tag{8.9}$$

where, as before, we assume scattering from spheres that are all the same size. The time-dependent intensity is proportional to the square modulus of the field

$$I_d(t) \propto |\boldsymbol{E}_s(\theta)|^2 \sum_{i,j}^{N} e^{i\boldsymbol{q}\cdot[\boldsymbol{r}_i(t)-\boldsymbol{r}_j(t)]}. \tag{8.10}$$

As expected, the scattered intensity of light depends on the *differences* between particle positions.

To obtain a quantitative measure of the temporal fluctuations of the scattered light, experiments typically measure the temporal autocorrelation function of the scattered light

$$g_I(t,\tau) = \frac{\langle I(t+\tau)I(t)\rangle}{\langle I(t)\rangle^2}. \tag{8.11}$$

The correlation function is calculated theoretically by taking an ensemble average, while it is measured experimentally by performing a time average. In most cases, these two methods yield identical results. There are notable exceptions, however, especially in glassy systems, where the equivalence of ensemble and time averages breaks down. This is discussed in Section 8.4.3.

Generally, it is not the intensity correlation function $g_I(t,\tau)$ that is calculated but the field autocorrelation

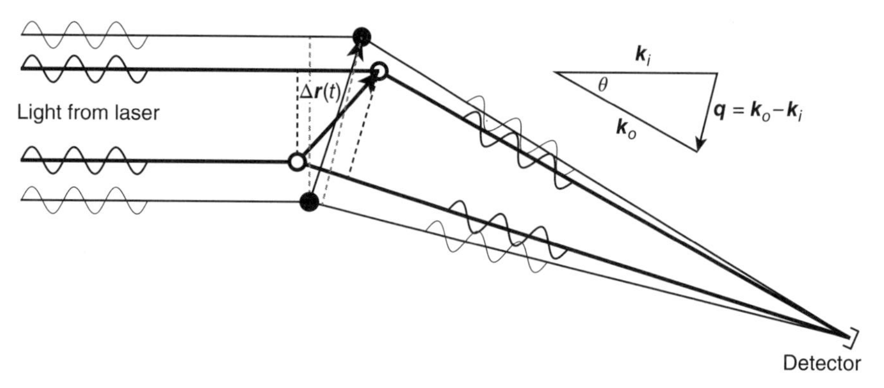

Figure 8.6 Light from a laser is scattered from two particles that move from their initial positions, shown as open circles, to their final positions, shown as closed circles.

function

$$g_E(t,\tau) = \frac{\langle E^*(t+\tau)E(t)\rangle}{\langle |E(t)|^2\rangle}. \quad (8.12)$$

In most cases, the scattered field obeys Gaussian statistics, as discussed above, in which case the two correlation functions are related by the Siegert relation [2, 4]

$$g_I(t,\tau) = 1 + |g_E(t,\tau)|^2. \quad (8.13)$$

The Siegert relation is typically used to relate the field autocorrelation function, which is generally what is calculated from theory, to the intensity autocorrelation function, which is generally what is measured experimentally.

The intensity autocorrelation function is determined experimentally by collecting a time series of the scattering intensity $I(t)$ from which $g_I(t,\tau)$ is calculated by averaging over all times t. To obtain a statistically meaningful result, the result should not depend on what time t is used, but only on the delay time τ, which is generally what is meant by a *stationary* process. For stationary processes, we can drop the t dependence and write $g_I(t,\tau) = g_I(\tau)$.

For delay times τ much greater than the longest relaxation time τ_{max} of the system being probed, $\langle E^*(t+\tau)\rangle$ and $\langle E(t)\rangle$ become statistically independent. On these timescales, the average in the correlation function factors such that $\langle E^*(t+\tau)E(t)\rangle = \langle E^*(t+\tau)\rangle\langle E(t)\rangle = 0$, since the electric field is equally likely to be positive or negative. Thus, we see that $g_E(\tau)$ decays from a maximum value of 1 at $t = 0$ to a value of 0 for $\tau \gg \tau_{max}$. Using the Siegert relation, this means that $g_I(\tau)$ decays from a value of 2 at $\tau = 0$ to a value of 1 for $\tau \gg \tau_{max}$.

The functional form of the decay of $g_I(t)$ depends on the dynamics of the scatterers. Therein lies the power of DLS. Different dynamical processes lead to different timescales and different functional forms for the decay of $g_I(t)$. For example, for a diffusive process

$$g_I(\tau) - 1 = \exp(-2Dq^2\tau). \quad (8.14)$$

This is the case, for example, of identical, spherical, and noninteracting Brownian particles of radius a, in which case D is the diffusion coefficient given by the Stokes–Einstein relation [2]:

$$D = \frac{k_B T}{6\pi\eta a} \quad (8.15)$$

with k_B Boltzmann's constant, T the absolute temperature, and η the solvent viscosity. If η is known, one can then calculate a from Equation 8.15, once D is obtained by fitting g_I to Equation 8.14. This is the principle underlying particle sizing, a popular application of DLS.

Note that in Equation 8.14, the relaxation time of g_I depends on both D and q. Physically, the q dependence stems from the fact that as q is decreased one probes the diffusive process on increasingly large length scales $\sim 1/q$, thus leading to a slower relaxation of g_I. In writing Equation 8.14 we assumed that interactions were negligible, which in experiments can be achieved at sufficiently small particle concentration. If this assumption is not fulfilled, Equation 8.14 often still applies, but D is replaced by an effective diffusion coefficient D_{eff} that depends on both interactions and q vector. Two notable cases are concentration fluctuations near the critical point in a binary liquid mixture, where the diffusivity at $q = 0$ vanishes [5], and highly charged colloidal liquids, where the diffusivity becomes very fast at low q [6].

The exponential decay and q^2 dependence of the decay rate are characteristic of a diffusive relaxation. Because we have ignored correlations between different scatterers, the diffusion coefficient appearing in Equation 8.14 is for self-diffusion. More generally, when interparticle correlations are taken into account, the decay of $g_I(\tau)$ pertains to the correlated relaxation of concentration fluctuations. The decay of these fluctuations varies substantially from system to system and depends sensitively on the interactions between particles. For scatterers of size a, this dependence is most pronounced, in general, when $qa \ll 1$, that is, when the light scattering probes length scales much larger than a typical scatterer. By contrast, when $qa \gg 1$, light scattering probes the dynamics of length scales smaller than a particle size, which is to say the dynamics of single particles.

8.4.2 Diffusing Wave Spectroscopy

In our discussion of the form factor (Section 8.3.2), we noted that for scatters whose size a is comparable to or larger than the wavelength λ of light, the scattering is typically too strong to do single light scattering experiments. To perform single scattering experiments, the mean free path ℓ must be much greater than the size of the sample, or more precisely, much greater than the distance that light travels through the sample. The mean free path is given by

$$\ell = \frac{1}{n_p \sigma_s}, \quad (8.16)$$

where n_p is the number density of particles and σ_s is the scattering cross section. For $a \gtrsim \lambda$, we noted earlier that $\sigma_s \simeq 2\pi a^2$. Noting that $n_p = 3\phi/4\pi a^3$, the mean free path in the limit that $a \gtrsim \lambda$ is given by

$$\ell \simeq \frac{2a}{3\phi}. \quad (8.17)$$

For a suspension of 1-μm-diameter particles at a volume fraction $\phi = 0.01$, this gives $\ell \simeq 33$ μm. Since samples are typically several millimeters or more in size, a mean free path of 33 μm means that single light scattering experiments are not practical. Various schemes have been proposed to reject multiply scattered light, thereby allowing DLS measurement on turbid samples. These methods generally rely on illuminating the samples with two beams and cross-correlating the signal from two detectors [7]. The optical layout is such that for single scattering the two detectors probe the same $\boldsymbol{q}$ vector (both in magnitude and orientation), while for multiply scattered light they collect photons issued from distinct paths. As a result, only the portion of the signal due to single scattering is identical for the two detectors, while that due to multiple scattering is fully uncorrelated and thus does not contribute to the cross-correlation. Commercial apparatuses implementing these methods are now available. Note, however, that for very turbid samples the single scattering component becomes vanishingly small as compared to the full detector signal, so that the method fails in practice. An alternative method applies to samples that multiply scatter light strongly enough for a statistical approach to be used to describe the transport of light through the sample. This powerful DLS method is called *diffusing wave spectroscopy* or DWS [8–10].

The scattering geometry for a DWS experiment is different from that of a single scattering DLS experiment. A typical scattering geometry is illustrated in Figure 8.7. Light from a laser is spread out into a beam that is considerably wider than the thickness L of the sample thickness. The incident light is scattered many times by the sample before exiting. Some of the multiply scattered light exits through the side opposite the entrance and, after passing through a collimating pinhole, impinges on a light detector that records the time-dependent intensity.

The qualitative picture for how DWS works is the same as for DLS and it is sufficient, once again, to consider a pair of light scattering paths through the sample at two different times, t and $t + \tau$. The two light paths enter the sample in phase from a laser. At time t, the two paths exit the sample with different phases because they have traveled different distances through the sample. At time $t + \tau$, the phase difference between the two paths will have changed compared to the phase at time t because the particle positions along each scattering path changed. This results in a changing intensity at the detector.

To make these ideas more quantitative, we write the temporal autocorrelation function as

$$g_E(\tau) = \frac{1}{N_p}\left\langle \sum_i^{N_p} e^{-i\Delta\phi_i(\tau)} \right\rangle, \tag{8.18}$$

where the sum is over different light *paths* through the sample, which we label as running from 1 to N_p [in the single scattering case, the number of light paths was equal to the number of particles that scattered light] and where $\Delta\varphi_i(\tau)$ is the change of phase of the light along the ith path, during the time interval τ. Each path consists of many scattering events from a sequence of particles. Our task is to figure out how the phase shift evolves in time as particles along any given path move.

Let's consider the ith path and designate n_i as the number of scattering events along this path. As the particles move, the change in phase for the entire path is just the sum of all the phase changes due to each particle's motion

$$\Delta\phi_i(\tau) = \sum_{j=1}^{n_i} \boldsymbol{q}_j \cdot \Delta\boldsymbol{r}_j(\tau) - \sum_{j=1}^{n_i} \boldsymbol{q}_j \cdot \boldsymbol{r}_j(t) = \sum_{j=1}^{n_i} \boldsymbol{q}_j \cdot \Delta\boldsymbol{r}_j(\tau), \tag{8.19}$$

where $\Delta\boldsymbol{r}_j(\tau) = \boldsymbol{r}_j(t+\tau) - \boldsymbol{r}_j(\tau)$. Different paths i through a sample will in general have a different number of scattering events n_i but only those paths with the same number of scattering events n can be considered to be statistically equivalent. Thus, the analysis of $\Delta\phi(\tau)$ proceeds in two steps: first we analyze all the paths that have the same number of scattering events n, and then we average over all possible n. To this end, we rewrite Equation 8.18 as a sum over paths of different lengths n

$$g_E(\tau) = \sum_n \frac{N_n}{N_p} e^{-i\Delta\phi_n(\tau)}, \tag{8.20}$$

where N_n is the number of paths of length n, and $\phi_n(\tau)$ is given by

$$\Delta\phi_n(\tau) = \sum_{j=1}^{n} \boldsymbol{q}_j \cdot \Delta\boldsymbol{r}_j(\tau). \tag{8.21}$$

Because $\Delta\phi_n(\tau)$ is the sum of many random variables, we expect $\Delta\phi_n(\tau)$ to have Gaussian statistics. Therefore,

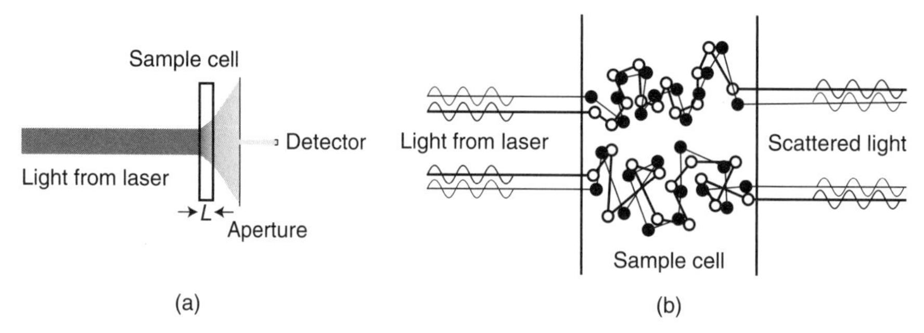

Figure 8.7 (a) Transmission scattering geometry for DWS. Light multiply scattered by a sample is collimated by an aperture and incident on a detector. (b) Two paths of light multiply scattered by the sample: open circles; paths at time t; closed circles; paths at time $t + \tau$.

Equation 8.20 becomes

$$g_E(\tau) = \sum_n \frac{N_n}{N_p} e^{-\frac{1}{2}\langle \Delta\phi_n^2(\tau)\rangle}. \tag{8.22}$$

Next we evaluate the average in the exponent for paths of length n:

$$\langle \Delta\phi_n^2(\tau)\rangle = \sum_{j=1}^{n} \langle [\boldsymbol{q}_j \cdot \Delta\boldsymbol{r}_j(\tau)]^2\rangle, \tag{8.23}$$

$$= n\frac{1}{3}\langle q^2\rangle\langle \Delta r^2(\tau)\rangle, \tag{8.24}$$

where we have ignored terms where $i \neq j$ consistent with our previous approximation that different particles are uncorrelated and where the factor of 1/3 comes from the angular average of $\cos^2\theta$ from the dot product over the unit sphere. For dense or strongly interacting systems, this approximation breaks down. The averages over $\boldsymbol{q}$ and $\boldsymbol{r}$ factorize as the light scattering direction and particle motion are uncorrelated. Substituting this result into Equation 8.22 gives

$$g_E(\tau) = \sum_n \frac{N_n}{N_p} e^{-\frac{1}{6}n\langle q^2\rangle\langle \Delta r^2(\tau)\rangle}. \tag{8.25}$$

The sum in Equation 8.25 is more conveniently calculated by passing to a continuum approximation. In this case, the number of scattering events is given by $n = s/\ell$, where s is the total length of a given path and ℓ is the mean free path introduced in Equation 8.17. From Equation 8.2, the average over q^2 is

$$\langle q^2\rangle = 2k^2\langle 1 - \cos\,\theta\rangle. \tag{8.26}$$

Thus, the argument of the exponential in Equation 8.25 becomes

$$\frac{1}{6}n\langle q^2\rangle\langle \Delta r^2(\tau)\rangle = \frac{1}{3}\frac{s}{\ell}k^2\langle 1 - \cos\,\theta\rangle\langle \Delta r^2(\tau)\rangle, \tag{8.27}$$

$$= \frac{1}{3}\frac{s}{\ell^*}k^2\langle \Delta r^2(\tau)\rangle, \tag{8.28}$$

where we have defined the *transport mean free path* as

$$\ell^* = \frac{\ell}{\langle 1 - cos\,\theta\rangle}. \tag{8.29}$$

Physically, the transport mean free path is the distance over which the direction of scattered light is randomized and can be regarded as the step or persistence length in a random walk. As pointed out previously, when the size of the scatterers is comparable to or larger than the wavelength of light, light is scattered preferentially in the forward direction (small θ) and thus several scattering events are required to randomize the direction of the scattered light. Physically, this is why $\ell^* \geq \ell$.

Passing to the continuum approximation, we define $P(s) = N_n/N_p$ as the fraction of paths of length s through the sample. Substituting this into Equation 8.25 and using Equation 8.28, we obtain

$$g_E(\tau) = \int_0^\infty P(s)e^{-\frac{1}{3}k^2\langle \Delta r^2(\tau)\rangle s/l^*}ds, \tag{8.30a}$$

$$= \int_0^\infty P(s)e^{-2Dk^2\tau s/l^*}ds, \tag{8.30b}$$

where we have used $\Delta r^2(\tau) = 6D\tau$ in passing from Equation 8.30a to Equation 8.30b.

To evaluate Equations 8.30a and 8.30b, the distribution of path lengths $P(s)$ must be determined. Note that the transport mean free path ℓ^*, and not the mean free path ℓ, appears in Equations 8.30a and 8.30b. This suggests that $P(s)$ need only be calculated on length scales of ℓ^* and greater. On these length scales, the transport of light is governed by a diffusion equation for light (not to be confused with the diffusion of the scatterers!) with a diffusion coefficient given by $D_\ell = c\ell^*/3$, where c is the speed of light in the sample. The path length distribution can thus be found by solving the diffusion equation for light using appropriate initial and boundary conditions. The procedure for calculating $P(s)$ is described elsewhere [10, 11]. For a typical sample, $\ell^* \sim 10^2$ µm, with a sample thickness L about 10 or more times greater.

The transmission geometry described in Figure 8.7a is often used. In this case, the temporal autocorrelation function is approximately

$$g_E(\tau) \simeq \frac{x}{\sinh x}, \tag{8.31}$$

where $x = (L/\ell^*)\sqrt{k^2\langle \Delta r^2(\tau)\rangle} = (L/\ell^*)\sqrt{6Dk^2\tau}$. The decay of $g_E(\tau)$ in Equation 8.31 is very nearly exponential with a decay rate of $(L/\ell^*)^2Dk^2$. For single scattering, the DLS correlation function decays with a decay rate of approximately Dk^2, or somewhat slower if measurements are performed at small angles. Thus, it is apparent that the DWS decay rate is accelerated by a factor of $(L/\ell^*)^2$ compared to DLS. This is easy to understand physically as $(L/\ell^*)^2$ is approximately the number of times light is scattered in a DWS transmission experiment. Each scattering event contributes, on average, a characteristic single scattering decay rate of Dk^2 to the correlation function, so the aggregate rate is $(L/\ell^*)^2Dk^2$.

In a typical DWS experiment, L/ℓ^* is about 10, so the decay rate is about 100 times faster than it would be for a DLS experiment. By using even thicker samples, L/ℓ^* can be increased even further so that the timescales probed can be 10^3–10^4 times faster than DLS. In DLS, the dynamics of the scatterers is probed on a length scale of the wavelength of light λ or about 500 nm. The acceleration of the decay of the correlation function in DWS by a factor of $(L/\ell^*)^2$ means that the dynamics can be probed on length scales of

$\sim 10^{-2} - 10^{-3}\lambda$, that is on subnanometer to nanometers length scales. A good example of this capability of DWS is provided by experiments that probe the ballistic motion of Brownian particles on submicrosecond timescales [11].

8.4.3 Dynamic Light Scattering from Nonergodic Media

A practical point of great importance in measuring the intensity autocorrelation function g_I is the time average indicated by the brackets in Equation 8.11. As a rule of thumb, one typically needs to average $g_I(\tau)$ over a duration at least 3–4 orders of magnitude larger than the characteristic time of the fluctuations of the intensity shown in Figure 8.2c. For most colloidal suspensions at moderate volume fraction and in the absence of strong interparticle interactions, this is not a severe limitation. The fluctuation time is close to that expected at infinite dilution, $(2q^2D)^{-1}$ (see Eq. 8.14), which, depending on the scattering angle and the particle diffusion coefficient, typically ranges from 10^{-6} to 10^{-1} s. A measurement then lasts a few minutes at most.

However, there are several instances where the relaxation time can exceed hundreds or even thousands of seconds. This may occur in small-angle experiments, where q^{-2} is very large. More importantly, the dynamics of glassy or gelled systems is generally slowed down by orders of magnitude due to particle crowding and the formation of space spanning structures, respectively. In these cases, the time average required by Equation 8.11 becomes practically unfeasible, since a measurement should last several days or even months. At a more fundamental level, the time average of Equation 8.11 is supposed to allow one to probe a sufficiently large number of system configurations, thereby performing an ensemble average. For systems with slow dynamics, this equivalence between time and ensemble averages breaks down; the system is then said to be nonergodic.

Note that if we still try to measure $g_I(\tau)$ by performing an insufficiently long time average, the resulting intensity correlation function will tend to have a lower-than-expected intercept (the intercept is the $\tau \to 0$ limit of g_I). This can be readily understood in the limit of perfectly frozen dynamics, where $I(t)$ is constant and Equation 8.11 yields $g_I(\tau) = 1$ for all τ. Indeed, erratic, lower-than-expected values of the intercept are often the signature of dynamics that are too slow to be captured by a traditional DLS experiment [12].

Two methods have been proposed to overcome this difficulty when using a traditional DLS apparatus. The first approach is a "brute force" method, where the ensemble average is obtained by sampling $g_I(\tau)$ for several independent speckles. This can be done by acquiring a number of separate runs, where the sample is slightly displaced in between runs, so that the detector is sequentially illuminated by different speckles [13]. Provided that the number of runs is sufficiently large, typically several hundred runs, the average of the acquired correlation functions correctly captures the dynamics of the ensemble. Alternatively, the sample can be continuously displaced, for instance, by rotation while measuring a single, long correlation function. In this case, a spurious decay is introduced on the timescale it takes a speckle to move across the detector, τ_{rot}. However, for time delays $\tau \ll \tau_{\text{rot}}$ the correlation function is correctly averaged and can be exploited. A simpler version of this method is often used in DWS experiments. The speckle pattern generated by the sample is renewed continuously by interposing an auxiliary sample with a fast dynamics or a rotating ground glass in between the main sample and the detector [14]. In this case, there is no need to move the sample, which avoids mechanical perturbations that may influence the sample dynamics. Similarly to the case where the sample is rotated, a spurious decay time is introduced in $g_I(\tau)$, dictated by the dynamics of the auxiliary sample or the rotation speed of the ground glass. Only data for time delays shorter than this spurious relaxation are considered.

The second approach has been proposed by Pusey and van Megen [12], who have shown that the correct, ensemble-averaged correlation function, $g_I^{(\text{e.a.})}(\tau)$, may be reconstructed from the intensity correlation function measured in a single (insufficiently long) run, $g_I^{(\text{1run})}(\tau)$, and the ensemble-averaged *static* intensity. In practice, one measures the intensity correlation function $g_I^{(\text{1run})}(\tau)$ for a given sample position, together with the time-averaged scattered intensity for that position, $I^{(\text{1run})}$. The sample is then rotated or translated while recording the instantaneous scattered intensity, in order to measure the ensemble-averaged static intensity, $I^{(\text{e.a.})}$. Note that this second step is far less time-consuming than in the brute force approach, since one only needs to measure the average of the instantaneous intensity, not that of the whole correlation function. Finally, the correct intensity correlation function is obtained from $g_I^{(\text{1run})}(\tau)$ and the ratio $I^{(\text{e.a.})}/I^{(\text{1run})}$ [12]. The correction formula is particularly simple if one chooses the sample position in such a way that $I^{(\text{e.a.})} = I^{(\text{1run})}$. In this case,

$$g_I^{(\text{e.a.})}(\tau) - 1 = [1 + g_I^{(\text{1run})}(\tau) - g_I^{(\text{1run})}(0)]. \qquad (8.32)$$

The general case is discussed in Ref. [12] (see Eq. 3.12 therein), together with the effects of the detector size as compared to the speckle size. In using this correction scheme, one should keep in mind the assumptions under which it was derived; namely the notion that the scatterers perform restricted motion about fixed positions and that the scattering volume is big enough to sample a sufficiently large ensemble of particle configurations. For colloidal glasses and gels, this is usually the case and this approach has been shown to give results similar to those obtained by the brute force method.

8.4.4 Multispeckle Methods

Multispeckle methods [15, 16] are a family of techniques designed to probe the dynamics of systems too slow to be measured by traditional DLS. The term was first introduced in Ref. [15] and the methods are equally well applicable to single or multiple scattering experiments. These methods use an approach conceptually similar to the brute force method discussed in Section 8.4.3 but take advantage of multielement detectors, such as CCD or complementary metal–oxide–semiconductor (CMOS) chips, to implement it in a parallel rather than sequential way. The idea is to illuminate simultaneously a large number of detectors, for example, the pixels of a CCD camera, with the speckle field scattered by a sample, such that the intensity fluctuations of a large number of independent speckles are recorded by the pixel array. For each pixel, an intensity time autocorrelation is then calculated, and the final $g_I(\tau)$ is obtained by averaging over the contributions of all pixels. Because the dynamics depend on the q-vector, the pixels used to form the average must correspond to nearly the same scattering vector. For small-angle scattering and isotropic samples, this can be achieved by averaging over rings of pixels associated to the same magnitude of $\mathbf{q}$, but with a different azimuthal orientation [16, 17], as shown in Figures 8.1b and 8.2a. Alternatively, the collection optics can be designed in such a way that the detector collects light in a small interval of $\mathbf{q}$ centered around a well-defined scattering vector [15], for example, by replacing the photomultiplier tube of Figure 8.1a by a multielement detector. These constraints are less tight for highly turbid samples probed by DWS. In this case, the scattered intensity depends very weakly on the direction of propagation after exiting the sample and one can safely assume that all CCD pixels are illuminated by light with the same temporal properties.

The outcome of a multispeckle experiment is a series of speckle images, recording the intensity $I_p(t)$ for each pixel p and time t. The intensity correlation function is then obtained from

$$g_I(\tau) = \left\langle \frac{\langle I_p(t) I_p(t+\tau)\rangle_p}{\langle I_p(t)\rangle_p \langle I_p(t+\tau)\rangle_p} \right\rangle_t, \qquad (8.33)$$

where $\langle \cdots \rangle_p$ denotes an average over an appropriate set of pixels and $\langle \dots \rangle_t$ denotes the average over t. The order of the averages in Equation 8.33 is important: if the reverse order was taken, the correlation function would not be correctly averaged. This can be readily seen in the limiting case of a sample whose dynamics are fully frozen, for which one expects $g_I(\tau) = g_I(0) = 2$ for all τ. In this case, I is constant in time but varies from pixel to pixel, so that one obtains the incorrect result $g_I(\tau) = 1$ if the average is performed first on time and then on pixels, while the correct result is found by performing the averages in the opposite order, as in Equation 8.33. Because the average over pixels already provides good-quality data, the time average can be quite short. In most cases, one simply averages over a time window comparable to the longest time delay τ of interest. For example, for a system with a relaxation time of 10^2 s, good data can be obtained by averaging over 10^4 pixels and a few hundred seconds, a tremendous gain as compared to conventional DLS, where a measurement time of about 1 day would be required. Not only this opens the way to measure very slow dynamics, but it also allows one to precisely characterize time-dependent dynamics, because "snapshots" of the dynamics can be taken, without "blurring" the evolution of dynamics by taking long time averages, as required by conventional methods. Examples include the slowing down of dynamics during aggregation, or aging phenomena in glassy and jammed materials (see, e.g., Ref. [18] and references therein).

Together with these very appealing features, multispeckle methods present also some disadvantages due to the limitations of multielement detectors. With typical acquisitions rates ranging from a few tens of hertz up to 10^4–10^5 Hz (for expensive cameras or 1D detectors) multielement detectors are much slower than phototubes or avalanche photodiodes that have acquisition rates up to 10^8 Hz. A solution to this problem is to combine both kinds of detectors in the same apparatus, as in Ref. [9]. A different approach is provided by Speckle Visibility Spectroscopy (SVS) [19], where one measures the blurring of the speckles in individual images due to the motion of the scatterers over the duration of the exposure time of the detector. Since for CCD and CMOS cameras the exposure time can be as short as a few microseconds, SVS greatly improves the range of dynamics accessible to multispeckle methods. Multielement detectors are also noisier than phototubes and have smaller sensitivity and dynamic range. Correction schemes that can alleviate these problems are discussed in Refs [17, 20]. Finally, we mention a mixed approach where one uses a phototube detector, but still records the intensity for many speckles by rotating the sample in a precisely controlled way. Each full turn of the sample generates a set of intensity values that can then be treated quite like the set of pixel intensity values of a multielement detector. Great mechanical stability and precisions are required, since one must make sure to sample exactly the same set of speckles at each turn. Details on these so-called interleave or echo approaches can be found in Refs [21, 22].

8.4.5 Time-Resolved Correlation

There is currently a growing interest in techniques that can characterize not only the average dynamics but also temporal and spatial fluctuations of the dynamics. This is motivated by the fact that many interesting systems have highly heterogeneous dynamics. Examples range from coarsening-driven bubble rearrangements in foams [23] to localized, intermittent rearrangements in a variety of glassy and jammed soft materials [24]. At first sight, scattering methods are not well suited for space- and time-resolved measurements, because (i) the detector is illuminated by light issued from the whole scattering volume, regardless of the location of the scatterers, and (ii) the intensity correlation function is averaged over time, thus "washing out" any potential signature of temporally heterogeneous dynamics. Multielement detectors, however, provide a solution to these problems. We describe here a time-resolved method (time-resolved correlation, TRC [20, 25]), while a version of this method with both temporal and spatial resolution, termed photon correlation imaging (PCI) [26], is introduced later in Section 8.5.1. Both methods are applicable to single and multiple scattering experiments alike.

In a TRC experiment, a multielement detector is placed in the far field, for example, as shown in Figure 8.1a or b. A series of speckle images is recorded and the change of the speckle pattern, which mirrors the change of the sample configuration, is quantified by calculating the degree of correlation between pairs of images taken at time t and $t + \tau$:

$$c_I(t, \tau) = \frac{\langle I_p(t) I_p(t + \tau)\rangle_p}{\langle I_p(t)\rangle_p \langle I_p(t + \tau)\rangle_p} - 1. \qquad (8.34)$$

A comparison with Equation 8.33 shows that $g_I(\tau) - 1 = \langle c_I(t, \tau)\rangle_t$. Thanks to the pixel average, however, the two-time correlation c_I is a "clean" enough quantity even when no time average is performed. Indeed, valuable information on temporally heterogeneous dynamics can be extracted from $c_I(t, \tau)$. To illustrate this point, we show in Figure 8.8 $g_I(\tau) - 1$ and $c_I(t, \tau)$ for two very different systems, a diluted suspension of Brownian particles and a shaving cream foam, both measured in a DWS transmission experiment. The intensity correlation function is close to an exponential decay for both systems. Note that the average dynamics is completely insensitive to the microscopic origin of the system reconfiguration: intermittent, localized bubble rearrangements due to internal stress imbalances that build up as a consequence of coarsening for the foam [23], particle diffusion for the colloidal suspension. By contrast, the time dependence of $c_I(t, \tau)$ at fixed τ highlights dramatically the temporally heterogeneous character of the foam dynamics, as opposed to the homogeneous dynamics for the colloids. For the foam, large fluctuations of c_I are observed; this is because the number and extent of bubble rearrangements over a given time lag is a stochastic quantity, leading to fluctuations in the fraction of the sample that has been reconfigured and hence of the degree of correlation [27].

The amplitude of the fluctuations may be characterized by the temporal variance of c_I, $\chi(\tau) = \langle c_I(t, \tau)^2\rangle_t - \langle c_I(t, \tau)\rangle_t^2$. Physically, we expect $\chi(\tau)$ to depend on the volume, V_r, of the regions that undergo a rearrangement, as compared to the scattering volume, V_s. If $V_r \ll V_s$, a large number of events N must occur in order to reconfigure the whole

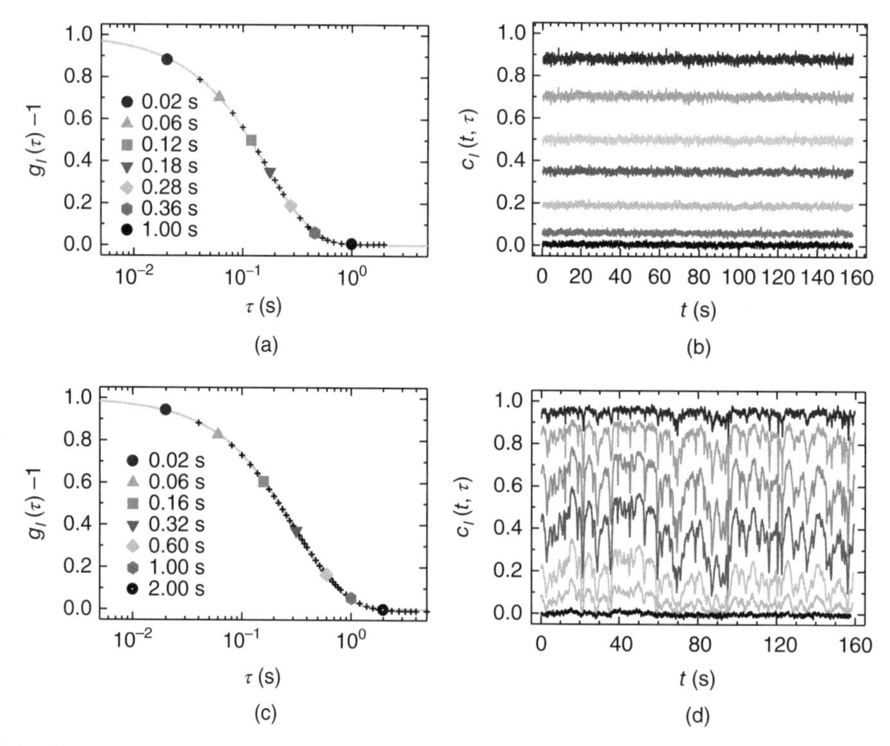

Figure 8.8 (a, c) $g_I(\tau) - 1$ as obtained by DWS for a dilute Brownian suspension (a) and a shaving cream foam (c). (b, d) Corresponding time-resolved degree of correlation as a function of time t. The large points in the left panels correspond to the time average of the traces shown in the right panels.

scattering volume (in single scattering) or all photon paths (in DWS) and thus fully decorrelate the scattered light. If N is large the relative number fluctuations will be small, such that χ will be modest. Conversely, χ will be large if the rearranged regions are a sizable fraction of the scattering volume. A more rigorous relation between χ and the size of the rearranged regions is discussed in Chapters 3, 4, and 6 of Ref. [24], where it is also shown that χ is analogous to the so-called dynamical susceptibility χ_4 used to characterize dynamic heterogeneity in many computer simulations of glassy and jammed systems. Other statistical characterizations of $c_I(t, \tau)$ are described in Ref. [20]; they include calculating its probability distribution function and its time autocorrelation, which provides information on the duration and separation between events. In some cases, individual rearrangements may be directly detected.

8.5 IMAGING AND SCATTERING

For many years, imaging and scattering methods have been well-distinct experimental tools, in terms of both the quantities that are measured and the optical layout that is used. These distinctions are, however, becoming increasingly blurred. Scattering methods based on imaging geometries have been developed, such as Photon Correlation Imaging (Section 8.5.1) and Near Field Scattering (Section 8.5.2). Similarly to conventional imaging, these scattering methods can provide spatially resolved information on the structure and dynamics of a system. On the other side, ordinary video microscopy has been demonstrated to be suitable for the determination of the q-dependent dynamic structure factor, as usually determined in scattering experiments (Section 8.5.3).

8.5.1 Photon Correlation Imaging

PCI [26] is an extension of TRC that allows one to measure a space- and time-resolved intensity correlation function. With respect to the TRC scheme described in Section 8.4.5, the major changes concern the collection optics. A scheme of a PCI apparatus for single scattering at 90° is shown in Figure 8.9; similar schemes are also applicable to other scattering angles and to DWS experiments in transmission or backscattering. Instead of collecting light scattered in the far field, one forms an image of the scattering volume onto a multielement detector, for example, a CCD chip, typically at a relatively low magnification $M \sim 1$. Note that the image is formed only using light scattered in a narrow cone centered around a well-defined scattering angle. Thus, PCI allies features typical of imaging methods with the sensitivity to a well-defined length scale $\sim 1/q$, as in conventional light scattering.

A typical image recorded in a PCI experiment is shown in Figure 8.9. Due to the low magnification, the coherent illumination, and the restricted range of q vectors accepted by the CCD, the scatterers are not individually resolved. Instead, one observes a speckle pattern, apparently similar to that obtained in conventional, far field multispeckle scattering. In contrast to far field speckles that are formed by the light issued from the whole scattering volume, however, each speckle in a PCI image receives only the contribution of scatterers located in a small volume, centered about the corresponding object point in the sample. The lateral size of this volume is $\sim \lambda z/(MD_{\text{ep}})$ [4] and its depth is equal to the beam dimension along the line of sight, for example, the beam diameter in Figure 8.9. Here, z is the lens-detector distance and D_{ep} is the diameter of the exit pupil of the collection optics, for example, the lens size in Figure 8.9. As a result of the imaging geometry, the fluctuations of the intensity of a given speckle are thus related to the dynamics of a well-localized, small portion of the illuminated sample.

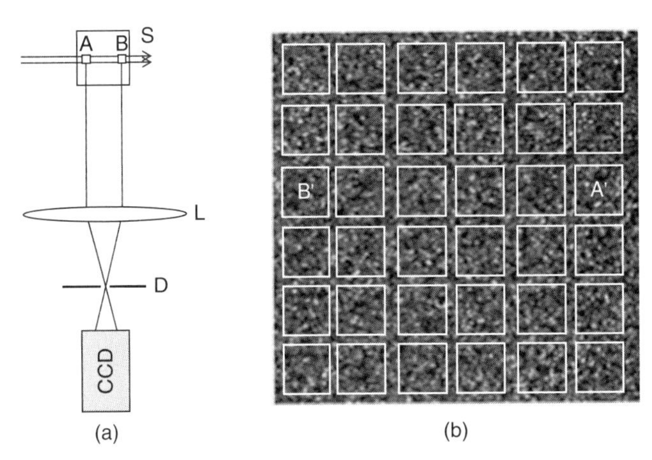

Figure 8.9 (a) Schematic view of an apparatus for space-resolved light scattering (PCI). The lens L makes an image of the sample S onto a multielement detector (CCD). The diaphragm D controls the range of scattering angles of the light accepted by the detection. In this case, only light scattered at $\theta \approx 90°$ contributes to the image formed onto the CCD. Regions A and B in the sample correspond to distinct portions A′ and B′ of the speckle image (b). The boxes overlaid to the speckle image indicate the ROIs used for the calculation of the local degree of correlation (see the text for more details).

The imaging processing for PCI is schematized in Figure 8.9. One divides the full image in regions of interest, ROIs, whose side corresponds typically to 15–50 μm in the sample. A local degree of correlation, $c_I(\mathbf{r}, t, \tau)$, is then calculated for each ROI, using Equation 8.34, where now the average is over the pixels belonging to the ROI at position $\mathbf{r}$. Figure 8.10 shows the temporal evolution of $c_I(\mathbf{r}, t, \tau)$ for a fixed time lag and three locations, measured at low angle for a quasi-2D shaving cream foam [26]. Regions A and B, separated by 0.05 mm, exhibit highly correlated traces, as indicated by concomitant drops of the degree of correlation. By contrast, region C, located 3.1 mm from A, shows uncorrelated dynamical events. This suggests that the

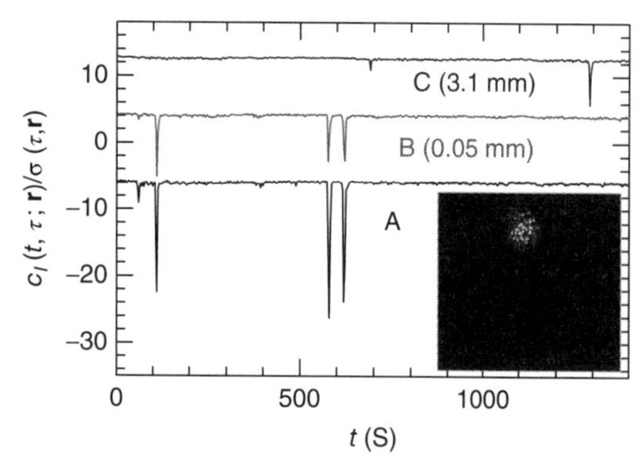

Figure 8.10 Degree of correlation measured for three distinct ROIs in a photon correlation imaging experiment on a quasi-2D foam. The time delay is $\tau = 3$ s; the sharp drops of c_I signal localized bubble rearrangements. Curves B and C correspond to ROIs at 0.05 and 3.1 mm from the ROI of A, as indicated by the labels. The degree of correlation is normalized by its temporal variance over the full duration of the experiment (about 7 h). For the sake of clarity, the curves have been offset vertically and only data for a shorter time window are shown here. Inset: Dynamic Activity Map (DAM), showing the spatial variation of the degree of correlation between two images separated by $\tau = 3$ s. Each pixel of the DAM corresponds to one ROI of the speckle images and is color-coded according to the value of c_I. The field of view is $5 \times 5\,\text{mm}^2$. The bright spot on the top of the DAM indicates a rearrangement event spanning about 10 bubbles.

spatial extent of the rearrangement events must be larger than 0.05 mm but smaller than 3.1 mm. Visual inspection of a map of the dynamical activity, shown in the inset of Figure 8.10, confirms that the region affected by a rearrangement extends over about 0.6 mm, corresponding to a few bubble sizes. Movies obtained by playing a sequence of dynamical activity maps can be viewed in the Supplementary Material of Ref. [26]. From these movies, the full temporal and spatial statistics of rearrangement events can be obtained. For example, in Ref. [27] it is shown that rearrangements in foams tend to persist in the same location over several events. This suggests that individual events do not fully relax the local internal stress accumulated during coarsening.

By comparing the temporal evolution of c_I at different locations, one can calculate a spatial correlation of the dynamics, $g_4(\Delta\mathbf{r}, \tau)$, whose space-varying part is proportional to $\langle\langle c_I(\mathbf{r}, t, \tau) c_I(\mathbf{r} + \Delta\mathbf{r}, t, \tau)\rangle_t\rangle_{\Delta\mathbf{r}}$ [26]. Surprisingly, the dynamics of a variety of jammed materials were found to be correlated over distances as large as several millimeters, presumably because the strain field set by a rearrangement can propagate very far in these predominantly solid-like materials [28].

8.5.2 Near Field Scattering

Near field scattering techniques have been recently demonstrated to be interesting alternatives to traditional small-angle light scattering techniques (for a review, see Ref. [29]). As mentioned in Sections 8.1 and 8.3, we generally exploit the scattered fields in the far field by either placing our detectors far from the sample or using a suitable set of optical elements. In this configuration, the light detected at a given angle from the primary beam contains the angular-specific scattering contributions of all particles within the scattering volume; the Fourier components of the scattered field are well separated in space and the speckle size d_{sp} depends solely on the beam diameter D and the sample-detector distance z, $d_{sp} = z \cdot (\lambda/D)$ [4]. In the near field configuration, this is not the case [30].

To understand this, let us consider a coherent beam of diameter D illuminating a sample of large particles with diameter d. As sketched in Figure 8.11a, any of these particles will mainly scatter the light in a cone with an angular width $\Theta \approx \lambda/d$. A sensor placed in the vicinity of the sample will only detect the scattered light of the subset of particles that are within $D^* \approx z \cdot \Theta \approx z \cdot (\lambda/d)$, as shown in Figure 8.11b. Remarkably, the size of the speckle depends here on the particle dimensions $d_{sp} = z \cdot (\lambda/D^*) \approx d$ and is independent of z as long $D^* < D$. The intensity measured at different detector positions (different observation angles) now mirrors the spatial distribution of the light scattered by the particles, rather than its Fourier transform. By simply taking the Fourier power spectrum of this intensity distribution, we can recover $I(q)$ as a function of q as usually measured in far field scattering experiments. In practice, most near field experiments are performed in the heterodyne mode, which entails the superposition of the primary beam to the scattered light. In this case, the Fourier transform of the near field image may also contain a q-dependent transfer function, which in principle can be calculated [31]. Alternatively, one can obtain the transfer function by performing a reference experiment using a sample, for which the scattering intensity is q-independent in the accessible q-range. The Fourier transform of the image taken in the near field then yields the q-dependence of the transfer function that can be used to isolate $I(q)$ [32].

In comparison to traditional small-angle scattering, the main advantage of near field techniques is that some of the strong experimental constraints of the far field experiments are relaxed. A typical setup for heterodyne near field scattering experiments is shown in Figure 8.12 [33]. In contrast to the small-angle setup shown in Figure 8.1b the primary beam is not blocked and the CCD detector can be placed off the optical axis, which also reduces the effect of stray light. Moreover, the distance b between lens and detector does not need to be carefully chosen as long as the plane imaged on the detector is within the near field regime, $z < D \cdot (d/\lambda)$. In addition to this simplified optical layout the correction scheme for stray light is simplified as well. Stray light scattered by imperfections or dust on the optical elements is a major problem in traditional small angle light scattering

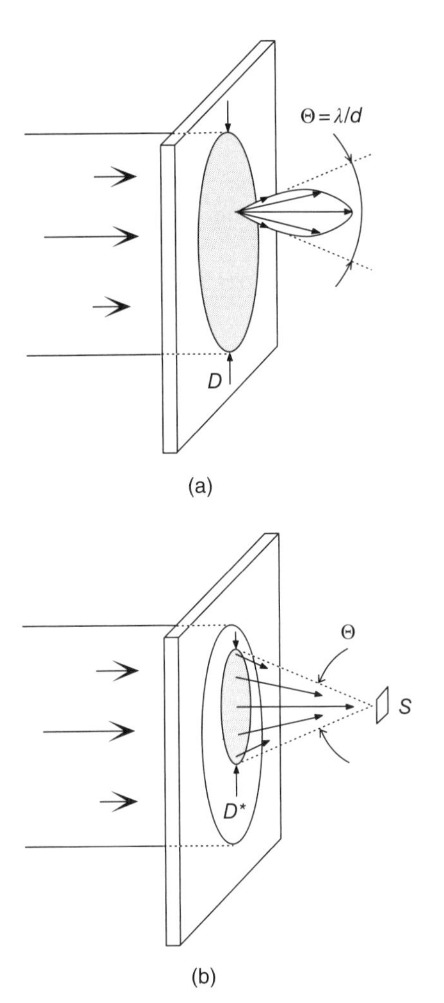

Figure 8.11 Illustration of scattering contributions in the near field plane. Source: From Ref. [30]. (a) A beam of diameter D illuminates a sample of colloidal particles of diameter d. Any of the particles within the illuminated volume scatters the light into a cone with an angular width of $\Theta \approx \lambda/d$. (b) The sensor S placed close to the sample receives the light from the portion of the sample with a diameter $D^* < D$, where again $\Theta \approx \lambda/d$.

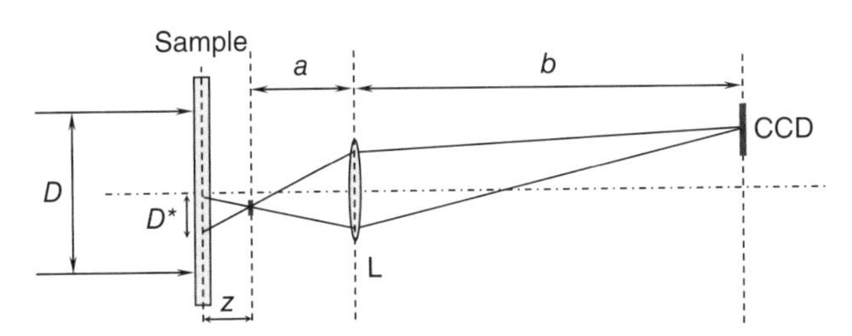

Figure 8.12 Schematic of a heterodyne near field setup. Source: Reprinted figure with permission from Ref. [33]. Copyright 2004 by the American Physical Society. The sample is illuminated with a collimated beam of size D and the near field speckle pattern formed in the plane located at a distance z from the cell is imaged onto a camera using a lens L.

(SALS) measurements. An example is shown in Figure 8.2, where the upturn of the averaged intensity for $q < 0.3°$ visible in (b) is due to stray light. One generally accounts for stray light contributions by performing a reference experiment, where the sample cell is filled with the background medium of the sample, that is, the solvent. After this reference experiment, the solvent is carefully removed, the cell is dried while maintained in the setup and then filled with the sample. The scattered intensities taken with and without the sample are then subtracted from each other, which corrects for stray light contributions [34]. In near field scattering, by contrast, one can use experimentally less demanding schemes provided the sample is reconfiguring in time. In this case, two images acquired with a delay that is sufficiently long to allow for the system to completely reconfigure are statistically independent. Subtracting these images from each other generates an image, where the contributions of the static stray light cancel out. This image is then processed to obtain $I(q)$ as a function of q, as outlined above [33].

Dynamic near field experiments that yield information equivalent to that obtained in DLS are based on the same principle. In this case, however, the delay time is chosen to be short compared to the typical reconfiguration time. Subtracting two images that have been acquired at t and $t + \tau$ generates an image containing only the nonstatic contributions; any reconfiguration of the system during τ will lead to a signal in the difference image. By taking the Fourier transform of such difference image, we obtain the q-dependent values of the power spectrum at the chosen τ. Performing this procedure for different τ enables us to construct the full lag-time dependence of the power spectrum at different q, yielding essentially the field autocorrelation function $g_E(\mathrm{q}, \tau)$ introduced in Section 8.4.

8.5.3 Differential Dynamic Microscopy

As typical interferometric methods, DLS techniques are usually based on the use of coherent light sources. However, it has been shown that the determination of the field temporal autocorrelation is not restricted to the use of coherent illumination and that q-dependent dynamics can also be retrieved from images taken with a microscope using an ordinary white light illumination with quite limited coherence [35, 36]. This method has been termed differential dynamic microscopy (DDM).

In DDM, one takes again advantage of a differential algorithm. A sequence of images is acquired in time and the difference between two images taken at different delays generates images that reflect by how much the system has reconfigured during a given delay time τ. Taking the Fourier transform of these difference images gives access to the Fourier power spectrum $|F(\mathbf{q}, \tau)|^2$ as a function of $\mathbf{q}$ and τ. Considering that the intensity distribution of the original images is related to the density distribution in the sample, one can then show that

$$|F(\mathbf{q},\tau)|^2 = A(\mathbf{q})[1 - g_E(\mathbf{q},\tau)] + B(\mathbf{q}), \qquad (8.35)$$

where $B(\mathbf{q})$ accounts for camera noise, while $A(\mathbf{q})$ depends on the q-dependent scattered and transmitted light [35, 36]. For the characterization of the dynamics of a system $A(\mathbf{q})$ and $B(\mathbf{q})$ can simply be treated as fitting parameters, the dynamics of the system determining the actual τ-dependence of $g_E(\mathbf{q},\tau)$ and thus $|F(\mathbf{q},r)|^2$. This fairly recent technique has been successfully applied for the characterization of the dynamics of colloids [35, 37] and bacteria [38] and presents a valuable addition to the tracking techniques normally used to characterize the dynamics of mesoscopic systems in microscopy.

REFERENCES

[1] Kerker M. *The Scattering of Light, and Other Electromagnetic Radiation*. New York: Academic; 1969.

[2] Berne BJ, Pecora R. *Dynamic Light Scattering*. New York: Wiley; 1976.

[3] Ashcroft NW, Lekner J. Phys Rev 1966;145:83.

[4] Goodman JW. *Speckle Phenomena in Optics: Theory and Applications*. Englewood: Roberts and Company; 2007.

[5] Burstyn HC, Sengers JV. Phys Rev A 1982;25:448.

[6] Brown JC, Pusey PN, Goodwin JW, Ottewill RH. J Phys A – Math Gen 1975;8:664.

[7] Schaetzel K. J Mod Opt 1991;38:1849.

[8] Maret G, Wolf PE. Z Phys B Condens Matter 1987;65:409.

[9] Pine DJ, Weitz DA, Chaikin PM, Herbolzheimer E. Phys Rev Lett 1988;60:1134.

[10] Weitz DA, Pine DJ. Diffusing-wave spectroscopy. In: Brown W, editor. *Dynamic Light Scattering*. Oxford: Clarendon Press; 1993.

[11] Zhu JX, Pine DJ, Weitz DA. Phys. Rev. A 1991;44:3948.

[12] Pusey PN, van Megen W. Physica A 1989;157:705.

[13] Xue JZ, Pine DJ, Milner ST, Wu XL, Chaikin PM. Phys Rev A 1992;46:6550.

[14] Zakharov P, Cardinaux F, Scheffold F. Phys Rev E 2006;73:011413.

[15] Kirsch S, Frenz V, Schärtl W, Bartsch E, Sillescu H. J Chem Phys 1996;104:1758.

[16] Wong APY, Wiltzius P. Rev Sci Instrum 1993;64:2547.

[17] Cipelletti L, Weitz DA. Rev Sci Instrum 1999;70:3214.

[18] Cipelletti L, Ramos L. J Phys Condens Matter 2005;17:R253.

[19] Bandyopadhyay R, Gittings AS, Suh SS, Dixon PK, Durian DJ. Rev Sci Instrum 2005;76:093110.

[20] Duri A, Bissig H, Trappe V, Cipelletti L. Phys Rev E 2005;72:051401.

[21] Muller J, Palberg T. Prog Colloid Polym Sci 1996;100:121.

[22] Pham KN, Egelhaaf SU, Moussaid A, Pusey PN. Rev Sci Instrum 2004;75:2419.

[23] Durian DJ, Weitz DA, Pine DJ. Science 1991;252:686.

[24] Berthier L, Biroli G, Bouchaud J-P, Cipelletti L, van Saarloos W, editors. *Dynamical Heterogeneities in Glasses, Colloids, and Granular Media*. Oxford: Oxford University Press; 2011.

[25] Cipelletti L, Bissig H, Trappe V, Ballesta P, Mazoyer S. J Phys Condens Matter 2003;15:S257.

[26] Duri A, Sessoms DA, Trappe V, Cipelletti L. Phys Rev Lett 2009;102:085702.

[27] Sessoms D, Bissig H, Duri A, Cipelletti L, Trappe V. Soft Matter 2010;6:3030.

[28] Maccarrone S, Brambilla G, Pravaz O, Duri A, Ciccotti M, Fromental JM, Pashkovski E, Lips A, Sessoms D, Trappe V, Cipelletti L. Soft Matter 2010;6:5514.

[29] Cerbino R, Vailati A. Curr Opin Colloid Interface Sci 2009;14:416.

[30] Giglio M, Carpineti M, Vailati A, Brogioli D. Appl Opt 2001;40:4036.

[31] Trainoff SP, Cannell DS. Phys Fluids 2002;14:1340.

[32] Vailati A, Cerbino R, Mazzoni S, Takacs CJ, Cannell DS, Giglio M. Nat Commun 2011;2:290.

[33] Ferri F, Magatti D, Pescini D, Potenza MAC, Giglio M. Phys Rev E 2004;70:041405.

[34] Tamborini E, Cipelletti L. Rev Sci Instrum 2012;83:093106.

[35] Cerbino R, Trappe V. Phys Rev Lett 2008;100:188102.

[36] Giavazzi F, Brogioli D, Trappe V, Bellini T, Cerbino R. Phys Rev E 2009;80:031403.

[37] Lu PJ, Giavazzi F, Angelini TE, Zaccarelli E, Jargstorff F, Schofield AB, Wilking JN, Weitz MBDA, Cerbino R. Phys Rev Lett 2012;108:218103.

[38] Wilson LG, Martinez VA, Schwarz-Linek J, Tailleur J, Bryant G, Pusey PN, Poon WCK. Phys Rev Lett 2011;106:018101.

9

RHEOLOGY OF SOFT MATERIALS

Hans M. Wyss[1]
Department of Mechanical Engineering and ICMS, Eindhoven University of Technology, 5612AJ Eindhoven, the Netherlands

9.1 INTRODUCTION

Rheology is the study of the flow and deformation of materials. Unlike the ideal concepts of elastic solids and viscous liquids, real materials generally behave intermediate between these two theoretical extremes of mechanical behavior. In particular, soft materials exhibit not just elastic or viscous properties, but their mechanical behavior is *viscoelastic* and depends sensitively on the typical timescales at which the material is deformed. Polymers and soft materials often show significant variations in their behavior in a range of timescales between milliseconds and minutes, a regime that we can directly experience and observe. This becomes apparent in many everyday examples of soft materials,

Fluids, Colloids and Soft Materials: An Introduction to Soft Matter Physics, First Edition. Edited by Alberto Fernandez Nieves and Antonio Manuel Puertas.

where the richness in mechanical behavior at timescales compatible with the human sensory system enables us to experience their viscoelastic nature.

As an example, the difference in mechanical properties between honey and mayonnaise is not so apparent when poking these materials with a knife, as their viscosities are similar. However, over a timescale of minutes, the honey in the jar will flow to form a flat surface again, while the mayonnaise will remain in the deformed state inscribed by the knife for an indefinite period. The two materials thus exhibit a fundamentally different rheological response, which can only be fully characterized by probing the material at different magnitudes of deformation and at different timescales.

9.2 DEFORMATION AND FLOW: BASIC CONCEPTS

In rheology, we probe the response of a material as a strain deformation or stress is applied. A shear deformation, which is most frequently used in such measurements, can be schematically depicted by considering a piece of material sandwiched between two plates. When the upper plate is laterally displaced by a distance Δx, a pure shear deformation occurs, as schematically shown in Fig. 9.1. The **shear strain** $\gamma(t)$ is then defined as $\gamma(t) = \frac{\Delta x(t)}{h}$, with h the distance between the two plates. To displace the upper plate, a time-dependent lateral force $F(t)$ is required. The **shear stress** $\sigma(t)$ is then given by this force divided by the surface area of the upper plate, $\sigma(t) = \frac{F(t)}{A}$. In general, the force $F(t)$ also strongly depends on the rate of deformation of the material. This is characterized by the **shear rate** $\dot{\gamma}(t)$, defined as the velocity of the upper plate $v(t) = \frac{\partial \Delta x}{\partial t}$ divided by the plate separation h, $\dot{\gamma}(t) = \frac{\frac{\partial \Delta x(t)}{\partial t}}{h} = \frac{v(t)}{h}$.

The general three-dimensional stresses, deformations, and the flow of a material are often theoretically described in tensorial form as second-order tensors for the stress, the deformation, and the rate of deformation. However, this treatment is beyond the scope of this chapter, and is also not necessary to grasp the essential ideas about the viscoelastic response of soft materials. While we will limit ourselves here to the discussion of shear deformations, the concepts and methods used are directly applicable also to extensional flows and deformations.[1] For a comprehensive discussion of the tensorial description of rheology, we refer to the excellent books by Macosko [1] and Larson [2].

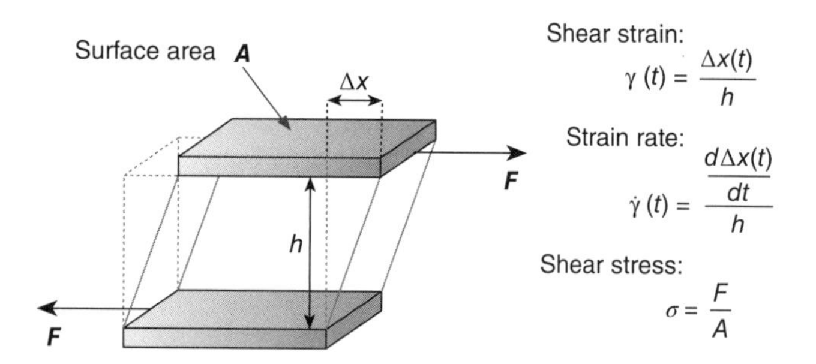

Figure 9.1 *Deformations and stresses.* Definition of strain, strain rate, and stress in a pure shear deformation.

9.2.1 Importance of Timescales

When considering the response of a material to deformations, both the characteristic timescales of the material and the timescale of the deformation are of key importance. The characteristic timescale of a material is the time it takes the material to adjust to an equilibrium stress state after applying a deformation. These timescales characterize the relaxation of stress in the material and are therefore referred to as *relaxation times*.[2] The rheological behavior becomes particularly rich if the timescale of deformation and the typical relaxation time are comparable.

Indeed, for many soft materials the typical relaxation times lie within a range that is directly observable to us, often between milliseconds and minutes; and most deformations that are applied to materials in practical applications exhibit a similar range of typical timescales. This is one reason for the particular richness observed in the rheological behavior of soft matter; because practically relevant and directly observable deformations occur on similar timescales as do the characteristic relaxation times of these materials, their behavior is highly sensitive to the applied deformation timescale. For slow deformations, at timescales longer than their relaxation times, they exhibit a liquid-like behavior, while their response becomes increasingly solid-like for faster deformations.

The effect of these timescales on the mechanical response of a material is conveniently characterized by the dimensionless *Deborah number De*

$$De = \frac{\tau_0}{\tau_d}, \tag{9.1}$$

where τ_0 is the typical relaxation time of the material, so the timescale over which the material adjusts to a new equilibrium stress state, and τ_d is the characteristic timescale of the deformation process, typically $\tau_d \approx 1/\dot{\gamma}$.

[1]However, what we do not consider in this treatment is the coupling between different modes of deformations; for instance, in most materials the application of a shear deformation will result in normal stress differences. We cannot adequately capture these effects using a one-dimensional description.

[2]The response of a material is usually not governed just by one single relaxation process; instead, most materials exhibit a range of relaxation processes, corresponding to a distribution of relaxation times that can cover a wide range of frequencies. However, the slowest relaxation process will usually dominate the mechanical response as it sets the characteristic timescale for a material to adjust to an equilibrium stress state following a deformation.

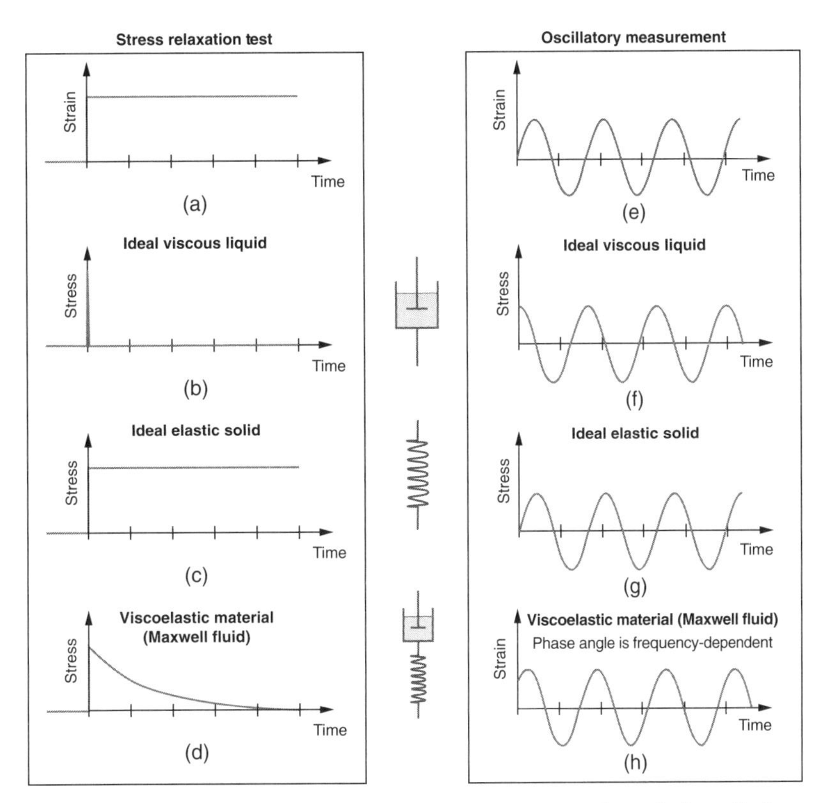

Figure 9.2 Timescale-dependent material response In a **stress relaxation test**, a step in strain is applied to a material a. Resulting stress response for a *viscous liquid* b, an *elastic solid* c, for, and for a *Maxwell fluid* (D). In **oscillatory measurements**, a sinusoidal strain deformation is applied e. For a linear response, the resulting stress response remains sinusoidal with a phase shift and amplitude that depends on the material properties. For a *viscous liquid* the response is proportional to the strain *rate* and thus *out of phase* with the applied strain f, for an *elastic solid* the response is *in phase* with the applied strain g, and for a *Maxwell fluid* the phase shift is between 0 and 90 degrees, depending on the oscillation frequency h.

For large Deborah numbers, $De \gg 1$ the material behaves like an elastic solid, while for $De \ll 1$ the material flows like a liquid.

9.3 STRESS RELAXATION TEST: TIME-DEPENDENT RESPONSE

The simplest rheological experiment for testing this time-dependent material response in detail is the so-called *stress relaxation test*, which also naturally exemplifies the basics of *elastic*, *viscous*, and *viscoelastic response*. In a stress relaxation test, an instantaneous strain deformation[3] γ_0 is applied to the material at time $t = 0$ and the resulting stress response is monitored as a function of time, as shown schematically in Figure 9.2a and b.

For a purely *elastic material*, the stress is by definition always *proportional to the applied strain deformation* $\gamma(t)$ and thus, as shown in Figure 9.2c, the stress remains at a fixed level, proportional to the applied strain $\sigma(t > 0) = G\gamma_0$, where the proportionality factor is the *elastic shear modulus G*.

For a purely *viscous material*, the stress is *proportional to the applied strain rate* $\dot{\gamma}(t)$ and thus for such an ideal liquid the stress vanishes immediately after the step in strain has been applied $\sigma(t > 0) = 0$, and stress occurs only at the points in time where the step in strain is applied, as shown in Figure 9.2b. In practice, a step in strain cannot be applied instantaneously but will have to occur with a finite strain rate $\dot{\gamma}_{\text{step}}$. Whenever a strain rate is applied, the stress in the viscous liquid is $\sigma(t) = \eta\dot{\gamma}(t)$, where the proportionality factor is the *shear viscosity* η.

Soft materials generally exhibit a behavior that lies between that of a purely elastic solid and an ideal viscous liquid; they thus exhibit a *viscoelastic* response. The simplest model for a viscoelastic material is the so-called *Maxwell model*, which possesses a single relaxation time τ. This model, similar to a shock absorber in a car, consists of a viscous damping element, a "dashpot" with a characteristic viscosity η_0, and an elastic spring with a characteristic elastic shear modulus G_∞ in series, as schematically shown in Figure 9.2. The relaxation time of such an ideal Maxwell

[3] A stress relaxation test is therefore also often referred to as a *step-strain* test.

fluid is given by the ratio of the characteristic viscosity and the characteristic shear modulus: $\tau = \frac{\eta_0}{G_\infty}$. At slow enough deformations ($\dot{\gamma} \ll 1/\tau$), the material behaves like a liquid of viscosity η_0, while at very fast deformations ($\dot{\gamma} \gg 1/\tau$) the response is that of an elastic solid with a shear elastic modulus G_∞.

In a stress relaxation test on such an *ideal Maxwell fluid*, just after the initial rapid strain step occurs, the stress in the material is determined by the elastic modulus G_∞ as $\sigma(t = 0) = G_\infty \gamma_0$, after which the *stress decays as a single exponential function* $\sigma(t) = G_\infty e^{-\frac{t}{\tau}}$, as shown in Figure 9.2d. The characteristic time of this decay is the relaxation time $\tau = \frac{\eta_0}{G_\infty}$ of the material.

Some soft materials, such as wormlike micelles [3–5] or supramolecular polymers [6, 7] actually exhibit a response remarkably close to this simple model. However, most viscoelastic materials show a more complex response, as a result of a broad range of different relaxation processes and characteristic timescales that govern their behavior.

9.3.1 The Linear Response Function $G(t)$

The simple stress relaxation test described in the previous section enables us to measure the *linear response function* $G(t)$ of a material, defined as

$$G(t) = \frac{\sigma(t)}{\gamma_0}, \tag{9.2}$$

where γ_0 is the magnitude of the step strain applied at $t = 0$. The response function $G(t)$ is also often referred to as the *(time-dependent) relaxation modulus*, as it captures the full linear viscoelastic response of a material and it is accessed via a stress relaxation measurement. Because the response is linear and any time-dependent strain deformation $\gamma(t)$ can be seen as a series of infinitesimal steps of magnitude $d\gamma = \dot{\gamma} dt$, the stress response for an arbitrary deformation history can be obtained by integrating over these infinitesimal strain steps as

$$\sigma(t) = \int_{-\infty}^{t} G(t - t')\dot{\gamma}(t')dt'. \tag{9.3}$$

The linear response function $G(t)$ can thus be interpreted as a *memory function* for the material response; if it vanishes at long times, the material "loses its memory" of deformations that occurred further than these timescales in the past, while more recent deformations still contribute to the material response. This occurs for all fluids and fluid-like materials, where relaxation times are finite and the stress response thus always eventually decays to zero when no further deformations occur. For elastic-like materials, however, all deformations that occurred in the past cumulatively contribute to the stress in the material. For an *ideal* elastic solid the response function is a constant, $G(t) = G$, and thus all past deformations contribute to the current stress state, as $\sigma(t) = G \int_{-\infty}^{t} \dot{\gamma}(t')dt' = G\gamma(t)$.

The response function $G(t)$ is directly analogous to other linear response functions encountered in spectroscopic techniques such as dielectric spectroscopy or infrared spectroscopy; in fact, rheology can be regarded as *mechanical spectroscopy*. The behavior of the ideal Maxwell fluid is the simplest possible response for a viscoelastic fluid; this material exhibits *one single relaxation time*. In fact, any valid linear response function can be described as a *superposition of single exponential functions* and thus as a superposition of linear Maxwell models as

$$G(t) \approx \sum_{k=1}^{N} G_k e^{t/\tau_k}, \tag{9.4}$$

provided the number of modes N is large enough. For the case $N \to \infty$ we can also rewrite Equation (9.4) as an integral

$$G(t) = \int_0^\infty \frac{H(\tau)}{\tau} e^{-t/\tau} d\tau, \tag{9.5}$$

where $H(\tau)$ is referred to as the *relaxation spectrum*, which describes a continuous distribution of relaxation times, rather than a finite number of individual timescales. However, usually a sum over a finite number of single exponential relaxation modes τ_k with associated magnitudes G_k can adequately describe the response of a material. Including a purely elastic component, described by an elastic modulus G_e, the relaxation modulus can then be written as

$$G(t) \approx G_e + \sum_{i=k}^{N} G_{(k)} e^{-\frac{t}{\tau_k}}. \tag{9.6}$$

The elastic modulus G_e accounts for the long-time elastic response of the material.[4] It describes the part of the stress that does not relax even at infinite times. The value of the elastic modulus G_e implies a definition of truly solid-like and liquid-like behavior; for solid materials $G_e > 0$, while for liquid-like materials G_e vanishes.[5]

Equation (9.4) is often used to fit experimental data; already for a small number of modes, $N < 10$, in most cases an adequate description of the data is obtained; however, the relaxation times τ_k extracted from such fitting procedures are usually not physically meaningful, unless the relaxation times in the material are clearly separated [8]. For instance, a commonly observed and theoretically expected functional form of $G(t)$ is a power-law decay. However, the very nature of such a decay is the absence of any characteristic timescale. A power law can be well approximated by a

[4] This would correspond to a mode with infinite relaxation time $\tau_e \to \infty$.

[5] $G_e = 0$ implies that if the material is deformed slowly enough (slower than the longest relaxation time), the material will behave like a liquid.

sum over single exponentials, but the individual extracted timescales clearly do not have any physical significance in this case.

9.4 OSCILLATORY RHEOLOGY: FREQUENCY-DEPENDENT RESPONSE

Another very common and useful way of probing the timescale-dependent viscoelastic response of soft materials is the use of oscillatory measurements, which characterizes the material response in frequency space. Instead of probing the relaxation of stresses in the material as a function of time, here we probe the viscoelastic behavior at a frequency ω by applying a sinusoidal shear deformation

$$\gamma(t) = \gamma_0 \sin(\omega t) \tag{9.7}$$

and measuring the resulting stress response $\sigma(t)$, thereby characterizing the viscoelastic behavior of a material at a characteristic frequency ω, which corresponds to a timescale $\tau = 2\pi/\omega$.

Performing such a measurement on an *ideal elastic material*, as shown schematically in Figure 9.2g, the stress must always be proportional to the applied strain

$$\sigma_{\text{elastic}}(t) = G \cdot \gamma(t), \tag{9.8}$$

where the proportionality constant G is the *shear modulus* of the material. For such a material, the stress response is thus completely *in phase* with the applied strain deformation $\sigma(t) = \sigma_0 \cdot \sin(\omega t)$.

In contrast, for an *ideal viscous liquid*, the stress is always proportional to the applied strain *rate*,

$$\sigma_{\text{viscous}}(t) = \eta\dot{\gamma}(t) = \eta\gamma_0\omega \cos(\omega t). \tag{9.9}$$

The stress response is thus out of phase with respect to the applied strain $\gamma(t)$, as shown schematically in Figure 9.2f; the phase angle δ between $\sigma(t)$ and $\gamma(t)$ is in this case 90°, or $\delta = \frac{\pi}{2}$.

For *viscoelastic* materials, such as the Maxwell fluid introduced above, the phase lag is between 0 and 90°, $0 < \delta < \frac{\pi}{2}$, with δ depending on the oscillation frequency ω, as shown schematically in Figure 9.2h. The general response of a linear viscoelastic material thus contains both in-phase and out-of-phase components, reflecting the elastic and viscous contributions, respectively.

9.4.1 Storage Modulus G' and Loss Modulus G''

It thus makes physical sense to split up the stress response into two parts, the in-phase response and the out-of-phase response, as follows:

$$\sigma(t) = G'\gamma_0 \sin(\omega t) + G''\gamma_0 \cos(\omega t). \tag{9.10}$$

Alternatively, if we write the applied oscillatory strain deformation as a complex function $\gamma(t) = \gamma_0 \cdot e^{i\omega t}$, then the in-phase response is determined by the real part and the out-of-phase response by the imaginary part of the stress response, as

$$\sigma(t) = G^*(\omega) \cdot \gamma_0 e^{i\omega t}, \tag{9.11}$$

where $G^*(\omega)$ is the *complex shear modulus*; G' and G'' are then, respectively, the real and imaginary parts of the complex modulus,

$$G^*(\omega) = G'(\omega) + i \cdot G''(\omega), \tag{9.12}$$

where $G'(\omega)$ is referred to as the *storage modulus* or the *elastic modulus*, and $G''(\omega)$ is called the *viscous modulus* or the *loss modulus*.

Generally, the storage and the loss moduli are frequency dependent. We say that a material is *liquid-like* at a given frequency or timescale if $G''(\omega) > G'(\omega)$; however, at a different frequency the material could be *solid-like*, with $G'(\omega) > G''(\omega)$.

A typical example of this behavior is shown in Figure 9.3, where G' and G'' are plotted as a function of frequency for a typical soft glassy material, a concentrated suspension, or paste, of soft microgel particles in water. At high frequencies, the storage modulus is clearly higher than the loss modulus; at frequencies even higher than those probed here G' is expected to reach a plateau, where it depends only weakly on frequency. In contrast, at the lowest frequencies probed

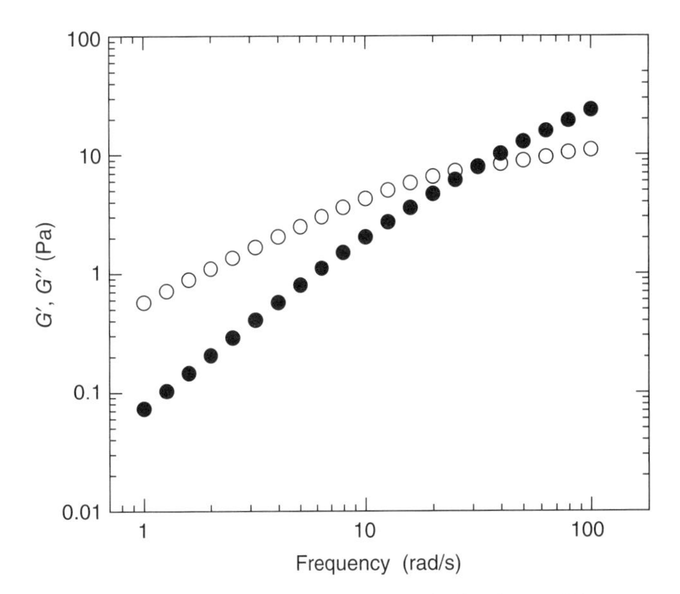

Figure 9.3 Viscoelastic response for a typical soft material, a concentrated suspension or paste of microgel particles. The storage modulus (solid symbols) and the loss modulus (open symbols) are plotted as a function of frequency, revealing a solid-like response at high frequencies and a liquid-like response at low frequencies.

the material is clearly liquid-like, with $G'' > G'$. Moreover, the loss modulus in this regime scales sensitively with frequency, similar to the case of a Newtonian liquid, albeit with a smaller slope, $G''(\omega) \propto \omega^n$, with $0 < n < 1$. In accord with this, the storage modulus falls off even more rapidly toward lower frequencies, with a power-law exponent of $\approx 2n$.

9.4.2 Relation Between Frequency- and Time-Dependent Measurements

As discussed in the previous sections, the linear viscoelastic behavior of soft materials can be characterized by either a stress relaxation experiment, which measures the time-dependent relaxation function $G(t)$, or by a frequency-dependent oscillatory experiment, which characterizes the complex modulus $G^*(\omega)$. Both measurements fully characterize the material response in the linear viscoelastic regime, within the time or frequency range accessed in the experiment. Thus, if $G(t)$ is known, we should be able to calculate the complex modulus $G^*(\omega)$ and the corresponding storage modulus $G'(\omega)$ and loss modulus $G''(\omega)$ as a function of frequency.

Indeed, as described above, the stress at any point in time is obtained from the shear rate history and the memory function $G(t)$; for a steady-state oscillatory measurement[6] we thus obtain

$$\sigma(t) = \int_{-\infty}^{t} G(t - t')\gamma_0 i\omega e^{i\omega t'}\, dt', \tag{9.13}$$

which, by a change of variables to $s = t - t'$, yields

$$\sigma(t) = \gamma_0 i\omega \int_0^{\infty} G(s)e^{i\omega(t-s)}\, ds = \gamma_0 e^{i\omega t} i\omega \int_0^{\infty} G(s)e^{-i\omega s}\, ds. \tag{9.14}$$

[6]We can prescribe a steady oscillatory state by assuming a deformation history with an oscillatory deformation $\gamma(t) = \gamma_0 e^{i\omega t}$ that started in the infinite past.

Thus, the two kinds of measurements are directly related to each other via a unilateral Fourier transform, as seen on the right-hand side of Equation (9.14). As a result, if $G(t)$ is known for all times $t > 0$, the complex modulus $G^*(\omega)$ is obtained as

$$G^*(\omega) = G'(\omega) + iG''(\omega) = i\omega \int_0^{\infty} G(t)e^{-i\omega t}\, dt. \tag{9.15}$$

9.5 STEADY SHEAR RHEOLOGY

In *steady shear rheology*, the viscosity of a material as a function of shear rate is measured. In these measurements, one is usually interested in steady-state flow, where the flow and the stress response of the material have equilibrated to an imposed constant shear rate. For *Newtonian fluids*, the shear stress is always proportional to the applied shear rate and the viscosity is thus constant. For viscoelastic materials, deviations from this ideal Newtonian behavior become apparent through variations in the measured viscosity as a function of the applied shear rate.

Most soft materials such as suspensions, foams, emulsions, pastes, polymer solutions, and many surfactant systems exhibit a regime of so-called *shear thinning* behavior, where the stress increases sublinearly with the applied strain rate, as shown in Figure 9.4a. The viscosity of these materials thus decreases with increasing shear rate, as shown in Figure 9.4b. This is often related to a structural change in the material, where building blocks of the material, such as colloidal particles or polymer chains, become aligned or organized, thereby facilitating macroscopic shear flow.

At higher shear rates, materials such as colloidal suspensions exhibit the opposite behavior, so-called *shear thickening*, where the shear stress increases stronger than linearly with increasing shear rate, as shown schematically

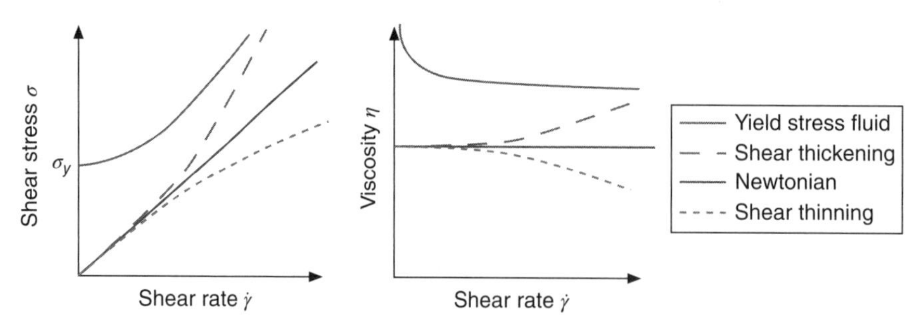

Figure 9.4 Main response types for soft materials under steady shear: *Newtonian fluid*: The shear stress σ increases linearly with $\dot{\gamma}$ (left-hand side); as a result the viscosity is constant at all shear rates (right-hand side). *Shear thickening*: σ increases stronger than linear with $\dot{\gamma}$; the viscosity increases with increasing shear rate. *Shear thinning*: σ increases sublinearly with increasing $\dot{\gamma}$; the viscosity thus decreases with increasing shear rate. *Yielding material:* Steady flow requires a minimum stress, the so-called yield stress σ_y. The stress increases from this low shear limit σ_y as the shear rate is increased. The apparent viscosity approaches infinity for very low shear rates, as $\eta = \sigma_y/\dot{\gamma}$.

in Figure 9.4a. This implies that the viscosity of the material increases as the shear rate increases, as shown in Figure 9.4b.

Many *solid-like soft materials* exhibit a so-called *yield stress*, a critical level of applied stress, σ_y, above which the material will start to flow. For values of the applied shear stress below this critical yield stress σ_y, the induced shear rate will thus vanish, $\dot{\gamma} = 0$. If a fixed shear rate is applied, the steady-state shear stress at low $\dot{\gamma}$ thus approaches the yield stress σ_y, as shown in Figure 9.4a. The apparent viscosity at low shear rates is then given by $\eta = \sigma_y / \dot{\gamma}$, which means it diverges in the zero shear rate limit. Such yielding behavior is frequently observed for polymers, pastes, solid-like emulsions and foams, and also in concentrated colloidal suspensions.

9.6 NONLINEAR RHEOLOGY

In the previous sections, we have considered the linear viscoelastic response of materials at small stresses and small strain deformations. If the deformation of the material is small enough, most materials exhibit a linear response, where the stress is directly proportional to the magnitude of the applied strain deformation. In this so-called *linear viscoelastic regime*, the material response can be directly linked to material properties such as the frequency-dependent storage and loss modulus. However, at higher degrees of deformation, most materials no longer respond in a linear manner; their stress response is no longer proportional to the magnitude of the applied strain deformation. The reason for this *nonlinear viscoelastic response* is often a structural change within the material itself that is induced by large deformations; for instance, this could be the breakage of bonds in a physical or chemical gel, shear-induced stretching and orientation of macromolecules, the stretching of chains in flexible and semiflexible polymer networks, or a change in the structural arrangement of particles in a colloidal suspension.

9.6.1 Large Amplitude Oscillatory Shear (LAOS) Measurements

When the response of a material is no longer in the linear viscoelastic regime, the concepts of storage and loss modulus G' and G'' become meaningless, as they are based on the assumption of a purely harmonic stress response to an applied sinusoidal strain deformation. Nevertheless, a wealth of useful information can be deduced from measurements in the nonlinear viscoelastic regime. These measurements are often referred to as large amplitude oscillatory shear (LAOS) measurements, as opposed to small amplitude oscillatory shear (SAOS) measurements, where the deformations are small enough to remain within the linear viscoelastic regime. Recent advances in the area of LAOS measurements deal with the interpretation of the data acquired in this regime [9]; they include the geometrical interpretation of the response when plotted as so-called Lissajous curves [10], Fourier-Transform rheology, which describes the shape of the stress response as a Fourier series [11, 12], as well as extensions of the concepts of the storage and the loss modulus to the nonlinear regime [9, 13].

9.6.2 Lissajous Curves and Geometrical Interpretation of LAOS Data

To illustrate the types of nonlinear behavior that can occur in oscillatory rheological measurements, the data obtained in LAOS measurements are often displayed in the so-called Lissajous–Bowditch curves, where the momentary stress $\sigma(t)$ is plotted as a function of the applied momentary strain deformation $\gamma(t) = \gamma_0 \sin(\omega t)$ or as a function of the applied momentary strain rate $\dot{\gamma}(t)$.

If the measurement is in the linear viscoelastic regime, then the response is harmonic with $\sigma(t) = G'(\omega)\gamma_0 \sin(\omega t) + G''(\omega)\gamma_0 \cos(\omega t)$ and, as a consequence, the Lissajous curve will be of elliptical shape, as shown in Figure 9.5a. This curve as a whole is tilted, with a slope corresponding to the elastic modulus $G'(\omega)$ of the material. For a linear material, this slope can either be taken from the line passing through the points of minimum and maximum strain, shown as a dotted line in Figure 9.5a, or as the tangential curve to the Lissajous curve at zero strain, shown as a dashed line in Figure 9.5a. Both slopes can be used as a definition of a characteristic elastic modulus, as suggested by Ewoldt et al. [9, 13]; these moduli are referred to as the large-strain elastic modulus,

$$G'_L = \frac{\partial \sigma}{\partial \gamma}(\gamma = 0) \tag{9.16}$$

and the minimum-strain elastic modulus,

$$G'_M = \frac{\sigma(\gamma = \gamma_0) - \sigma(\gamma = -\gamma_0)}{2\gamma_0}, \tag{9.17}$$

respectively. For the case of linear response, both are equal to the linear storage modulus $G'(\omega)$ of the material, but they offer a useful and physically meaningful way of extending the concept of storage modulus to the nonlinear regime. Typically, these definitions can also serve as an indication for the presence of strain-stiffening or strain-softening behavior based on the in-cycle response. The Lissajous curve for a strain-stiffening material will exhibit an upturn in stress at the highest strains within a cycle, as schematically shown in Figure 9.6a, while for strain-softening materials a downturn in stress is observed at the largest strains, as shown in Figure 9.6b. The moduli G'_L and G'_M can thus offer a useful way of indicating the presence of strain-stiffening or strain-softening response and for quantifying the extent of this nonlinear behavior. To do so, a strain-stiffening

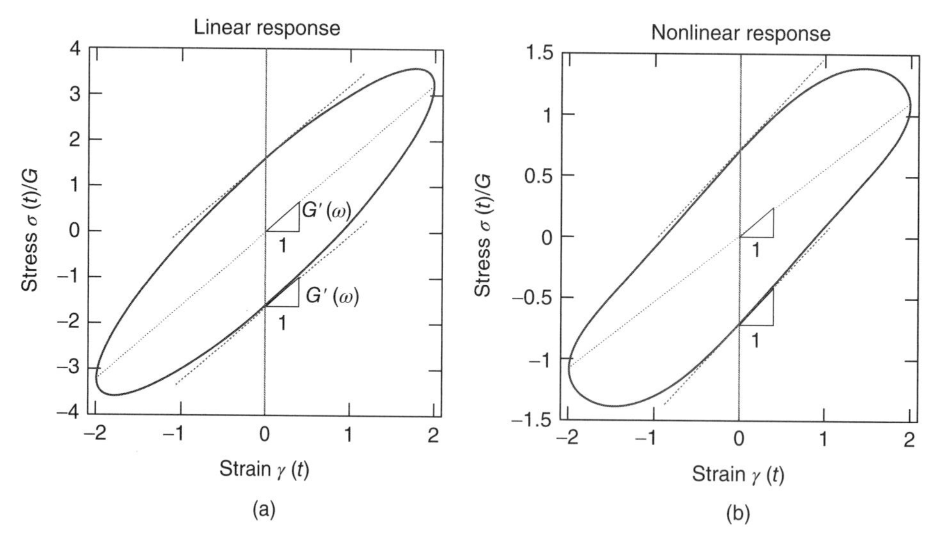

Figure 9.5 Lissajous curves, where the stress $\sigma(t)$ is plotted as a function of the applied strain $\gamma(t)$. (a) for the case of a *linear viscoelastic response*. The response is harmonic, and, as a consequence, the curve has an elliptical shape. The storage modulus is deduced from the slope of the curve at zero strain or from the slope of the line between the two points of maximum and minimum strain. The area under the Lissajous curve corresponds to the dissipated energy per cycle and volume and is directly related to the loss modulus $G''(\omega)$ (b) For the case of a nonlinear response, the stress response is anharmonic; as a consequence, the Lissajous curve deviates from an elliptical shape. The slopes G'_M and G'_L, defined analogous to the linear case are generalizations of a storage modulus to the nonlinear regime. The area under the curve still corresponds to the dissipated energy per cycle and is completely characterized by the first harmonic of the stress response.

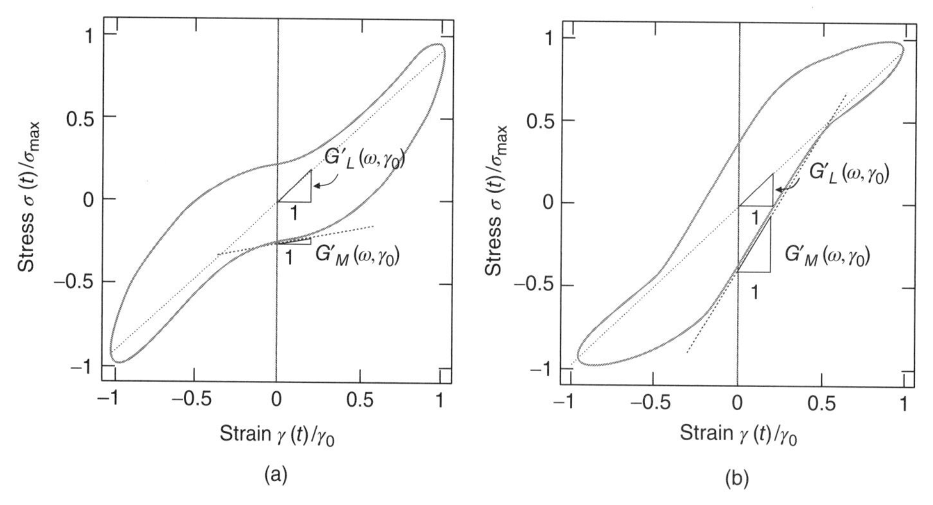

Figure 9.6 In-cycle strain-stiffening and strain-softening behavior: Schematic representation of Lissajous curves that exhibit in-cycle strain stiffening (a) and in-cycle strain softening (b), respectively.

ratio S can be defined [13], based on these two moduli, which indicates the degree of in-cycle strain stiffening in the material:

$$S = \frac{G'_L - G'_M}{G'_L}. \tag{9.18}$$

For a strain-softening response, the ratio is generally negative, $S < 0$, where the magnitude quantifies the degree of in-cycle strain softening. For a linear response, the strain-stiffening ratio vanishes, as expected. However, some caution should be used in applying these strain-stiffening/-softening ratios, as it has been shown that they can fail to correctly reflect the stiffening/softening for certain material models such as the Bingham model [14].

Similar useful measures can also be defined for the dynamic viscosity, based on Lissajous curves of stress as a function of strain rate. In this manner, the minimum-rate dynamic viscosity η'_M and the large-rate dynamic viscosity

η'_L can be defined, which, in analogy to the strain-stiffening ratio, enables the introduction of a shear-thickening ratio T as

$$T = \frac{\eta_L - \eta_M}{\eta_L}, \tag{9.19}$$

where $T = 0$ for a linear response, $T > 0$ for in-cycle shear thickening, and $T < 0$ for in-cycle shear thinning.

9.6.3 Fourier Transform Rheology

In *Fourier Transform Rheology* (FTR) [9, 11, 12, 15], the anharmonic stress response in LAOS is analyzed in terms of Fourier series. This analysis is based on time-dependent stress data in steady-state oscillation, where the response of the material within each cycle, and the shape of the corresponding Lissajous curves, remains fixed. In this steady-state oscillation, the stress response $\sigma(t)$ is *periodic in time* with the same time periodicity $T = \frac{2\pi}{\omega}$ as the applied strain deformation; the shape of the stress response and the corresponding Lissajous curves can therefore be accurately described by a *Fourier series*

$$\sigma(\omega, \gamma_0, t) = \sum_{n \text{ odd}}^{\infty} \left[G'_n(\omega, \gamma_0) \sin(n\omega t) + G''_n(\omega, \gamma_0) \cos(n\omega t)\right], \tag{9.20}$$

with Fourier coefficients

$$G'_n\left(\omega, \gamma_0\right) = \int_t^{t+T} \sigma\left(\omega, \gamma_0, t'\right) \cos\left(\omega n t'\right) dt', \tag{9.21}$$

and

$$G''_n\left(\omega, \gamma_0\right) = \int_t^{t+T} \sigma\left(\omega, \gamma_0, t'\right) \sin\left(\omega n t'\right) dt'. \tag{9.22}$$

Thus, the first-order coefficients we can express directly as a function of the stress and strain only, as $G'_1\left(\omega, \gamma_0\right) = \int_t^{t+T} \frac{\sigma(\omega,\gamma_0,t')\ \dot{\gamma}(t')}{\omega\gamma_0}\, dt'$, and $G''_1\left(\omega, \gamma_0\right) = \int_t^{t+T} \frac{\sigma(\omega,\gamma_0,t')\ \gamma(t')}{\gamma_0}\, dt'$.

The Fourier coefficients obtained from FTR often serve as a convenient basis for the further analysis of in-cycle stress–strain data. Quantities derived from the in-cycle response, such as the generalized moduli discussed above, can be readily expressed in terms of these Fourier coefficients; however, the individual Fourier coefficients themselves have only a limited physical meaning.

An exception to this are the first-order coefficients G'_1 and G''_1, which, in fact, remain physically meaningful even outside the linear regime.

Considering the role of G''_1, we note that the area under the Lissajous curve plotting $\sigma(t)$ versus $\gamma(t)$ has units of energy per volume; indeed, it corresponds to the *energy dissipated per oscillation cycle per unit of volume*. Thus, the loss modulus should be directly related to this dissipated energy. For a linear viscoelastic material, the dissipated energy per cycle can be written as

$$E_{\text{diss.}} = \int_0^{\frac{2\pi}{\omega}} \sigma(t)\, \dot{\gamma}(t)\, dt = \int_0^{\frac{2\pi}{\omega}} \dot{\gamma}^2 \eta\, dt, \tag{9.23}$$

which is equal to the area under the Lissajous curve $\sigma(t)$ versus $\gamma(t)$. This relation also holds for nonlinear response; in fact, it has been shown that the dissipated energy is always equal to the area under the Lissajous curve and that even for nonlinear response, only the first harmonic of the stress response contributes to the dissipated energy [15]. This implies that $E_{\text{diss.}} = \pi\gamma_0^2 G''_1$, and thus

$$G''_1 = \frac{E_d}{\pi\gamma_0^2} \tag{9.24}$$

can also be employed as a physically meaningful extension of the loss modulus to the nonlinear regime.

In summary, the detailed analysis of in-cycle stress–strain data offers meaningful ways of characterizing material response in the nonlinear regime, which sheds light on aspects of their behavior that could not otherwise be readily obtained. The in-cycle data is also a useful way of testing constitutive models describing the response of a material, as they are often highly sensitive to model parameters.

9.7 EXAMPLES OF TYPICAL RHEOLOGICAL BEHAVIOR FOR DIFFERENT SOFT MATERIALS

9.7.1 Soft Glassy Materials

A surprisingly wide range of soft materials exhibit a glass-like behavior and they are often termed *Soft Glassy Materials* [16, 17]. Essentially, this class of materials includes all systems that contain particle-like building blocks at high concentrations; examples include dense colloidal suspensions [18], emulsions [19], foams [20], microgel suspensions and pastes [21], as well as a wide range of foods, biological systems, and industrial products.

In many aspects, their behavior is remarkably close to that of molecular or polymeric glasses. However, as molecular glass formers undergo a glass transition upon decreasing their temperature toward the glass transition temperature, for soft glassy materials a glass transition occurs as their density or concentration is increased. As the concentration is increased, the free volume available to each particle decreases, which eventually leads to structural and dynamic arrest in the material, where the material gets trapped in a disordered solid-like state, a glass. The relaxation of shear stress in a soft glassy material is only possible via

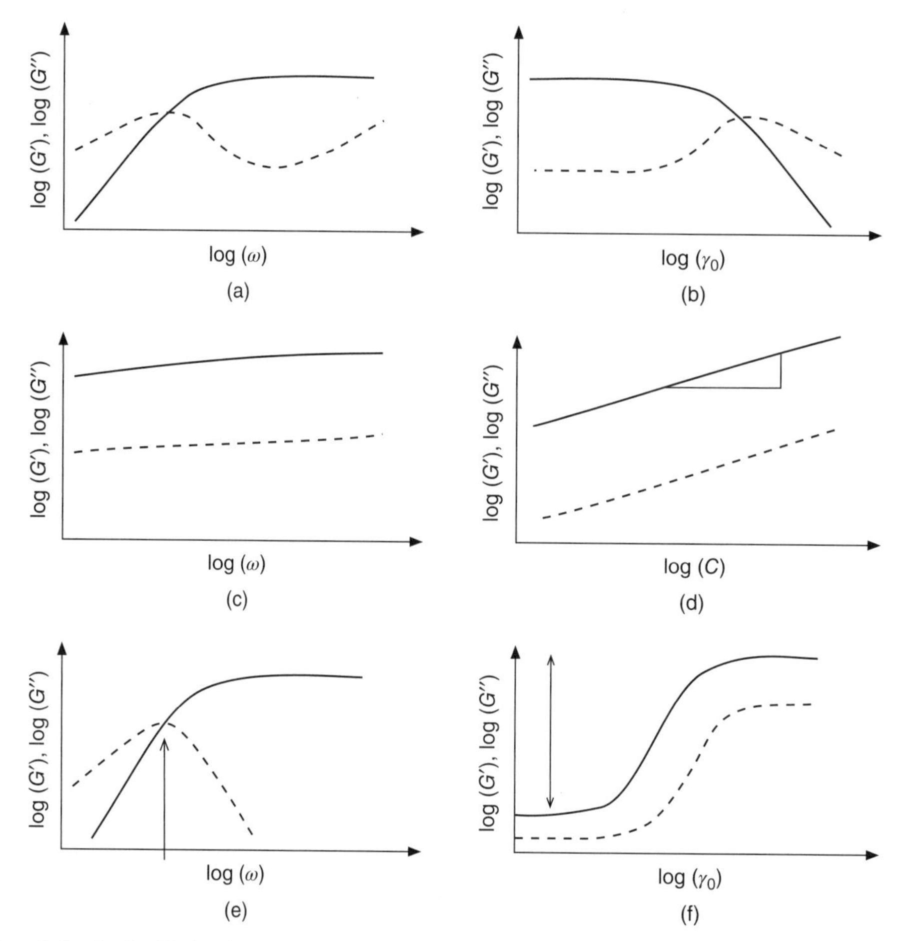

Figure 9.7 Examples of rheological behaviors for different types of soft materials. *Soft glassy materials*: The frequency dependence of the viscoelastic moduli exhibits peak in the loss modulus at the structural relaxation timescale and an increase toward the highest frequencies (a). The strain-dependent response also shows a peak in the loss modulus, a highly characteristic feature of glassy soft materials (b). *Gel networks*: The storage modulus of gels dominates over the loss modulus and often depends only weakly on frequency (c). The storage modulus of gels often scales as a power law with concentration (d). *Transient networks*: Some transient materials such as supramolecular systems or wormlike micelles exhibit a frequency-dependent behavior remarkably close to a single timescale Maxwell model (e). *Strain-stiffening materials*: Many biopolymers show a dramatic strain-stiffening behavior, where the viscoelastic moduli increase by orders of magnitude as the applied strain or stress is increased (f).

particle rearrangement events, where particles must escape from the "cage" formed by surrounding particles and slip past some of their immediate neighbors. The timescale for this process is called the structural relaxation time or the alpha relaxation time τ_α of the material; it increases dramatically with increasing concentration, as rearrangements become increasingly difficult and require cooperative rearrangements of many particles.

These dynamical hallmarks of Soft Glassy Materials are directly reflected in their typical rheological response. As schematically shown in Figure 9.7a, the typical frequency-dependent response of these materials shows a storage modulus G' (solid line) that at high frequencies depends only weakly on frequency and dominates the viscoelastic response, whereas it falls off like a power law at low frequencies with a power-law exponent roughly twice of that found for the loss modulus. At intermediate frequencies, the loss modulus G'' (dashed line) exhibits a highly typifying minimum in its frequency-dependent response. Toward lower frequencies, there is a pronounced peak, with a position that corresponds to the typical structural relaxation time τ_α.

The typical nonlinear response of these materials also shows highly pronounced and typifying features: compared to other soft materials, the LAOS response remains remarkably *harmonic* even at large deformations. A likely cause for this behavior is the fact that in densely packed systems no significant changes in microstructure can occur. This implies that anharmonic contributions to the stress response are small, thus providing a justification for interpreting the data in terms of linear viscoelasticity, with $G' \approx G'_1$ and $G'' \approx G''_1$.

A robust hallmark of these systems is that as the strain increases at fixed ω, the loss modulus initially increases to a peak, before ultimately decreasing at the largest strains, as shown in the example in Figure 9.7b. In fact, this remarkably robust behavior can be explained as a natural consequence of a slow structural relaxation process that is expected in these materials at very low frequencies. Importantly, the timescale of such structural relaxations must depend on the strain rate of deformation; the relaxation speeds up gradually as the system is subjected to shear [22]. As the strain amplitude of deformation is increased in a strain-sweep measurement, the strain rate increases and the characteristic frequency of the slow relaxation process moves toward higher frequencies, driven by the applied shear. As a consequence, a peak in the loss modulus is observed at the point where this characteristic frequency becomes comparable to the oscillation frequency of the strain deformation, ω. It thus appears that the peak in G'' seen in the frequency-dependent measurement in Figure 9.7a and the peak seen in the strain-dependent measurement shown in Figure 9.7b are directly related in that they both originate from the structural relaxation of the material.

This behavior is exploited in a recently introduced approach to oscillatory rheology, termed Strain-Rate Frequency Superposition (SRFS) [23], where the amplitude of the applied strain rate $\dot{\gamma}_0 = \gamma_0\omega$ is kept constant as ω is varied, thus enabling a systematic study of the effects of strain rate on the material response. Surprisingly, it is observed for a range of soft glassy materials that the shape of this response remains similar, independent of the applied strain rate amplitude. This suggests that the observed strain-rate-driven relaxation at large strains remains surprisingly similar to the thermally activated structural relaxation of the material [23]. Such a similarity is not at all obvious; indeed, significant differences in the shape of the observed response have been identified in experiments on star polymers, concentrated emulsions [24], or wormlike micelles [25], while in other experiments on polymeric thickener solutions [25], concentrated water-in-oil emulsions [25], or triblock copolymer systems [26], good agreement was observed. A full understanding of the role of strain rate in the relaxation mechanism of soft glassy materials is thus still lacking.[7] Detailed future investigations, including the direct imaging of structural relaxation at the single particle level, should give more insights into the physical mechanisms underlying this behavior.

[7]It has also been observed [24] that the viscoelastic response extracted from SRFS measurements can deviate significantly from the Kramers–Kronig relations, linking the ω-dependence and magnitudes of $G'(\omega)$ with those of $G''(\omega)$, which should hold for a purely linear viscoelastic response.

9.7.2 Gel Networks

Similar to glasses, gels are materials in an out-of-equilibrium solid-like state. However, the gel state is distinctly different from the glassy state in that the physical origin of the elasticity and the kinetic and structural arrest is fundamentally different. While for glasses it is the local crowding of molecules or particles that leads to arrest, for gels it is the formation of a long-range network of molecules or particles that leads to the formation of a solid-like state [27–29]. Therefore, as their formation does not require local crowding, gels can be formed at much lower concentrations than glasses.

The typical frequency-dependent rheological behavior of gels, schematically shown in Figure 9.7c, shows a storage modulus G' that dominates the response and depends only weakly on frequency over the entire frequency range. If the cross-links in the network are permanent, these materials are true solids, where stresses never fully relax upon applying a strain deformation. This implies that in the zero-frequency limit, the storage modulus remains finite. Alternatively, in a stress relaxation test a truly elastic component $G_e > 0$ would be present in the response function (see Eq. 9.6).

A simple estimate for the elastic modulus is to consider the thermal energy multiplied with the number density of cross-links in the system. An elastic modulus has units of energy per volume, so a dimensional analysis suggests we should look for a typical energy scale and for a typical volume in our material. For colloids, the typical volume would be the volume occupied by each particle and for polymers it would be the average volume per cross-link, thus $G'_0 \approx \frac{k_BT}{\zeta^3}$, where ζ is the mesh size of the polymer network [30]. Such a simple estimate often gives a good approximation of the typical magnitude of the elastic modulus for both polymeric and colloidal gels.

If the links in a gel network are very weak, then a so-called transient gel network can be formed. This implies that bonds are reversible and nonpermanent. At short timescales, such a system can still show an elastic modulus comparable to that of a permanently bound gel of similar mesh size. However, at long timescales, these systems will be fluid-like and stresses in the material can relax when given enough time. Some of these materials can display a frequency-dependent behavior that is remarkably close to that of the simple Maxwell model discussed above and schematically shown in Figure 9.7e. Examples include telechelic polymers [7], wormlike micelles [5], and supramolecular polymer solutions [6].

Another interesting feature of gels is that simple scaling arguments based on the properties of the network structure often yield surprisingly accurate results, which exemplifies the tight connection between the structure and the rheological properties of materials. While the structure of gels is disordered, their structure often can be described in scaling arguments. For instance, even a single polymer chain has scale-free properties; looking at it at different length scales, the morphology is similar. This is also the case for gel networks, which often exhibit fractal properties, implying that they are self-similar and possess no characteristic length scale. As a result, their structures are governed by power laws, a functional form that exhibits no characteristic scales [27]. For instance, the fractal dimension of a fractal cluster is defined by a power law that describes the mass as a function of length scale, $M \propto L_f^d$, where d_f is the fractal dimension. As a result of power laws describing the structural features of these materials, their rheological properties are also governed by power laws, as schematically shown in Figure 9.7d, where a typical dependence of the storage modulus and the loss modulus on concentration is shown; the exponent m can often be directly linked to the structural features of the system via scaling arguments [27, 30].

9.7.3 Biopolymer Networks: Strain-Stiffening Behavior

Some of the most dramatically nonlinear viscoelastic behavior is exhibited by biopolymer networks and biological tissues. Many of these materials show a remarkable strain-stiffening behavior, schematically shown in Figure 9.7f, where the viscoelastic moduli can increase by more than two orders of magnitude as the applied strain deformation increases [31–34]. One possible physical origin for this behavior is the semiflexible nature of the chains that make up these networks [31]. The flexibility of these chains is characterized by the so-called *persistence length* l_p, which quantifies the contour length over which directional correlations are lost within the chain; stiff chains thus exhibit a long l_p, while flexible chains have short l_p values. For semiflexible networks, the persistence length is comparable to the network mesh size; the contour length between cross-links is thus larger but comparable to the mesh size. As a result, these chains show a highly asymmetric stress response in terms of stretching versus compressing; a significant stretching out of chains and the accompanying chain stiffening is achieved at much lower levels of strain than in networks of flexible chains. While this asymmetry is one important reason for the observed strain-stiffening behavior, the detailed physical origins of the nonlinear mechanical response of these materials is still not fully understood; effects such as bundle formation or the nonpermanent cross-links present in biopolymer systems also have to be considered to arrive at a reasonable description of the mechanical behavior of these materials [32, 35, 36]. Adequately probing the viscoelastic response of these highly nonlinear materials is also a challenge that is the topic of intense research [37, 38].

9.8 RHEOMETERS

The instruments for measuring the rheological properties of a material are called *rheometers*, or, if only the viscosity of materials is tested, *viscometers*. Below we introduce the most commonly used rheometers and discuss their advantages and limitations.

9.8.1 Rotational Rheometers

The instruments most commonly used for rheological measurements are *rotational rheometers*. In these instruments, the sample is placed in a rotationally symmetric measuring geometry. The sample is, for instance, placed in the gap between two parallel, circular plates. As a time-dependent torque or an angular displacement is applied to one of the plates, the resulting deformation or stress in the material is then simultaneously probed. Locally in the sample, the deformation is very close to a linear shear deformation. The reason why a true linear shear is generally not used is the fact that in such an instrument the maximum strain deformation would always be limited, while arbitrarily large strains can be applied to a material in a rotational rheometer. This is especially important for steady shear measurements, where a constant shear rate is applied and the maximum strain thus grows linearly with the time of the measurement. All rotational rheometers can therefore also be used as a viscometer.

9.8.2 Measuring Geometries

Different measuring geometries can be used in rotational rheometers, depending on the properties of the material to be measured, the available sample volume and the torque sensitivity of the instrument.

The simplest measuring geometry is the *parallel plates geometry*, schematically shown in Figure 9.8a. The sample is held between two parallel plates in the shape of a circular disc. This geometry has the advantage that samples are conveniently loaded even if they are solid-like. Moreover, the gap height h is homogeneous across the sample and can be freely chosen depending on the measuring requirements. The sample volume can thus be varied in this geometry, which is convenient for samples that are available only in small quantities. The strain deformation is *not homogeneous* in this geometry; in the middle of the sample, on the rotation axis, the strain deformation is zero, increasing linearly toward the edge of the plates, where it reaches a maximum. However, for a linear viscoelastic response, the stress is always proportional to the applied strain and the stress–strain curve can thus be obtained from the applied (maximum) strain and the measured torque T, as the torque is given by

$$T = \int_0^R \sigma(r)\, 2\pi r^2\, dr, \qquad (9.25)$$

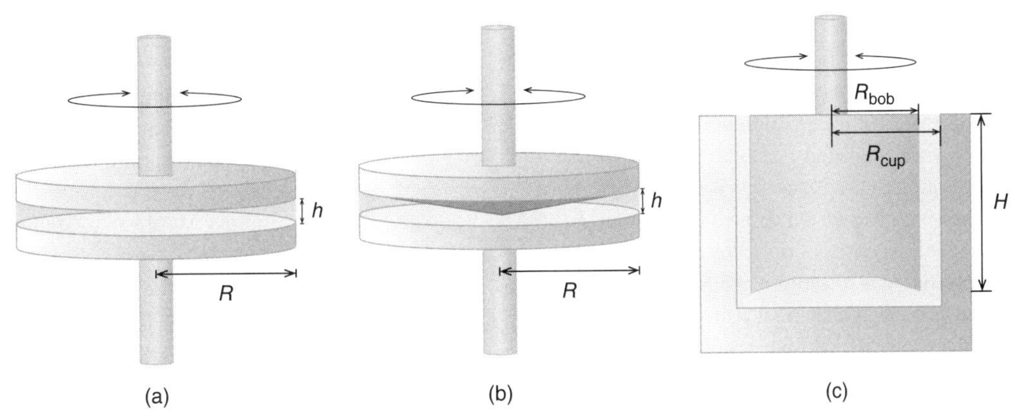

Figure 9.8 Measuring geometries used in rotational rheometers. (a) Parallel plates geometry. (b) Cone-plate geometry. (c) Bob-cup or *Couette* geometry.

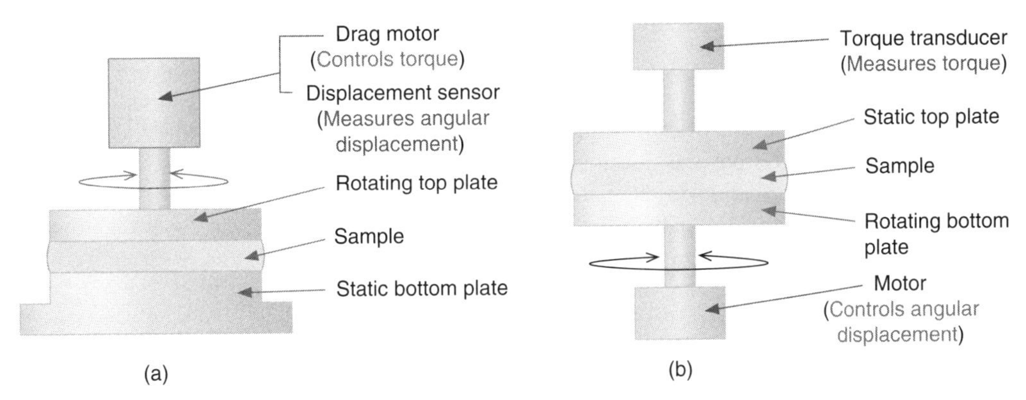

Figure 9.9 Types of rotational rheometers. (a) *Stress-controlled rheometer*: A time-dependent torque is applied to the upper tool of the measuring geometry. Simultaneously, the resulting angular displacement of the tool is measured. The lower tool remains fixed and has no active role in the measurement. (b) *Strain-controlled rheometer*: The angular displacement of the lower tool is controlled as a function of time by a motor, while simultaneously the resulting torque is measured at the upper tool by a torque transducer.

where $\sigma(r)$ is the local stress at a distance r from the center of the tool and R is the radius of the tool. For linear response, we have $\sigma(r) = \frac{r}{R}\sigma_R$, with σ_R the stress at the edge of the geometry. Other frequently used measuring geometries include the cone-plate geometry (shown schematically in Fig. 9.8b), and the bob-cup or Couette geometry (Fig. 9.8c). These geometries are designed to avoid variations of the shear stress across the sample, thus simplifying the interpretation of the acquired data in both linear and nonlinear measurements.

In a *stress-controlled rotational rheometer*, schematically shown in Figure 9.9a, a time-dependent stress is applied to the sample by controlling the torque on one side of the measuring geometry. The resulting strain deformation of the material is recorded simultaneously by following the angular displacement of the same side of the tool as a function of time. Usually, the detection of displacement is achieved optically, and thus contact-free, for instance, by directly detecting the laser light reflected from a micropatterned disc, attached to the shaft of the measuring geometry. Such optical encoders can achieve very high angular resolution, typically smaller than 100 nrad ($< 10^{-7}$rad).

9.8.3 Stress- and Strain-Controlled Rheometers

There are two main types of design for a rotational rheometer, stress-controlled and the strain-controlled rheometers, as schematically shown in Figure 9.9.

In a *strain-controlled rotational rheometer*, schematically shown in Figure 9.9b, the application of a strain deformation and the detection of the resulting stress occur separately at the top and bottom part of the measuring tools. For instance, in a cone-plate tool, the rotation of the lower plate is controlled by a motor, resulting in a time-dependent angular displacement $\phi(t)$. The stress response of the sample is followed by measuring the time-dependent torque $\tau(t)$ exerted by the sample on the upper cone tool, while keeping the angular displacement of the upper tool fixed.

Because strain-controlled rheometers require a torque transducer on one side and a precision motor on the other side, they are generally more expensive to build than stress-controlled rheometers. However, strain-controlled rheometers have a number of advantages over stress-controlled rheometers, which can make them the instruments of choice at least for certain types of measurements. The first advantage is the fact that the friction of the bearing does not need to be calibrated as a function of tool position and speed. In stress-controlled rheometers, the upper part of the geometry is rotating in a bearing of very low friction. The friction in this bearing is an important factor that limits the resolution of the instrument; unfortunately, even for the most advanced bearings available, this friction can vary significantly for different angular positions of the tool and for different angular velocities.[8] To account for the effects of this bearing friction, stress-controlled rheometers rely on separate calibration measurements that quantify the friction as a function of angular position and angular velocity. A similar calibration is performed to quantify the moment of inertia of the tool and motor shaft. The contributions of both the resulting inertial torques and the bearing friction are then subtracted from the raw torque, thereby isolating the torque that is applied to the sample. Modern stress-controlled rheometers have advanced low friction bearings, where also the variation of friction with the angular position is minimized. As a result, these instruments can reliably resolve torques on the order of 0.1 μN m. To give some perspective, if we were to drop a grain of sand into our outstretched arm, the extra torque exerted by the weight of the grain would correspond to roughly 10 μN m.

9.9 CONCLUSIONS

Soft materials exhibit a remarkably rich viscoelastic response, due to the fact that the typical relaxation times in these materials often lie in a regime that is directly observable. The timescales at which these materials are deformed and probed are therefore of key importance.

To gain full insight into the often complex behaviors of soft materials, a range of experimental methods covering their structure, dynamics, and their mechanics need to be combined. True insight into their behavior can only be gained through the combination of methods quantifying these different aspects of a material. As the structure and dynamics of soft matter is directly reflected in their viscoelastic response, rheology plays a key role in this –it is an indispensable tool for understanding and studying this fascinating class of materials.

Both stress relaxation tests and oscillatory rheological measurements are suited for studying the material response at different characteristic timescales. Which method to use depends on the material properties of interest; for instance, when interested in the behavior at very long timescales, stress relaxation is preferred as the time it takes to conduct the measurement equals the characteristic timescale accessed in the experiment. However, the torque response in stress relaxation measurements is often too small and at small times the measurement is inaccurate, as a truly instantaneous step in strain cannot be realized in an experiment.

The nonlinear viscoelastic response of soft materials is an area of growing interest, largely due to the presence and importance of mechanical nonlinearities in biological materials as well as in a wide range of industrial products. New approaches for experimentally characterizing such a nonlinear response and for analyzing and interpreting this response are currently being investigated.

Rheology of soft materials is a fascinating field of study. The extremely rich phase behavior of soft materials leads to a wide range of structural and dynamical features observed in these materials; this richness is directly reflected in the resulting viscoelastic response and thus rheological studies are often of key importance for gaining information on the physical mechanisms that govern their behavior.

[8]The normal force acting on an air bearing can also significantly influence the friction. This effect is difficult to correct by measuring a calibration curve.

REFERENCES

[1] Macosko CW. *Rheology: Principles, Measurements, and Applications*. New York: Wiley-VCH; 1994.

[2] Larson RG. *The Structure and Rheology of Complex Fluids*. New York: Oxford University Press; 1999.

[3] Cates ME. Reptation of living polymers: dynamics of entangled polymers in the presence of reversible chain-scission reactions. Macromolecules 1987;20(9):2289.

[4] Turner MS, Cates ME. Linear viscoelasticity of living polymers: a quantitative probe of chemical relaxation times. Langmuir 1991;7(8):1590.

[5] Cates ME, Fielding SM. Rheology of giant micelles. Adv Phys 2006;55(7–8):799.

[6] Shikata T, Ogata D, Hanabusa K. Viscoelastic behavior of supramolecular polymeric systems consisting of N,N ',N "-tris(3,7-dimethyloctyl)benzene–1,3,5-tricarboxamide and N-alkanes. J Phys Chem B 2004;108(2):508.

[7] Seiffert S, Sprakel J. Physical chemistry of supramolecular polymer networks. Chem Soc Rev 2012;41(2):909.

[8] Baumgaertel M, Winter HH. Determination of discrete relaxation and retardation time spectra from dynamic mechanical data. Rheol Acta 1989;28(6):511.

[9] Hyun K, Wilhelm M, Klein CO, Cho KS, Nam JG, Ahn KH, Lee SJ, Ewoldt RH, McKinley GH. A review of nonlinear oscillatory shear tests: analysis and application of large amplitude oscillatory shear (LAOS). Prog Polym Sci 2011;36(12):1697.

[10] Cho KS, Hyun K, Ahn KH, Lee SJ. A geometrical interpretation of large amplitude oscillatory shear response. J Rheol 2005;49(3):747.

[11] Wilhelm M, Maring D, Spiess HW. Fourier-transform rheology. Rheol Acta 1998;37(4):399.

[12] Wilhelm M. Fourier-transform rheology. Macromol Mater Eng 2002;287(2):83.

[13] Ewoldt RH, Hosoi AE, McKinley GH. New measures for characterizing nonlinear viscoelasticity in large amplitude oscillatory shear. J Rheol 2008;52(6):1427.

[14] Rogers SA, Lettinga MP. A sequence of physical processes determined and quantified in large-amplitude oscillatory shear (LAOS): application to theoretical nonlinear models. J Rheol 2012;56(1):1.

[15] Ganeriwala SN, Rotz CA. Fourier-transform mechanical analysis for determining the nonlinear viscoelastic properties of polymers. Polym Eng Sci 1987;27(2):165.

[16] Sollich P, Lequeux F, Hebraud P, Cates ME. Rheology of soft glassy materials. Phys Rev Lett 1997;78(10):2020.

[17] Sollich P. Rheological constitutive equation for a model of soft glassy materials. Phys Rev E 1998;58(1):738.

[18] Mason TG, Weitz DA. Linear viscoelasticity of colloidal hard-sphere suspensions near the glass-transition. Phys Rev Lett 1995;75(14):2770.

[19] Mason TG, Bibette J, Weitz DA. Yielding and flow of monodisperse emulsions. J Colloid Interface Sci 1996;179(2):439.

[20] Mayer P, Bissig H, Berthier L, Cipelletti L, Garrahan JP, Sollich P, Trappe V. Heterogeneous dynamics of coarsening systems. Phys Rev Lett 2004;93(11):115701.

[21] Mattsson J, Wyss HM, Fernandez-Nieves A, Miyazaki K, Hu Z, Reichman DR, Weitz DA. Soft colloids make strong glasses. Nature 2009;462(7269):83.

[22] Miyazaki K, Wyss HM, Weitz DA, Reichman DR. Nonlinear viscoelasticity of metastable complex fluids. Europhys Lett 2006;75(6):915.

[23] Wyss HM, Miyazaki K, Mattsson J, Hu Z, Reichman DR, Weitz DA. Strain-rate frequency superposition: a rheological probe of structural relaxation in soft materials. Phys Rev Lett 2007;98(23):238303.

[24] Erwin BM, Rogers SA, Cloitre M, Vlassopoulos D. Examining the validity of strain-rate frequency superposition when measuring the linear viscoelastic properties of soft materials. J Rheol 2010;54(2):187.

[25] Kowalczyk A, Hochstein B, Staehle P, Willenbacher N. Characterization of complex fluids at very low frequency: experimental verification of the strain rate-frequency superposition (SRFS) method. Appl Rheol 2010;20(5):U19.

[26] Mohan PH. Bandyopadhyay R. Phase behavior and dynamics of a micelle-forming triblock copolymer system. Phys Rev E 2008;77(4):041803.

[27] Kantor Y, Webman I. Elastic properties of random percolating systems. Phys Rev Lett 1984;52(21):1891.

[28] Krall AH, Weitz DA. Internal dynamics and elasticity of fractal colloidal gels. Phys Rev Lett 1998;80(4):778.

[29] Trappe V, Sandkühler P. Colloidal gels – low-density disordered solid-like states. Curr Opin Colloid Interface Sci 2004;8(6):494.

[30] de Gennes PG. *Scaling Concepts in Polymer Physics*. Ithaca (NY): Cornell University Press; 1979.

[31] Mackintosh FC, Kas J, Janmey PA. Elasticity of semiflexible biopolymer networks. Phys Rev Lett 1995;75(24):4425.

[32] Gardel ML, Shin JH, MacKintosh FC, Mahadevan L, Matsudaira P, Weitz DA. Elastic behavior of cross-linked and bundled actin networks. Science 2004;304(5675):1301.

[33] Storm C, Pastore JJ, MacKintosh FC, Lubensky TC, Janmey PA. Nonlinear elasticity in biological gels. Nature 2005;435(7039):191.

[34] Gardel ML, Nakamura F, Hartwig J, Crocker JC, Stossel TP, Weitz DA. Stress-dependent elasticity of composite actin networks as a model for cell behavior. Phys Rev Lett 2006;96(8):088102.

[35] Piechocka IK, Bacabac RG, Potters M, MacKintosh FC, Koenderink GH. Structural hierarchy governs fibrin gel mechanics. Biophys J 2010;98(10):2281.

[36] Wolff L, Fernández P, Kroy K. Resolving the stiffening-softening paradox in cell mechanics. PLoS ONE 2012;7(7):e40063.

[37] Semmrich C, Larsen RJ, Bausch AR. Nonlinear mechanics of entangled F-actin solutions. Soft Matter 2008;4(8):1675.

[38] Broedersz CP, Kasza KE, Jawerth LM, Münster S, Weitz DA, MacKintosh FC. Measurement of nonlinear rheology of cross-linked biopolymer gels. Soft Matter 2010;6(17):4120.

10

OPTICAL MICROSCOPY OF SOFT MATTER SYSTEMS

Taewoo Lee[1], Bohdan Senyuk[1], Rahul P. Trivedi[1], & Ivan I. Smalyukh[1,2]

[1]*Department of Physics and Liquid Crystal Materials Research Center, University of Colorado, Boulder, CO, USA*

[2]*Renewable and Sustainable Energy Institute, National Renewable Energy Laboratory and University of Colorado, Boulder, CO, USA*

10.1 INTRODUCTION

The fast-growing field of soft matter research requires increasingly sophisticated tools for experimental studies. One of the oldest and most widely used tools to study soft matter systems is optical microscopy. Recent advances in optical microscopy techniques have resulted in a vast body of new experimental results and discoveries. New imaging modalities, such as nonlinear optical (NLO) microscopy techniques that were developed to achieve higher resolution, enable soft matter research at length scales ranging from the molecular to the macroscopic.

The aim of this chapter is to introduce a variety of optical microscopy techniques available to soft matter researchers, starting from basic principles and finishing with a discussion of the most advanced microscopy systems. We describe traditional imaging techniques, such as bright field and polarizing microscopy (PM), along with state-of-the-art three-dimensional (3D) imaging techniques, such as fluorescence confocal and NLO microscopies. Different approaches

Fluids, Colloids and Soft Materials: An Introduction to Soft Matter Physics, First Edition. Edited by Alberto Fernandez Nieves and Antonio Manuel Puertas.

are discussed along with their applications in the study of soft matter systems by providing typical examples. We note that this chapter gives an overview of what can be done in the studies of soft matter systems with different optical microscopy modes available. For more comprehensive and detailed description of microscopy techniques and sample studies in different imaging modes as well as the physics behind different studies of soft matter phenomena, readers can refer to the original works [1–14]. The introduction into imaging basics and a detailed description of optical imaging techniques can be found in recent textbooks and reviews [15–22].

The chapter is organized as follows. Section 10.2 deals with basic principles and parameters important to all optical microscopy modalities. Sections 10.3–10.5 describe traditional optical microscopy techniques. Fluorescence and confocal microscopies are discussed in Sections 10.6–10.8. Section 10.9 provides an extended account of NLO microscopy techniques. An example of a particle-tracking technique that utilizes a pre-engineered point spread function (PSF) and its application in imaging is detailed in Section 10.10. Since many experiments in soft matter systems require simultaneous noncontact optical manipulation and 3D imaging, integration of these techniques is immensely useful and is discussed in Section 10.11. We conclude our review on optical microscopy and discuss future perspectives of its use in the study of soft matter systems in Section 10.12.

10.2 BASICS OF OPTICAL MICROSCOPY

The history of optical microscopy began centuries ago. It is difficult to say who invented the compound optical microscope, but it has been used extensively for research in different branches of science starting from the beginning of the 17th century. Since that time, optical microscopy has developed into its own research field and industry at the forefront of science and engineering. The developments of new techniques in optical microscopy have often led to significant breakthroughs in many different scientific fields.

The simplest optical imaging system that can be used as a microscope consists of two converging lenses: an objective lens and an eyepiece. The optical train of this microscope and its corresponding ray diagram are shown in Figure 10.1a. Illumination light transmitted through the sample is collected by the objective lens and transferred to the eyepiece, forming an image on the retina of the observer's eyes. Various types of light-detecting devices, such as charge-coupled device (CCD) cameras, photodiodes, avalanche photodiodes, photomultiplier tubes (PMTs), and other optical sensors are nowadays widely used for collecting the image. Electronic scanning systems such as galvano-mirrors and acousto-optic deflectors or fast confocal illumination systems such as Nipkow disks are often utilized in modern optical microscopes and imaging systems.

Objective lenses are perhaps the most essential optical components and are characterized by parameters such as numerical aperture (NA) and working distance (WD), which largely define their performance. The NA of the objective determines the range of angles over which the objective lens can accept light and is defined as

$$\mathrm{NA} = n \cdot \sin\theta,$$

where n is the refractive index of the medium between the objective lens and the sample, and θ is the half-angle of the cone of light accessible for collection by the objective (Fig. 10.1b). While NA is less than unity for air objectives, oil immersion objectives can have NA close to 1.5. Another important parameter characterizing the objective is its WD, defined as the distance from the front lens of the objective to the surface of the sample when the inspected area is in focus (Fig. 10.1b). The depth up to which the sample can be imaged is often restricted by the available WD. Objectives with larger NA usually have smaller WD and vice versa.

Numerical aperture determines important microscopic characteristics such as the PSF and, consequently, the resolution. The PSF (Fig. 10.1c) can be thought of as the light-intensity distribution in the image acquired by the microscope from a point source and is given by [23]

$$\mathrm{PSF}(r) = \left[\frac{2J_1(ra)}{ra}\right]^2,$$

where $a = 2\pi\mathrm{NA}/\lambda$, J_1 is the Bessel function of the first kind, λ is the wavelength of light, and r is the distance from the center of the peak in light intensity (Fig. 10.1c). In the image plane of a standard optical system, the PSF is shaped as the Airy diffraction pattern with the first minimum at $r = W/2$ (Fig. 10.1d). Two Airy disks separated by the distance $R \geq W/2$ (Fig. 10.1d) can be resolved into separate entities, but not at smaller R [21]. This limit is often called the Rayleigh criterion or the diffraction limit and defines the lateral resolution of the objective as $r_{\mathrm{lateral}} = 0.61\lambda/\mathrm{NA}$, that is, the resolution in the plane orthogonal to the microscope's optical axis. From the definition of r_{lateral}, it follows that objectives with higher NA can resolve finer details; however, even for high-NA objectives, the lateral resolution can be only slightly smaller than the wavelength of light used for imaging. Although conventional optical microscopes have poor resolution along the optical axis of a microscope, recently introduced optical imaging techniques, such as confocal and NLO microscopies, allow for optical imaging with high axial resolution. The axial resolution of an optical system will be introduced later when discussing fluorescence confocal microscopy (FCM) in Section 10.7.

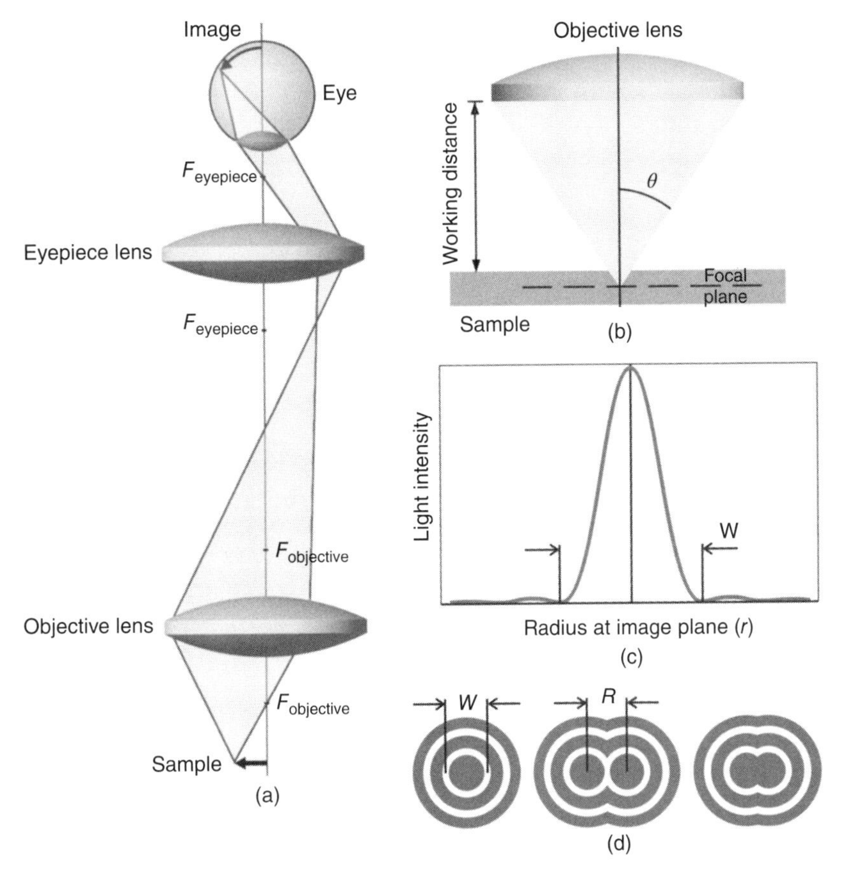

Figure 10.1 Principles of imaging with an optical microscope: (a) ray diagram of the simplest two-lens microscope; (b) definition of parameters of an objective lens; (c) point spread function with the first minimum at $r = W/2$; (d) Airy patterns in the image plane. "$F_{\text{objective}}$" and "F_{eyepiece}" mark the foci of objective and eyepiece lenses, respectively. "R" shows the separation distance between the centers of Airy disks. (*See color plate section for the color representation of this figure.*)

Modern imaging approaches allowing one to overcome the diffraction limit for both radial and axial directions will be discussed in Section 10.12.

10.3 BRIGHT FIELD AND DARK FIELD MICROSCOPY

Bright field imaging is perhaps the simplest of all optical microscopy methods. A sample is illuminated by unpolarized white light, and the contrast in the image results from direct interaction of the probing light with the sample (absorption, refraction, scattering, reflection, etc.). Figure 10.2a shows a simple schematic diagram of transmission mode (dia-illumination) bright field microscopy, where the illumination light is focused onto the sample by a condenser lens with numerical aperture NA_{cond}, and the transmitted light is collected by the objective lens with numerical aperture NA_{obj}. High-quality imaging can be performed at optimum illumination of the sample using the so-called Köhler illumination [22]. The ratio between the NAs of the objective and condenser lenses affects the resolution and contrast of imaging. The resolution in the microscope image is optimized when the NAs of condenser and objective are equal. Using objectives with NA_{obj} higher than NA_{cond} increases the image contrast but decreases resolution [21, 22, 24].

The example images obtained with bright field microscopy show the texture of toroidal focal conic domains in a thin film of a smectic A liquid crystal spread over a glycerol surface (Fig. 10.2b) and an array of *Pseudomonas aeruginosa* cells trapped by an array of infrared (invisible in the image) laser beams (Fig. 10.2c). The contrast in these images (Fig. 10.2b and c) results mostly from refraction and scattering of light at the boundaries of objects having refractive index different from that of the surrounding medium. One of the main limitations of bright field imaging is that samples with weak spatial variation of the refractive index or absorption produce images with poor contrast.

Dark field microscopy is a method in which the image is formed by collecting only the light scattered by the sample. All unscattered light coming directly from the illumination source is excluded from the image, thus forming a dark

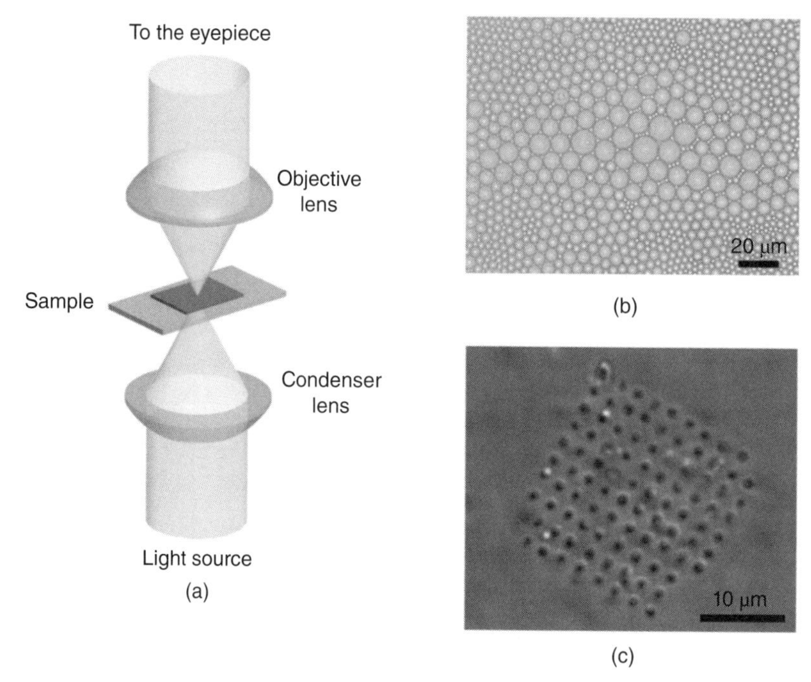

Figure 10.2 Bright field microscopy diagram (a) and textures: (b) toroidal focal conic domains in a smectic liquid crystal (8CB) film spread over a glycerol surface; (c) an array of *P. aeruginosa* cells trapped by an array of infrared laser beams. (*See color plate section for the color representation of this figure.*)

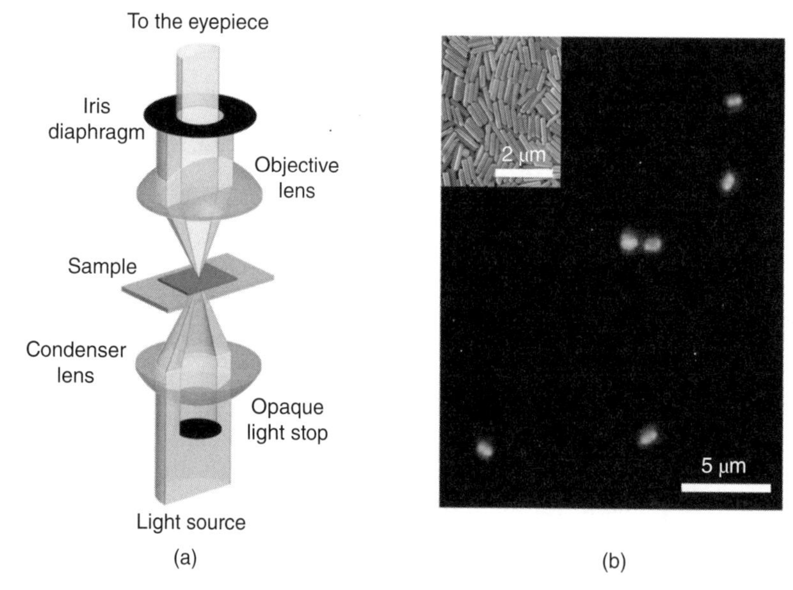

Figure 10.3 Dark field microscopy diagram (a) and texture of gold rod-like colloids (b) dispersed in a smectic (8CB) homeotropic liquid crystal cell. Inset in (b) shows the transmission electron microscopy image of gold rod-like colloids. Source: TEM image courtesy of Nanopartz Inc. (*See color plate section for the color representation of this figure.*)

background (Fig. 10.3a). To achieve this, the specimen is illuminated with a hollow cone of light (Fig. 10.3a) formed by a special condenser with an opaque light stop blocking the direct light. The oblique illumination light is scattered by objects in the field of view and only the scattered light is collected by the objective, while unscattered direct light is blocked. For optimized imaging in this optical arrangement, NA_{cond} should be larger than NA_{obj} [22, 24].

Dark field microscopy is typically used for imaging of samples with sparse scattering objects dispersed in a nonscattering medium and is especially useful in the study of objects having size below the diffraction limit. For example, this technique is employed when investigating systems ranging from red blood cells to plasmonic metal nanoparticles [1]. Figure 10.3b shows a dark field image of individual gold rod-like colloidal particles dispersed in a smectic liquid crystal. The color and contrast in the image are formed due to absorption and scattering of the white illumination light by rod-shaped gold nanoparticles shown in the inset of Figure 10.3b.

10.4 POLARIZING MICROSCOPY

The polarizing microscopy is an imaging method that uses polarized light and is usually utilized for imaging birefringent and optically active materials, such as liquid crystals, minerals, and crystals [22, 24]. Typically, the illumination is linearly polarized by a "polarizer," and upon transmitting through the sample, it is passed to another polarizer (called "analyzer") (Fig. 10.4a). The analyzer is usually kept parallel or orthogonal to the polarizer. The intensity of the light transmitted by the sample between two crossed polarizers depends on the angle β between the polarization of light and the optic axis of birefringent sample and phase retardation $2\pi\Delta nd/\lambda$ introduced by a uniform sample of thickness d [24]:

$$I = I_0 \sin^2(2\beta)\sin^2\left(\frac{\pi\Delta nd}{\lambda}\right),$$

where I and I_0 are the intensities of transmitted and incident light, respectively, and Δn is the birefringence. If the sample is viewed in the same position between parallel polarizers instead, the intensity of transmitted light becomes $I = I_0[1 - \sin^2(2\beta)\sin^2(\pi\Delta nd/\lambda)]$. In both cases, linearly polarized incident light is split by the birefringent sample into two components, extraordinary and ordinary, propagating within the sample at different speeds, which generally results in elliptical polarization of light upon exiting the sample (Fig. 10.4a). If $\beta = 0$ or $\pi/2$, only ordinary or extraordinary waves propagate so that polarization of light remains linear. The analyzer probes the polarization state of the transmitted light as altered by the birefringent medium.

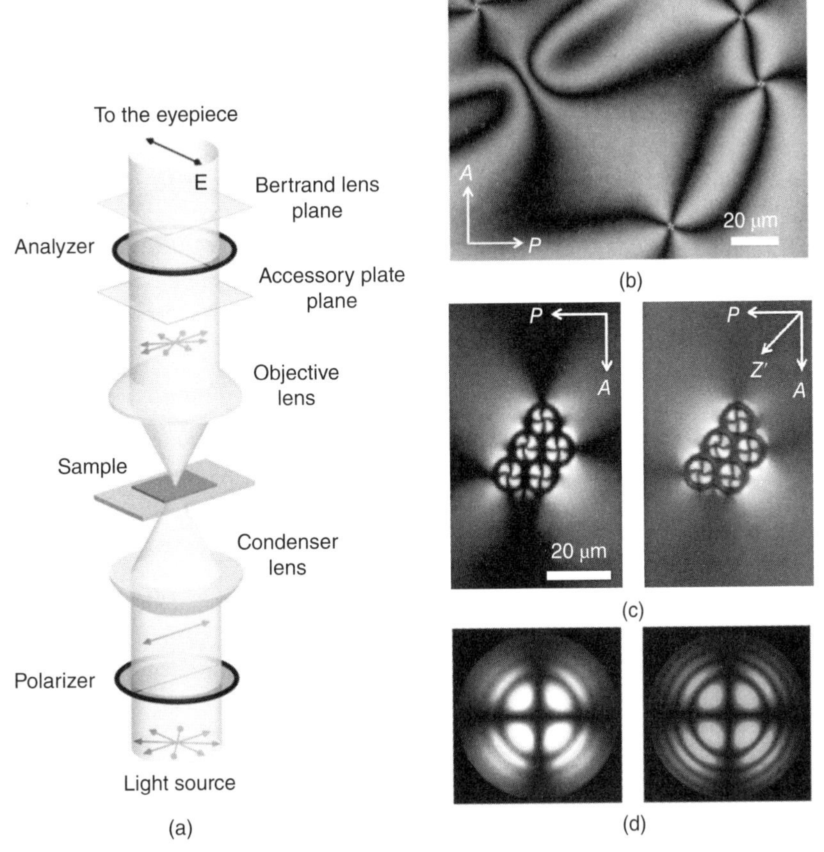

Figure 10.4 Polarizing microscopy: (a) diagram; (b) schlieren texture observed in the thin film of a nematic liquid crystal spread over glycerol surface; (c) texture of a nematic liquid crystal around a colloidal particle of nonzero genus [2] observed without (left) and with (right) the red plate inserted; (d) conoscopic images obtained for a homeotropic uniaxial smectic A (8CB) sample with a Bertrand lens inserted under the white (left) and green (right) light. Arrows P, A, and Z' show a direction of polarizer, analyzer, and "slow axis" of the red plate, respectively. (*See color plate section for the color representation of this figure.*)

The intensity pattern in the image is then used to deduce the spatial variations of the orientation of the optic axis and/or the value of the birefringence in the lateral plane of an optically anisotropic sample.

An example image acquired with PM between crossed polarizers (Fig. 10.4b) shows the texture of a thin film of a nematic liquid crystal [25] spread over a liquid substrate such as glycerol. The contrast and color in the image result from the phase shift between the two propagating modes. The dark brushes in the image correspond to the locations where the local orientation of the optic axis in the sample is along the polarizer ($\beta = 0$) or analyzer ($\beta = \pi/2$). Vivid colors are produced as a result of interference when the phase retardation of visible light is on the order of 2π. The color pattern can be utilized to deduce the values of optical anisotropy, orientation of the optic axis, and the film thickness [22, 24].

The capabilities of the PM technique in the study of birefringent samples can be greatly expanded by using accessory birefringent plates and wedges in the optical path. The orientation of the optic axis of a birefringent material can be determined by introducing a full wavelength (typically ~530 nm) retardation plate (the so-called red plate) into the optical train [22, 24] after the sample and objective (Fig. 10.4a). In the example shown in Figure 10.4c, the in-plane orientation of the local liquid crystal optic axis is parallel to the "slow axis" of the red plate in the bluish regions and perpendicular to it in the yellowish regions. The birefringence of the sample and its sign can be determined using a quartz wedge, quarter-wave retardation plate (Senarmont method), or the Berek compensator [22, 24]. The principles behind the use of all of these accessory plates are based on probing the addition or subtraction of the phase retardations due to the retardation plate or wedge (with known orientation of the optic axis and known phase retardation) and the studied sample. The analysis of the resulting phase retardation patterns yields information complementary to that obtained by imaging the sample between a pair of polarizers alone.

The introduction of a Bertrand lens [22] into the optical train of the microscope enables an imaging mode known as "conoscopy" (Fig. 10.4d). The Bertrand lens is inserted in the optical train after the analyzer (Fig. 10.4a) and brings the objective's back focal plane into the focus [22]. In this arrangement, the sample is illuminated by a strongly convergent cone of light, which provides an interference image (Fig. 10.4d) formed by light passing through the sample at different angles. Conoscopic arrangement of a polarizing microscope allows determination of the type of birefringent materials (i.e., uniaxial or biaxial), orientation of the optic axes, and the sign of birefringence [22, 24].

10.5 DIFFERENTIAL INTERFERENCE CONTRAST AND PHASE CONTRAST MICROSCOPIES

Most common optical microscopy techniques used to enhance the contrast in transparent nonbirefringent samples with weak spatial variation of the refractive index are differential interference contrast (DIC) and phase contrast microscopies. In the DIC technique, image contrast results from the gradient of the refractive index within the sample. A sample is placed between crossed polarizers and two beamsplitters, the Nomarski-modified Wollaston prisms (Fig. 10.5a), such that the prism's optic axis is kept at 45° with respect to the crossed polarizers. The polarized illumination light is separated by the first beamsplitter into two orthogonally polarized and spatially displaced (sheared) components (red and blue in Fig. 10.5a). Each component propagates along a different path, giving rise to an optical path difference upon exiting the sample. After being collected by an objective, the two components are recombined by the second Nomarski prism and transmitted through the analyzer to the detector. As a result, the contrast in DIC images depends on the path length gradient along the shear direction; this enhances the contrast due to edges or interfaces between sample areas with different refractive indices. Figure 10.5b shows a DIC image of an oil-in-water emulsion [26]. In this example, the interface between water and oil droplets is highlighted due to the difference in their refractive indices and the resultant gradient of optical paths along the shear direction. The image contrast increases as this gradient becomes steeper.

The phase contrast microscopy technique transforms small spatial variations in phase into corresponding changes in the intensity of transmitted light. The diagram of the phase contrast microscope's optical train is presented in Figure 10.6a. The illumination light passes through an annular ring placed before the condenser lens and is focused on the sample. It either passes as undeviated (yellow in Fig. 10.6a) or is diffracted with changed phase (violet in Fig. 10.6a) depending on the composition of the sample. Both undeviated and diffracted light is collected by the objective and transmitted to the eyepiece through the phase ring, which introduces an additional phase shift to the undeviated reference light. Parts of the sample having different refractive indices appear darker or brighter compared to the uniform background, forming the phase contrast image. Phase contrast imaging is insensitive to polarization and birefringence effects. This is a major advantage when examining living cells and a number of other soft matter systems. Figure 10.6b shows the texture of elongated mycelium hyphae of a common *Aspergillus* mold confined within microfluidic channels [27] obtained with the phase contrast technique.

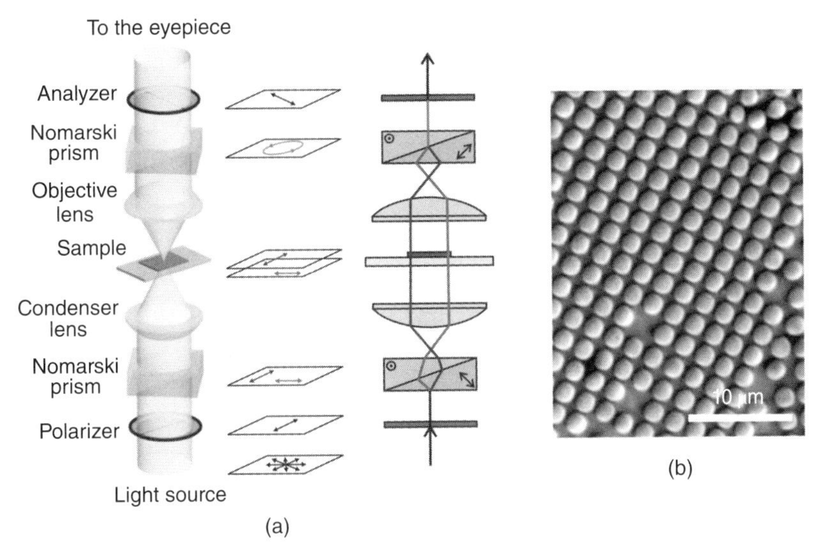

Figure 10.5 Differential interference contrast microscopy (a) diagram and (b) texture of an oil-in-water emulsion. Source: Image courtesy of V.N. Manoharan; see Ref. [26] for detailed description of the image in (b). (*See color plate section for the color representation of this figure.*)

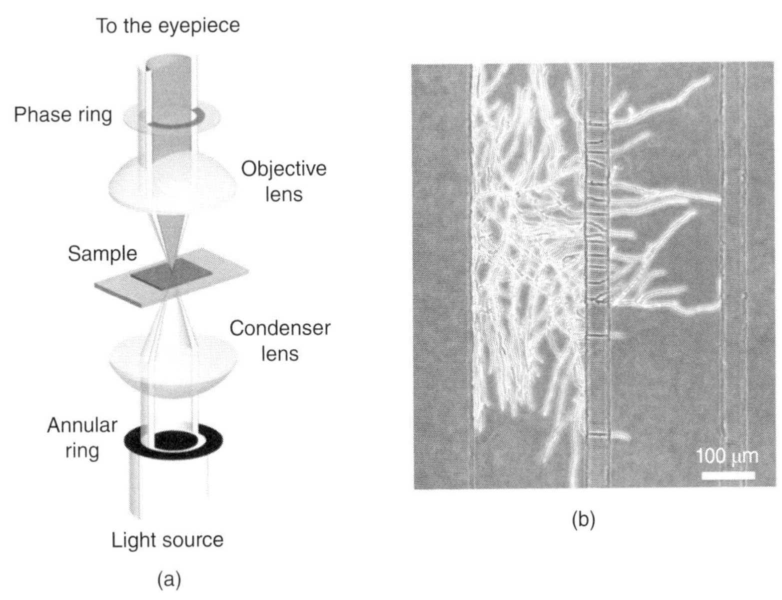

Figure 10.6 Phase contrast microscopy: (a) diagram and (b) the image of elongating mycelium hyphae of a common *Aspergillus* mold confined within polydimethylsiloxane-based microfluidic channels. Source: Image courtesy of L. Millet ; see Ref. [27] for detailed description of the image in (b). (*See color plate section for the color representation of this figure.*)

10.6 FLUORESCENCE MICROSCOPY

The optical microscopy techniques described so far are based on transmission, absorption, refraction, or scattering of light. Fluorescence microscopy is a method based on the fluorescence phenomenon, where absorption of light by fluorescent molecules of dyes or fluorophores results in the emission of light at a longer wavelength (Fig. 10.7a and b). The dye molecules absorb light of a specific wavelength (Fig. 10.7b), which results in a transition to a higher energy level known as the excited state (Fig. 10.7a). After a short time delay (determined by the lifetime of molecules in the excited state, typically in the range of nanoseconds), the molecule returns to the initial ground state (Fig. 10.7a) with the emission of light with a longer wavelength than that of the absorbed light (Fig. 10.7b). The difference between the wavelengths of maximum absorption and fluorescence (called the Stokes shift) is caused by the loss of part of

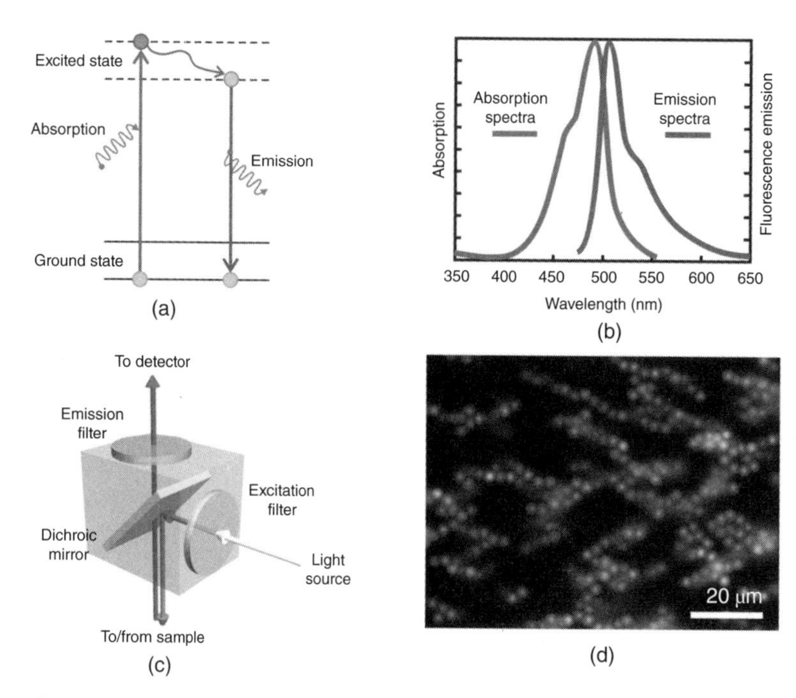

Figure 10.7 Fluorescence microscopy: (a) Jablonski diagram; (b) typical absorption and emission spectra of a fluorescent dye; (c) filters cube; (d) fluorescence texture of spherical microparticles tagged with a dye and dispersed in a nematic liquid crystal. (*See color plate section for the color representation of this figure.*)

the absorbed energy due to nonradiative processes. This wavelength difference makes it possible to effectively separate the excitation and emission signals by the use of optical filters.

The basic function of a fluorescence microscope is to irradiate the sample with light of a specific wavelength and then to separate the fluorescence emission signal from the excitation light of much higher intensity. In the fluorescence microscope setup, this is typically achieved with a filter cube consisting of excitation and emission filters and a dichroic mirror (Fig. 10.7c). The constituents of interest in a sample are labeled with one or more fluorescent dyes. The excitation wavelength is selected by the excitation filter, reflected by the dichroic mirror to the sample, and absorbed by the dye molecules in the sample. The fluorescent light emitted by the dye molecules transmits through the dichroic mirror and, after being separated from the excitation light by the emission filter (Fig. 10.7c), is collected by the photodetector. This forms an image with the bright areas corresponding to the sample regions marked with the fluorescent dye and the dark areas corresponding to the regions without the dye (Fig. 10.7d). A fluorescence microscopy image in Figure 10.7d shows colloidal structures formed in a nematic liquid crystal by melamine resin microspheres labeled with a fluorescent dye that emits green fluorescence light [3].

10.7 FLUORESCENCE CONFOCAL MICROSCOPY

A more advanced technique, which offers 3D imaging capability by combining the features of fluorescence and confocal microscopies, is FCM as shown in Figure 10.8. The main feature of confocal microscopy is that the inspection region at a time is a small voxel (volume pixel), and the signal arising from the neighboring region is prevented from reaching the detector by having a pinhole in the detection image plane (Fig. 10.8a). The inspected voxel and the pinhole are confocal, so that only the light coming from the probed voxel reaches the detector. This enables diffraction-limited imaging resolution not only in the lateral plane but also along the optical axis of the microscope. It is thus possible to construct a 3D image of the sample by scanning (typically achieved using galvano-mirrors or acousto-optic deflectors [21]) voxel by voxel within the volume of interest. In FCM, the contrast and sensitivity of confocal microscopy are greatly enhanced as the fluorescent dye used to tag the sample absorbs light at the wavelength of the excitation laser beam and emits at a longer wavelength. This allows for imaging of fluorophore-tagged samples with diffraction-limited resolution in the axial plane of the microscope, a capability not provided by the conventional fluorescence microscopy.

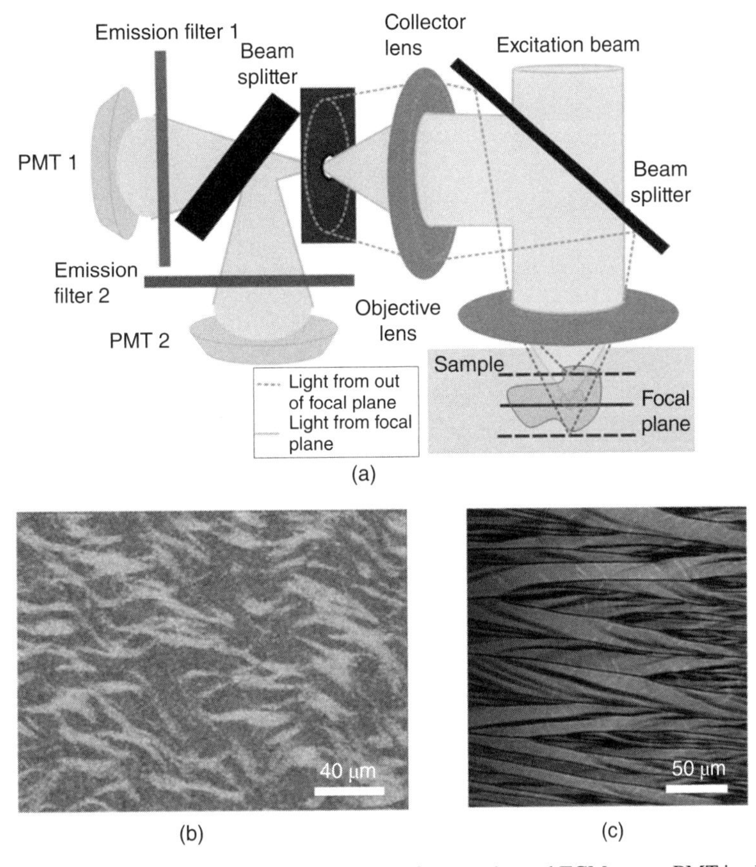

Figure 10.8 Fluorescence confocal microscopy: (a) principal diagram of a two-channel FCM setup: PMT is photomultiplier tube; (b) image of phase-separated domains of F-actin (green) and DNA (red) labeled by two different dyes [5]; (c) polarizing microscope texture of the pattern formed in a drying droplet of aqueous DNA and colocalized with the fluorescence confocal signal from a small number of molecules marked by fluorescent dye (green); see Refs [5, 6] for the detailed description of samples. (*See color plate section for the color representation of this figure.*)

Figure 10.8a illustrates the basic principles of FCM. The combined features of fluorescence and confocal microscopies offer enhanced lateral resolution in isotropic samples, $r_{\text{lateral}} = 0.44\lambda/\text{NA}$ [21]. The axial resolution of FCM is $\Delta z = 1.55n\lambda/\text{NA}^2$, where n is the refractive index of the medium between the objective lens and the sample [4, 17, 20, 21]. These theoretical limits of resolution, however, are rarely achieved in experiments because of different kinds of sample inhomogeneities, light scattering, spherical and chromatic aberrations, sample birefringence, and index mismatches at dielectric interfaces along the light path. These and other effects can significantly worsen both axial and lateral resolutions of FCM. For example, birefringent samples such as liquid crystals cause the excitation beam to split into two separate beams that are focused at different spots, thus degrading the spatial resolution and making it dependent on the depth of imaging and sample birefringence. Photobleaching, temporary loss of the ability to fluoresce due to photon-induced chemical damage, is another reason for worsening the imaging quality [17, 21].

Similar to conventional fluorescence microscopy (although we did not provide examples of this in the previous section), FCM enables multicolor imaging of sample composition patterns. Different constituents of the studied sample can be labeled with different dyes tailored to have different excitation and emission wavelengths, and the resultant fluorescence patterns from each of them can be collocated to form separate images or overlaid to form a single multicolor image. For example, Figure 10.8b shows an image of phase-separated domains in a DNA/F-actin biopolymer mixture with the DNA and F-actin each labeled with a different dye [5]. Likewise, to obtain complementary information, one can use transmission mode bright field microscopy, polarizing microscopy, or a number of other optical imaging modalities to obtain images that can then be overlaid with the FCM images. This is demonstrated by an

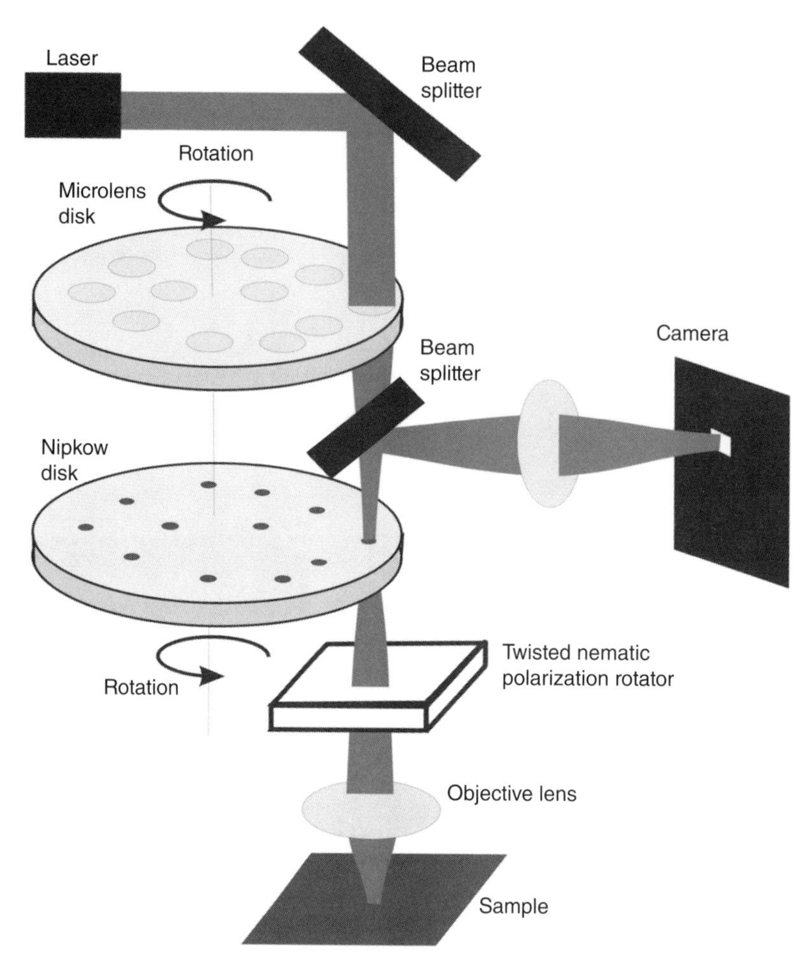

Figure 10.9 A schematic of Nipkow disk fluorescence confocal microscopy. Twisted nematic polarization rotator is typically used only in studies of samples with orientational ordering. (*See color plate section for the color representation of this figure.*)

example of FCM textures (Fig. 10.8c) of a DNA pattern at the perimeter of a dried drop overlaid with the PM image of the same sample area [6]. Multicolor imaging is especially useful in the study of composite soft matter systems.

Since imaging with a conventional confocal microscope is relatively slow, it is most often utilized in the study of stationary structures. To study dynamic processes in soft matter systems, researchers have recently started to utilize fast confocal microscopy systems, which can be implemented by the use of the fast laser scanning with acousto-optic deflectors (instead of galvano-mirrors) or using the Nipkow disk. One example of implementation of the fast confocal imaging system, shown in Figure 10.9, employs a rotating Nipkow disk [17, 21] having thousands of pinholes, supplemented by a coaxial disk with microlenses. The two disks are mechanically connected and rotated together by an electric motor. The sample is scanned by thousands of excitation beams at once, and the resultant speed of FCM imaging is higher by orders of magnitude as compared to that of a conventional confocal microscope. The vertical refocusing is performed by a fast piezo z-stepper drive that is capable of an accurate (typically ~ 50 nm or better) vertical position setting. The speed of imaging can reach 100–1000 frames/s, although it depends on many factors, such as the needed contrast (i.e., the integration time of the fluorescence signal), the size of the scanned area, and the frame rate of the used CCD camera. Thus, Nipkow disk-based FCM allows one to decipher the dynamics of colloidal structures and other soft matter systems with up to millisecond temporal resolution. For example, fast FCM has been widely used to simultaneously localize multiple (millions) dye-marked particles and has provided deep insights into the physics of colloidal aggregation and phase transition in colloidal fluids, gels, glasses, and so on [28].

10.8 FLUORESCENCE CONFOCAL POLARIZING MICROSCOPY

The FCM technique described earlier visualizes the distribution of fluorescent dye, providing insight into the spatial distribution of different dye-tagged constituents throughout

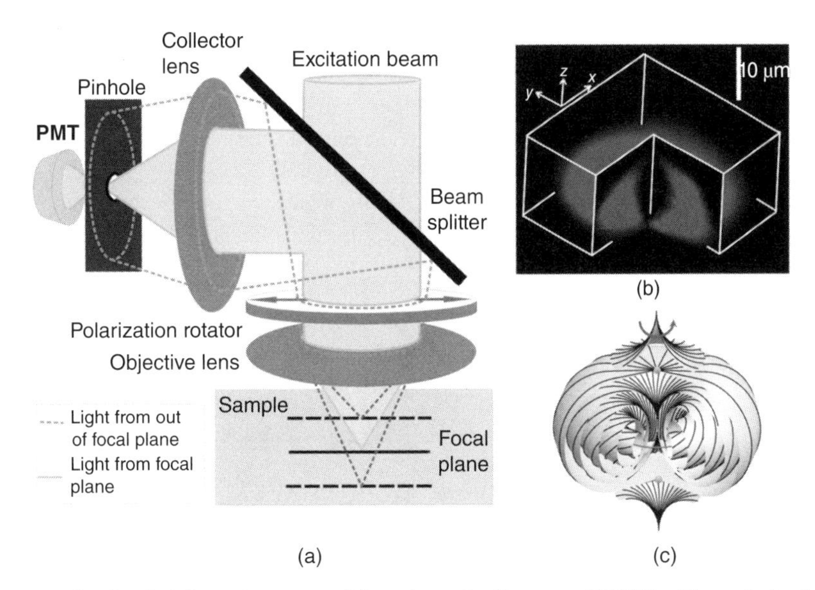

Figure 10.10 Fluorescence confocal polarizing microscopy: (a) a schematic diagram of FCPM with a polarization rotator; (b) 3D image of the T3-1 toron director configuration obtained using FCPM with circularly polarized probing light (see Ref. [7] for details); (c) a schematic representation of the T3-1 configuration with the double-twist cylinder looped on itself and accompanied by two hyperbolic point defects (blue dots). (*See color plate section for the color representation of this figure.*)

the sample. The orientational order of anisotropic soft matter systems can be probed by combining conventional FCM with the capability of polarized light excitation and detection. In fluorescence confocal polarizing microscopy (FCPM) [4], this is commonly achieved by annexing the FCM setup with a polarization rotator introduced before the objective lens (Fig. 10.10a), which enables controlled polarized excitation. In addition, FCPM requires that the specimen be stained with anisometric dye molecules which, on average, align parallel or perpendicular to the molecules of the studied "host" material. FCPM signal strongly depends on the angle between the transition dipole moment of the dye molecules and the polarization of excitation light. The intensity of fluorescence is maximized when the linear polarization of excitation light is parallel to the transition dipoles of excitation and fluorescence of the dye molecules in the sample, and it is at a minimum when the polarization is perpendicular to the transition dipoles. This strong orientational dependence of the measured fluorescence signal allows one to decipher the 3D molecular orientation patterns and liquid crystal director fields directly from the FCPM images.

The basic principle of FCPM is illustrated with the help of a schematic shown in Figure 10.10a. Let us assume that the transition dipoles of both excitation and fluorescence are parallel to the long axis of the dye molecules. The linearly polarized light incident on the sample causes fluorescence of the dye molecules. The efficiency of light absorption and the intensity of the detected fluorescence are determined by the angle β between the polarization of incident light and the long axis of the dye molecules. Thus, the intensity of the detected fluorescence signal is proportional to $\cos^4\beta$ (in the case of collinear polarized detection) [4]. Fluorescent light emitted from the focal spot passes through the pinhole located in the focal plane of the collector lens that is conjugate to the focal plane of the objective. Signals from out-of-focus regions are prevented from reaching the detector, similar to the case of FCM. To obtain a 3D image of the sample, the tightly focused laser beam raster-scans in the lateral plane (perpendicular to the microscope's optical axis), and then, by moving the objective (or the sample-stage) stepwise along the axial direction, the scan is repeated at each axial depth. As a result, the resultant 3D image comprises a stack of thin (submicron) horizontal optical slices stored in the computer memory that can be then software-processed and presented in a variety of different formats.

FCPM offers a detailed 3D visualization of orientational ordering in soft matter systems like liquid crystals. The example of 3D FCPM texture shown in Figure 10.10b visualizes the structure of the so-called toron (Fig. 10.10c) [7] generated by an infrared laser beam in a cholesteric liquid crystal sandwiched between two glass substrates with vertical boundary conditions. A schematic representation in Figure 10.10c shows the reconstructed director configuration of the toron with a double-twist cylinder looped on itself and accompanied by two hyperbolic point defects. FCPM imaging can thus visualize the static equilibrium structures of long-range molecular alignment in soft matter systems. On the other hand, the fast FCPM setup, such as the one based on a Nipkow disk confocal microscopy and an achromatic polarization rotator (Fig. 10.9), can be used to visualize dynamics of 3D director fields in liquid crystals.

10.9 NONLINEAR OPTICAL MICROSCOPY

NLO techniques utilize intrinsic and/or extrinsic NLO responses of materials such as biological systems and a variety of soft matter systems [29]. These imaging techniques are based on nonlinear light-matter interactions that result in either emission, such as multiphoton excitation fluorescence/luminescence, or scattering, such as second harmonic generation (SHG) [30], sum frequency generation (SFG) [31], third harmonic generation (THG) [32], and coherent Raman scattering, namely, coherent anti-Stokes Raman scattering (CARS) [8–10, 33] and stimulated Raman scattering (SRS) [34, 35].

NLO microscopy has recently emerged as a powerful tool for noninvasive, label-free imaging with high 3D resolution capable of probing highly scattering thick biological and soft matter samples. In contrast to conventional single-photon excitation schemes (such as those used in fluorescence microscopy and ordinary Raman microscopy), the NLO signal is generated from NLO interactions involving multiple excitation photons. Hence, there are several advantages of using NLO microscopy as compared to conventional (linear) optical imaging [21, 29, 33]: (i) low out-of-focus photobleaching (multiphoton absorption occurs only at the focus), (ii) low photodamage (e.g., biological tissues absorb less in the near infrared), (iii) ability to excite ultraviolet (UV)-excitable fluorophores with visible or near-infrared light sources via two- or three-photon absorption (since UV microscope optics and UV laser sources that would be needed for a single-photon excitation are not easily accessible.), (iv) inherent optical sectioning (no pinhole required because of a small excitation volume at the focal spot), (v) ability to work with thick specimens (larger penetration depth), and (vi) chemical bond selectivity in coherent Raman imaging. However, NLO microscopy typically requires more expensive pulsed lasers (typically a femtosecond pulsed laser) and more complicated microscopy setups that are rarely available commercially.

Multiphoton excitation is a nonlinear process typically associated with absorbing two, three, or more photons of near-infrared light and emitting a single photon at a wavelength shorter than the excitation wavelength. The comparison of energy diagrams of the single- and multiphoton processes is shown in Figure 10.11a. Nonlinear processes in an optically excited nonlinear medium can be described by the induced polarization $P(t)$ expressed as a power series in an electric field $E(t)$:

$$P(t) \propto \chi^{(1)}E(t) + \chi^{(2)}E^2(t) + \chi^{(3)}E^3(t) + \cdots,$$

where the coefficients $\chi^{(n)}$ are the nth order susceptibilities of the medium [29]. This expression can be used to describe a number of NLO processes discussed later from the standpoint of their use for imaging purposes.

In this section, we first describe multiphoton excitation fluorescence microscopy and multiharmonic generation microscopy and then proceed to the discussion of molecular imaging techniques utilizing coherent Raman scattering, such as recently introduced CARS microscopy, CARS polarizing microscopy, and SRS microscopy.

10.9.1 Multiphoton Excitation Fluorescence Microscopy

Compared to the single-photon fluorescence technique, multiphoton excitation fluorescence microscopy requires higher peak power of excitation. A femtosecond or picosecond pulsed laser is typically used as an excitation light source with a wavelength in the near-infrared region. The intensity of the excitation light falls off inversely as the square of the axial distance from the focal plane, and the efficiency of multiphoton absorption away from the focal point is extremely low because the probability of exciting a fluorophore falls off inversely as the fourth power of the axial distance [29]. Unlike in FCM, the detection of fluorescent light does not require a pinhole, because the excitation volume is inherently small enough to enable high-resolution imaging in 3D. The use of a near-infrared excitation laser typically helps to image thick samples with deep penetration and less photodamage to the sample because spectral range (600–1300 nm) is known as a "tissue optical window" with low light losses due to absorption and scattering [20, 21].

A simple schematic diagram of multiphoton excitation fluorescence microscopy setup is shown in Figure 10.11c. The setup utilizes a tunable pulsed laser, an xy scanning mirror, an inverted microscope with multiple detection channels, and a rotating polarizer. Two different NLO imaging modes can be implemented simultaneously, for example, by collecting the fluorescence signal in the backward detection (also known as epi-detection) and the SHG signal in the forward detection, as shown in Figure 10.11c.

Polarization-sensitive excitation and detection of NLO signals is very useful in the study of anisotropic materials such as liquid crystals for imaging of 3D patterns of long-range molecular orientation. An example of multiphoton excitation fluorescence polarizing microscopy in Figure 10.12 shows a three-photon excitation fluorescence image of a colloidal particle in a smectic liquid crystal [10] (note that the liquid crystal molecules in this particular case serve as fluorophores themselves). The intensity of detected NLO polarizing microscopy signals depends on the angle β between the polarization of excitation pulses and the liquid crystal director $\mathbf{n}(\mathbf{r})$ as $\sim \cos^{2m}\beta$ for the detection with no

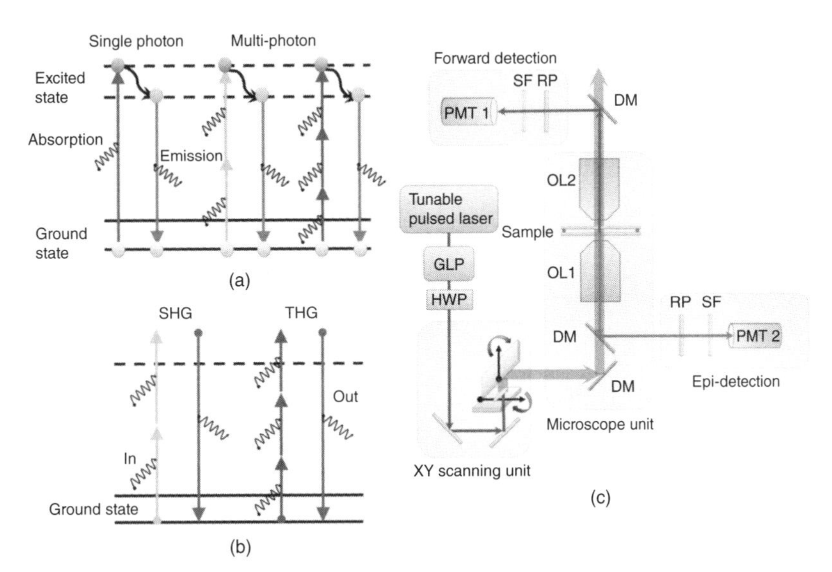

Figure 10.11 Multiphoton excitation fluorescence and multiharmonic generation microscopy. Energy diagram of (a) single and multiphoton processes and (b) second and third harmonic generation processes. (c) A schematic diagram of nonlinear optical imaging setup based on a tunable pulsed laser, an *xy* scanning mirror, and an inverted microscope with multiple detection channels. GLP, HWP, and RP are used to control light polarization. DM, dichroic mirror; GLP, Glan laser polarizer; HWP, half wave plate; OL, objective lens; PMT, photomultiplier tube; RP, rotating polarizer; SF, selection filters. (*See color plate section for the color representation of this figure.*)

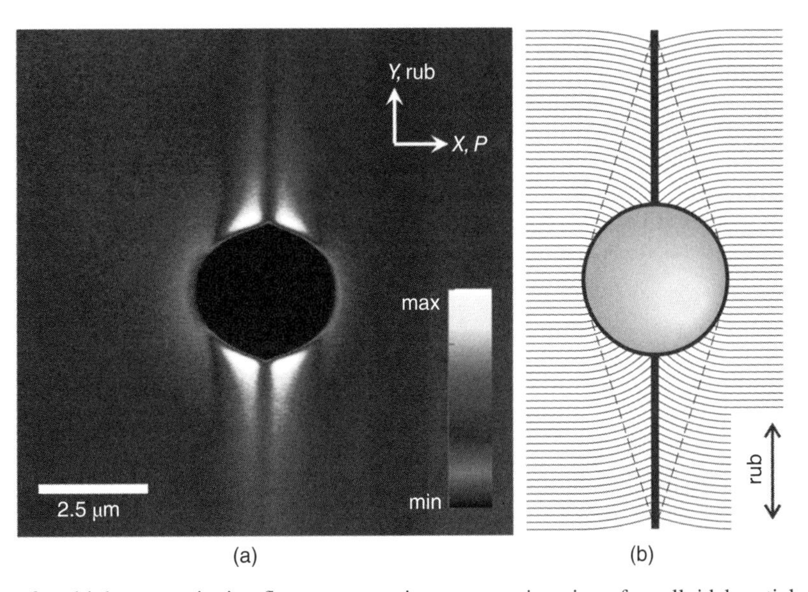

Figure 10.12 Application of multiphoton excitation fluorescence microscopy to imaging of a colloidal particle in a smectic liquid crystal (see Ref. [10] for detailed description). (a) Three-photon excitation fluorescence image of smectic liquid crystal layer deformations around a melamine resin sphere obtained with excitation at 870 nm and detection within 390–450 nm. (b) Reconstructed deformations of smectic layers (thin solid lines) around a spherical inclusion. Liquid crystal molecules are anchored tangentially at the surface of the particle. Arrow marked with "rub" shows the direction of substrates rubbing. Thick solid line shows the singular defect line, and dashed lines enclose the regions of strong layer deformations corresponding to the areas of maximum signal in (a). (*See color plate section for the color representation of this figure.*)

polarizer and as $\sim \cos^{2(m+1)}\beta$ for imaging with the polarizer in the detection channel collinear with the polarization of the excitation beam, where m is the order of the nonlinear process (e.g., $m = 2$ for two-photon excitation and $m = 3$ for three-photon excitation [8–10]).

10.9.2 Multiharmonic Generation Microscopy

SHG and THG microscopies derive contrast from variations in a specimen's ability to generate the respective harmonic signal from the incident light [30–32]. These multiharmonic generation microscopies also require an

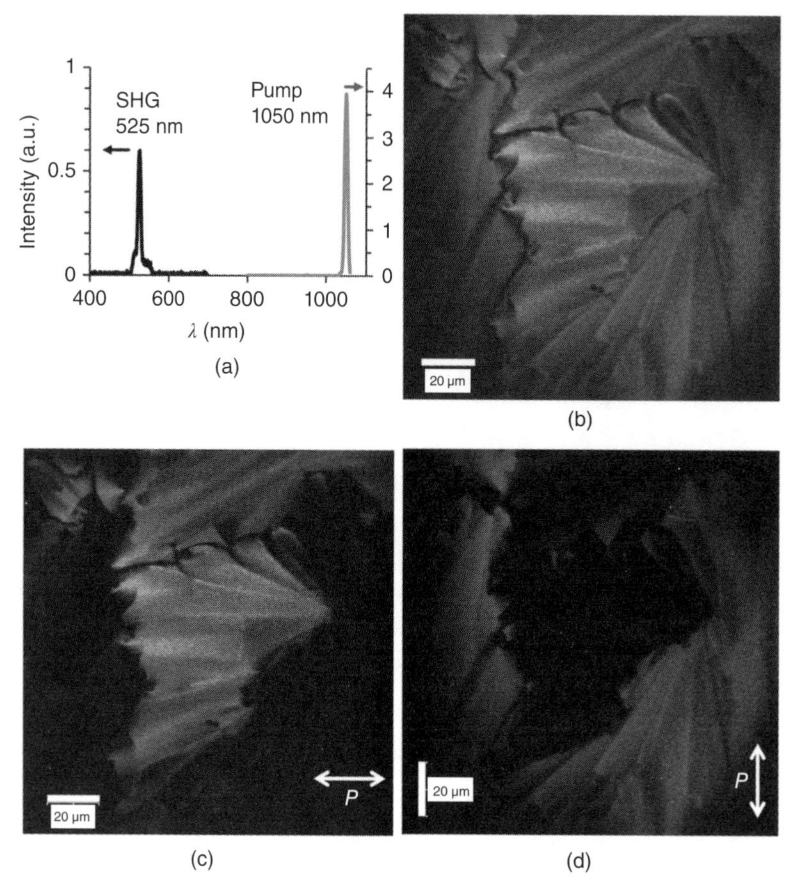

Figure 10.13 Second harmonic generation imaging of a smectic C* liquid crystal. (a) A spectrum showing the excitation pulse and the generated SHG signal. (b) In-plane colocalized superimposed texture of two SHG images of (c) and (d) that were obtained separately for two orthogonal polarizations (see Ref. [10] for a detailed description). (*See color plate section for the color representation of this figure.*)

intense laser light source (i.e., a pulsed laser). SHG signal emerging from the material is at exactly half the wavelength (frequency doubled) of the incident light interacting with the material via the second-order nonlinear process. Although two-photon excitation fluorescence is also a second-order nonlinear process, it loses some energy during relaxation from the excited state and, therefore, fluorescence occurs at wavelengths longer than a half of excitation wavelength. In contrast, SHG process is energy conserving. The energy diagram of SHG and THG processes (Fig. 10.11b) shows that there is no energy loss in the scattered light after multiple photons combine into a single photon of higher energy. In general, the setup for multiharmonic generation microscopy is similar to that of multiphoton excitation fluorescence microscopy. Therefore, multiharmonic generation imaging can be implemented with detection in the forward channel, and simultaneously multiphoton excitation fluorescence imaging can be collected in the epi-detection channel, as shown in Figure 10.11c. SHG does not depend on excitation of fluorescent molecules. Hence, it is not required to tag the specimen with a dye, and the effects of photobleaching and photodamage are avoided. The second-order term $\chi^{(2)}$ in nonlinear process is nonzero only in media with no inversion symmetry while the $\chi^{(3)}$ term is nonzero for all media. Thus, imaging can also provide information about the symmetry of the studied materials by using these nonlinear processes.

As an example of the use of SHG microscopy in the study of soft matter, Figure 10.13 shows SHG images of a smectic C* liquid crystal obtained by using excitation at 1050 nm and detection at 525 nm for two orthogonal polarizations of excitation light. The spectra and selection filters corresponding to SHG images are shown in Figure 10.13a. A strong SHG signal reveals the polar ordering of the smectic C* phase with focal conic domains [10].

10.9.3 Coherent Anti-Stokes Raman Scattering Microscopy

CARS microscopy is a noninvasive, label-free nonlinear imaging technique that utilizes molecular vibrations to obtain imaging contrast [8–10, 33]. CARS microscopy is often compared to conventional Raman microscopy,

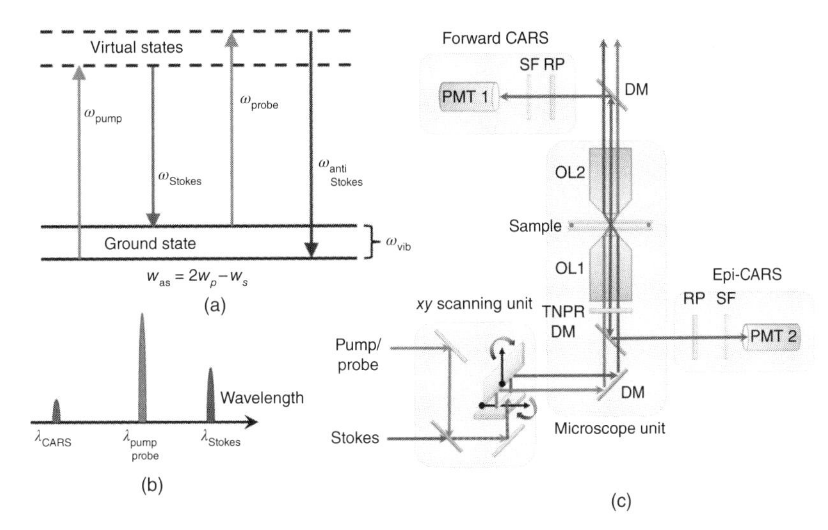

Figure 10.14 Coherent anti-Stokes Raman scattering microscopy. (a) The energy diagrams of CARS at $\omega_{as} = 2\omega_p - \omega_s$ when $\omega_{\text{vib}} = \omega_p - \omega_s$. (b) CARS signal generated at shorter wavelength than the pump and Stokes wavelengths. (c) A schematic diagram of CARS polarizing microscopy setup utilizing the synchronized pump/probe and Stokes pulses, *xy* scanning galvano-mirrors, and inverted microscope with forward and epi-detection channels. DM, dichroic mirror; OL, objective lens; PMT, photomultiplier tube; RP, rotating polarizer; TNPR, twisted nematic polarization rotator; SF, selection filter. (*See color plate section for the color representation of this figure.*)

as both techniques probe the same Raman active modes. Spontaneous Raman scattering can be induced by a single continuous wave laser, whereas CARS requires at least two pulsed laser sources at different frequencies. The spontaneous Raman signal is typically detected on the red side of the spectrum compared to the excitation radiation, where it might be difficult to discriminate it from fluorescence signal. The CARS signal is detected on the blue side, which is free from fluorescence interference, but it typically comes with a nonresonant background contribution.

The CARS technique utilizes a third-order nonlinear process, which involves three photons at two different frequencies called pump/probe ($\omega_p = \omega_{\text{pump}} = \omega_{\text{probe}}$) and Stokes ($\omega_s = \omega_{\text{Stokes}}$) beams as shown in Figure 10.14a. When the frequency difference between the pump/probe and the Stokes beams matches a certain molecular vibrational frequency ($\omega_{\text{vib}} = \omega_p - \omega_s$) and the phase-matching condition of three input photons is fulfilled, a strongly enhanced blue-shifted anti-Stokes ($\omega_{as} = 2\omega_p - \omega_s$) resonance signal is generated in the sample. The resultant CARS signal is at a wavelength shorter than that of the pump/probe and Stokes beams, as shown in Figure 10.14b. The CARS intensity scales with the intensities of the excitation beams as follows:

$$I_{\text{CARS}}(2\omega_p - \omega_s) \propto (\chi^{(3)}_{\text{CARS}})^2 I_p^2(\omega_p) I_s(\omega_s),$$

where $\chi^{(3)}_{\text{CARS}}$ is the third-order susceptibility and I_p and I_s are the intensities of the pump/probe and the Stokes beams, respectively.

A simple schematic diagram of CARS microscopy setup based on synchronized pump/probe and Stokes beams, an *xy* scanning mirror, and an inverted microscope with forward and epi-detection channels is shown in Figure 10.14c [10]. There are several options for a light source in CARS microscopy: (i) two tightly synchronized pulsed lasers, (ii) a synchronously pumped intracavity-doubled optical parametric oscillator, (iii) a single femtosecond laser pulse spectrally shaped to select two frequencies ω_p and ω_s, or (iv) a synchronously generated supercontinuum (by using highly nonlinear fiber) with filter-selected ω_p and/or ω_s from a single femtosecond laser. CARS microscopy has several unique benefits in the study of soft matter and biological samples: (i) intrinsic vibrational contrast (no labeling needed), (ii) a strong, coherently enhanced signal (CARS is more sensitive than conventional vibrational microscopy), and (iii) 3D sectioning capability with less photodamage, no photobleaching, and deeper penetration in thick turbid media, similar to other NLO microscopy techniques.

10.9.4 Coherent Anti-Stokes Raman Scattering Polarizing Microscopy

As discussed earlier, polarization-sensitive imaging is an essential tool for imaging soft matter systems possessing long-range orientational order. The CARS polarizing microscopy setup (Fig. 10.14c) has a polarization control based on a twisted nematic achromatic polarization rotator for excitation and epi-detection of CARS signals and an optional rotating polarizer in the forward-detection channel. Since CARS is a three-photon process, the dependence of the CARS signal on the angle β between the polarization of excitation light and the director of the sample is $\propto \cos^6\beta$

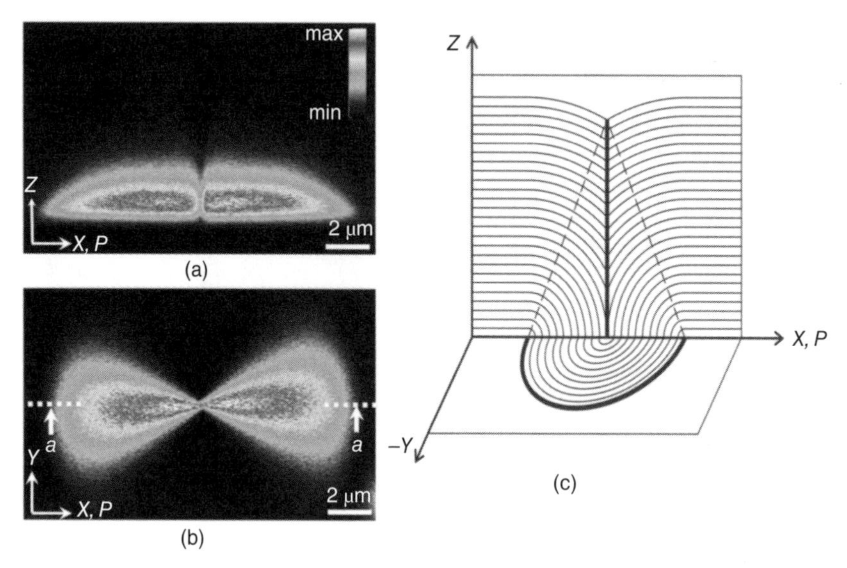

Figure 10.15 Coherent anti-Stokes Raman scattering images of toric focal conic domain in a smectic liquid crystal thin film on a solid substrate. (a) Vertical cross-sectional and (b) in-plane images. (c) Reconstructed structure of smectic layers (thin solid lines) in the toric focal conic domain. (*See color plate section for the color representation of this figure.*)

with no polarizer in the detection channel and $\propto \cos^8\beta$ with a polarizer collinear to excitation polarization in the detection [8–10]. Similar to other NLO polarizing microscopy techniques, CARS polarizing microscopy exhibits a stronger sensitivity to spatial variations of the director $\mathbf{n}(\mathbf{r})$ as compared with the single-photon FCPM imaging.

As an example of the use of the CARS polarizing microscopy technique, Figure 10.15a and b show a vertical cross section and in-plane images of a toric focal conic domain in a thin film of smectic liquid crystal on a solid substrate. The signal due to CN-triple-bond vibration of the smectic liquid crystal (4-cyano-4′-octylbiphenyl, 8CB) provides information about the spatial pattern of the liquid crystal director. When the director is parallel to the polarization of excitation light, the intensity of the CARS signal is maximized, whereas the CARS intensity is minimized when the director is perpendicular to the polarization of incident light. The reconstructed structure of smectic layers in the toric focal conic domain is shown in Figure 10.15c, which was obtained using the 3D CARS images (Fig. 10.15a and b).

10.9.5 Stimulated Raman Scattering Microscopy

SRS is a technique in which two beams, pump (at frequency ω_p) and Stokes (at frequency $\omega_s < \omega_p$), combine to amplify the Stokes Raman signal when the difference between ω_p and ω_s equals the vibrational frequency of a certain chemical bond in the molecules comprising the material as shown in Figure 10.16a. Since there are two photons involved, SRS is a second-order NLO process and the intensity of the SRS signal is proportional to the product of the intensities of the pump and the Stokes beams. As a result of the amplification of the Raman signal, the pump beam experiences a loss in its intensity (stimulated Raman loss) while the Stokes beam experiences a gain in its intensity (stimulated Raman gain). Since the gain/loss mechanism occurs only when the difference between the frequencies of the pump and Stokes beam equals the molecular vibration frequency, the nonresonant background is significantly reduced compared to that of CARS.

The relative Raman gain/loss is a very small fraction of the excitation signal intensities. Therefore, a typical implementation of SRS microscopy involves modulation of one of the excitation beams (e.g., Stokes), as shown in Figure 10.16b. The pump beam experiences Raman loss only when the Stokes pulse is present; thus, the Raman loss signal in the pump occurs at the Stokes modulation rate, which is then selectively detected by a photodiode with the help of a lock-in amplifier. A simplified representative setup is shown in Figure 10.16c. Similar to CARS, SRS can also be implemented to image multiple molecular functionalities simultaneously. This is demonstrated in the example in Figure 10.16d, where a drug penetration enhancer dimethyl sulfoxide (DMSO) is imaged (shown in green) together with lipids (shown in red), demonstrating that DMSO is insoluble in the lipids [34, 35]. Similar to CARS and other NLO imaging techniques, SRS can be extended to enable orientation-sensitive imaging by means of polarized excitation and detection [11].

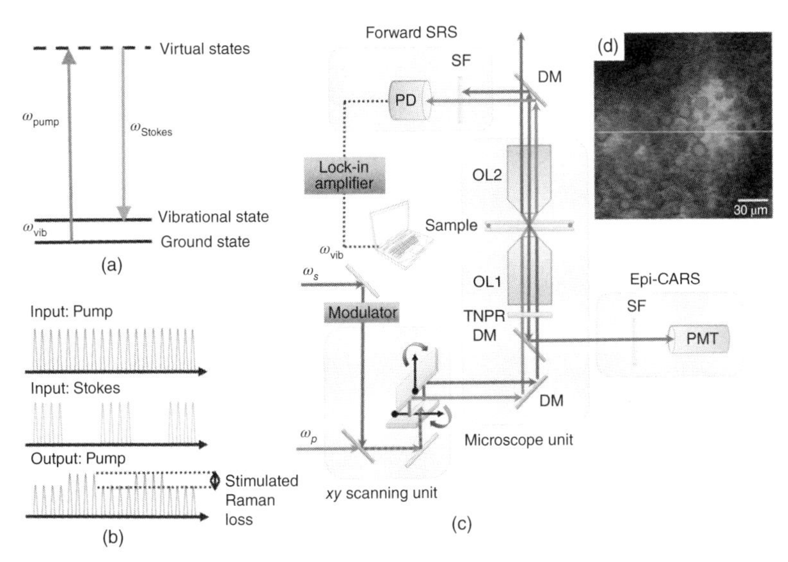

Figure 10.16 Stimulated Raman scattering microscopy. (a) Energy diagram and (b) detection scheme for SRS. Stokes beam is modulated at high frequency, and the resulting amplitude modulation of the pump pulse (stimulated Raman loss) can be detected. (c) A schematic diagram for SRS microscope with forward SRS and epi-CARS detection. (d) Two-color SRS image of DMSO (green) and lipids (red) in the subcutaneous fat layer. DM, dichroic mirror; OL, objective lens; PD, photodiode; PMT, photomultiplier tube; SF, selection filter. Source: Image reprinted from [34] with permission from AAAS. (*See color plate section for the color representation of this figure.*)

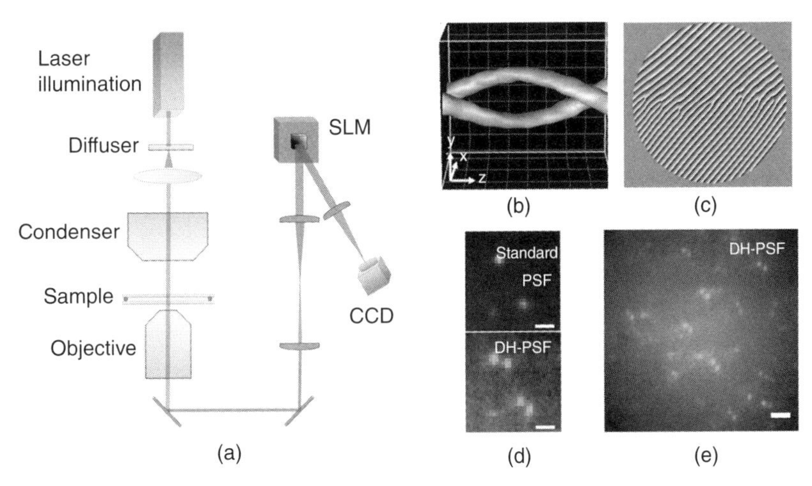

Figure 10.17 Optical imaging by the use of engineered PSF. (a) Optical setup for imaging with pre-engineered PSF. (b) Schematic representation of double-helix PSF. (c) Typical phase mask used to generate double-helix PSF. (d–e) Single molecule detection in 3D using double-helix PSF technique. Source: Images from [36] copyright 2009 National Academy of Sciences, USA. (*See color plate section for the color representation of this figure.*)

10.10 THREE-DIMENSIONAL LOCALIZATION USING ENGINEERED POINT SPREAD FUNCTIONS

Along with imaging of materials, high-resolution tracking and localization of molecules and particles is helpful for understanding their self-assembly in soft matter systems. For instance, the image blurring caused by displacing the object away from the focal plane can be used as a rough measure in determining the axial position of an object in the image. One approach to localize an object in 3D with much finer resolution is to design the PSF of the imaging system, as opposed to relying on the standard Airy disk PSF (Fig. 10.1c). Figure 10.17a shows an example of the implementation of such an imaging system where the PSF is made up of two lobes that rotate as they traverse along the microscope's optical axis, forming a "double-helix" near the focal point (Fig. 10.17b). The setup is built around a phase mask at the Fourier plane in the imaging path, which can be programmed by the use of a spatial light modulator (SLM), as shown in Figure 10.17c [12]. The SLM modulates the phase of the imaging beam at its Fourier plane, so that the resultant image of a particle captured by the CCD camera is presented in the form of two lobes. The axial position of an imaged particle is encoded as the angle of the line joining the lobes (Fig. 10.17d). This technique can be used to localize scatterers (e.g., colloids) or fluorescent emitters (e.g., single fluorescent molecules) with the help of appropriate dichroic mirror and filters (Fig. 10.17e) achieving nanoscale resolution (typically ~10 nm) in this spatial localization technique [36].

10.11 INTEGRATING THREE-DIMENSIONAL IMAGING SYSTEMS WITH OPTICAL TWEEZERS

Optical manipulation has proved to be of great importance in the study of colloidal systems, liquid crystals, and biological materials. Optical tweezers have been employed to study various liquid crystal systems, colloidal assemblies, and interactions, as well as topology and structure of defects [13]. Manipulation of foreign inclusions and defects gives rise to a variety of director patterns in liquid crystals, which requires 3D imaging to fully understand their structures [7].

An integrated optical imaging and manipulation system, such as the one schematically depicted in Figure 10.18a, allows for simultaneous noncontact optical manipulation and imaging [14]. The optical manipulation employs a liquid crystal-based SLM, which spatially modulates the phase of the incident trapping laser beam. The phase modulation is controlled in real time by programming the SLM to display holograms corresponding to the specified trap pattern at video rate. The lenses in the optical train form a 4*f* telescope, so as to image the plane of the SLM at the back aperture of the trapping objective. The main advantage of SLM-based holographic optical tweezers is that it allows simultaneous optical manipulation of multiple particles in the lateral as well as the axial directions. The SLM also allows generation of non-Gaussian beams (e.g., Laguerre–Gaussian beams), which can be used to generate liquid crystal director patterns under the effect of various unconventional light-intensity profiles [7].

Simultaneous 3D optical manipulation and imaging capabilities of the system are demonstrated in Figure 10.18b–d, where colloidal particles immersed in a cholesteric liquid crystal are moved axially perpendicular to the cholesteric layers, and the layer configurations are imaged in vertical cross sections with FCPM [14]. The cholesteric layer deformations caused by the presence of particles can be

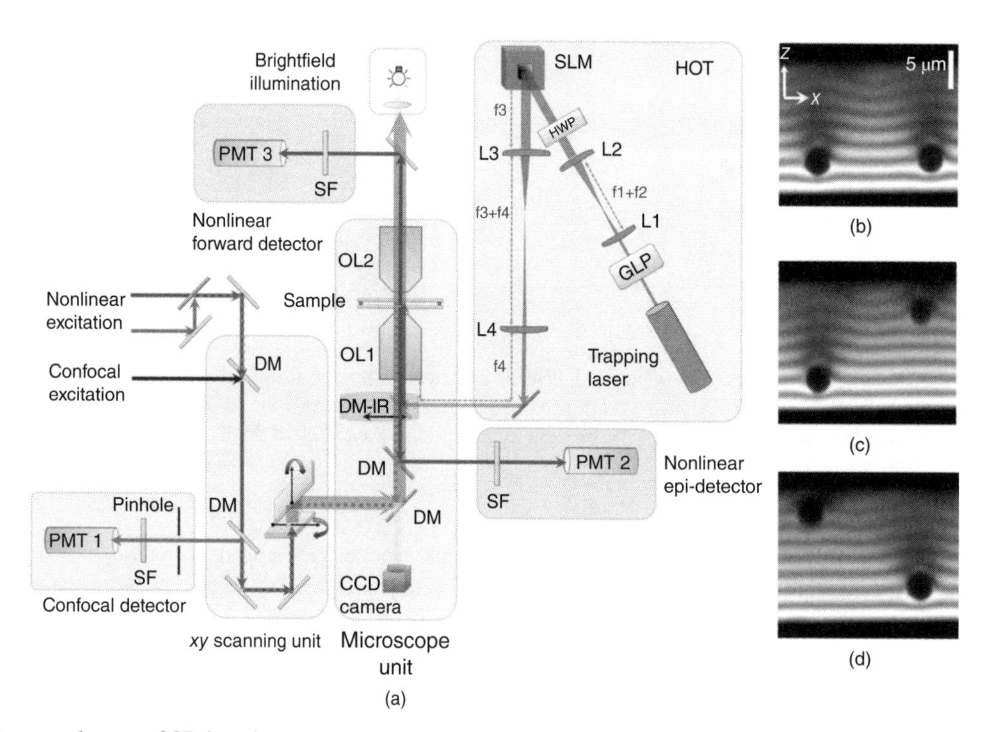

Figure 10.18 Integrated setup of 3D imaging and holographic optical trapping. (a) A schematic diagram of 3D optical manipulation with holographic optical trapping (HOT) and simultaneous imaging with FCPM and nonlinear optical microscopy. (b–d) FCPM vertical cross sections showing spherical particles in cholesteric liquid crystal moved axially and imaged using the integrated setup. The periodicity of bright stripes roughly corresponds to the half pitch of the cholesteric liquid crystal. DM, dichroic mirror; DM-IR, dichroic mirror for trapping; OL, objective lens; PMT, photomultiplier tube; SF, selection filters. (*See color plate section for the color representation of this figure.*)

clearly seen in the FCPM images. The integration of 3D noncontact manipulation with 3D imaging enables the design and characterization of soft matter composites comprising colloids, nanoparticles, liquid crystals, polymers, and so on. The integrated setup allows one to uncover the nature of interactions between various components of the materials system as well as control of their assembly and organization. In addition, integration of an engineered PSF tracking and localization technique can expand functionality of the system to probe molecular and colloidal interactions with higher precision.

10.12 OUTLOOK AND PERSPECTIVES

Some of the grand challenges from the standpoint of the use of optical microscopy in the study of soft matter systems include the need of nanoscale spatial resolution (to directly probe molecular and colloidal self-assembly in material systems), sub-millisecond temporal resolution (to probe fast dynamic processes), sensitivity to different chemical bonds of molecules comprising soft matter systems, and the ability to visualize orientations of molecules and the chemical bonds comprising them. Out of all of these challenges, achieving nanoscale resolution of optical imaging is, perhaps, the most important. However, in terms of resolution, optical imaging has undergone only incremental improvements over centuries of its use, even though imaging techniques are most frequently used not only in soft matter research but also in many other branches of science. Novel imaging approaches with improved resolution are especially needed to further the collective understanding of soft matter systems on the nanoscale. Several recently introduced approaches show the promise of providing significant breakthroughs in terms of overcoming resolution limits and may also be applied to the study of soft matter systems. In the future, their further development may allow for direct imaging of the self-assembly of molecules and nanoparticles and their condensed phase behavior.

Recently, a bulk of theoretical and experimental research has shown the feasibility of diffraction-unlimited optical imaging by the use of simple optical elements based on metamaterials with a negative index of refraction. Theoretical works by Veselago [37] and Pendry [38], followed by numerous theoretical and experimental explorations, have demonstrated that the diffraction limit of conventional lenses is not a limiting factor for focusing of light by a lens made of a flat slab of metamaterial with a negative index of refraction, often referred to as a "superlens" or a "perfect lens." The improved resolution of the negative-index superlens is due to the transmission of the evanescent surface waves, which are not lost (unlike in the case of conventional lenses). Being first considered theoretically over four decades ago by Veselago [37], metamaterials (also called "left-handed materials") have never been encountered in nature or fabricated until the theoretical work of Pendry [38] provided important physical insights into how these unusual materials can be realized. Metamaterials are composed of structural units much smaller than the wavelength of incident light so that they appear homogeneous to the electromagnetic waves. To obtain metamaterials with a negative refractive index at optical frequencies, one has to achieve simultaneous "negative" electric and magnetic responses in the artificial nanofabricated composite material [37–39]. Although the practical uses of metamaterials are still hindered by a number of technical challenges, such as losses, lenses and other optical elements made of metamaterials may one day revolutionize optical microscopy and enable nanoscale diffraction-unlimited imaging [39].

Several new techniques enabling sub-diffraction-limited high-resolution optical imaging include near-field scanning optical microscopy (NSOM) [40], photoactivated localization microscopy (PALM) [41, 42], and stochastic optical reconstruction microscopy (STORM) [43]. They can be extended for the study of soft matter systems, including the orientation-sensitive imaging of long-range molecular alignment patterns in liquid crystals. NSOM is a type of scanning probe microscopy for nanoscale investigation by observing the properties of evanescent waves. This is done by placing the probe very close to the sample surface. Light passes through a subwavelength diameter aperture and illuminates a sample that is placed within its near field, at a distance smaller than the wavelength of the light. With this technique, the resolution of the image is limited by the size of the detector aperture and not by the wavelength of the illuminating light. In particular, lateral resolution of 20 nm and vertical resolution of 2–5 nm have been demonstrated [40]. However, NSOM is difficult to operate in a noninvasive mode and has a limited imaging depth.

The basic concept of PALM and STORM techniques is to fill the imaging area with many nonfluorescing fluorophores that can be photoactivated into a fluorescing state by a flash of light [41–43]. Since this photoactivation is stochastic, only a few, well-separated molecules can be detected and then localized by fitting the PSF with Gaussians with high precision. This process is repeated many times while photoactivating different sets of molecules and building up an image molecule by molecule. The resolution of the final reassembled image can be much higher than that limited by diffraction. The demonstrated imaging resolution is 20–30 nm in the lateral dimensions and 50–60 nm in the axial dimension [44]. However, the major problem with these emerging techniques is that it takes time on the order of hours to collect the data to get these high-resolution images. Study of soft matter systems such as biological structures and colloidal nanoparticle dispersions will tremendously benefit from further development and use of these techniques.

In conjunction with high spatial resolution, imaging systems with high temporal resolution are required to study dynamic processes in soft matter systems, for instance, the time evolution of self-assembled structures, the effect of applied external fields, and so on. Nowadays, high-speed cameras with capture rates of up to a million frames per second are commercially available. There has been significant development on the front of enabling nonlinear, noninvasive optical imaging modalities such as CARS and SRS to operate at video rate, which will immensely benefit research on biological and soft matter systems. This will eventually allow for *in vivo* optical imaging with molecular selectivity [45] and direct characterization of self-assemblies in composite systems.

A variety of optical microscopy techniques discussed here has driven the research in the field of soft matter systems to new frontiers. The older microscopy techniques have only allowed 2D imaging, and hence the actual 3D configurations of material components composing soft matter systems was often left to an educated guess [25]. The introduction of 3D imaging techniques has facilitated unambiguous determination of the molecular and colloidal arrangement with precise localization of other elements. The advent of pulsed lasers has enabled the implementation of NLO microscopies, furthering this trend and rendering the use of labeling agents unnecessary, while at the same time giving similar or better spatial resolution. NLO microscopy has also made it more convenient to image composite soft matter systems by allowing imaging of different constituents in different nonlinear modalities simultaneously, instead of going through the difficult and often impossible process of finding appropriate dyes that would bind to each of the constituents selectively. Further general development of optical microscopy techniques is an ongoing quest that will continue to contribute to the body of knowledge of soft matter systems.

REFERENCES

[1] Liu Q, Cui Y, Gardner D, Li X, He S, Smalyukh II. Nano Lett 2010;10:1347.

[2] Senyuk B, Liu Q, He S, Kamien RD, Kusner RB, Lubensky TC, Smalyukh II. Nature 2013;493:200–205.

[3] Smalyukh II, Lavrentovich OD, Kuzmin AN, Kachynski AV, Prasad PN. Phys Rev Lett 2005;95:157801.

[4] Smalyukh II, Shiyanovskii SV, Lavrentovich OD. Chem Phys Lett 2001;336:88.

[5] Lai GH, Butler JC, Zribi OV, Smalyukh II, Angelini TE, Purdy KR, Golestanian R, Wong GCL. Phys Rev Lett 2008;101:218303.

[6] Smalyukh II, Zribi OV, Butler JC, Lavrentovich OD, Wong GCL. Phys Rev Lett 2006;96:177801.

[7] Smalyukh II, Lansac Y, Clark NA, Trivedi RP. Nat Mater 2010;9:139.

[8] Kachynskii A, Kuzmin A, Prasad PN, Smalyukh II. Appl Phys Lett 2007;91:151905.

[9] Kachynski AV, Kuzmin AN, Prasad PN, Smalyukh II. Opt Express 2008;16:10617.

[10] Lee T, Trivedi RP, Smalyukh II. Opt Lett 2010;35:3447.

[11] Lee T, Mundoor H, Gann DG, Callahan TJ, Smalyukh II. Opt Express 2013;21:12129.

[12] Conkey DB, Trivedi RP, Pavani SRP, Smalyukh II, Piestun R. Opt Express 2011;19:3835.

[13] Trivedi RP, Engström D, Smalyukh II. J Opt 2011;13:044001.

[14] Trivedi RP, Lee T, Bertness K, Smalyukh II. Opt Express 2010;18:27658.

[15] Bradbury S, Bracegirdle B. *Introduction to Light Microscopy*. United Kingdom: Taylor Francis Ltd; 1998.

[16] Levine S, Johnstone L. *The Ultimate Guide to Your Microscope*. Sterling Publishing; 2008.

[17] Müller M. *Introduction to Confocal Fluorescence Microscopy*. 2nd, in Tutorial texts series ed. Vol. 69. Bellingham, WA, United States: SPIE Press; 2006.

[18] Mertz J. *Introduction to Optical Microscopy*. Roberts and Company Publishers; 2009.

[19] Murphy DB, Davidson MW. *Fundamentals of Light Microscopy and Electronic Imaging*. 2nd ed. John Wiley & Sons, Inc.; 2012.

[20] Kubitscheck U. *Fluorescence Microscopy: From Principles to Biological Applications*. Weinheim: Wiley-VCH; 2013.

[21] Pawley JB. *Handbook of Biological Confocal Microscopy*. 3rd ed. New York: Springer; 2006.

[22] Nesse WD. *Introduction to Optical Mineralogy*. 3rd ed. Oxford: Oxford University Press; 2004.

[23] Jue T. *Fundamental Concepts in Biophysics*. New York: Humana Press; 2009.

[24] Bloss FD. *An Introduction to the Methods of Optical Crystallography*. New York: Holt, Rinehart and Winston, Inc.; 1961.

[25] de Gennes PG, Prost J. *The Physics of Liquid Crystals*. 2nd ed. Oxford: Oxford University Press; 1995.

[26] Manoharan VN, Imhof A, Pine DJ. Adv Mater 2001;13:447.

[27] Millet LJ, Stewart ME, Sweedler JV, Nuzzo RG, Gillette MU. Lab Chip 2007;7:987.

[28] Dinsmore AD, Weeks ER, Prasad V, Levitt AC, Weitz DA. Appl Optics 2001;40:4152.

[29] Boyd RW. *Nonlinear Optics*. 3rd ed. Academic Press; 2008.

[30] Yoshiki K, Hashimoto M, Araki T. Jpn J Appl Phys 2005;44:L1066.

[31] Fu Y, Wang H, Shi R, Cheng J-X. Biophys J 2007;92:3251.

[32] Pillai RS, Oh-e M, Yokoyama H, Brakenhoff CJ, Muller M. Opt Express 2006;14:12976.

[33] Potma EO, Xie XS. Coherent anti-Stokes Raman scattering (CARS) microscopy: instrumentation and applications. In: Masters BR, So PTC, editors. *Handbook of Biomedical Nonlinear Optical Microscopy*. New York, NY: Oxford University Press; 2008. p 164–186.

[34] Freudiger CW, Min W, Saar BG, Lu S, Holtom GR, He C, Tsai JC, Kang JX, Xie XS. Science 2008;322:1857.

[35] Min W, Freudiger CW, Lu S, Xie XS. Annu Rev Phys Chem 2011;62:507.

[36] Pavani SRP, Thompson MA, Biteen JS, Lord SJ, Liu N, Twieg RJ, Piestun R, Moerner WE. Proc Natl Acad Sci U S A 2008;106:2995.

[37] Veselago VG. Soviet Phys Uspekhi 1968;10:509.

[38] Pendry JB. Phys Rev Lett 2000;85:3966.

[39] Zhang X, Liu Z. Nat Mater 2008;7:435.

[40] Oshikane Y, Kataoka T, Okuda M, Hara S, Inoue H, Nakano M. Sci Tech Adv Mater 2007;8:181.

[41] Betzig E, Patterson GH, Sougrat R, Lindwasser OW, Olenych S, Bonifacino JS, Davidson MW, Lippincott-Schwartz J, Hess HF. Science 2006;313:1642.

[42] Hess ST, Girirajan TPK, Mason MD. Biophys J 2006;91:4258.

[43] Rust MJ, Bates M, Zhuang X. Nat Methods 2006;3:793.

[44] Huang B, Wang W, Bates M, Zhuang X. Science 2008;319:810.

[45] Saar BG, Freudiger CW, Reichman J, Stanley CM, Holtom GR, Xie XS. Science 2010;330:1368.

SECTION IV

COLLOIDAL PHASES

11

COLLOIDAL FLUIDS

JOSÉ LUIS ARAUZ-LARA
Instituto de Física, Universidad Autónoma de San Luis Potosí, San Luis Potosí, S.L.P., Mexico

11.1 INTRODUCTION

Colloidal fluids play an essential role in understanding the fundamental properties of soft materials and in designing some of their applications. For the sake of simplicity, let us consider here a colloidal fluid as consisting of "particles" dispersed in a fluid. The main characteristic of the particles being their size, which should be in the range of nanometers to about 1 μm. In other words, they should be much larger than the solvent's molecules but small enough to remain dispersed in the fluid, against gravity, purely by thermal agitation of the solvent. A great variety of systems, both natural and man-made, fall within this category. For instance, proteins, virus or DNA in solution, crude oil, polymeric solutions, some foods, cosmetics, and pharmaceuticals are colloidal fluids. Thus, the interest in such systems arises from both the scientific and the engineering point of view. The description of physical properties of colloidal fluids, such as stability, structure, dynamics, rheology, and thermodynamics, in terms of fundamental quantities is, in general, a complex task. Most systems contain different colloidal species, with different interparticle interactions. Those interactions depend not only on the nature of the particles themselves but also on the solvent properties, the amount of electrolytes, and temperature. The solvent can also be complex, with diverse molecular species, dielectric, conductivity, and viscosity properties. Nevertheless, significant progress has been done in the development of models of effective interparticle interactions; some of them are discussed below.

The aim of this chapter is to introduce some concepts and techniques currently used in the study of colloidal fluids, particularly those concerning the static and dynamic structural properties. Those methods are better understood when they are discussed for simple cases. Thus, we consider here monodisperse homogeneous suspensions of spherical particles, with spherically symmetric interactions, dispersed in a continuum three-dimensional (3D) viscous fluid.

Fluids, Colloids and Soft Materials: An Introduction to Soft Matter Physics, First Edition. Edited by Alberto Fernandez Nieves and Antonio Manuel Puertas.

Naturally, colloidal particles can be of any shape and can be in quite complex environments, for instance, in capillaries, in fracture rocks, and inside biological cells. Although there is also progress in the development of methods to address the study of more complex situations, the discussion of them is not carried out within this chapter. However, we consider here in some detail one interesting case, namely, the case of quasi-two-dimensional (Q2D) colloidal fluids. Those systems consist of colloidal suspensions confined between plane parallel walls. Such systems will be presented here as model systems, where the definition and calculation of various physical properties can be illustrated with images in real space and time. Thus, in the next section, the Q2D system considered here is presented.

In Section 11.3, the concept of static structure is introduced. In Section 11.4, the effective DLVO pair potential for colloidal interactions is discussed. In Section 11.5, the integral equation of Ornstein and Zernike for the static structures is presented. Section 11.6 concerns the static structure factor and Section 11.8 the dynamic structure factor. Finally, the conclusion is presented in Section 11.9.

11.2 QUASI-TWO-DIMENSIONAL COLLOIDAL FLUIDS

Understanding of the physical properties of colloidal systems under conditions of confinement is an area of current interest from different perspectives [1–16]. Here, we discuss the case of Q2D colloidal fluids, that is, colloidal fluids confined between two plane parallel walls, with the interwalls distance being only slightly larger than the particle's size. Therefore, the motion of the particles in the direction perpendicular to the walls is highly restricted and the main motion is along the unbounded parallel plane. Thus, the systems can be regarded as effectively two dimensional. Although the effects of confinement are really interesting, Q2D colloidal fluids are considered here mainly for illustrating the concepts and the physical quantities of interest discussed in this chapter. The Q2D colloidal fluids presented here are prepared following a standard procedure [12]. Briefly, monodisperse water suspensions of polystyrene spheres, carrying negative-charged sulfate end groups on the surface, are extensively dialyzed against ultrapure water to eliminate the surfactant added by the manufacturer (Duke scientific). In a clean atmosphere of nitrogen gas, the suspension of particles of diameter σ is mixed with a small amount of larger particles of diameter h. A small volume of the mixture (≈ 1 μl) is confined between two clean glass plates (a slide and a cover slip), which are uniformly pressed one against the other until the separation between the plates is h. Thus, the larger particles scattered across the sample serve as spacers with an average distance between them being ~ 100 μm. The system is then sealed with epoxy resin, and the species of mobile particles allowed to equilibrate in this confined geometry at room temperature. The samples are placed on the stage of an optical microscope and observed from the top view (perpendicular to the walls plane). The motion of the particles is recorded using standard video equipment coupled to the microscope. From the analysis of the images, the 2D trajectories $\mathbf{r}_j(t)$ of the particles are determined with a time resolution $\Delta t = 1/30$ s and a spatial resolution of 25 nm [17]. As discussed below, from $\mathbf{r}_j(t)$ various physical quantities of these systems can be calculated. Figure 11.1 shows the schematics of the side view of the sample cell, with the open circles representing the mobile colloidal particles and the shadow circles representing the spacer particles. It is also shown here, a top view image of a portion of the field of view of an actual system, where $\sigma = 2.05 \pm 0.06$ μm and $h = 2.92 \pm 0.09$ μm. In this system, and the others shown below, small ions dissociate from the glass walls in sufficient amount to screen the electrostatic repulsion between particles, allowing them to approach each other close to contact but without aggregating, that is, these systems behave as effective hard-sphere suspensions.

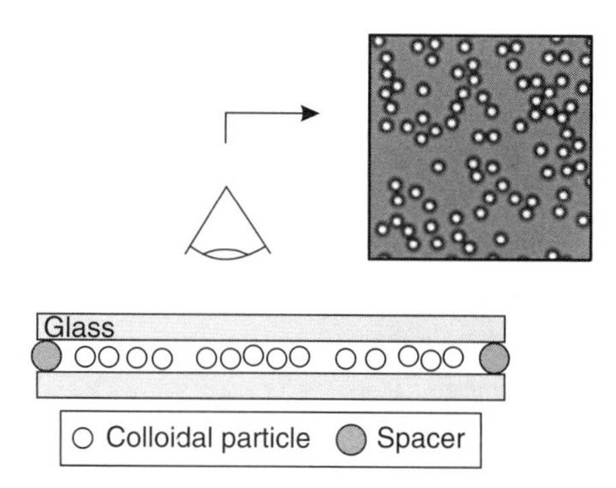

Figure 11.1 Image of 2.05 μm polystyrene spheres confined between two glass plates and a schematic side view of the sample cell.

11.3 STATIC STRUCTURE

Colloidal particles dispersed in a fluid move randomly due to the collisions with the solvent's molecules. Nevertheless, their positions are correlated due to the interparticle interactions, that is, the particles cannot take any arbitrary configuration but only those allowed by the direct interactions, producing in this way a local structure of the particles referred to as the static structure. The physical quantities describing the static structure, such as the static structure factor and the pair correlation function, can be determined experimentally by different techniques, and they can be calculated from their statistical mechanics expressions when the interparticle direct interactions are known. Thus, the precise determination of those interactions is one of the main issues in characterizing a colloidal suspension since they determine properties of the system such as stability, phase transitions, dynamics, and rheology. An indirect way to obtain insight into the interparticle interactions is the experimental determination of the static structure. Then, let us proceed to develop a quantitative description of the static structure.

Let us consider a system having N identical particles, volume V, and temperature T. We then ask the following question: what is the conditional probability density $\rho^{(2)}(\mathbf{r}_1, \mathbf{r}_2)$ of finding a particle at the position $\mathbf{r}_1$ and a second particle at $\mathbf{r}_2$, at the same time? See Figure 11.2 for a 2D representation of such systems. In the absence of interparticle interactions, the particles can be in any position at any time, regardless of the location of the other particles, that is, the system behaves as an ideal gas and $\rho^{(2)}(\mathbf{r}_1, \mathbf{r}_2) = \rho^2$, where $\rho \equiv N/V$ is the average particle concentration. However, when the particles do interact they cannot take configurations arbitrarily, but their positions are correlated. In this case, the system deviates from the ideal gas behavior and one can write in general

$$\rho^{(2)}(\mathbf{r}_1, \mathbf{r}_2) = \rho^2 g(\mathbf{r}_1, \mathbf{r}_2), \tag{11.1}$$

where the factor $g(\mathbf{r}_1, \mathbf{r}_2)$ in Equation 11.1 is introduced here to account for the correlation between pairs of particles. This quantity is referred to as the two-particle correlation function, or pair correlation function, and it depends on the interparticle interactions. From Equation 11.1, one can see that $g(\mathbf{r}_1, \mathbf{r}_2)$ should have the following general properties. At sufficiently large interparticle distances, the interparticle correlation should be lost, that is, $\rho^{(2)}(\mathbf{r}_1, \mathbf{r}_2) = \rho^2$. Then, the pair correlation function should equal 1 in the limit of long interparticle distances $|\mathbf{r}_2 - \mathbf{r}_1|$. On the other hand, in the case of particles with hard-core or repulsive interactions, the pair correlation should vanish at distances shorter than an exclusion zone. However, for arbitrary distances, in order to determine the actual pair correlation function for a specific system, one needs to develop a statistical mechanics model for this quantity.

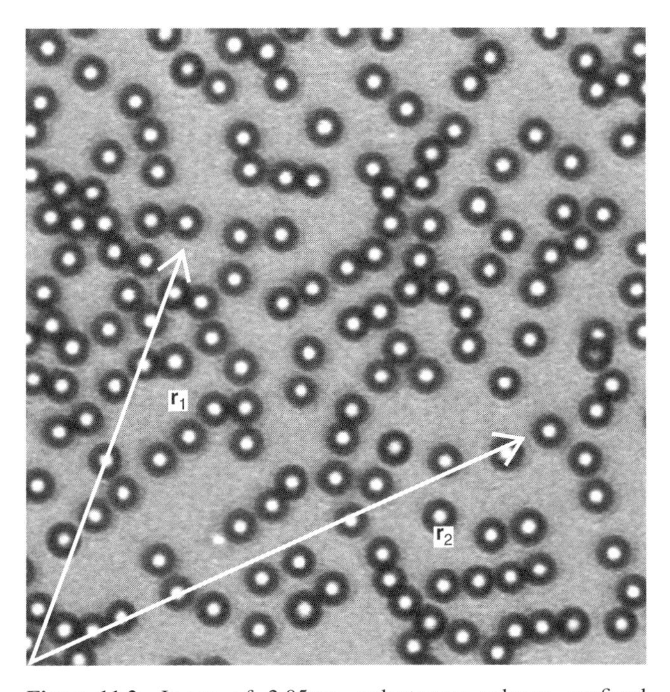

Figure 11.2 Image of 2.05 μm polystyrene spheres confined between two glass plates. Positions $\mathbf{r}_1$ and $\mathbf{r}_2$ in the sample are pointed out by arrows.

Colloidal fluids resemble simple liquids, with the colloidal particles playing the role of "atoms" and the dispersing medium being the "vacuum." Thus, a fundamental, statistical mechanics description of the static structural properties can be formulated on the same basic principles. The differences in approaching the description of colloidal and simple liquids arise when one is concerned with their dynamic properties or other transport properties. Thus, let us now briefly review the important physical concepts and quantities describing the static structure of classical liquid systems [18]. For a system of N identical interacting particles in a volume V, in thermal equilibrium at the temperature T, we can introduce a set of configurational functions, namely, the n-particle distribution function, which provide a quantitative measure of the correlations between the positions of subsets of n-particles. For a system in thermal equilibrium, the normalized configuration probability density $P(\tilde{r})$ is given by [18],

$$P(\tilde{r}) = Z^{-1} e^{-\beta U(\tilde{r})}, \tag{11.2}$$

where $\tilde{r} \equiv (\mathbf{r}_1, \mathbf{r}_2, \ldots, \mathbf{r}_N)$ is the instantaneous configuration of the N particles, $U(\tilde{r})$ is the total potential energy of the system, and the constant $Z = \int d\tilde{r} e^{-\beta U(\tilde{r})}$ with $d\tilde{r} = d\mathbf{r}_1 d\mathbf{r}_2 \cdots d\mathbf{r}_N$. Then, the single-particle density $\rho^{(1)}(\mathbf{r}_1)$, that is, the probability per unit volume of finding one of the N particles in the position r_1, irrespective of the configuration of the other $N-1$ particles, is obtained by integrating $P(\tilde{r})$ on all the configuration space, except on the coordinate $\mathbf{r}_1$,

$$\rho^{(1)}(\mathbf{r}_1) = N \int P(\tilde{r}) d\mathbf{r}_2 dr_3 \cdots d\mathbf{r}_N, \tag{11.3}$$

where the factor N in this equation follows from the fact that particle 1 can be chosen in N equivalent ways since all the particles in the system are identical. From Equations (11.2) and (11.3) one can see that the single-particle density satisfies the condition

$$\int \rho^{(1)}(\mathbf{r}_1) d\mathbf{r}_1 = N. \tag{11.4}$$

For homogeneous systems, $\rho^{(1)}(\mathbf{r})$ is independent of $\mathbf{r}$, that is, $\rho^{(1)}(\mathbf{r}) = \rho = N/V$.

As the next step, we consider the two-particle distribution function $\rho^{(2)}(\mathbf{r}_1, \mathbf{r}_2)$, defined as

$$\rho^{(2)}(\mathbf{r}_1, \mathbf{r}_2) = N(N-1) \int P(\tilde{r}) d\mathbf{r}_3 d\mathbf{r}_4 \cdots d\mathbf{r}_N. \tag{11.5}$$

Then, $[N(N-1)]^{-1}\rho^{(2)}(\mathbf{r}_1,\mathbf{r}_2)d\mathbf{r}_1\, d\mathbf{r}_2$ is the probability of finding one particle in the volume $d\mathbf{r}_1$ at $\mathbf{r}_1$, and a second particle in the volume $d\mathbf{r}_2$ at $\mathbf{r}_2$, irrespective of the positions of the remaining $(N-2)$ particles. The factor $N(N-1)$ is the number of equivalent ways that two particles can be chosen from a total of N. The normalization of the two-particle density function is given by

$$\int \rho^{(2)}(\mathbf{r}_1,\mathbf{r}_2)d\mathbf{r}_1\, d\mathbf{r}_2 = N(N-1). \tag{11.6}$$

One can go on and write down the expression for the n-particle density function $\rho^{(n)}(\mathbf{r}_1,\mathbf{r}_2,\ldots,\mathbf{r}_n)$ as the integral of $P(\tilde{r})$ over the coordinates of the $(N-n)$remaining particles. However, for the purpose of our discussion, it is sufficient to consider only up to $\rho^{(2)}(\mathbf{r}_1,\mathbf{r}_2)$. Then, according to Equations (11.1) and (11.5), the formal expression for the two-particle correlation function is given by

$$g^{(2)}(\mathbf{r}_1,\mathbf{r}_2) = V^2 Z^{-1}\int e^{-\beta U(\mathbf{r}_1,\mathbf{r}_2,\ldots,\mathbf{r}_N)}\, d\mathbf{r}_3 d\mathbf{r}_4\cdots d\mathbf{r}_N. \tag{11.7}$$

For homogeneous and isotropic systems, the two-particle correlation function depends only on the distance between the particles $r = |\mathbf{r}_1 - \mathbf{r}_2|$, that is,

$$g^{(2)}(\mathbf{r}_1,\mathbf{r}_2) = g^{(2)}(|\mathbf{r}_1-\mathbf{r}_2|) \equiv g(r). \tag{11.8}$$

The function $g(r)$, referred to as the radial distribution function, is then the conditional probability of finding a particle at a distance r from a particle located at the origin. This function satisfies the normalization condition

$$\int_0^\infty 4\pi\rho g(r) r^2\, dr = N-1. \tag{11.9}$$

In order to have a concrete example of the behavior of the radial distribution function of an actual colloidal suspension, let us consider here a Q2D system. Figure 11.3 shows an image of a Q2D suspension of polystyrene spheres of diameter 1.9 ± 0.05 μm at a particle's area fraction $\phi_a \equiv \pi\sigma^2\rho_a/4 = 0.48$. Here, $\rho_a \equiv N/A$ is the 2D particles concentration, with N being the number of particles in the area A. Let us first illustrate the way $g(r)$ is obtained from the images of the system. On top of the image in Figure 11.3 a pattern of concentric circles is drawn, centered on one of the particles. The number of particles in each circle of radius r and width Δr, normalized with the average number of particles in the corresponding ring, provides an account of $g(r)$. Since all particles are equivalent, in order to improve the statistics, the procedure is repeated for each of them in the field of view. As the system evolves with time, the particles take different equilibrium configurations. Then, the same calculation is performed for a number of images. The normalized averaged histogram, obtained by considering those from the different particles in different frames, produces a smoother radial distribution function. The resulting pair correlation function for this system is shown in Figure 11.4. The curve in this figure is quite smooth and shows the main features quite clearly; it was obtained by averaging the histograms from ~10^4 frames. As one can see here, $g(r)$ vanishes for $r < \sigma$. Since the particles are effective hard spheres, they can not penetrate the hard core and the exclusion zone is just that given by

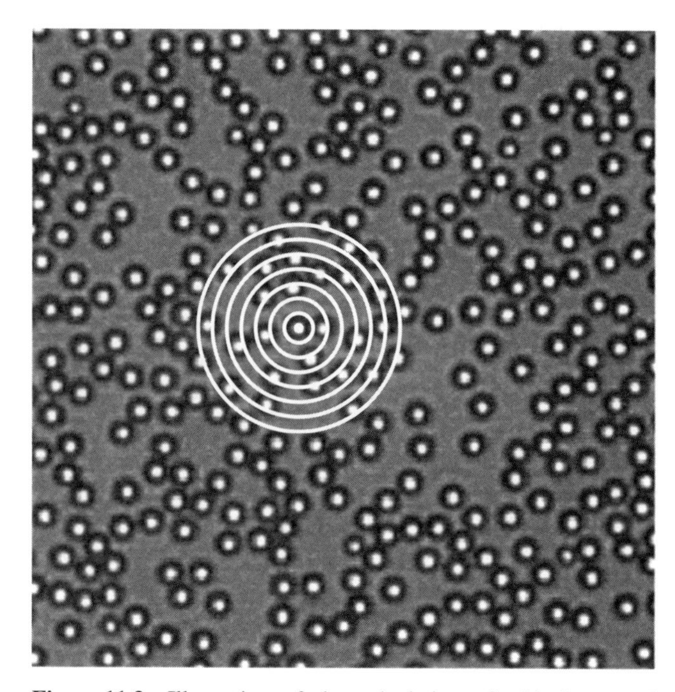

Figure 11.3 Illustration of the calculation of $g(r)$. Image of 1.9 μm polystyrene spheres confined between two glass plates, where concentric circles are drawn centered on one particle. See text for discussion.

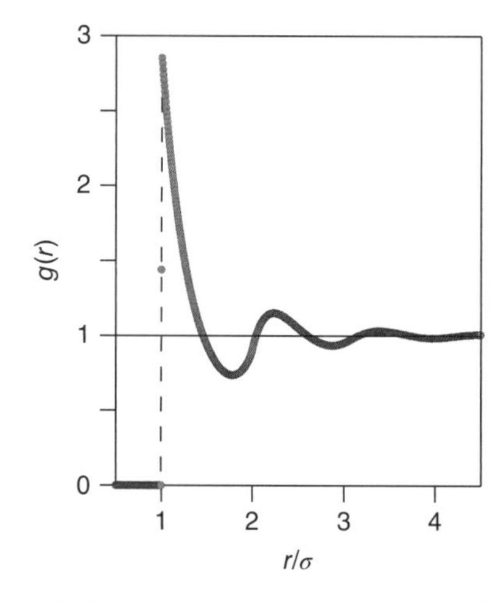

Figure 11.4 Radial distribution function for the Q2D system of effective hard spheres in Figure 11.3.

the excluded volume. At larger distances, $g(r)$ exhibits maxima and minima, above and below 1, respectively, with the main peak at contact. Thus, given a particle, the probability of finding a second one is highest at contact. The height of subsequent maxima and the depth of the minima decrease as the distance increases, and in the limit of large distances $g(r) \to 1$, that is, the particle's positions become uncorrelated. The maxima represent in the average, layers of nearest neighbors, first, second, and so on, respectively, and the minima represent exclusion zones between those layers produced by the particles in the layers. For systems of hard spheres, where the interaction is only that of excluded volume, the only relevant parameter is the particle's concentration, it determines the height and the number of peaks of $g(r)$, that is, the range of the correlation. In the case of long-range repulsive interactions, the radial distribution function would also exhibit maxima and minima, that is, layers of particles and exclusion zones, but in such cases the range of the correlation depends also on the strength of the interaction.

The radial distribution function plays an important role in the study of liquid systems. This quantity can be determined indirectly by a number of experimental techniques, for instance, X-ray and neutron scattering (for atomic and molecular fluids), light scattering and imaging techniques (for colloidal liquids and other complex fluids). From the theoretical point of view, $g(r)$ can be obtained by various approximations and from computer simulations when the interparticle pair potential is known. The static structure is also important in determining physical quantities such as the dynamic and other transport properties, as well as the thermodynamics of the system [19–24].

11.4 MODEL PAIR POTENTIAL

The evaluation of the pair correlation function, and other properties, requires to have an accurate theoretical model for the total potential energy of the system. The interactions between particles can be of different nature, electrostatic, magnetic, entropic, and so on. Since a rich variety of particles get charged when they are in solution, here we discuss in some detail the electrostatic interaction between colloidal particles. In a suspension of colloidal particles having an electric charge Q, besides the particles themselves, there are in the solution small ions in sufficient amount to balance the charge of the colloidal particles, that is, the total charge in the system should be zero due to the general condition of electroneutrality. Those ionic species are referred to as counterions since their charge is opposite in sign to that of the colloidal particles. Real systems, for instance biological systems, usually contain added electrolytes. Then, in a more general case, one should consider the presence of an arbitrary number of ionic species in the solution. Thus, the determination of the potential energy is a complex many-body problem with no general analytical solution. However, it has been possible to advance in the study of colloidal systems by performing some physically justified approximations. One of such approximations is the assumption that the total potential energy is pairwise additive, that is,

$$U(\tilde{r}) = \sum_{i} \sum_{j>i} u(\mathbf{r}_i, \mathbf{r}_j). \tag{11.10}$$

Equation 11.10 is called the superposition approximation, with $u(\mathbf{r}_i, \mathbf{r}_j)$ being an effective interparticle pair potential between particles i and j located at $\mathbf{r}_i$ and $\mathbf{r}_j$, respectively. The reduction in the description of a multicomponent system to a one-component system, implied in Equation 11.10, neglects contributions that become important in highly interacting systems [25, 26]. However, considering low intensity interactions, an analytical expression for the pair potential has been derived, which has been successful in the description of suspensions of interacting particles even in cases where they are highly charged [27]. Let us note here, that for central potentials such as the hard-sphere potential,

$$u(\mathbf{r}_i, \mathbf{r}_j) = u(|\mathbf{r}_i - \mathbf{r}_j|) \equiv u(r). \tag{11.11}$$

For charged particles in solution, the current model potential used extensively in the calculation of different properties of colloidal suspensions is the celebrated DLVO pair potential due to Derjaguin, Landau, Verwey, and Overbeek [27]. The DLVO pair potential $u_{DLVO}(r)$ is the sum of three terms, that is,

$$u_{DLVO}(r) = u_{HS}(r) + u_Y(r) + u_W(r). \tag{11.12}$$

The first component is an excluded volume or hard-sphere term $u_{HS}(r)$, expressing the fact that particles cannot interpenetrate. The hard-sphere condition is given by

$$u_{HS}(r) = \begin{cases} \infty, & r < \sigma \\ 0, & r > \sigma. \end{cases} \tag{11.13}$$

The second term in Equation 11.12 is a repulsive Yukawa (or screened Coulomb) potential due to the repulsive electrostatic interaction between particles. Particles in solution can acquire a net electrical charge by different mechanisms, the most common is that by releasing small ions into the solution [28, 29]. These ions, together with the ions of any added electrolytes in the solution, form an ionic cloud (referred to as the electrical double layer) around the colloidal particles, which screens the electrostatic interaction between the particles. For two identical, isolated, electrically charged spheres, surrounded by a cloud of point-like ions, an analytical expression for the effective electrostatic interaction potential can be obtained within the linear approximation of the Poisson–Boltzmann equation, that is, for the case of weakly interacting particles [29]. Explicitly,

$$u_Y(r) = \frac{Q^2}{\epsilon(1+\kappa\sigma/2)^2}\frac{e^{-\kappa(r-\sigma)}}{r}, \quad r > \sigma. \tag{11.14}$$

Here, r is the distance between the centers of the two particles, Q and σ are as earlier the particles' electric charge and diameter, respectively, ϵ is the solvent's dielectric constant, and the screening constant κ is given by

$$\kappa^2 = \frac{\beta 4\pi e^2}{\epsilon}\sum_i n_i z_i^2, \tag{11.15}$$

where e is the electron's charge, n_i and z_i are the concentration and valency of the small ions of species i, and $\beta^{-1} = k_B T$ is the thermal energy, that is, T is the absolute temperature and k_B Boltzmann's constant. The inverse of κ determines the spatial extent of the electric double layer and the range of the electrostatic interactions, and it is generally referred to as the Debye length.

The direct measurement of the interaction potential in actual systems is a rather complex matter. Thus, the experimental test of the accuracy of Equation 11.14 is usually done indirectly by comparing measurements of properties, such as the static and the dynamic structure, in model experimental systems with theoretical and/or computer simulations results obtained by using Equation 11.14 [19–21, 24, 30, 31]. Such comparisons have been possible due to the availability of model systems of charged particles in solution with very low size polydispersity. Those systems are now simple to produce and they are also available commercially. The typical systems studied have been aqueous suspensions of either polystyrene latex spheres or silica particles. In both cases, the particles' surface can carry ionizable polar groups that dissociate when immersed in a polar solvent such as water, leaving on the surface either a negative or a positive net charge. Nevertheless, the calculated quantities are brought to a very good agreement with the corresponding measured quantities when an effective value for the electric charge of the particles is used, instead of the bare charge value [19–21, 24, 30, 31]. The effective charge is usually one or two orders of magnitude smaller than the value of the bare charge, which for submicron size polystyrene particles in water at room temperature can be as high as 10^3–10^5 electronic charges [29]. Then, Q enters in the calculations as a free parameter. Although Equation 11.14 should be valid only when the strength of $u(r)$ allows the linear approximation of the Poisson–Boltzmann equation, that is, at large particle–particle separations, low charge and/or high ionic concentrations, this equation provides the correct qualitative functional form for the pair potential of charged colloidal particles in the bulk. Nevertheless, one should keep in mind that issues such as the physical mechanism of charge renormalization [32, 33], effects such as volume, the finite size of the small ions, and their correlations, and the range of applicability of Equation 11.14, in terms of Q, κ, and σ, still need to be elucidated [25, 26].

The third term of the DLVO pair potential, referred to as van der Waals or dispersion forces, arises from the interaction between induced dipoles due to quantum fluctuations in the charge distribution of the molecules of the particles [28, 29]. This term depends on the shape of the particles, the material of the particles, and the solvent's polarizability. Explicit expressions for different geometries have been derived and they are available in the literature [28, 29]. Such expressions are given as the product of two factors, one depending on the geometry and the other on the materials. For two spherical particles of the same size, the van der Waals interparticle potential $u_W(r)$ is given by

$$u_W(r) = \frac{-A}{12}\left[\frac{\sigma^2}{r^2-\sigma^2} + \frac{\sigma^2}{r^2} + 2\ln\left(\frac{r^2-\sigma^2}{r^2}\right)\right], \quad r > \sigma. \tag{11.16}$$

The factor in square parenthesis is the explicit geometric factor for two spheres. As one can see here, this potential diverges at contact. The quantity A is a constant, the Hamaker constant, depending on the particle's material and on the properties of the solvent. For particles of the same material, the Hamaker constant is positive with values in the order of 10^{-19}–10^{-21} J. Thus, for two isolated identical particles, immersed in a solvent of different index of refraction, $u_W(r)$ is attractive and diverges at contact.

A schematic representation of the DLVO pair potential is shown in Figure 11.5. The shadow area is the hard-sphere exclusion zone. The dashed line represents the screened electrostatic potential, the dot-dashed line corresponds to the attractive van der Waals interaction, and the solid line is the full DLVO pair potential. The Yukawa term has a finite maximum at contact, whereas the van der Waals potential diverges as r approaches σ, from $r > \sigma$. Thus, the DLVO pair potential has a deep minimum at contact. At larger distances, the combination of the van der Waals and the electrostatic repulsive force produces an energy barrier, and

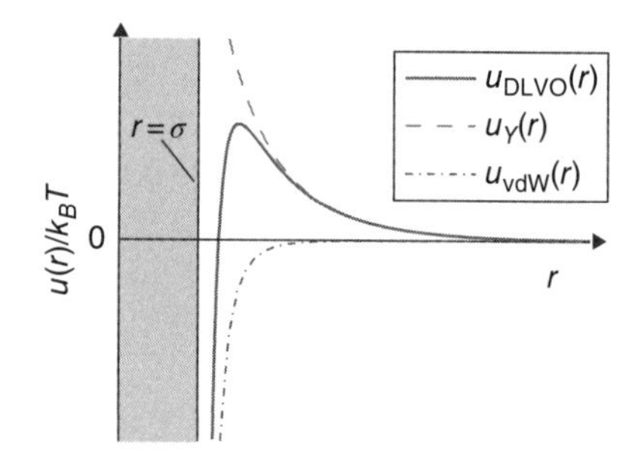

Figure 11.5 Schematic of the DLVO pair potential.

at even larger distances a shallow secondary minimum, now due to the fact that the van der Waals interaction decays much more slowly than the repulsive Yukawa interaction. For a given system, the height of the barrier depends on the amount of ions present in the solution. For low ionic concentrations, the double- layer extension is large and the repulsive electrostatic interaction is dominant; the energy barrier can be much higher than $k_B T$. In such cases, the particles are kept far apart from each other by the electrostatic repulsion, that is, they remain dispersed in the solvent and the suspension is said to be stabilized by the electric charge. Since the particles do not reach contact, the van der Waals and hard-core interactions do not play an explicit role. Therefore, for highly interacting colloidal particles, it is sufficient to consider only the screened Coulomb potential $u_Y(r)$ as the effective pair potential. As the ionic concentration is increased, the strength of the repulsion decreases and the particles can approach closer to each other. For higher ionic concentration, the electrostatic repulsion will be sufficiently screened, and the particles would be able to overcome the potential barrier by thermal fluctuations falling into the first minimum of the DLVO potential, that is, they will be irreversibly bound to each other by the effect of the attractive van der Waals forces. In such cases, the system is destabilized and the particles form aggregates [34, 35].

11.5 THE ORNSTEIN–ZERNIKE EQUATION

From the theoretical point of view, the current method employed to compute the static structure of a fluid system is the Ornstein–Zernike integral equation (OZ). Introduced by Ornstein and Zernike in 1914 [18, 36], this equation splits formally the total pair correlation $h(r)$ in the direct correlation between pairs and the indirect correlation mediated by the neighbor particles, that is,

$$h(r) = c(r) + \rho \int d\mathbf{s}\, c(s) h(|\mathbf{r} - \mathbf{s}|). \qquad (11.17)$$

Equation 11.17 is actually the definition of the direct correlation function $c(r)$, where the total correlation is given by $h(r) \equiv g(r) - 1$. The indirect part is the correlation between a particle at the origin and a third particle, via the correlations, direct and indirect, of that third particle with the second. Since in the Ornstein–Zernike equation both functions $h(r)$ and $c(r)$ are unknown, it is clear that in order to calculate $g(r)$ for a specific system, we require an additional relation between these two functions. It should also be clear that the additional relation must include information about the system being considered. Different relations between the interaction pair potential $u(r)$ and the functions $h(r)$ and/or $c(r)$ have been introduced in the literature as closure relations to Equation 11.17. Here we present some of the most frequently employed closures.

The simplest closure relation is the so-called mean spherical approximation (MSA). This approximation is defined by the conditions [18, 36]

$$h(r) = -1, \quad r < \sigma, \qquad (11.18)$$

and

$$c(r) = -\beta u(r), \quad r > \sigma. \qquad (11.19)$$

Equation 11.18 is the hard-sphere condition and Equation 11.19 is the asymptotic relation between $c(r)$ and $u(r)$. Thus, Equation 11.19 is an approximation for $c(r)$ at arbitrary r. Other closure relations are the hypernetted chain approximation (HNC), defined as [18, 36]

$$c(r) = e^{-\beta u(r)} e^{\gamma(r)} - \gamma(r) - 1, \qquad (11.20)$$

where $\gamma(r) = h(r) - c(r)$. The Percus–Yevick approximation (PY), defined as [18, 36]

$$c(r) = e^{-\beta u(r)} [\gamma(r) + 1] - \gamma(r) - 1, \qquad (11.21)$$

and the Rogers–Young approximation (RY), defined by Rogers and Young [37]

$$c(r) = e^{-\beta u(r)} \left[1 + \frac{e^{\gamma(r) f(r)} - 1}{f(r)} \right] - \gamma(r) - 1, \qquad (11.22)$$

where the function $f(r)$ is defined as

$$f(r) = 1 - e^{-\lambda r}, \qquad (11.23)$$

with λ being a mixing parameter. It should be noted that the RY closure relation is an interpolation between the HNC and PY approximations. The HNC and PY approximations are recovered taking the limit $\lambda \to \infty$ and $\lambda \to 0$, respectively, in Equations 11.22 and 11.23. The value of the free parameter λ is adjusted to satisfy consistency between thermodynamic quantities calculated by different routes [37].

The MSA has the beauty of being simple, and the advantage that for pair potentials such as those given by Equations 11.13 and 11.14, the solution is analytical [38]. The other approximations HNC, PY, and RY, have to be solved numerically, except for the case of a system of hard spheres, where the PY approximation has analytical solution. For hard spheres, MSA and PY coincide.

11.6 STATIC STRUCTURE FACTOR

In Section 11.3, it was shown that measuring the pair correlation function requires the simultaneous determination of the positions of the N particles in a statistically significant portion of the system. As discussed above, this is simple to do in the case of Q2D systems. In the three-dimensional (3D) case, that is possible using confocal microscopy only when the dynamics of the systems is slow enough, so that it is possible to scan a representative volume of the system without any appreciable change in the configuration. In the more general 3D case, the structure is usually measured using scattering techniques: light, neutrons, or X-rays, depending on the space scale of interest. Static light scattering by colloidal suspensions measures the quantity referred to as the static structure factor $S(k)$, which describes the system structure in the reciprocal space. The radial distribution function is directly related to the Fourier transform of $S(k)$, as it is explained below. Thus, let us consider a system of N particles in a volume V. The local particle concentration $\rho(\mathbf{r})$ at the position $\mathbf{r}$ is given by

$$\rho(\mathbf{r}) = \sum_{i=1}^{N} \delta(\mathbf{r} - \mathbf{r}_i). \tag{11.24}$$

In systems in thermal equilibrium, statistical (or thermal) fluctuations produce deviations of the local concentration of particles from its average value ρ. This fact is illustrated in Figure 11.6, where it shows the Q2D system in Figure 11.3

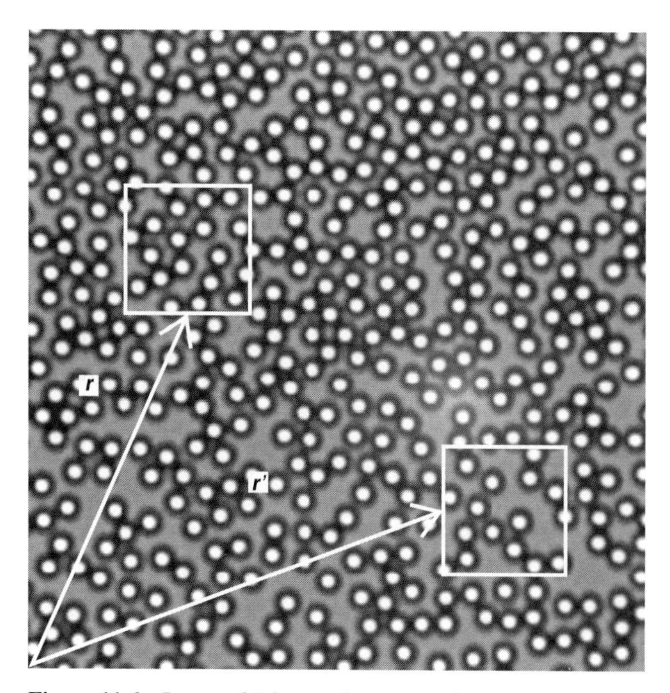

Figure 11.6 Image of 1.9 μm polystyrene spheres at an area fraction of 0.48 confined between two glass plates. Small areas at $\mathbf{r}$ and $\mathbf{r}'$ are enclosed in squares.

but enclosing within boxes small areas at positions $\mathbf{r}$ and $\mathbf{r}'$. Although the system is closed, with a constant average particle concentration, locally the number of particles within the boxes is different at both locations, that is, $\rho(r)$ exhibits the fluctuations just mentioned above. A simple question to ask here is whether the local concentration at $\mathbf{r}$ is related to that at $\mathbf{r}'$. In order to answer this question, let us calculate the autocorrelation function $\langle\rho(\mathbf{r})\rho(\mathbf{r}')\rangle$ of the local particle concentration at two different locations $\mathbf{r}$ and $\mathbf{r}'$. The angular parenthesis here mean an average over an equilibrium ensemble of configurations, that is, $\langle A(\tilde{r})\rangle \equiv Z^{-1} \int d\tilde{r}\, e^{-\beta U(\tilde{r})} A(\tilde{r})$. Then,

$$\begin{aligned}\langle\rho(\mathbf{r})\rho(\mathbf{r}')\rangle &= \sum_j \sum_{i\neq j} \int d\tilde{r}\; P(\tilde{r})\delta(\mathbf{r} - \mathbf{r}_i)\delta(\mathbf{r}' - \mathbf{r}_j) \\ &= N(N-1) \int d\tilde{r}\; P(\tilde{r})\delta(\mathbf{r} - \mathbf{r}_1)\delta(\mathbf{r}' - \mathbf{r}_2) \\ &= \int d\mathbf{r}_1\; d\mathbf{r}_2 \delta(\mathbf{r} - \mathbf{r}_1)\delta(\mathbf{r}' - \mathbf{r}_2)\rho^{(2)}(\mathbf{r}_1, \mathbf{r}_2) \\ &= \rho^{(2)}(\mathbf{r},\mathbf{r}'). \end{aligned} \tag{11.25}$$

Thus, the autocorrelation function of the local particle-concentration is nothing else but the pair correlation function.

Fluctuations in the local number of particles $\rho(r)dV$ depend on the size of the cell considered, that is, if the cell's volume dV increases or decreases, quantities such as the local concentration $\rho(\mathbf{r})$, and its autocorrelation function $\rho^{(2)}(\mathbf{r}, \mathbf{r}')$, will vary with the cell's size. In fact, one may ask, for instance, what is the dependence of quantities such as the correlation function of particle concentration on the wavelength of the fluctuations. The quantity containing such information is the autocorrelation function of the Fourier components of $\rho(r)$ referred to as the static structure factor $S(k)$. The static structure factor is written as

$$S(\mathbf{k}) = \frac{1}{N}\langle\rho(\mathbf{k})\rho(-\mathbf{k})\rangle, \tag{11.26}$$

where $\rho(\mathbf{k})$ is the Fourier transform of $\rho(\mathbf{r})$, that is,

$$\begin{aligned}\rho(\mathbf{k}) &= \int d\mathbf{r}\; \exp(-i\mathbf{k}\cdot\mathbf{r})\rho(\mathbf{r}) \\ &= \sum_{j=1}^{N} \exp(-i\mathbf{k}\cdot\mathbf{r}_j). \end{aligned} \tag{11.27}$$

Thus,

$$\begin{aligned}S(\mathbf{k}) &= \frac{1}{N}\left\langle \sum_i \sum_j \exp(-i\mathbf{k}\cdot\mathbf{r}_i)\exp(+i\mathbf{k}\cdot\mathbf{r}_j) \right\rangle \\ &= 1 + \frac{1}{N}\left\langle \sum_i \sum_{i\neq j} \exp[-i\mathbf{k}\cdot(\mathbf{r}_i - \mathbf{r}_j)] \right\rangle. \end{aligned} \tag{11.28}$$

For isotropic and homogeneous systems, the static structure factor can be written as

$$S(\mathbf{k}) = 1 + \rho \int d\mathbf{r} \exp(-i\mathbf{k}\cdot\mathbf{r})g(\mathbf{r})$$
$$= 1 + \rho \int d\mathbf{r} \exp(-i\mathbf{k}\cdot\mathbf{r})[g(\mathbf{r}) - 1] + \rho\delta(\mathbf{k}). \tag{11.29}$$

As one can see here, the static structure factor is related directly to the Fourier components of the radial distribution function. For isotropic systems, $S(\mathbf{k})$ and $g(\mathbf{r})$ depend only on $k = |\mathbf{k}|$ and $r = |\mathbf{r}|$, respectively. Then

$$S(k) = 1 + \rho h(k) + (2\pi)^3 \rho\delta(r), \tag{11.30}$$

where the function $h(k)$ is the Fourier transform of the total correlation function. The Dirac delta function appearing in Equations 11.29 and 11.30 is the Fourier transform of a constant and do not contain any information on the particles' correlations. Therefore, that term is usually ignored and the relation between the static structure factor and the Fourier transform of the total correlation function is assumed to be

$$S(k) = 1 + \rho h(k). \tag{11.31}$$

Figure 11.7 shows the static structure factor corresponding to the Q2D system of effective hard spheres whose radial distribution function is shown in Figure 11.4. As one can see here, $S(k)$ has maxima and minima corresponding to the relevant scales of the system's structure. For instance, the first maximum occurs at $\kappa\sigma = 5.8$, corresponding to a wavelength $\lambda \equiv 2\pi/k = 1.08\sigma$, a value very close to one of the relevant scales in the system, namely, the size of the particles. For systems with long-range repulsive interactions between the particles, the position of the first peak in $S(k)$corresponds to a wavelength of the size of the interparticle distance.

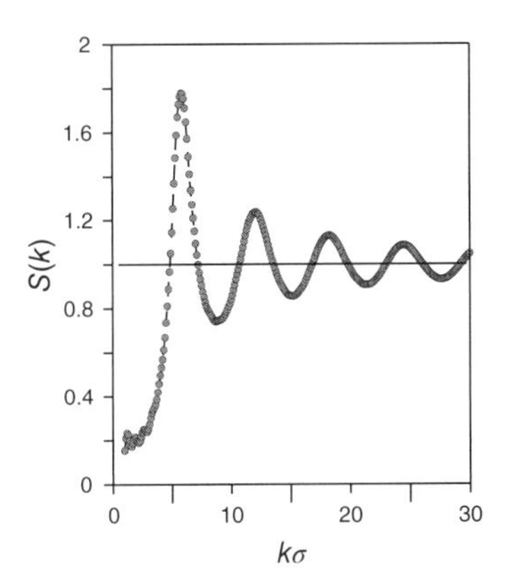

Figure 11.7 Static structure factor obtained by Fourier transforming $g(r)$ in Figure 11.4.

For 3D systems, the static structure factor is experimentally determined by scattering techniques, such as static light scattering (SLS). A detailed description of this technique can be found in Chapter 8.

11.7 SELF-DIFFUSION

Colloidal particles suspended in a fluid are observed to move randomly. In a monodisperse system, two types of diffusion phenomena are observed. On one hand, one can follow the motion of individual particles, referred to as self-diffusion or Brownian motion, and on the other hand, one can look at the collective motion of particles, that is, at the relaxation of fluctuations in the local concentration. Let us first describe the motion of individual particles, which appears random in the timescale of the diffusive regime. For isolated particles, that is, in the limit of infinite dilution, such erratic motion is due to collision of the particles with the fluid's molecules. In this regime, the motion of single particles is described by the Langevin equation

$$m\frac{d^2\mathbf{r}(t)}{dt^2} = -\gamma\frac{d\mathbf{r}(t)}{dt} + \mathbf{f}(t), \tag{11.32}$$

where $\mathbf{r}(t)$ and m are the particle's position at time t and mass, respectively, γ is the translational friction coefficient and $\mathbf{f}(t)$ is a random function representing the force on the particle due to the collisions with the solvent's molecules. Thus, Equation 11.32 is a stochastic equation whose solution is the normalized probability distribution function $P(\Delta\mathbf{r}, t)$ of single particle displacements $\Delta\mathbf{r}$, during a time t [36]. This distribution function is the fundamental quantity describing Brownian motion. However, single particle motion is usually discussed in terms of a more simple quantity, namely, the mean squared displacement $\langle[\mathbf{r}(t) - \mathbf{r}(0)]^2\rangle$, where the angular parenthesis indicate an equilibrium ensemble average. Thus, the mean squared displacement is nothing but the second moment of $P(\Delta\mathbf{r}, t)$, that is,

$$\langle[\mathbf{r}(t) - \mathbf{r}(0)]^2\rangle = \int d(\Delta\mathbf{r})(\Delta\mathbf{r})^2 P(\Delta\mathbf{r}, t). \tag{11.33}$$

For isolated particles, the normalized distribution function $P(\Delta\mathbf{r}, t)$ is a Gaussian function [36], that is,

$$P(\Delta\mathbf{r}, t) = [4\pi W(t)]^{-d/2} \exp\left[-\frac{\Delta\mathbf{r}^2}{4W(t)}\right], \tag{11.34}$$

where $W(t) \equiv \langle[\mathbf{r}(t) - \mathbf{r}(0)]^2\rangle/2d$, with d being the system's dimensionality. Thus, the mean squared displacement is the width of the steps distribution function.

At finite concentrations, the single particle motion is still random, but it also depends on the interparticle interactions both direct and hydrodynamic. In order to illustrate the effect of concentration, in Figure 11.8 are shown trajectories of particles in Q2D systems at finite concentrations, consisting of 10^4 time steps of 1/30 s. The more and the less extended trajectories corresponding to a particle in systems at area fractions of about 10% and 50%, respectively. Then, the effect of the interparticle interactions (excluded volume and hydrodynamic in these systems) is the hindering of the particle's motion but, as one can see here, the random nature is preserved. Examples of the distribution of steps are presented in the following section.

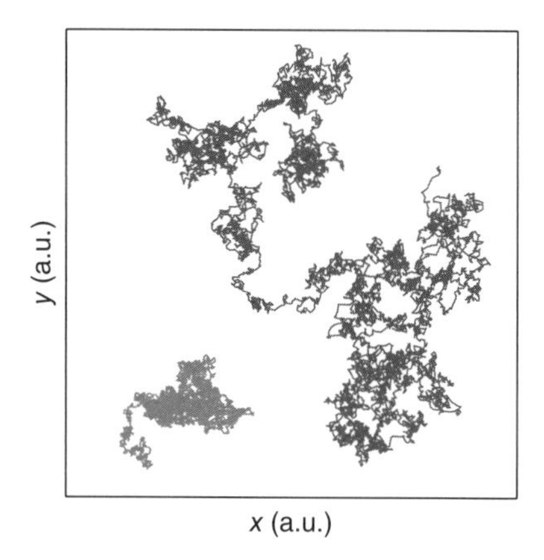

Figure 11.8 Single particle trajectories of particles in Q2D systems at area fractions of 10% and 50%, the more and the less extended trajectory, respectively.

11.8 DYNAMIC STRUCTURE

The collective motion, on the other hand, appears only at finite concentrations arising from the interparticle interactions. The description of both dynamic properties, single particle and collective motions, is provided by the time correlation function of the local particles concentration $G(r,t) \equiv \frac{1}{N}\langle\rho(\mathbf{r}', t=0)\rho(\mathbf{r}'', t)\rangle$ with $r \equiv |\mathbf{r}'' - \mathbf{r}'|$. For illustration, Figure 11.9 shows images of the Q2D system in Figure 11.6 taken at time steps of 5 s. In Figure 11.9a, considered here as $t = 0$, the local concentration of particles is highlighted by a box at positions $\mathbf{r}'$ and $\mathbf{r}''$. As one can see in Figure 11.9b–d, the local concentration at $\mathbf{r}''$ changes with time. The questions here are to what extent and for how

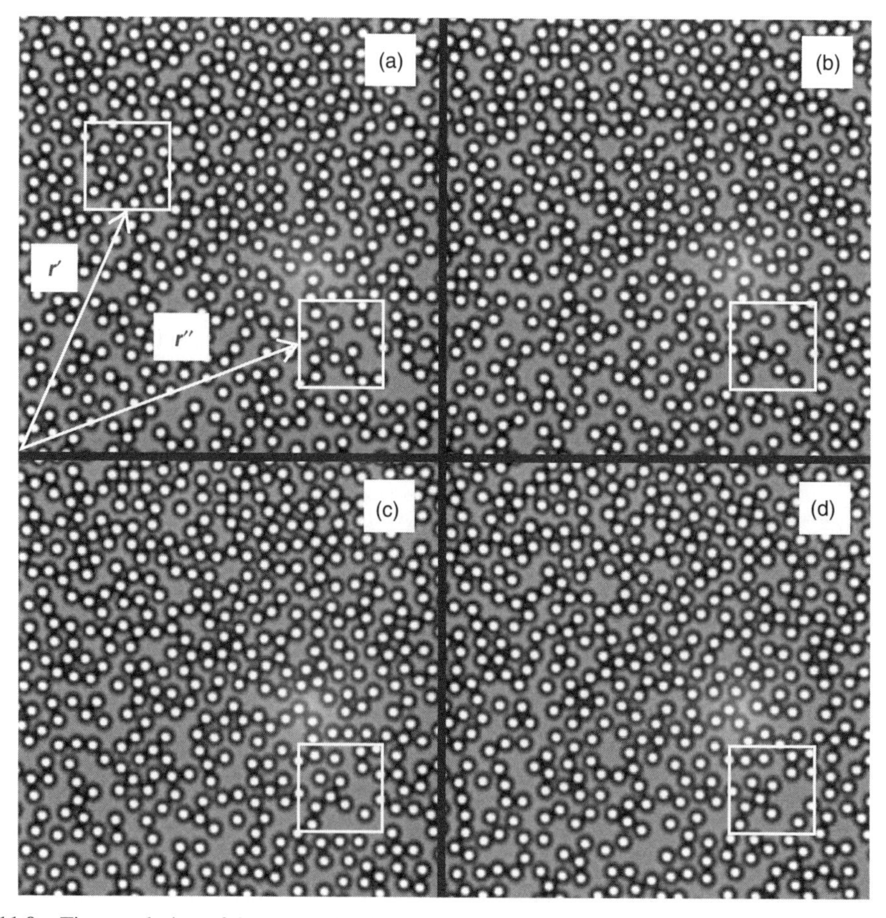

Figure 11.9 Time evolution of the system in Figure 11.6. Images (a)–(d) were taken 5 s apart from each other.

long the evolution of the concentration at $\mathbf{r}''$ depends on that at the position $\mathbf{r}'$ and $t = 0$. The quantity $G(r, t)$ measures precisely such dependence.

Let us consider here some properties of the correlation function $G(r, t)$, referred to as the van Hove function. Using the expression for the local concentration given in Equation 11.24, we obtain for a system of N identical particles,

$$G(r,t) = \frac{1}{N}\left\langle \sum_{j,l=1}^{N} \delta(\mathbf{r} - \mathbf{r}_j(t) + \mathbf{r}_l(0)) \right\rangle. \tag{11.35}$$

The correlation function $G(r, t)$ can be split into two terms, the self- and the distinct part, $G_s(r, t)$ and $G_d(r, t)$, respectively. The former containing the terms $j = l$ and the latter the terms $j \neq l$, that is,

$$G(r,t) = G_s(r,t) + G_d(r,t), \tag{11.36}$$

with

$$G_s(r,t) = \langle \delta(\mathbf{r} - \mathbf{r}_1(t) + \mathbf{r}_1(0)) \rangle, \tag{11.37}$$

and

$$G_d(r,t) = \frac{1}{N}\left\langle \sum_{j\neq l}^{N} \delta(\mathbf{r} - \mathbf{r}_j(t) + \mathbf{r}_l(0)) \right\rangle. \tag{11.38}$$

The quantity $G_s(r, t)$ describes the process of self-diffusion, that is, the self-correlation of individual particles, which is nothing but the distribution function of steps introduced in the previous section. The function $G_d(r, t)$ describes the time-correlation between different particles. Initially, a particle is correlated with itself only at the particle's initial position, that is, $G_s(r, 0) = \delta(r)$. On the other hand, the correlation between different particles is provided by the radial distribution function, that is, $G_d(r, 0) = n^* g(r)$, where $n^* = \rho\sigma^3$ is the reduced concentration. As one can see from Equation 11.35, the calculation of the van Hove function requires to determine the trajectories $\mathbf{r}_j(t)$ of the N particles in the system. In the case of three-dimensional systems, such information is not available except in very few situations. Therefore, the discussion of the colloidal dynamic properties is usually carried out not in real space but in the reciprocal space, where the correlation function can be measured by scattering techniques. Nevertheless, we can have again some insight into the behavior of $G(r, t)$ by resorting to Q2D systems. Thus, Figure 11.10 shows $G(r, t)$ versus r measured at different times in a Q2D system of area fraction $\phi_a = 0.48$ and $\sigma = 1.9$ μm. At $t = 0$, the self-part is a peaked function at $r = 0$ (not shown) and $G_d(r, t)/n*$ is $g(r)$ (dots with solid line). As time increases, the self-part of the van Hove function spreads out as a Gaussian function, with its width being directly the mean squared displacement

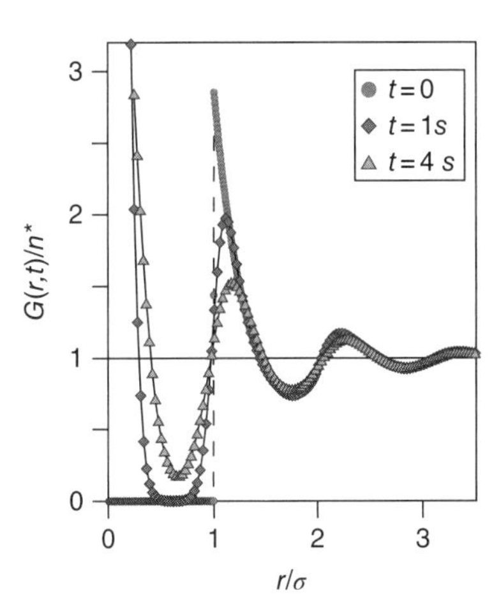

Figure 11.10 Measured van Hove function, at different times, corresponding to system in Figure 11.9.

of single particles, whereas the initial structure of the distinct part smears down due to the loss of interparticle correlation. As one can see here, at short times both components are clearly differentiated from each other, but at longer times they merge together more for longer times. In the limit of long times, the motion of the particles is expected to become uncorrelated, that is, the self-part should vanish and the distinct part should become flat at the value n^*.

As mentioned above, experimental studies of colloidal dynamic processes are usually carried out by using different light scattering techniques, in particular by the so-called dynamic light scattering (DLS), described in more detail in Chapter 8. In DLS, one measures the dynamic structure factor $F(k, t)$, which is the k-Fourier component of the van Hove function, that is,

$$F(k,t) = \frac{1}{N}\langle \rho(\mathbf{k}, t)\rho(-\mathbf{k}, 0) \rangle, \tag{11.39}$$

where $\rho(k, t)$ is the Fourier transform of the local concentration, that is,

$$\rho(k,t) = \sum_{i=1}^{N} \exp[\mathbf{k} \cdot \mathbf{r}_i(t)]. \tag{11.40}$$

Figure 11.11 shows the dynamic structure factor of the system in Figure 11.10, that is, the space Fourier transform of the corresponding $G(r, t)$. At $t = 0$ the dynamic structure factor is, by definition, the static structure factor, that is, $F(k, 0) = S(k)$. At later times, $F(k, t)$ describes the dynamics of the system as it occurs at different wavelengths. As one can see here, the initial structure $S(k)$ fades out as the system evolves and the decay is faster for larger wavevector k, that is, the dynamics of particle fluctuations of shorter wavelengths is faster. In the limit of long k, that is, short wavelength, one is looking at the dynamics inside a cell with only one particle. Thus, in this limit, $F(k, t)$ describes single particle motion. On the other

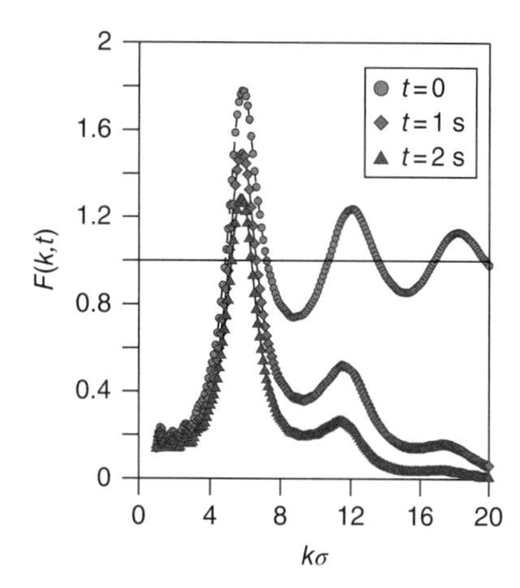

Figure 11.11 Dynamic structure factor of the systems in Figure 11.10.

hand, in the long wavelength limit, $k \to 0$, one is looking at macroscopic scales and it is referred to as the thermodynamic limit. The dynamic structure factor can also be split in the self- and distinct part, that is, $F(k,t) = F_s(k,t) + F_d(k,t)$, where $F_s(k,t)$ and $F_d(k,t)$ describe single particle dynamics and the dynamic correlation between different particles, respectively. Those quantities are related throughout Fourier transformations with the self- and distinct part of the van Hove function discussed above.

11.9 CONCLUSIONS

In this chapter, a general description of the structural properties of colloidal dispersions is presented. Both static and dynamic properties are shown to arise from the correlations between particle concentration fluctuations at different wavelengths. For a more visual presentation, the basic concepts are discussed using optical images of Q2D systems, where the particle's configurations can be observed directly in real space. Since colloidal liquids resemble atomic liquids, those images also allows one to visualize the properties of atomic liquids. The concepts and methodologies presented here focus on the static structure, but we also discuss briefly dynamic properties such as self- and collective diffusion. For simplicity, we only discuss the structural properties of monodisperse systems of spheres interacting throughout central potentials. However, one can extend concepts and methodologies to treat effects such as polydispersity, shape of the particles, and rheology of dispersing fluid. Those issues are quite important in determining different properties of systems and they are currently subjects of intense research. From the theoretical and the computer simulation point of view, there is also a vast literature and efforts dealing with the description of different properties. The discussion of those models is not within the aim of this chapter, but we provide here reference to some interesting works [39–42].

ACKNOWLEDGMENTS

The collaboration of M.A. Ramírez-Saito in the preparation of the figures, is greatly acknowledged, as well as financial support from Consejo Nacional de Ciencia y Tecnología, Mexico.

REFERENCES

[1] Sahimi M. *Flow Transport in Porous Media and Fractured Rock: From Classical Methods to Modern Approaches*. Weinheim: VCH; 1995.

[2] Lobry L, Ostrowsky N. Phys Rev B 1996;53:12050.

[3] Carbajal-Tinoco MD, Cruz de León G, Arauz-Lara JL. Phys Rev E 1997;56:6962.

[4] Zhan K, Méndez-Alcaraz JM, Maret G. Phys Rev Lett 1997;79:175.

[5] Acuña-Campa H, Carbajal-Tinoco MD, Arauz-Lara JL, Medina-Noyola M. Phys Rev Lett 1998;80:5802.

[6] Dufresne ER, Squires TM, Brenner MP, Grier DG. Phys Rev Lett 2000;85:3317.

[7] Pesché R, Nägele G. Phys Rev E 2000;62:5432.

[8] Santana-Solano J, Arauz-Lara JL. Phys Rev Lett 2001;87:038302.

[9] Cui B, Lin B, Rice SA. J Chem Phys 2001;114:9142.

[10] Cui B, Diamant H, Lin B, Rice SA. Phys Rev Lett 2004;92:258301.

[11] König H, Hund R, Zahn K, Maret G. Eur Phys J 2005;18:287.

[12] Santana-Solano J, Ramírez-Saito A, Arauz-Lara JL. Phys Rev Lett 2005;95:198301.

[13] Löwen H. J Phys Condens Matter 2009;21:474203.

[14] Leng J. Phys Rev E 2010;82:021405.

[15] Cervantes-Martínez AE, Ramírez-Saito A, Armenta-Calderón R, Ojeda-López MA, Arauz-Lara JL. Phys Rev E 2011;83:030402.

[16] Bonilla-Capilla B, Ramírez-Saito A, Ojeda-López MA, Arauz-Lara JL. J Phys Condens Matter 2012;24:464126.

[17] Crocker JC, Grier DG. J Colloid Interface Sci 1996;179:298.

[18] Hansen JP, McDonald IR. *Theory of Simple Liquids*. 2nd ed. New York; Academic press; 1986.

[19] Nägele G, Medina-Noyola M, Klein R, Arauz-Lara JL. Physica A 1988;149:123.

[20] Krauze R, Nägele G, Karrer D, Schneider J, Klein R, Weber R. Physica A 1988;153:400.

[21] Krause R, Arauz-Lara JL, Nägele G, Ruiz-Estrada H, Medina-Noyola M, Weber R, Klein R. Physica A 1991;178:241.

[22] Hess W, Klein R. Adv Phys 1983;32:173.

[23] Arauz-Lara JL, Medina-Noyola M. J Phys A Math Gen 1986;19:L117.

[24] Nägele G. Phys Rep 1996;272:215.

[25] Denton AR. Phys Rev E 2000;62:3855.

[26] Denton AR. Phys Rev E 2004;70:031404.

[27] Verwey EJW, Overbeek JTG. *Theory of the Stability of Lyophobic Colloids*. Amsterdam: Elsevier; 1948.

[28] Israelachvili JN. *Intermolecular and Surface Forces*. 2nd ed. London: Academic Press; 1992.

[29] Hiemenz PC, Rajagopalan R. *Principles of Colloid and Surface Chemistry*. 3rd ed. New York: Marcel Dekker, Inc.; 1997.

[30] Gaylor K, Snook I, van Megen W. J Chem Phys 1981; 75:1682.

[31] Pusey PN. In: Hansen JP, Levesque D, Zinn-Justin J, editors. *Liquids, Freezing and Glass Transition*. Amsterdam: Elsevier; 1991.

[32] Alexander S, Chaikin PM, Grant P, Morales GJ, Pincus P. J Chem Phys 1984;80:5776.

[33] Gisler T, Schulz SF, Borkovec M, Sticher H, Schurtenberger P, D'Aguanno B, Klein R. J Chem Phys 1994;101:9924.

[34] Sonntag H, Strenge K. *Coagulation Kinetics and Structure Formation*. New York: Plenum Press; 1987.

[35] Russel WB, Saville DA, Schowalter WR. *Colloidal Dispersions*. Cambridge: Cambridge University Press; 1989.

[36] McQuarrie DA. *Statistical Mechanics*. New York: Harper and Row; 1976.

[37] Rogers FJ, Young DA. Phys Rev A 1984;30:999.

[38] Hoye JS, Blum L. J Stat Phys 1977;16:399.

[39] Ramírez-González PE, Juárez-Maldonado R, Yeomans-Reyna L, Chávez-Rojo MA, Chávez-Páez M, Vizcarra-Rendón A, Medina-Noyola M. Rev Mex Física 2007;53:327.

[40] Yeomans-Reyna L, Chávez-Rojo MA, Ramírez-González PE, Juárez-Maldonado R, Chávez-Páez M, Medina-Noyola M. Phys Rev E 2007;76:041504.

[41] Juárez-Maldonado R, Medina-Noyola M. Phys Rev E 2008;77:051503.

[42] Juárez-Maldonado R, Medina-Noyola M. Phys Rev Lett 2008;101:267801.

12

COLLOIDAL CRYSTALLIZATION

ZHENGDONG CHENG
Artie McFerrin Department of Chemical Engineering, Materials Science and Engineering, The Professional Program in Biotechnology, Texas A&M University, College Station, TX, USA

The crystallization of a metastable melt is one of the most important nonequilibrium phenomena in condensed matter physics. Colloidal hard spheres offer the simplest and direct realization of crystallization. Together with computer simulations, quantitative level understanding of crystallization has been achieved using colloidal hard spheres. The discrepancy between the simulation and experimental nucleation rate densities of hard spheres calls for further advancement on experiment, theory, and simulation. Charged sphere crystallization puts back the enthalpy into the crystallization process and opens new dimensions for control of the crystallization. Currently, we only qualitatively understand the nucleation of charged sphere crystallization. Microgels offer unique ways to manipulate the colloidal size, through which our ultimate understanding of nucleation might be accomplished.

12.1 CRYSTALLIZATION AND CLOSE PACKING

The first attempt to explain crystal structure on the basis of interaction between atoms was made by Barlow (1898) [1]. He developed the *principle of close packing* by assuming mutual repulsion of solid elementary particles that are held

Fluids, Colloids and Soft Materials: An Introduction to Soft Matter Physics, First Edition. Edited by Alberto Fernandez Nieves and Antonio Manuel Puertas.

together by a general pressure. Two types of contacts were allowed by Barlow between spherical atoms: touching and interpenetrating. The first type of contact one would link today to weak ionic or metallic bonds. The second type is to covalent bonds. Barlow derived so many different crystal structures by packing that their number was comparable to the number actually found, which he took as an indication that close packing is the main principle governing crystal structure.

Today, the macroscopic reasoning requires equilibrium structure to have the lowest free energy, G. The difference in free energy between two structures ΔG is determined by their entropy difference ΔS and enthalpy difference ΔH:

$$\Delta G = \Delta H - T\Delta S.$$

On inspection of this equation, one can see that more order (which intuitively means lower entropy according to conventional wisdom) at any temperature leads to a more stable crystal *when accompanied by sufficient lower enthalpy* (larger interaction). However, it is

$$\Delta G = -T\Delta S,$$

that determines the hard-sphere disorder–order transition and the structure of the ordered phase, which is a very *intriguing* observation. *At first thought almost everybody cannot believe the conclusion they obtained for a collection of hard spheres.* At a given temperature, all that can be done to minimize the free energy is to maximize the entropy, that is, to maximize the total number of accessible states. One would seek to write down expressions for the entropy of the ordered state and the disordered state in order to investigate the disorder–order transition at all volume fractions and then choose the highest (entropy) of the two for any given volume fraction. *Counterintuitively*, it is the *ordered* phase has higher entropy than the disordered phase does at large volume fractions of hard spheres. The reduction in entropy associated with the formation of long-range order is offset by the increased "packing" contribution to the entropy associated with the particles' greater freedom for local motions in the ordered state [2]. This *disorder–order transition* problem has not been completely solved analytically to date [3] (analytically crystal has higher entropy at random close-packing volume fraction, which is about 63%) but has been solved by Monte Carlo simulations [4].

One may easily accept close packing as the principle of crystal structure for spheres or even irregularly shaped but rigid structure units. Macromolecules, however, are not rigid. Their shape in a crystal is determined by intramolecular free energy, which arises from rotation around bond angles. For a full description, intramolecular and intermolecular free energies must be minimized simultaneously. Fortunately, intramolecular free energy changes are frequently larger than the intermolecular free energy changes. For this reason, the low free energy shape of an isolated polymer chain can usually be evaluated as a separate problem. Close-packing considerations are then applied to the low-energy shape, which is treated as a rigid structure. The resulting approximate crystal structure will usually be close to the one found experimentally. The experimentally evaluated structure can in turn often be rationalized in terms of close-packing considerations of the lowest free energy molecular chain conformation.

Phase transitions observed in colloidal systems can be divided into two classes [5, 6]: (i) Disorder–order transition driven by entropy in systems dominated by repulsive interparticle potential and (ii) fluid–fluid or fluid–solid transitions caused by weak attractions. The first class corresponds to the liquid–solid or freezing transition in molecular systems and is distinguished by the small difference in density between the two coexisting phases and the crystalline order of the denser phase. The second, akin to gas–liquid and gas–solid transitions, involves large density differences.

Disorder–order transition for hard spheres was predicted quantitatively in the 1950s [7–9] but not quantified experimentally till the work of Hachisu and Kobayashi (1974) [10] with aqueous latexes at high electrolyte concentrations. Monodisperse charged spheres develop similar crystalline order at low ionic strengths, owing to long-range electrostatic repulsions. Hachisu and Takano [11] argued the latter is an effective hard-sphere phase transition with the effective radius including the particle and its electrical double layers. Experiments on better hard-sphere-like PMMA spheres, which are sterically stabilized, were performed by the Bristol group in the United Kingdom [12] and Princeton group in the United States [13–17].

Since the range of concentrations for which two phases coexist is rather narrow, attention has focused primarily on the structure and properties of the ordered solid, and more recently the dynamics of the phase transition and the possibility of metastable glassy states (e.g., Ref. [18]).

12.1.1 van der Waals Equation of State and Hard Spheres as Model for Simple Fluids

The celebrated van der Waals equation of state is

$$p = \frac{k_B T}{v - b} - \frac{a}{v^2},$$

where p is the pressure, $k_B T$ is the thermal energy (Boltzmann constant k_B times the temperature T), v is the volume per molecule ($v = V/N$ with V the volume and N the number of molecules in the system), while a and b are two constants characteristic of the fluid. The second term in the equation of state (EOS), a/v^2, accounts for the attractions between molecules in a mean field fashion. The potential energies of attraction felt by each molecule are proportional to the number of molecules contributing to it and hence to the number density N/V. Thus, the configurational Helmholtz free energy for a van der Waals fluid is

$$A = -TS_{\text{hs}} - aN \cdot N/V,$$

where S_{hs} is the configurational entropy of a fluid of nonattracting hard spheres at the same density as the real fluid and a is the constant of proportionality in the energy of attraction, which is given in terms of the potential energy of attraction U between pairs of molecules, which is a function of the location of the two molecules:

$$a = -\frac{1}{2}\int U \, d\tau,$$

where the integration is over all relative separations of the molecules and $d\tau$ is the element of volume in the integration. The pressure is then

$$p = -\left(\frac{\partial A}{\partial V}\right)_{N,T} = p_{\text{hs}} - \frac{a}{v^2},$$

where p_{hs} is the pressure of the hard-sphere fluid without attraction at the same density and temperature as the real fluid. One can then obtain the van der Waals equation or an improved version of it according to how p_{hs} is represented.

Thus, the first term on the right-hand side of the van der Waals equation of state is an approximation to the pressure p_{hs} of a fluid of nonattracting hard spheres. It is exact in one dimension (hard spheres or hard rods on a line), but it is a poor approximation in three dimensions.

The EOS is considerably improved when $1/(v-b)$ is replaced by a more accurate representation of p_{hs}/k_BT for a hard-sphere fluid in three dimensions. Though originally developed to describe the liquid–gas critical point, van der Waals mean field approach is now known to be most useful and accurate at the high densities that characterize a liquid away from the critical point, where the nearly incompressible nature of the fluid tends to inhibit the fluctuations that invalidate a mean field theory.

The fundamental physical idea behind van der Waals theory is that *the configuration of the molecules in the fluid are determined largely by the hard-sphere repulsions and very little by the attractions*, since the attraction forces exerted by each molecule's many neighbors largely cancel each other and the average attraction just sets particle density. The successful exploitation of this picture can be found in textbook [19] descriptions of perturbation theories of simple atomic liquids. But the range of utility of the van der Waals picture is far broader. Its variety and usefulness have been documented in numerous studies extending from computer simulations of condensed materials to analytical equilibrium theories of polyatomic organic liquids and mixtures to models of transport and even to vibrational relaxation in liquids [20].

12.1.2 The Realization of Colloidal Hard Spheres

Prominent examples of well-characterized hard spheres include polymethylmethacrylate (PMMA) spheres, sterically stabilized by polyhydroxy stearic acid (PHSA) in mixed solvent of decalin and tetralin, as well as sterically stabilized silica spheres. If sufficiently large amounts of screening electrolyte are added, charged spheres exhibit an effective hard-sphere behavior. Advanced methods to precisely control the preparation parameters are available.

Minimizing van der Waals Interaction by Refractive Index Matching The van der Waals attraction between two spheres of radius R, obtained by summing over all the interaction between pairs of atoms is

$$V_A(r) = -\frac{A}{6}\left[\frac{2R^2}{r^2 - 4R^2} + \frac{2R^2}{r^2} + \ln\left(1 - \frac{4R^2}{r^2}\right)\right],$$

where r is the center-to-center separation [5, 21]. The Hamaker constant A is determined by the material properties of the particles and suspension medium, in particular their frequency-dependent polarizability. If the particles and the liquid have equal polarizability, $A = 0$ (cf. Eq. 5.9.3 of Ref. [5]). An estimation by Israelachvili is [22]

$$A = \frac{3h\nu(n_1 + n_2)^2(n_1 - n_2)^2}{16\sqrt{2}(n_1^2 + n_2^2)^{3/2}},$$

where h is Planck's constant, ν is the characteristic frequency, and n_1, n_2 are the optical refractive indices of the colloids and the solvent, respectively. Thus, if the refractive indices of the particles and the liquid are matched at the frequency of visible light, van der Waals attractions are expected to be negligible (or at least small, as some difference in polarizability may still contribute at other frequencies). The typical value of the Hamaker constants for substances immersed in water ranges from 30×10^{-20} J for metals to 3×10^{-20} J for oxides and halides to 0.3×10^{-20} J for hydrocarbons [23].

The effect of van der Waals forces is to create a deep potential minimum near $r = 2R$ that can be many times greater than the thermal energy k_BT. Suspended particles that are unprotected will soon aggregate irreversibly under the influence of these strong attractions. It is therefore necessary to provide some stabilization mechanism by introducing a large positive potential barrier in the interparticle potential $V(r)$. Two approaches are common: steric and electrostatic stabilization.

Stabilization: Steric and Electrostatic Steric means that a process is inhibited due to the physical obstructions caused by a certain molecular arrangement. Steric stabilization is achieved by coating colloidal particles with layers of polymers. Many types of "polymer" may be used, ranging from relatively short alkane chains through various more complex structures to essentially random coils. The coating may be physically adsorbed on the particle or chemically bonded to it. In a good solvent, the polymer layer will swell and will preferably extend *beyond* the range of the van der Waals attraction, hence preventing the particle aggregation caused by van der Waals attraction. For the case of PMMA particles with grafted PHSA chains, the layer can extend to 10 nm, which can be beyond the grasp of the van der Waals attraction. When two polymer layers are brought into contact, they will resist compression because their osmotic pressure is a strong function of local macromolecular concentration. Hence, the spheres will experience a very deep repulsive interaction before the van der Waals force becomes significant. The effectiveness depends on various parameters. A solvent must have a relatively good match with the particle index of refraction, and the polymers must also readily dissolve in it. Silica [24, 25] and PMMA [26] particles are commonly treated in this fashion.

Costello *et al.* [27, 28] measured the force between two mica surfaces coated with a PMMA (backbone) PHSA (side chain) comb copolymer, with the PMMA backbone directly adsorbed on the mica and the PHSA side chains protruding into the solvent. The interaction thus mimics what between the two surfaces of PHSA-stabilized PMMA colloids.Bryant *et al.* [29] performed similar experiments, but the PHSA was directly adsorbed on the mica surfaces without the PMMA. According to the model proposed by Alexander and de Gennes [30, 31], which is valid for high polymer grafting density, each chain is assumed to be consisting connected semidilute blobs. The chains are stretched by osmotic repulsion between the blobs. This tendency is opposed by the increase in elastic free energy of the chain upon stretching. The resulting expression for the force per unit area between two parallel plates at a distance r is

$$F(r) = \frac{\alpha k_B T}{S^3}\left[\left(\frac{2L}{r}\right)^{9/4} - \left(\frac{r}{2L}\right)^{3/4}\right] \quad \text{for } 0 < r < 2L,$$

where S is the mean spacing between grafting, L is the thickness of the polymer layer, and α is a numerical prefactor. Integration of this equation yields the corresponding interaction energy density. At the distance where the interaction kicks in, a polymer thickness $L = 12.5$ nm can be estimated [27, 28], and a fit of the model to data gives $\alpha = 0.025$ and $S = 2.8$ nm [27, 28].

By using the Derjaguin approximation (cf. Eq. 5.7.2 of Ref. [5]), we can estimate the interaction potential between two spheres using the measurement between two parallel plates:

$$V_{\text{steric}}(r) = \pi R \int_{r-2R}^{\infty} F(h)dh - \frac{A}{6}\left[\frac{2R^2}{r^2 - 4R^2} + \frac{2R^2}{r^2} + \ln\left(1 - \frac{4R^2}{r^2}\right)\right].$$

Electrostatic stabilization is achieved by the charges on the surface of colloidal particles. Charged colloidal particles can be regarded as "macroions." When two macroions approach each other, overlap of their double layers causes a repulsive force, which can stabilize the particles against aggregation. In the celebrated DLVO (Derjaguin–Landau–Verwey–Overbeek) theory [32] treatment of the electrical double layer, the small ions in the suspension, regarded as point charges, are assumed to move rapidly enough in Brownian motion that, on any timescale of interest, their average spatial distribution in the potential field of the macroions can be taken to be the equilibrium Boltzmann distribution. Poisson's equation is then applied to calculate the potential. To make analytical progress, the resulting Poisson–Boltzmann equation is linearized, as in the Debye–Huckel theory of simple electrolytes. This approach leads to the celebrated DLVO expression for the interaction energy of an isolated pair of macroions in a bath of electrolytes. In the simplest case, it has the form of a *hardcore Yukawa* potential (also called *screened Coulomb* potential) [33]:

$$V_c(r) = \begin{cases} \infty & r \le \sigma, \\ \varepsilon \dfrac{e^{[-\kappa\sigma(r/\sigma - 1)]}}{r/\sigma} & r > \sigma, \end{cases}$$

where $\sigma = 2R$ and ε is the interaction strength, $\varepsilon / k_B T = \frac{Z^2 e^2}{8\pi\varepsilon_r\varepsilon_0 R} = \frac{B}{2R} Z^2$ for two colloids with size $2R$ and charge Ze, with $B = e^2/(4\pi\varepsilon_r\varepsilon_0 k_B T)$ the Bjerrum length, ε_r the relative dielectric constant of the liquid and ε_0 the vacuum permeability; $B = 0.7$ nm for water and κ is the inverse Debye length characterizing the screening of the interaction. In the limit of small particle size, the potential is called a point Yukawa potential [34]:

$$V_c(r) = \begin{cases} \infty & r \le 2R, \\ \dfrac{q^2}{4\pi\varepsilon_r\varepsilon_0 r} e^{-\kappa r} & r > 2R \end{cases} \qquad (R \to 0),$$

where q is the charge on the macroion $q = Ze$.

An equation proposed by Bell *et al.* [35] and adapted by Glendinning and Russel [36] derives the interparticle potential from a linear superposition of the electrostatic potentials surrounding each particle, assuming that either the effective surface potential, often taken as the ζ-potential (zeta potential), or the surface effective charge remains constant:

$$V_c(r) = \begin{cases} \left(\frac{k_B T}{Ze}\right)^2 \psi_s^2 \left(\frac{R}{\frac{e^2}{4\pi\varepsilon_r\varepsilon_0 k_B T}}\right)^2 \frac{1}{\left[1+\phi/\left(\frac{\kappa^2 R^2}{3(1+\kappa R)}\right)\right]^2} \\ \exp(2\kappa R)\frac{\exp(-\kappa r)}{r}, \\ \qquad \text{constant effective potential } \psi_s, \\ \frac{Z^{*2}e^2}{4\pi\varepsilon_r\varepsilon_0}\left(\frac{\exp(\kappa R)}{1+\kappa R}\right)^2 \frac{\exp(-\kappa r)}{r}, \\ \qquad \text{constant effective charge } Z^*, \end{cases}$$

where $\psi_s = Ze\varphi_s/k_B T$ is the dimensionless surface potential, with φ_s the zeta potential [37, 38], and ϕ is volume fraction of the colloids. For typical numbers, $R/B = 100$ and $\kappa R = 0.25$, the surface effective charge $Z^* = (R/B)(1+\kappa R) = 125\psi_s$ and the volume fraction at which the colloidal charge has decayed to half of its dilute-limit value Z^* is $\phi^* = \frac{\kappa^2 R^2}{3(1+\kappa R)} \approx 0.017$. With $\psi_s \approx 1-2$, which corresponds to a surface potential of 25–50 mV, one should expect a few hundred effective charges in the dilute limit and a significant charge reduction for $\phi > 10^{-2}$. The geometric factor $(\exp(\kappa R)/(1+\kappa R))^2$ in the above equation for the case of constant effective charge results from the finite particle size [39].

The Debye screening length κ^{-1} is defined through following equation:

$$\kappa^2 = \frac{z_c z_p e^2 n_p}{\varepsilon_r \varepsilon_0 k_B T} + \sum_{k=1}^{N} \frac{(ez^k)^2 n_b^k}{\varepsilon_r \varepsilon_0 k_B T}.$$

The first term accounts for the counterions, with charge ez_c, that are dissociated from the particle surface; the concentration is just the particle concentration n_p multiplied by the number of charges per particle ez_p. The terms in the sum account for any ionic species.

To relate ψ_s to the surface charge, we use an approximation formula of Loeb *et al.* [40]

$$\frac{q}{4\pi R^2} = \frac{\varepsilon\varepsilon_0 k_B T}{ez}\kappa\left[2\sinh\left(\frac{1}{2}\psi_s\right) + \frac{4}{\kappa R}\tanh\left(\frac{1}{4}\psi_s\right)\right].$$

The total *interaction potential* resulting from adding V_A and V_C typically contains a deep primary minimum, a secondary minimum and a primary maximum $\Phi_{\max}$. At moderate to low ionic strengths, the repulsive barrier is large, $\Phi_{\max}/k_B T \gg 1$, making diffusion of initially dispersed particles into the primary minimum very slow, hence stabilizing the dispersion. At high ionic strengths, the repulsive barrier disappears, permitting rapid and irreversible flocculation into the primary minimum.

We finally note that although charge is often thought to play no role in nonpolar media, there is ample [41–47], sometimes contradictory [48], evidence of surface charging in nonpolar colloidal suspensions. Much of the reason for this uncertainty is the extremely low level of charging in nonpolar environments. While a particle of radius $R = 500$ nm might carry a surface charge eZ of the order of 10^3 electrons in an aqueous dispersion with $\varepsilon_r \approx 80$, the charge in a nonpolar solvent will be some two or three orders of magnitude smaller. Even if small in absolute terms, this charge still produces surprisingly strong effects in low dielectric environments. For instance, using Coulomb's law, the contact value of the interaction potential $\varepsilon/k_B T$ between two colloidal spheres with radius R and charge eZ in a solvent with relative permittivity ε_r is

$$\frac{\varepsilon}{k_B T} = \frac{1}{k_B T}\frac{Z^2 e^2}{8\pi\varepsilon_r\varepsilon_0 R} = \frac{B}{2R}Z^2.$$

For dodecane, $B = 28.0$ nm the repulsive interaction between 500 nm particles with a typical charge of $10e$ is about $3\,k_B T$; this value is large enough to have a dramatic effect on the structure of a colloidal suspension.

Fighting Gravity Let us consider the sedimentation of a colloidal suspension in gravity. Since the thermal motion of the colloids opposes sedimentation, there is a nonzero gravitational length

$$h_g \sim \frac{k_B T}{\Delta\rho\left[\frac{4\pi}{3}\left(\frac{d}{2}\right)^3\right]g},$$

where $\Delta\rho$ is the density deference between the colloids and the solvent and g is gravity. For a colloid of radius 500 nm, $h_g \approx 20\,\mu$m. Therefore, for a sample height higher than this value, sedimentation effects and the associated convection will be important. The best way to avoid these effects is to have the colloids in space. The averaged gravity in a "quiet" Space Shuttle, referred to as microgravity, is $10^{-6}g$ [15, 49].

Another possibility is to increase the gravitational length by matching the density of the colloids to that of the continuous phase. Cycloheptyl bromide (CHB) has relative large density $\rho = 1.289$ g/cm^3 and refractive index $n = 1.505$. A mixture of decalin ($n = 1.4750$) and CHB can achieve a good density match with PMMA particle, $\rho_{\text{PMMA}} = 1.192$ g/cm^3 and in the meantime a relatively good refractive match ($n_{\text{solvent}} = 1.498$ g/cm^3). The bulk refractive index of the PMMA particles synthesized by Professor R.H. Ottewill and his group at Bristol University is $n = 1.503$. CHB is a nonsolvent for PMMA [50]. Following the initial applications of CHB in PMMA suspensions for density matching [50, 51], explosive use of this solvent system has been seen in the past decade. However, the PMMA spheres are slightly charged from the presence of water or from the dye incorporated in the particles [52–54]. In principle, three solvents should offer density match and refractive index match simultaneously to a colloid suspension.

Polydispersity All colloidal suspensions are polydisperse to some extent. In the best cases, polydispersity can be as small as a few percent. In many applications, particles with such narrow size distributions are generally denoted as "monodisperse." Monodisperse colloids with a variety of chemical compositions, shapes, and morphologies [55] have been synthesized [56–58]. The methods to obtain monodisperse colloids include (i) caged growth inside nanoporous structures or objects (micelles or inverse micelles) [59], (ii) seeded growth [60], (iii) burst nucleation flowed by growth [58, 61], (iv) Ostwald ripening of unstable nuclei [62], and surface reaction limited growth [63].

Polydispersity is an important parameter controlling phase transitions [64, 65]. It is known that subtle difference in particle size distributions can have a huge impact on hard-sphere crystallization [66–72]. Simulations indicate that crystal nucleation in polydisperse colloids is suppressed due to increase in the surface free energy, and vitrification at high supersaturations should yield colloidal glasses that are truly amorphous, rather than nanocrystalline [67]. Polydispersity affects the type of crystalline structure as well as the crystal morphology of the resulting material. It introduces bad spots, dislocations, stacking faults, and unregistered planes [71, 73–75]. Therefore, the polydispersity details are also critical for many applications; for example, the use of colloidal crystals as a starting material for photonic band gap materials will be limited by the polydispersity as imperfect crystal structures will reduce the width of the band gap.

To address the effects of small polydispersity changes on crystallization of hard-sphere suspensions, it has been necessary to develop methods to prepare and characterize suspensions with subtle differences in size distribution. It has been demonstrated [76] that polydispersity can be tuned by fractionation, a technique that has been applied to both submicron- and micron-sized particles. Scanning or transmission electron microscopy, as well as light scattering, can be used to determine the size distribution of a colloidal suspension.

12.2 CRYSTALLIZATION OF HARD SPHERES

12.2.1 Phase Behavior

What causes the hard-sphere fluid–crystal transition? A simple free volume argument is shown schematically in Figure 12.1. For the fluid state, the highest volume fraction achievable corresponds to random close packing (RCP) $\phi_{\rm rcp} \approx 0.64$, so the entropy for a liquid with volume fraction ϕ, when ϕ is large, is proportional to [77]:

$$\begin{aligned} S_{\text{Liquid},\phi} &\sim k_B \ln(\text{free volume for centers of hard spheres} \\ &\qquad \text{at a liquid state with } \phi) \\ &\sim k_B \ln(0.64/\phi - 1). \end{aligned}$$

On the other hand, for the crystalline state, the highest volume fraction corresponds to hexagonal close packing (HCP) $\phi_{\text{Hex c.p.}} \approx 0.74$, so the entropy for a crystal with volume fraction ϕ, when ϕ is large, is proportional to:

$$\begin{aligned} S_{\text{Crystal},\phi} &\sim k_B \ln(\text{free volume for centers of hard spheres} \\ &\qquad \text{at a crystalline state with } \phi) \\ &\sim k_B \ln(0.74/\phi - 1). \end{aligned}$$

Therefore, for ϕ reasonably high (for ϕ too small, the particle will move around in the whole sample space, not just in the cage formed by the neighbors as pictured in Fig. 12.1), $S_{\text{Crystal}(\phi)} > S_{\text{Liquid}(\phi)}$, just as Figure 12.1 tells us that the black area, representing the free volume for the center of hard sphere to move freely, in the crystal phase is larger than that in the liquid state. Hence, the crystal phase is thermodynamically more stable. So the hard-sphere disorder–order (Alder–Wainwright) transition occurs for purely entropy reasons: by crystallization, the system gains entropy via creation of more space within the ordered domain.

Even though there is a long history of studying hard-sphere models [4, 78–82], no rigorous statistical mechanical proof exists for the disorder–order transition in dimension higher than 2 [83, 84]. Wang [84] presented an interesting mean-field argument to locate the freezing point in hard spheres utilizing the observation that near the freezing point, the volume of the effective cage of a hard-sphere fluid is no less than $(2R)^d$ in dimension d if the spheres sit on a square lattice, where R is the particle radius. It means that any cage can only shrink to $(2R)^d + \varepsilon$ with $\varepsilon \to 0^+$. It was also observed that, when two spheres exchange positions (Fig. 12.3), they would occupy a deformed cage. When a cage deforms to increase volume, it has to gain volume from neighboring cages. Its maximum possible volume is $v + 2d[v - (2R)^d - \varepsilon]$ with v the volume occupied by each particle on a square lattice (i.e., volume of the lattice cell) because there are only $2d$ nearest neighbors. Since there are two particles in the deformed cage in the transient state, the local number density is $2/v + 2d[v - (2R)^d - \varepsilon]$. When this density equals the density at close packing that is the highest can be achieved,

$$\frac{2}{v + 2d[v - (2R)^d - \varepsilon]} = \rho_{\rm cp},$$

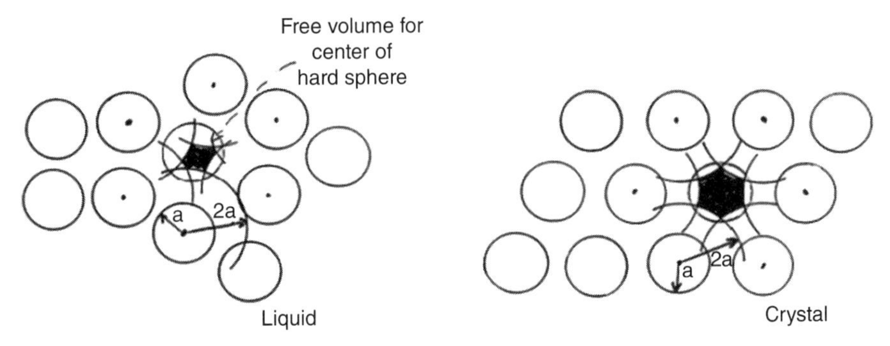

Figure 12.1 *Free volume argument for hard-sphere disorder–order transition.* Hard-sphere disorder–order (Alder–Wainwright) transition occurs for purely entropy reason: by separating into two phases the system gains entropy by creating more space within the ordered domain. Hand drawing by Prof. Paul Chaikin.

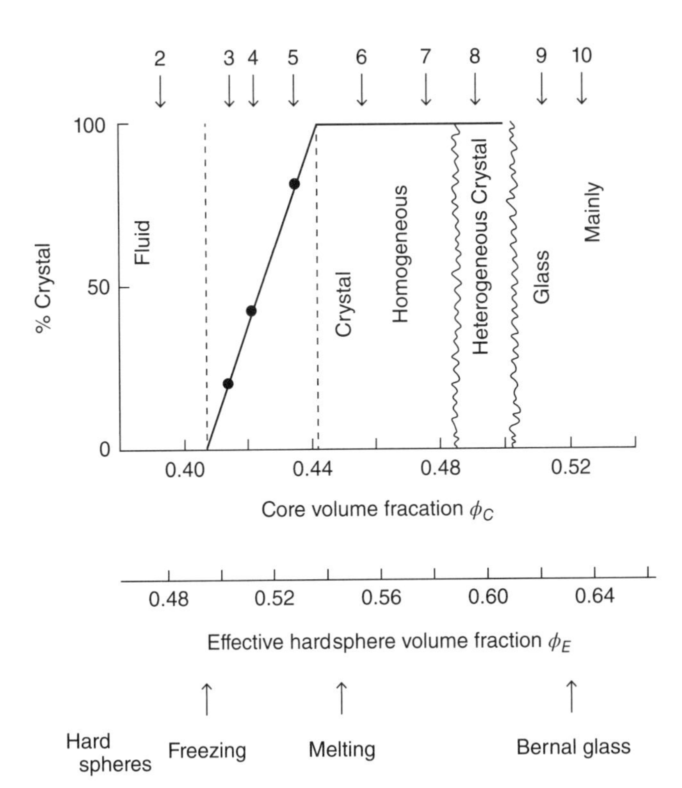

Figure 12.2 *Phase diagram for PMMA–PHSA dispersions* (614-nm-diameter spheres). Arrows at the bottoms indicate volume fractions of hard spheres obtained from computer simulation for freezing, melting, and random close packing (Bernal glass).

freezing occurs. Therefore, in the limit of $\varepsilon \to 0$,

$$\frac{2}{v_f + 2d[v_f - (2R)^d]} = \rho_{cp}.$$

In three dimensions, $d = 3$, $\rho_{cp} = \sqrt{2}(2R)^{-3}$, $\phi_{cp} = \pi\sqrt{2}/6$. Hence,

$$\phi_f = \frac{\pi}{6}\rho_f = \frac{\pi}{6}\frac{1}{v_f} = \frac{7\pi}{6(6+\sqrt{2})} = 0.4943.$$

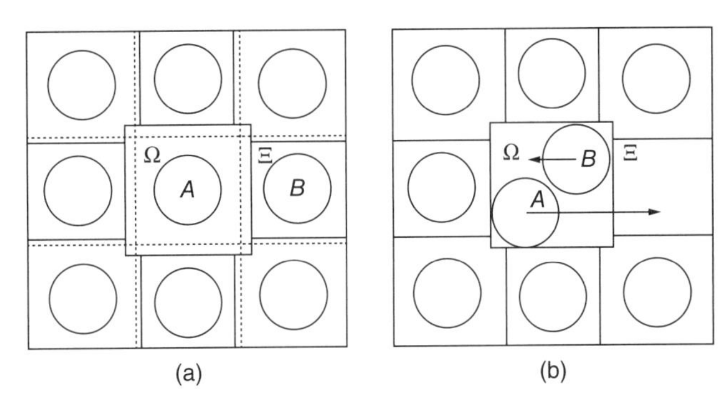

Figure 12.3 *Exchange of positions of two hard spheres.* The spheres are located on a cubic lattice, as illustrated on (a). One of the cells expands to accommodate two spheres (b) in a transient state in order to exchange the positions of the two.

This value is in a remarkable good agreement with simulation results [4, 8, 85].

In experiments, the crystal fractions in the samples in the fluid–crystal coexistence region are often used to determine the phase boundaries in the phase diagram. Figure 12.4 shows a photo of some PMMA crystal samples. The crystals settled to the bottom of the container because the density of the crystal is slightly larger than the coexisting fluid. However, a thin layer of crystals can also grow at the bottom by sedimentation. A long time, weeks or even months, is required for the measurement of this thin layer growth [15, 86]. Figure 12.2 shows the phase diagram determination for a PMMA suspension [12]. Rescale the freezing point to 0.494, the melting point was determined experimentally to be 0.536, slightly less than the simulation result of 0.545 for hard spheres. It indicated a slightly soft interaction between the spheres when they are very close to each other, due to steric or electrostatic stabilization.

Figure 12.5 shows the seminal work on PMMA spheres by Pusey and van Megen [12]. Starting from the fluid sample at the right (which is sample number 2 labeled in Figure 12.2, where a very thin layer of crystal is visible at the bottom of the vial), three samples in the coexistence region

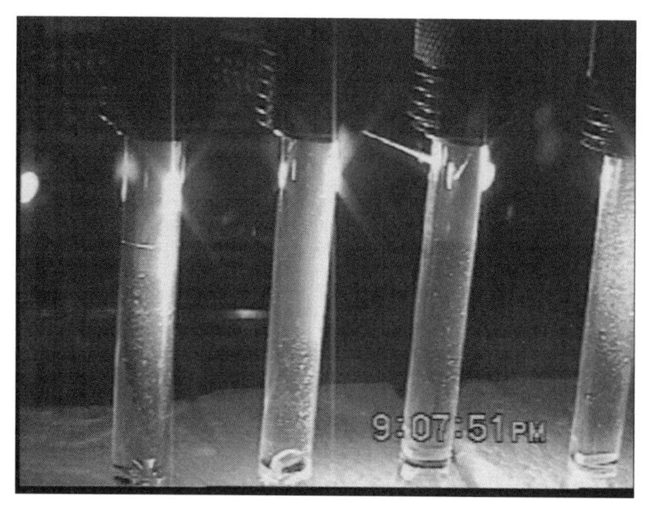

Figure 12.4 *Crystal fraction measurement by sedimentation.* Left three samples are in fluid–crystal coexistence region. Measurement of the fluid–crystal interface position as a function of time for 2 months and extrapolating back to time zero to subtract crystal growth due to sedimentation will yield the total amount of crystal phase in the sample due to homogeneous nucleation and growth.

followed. The iridescent color grains are Bragg scattering of visible light from the colloidal crystals. The crystals settle to the bottom half of the sample due to gravity. Moving further left, we see two samples with crystallites filling up the whole vial (cf. Fig. 12.2, samples 6 and 7, which are the fifth and sixth from the left in Fig. 12.5). The crystal morphology in sample number 8 (the seventh from the right in Fig. 12.5) is very different from the crystals in the samples of lower concentrations. Here, heterogeneous crystal growth takes over. When sample concentration is very close to RCP, the so-called "glass" transition comes to play a role (sample numbers 9 and 10, the eighth and ninth photographs from the right), no crystallites are visible any more.

12.2.2 Equation of State of Hard Spheres

Let us quantitatively characterize the hard-sphere fluid–crystal transition. The three branches of the hard-sphere EOS are sketched in Figure 12.6. For ideal gas, as Perrin measured using his monodisperse emulsions, the osmotic pressure Π is

$$\Pi = nk_BT,$$

where n is the particle number density. Hard-sphere fluids or crystals deviate from this by a volume fraction–dependent compressibility factor Z.

$$\Pi = nk_BTZ(\phi).$$

For fluids, the Carnahan and Starling approximation [87] captures the first seven virial coefficients (defined as the coefficients in the density expansion of the compressibility factor) and available results from computer simulations:

$$Z_{\text{fluid}}(\phi) = \frac{1 + \phi + \phi^2 - \phi^3}{(1 - \phi)^3}.$$

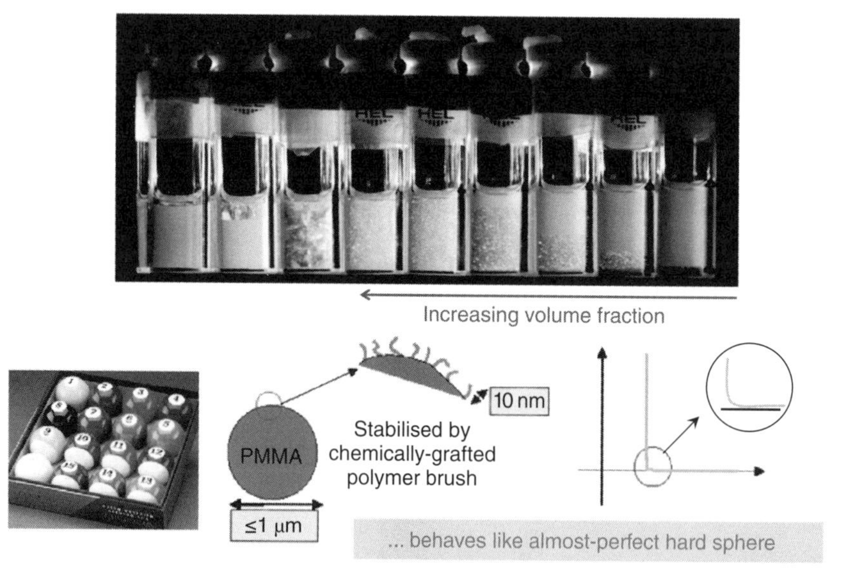

Figure 12.5 *Phase diagram of PMMA hard spheres.* Top row shows samples 4 days after melting. Bottom row: middle panel illustrates the micron size of the particles and their steric stabilization; right illustrates the near hard-sphere potential, highlights the softness at contact; left, the hard spheres in the pool game.

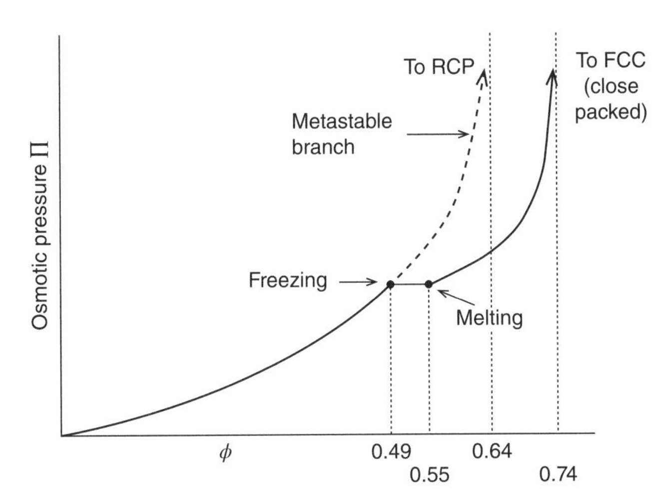

Figure 12.6 *Equation of state of the hard spheres.* The fluid and crystal phases are bridged by the coexistence region located between the freezing and melting points. The metastable state extends the stable fluid phase up to the random close packing at $\phi = 0.64$. The crystal phase extends to the face center cubic close packing $\phi = 0.74$.

For face-centered-cubic (FCC) crystals, Hall [88] constructed a modified Padé approximant from computer simulation results:

$$\begin{aligned} Z_{\mathrm{FCC}} = {} & \frac{12-3\beta}{\beta} + 2.557696 + 0.1253077\beta \\ & + 0.1762393\beta^2 - 1.053308\beta^3 \\ & + 2.818621\beta^4 - 2.921934\beta^5 + 1.118413\beta^6, \\ & \text{with } \beta = 4(1-\phi/\phi_{\max}), \text{ and } \phi_{\max} \approx 0.74. \end{aligned}$$

X-ray microdensitometry has been used to measure the density as a function of sample height for colloidal samples that have reached equilibrium in the gravitational field of Earth. The volume fraction profile in an equilibrium sediment reflects the balance between gravitational forces and the osmotic pressure gradients. Integration of this balance from the upper surface to any depth h determines the osmotic pressure from the total weight of the overlying particles,

$$\Pi(h) = \int_h^{\infty} g\Delta\rho\phi(z)dz,$$

where $\Delta\rho$ is the density difference between the particle and the solvent.

Results from experiments on PMMA hard spheres [13] and molecular dynamics simulations [89] are presented in Figure 12.7, together with the Carnahan–Starling and Hall equations.

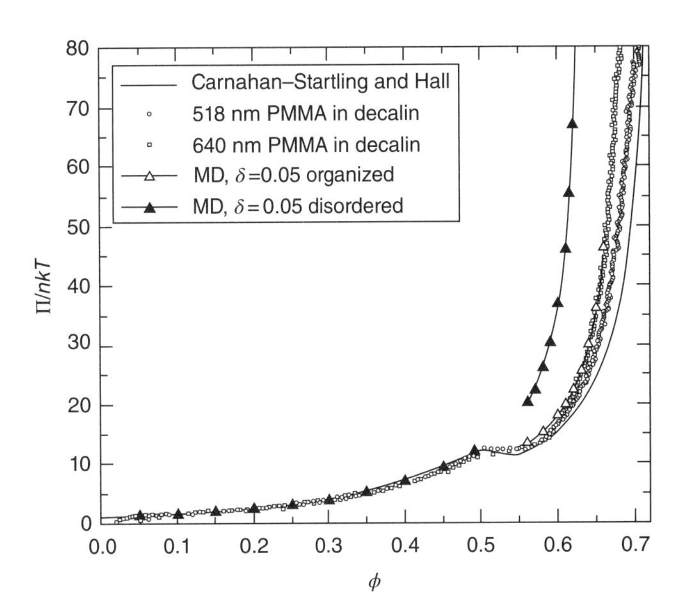

Figure 12.7 *The equations of state for hard spheres.* The molecular dynamics simulation results for the organized and disordered systems with 5% polydispersity determined from molecular dynamics simulations represent an upper and a possible lower bound for the X-ray densitometry data from 518- and 640-nm-diameter PMMA–PHSA in *cis/trans*-decalin. The solid line represents the Carnahan–Startling equation and Hall equation with $\phi_{\max} \approx 0.74$.

The Stable Crystal Branch of the Hard-Sphere Equation of State Ends at $\phi_{\mathrm{FCC}} = \pi/\sqrt{18}$: Kepler Conjecture
The curve above the melting point, shown in Figures 12.6 and 12.7, ends at the close-packed FCC volume fraction, $\phi_{\mathrm{FCC}} = \pi/\sqrt{18}$.

The Kepler conjecture, named after the 17th-century German astronomer Johannes Kepler, is a mathematical conjecture about sphere packing in 3D Euclidean space. It states that no arrangement of equally sized spheres filling space has a greater average density than that of the cubic close packing of the FCC lattice or HCP. The Kepler conjecture asserts that the volume fraction of a packing of congruent spheres in 3D is never greater than $\pi/\sqrt{18}$, or approximately 0.74048. This is the oldest problem in discrete geometry and is an important part of Hilbert's 18th problem. Hilbert's address in 1900 to the International Congress of Mathematicians in Paris might be regarded as the most influential speech ever given to mathematicians, given by a mathematician, or given about mathematics. In his speech, Hilbert outlined 23 major mathematical problems to be studied in the 20th century. The 18th problem asked three separate questions about lattices and sphere packing in Euclidean space.

The Kepler conjecture was proved by Sam Ferguson and Tom Hales in 1998 using 250 pages of notes and 3 GB of computer programs, data, and results. They followed an approach suggested by Laszlo Fejes Toth in 1953 that the maximum density of all arrangements could be found by minimizing a function with 150 variables and systematically applied the linear programming method to find a lower bound on the value of this function for each one of a set of over 5000 different configurations of spheres (i.e., *proof by exhaustion* checking many individual cases using complex computer calculations). Despite the unusual nature of the proof, the editors of the Annals of Mathematics agreed to publish it, provided it was accepted by a panel of 12 referees. In 2003, after 4 years of work, the head of the referee's panel Gabor Fejes Toth (son of Laszlo Fejes Toth) reported that the panel was "99% certain" of the correctness of the proof, but they could not certify the correctness of all of the computer calculations. The results were not published in full until 2006 [90–92]. An ongoing project, called *F*lys*p*e*c*k (the *F*ormal *P*roof of *K*epler), seeks to give a formal proof of the Kepler conjecture. It was estimated that it would take around 20 years to produce a complete formal proof. Therefore, the Kepler conjecture is now very close to being accepted as a theorem.

The Metastable Fluid Branch of the Hard-Sphere Equation of State Ends at the Maximally Random Jammed State The fluid branch of the hard-sphere EOS starts at $\phi = 0$ and ends at the freezing-point volume fraction, $\phi \approx 0.494$. The metastable extension of the fluid branch follows the stable fluid branch and is conjectured to end at RCP, where $\phi \sim 6/(6 + 2\sqrt{3}) \approx 0.634$ [93–97]. A large-scale molecular dynamics simulations of hard spheres in the metastable branch shows that near $\phi_{\rm rcp}$, the pressure scales as $(\phi_{\rm rcp} - \phi)^{-1}$ and $\phi_{\rm rcp} = 0.644 \pm 0.005$ [98].

The term RCP is widely used to designate the "random" packing with the highest achievable density. It is ill defined since a precise definition of randomness is lacking and random packings can also contain regions with certain degree of order. It has been suggested that RCP could be replaced by the *maximally random jammed* (MRJ) state, which can be made precise [99]. MRJ corresponds to the most disordered among all jammed (mechanically stable) packings. A jammed packing is one in which the particle positions and orientations are fixed by impenetrability constraints and boundary conditions [100].

Replacing RCP with MRJ inevitably leads to the necessity of distinguishing the MRJ state among the entire collection of jammed packings. While the ideal method of addressing this question would be to enumerate and classify all possible jammed hard-sphere configurations, practical limitations prevent such a method from being employed. Instead, a large number of representative jammed hard-sphere configurations are numerically generated and evaluated by several commonly employed order metrics [101]. Even in the large-system limit, jammed systems of hard spheres can be generated with a wide range of packing fractions from $\phi \approx 0.52$ to the fcc limit of $\phi \approx 0.74$. Moreover, at a fixed packing fraction, the variation in the order can be substantial, indicating that the density alone does not uniquely characterize a packing [101]. It has been established that $\phi_{\rm MRJ} \approx 0.637$, which has been obtained by a variety of different order metrics. The volume fraction is consistent with what has traditionally been associated with RCP in three dimensions. The MRJ state is the divergent point of the metastable branch and has minimal values of typical order parameters of crystals, including bond-orientational order and translational order. This concept has been debated and generalized [99, 102, 103]. There might be multiple metastable branches and RCP corresponding to the endpoint of the random jammed metastable state branch [103].

Replica theory of glass transition [104, 105] predicts that hard spheres remain in a liquid state up to a certain $\phi \in [\phi_d, \phi_K]$, where a dynamic transition to a glass phase takes places. Here, ϕ_d is the density where *many* metastable states first appear in the liquid phase and ϕ_K is the Kauzmann density of the ideal glass transition. The glassy states can become jammed at infinite pressure within a range of densities between the glass threshold, $\phi_{\rm th}$, and what is called the glass close packing, $\phi_{\rm GCP}$: $\phi_j(\phi) \in [\phi_{\rm th}, \phi_{\rm GCP}]$; RCP can be found everywhere in this range depending on the way that the sample is prepared.

A thermodynamic view of the sphere packing problem was proposed recently [97, 106, 107]. It states that RCP is the "freezing point" in a first-order phase transition between disordered and ordered packing phases. Despite the inherent out-of-equilibrium nature of jammed matter, the formation of jammed crystallites can be mapped to a thermodynamic process that occurs at a precise compactivity, where the volume and the entropy are discontinuous. For an equilibrium system of hard spheres, the freezing point is at $\phi_f \sim 0.49$ and melting point is at $\phi_m \sim 0.54$. For jammed hard spheres, the transition shifts to $\phi_f \sim 0.64$ and $\phi_m \sim 0.68$. The coexistence regime consists of a mixture of glass and crystal.

Experimental Measurement of the Phase Diagram of Hard Spheres in Microgravity The particles consist of uniform PMMA spheres, 508 or 518 nm in diameter, polydispersity ~5%, with a thin (10 nm) grafted layer of PHSA to prevent aggregation. The particles are suspended in a refractive index matching mixture of decalin and tetralin ($n = 1.511$). The volume fractions of the samples were determined on Earth by constructing the phase diagram using the sedimentation method and calibrating with respect to the freezing point by setting $\phi_f \sim 0.494$ as the lowest volume fraction that first showed crystallization.

Figure 12.8a shows the height of the crystal–liquid interface measured from the sample bottom (total length of the sample is 7.3 cm) for the Space Shuttle experiment "*C*olloidal *D*isorder and *O*rder *T*ransition *first* flight"

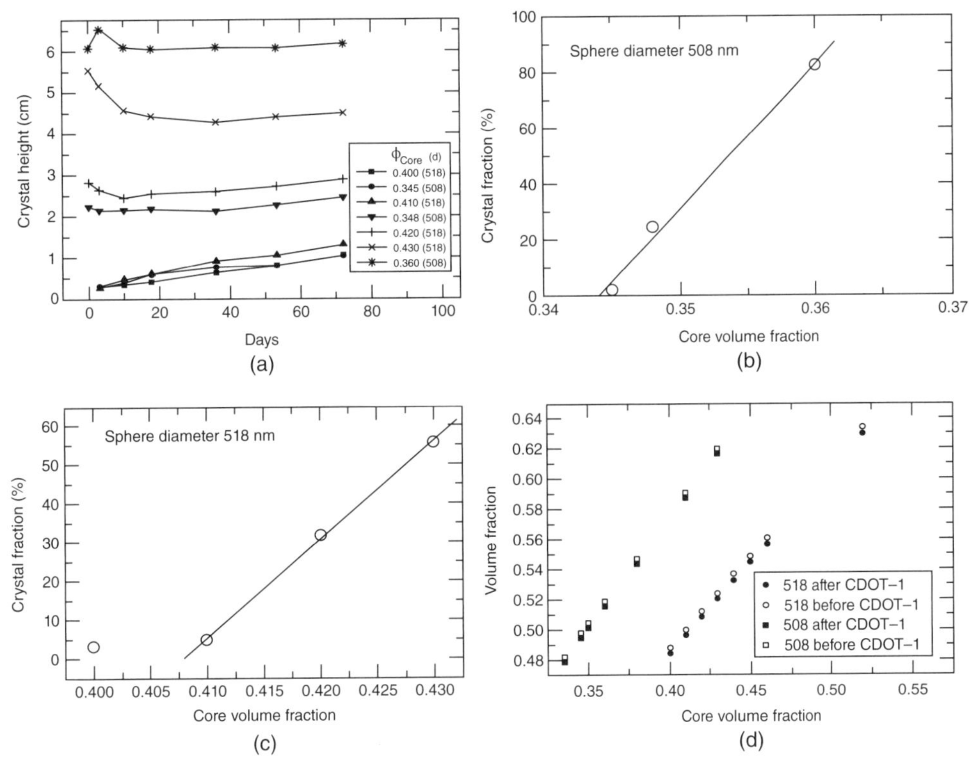

Figure 12.8 *Volume fraction determination of samples by sedimentation.* (a) Crystal height as a function of days after mix-melting. (b) and (c) Crystal fraction versus core volume fraction for coexistence region suspensions of spheres with diameters of 508 and 518 nm, respectively. (d) Volume fraction for CDOT-1 samples measured before and after the flight.

(CDOT-1) samples in the coexistence region. The temperature was $21 \pm 1\,^\circ$C. Within a few weeks, the samples separated into three distinct layers with the crystalline solid at the bottom, a clear supernatant on top, and a fluid dispersion in between. As the fluid phase settled and the crystalline layer grew, the height of each layer was carefully measured with a cathetometer. The linear fit to the data of height as a function of time for times longer than 30 days measured the growth speed of the crystalline due to sedimentation. The intercept of these fits with the vertical axis (height at time zero) gave the equilibrium crystal fractions of these samples. Notice that in Figure 12.8a, for samples below the phase transition, crystal layer grew due to sedimentation, while for samples in the coexistence region, crystallites nucleated over the sample volume, and observed crystalline height decreased first due to sedimentation of these crystallites and increased again at a later time due to the growth of the crystalline layer because of sedimentation of individual spheres from the liquid phase. Figure 12.8b and c is the crystal fraction versus core volume fraction for spheres with a diameter of 508 and 518 nm, respectively. The solid lines in (b) and (c), which are linear fits to the data, cross the core volume fraction axis at the corresponding effective hard-sphere freezing points. This information was used to rescale the core volume fractions (assuming the freezing point is $\phi_f \sim 0.494$) to obtain the effective hard-sphere volume fractions. Figure 12.8d shows the volume fractions calculated from two independent measurements before and after CDOT-1 flight.

With a stirring bar, the samples were mixed (shear melted) to set the starting time for nucleation and growth. The samples were photographed and observed at the beginning, the middle, and the end of the missions by the astronauts. Those that became crystals appear glittery; both liquid and glassy samples look transparent. By shining a light through each sample, the astronauts were able to determine whether the suspension remained a fluid, a glass or formed crystals. Digital images of the samples were taken and transmitted to Earth.

Figure 12.9 shows the sample photographs from the first (the Space Shuttle *Columbia* Mission USMSL-2/STS-73 in October 1995) and second flight (the Space Shuttle *Discovery*, STS-95 in November 1998; Senate Glenn on board) of CDOT. The pictures were taken approximately 4 days after shear melting. The close-up of sample 18 ($\phi = 0.516$), shown below samples 13-6, was from CDOT-2 (the second flight of CDOT). A cluster of dendritic (snow-flake-like) crystals is clearly visible. Similar crystal morphology was observed for the first time in the CDOT-1 samples ($\phi = 0.504,\ 0.512$) [14]. These findings demonstrate the instability of crystal growth in the liquid–crystal coexistence

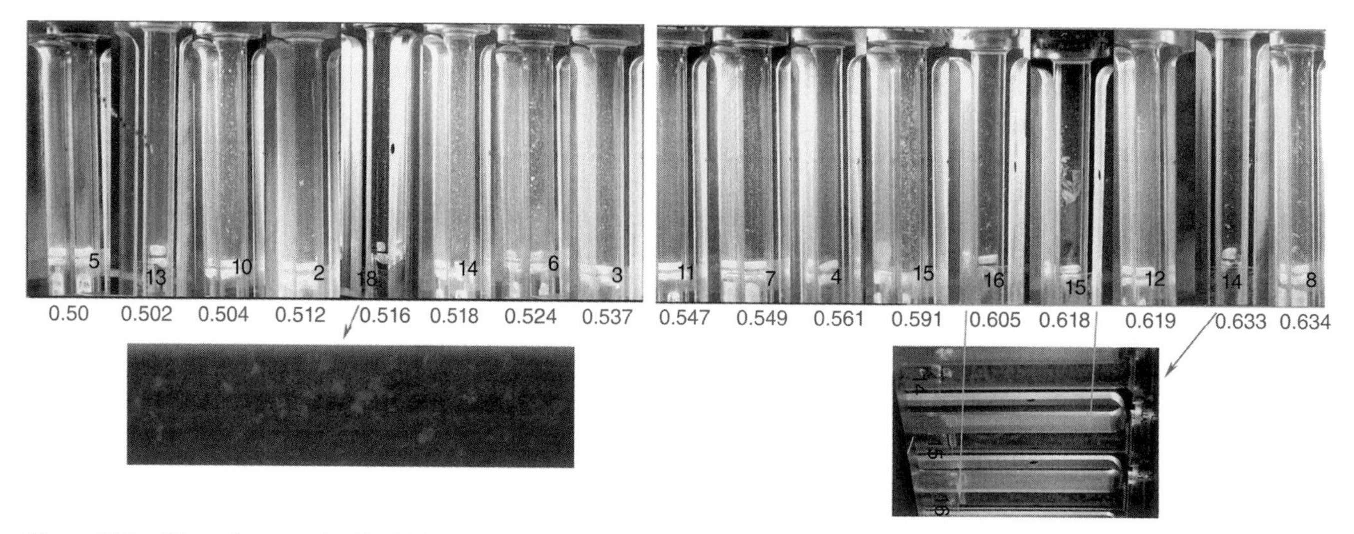

Figure 12.9 *Phase diagram of colloidal hard spheres.* The pictures were taken in microgravity at approximately 4 days after mix-melting. Left inset: Dendrites for CDOT-2 sample $\phi = 0.516$. Right inset: Picture taken 12 h after Space Shuttle Discovery landing for samples with $\phi = 0.605$, 0.618, and 0.633. CDOT-1 samples: $\phi = 0.50$; 0.504; 0.512; 0.518; 0.524; 0.537; 0.547; 0.549; 0.561; 0.591; 0.619; 0.634; CDOT-2 samples: $\phi = 0.502$; 0.516; 0.605; 0.618; 0.633. Crystal morphology was changed at 0.591 when increasing volume fraction ϕ.

region. Such dendritic crystals have not been reported for hard spheres in normal gravity [16]. As a crystallite grows, its mass increases and it sinks faster in normal 1g gravity. Convection from sedimenting crystallites of radius R is faster than diffusion over the distance R when R is larger than several times colloidal sphere size, which might alter the growth mechanism. Also, dendritic arms are easily sheared off. Therefore, gravity masks the crystal growth instability in the liquid–crystal coexistence region.

Interestingly, we do not observe formation of amorphous or glassy phases, even for samples near RCP (Fig. 12.9) [14], in contrast to experimental results on Earth (Fig. 12.5). For samples with volume fractions larger than 0.591, the nucleation rate is slower compared to what has been formerly referred to as the fully crystalline regime, $\phi \in [0.545, \sim 0.58]$. Therefore, the number of crystallites was less and the crystallites were larger. This is the reason for the change in crystallite morphology at $\phi = 0.591$ in Figure 12.9.

Dense Packing of Nonspherical Particles The phase behavior of nonspherical particles is of great interest for the understanding of natural materials such as clay, graphene, sand, and carbon nanotubes and for technological advancement, for example, for the application of liquid crystals of rod-like molecules in displays and clay for rheology enhancement in oil recovery. In this section, we briefly describe the dense packing of nonspherical particles. This section is not intended to be a complete account of the subject but rather an introduction.

Nonsphericity introduces rotational degrees of freedom not present in sphere packings and can dramatically alter the jamming characteristics from those of sphere packings. The densest known packings of all the Platonic and Archimedean solids (Fig. 12.10) and ellipsoids in 3D have packing fractions that exceed the optimal sphere packing value $\phi_{\max}^{\text{sphere}} = \phi_{\text{FCC}}^{\text{sphere}} = \pi/\sqrt{18} \approx 0.7408$, due to the break of the continuous rotational symmetry of the sphere [108–112]. A *Platonic solid* is a convex polyhedron that is regular in the sense of a regular polygon; particularly, the faces of a Platonic solid are congruent regular polygons,

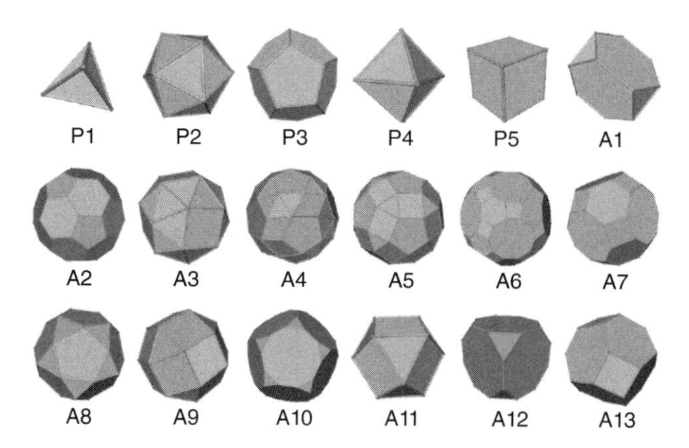

Figure 12.10 *The five Platonic solids and the 13 Archimedean solids.* The five Platonic solids are the tetrahedron (P1), icosahedron (P2), dodecahedron (P3), octahedron (P4), and cube (P5). The 13 Archimedean solids are the truncated tetrahedron (A1), truncated icosahedron (A2), snub cube (A3), snub dodecahedron (A4), rhombicosidodecahedron (A5), truncated icosidodecahedron (A6), truncated cuboctahedron (A7), icosidodecahedron (A8), rhombicuboctahedron (A9), truncated dodecahedron (A10), cuboctahedron (A11), truncated cube (A12), and truncated octahedron (A13). The cube (P5) and truncated octahedron (A13) are the only Platonic and Archimedean solids, respectively, that tile space.

with the same number of faces meeting at each vertex; thus, all its edges are congruent, as are its vertices and angles. An *Archimedean solid* is a highly symmetric, semiregular convex polyhedron composed of two or more types of regular polygons meeting at identical vertices.

Prolate and oblate spheroids (an ellipsoid in which two of the semiaxes are equal) can be obtained from a sphere by a linear stretch and shrinkage of the space along the axis of symmetry. This is an affine transformation that does not change the space occupation fraction by the particles. Stretching or shrinking a packing of spheres leads to an ellipsoid packing that keeps the original packing fraction but with all the particles in the same orientation. Exploiting the rotational freedom so that the ellipsoids are not all required to have the same orientations leads to higher packing fractions.

Figure 12.11 presents the MRJ volume fraction as a function of the ellipsoid aspect ratio [113]. The ellipsoid semiaxes have ratios $a : b : c = 1 : \alpha^{\beta} : \alpha$, where $\alpha > 1$ is the aspect ratio (for general particle shapes, α is the ratio of the radius of the smallest circumscribed sphere to the largest inscribed sphere), and $0 \leq \beta \leq 1$ is the "oblateness" or skewness ($\beta = 0$ corresponds to prolate and $\beta = 1$ to an oblate spheroid). ϕ_{MRJ} increases linearly initially with $\alpha - 1$ from its sphere value, $\phi_{\mathrm{MRJ}} \approx 0.64$, then to values as high as about 0.74 for the self-dual ellipsoids with $\beta = 1/2$. It eventually decreases again for higher aspect ratios. The contact number (or number of touching neighbors) also shows a rapid rise with $\alpha - 1$ and plateaus at values somewhat below isoconstrained (defined as the number of constraints or contacts is equal to the total number of degrees of freedom), $Z \approx 10$ for spheroids, and $Z \approx 12$ for nonspheroids (Fig. 12.11 inset). The existence of a cusp (i.e., nondifferentiable) minimum at the sphere point is at

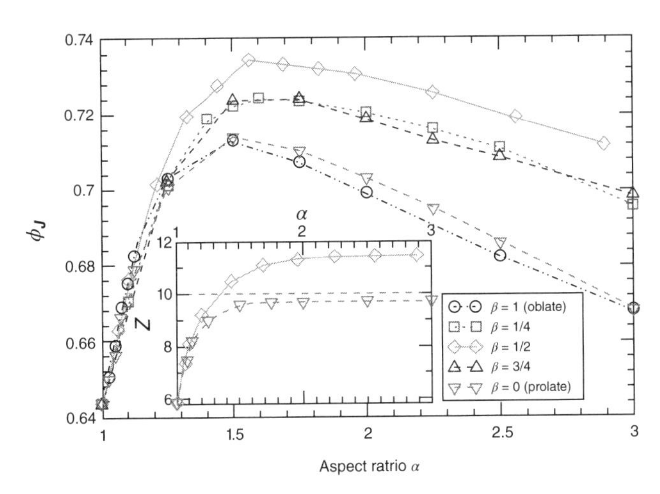

Figure 12.11 Jamming density and average contact number (inset) for simulated packings of $N = 10{,}000$ ellipsoids with ratios between the semiaxes of $1 : \alpha^{\beta} : \alpha$. The isoconstrained contact numbers of 10 and 12 are shown as a reference.

odds with the prevailing expectation in the literature that for "generic" (disordered) jammed frictionless particles, the total number of (independent) constraints equals the total number of degrees of freedom, d_f, implying [77] a mean contact number of $Z = 2d_f$. Recall that $d_f = 2$ for disks, $d_f = 3$ for ellipses, $d_f = 3$ for spheres, $d_f = 5$ for spheroids, and $d_f = 6$ for general ellipsoids [110]. This has been referred to as the *isostatic* conjecture [114] or *isocounting* conjecture [113]. The isostatic conjecture would suggest that an infinitesimal deviation from sphericity ($\alpha = 1$) would require a jump from 6 to 12 contacts, which is not the continuous change shown in the inset of Figure 12.11. The absence of this discontinuity in Figure 12.11 indicates that jammed ellipsoid packings are *hypostatic* $Z < 2d_f$ near the sphere point, and only become nearly isostatic for large aspect ratios. In fact, the isostatic conjecture is only rigorously true for amorphous sphere packings after removal of rattlers (defined as particles trapped in a cage of jammed neighbors but free to move in the cage). It has recently been rigorously shown that if the curvatures of nonspherical particles at their contact points are included in a second-order and higher order analysis, then hypostatic packings of such particles can indeed be jammed [113].

As an example for the phase behavior of nonspherical hard particles, the EOS for hard cubes (P5 of Fig. 12.10) is plotted in Figure 12.12 [115]. We can see a higher crystal close packing than that of hard spheres ($\phi_{\mathrm{fcc}} \approx 0.74$). There is an orientation-related additional phase (cubatic) compared to hard spheres between the fluid and crystal phases, and the fluid branch of the EOS ends a bit earlier than that of hard spheres.

12.2.3 Crystal Structures

Considering the close packing of hard spheres, if all spheres are identical in size, a planar hexagonal arrangement is the only way for the closest packing. Interesting effects arise from the fact that the centers of the spheres are at the sites of the triangular lattice (hexagonal Bravais lattice) while the empty spaces between spheres (interstitial sites) are on a honeycomb lattice, which is a Bravais lattice with a two-point basis. Crystal close packing can be obtained by stacking these hexagonal planes (Fig. 12.13). With a first layer as A, there are two equivalent placements B and C of the second layer above the interstitial sites of the first layer, corresponding to the two nonequivalent positions in the unit cell of the honeycomb lattice. The FCC lattice is the arrangement $ABCABC\ldots$ (ABC repeats), while the hexagonal close-packed (HCP) structure is $ABABAB\ldots$ (AB repeats). The random arrangement $ABACBACBCA\ldots$ with no repeating sequence corresponds to the random hexagonal close-packed (RHCP) structure. Any of these arrangements has the same volume fraction, $\phi = 0.7404$.

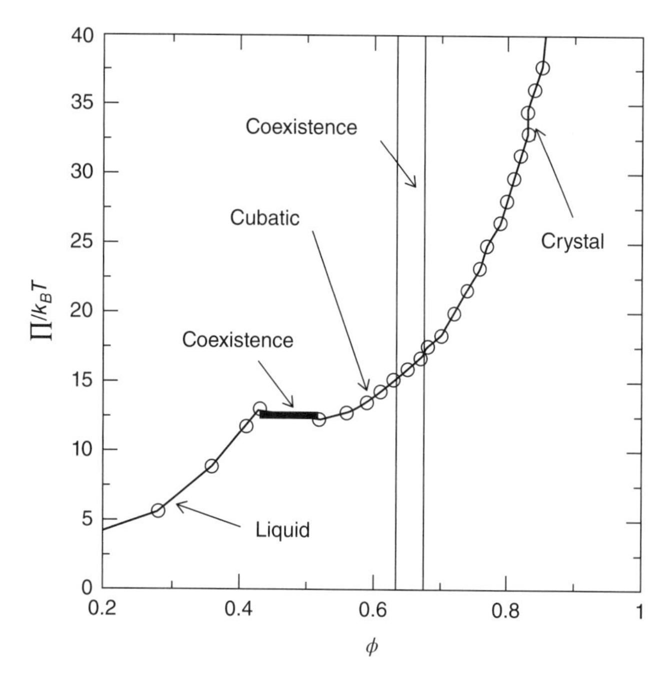

Figure 12.12 *Equation of state for hard cubes.* It is obtained by Monte Carlo for free rotating hard cubes. A cubatic-crystalline transition was suggested between $\phi = 0.634$ and $\phi = 0.674$, but the nature of the transition was not well characterized. The crystal phase is simple cubic structure.

The first numerical studies calculated the pressure from molecular dynamics simulations and obtained entropies by integrating pressure versus volume along reversible paths from some states with known entropy [82, 116–118]. These studies were not able to detect the entropy difference between FCC and HCP crystals. Later, Frenkel and Ladd [119] considered a path connecting the hard-sphere model to Einstein crystals of the same lattice structure, by adding to the model ideal springs tethering each ball to its lattice site. Using this approach, they integrated the derivative of free energy with respect to the spring constant, to obtain an upper and a lower bound for the entropy difference per sphere:$-0.001 < \Delta S^* < 0.002$ in units of k_B, where $\Delta S^* \equiv S_{\mathrm{FCC}} - S_{\mathrm{HCP}}$. Bolhuis *et al.* [120] used a new implementation of the multicanonical Monte Carlo (MCMC) method [121, 122], together with the Einstein crystal method, to accurately resolve an entropy difference of roughly $10^{-3}k_B$ per sphere between FCC and HCP crystals, with the FCC crystal having the higher entropy at the melting point. This quantitatively corrected the pressure-integration study of Woodcock [123], confirming his result that the FCC crystal has higher entropy. Bruce *et al.* found an apparently superior implementation of the multicanonical method for this problem, reducing the statistical error in ΔS^* down to $10^{-5}k_B$ [124].

Experimentally, Pusey *et al.* [125] observed a mixture of FCC and HCP crystallites with a bias toward FCC, whereas under microgravity RHCP structure is detected first [14, 16], with the FCC emerging at a later time in the crystallization process [17, 49]. Highly swollen ionic hydrogel particles were also found to form RHCP structures [126].

Scattering from Structures with Mixed FCC and HCP
Due to the small entropy difference between FCC and HCP, the structure of hard-sphere crystals is consistent with stacking of hexagonal planes of spheres with a stacking probability α for FCC stacking. Figure 12.14 shows the theoretical structure factors. For $\alpha = 0$, $\alpha = 0.5$, and $\alpha = 1$, the structure corresponds to HCP, random hexagonal close packing, and FCC, respectively. The calculations use crystals containing $100 \times 100 \times 100$ spheres randomly oriented; they then correspond to a powder diffraction experiment. Two operations are required: determination of its reciprocal-space

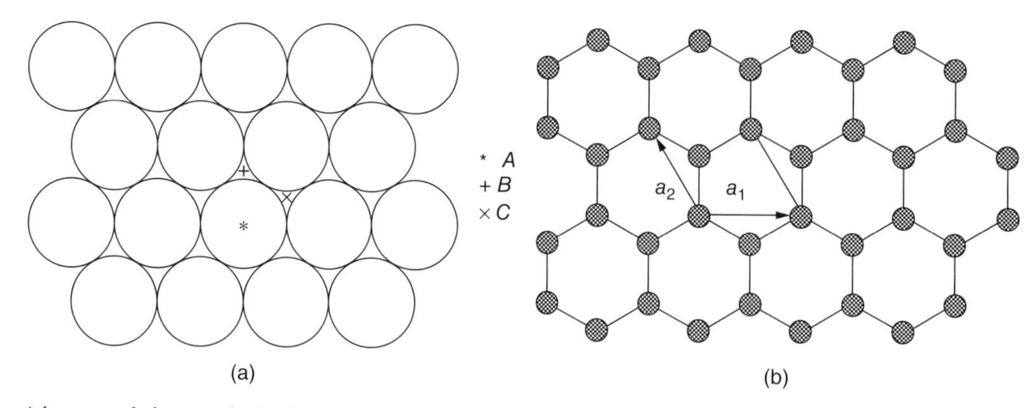

Figure 12.13 *A hexagonal close-packed sphere plane.* (a) The centers of the spheres are at the sites of the triangular lattice. The empty space between spheres lies on a honeycomb lattice. (b) The two-dimensional honeycomb lattice showing the primitive translation vectors and the two-atom unit cell. In (a), the two nonequivalent sites in the lattice are marked with + and ×. "*ABC*" notation: Current spheres' plane position sitting on "*" in (a) is in position class "*A*"; if we stack planes in three dimensions respecting to the "*A*" plane, the planes sitting on top of "+" are in position class "*B*," and the planes sitting on top of "×" is in position class "*C*." Then 3D close-packing structure can be represented by a sequence of letters *A*, *B*, and *C*.

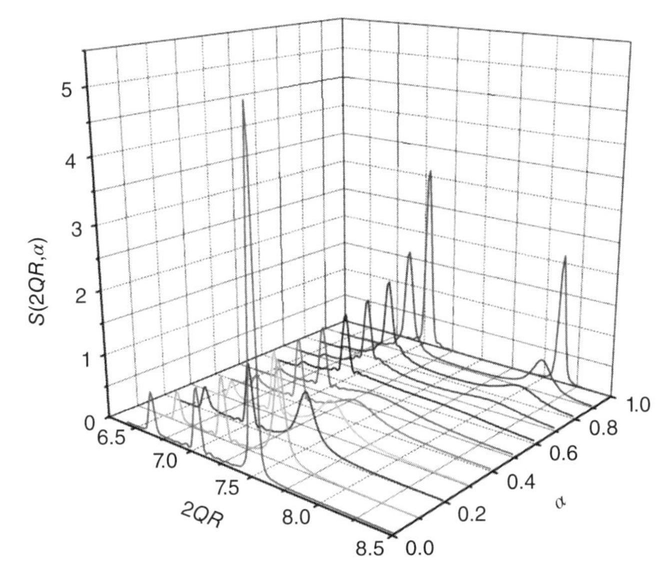

Figure 12.14 *The structure factors of hexagonal close-packed plane stacking structures.* α is the probability having an FCC stacking. $\alpha = 0$, HCP; $\alpha = 0{:}5$, RHCP; $\alpha = 1$; FCC. Q is the scattering vector and R is the radius of the spheres.

structure and construction of the orientational average. For stacking of hexagonally close-packed planes, the former was worked out many years ago using finite-difference equations by Wilson [127] and by Hendricks and Teller [128]. Loose and Ackerson derived a very compact and useful form using a transfer matrix method [129]. We calculated powder patterns using the expressions of Wilson [127] and the orientational-averaging procedure described by Brindley and Mering [130]. The thermal diffusion scattering can be included to account for the Brownian motion of the spheres in the crystallites [131].

To illustrate how to experimentally obtain the crystal structure factor measured by static light scattering, we will again use the microgravity experiments. These were part of "The Physics of Hard Sphere Experiments (PhaSE)" that flew on board the Space Shuttle's Microgravity Science Laboratory (MSL-1) in 1997 (STS-83 and STS-94).

The static structure factor of *metastable* hard-sphere *fluids* can be described quite accurately by the Percus–Yevick result, $S_{\mathrm{PY}}(Q, \phi)$, for hard-sphere fluids. An analytic expression was obtained by Baxter in 1968 in the Percus–Yevick approximation [132]:

$$\begin{aligned}\frac{1}{S_{\mathrm{PY}}(Q,\phi)} &= 1 + \frac{24\phi}{x^3}\{a(\sin x - x\cos x) + b \\ &\quad [(2/x^2 - 1)x\cos x + 2\sin x - 2/x] \\ &\quad + \frac{\phi a}{2}[24/x^3 + 4(1 - 6/x^2)\sin x \\ &\quad - (1 - 12/x^2 + 24/x^4 s\cos x)]\},\end{aligned}$$

where

$$x = 2QR, \quad a = (1 + 2\phi)^2(1 - \phi)^{-4},$$
$$b = -3\phi(\phi + 2)^2/2(1 - \phi)^4.$$

To allow for the known overestimation of the structure by the simple PY theory, the volume fraction used in these calculations should be given by

$$\phi' = \frac{\phi - \phi^2}{16},$$

an empirical correction as suggested by Verlet and Weis [133]. We first determine the product of the particle form factor, $P(Q)$, and the instrument factor (taking care of the scattering and absorption in the instrument), $\alpha(Q)$, in arbitrary units, by dividing the scattered intensity immediately after shear melting, $I(Q; t = 0)$, by $S_{\mathrm{PY}}(Q, \phi')$ at the hard-sphere volume fraction ϕ corresponding to that of the sample:

$$\alpha(Q)P(Q) = \frac{I(Q,0)}{S_{\mathrm{PY}}(Q,\phi')}.$$

The structure factor $S(Q; t)$ of the crystallizing suspension is then obtained from

$$S(Q,t) = \frac{I(Q,t)}{\alpha(Q)P(Q)}.$$

This quantity represents the structure factor of the colloidal crystal plus the colloidal fluid that has not crystallized. To increase the statistics of the data, $I(Q, 0)$ and $I(Q, t)$ are averaged azimuthally (i.e., intensity is averaged over the Bragg ring resulted from the scattering from the many crystallites). To subtract the fluid contribution from $S(Q, t)$, a scale factor $\beta(t)$ is decreased from unity (meaning fully in fluid phase) until

$$S_c(Q,t) = S(Q,t) - \beta(t)S_{\mathrm{PY}}(Q,\phi_L)$$

approaches zero at small and large Q (hence $S_c(Q, t)$ is the structure factor for crystallites only at time t). Here, ϕ_L is the volume fraction of the liquid region in the sample. As time t increases, more and more spheres are converted from fluid to crystal and the volume fraction of the liquid ϕ_L decreases. In the meantime, the structure of the crystallites $S_c(Q, t)$ changes and no longer falls to zero at the largest Q visible in the detector. The crystal structure factor extraction procedure is shown in Figure 12.15 in a specific example. The visible peaks corresponds to the $\{111\}$ peak, and the broadened $\{101\}$ and $\{10\bar{1}\}$ peaks of the FCC structure.

Figure 12.16 shows the growth of the FCC structure for a sample exhibiting phase coexistence regime at $\phi = 0.528$. The FCC $\{200\}$ peak appears at $2qR \sim 8$. The peak emerges after $t = 4642$ s. For a sample with $\phi = 0.552$, the FCC

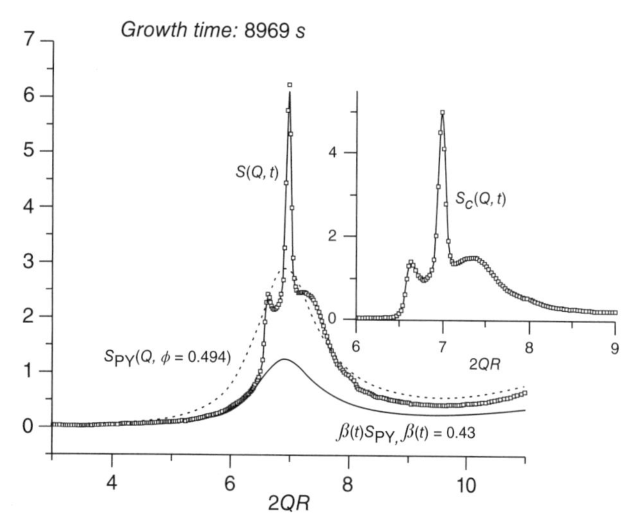

Figure 12.15 *Extraction of the crystal structure factor $S_c(Q;t)$ from the measured total structure of the sample $S(Q;t)$.* Shown is the structure factor of PHaSE sample with $\phi = 0.528$ after growth for 8969 s. The structure of the liquid is modeled by $S_{\text{PY}}(Q;\phi'_L)$. Since the sample is in the coexist regime and the growth time is very large, $\phi_L(t = 8969\,\text{s}) = 0.494$, the freezing volume fraction. Generally, ϕ_L is a function of time. Adjusting $\beta(t)$, $\beta(t)S_{\text{PY}}$ should well describe the small Q part of $S(Q;t)$, which is the scattering from the liquid part of the sample. $S(Q;t) - \beta(t)S_{\text{PY}}$ gives the structure of the crystallites (inset), which should be described by the hexagonal close-packed plane stacking structure with stacking probability α (cf. Fig. 12.14).

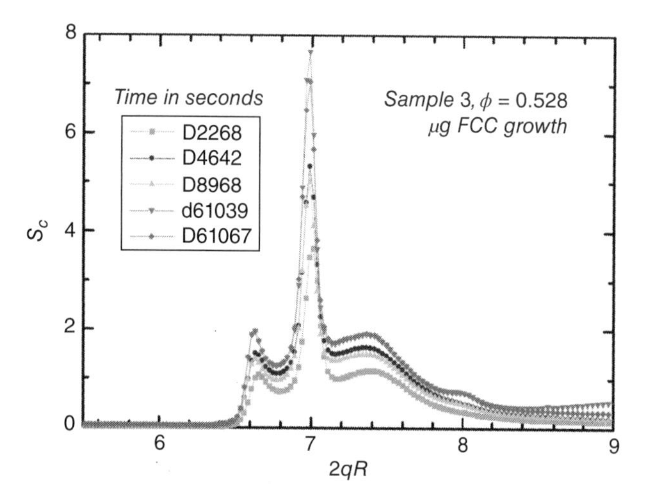

Figure 12.16 The growth of FCC structure of a sample with $\phi = 0.528$ in microgravity.

{200} peak emerges at $t = 3650\,\text{s}$, for a sample near the glass transition, $\phi = 0.575$, FCC peaks appear during the nucleation stage ($t \sim 600\,\text{s}$) and remain for later times. Therefore, we always see the gradual growth of the FCC structure. With increasing volume fraction, formation of the FCC structure appears sooner. We thus conclude that the FCC structure is the equilibrium stable structure for hard-sphere crystals.

12.2.4 Crystallization Kinetics

For the study of crystallization kinetics, the major problems in atomic systems are the high speed of crystal nucleation and growth, as well as the difficulty of preventing heterogeneous nucleation in conditions of large supercooling [134]. These are circumvented when colloids are used as model systems. As a result of the ratio of the dimension of colloidal particles to atomic dimensions (typically a factor of $\sim 10^3$), the timescales of particle diffusion and all phenomena governed by diffusion are scaled up. The relevant timescales in the case of colloids are therefore milliseconds. The use of colloids as model many-body systems allows to probe in real-time processes that are inaccessible in atomic systems. On the other hand, the typical colloidal length scale facilitates the use of comparably simple yet powerful optical methods such as light scattering and microscopy, thus yielding complementary information from reciprocal and real space. They also allow for controlled experiments in restricted geometries. Therefore, in colloidal crystallization not only are the kinetics accessible by scattering with higher temporal resolution than in their atomic counterparts, but application of complementary methods allows the study of the same phenomena on smaller length scales even down to the single-particle level. Further, a comprehensive set of ingredients needed to interpret the data, like the EOS or the diffusion dynamics, became readily available from independent experiments under reproducible conditions. Finally, theoretical calculations and simulations have also supported and advanced the understanding of the phase transition kinetics.

Theoretical Adaption of the Classical Nucleation and Growth Theory of Crystallization Significant progress was made in adapting the classical nucleation and growth theory for molecular phase transitions to colloidal systems [135–137]. Models can predict the nucleation rate and stable nucleus size as a function of concentration. The nucleation rate J has the general form:

$$J = J_0 \exp\left[\frac{-\Delta G^*}{k_B T}\right] = \left(\frac{\beta n D}{l^2}\right) \exp\left[\frac{-\Delta G^*}{k_B T}\right],$$

where the nucleation barrier,

$$\Delta G^* = \frac{16\pi\gamma^3}{3(\Delta\mu n)^2},$$

is connected to the macroscopic surface tension γ, the difference in chemical potential $\Delta\mu = \mu_{\text{solid}} - \mu_{\text{liquid}}$ between solid and liquid per particle and the particle density n. The frequency factor, J_0, is conveniently expressed both in terms of n and the ratio between a diffusion constant D and a typical length scale l (such as the particle radius) squared. β is a proportionality constant assumed to be of order unity. The theory predicts a slower nucleation rate near the freezing transition than observed experimentally [138, 139]. If a suspension is quickly compressed to close packing, the particles have no time to order themselves and the suspension becomes glassy. These glassy samples eventually crystallize by slow growth of large and irregularly shaped crystals on secondary nucleation regions, such as the meniscus, container walls or regions of shear alignment remaining from mixing. Pusey and van Megen [18] examined the nucleation and crystallization of PMMA–PHSA particles in decalin and carbon disulfide. For $0.535 < \phi < 0.585$, nucleation appears to be homogeneous, randomly distributed throughout the bulk. In the sample at $\phi = 0.595$, however, heterogeneous nucleation begins first at the meniscus and container walls and then proceeds into the bulk. Samples at $\phi = 0.62$ and 0.63 also nucleate heterogeneously but start from the meniscus and not the walls of the container. These latter samples had slower growth rates than the sample at $\phi = 0.595$. Studies on charge-stabilized silica exhibit a similar transition from homogeneous to heterogeneous nucleation with concentration [140]. In contrast, in microgravity environment, samples that stayed glassy on ground for 1 year crystallize quite quickly and very large crystallites were produced [14, 16].

The simplest law describing the velocity v of a planar interface between a metastable liquid and a crystal is due to Wilson–Frenkel [141], which states

$$v = v_\infty \left(1 - \exp\left[\frac{-\Delta\mu}{k_B T}\right]\right),$$

where v_∞ is the limiting growth velocity, estimated by $v_\infty = Dd/l^2$ via the self-diffusion, through an interface of thickness d, of a particle to their target places, or the diffusional registering of adjacent layers of particle planes in the suspension near the crystallites [142].

For hard spheres, growth is considerably complicated by transient *compression* of crystals and the formation of depletion zones, because that the crystal phase has a higher density than the liquid phase in osmotic equilibrium. Ackerson and Schatzel [136] developed a numerical model for the growth of an isolated spherical crystal. Growth exponents were observed to first undergo long transients before approaching 1/2 and 1, the long-time asymptotic limits expected for diffusion (i.e., rate of spheres diffuse to the interface) and surface-limited (i.e., rate of the spheres to be integrated into the corrected lattice positions) growth, respectively. Consequences of this model were instructively illustrated by Russel for crystal growth instability analysis presenting results from microgravity experiments [16].

Crystallization Kinetics via Bragg Scattering Bragg scattering monitors the intensity of laser light scattered off one or more set of lattice planes through a Debye–Scherrer-like experiment, thus probing the crystal structure. Time evolution at several scattering angles can be monitored simultaneously through the use of fiber optics or photodiode arrays.

Harland *et al.* [143, 144] integrated the Bragg peak of the structure factor to calculate the fraction of sample that had converted from fluid to crystal. The time-dependent crystal fraction, crystallite size, and nucleation rate indicated a change in crystallization mechanism around the melting transition. Below melting, the observed crystallization follows classical theory where nuclei form throughout the sample and grow. Above melting, however, crystal growth appears to be suppressed by very high nucleation rates. Moreover, the nucleation at very high concentrations accelerates with time rather than remaining constant. Pusey and van Megen [18] also found that nuclei number density increases with particle concentration, and the subsequent growth led to progressively smaller crystals. But in the coexistence region, the crystallite size appears to be independent of concentration.

Figure 12.17 displays the status of a sample at $\phi = 0.528$ at different nucleation and growth stages in normal and microgravity as captured by a color CCD camera. To analyze the crystallization kinetics, data from Bragg scattering, such as that shown in Figure 12.16 is used. A Gaussian function and a constant background (taking care of the broad scattering from the RHCP structure, cf., the $\alpha = 0.5$ line in Fig. 12.14) fit the {1 1 1} peak and locate the maximum $Q_m(t)$ and the width $\Delta Q(t)$ at half-maximum. The fraction $X(t)$ of the sample converted from fluid to crystal is calculated from

$$X(t) = c\int_{q_1}^{q_2} S_c(Q,t)dQ,$$

where c is a constant (which will adjust X to the correct percentage of sample that occupied by crystallites) and the constant background (from the RHCP scattering) was excluded.

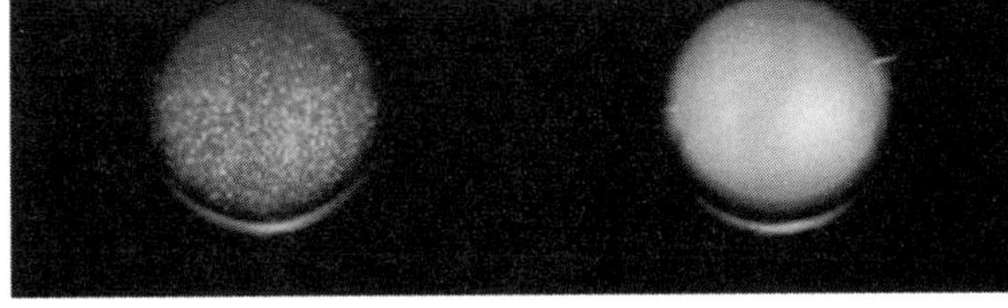

Figure 12.17 Crystallization of a PHaSE sample ($\phi = 0.528$) in 1g and μg.

The fits are largely insensitive to the cutoffs q_1 and q_2, which defines the $\{1\,1\,1\}$ scattering peak region. The average linear dimension of the crystal, in units of the particle diameter $2R$, is given by

$$L(t) = \frac{\pi\kappa}{\Delta Q(t)R},$$

where $\kappa = 1.155$ is the Scherrer constant for a crystal of cubic shape. The number density of crystallites with the average size is then estimated as

$$N_c(t) = \frac{X(t)}{L^3(t)}.$$

Finally, it is easily shown that the volume fraction $\phi_c(t)$ of an FCC crystal is related to the location of $Q_m(t)$ of the $\{1\,1\,1\}$ reflection as follows:

$$\phi_c = \frac{2}{9\pi^2\sqrt{3}}[Q_m(t)R]^3.$$

Crystallization Kinetics for Samples in the Crystalline Regime Figure 12.18 presents the change in the average crystallite size and the number of crystallites for a sample in microgravity and on Earth with $\phi = 0.552$ [17]. The crystallization on Earth stopped after $\sim$ 1000 s as a result of the sedimentation of the crystals. We observed enhanced coarsening and growth in microgravity. The crystallites are smaller but more numerous on Earth than in microgravity, suggesting an enhancement of nucleation in ground experiments. We observed larger exponents for the increase in the average crystallite size and a decrease in crystallite number in microgravity. Therefore, gravity alters the nature of crystallization. Sedimentation limits the growth of individual crystallites and affects the long-range interaction between the crystallites that is transported through diffusion in the surrounding liquid, hence affecting the coarsening and growth processes.

Let us now focus on the physics revealed by microgravity data. For $t < 300$ s, we observed $N_c \propto t$, which is the nucleation stage. At this stage, we thus assume

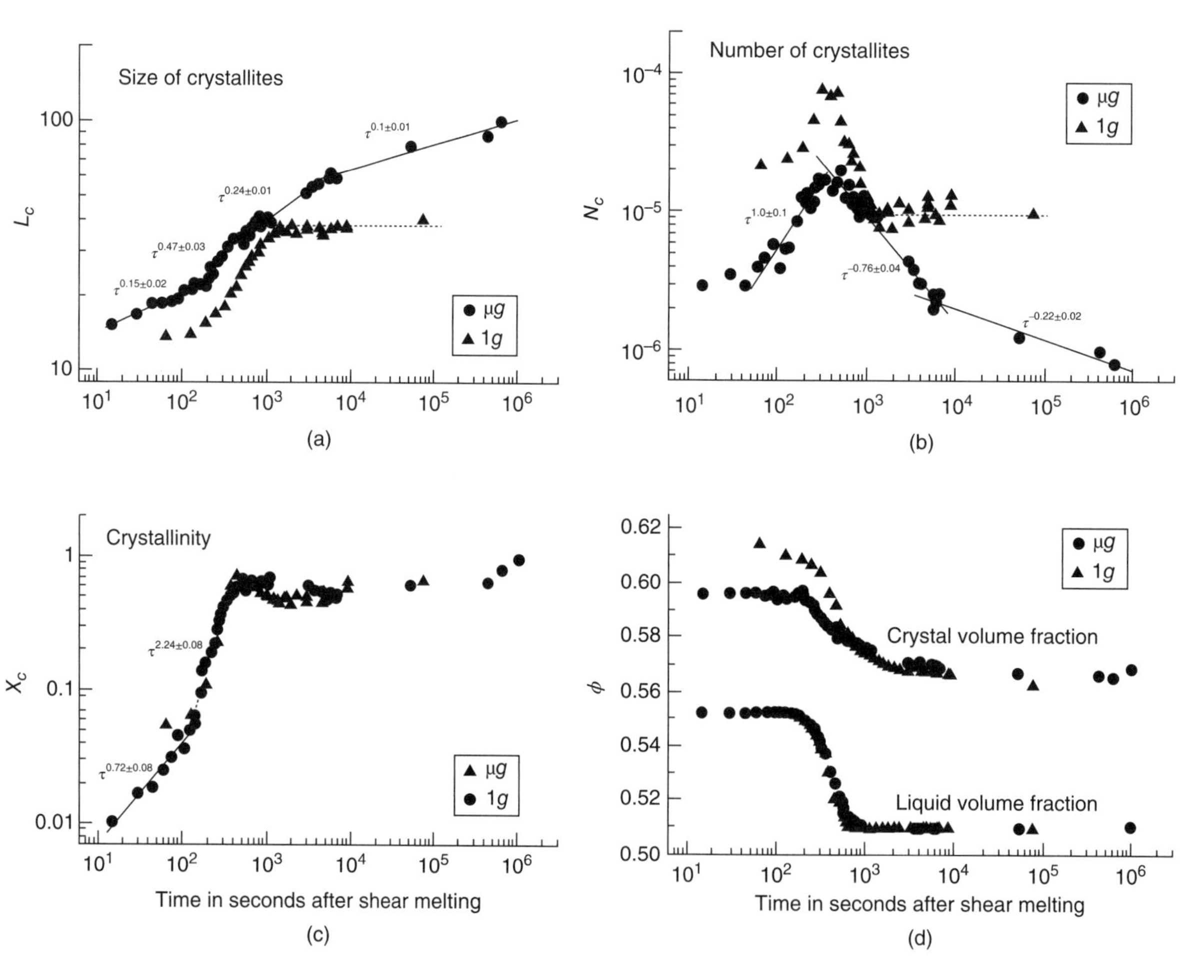

Figure 12.18 The crystallization kinetics of hard spheres at $\phi = 0.552$. L and N_c are measured in $2R$ and $(2R)^{-3}$, respectively.

that $N_c = (dN_c/dt)t$, where dN_c/dt is a constant and $dL_c^3/dt = L_c L_{\text{nucleus}}^2/\tau$, where $\tau = L_{\text{nucleus}}^2/D_0$ is a characteristic diffusion time with D_0 being the Stokes–Einstein diffusion coefficient. In addition,

$$X_c(t) = \int_0^t \frac{dN_c}{dt'} L_{\text{nucleus}}^3 d' + \int_0^t \frac{dN_c}{dt'}(t') [L_c^3(t-t') - L_{\text{nucleus}}^3] d'$$

$$= \int_0^t \frac{dN_c}{dt'}(t') \left[L_{\text{nucleus}}^2 + \frac{2}{3}\frac{L_{\text{nucleus}}^2}{\tau}(t-t') \right]^{3/2} dt'$$

$$= \frac{3}{5} L_{\text{nucleus}}^3 \tau \frac{dN_c}{dt'} \left[\left(1 + \frac{2}{3}\frac{t}{\tau}\right)^{5/2} - 1 \right],$$

$$L_c(t) = \frac{\int_0^t \frac{dN_c}{dt'}(t') \left[L_{\text{nucleus}}^2 + \frac{2}{3}\frac{L_{\text{nucleus}}^2}{\tau}(t-t') \right]^{1/2} dt'}{\frac{dN_c}{dt'} t}$$

$$= \frac{L_{\text{nucleus}} \tau \left[\left(1 + \frac{2}{3}\frac{t}{\tau}\right)^{3/2} - 1 \right]}{t}.$$

In the short-time limit, $t/\tau \ll 1$,

$$X_c(t) = L_{\text{nucleus}}^3 \frac{dN_c}{dt'} t,$$

$$L_c(t) = L_{\text{nucleus}} \left(1 + \frac{4}{9}\frac{t}{\tau}\right).$$

This short-time $X_c(t)$ behavior is in good agreement with the experimental result, $X_c(t) \sim t^{0.72}$, and L_c is almost independent of t, $L_c(t) \sim t^{0.15}$. The long-time limits, $t/\tau \gg 1$,

$$X_c(t) = \left(\frac{2}{3}\right)^{1/2} \frac{4}{45} L_{\text{nucleus}}^3 \frac{dN_c}{dt'} \frac{t^{5/2}}{\tau^{3/2}},$$

$$L_c(t) = \left(\frac{2}{3}\right)^{2/3} \left(\frac{t}{\tau}\right)^{1/2} L_{\text{nucleus}},$$

which also agree with the experimental data: $X_c(t) \sim t^{2.24}$ and $L_c(t) \sim t^{0.47}$. Figure 12.18 shows that the growth exponent of $L_c(t)$ changes from 0.5 (nucleation and growth stage) to 0.25 (crossover stage) and to 0.1 (coarsening stage) at later times. We attribute this to interactions between crystallites, which are present at high concentration. The crossover exponent 0.25 is predicted by liquid–liquid phase-separation theory [145] and is considered to be a result of drop–drop interactions at high concentration. In simulations on phase separation that now combine steady-state homogeneous nucleation theory with the classical Lifschitz–Slyozov mechanism for Ostwald ripening, modified to include the spatial correlation among droplets and the interdroplet diffusional interactions [146], a diffusive stage (the crossover stage) generally observed between nucleation and Ostwald ripening.

In conclusion, the general picture outlined by classical nucleation and growth theory [135] is correct. However, *the high density of crystallites* could affect the crystallization process quantitatively and qualitatively. The data on the whole lifetime of hard-sphere metastable states provide a solid foundation for completing the theory of crystallization kinetics in particular and phase transformations in general.

Crystallization Kinetics for Samples in the Fluid–Crystal Coexistence Region Figure 12.19 presents the change in the average crystallite size and the number of crystallites for a sample in the fluid–crystal coexistence region with $\phi = 0.528$. Let us discuss here two important observations for this sample [49]. We can conclude that a picture with single crystallites growing in a metastable liquid significantly misses much of the kinetics that we observe. A complete picture must involve the interaction between the growing crystallites.

Interaction Between Two Crystallites: Growth and Coarsening One of the salient features, which has not been emphasized in previous experiments, is the existence of strong interactions between individual crystallites. In Figure 12.19, from $t = 150$ s to about 800 s, the size of the crystallites increases as $L_c(t) \sim t^{\delta}, \delta = 0.99 \pm 0.05$, while X over this interval increases at a rate considerably smaller than L_c^3. Thus, the number of crystallites decreases roughly as $N_c(t) \sim t^{-\gamma}, \gamma = 0.99 \pm 0.05$. This is unusual and in contrast with the classical theory, which focuses on the nucleation and growth of *isolated* crystallites. We identify this process as "simultaneous coarsening and growth" in which small crystallites shrink and eventually disappear, causing the number of crystallites to decrease, while large crystallites keep growing, causing the measured average crystallite size to increase with time. The growth exponent for the size is about unity, resulting from the combination of normal diffusion-limited growth and coarsening, apparently due to direct transport from small to large crystallites [146]. Because of the different curvatures of these crystallites, the smaller crystallites are under higher pressure due to surface tension and hence have a high internal volume fraction. Therefore, the corresponding liquid volume fraction surrounding a small crystallite is relatively higher than that surrounding a large crystallite. Mass transport, via density gradients in the intervening fluid dispersion, moves particles from the small crystallites to the large crystallites.

Growth Instability: Dendrites The most distinctive feature of the sample at phase coexistence is that, at a later stage ($t > 800$ s), N_c increases after rapidly decreasing from 100 to

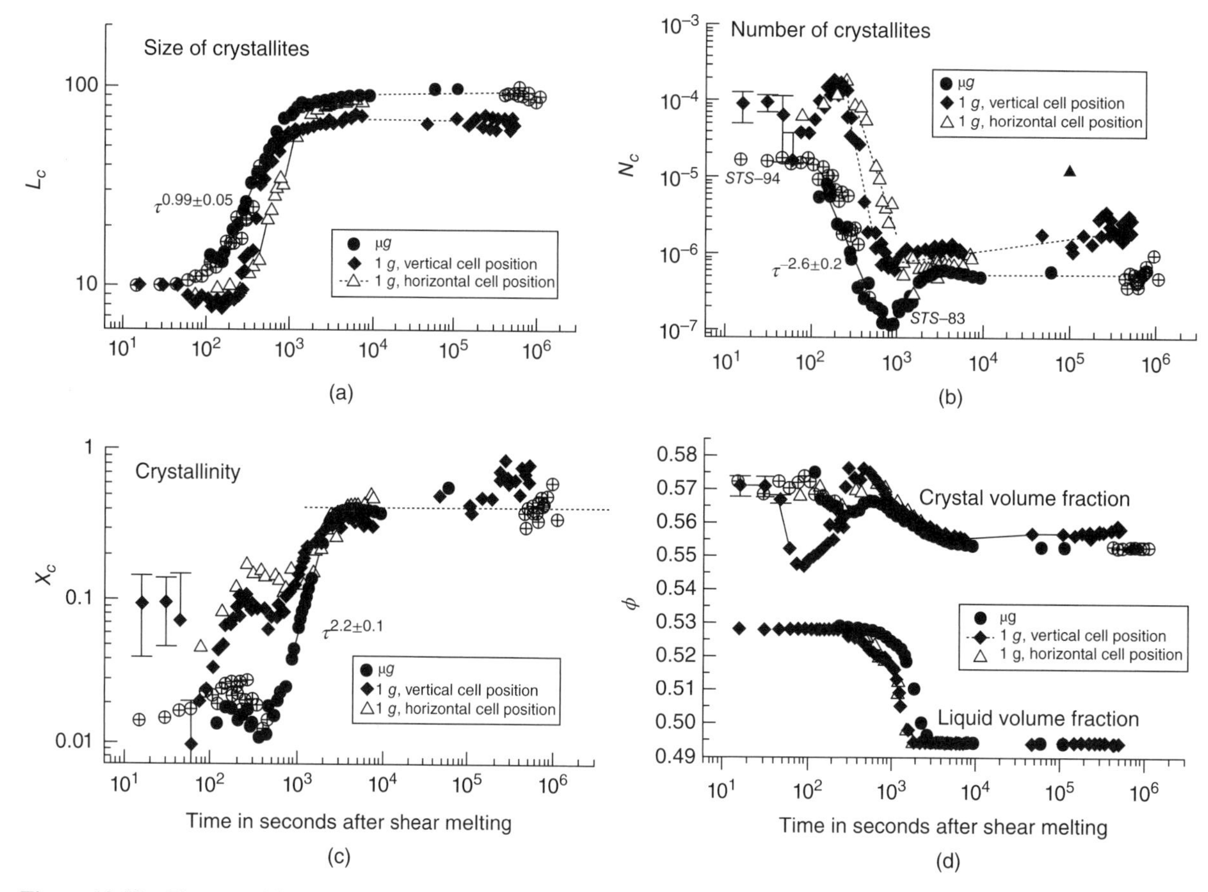

Figure 12.19 The crystallization kinetics of hard spheres at $\phi = 0.528$. L and N_c are measured in $2R$ and $(2R)^{-3}$, respectively.

800 s as a consequence of the constancy of L and an increasing X. In comparison, this phenomenon is totally absent in gravity and cannot be interpreted as a nucleation process. We identify it as an instability in the diffusion-limited growth: dendritic crystal growth. Dendrites, the primary features of metal alloys, are of great technological importance and scientific interest. In colloidal systems, dendrites were first reported for charged colloids near a surface [147]; the CDOT experiment was the first observed hard-sphere dendrites [14, 16]. Figure 12.9 shows that dendrites grown in all CDOT coexistence samples. The dendrites grown at $\phi = 0.52$ in another experiment on the Mir Space Station remained intact with negligible or, at least, incomplete annealing after 2.5 months. In normal gravity, the fragile dendritic arms are prevented from growing by sedimentation due to either viscous stresses on the crystallite or fluid flow altering the diffusion field. Experiments on Earth (without density matched samples) seldom observe the growth of dendrites.

A linear stability analysis [16] of hard-sphere crystal growth indicates a dendritic instability at $\sim 13r_c$, with r_c the critical nucleus, for samples in microgravity at $\phi = 0.504$, similar to the instability at $\sim 7r_c$ predicted for molecular systems [148]. Figure 12.19 suggests the *critical* nucleus (nucleus that will continuously grow) to be about 10 times the diameter of the sphere, that is, $r_c = 10 \times 2R$. At $t \approx 800$ s, the average crystallite size is about 70 times the diameter of the sphere, that is, $r_{\text{crit}} = 70 \times 2R$, being the diameter at which the dendritic instability takes place. Therefore, the growth of this instability takes place at about $r_{\text{crit}} = 7r_c$, which is in reasonable agreement with the above-mentioned theoretical and experimental results.

We can also estimate other properties of the dendrites. Figure 12.19 shows that the apparent number density of the crystallites increases from $10^{-7}(2R)^{-3}$ to $6 \times 10^{-7}(2R)^{-3}$ from 800 to 3000 s. During this interval, L changes only slightly while X changes from 0.02 to 0.4, that is, a 20-fold increase. In a crude model, we imagine that once the size surpasses r_{crit} further growth is predominantly from the dendritic arms. We then have

$$X \propto N_a N_c \rho^2 L_a \propto t^{2.2},$$

with the number of arms N_a, the arm radius ρ, and the arm length L_a. In this interpretation, the width of the azimuthally averaged Bragg peak, corresponding to $1/L$, is $\approx 1/\rho$. Since ρ grows as $\rho \sim L \sim t^{1/12}$, the maximum dendrite growth rate, assuming *marginal stability* (the margin stability hypothesis is, when a local gradient exceeds the critical value set by a stability criterion, the fluctuation-driven flux increases rapidly, thus driven the gradient back to marginality), is $dL_a/dt \propto 1/\rho$, and then $L_a \propto t^{11/12}$. This can also be obtained from mass conservation for the growing arm [16]: $(\phi_c - \phi_i)\frac{d}{dt}\rho^2 L_a = D\rho^2 \frac{(\phi_f - \phi_i)}{\rho}$, where ϕ_c, ϕ_f, ϕ_i are the volume fractions in the crystal, liquid, and at the interface, respectively, and D is the diffusion coefficient. Combined with the equation above for X, we have $N_a \propto t^{1.1}$. Then, from $t_0 = 800\,\text{s}$ to $t_1 = 3000\,\text{s}$, $X \approx (\rho_0^3 + N_a \rho^2 L_a)/\rho_0^3$ increases by a factor 20, which gives a number of arms of $N_a \sim 6$ at long times.

Crystallization Kinetics for Samples Near the "Glass Transition" Are Dominated by Nucleation Close to the glass transition, the dynamics become very slow. Hence crystallization is slow. Figure 12.20 reveals that the increase in X is companied (from $t = 1000$ to 3000 s) by a stage of *accelerated nucleation* [144], where $N_c \sim t^{1.4}$. When a colloidal fluid is quenched into the crystal region, nucleation results in a reduction in the concentration and a concomitant increase in the particle diffusivity of the remaining fluid. If the concentration of the quenched fluid exceeds that at which the nucleation has a maximum, then the increase in diffusivity following any nucleation process would be greater than the increase of the nucleation barrier, and as a consequence, nucleation is *accelerated*.

From the {1 1 1} peak position, we can determine the volume fraction of the crystal phase. The volume fraction of the liquid phase can also be determined from the data reduction procedure for determining the structure factor of the sample. The volume fraction of the liquid phase in samples slightly above freezing and within phase coexistence decreases monotonically with time. For a sample with $\phi = 0.575$, the change in volume fraction of the liquid phase is not needed to give a reasonable interpretation of the changing total structure of the sample $S(Q; t)$, probably due to the slow growth of the crystal.

For $\phi = 0.528$, during the nucleation stage ($t < 100\,\text{s}$), the volume fraction of the crystal is almost constant, $\phi_c = 0.572$, but significantly larger than that of the surrounding liquid phase, that is, nuclei are compressed by the metastable liquid. In the next coarsening stage ($100 < t < 800$), while the volume fraction of the surrounding liquid still does not change much, the volume fraction of the crystal phase decreases from 0.572 to about 0.562 as the crystallites

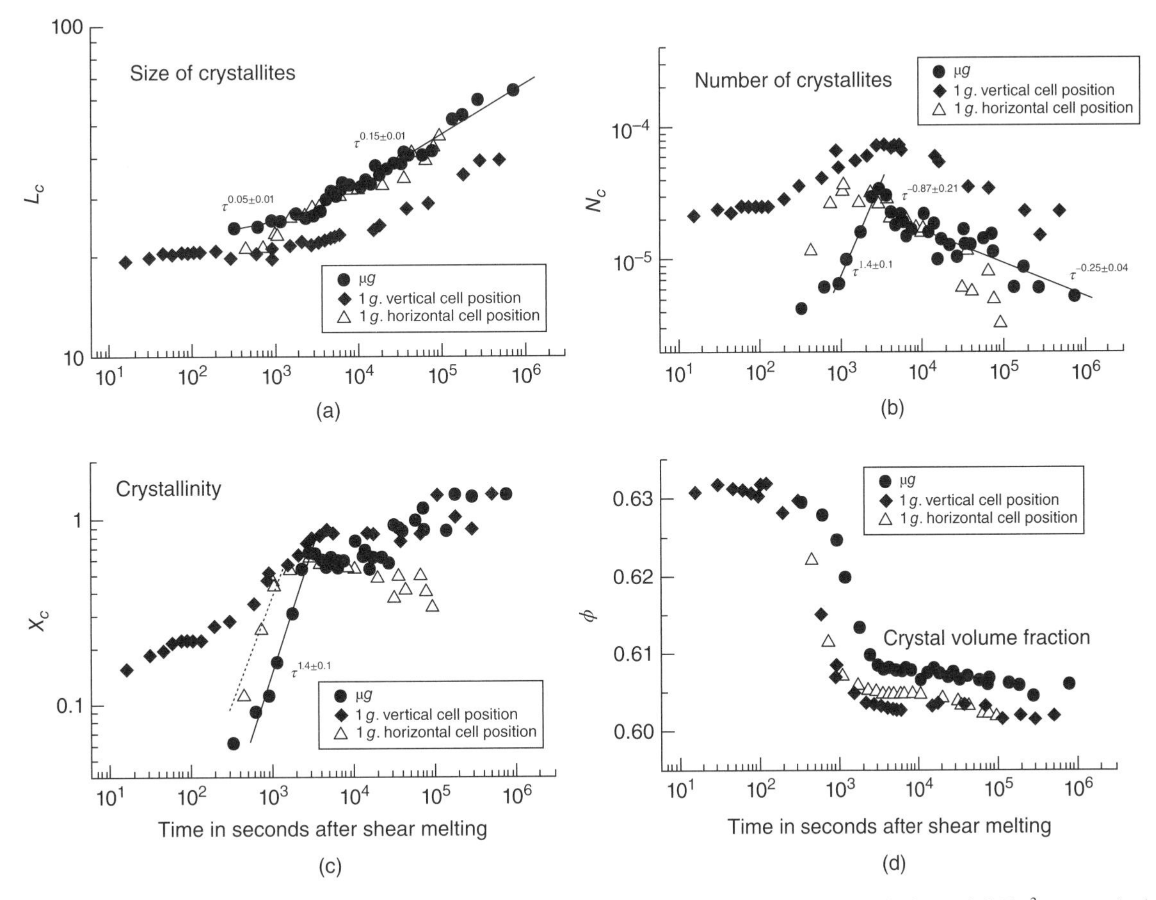

Figure 12.20 The crystallization kinetics of hard sphere at $\phi = 0.575$. L and N_c are measured in $2R$ and $(2R)^{-3}$, respectively.

change in size. In the initial time of the dendritic growth, the average volume fraction increases a little bit to about 0.567, which can be attributed to the small size of the arms that have begun to grow. This reflects *the Gibbs–Thompson effect*: the surface tension leads to larger pressure exerted on small crystals compared to larger ones in the same metastable fluid. Gibbs noted that the surface tension on a small crystal in equilibrium will lead to a lower melting temperature than for a bulk crystal [149]. This effect is included in many theories of crystal growth [136, 148]. During the dendritic growth of the crystal, as the spheres convert quickly from the liquid to the crystal phase, the volume fraction of the crystal decreases very fast because less and less pressure is exerted on the crystal by the liquid phase as the liquid volume fraction decreases. The final volume fraction of the liquid is 0.494 and the final volume fraction for the crystal phase is 0.552, which is the same value measured previously [143] and consistent with hard-sphere expectations.

For $\phi = 0.552$, the nuclei have a higher volume fraction, $\phi_c = 0.597$. During the nucleation stage, the volume fraction of the nuclei decreases just slightly, as the average crystal size does not change much. Much of the decrease in the crystal volume fraction is associated with the decrease in the surrounding liquid volume fraction due to the conversion of the liquid phase to the crystal phase. At the end of the experiment, the crystallization process was not quite finished, leaving 0.563 for the crystal and 0.51 for the not crystallized liquid.

For $\phi = 0.575$, nucleation dominates the crystallization process. We still observed the decrease in the volume fraction of the nuclei, from 0.662 to 0.605. This can be interpreted as the nuclei being in equilibrium with the depletion zone instead of the bulk liquid phase at some distance away.

We can identify $L_c \sim t^{0.15}$ and $N_c \sim t^{-0.25}$ for the coarsening stage of the sample at $\phi = 0.575$. Possibly, also an intermediate stage with $N_c \sim t^{-0.87}$ can be identified. No theoretical accounts have been put forward for the observed growth exponents.

Effect of Gravity on Crystallization Kinetics Besides the absence of dendritic growth in normal gravity and the absence of glass phase in microgravity, as already discussed, we summarize here other observations that can be drawn from Figures 12.18–12.20.

Coarsening First of all, let us note that experiments in normal gravity measure an effective average crystallite size and an effective nucleation rate since crystallites settle out of the beam. The effect of gravity on the coarsening process was clearly demonstrated on sample at $\phi = 0.552$ (Fig. 12.18), where much of the features of the coarsening have shown up in microgravity. On ground, the "coarsening and growth I" was enhanced. The decrease in the number density at this stage has a larger magnitude. The reason is that there are a larger number of crystallites on ground. However, although there is a "coarsening and growth II" in microgravity, the crystallization on ground actually stops (Fig. 12.18 L_c panel). There is no decrease in number density and the average crystal size does not change after about 1000 s. This is due to the *sedimentation* of the crystals.

We also observed enhanced coarsening and growth in gravity for the sample at coexistence $\phi = 0.529$ (Fig. 12.19). We observed a larger magnitude of the growth exponent for the average crystallite size and a larger magnitude of the decrease exponent for the crystallite number. Of the observed dendritic growth in the sample at $\phi = 0.528$ in microgravity, we could only see a little trace on the ground. So *sedimentation eliminated the growth instability*. The depletion zone between the growing crystals and the bulk liquid may disappear on the ground due to sedimentation. There is no liquid between crystallites at later times ($t > 800$ s), so *dendritic growth is eliminated*. Therefore, gravity affects the "nature" of crystallization.

For the sample at $\phi = 0.575$ (Fig. 12.20), the effect was hard to observe because the "coarsening and growth II" observed at $\phi = 0.552$ (Fig. 12.18) was absent due to the dominance of nucleation.

Overall Crystallization: Gravity Does Not Affect the Final Crystallinity and Overall Crystallization Speed of the Samples An interesting phenomena can be noticed from the X versus time plots. Overall, gravity does not affect the conversion from liquid to crystal. The $X(t)$ curves for all three samples can be normalized by a single constant $C_{x(t),\mathrm{Norm}} \sim 0.75 \pm 0.2$. The crystallinity of the sample at $\phi = 0.552$ at $t = 10^6$ s reaches unity. This same normalization constant sets the final crystallinity of the sample at $\phi = 0.528$ to ~ 0.6, which is in good agreement with the calculation from the phase diagram: $X(\mathrm{Final}) = \frac{0.528-0.494}{0.545-0.494} = 0.67$. The normalization also gives a reasonable crystallinity for the sample at $\phi = 0.575$: $X(t = 3000\,\mathrm{s}) \approx 1$.

A characteristic time T_ζ can be defined as the time when the conversion from liquid to crystal has virtually finished. The inverse of T_ζ characterizes the crystallization speed. $T_\zeta(\phi = 0.528) \sim 3000$ s; $T_\zeta(\phi = 0.552) \sim 400$ s; $T_\zeta(\phi = 0.575) \sim 3000$ s. We see that the crystallization speed is maximized around the melting point $\phi_m = 0.552$. At "lower supercooling," the thermodynamic driving force is weaker; at "higher supercooling," the transport of spheres is slowed down by the dynamics. Our experiments show that T_ζ is independent of the strength of gravity. Since that the thermodynamic driving force either produces new nuclei or grows larger crystallites, ground experiments produce more crystallites than microgravity experiments. Therefore, *gravity limits growth and enhances nucleation*; the combination of these two effects makes $X(\phi = 0.552)$ independent of gravity strength. For $\phi = 0.575$, crystallization is dominated by nucleation, so T_ζ is not affected by gravity. For $\phi = 0.528$, crystallization is driven by the presence of the nucleation barrier. Sedimentation breaks up initial "cluster" structures in the "supercooled" liquid state. Hence, crystallization on ground lags behind with respect to microgravity. Crystallization speeds up after that and finally catches up at the same T_ζ observed in microgravity.

Nucleation First of all, the nuclei are compressed to a higher volume fraction on the ground due to the sedimentation, as shown in the $\phi(t)$ curve of the sample at $\phi = 0.552$ (Fig. 12.18). The primary reason is that the *size of the nuclei* (which is, to a large extent, equal to the early time crystallites) *on ground is smaller than in microgravity*, cf. the $L_c(t)$ plot at early times. This is because sedimentation intervenes the formation process of a cluster in the nucleation stage, preventing the growth of large clusters. Since the crystallites have initially a smaller size, ground growth lags behind microgravity growth in all the samples.

A very interesting phenomenon is that nucleation depends on the orientation of the sample cell on the ground experiments. We conducted experiments with samples at $\phi = 0.528$ and $\phi = 0.575$, with the axis of the cell being either vertical, where the crystallites at the bottom are seen by the Bragg beam, or horizontal, where the crystallites at the bottom are not seen by the Bragg beam. The difference in the number of crystallites for these two orientations can be observed clearly from the measurement at $\phi = 0.575$. For the vertical orientation, the Bragg beam saw more crystallites and more crystallites were produced at early times. This means more crystallites have been nucleated at the bottom.

The additional information from the experiments for $\phi = 0.528$ is that (i) the initial clusters might be destroyed due to sedimentation (N_c decreases and X_c decreases); (ii) the crystallites at the bottom (seen in the vertical orientation experiment) were compressed by sedimentation: ϕ_c increases during 100–200 s; (iii) as time increases, more and more small crystallites were produced at the bottom (L_c decreases, N_c increases, $N_c(t)$(vertical) > $N_c(t)$(horizontal); (iv) there were more crystallites in the vertical orientation. Hence, the coarsening and growth was faster in that orientation (N_c decreases more rapidly, so that L_c increases).

These results may be interpreted as the effect of the volume fraction gradient built up in gravity and nucleation taking place at the bottom of this gradient [150].

Comparison of Hard-Sphere Experiments on Crystallization Kinetics Figure 12.21 compares the results from previous experiments [136, 143] with microgravity results [17, 49]. There is a reasonable agreement between these experiments for samples at coexistence with $\phi \sim 0.52$. Harland and van Megen [143] essential measured the long-time kinetics, while small-angle scattering experiments by Ackerson and Schatzel [136] measured the short-time kinetics. The PHaSE instrument, having a better resolution and higher statistics, measured the whole sample history, revealing the coarsening of the crystallites at short times ($\tau < 10^{3.2}$) and the growth instabilities right after ($10^{3.2} < \tau < 10^4$).

There is large disagreement on the crystalline samples with $\phi \sim 0.553$. Crystallites were larger and crystallization was faster in microgravity than in normal gravity. Sedimentation limits crystal growth, hence apparently enhances nucleation (more crystallites are produced in normal gravity than in microgravity). Gravity compresses the crystallites as shown by the $\phi_c(t)$ curves of Figure 12.21. Gravity also produces a nonuniform volume fraction distribution. The $L(t)$, $N_c(t)$ and $X_c(t)$ curves in normal gravity are similar to the corresponding curves in microgravity for $\phi = 0.575$.

Confocal Imaging: Catch the Critical Nucleus Direct imaging in three dimensions allowed the identification and observation of both nucleation and growth of crystalline regions. By following their evolution (Fig. 12.22), the critical nuclei can be identified and nucleation rates can be measured [52]. The structure of the nuclei was found to be a random hexagonal close-packed structure, and their average shape was rather nonspherical, with rough rather than faceted surfaces. It can be seen from Figure 12.22 that the crystallites are very close to each other, so strong interactions among them are possible.

Figure 12.23 illustrates the ability to locate each particle (around 1 μm in diameter) and separate the fluid phase and the crystal phase. The heterogeneous growth of the hard-sphere crystals in the coexistence region are demonstrated in Figure 12.24. Notice that the crystallites are not very ordered at the wall.

How Well Do We Understand Nucleation? Even after centuries of investigation of nucleation, the experimental measurement of the nucleation rate of hard spheres remains many orders of magnitude larger than the theoretical predictions [151]. Figure 12.33 summarizes our current understanding of the nucleation rates in hard spheres.

The dimensionless nucleation rate for hard spheres are defined as

$$I = \frac{\text{Nucleation rate}}{\text{Crytallization volume}} \times (2R_h)^5/D_0,$$

where R_h is the hydrodynamic radius of the spheres, D_o the self-diffusion coefficient at infinite dilution, which can be calculated via Stokes–Einstein Equation $D_0 = (k_B T)/(6\pi\eta R_h)$. Hard-sphere nucleation rates are compared here. Computer simulation [151] agrees with the measurements on PMMA–PHSA spheres in microgravity [152] and gravity [143] at $\phi = 0.56$; but for most of the volume fractions, discrepancies of many orders of magnitude

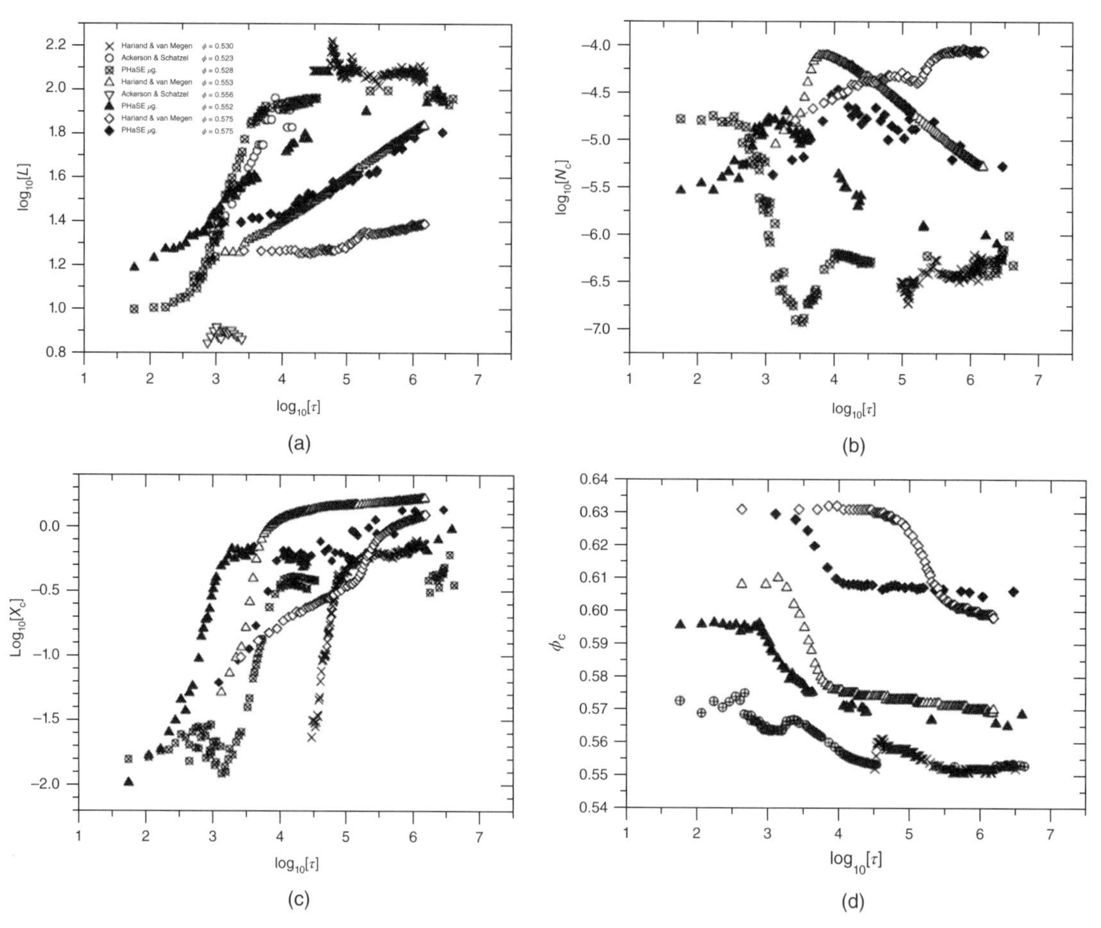

Figure 12.21 A comparison of three experiments: Harland and van Megen, small-angle scattering by Schatzel and Ackerson, and PHaSE μg experiment. The elapse time τ is in units of R^2/D_0. L and N_c are measured in $2R$ and $(2R)^{-3}$, respectively.

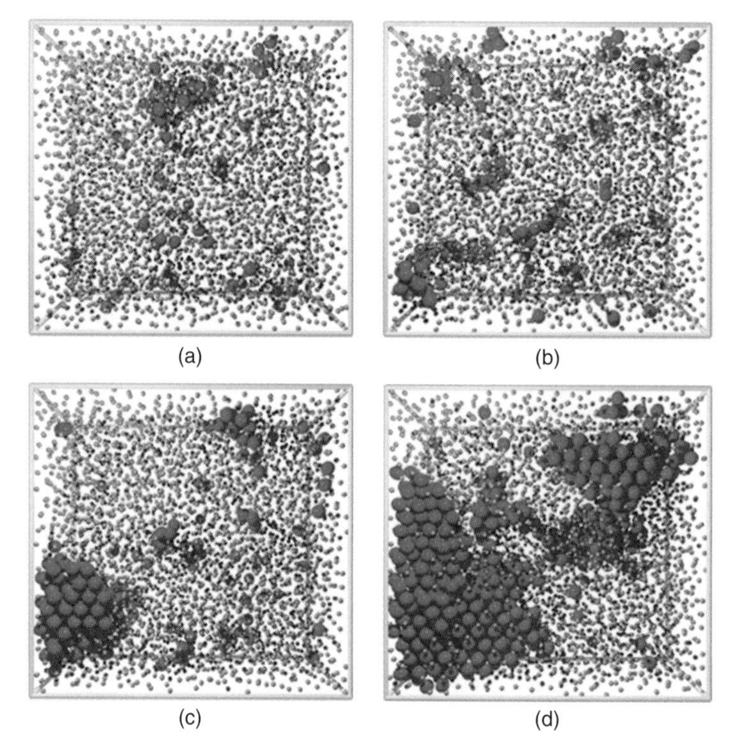

Figure 12.22 The confocal imaging of the nucleation process.

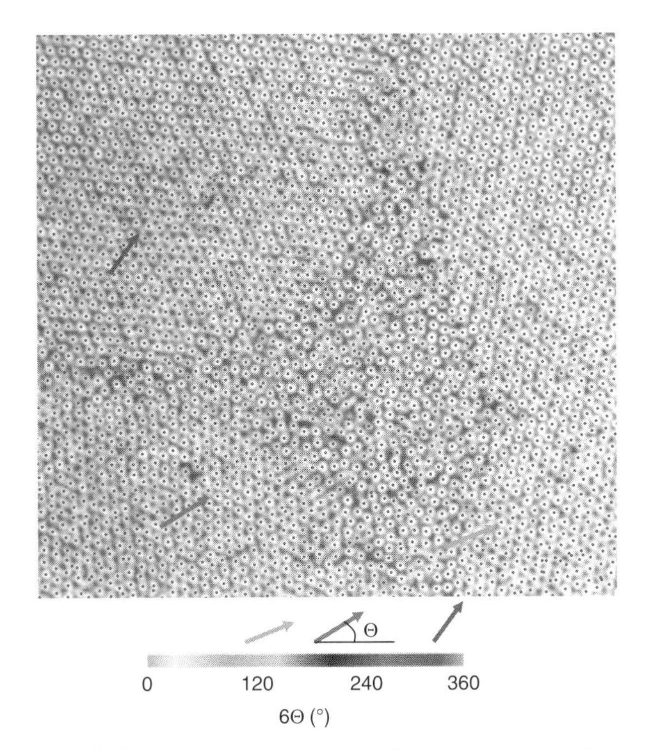

Figure 12.23 *Confocal imaging of colloidal phases.* After locating the positions of each colloid (the dots), the crystallites are identified by local correlation and angularly coded in their orientation.

are observed. The accuracy of both experiments and simulations are expected to improve. Advanced theories are expected to be established to interpret experimental and simulation results.

The crystal nucleation rate depends on the probability that a critical nucleus spontaneously forms from a liquid, P_{crit}, spontaneously, and on a kinetic factor κ that describes the rate at which the critical nuclei subsequently grow. P_{crit} can be written in terms of the free energy required for nucleus formation, ΔG_{crit}, the temperature as $P_{\text{crit}} = \exp(-\Delta G_{\text{crit}}/k_B T)$. The nucleation rates $I = \kappa \exp(-\Delta G_{\text{crit}}/k_B T) = \kappa \exp[-(16\pi/3)\gamma^3/(n|\Delta\mu|)^2]$, where n is the number density of particles in the crystal, γ is liquid–crystal surface tension, and $\Delta\mu$ is the chemical potential difference of the crystal and liquid. Classical nucleation theory (CNT) uses κ, and frequently also γ, as fitting parameters to interpret the experiments. Recent computer simulations on hard spheres [151], predicted absolute nucleation rates with no adjustable parameters or assumptions. Compared to these simulation results, CNT-fitted experimental data yield a ΔG_{crit} that is 30–50% too low; therefore, P_{crit} is about five orders of magnitude smaller if we calculate using a barrier height of about $40k_BT$, which is the barrier height at $\phi = 0.52$ as obtained in the simulations [151]. In addition, simulation [151] and experiments [52] (Fig. 12.22) invalidate the spherical morphology assumption in CNT for the nucleating clusters. Therefore, a theory other than simple CNT needs to be developed. Such a theory needs to account for the interaction among nuclei and for the nonspherical morphology of the nuclei [153, 154].

On the experimental side, the accuracy of the measurements of the nucleation rate might not be good enough. For atoms, nucleation takes place on the order of picoseconds,

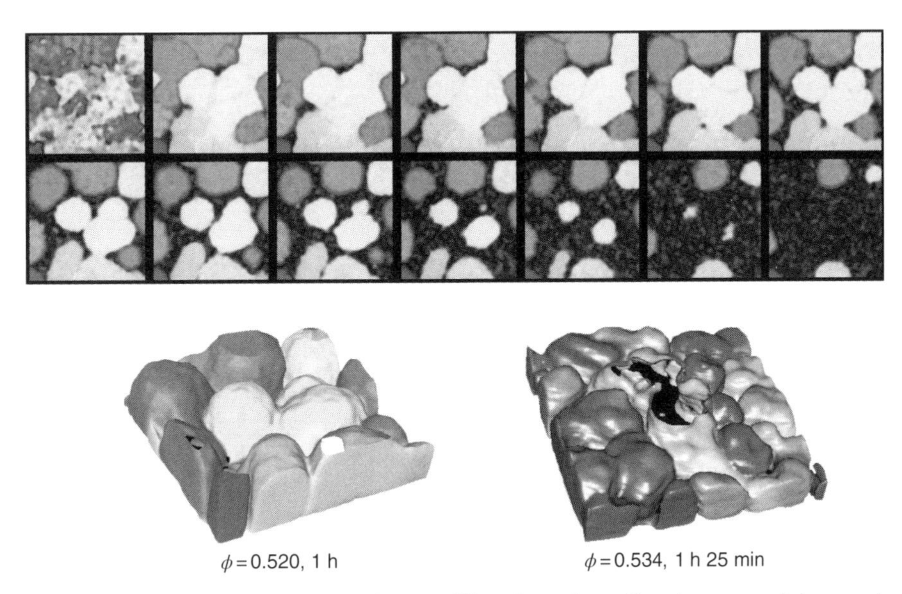

Figure 12.24 *Heterogeneous growth of hard-sphere crystals.* Top: Slices from the wall to the center of the sample (order of reading, left to right, top to bottom). Bottom: Three-dimensional representations of the crystallites only. Color codes the orientation of the crystal (cf Fig. 12.23).

which is *too fast* for scientists to investigate the details with the current experimental techniques. Colloidal systems with particles of 200–600 nm crystallize in tens of seconds to minutes, allowing the critical nucleus to be directly visualized by confocal microscopy [52] and the crystallization kinetics of hard spheres to be investigated using Bragg light scattering [49, 143]. However, CCD cameras take minutes to capture an individual Bragg scattering image of the colloidal crystals, which might *still not be fast enough*. Because of this last problem, UV–Vis spectroscopy has been used to capture the turbidity peak [155] (corresponding to Bragg scattering) with time resolutions of 0.7 s. The interparticle potential of the poly-*N*-isopropylacrylamide (PNIPAM) particle suspensions at pH = 2.8 was confirmed to be hard-sphere-like by measuring the relative width of the coexistence region; it was found $(\phi_{\text{melting}} - \phi_{\text{freezing}})/\phi_{\text{freezing}} = 0.07$, consistent with 5% polydisperse hard spheres [86]. The nucleation rates obtained are compared in Figure 12.25 showing better agreement with simulations than the results for PMMA spheres for full crystalline samples with $\phi > 0.545$.

Therefore, the focus of the debate is shifting to the coexistence regime, where the nucleation rate data are very limited and simulation indicates a very low nucleation rate (about 1 nucleus/month/mm^3 for samples consisting of 200 nm hard spheres) [151, 156, 157] and is hence a challenging task for experimental measurement.

Measure the Nucleation Rate Using Microfluidic Droplets Crystallization of molecular liquid can easily be induced by heterogeneous nucleation on the surface of dust particles; hence, homogeneous nucleation cannot be measured since they are much slower than heterogeneous nucleation. In the 1950s, Turnbull isolated dust particles to a small fraction of the emulsions, so that a homogeneous nucleation rate has been measured [134]. This *emulsion crystallization* method has been attempted in the case of colloids [158–160]. However, polydispersity in emulsion droplet size can lead to the variation in nucleation volume of individual nucleation event (which takes place in the volume of the emulsion droplet), and hence can affect the nucleation rate measurement dramatically [161]. A microfluidic-based droplet system will drastically reduce the droplet polydispersity and, therefore, improve the accuracy of the nucleation rate measurement. Crystallization and self-assembly in such a microfluidic-based droplet system have been widely studied for photonic crystal applications [162, 163], high-throughput screening and diagnostics [164], and protein crystallization [165–167]. To test the feasibility of applying the emulsion crystallization approach to colloidal systems, the thermal-responsive PNIPAM microgel particles were used, the volume concentration of which can be readily tuned by simply raising (lowing) temperature to shrink (swell) the particle gel network.

PDMS or glass capillary microfluidic devices were used to make uniform droplets with two different sizes: 500 and 100 μm in diameter. The multiple crystallites in the large drops indicate correlated nucleation[144]: once a nucleus is produced, it reduces the concentration of colloids in its vicinity and this depletion enhances diffusion locally. Therefore, additional nuclei tend to be produced in the neighborhood (Fig. 12.26a). Similar phenomena could happen in the bulk sample (Fig. 12.26b). So, smaller drops are required to avoid multinuclei production when measuring the nucleation rate. Figure 12.27 illustrates the nucleation rate measurement with 100 μm size droplets. The fraction of drops containing crystals was plotted against time, so the slope determines the steady-state nucleation rate. Figure 12.27d compares the normalized nucleation rates of emulsion crystallization with bulk experiments. One can conclude that the bulk experiments [155] *overestimated* the normalized nucleation rates by four orders of magnitude due to strong correlations between nuclei.

With a better control of the interparticle potential of the PNIPAM spheres [86], emulsion crystallization will contribute to a profound understanding of colloidal nucleation and growth, especially with the ability to locate individual spheres (Fig. 12.28, cf. Fig. 12.23). Homogeneous and heterogeneous nucleation could be investigated. Emulsion crystallization is very promising as a method to achieve accurate nucleation rate measurements by eliminating internuclei interactions. Furthermore, it can be used to measure log(I) down to −19 (Fig. 12.27, cf. Fig. 12.25).

Single Crystal Growth in a Temperature Gradient Large hard-sphere crystals are of special interest to the understanding of the ground state of the hard-sphere crystalline state (HCP vs FCC) and to the investigation of lattice dynamics [168] in addition to their technological interests. Counterintuitively, temperature gradients can control nucleation and growth of hard-sphere crystals, even though the phase diagram of hard spheres is independent of temperature [51].

The effect of a temperature gradient on a hard-sphere suspension is shown schematically in Figure 12.29. Each curve describes the osmotic pressure versus volume fraction at a uniform temperature, with $T_1 > T_2$. At each uniform temperature, the coexistence region is $0.494 < \phi < 0.545$. A sample with an initial volume fraction in this region will have small crystals with $\phi = 0.545$ randomly distributed over the whole sample after the nucleation and growth processes have finished if sedimentation is not very important. The crystallization takes place spontaneously, allowing no control over the position of the nuclei, growth rate, and size of the final crystals. The situation is different for a

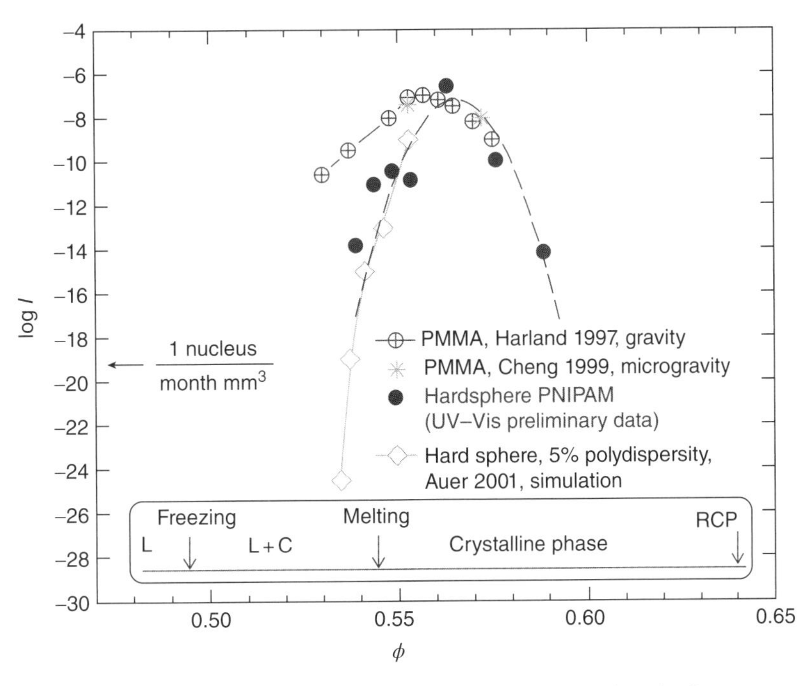

Figure 12.25 Comparison of normalized hard-sphere crystal nucleation rates.

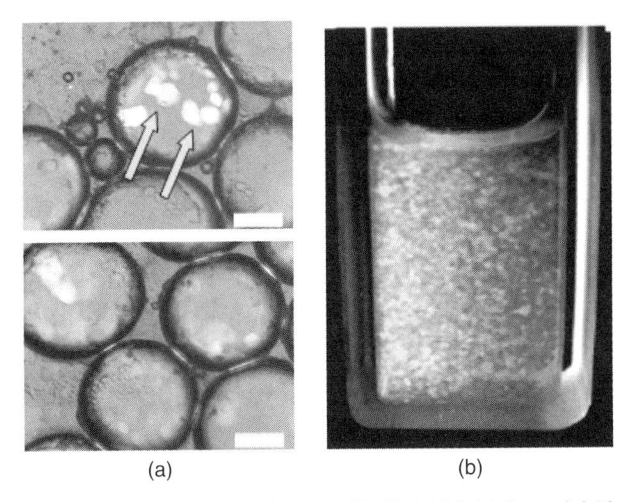

Figure 12.26 *Crystals in drops and bulk.* (a) Multiple nuclei (the color flecks such as pointed out by the arrows) exist in large droplets ($T = 23.6\,°C$, crossed polarizer images, scale bars: 200 μm). (b) Similar phenomena might exist in the bulk sample (22 °C, the diameter of the vial is 2 cm).

sample under a temperature gradient. Figure 12.29 shows that the high-temperature fluid with volume fraction slightly below 0.494 will coexist in mechanical equilibrium with the low-temperature crystal with volume fraction 0.545. Suppose we have a sample with ϕ just below the freezing volume fraction 0.494. The application of a temperature gradient will drive particles from the high-temperature region to the low-temperature region with the help of convection [51], and nucleation and growth of crystals will take place there as long as the local particle concentration becomes larger than the volume fraction for freezing. The speed of the nucleation and growth process depends on the local concentration driven by particle transport under the temperature gradient. Technically, by adjusting the strength of the temperature gradient, one may generate only a single nucleus in the low-temperature region by imposing a nucleation speed as low as possible and allow this nucleus to grow into a large single crystal (Fig. 12.30) by moving slowly the temperature gradient to slowly advance the crystallization interface. If the growth is slow enough, the single crystal may have the equilibrium thermodynamic structure.

12.3 CRYSTALLIZATION OF CHARGED SPHERES

Charged colloidal particles in liquid electrolytes have vast industrial applications [169]. They can also be used as a model system for metals and alloys [170]. It is important to understand the stability and phase behavior as a function of colloid concentration and ionic strength. Charged colloids exhibit a variety of order–disorder and structural transitions [33, 37, 171–173].

12.3.1 Phase Behavior

At *low* ionic strengths, the long-range electrostatic repulsion between charged spheres induces a phase transition at

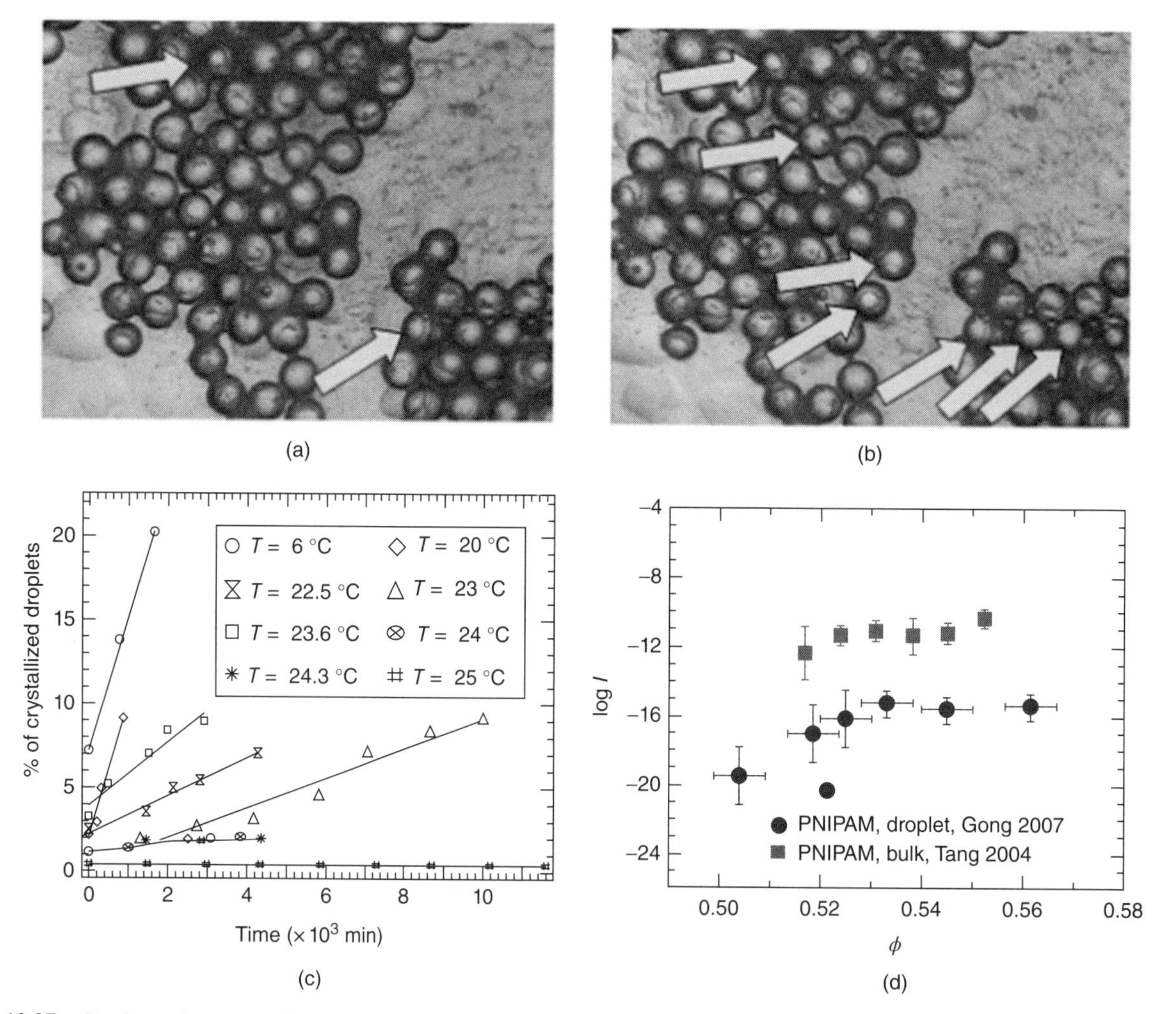

Figure 12.27 *Single-nucleus crystallization events can be observed in small droplets.* (Drop diameter 100 μm, 23.6 °C, colloids are charged PNIPAM spheres at pH = 7.4.) (a) and (b) Two images taken at the same location using a programmable scanning stage at 8 h interval. Arrows point at the crystallized droplets. (c) The slopes in the crystallized fraction of the droplets versus time plot give the nucleation rates. (d) Dimensionless nucleation rate for charged spheres. Results from emulsion crystallization (circles) are four orders of magnitude smaller than the bulk measurements.

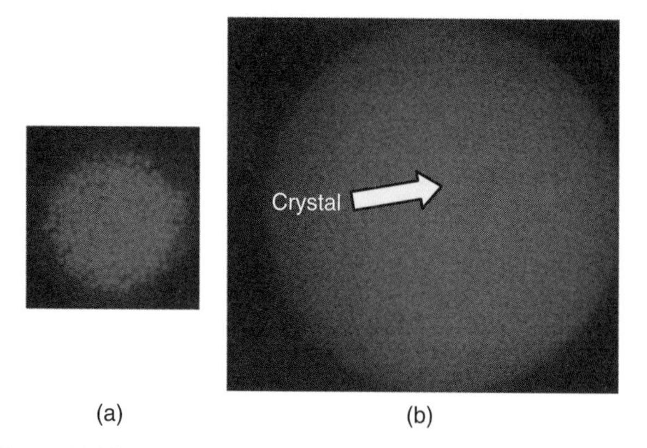

Figure 12.28 *Confocal images of 700 nm PNIPAM spheres in droplets.* Nucleation is homogeneous inside a large droplet (b); surface effect is important in small droplet (a).

volume fraction as low as 10^{-3} with characteristics similar to that of hard spheres [10, 174, 175]: (i) The order phase can diffract visible light via Bragg scattering showing brilliant opalescence. (ii) The disordered phase can be obtained from the ordered phase by dilution or addition of excess electrolyte. (iii) The width of the fluid-coexistence regime is narrow. Such a disorder–order transition can be explained using an *effective hard-sphere model* [5, 10, 176, 177]. At relatively high ionic strength, the Debye length κ^{-1} is much shorter than the interparticle distance.

$$\kappa^{-1} = \frac{1}{\left[\frac{z_c z_p n_p}{\varepsilon_r \varepsilon_0 k_B T} + \sum_{k=1}^{N} \frac{\left(ez^k\right)^2 n_b^k}{\varepsilon_r \varepsilon_0 k_B T}\right]} \approx \left(\frac{\left(ez^k\right)^2 n_b^k}{\varepsilon_r \varepsilon_0 k_B T}\right)^{-1/2}$$
$$= (8\pi B n_s)^{-1/2},$$

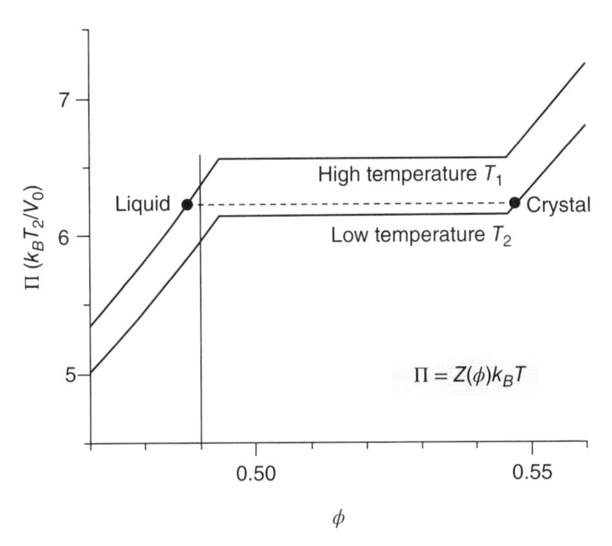

Figure 12.29 *The performance of a temperature gradient on a hard-sphere suspension.* Shown here are the calculations of equation of state using Carnahan–Starling equation and Hall ($\phi_{max} = 0.74$) equation for $T_2 = 300$ K and $T_1 = 320$ K. V_0 is the volume of the sphere. Under each uniform temperature field, the fluid and crystal coexistence region is $0.494 < \phi < 0.545$. While under a temperature gradient, high-temperature liquid coexists with low-temperature crystal in a steady-state fashion. The temperature gradient can control the strength of "supercooling" of metastable state, the site of nucleation and the speed of crystallization. It is in principle a powerful tool to engineer large hard-sphere crystals and to investigate the equilibrium hard-sphere crystal structure.

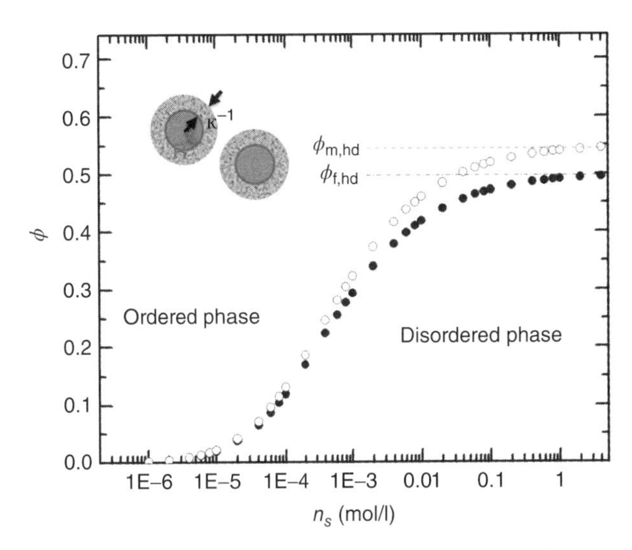

Figure 12.31 Order–disorder phase diagram of charged spheres as predicted by effective hard-sphere model (suspension of $R = 50$ nm particles in water, $B = 0.7$ nm). The phase transition depends on ionic strength of the solvent and volume fraction of the particles.

where we can ignore the small the ion de-association of the particle, described by the first term, and that the valence of the positive and negative ions is $z = 1$. n_s is the concentration of ions. In such case, each particle is surrounded by its own double layer and behaves as it was electrically neutral unless it approaches another particle so that their electric double layers interact (Fig. 12.31). Introducing an effective hard-sphere radius $R^* = R + \kappa^{-1}$ and an associated effective volume fraction, $\phi^* = n \times (4\pi/3)R^{*3} = \phi(1 + \kappa^{-1}/R)^3$, the crystallization and melting phase transitions take place at $\phi_f^* = 0.494$ and $\phi_m^* = 0.545$, consistent with the hard-sphere model [5, 10, 176]. The phase diagram produced by Hachisu *et al.* [10] for aqueous polystyrene (PS) suspensions is the first most complete investigation on the order–disorder transition that was available before the PMMA hard spheres were studied.

The obvious weakness of the effective hard-sphere model is that it oversimplifies the interparticle interactions and excludes the possibility of crystallization into other than close-packed structures. Charged colloids do crystallize into body-centered-cubic (BCC) crystals at low n_s for highly charged colloids. Figure 12.32 presents

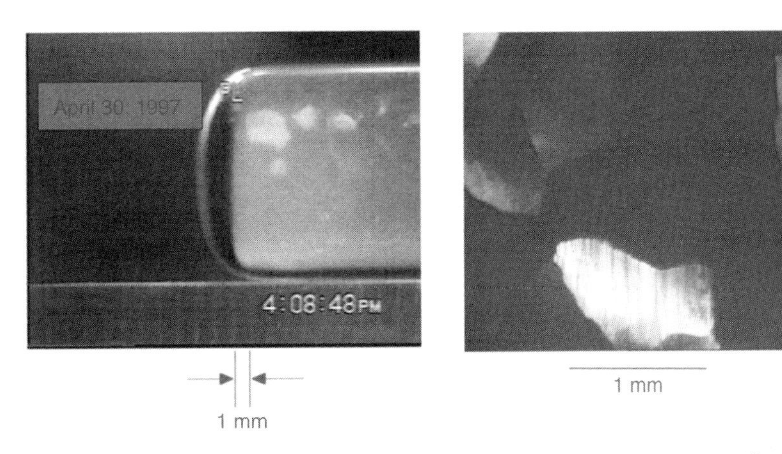

Figure 12.30 Hard-sphere crystals grown under a temperature gradient.

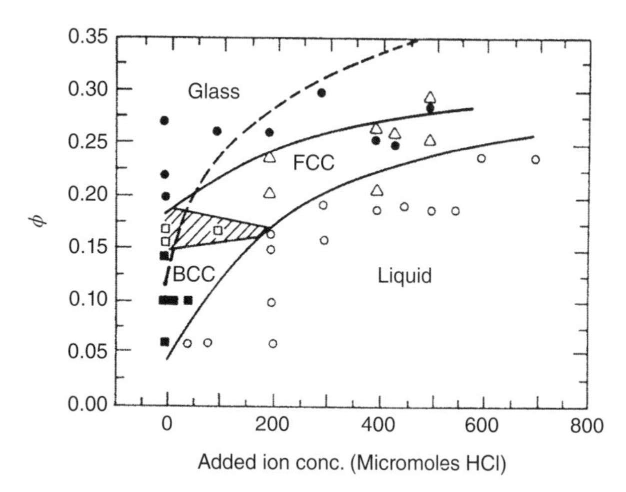

Figure 12.32 Experimental phase diagram of charged polystyrene colloidal suspension: ■, bcc; △, fcc; ○, liquid, and ●, glass. The open squares (□) in the hatched area mark the coexistence of bcc and fcc crystals.

the experimental phase diagram of PS spheres, diameter 91 nm, suspended in 90% methanol–10% water solution [172]. The particles have a surface charge of ~500*e*. The observed FCC–BCC–fluid triple point is located at an acid concentration of $n_{s,\mathrm{HCl}} = 200\,\mu\mathrm{M}$ and particle volume fraction of $\phi = 0.17$. Monovoukas and Gast obtained an FCC–BCC–fluid triple point for $n_{s,\,\mathrm{HCl}} = 2.65\,\mu\mathrm{M}$ and $\phi = 0.083$ for aqueous suspensions of PS spheres of 66.7 ± 0.5 nm in diameter and a surface charge of $1200 \pm 40e$ per particle [37].

The phase diagram for spheres interacting with a hardcore repulsive Yukawa potential has been determined using computer simulations (Fig. 12.33) [33]. For sufficiently high contact values of the pair potential (e.g., $\varepsilon/K_BT > 20$), BCC crystals form in the low salt, weak screening regime ($1/\kappa\sigma$ close to 0.5 in Fig. 12.33), while FCC crystals form for high screening. The coexistence of fluid and BCC, BCC, and FCC for volume fraction $\phi < 0.5$ is well described by the phase boundaries of point Yukawa particles. Therefore, similar to the formation of Wigner crystals in the case of electrons [178], charged spheres crystallize and make transitions from BCC to FCC in the weak screening regime to minimize electrostatic repulsion energy. They constitute an example of *enthalpy-driven phase transitions* [34, 179]. Figure 12.33 also shows that the hardcore favors the FCC for $\phi > 0.5$, independent of the screening. This transition, in contrast, is an example of an *entropy-driven phase transition*. Consequently, a second triple point appears in the phase diagram in the weak screening regime, as shown clearly in Figure 12.33a. This triple point sets a lower limit for the strength of the Yukawa interaction for which a BCC phase exists. All the phase coexistence regions in the phase diagrams for hardcore repulsive Yukawa system are very narrow with only a small density jump for the coexisting phases.

Generally, the effective charge and the salt concentration are treated as the fitting parameters in calculations for comparison with experiments [178]. The total charge and the effective charge on the particle can be estimated experimentally from conductometric titration and conductivity experiments, respectively [180]. The actual charge is quite low as compared to the titratable charge because the degree of dissociation of the ionizable groups is always much less than unity. The charge estimated from conductivity experiments is often renormalized so that the calculated values can fit the experimental ones.

Using blinking optical tweezers to drive isolated particle-pairs out of equilibrium, the interparticle forces can be extracted from the statistics of trajectories of the particles [181–183]. The screening length κ^{-1}, the effective surface potential, and the hydrodynamic radius of the particle R can be obtained in a single measurement. It was found that the electrostatic interactions of the particles depend sensitively on surface composition as well as on the concentration and chemistry of the charge control agent in the solvent [183]. As an example for charge control agent, polyisobutylene succinimide (OLOA-1200), which is a commercial dispersant, has long been known to charge carbon black in oil [184]. Nonpolar colloids have been used as electrophoretic ink in flexible electronic displays [185].

The effective charge can also be measured by optical tracking of a single colloid in an optical tweezer trap and driven by a sinusoidal electric field [186]. The trapped particle forms a strongly damped harmonic oscillator whose fluctuations are a function of the ratio between the root-mean-square average of electric and thermal forces on the particle ϖ At low applied fields ($\varpi < 1$) the particle is confined to the optical axis, while at high fields ($\varpi > 1$) the probability distribution of the particle has a double peak. The periodically modulated thermal fluctuations are measured with nanometer sensitivity using an interferometric position detector. Charges, as low as a few elementary charges, can be measured with an uncertainty of about 0.25*e*.

Experiments on deionized aqueous suspensions of highly charged spherical latex colloids with ionizable sulfate groups and silica particles [187–190] showed evidences for an effective charge that is independent of volume fraction [191, 192]. However, increasing evidence suggests the breakdown of the constant-charge assumption [193–196]. It has been shown that the constant potential condition mimics charge-regulation on the colloidal surface (through an association–dissociation equilibrium of chargeable groups on the surface [37, 197, 198]) fairly well [38]. The phase diagram of charged spheres with a *constant Zeta potential* has been calculated [38]. Significant deviations from Figure 12.33 are revealed at large ϕ, especially at high zeta potentials. Although the resulting phase diagrams

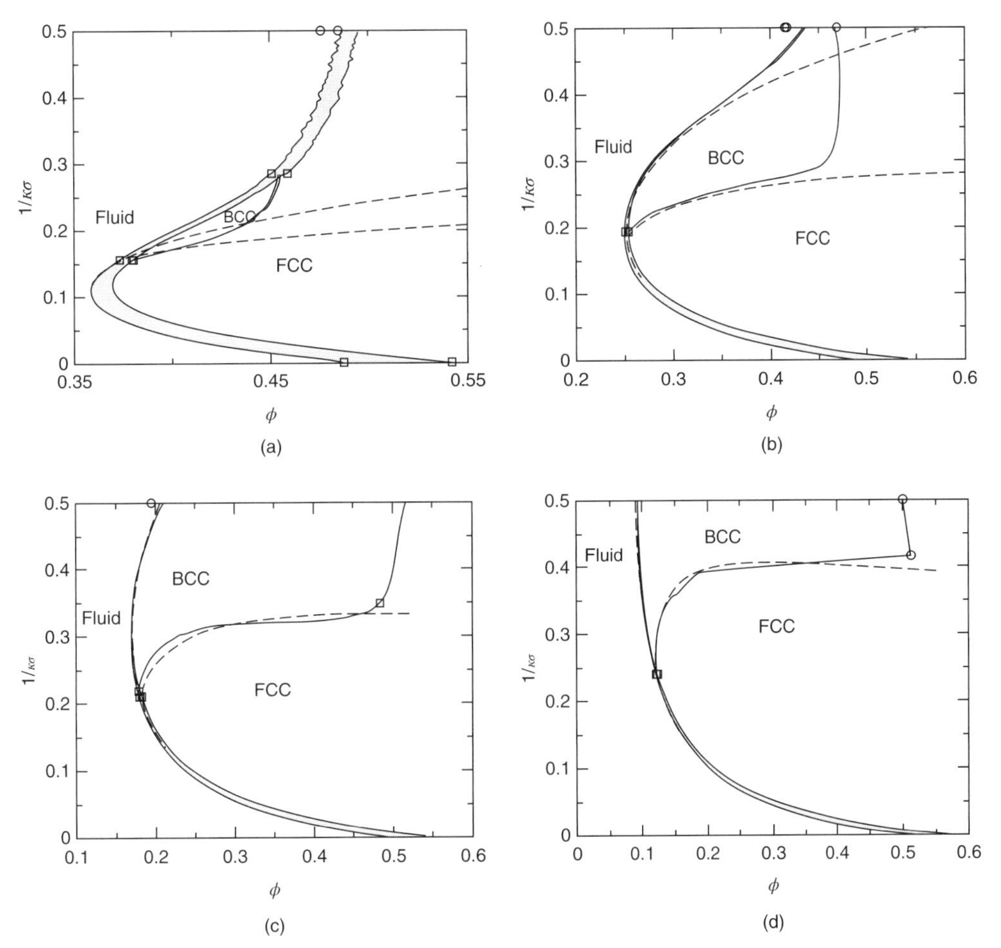

Figure 12.33 Phase diagram of charged spheres with a hardcore repulsive Yukawa pair potential. The phase diagram can be varied by ϕ, κ (varies by salt concentration n_s), and ε (varies by surface charges q). (a) $\varepsilon/k_BT = 8$, (b) $\varepsilon/k_BT = 20$, (c) $\varepsilon/k_BT = 39$ and (d) $\varepsilon/k_BT = 81$. The solid lines are coexistence lines with tie lines horizontal, and the grey areas are coexistence regions. Dashed lines are phase boundaries for point Yukawa particles. In charge-stabilized colloidal suspensions, the lower part of the diagram ($1/\kappa\sigma = 0$) is a high-salt regime and the upper part ($1/\kappa\sigma = 0.5$) is a low-salt regime.

do feature FCC and BCC phases, they are dominated by the (reentrant) fluid phase due to the colloidal discharging with increasing colloid concentration and decreasing salt concentration (Fig. 12.34).

Now recall that the Bjerrum length is 7 nm for water and 28 nm for dodecane. The interaction potential at contact is $\varepsilon/k_BT = BZ^2/2R$. For 500 nm particles carrying a typical charge of 10e, the contact potential value is about $3k_BT$ in dodecane, which is large enough to have a dramatic effect on the structure of a colloidal suspension. In fluorescently labeled and sterically stabilized dispersion of PMMA spheres dispersed in index and density-matched solvent with a relative dielectric constant $5 < \varepsilon_r < 10$, the Debye screening length is comparable to the particle size σ, even for particles with sizes of several micrometers. For such systems, quantitative 3D real-space measurements can be performed using confocal microscopy [199, 200]. Moreover, by addition of salt (tetrabutylammonium chloride), κ^{-1} can be varied and the surface potential of the particle can be set to a value in between +100 and −100 mV. Comparison of the radial distribution functions and displacements from the lattice position with Monte Carlo simulations revealed that the structure in the liquid and the crystallization volume fraction could be described with a Yukawa potential. Phase behaviors that were previously observed only for deionized dispersions in water using ionic exchange resins, such as extreme long-range repulsions, coexistence of high-density and low-density colloidal crystals and void formation, were also observed without using ion exchange resins [201–203].

Since the *sign* of the surface potential of the fluorescently labeled spheres in apolar suspensions can be tuned at ease by salt concentration (or in aqueous suspension by adjusting pH [204]), large crystals from oppositely charged particles have been fabricated [173, 205, 206]. In contrast to atomic systems, the stoichiometry of these colloidal crystals is not dictated by charge neutrality of the colloids (colloids

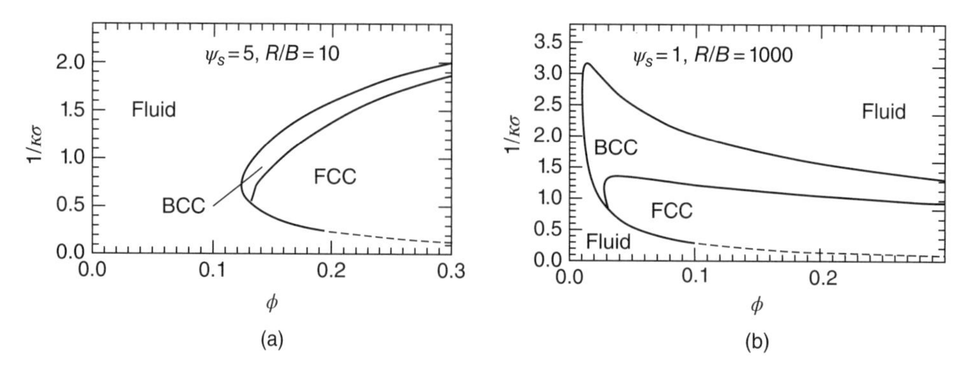

Figure 12.34 *Phase diagram of charged spheres at constant zeta potentials.* For large spheres ($R/B = 1000$) reentrance of the fluid phase (fluid–crystal–fluid transition sequence as increasing volume fraction ϕ) is easy to realize, which is absent for small spheres ($R/B = 10$) and large zeta potential ($\psi_s = 5$).

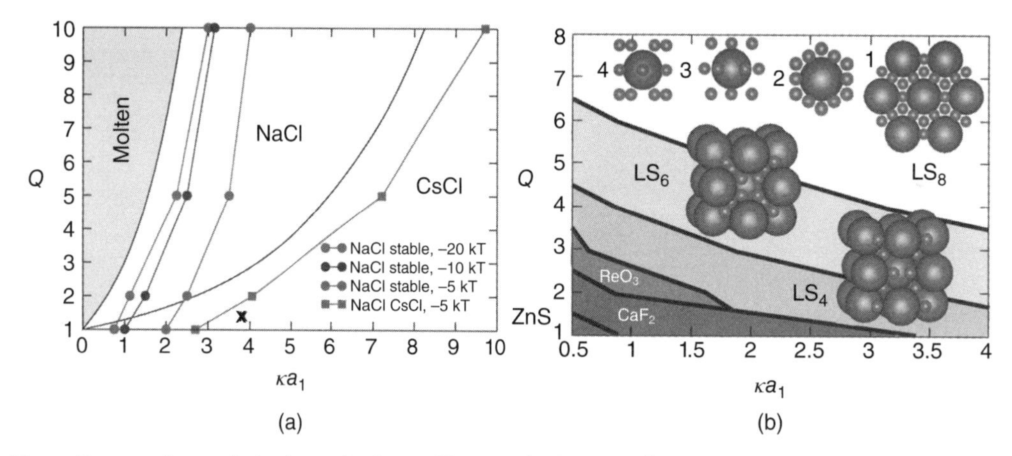

Figure 12.35 *Phase diagram of oppositely charged spheres.* Charge ratio $Q = -Z_1/Z_2$ and size ratio a_2/a_1 of 1 (a) and 0.31 (b). Full curves indicate crystal–crystal transitions from Madelung energy calculations (zero-temperature; black) and finite-temperature, zero-pressure MC simulations with a contact energy V_{12} of $-20k_BT$ (green), $-10k_BT$ (blue) and $-5k_BT$ (red). (a) The NaCl-structure melting line and NaCl–CsCl phase boundary. (b) Crystal phases coexisting with a dilute gas of small colloids, from Madelung energies. Insets: 1, projection of the three basic planes of LS_8; 2, configuration around a large colloid; 3, side view; 4, side view rotated by 90°. (*See color plate section for the color representation of this figure.*)

with their surrounding diffuse layer of counterions are charge-neutral objects); instead, the equilibrium structure of the crystal is determined by geometric requirements, stemming from short-range repulsions, together with the potential energy of the lattice and entropy associated with thermal effects [173]. This allows to obtain a remarkable diversity of new binary structures (Fig. 12.35) [173]. An external electric field melts the crystals, confirming that the constituent particles are indeed oppositely charged. Colloidal model systems can thus be used to study the phase behavior of ionic species [207–209].

The influence of *temperature* on the phase behavior of charged spheres is complicated [33, 39, 171, 210]. Figure 12.36 shows that the phase diagram of charged spheres depends on temperature through the dimensionless temperature $k_BT/\varepsilon = 2R/BZ^2 \propto \varepsilon_r T$ and $1/\kappa\sigma \propto \varepsilon_r T$. So temperature enters the problem as $\varepsilon_r T$. If the dielectric constant of the solvent arises from free dipoles and follows a curie law, the partition function is temperature independent and the system is athermal. If the dielectric constant of the solvent increases faster than linearly with inverse temperature, which is the case for water, then the high entropy phase (FCC crystal are more ordered than BCC crystal, solid is more ordered than liquid) occurs upon either heating or cooling from the more ordered phase (BCC–FCC–BCC transition or fluid–solid–fluid transition) [171]. Upon increasing temperature, freezing was observed for low-charge and low-salt colloids and melting was observed for high-charge and high-salt colloids. These transitions were thermoreversible [39].

Growing evidence suggests that protein crystallization can be understood in terms of an order–disorder transition between weakly attractive (charged) particles [211–214]. The phase diagrams are presented in Figure 12.37 (ρ is

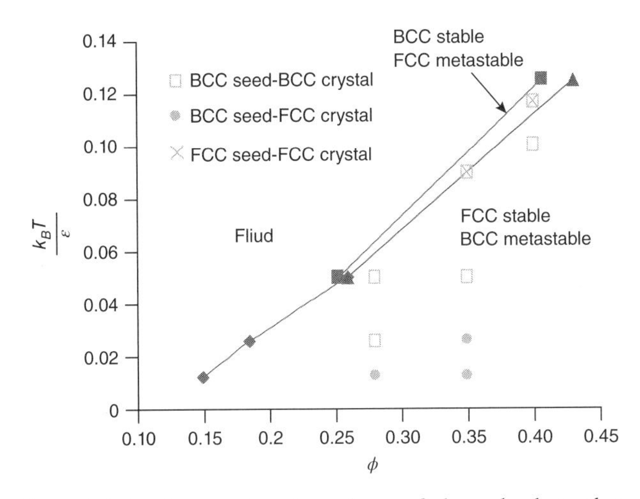

Figure 12.36 Temperature dependence of charged sphere phase behaviors. Plot is for the hardcore repulsive Yukawa fluid for $1/\kappa\sigma = 0.2$.

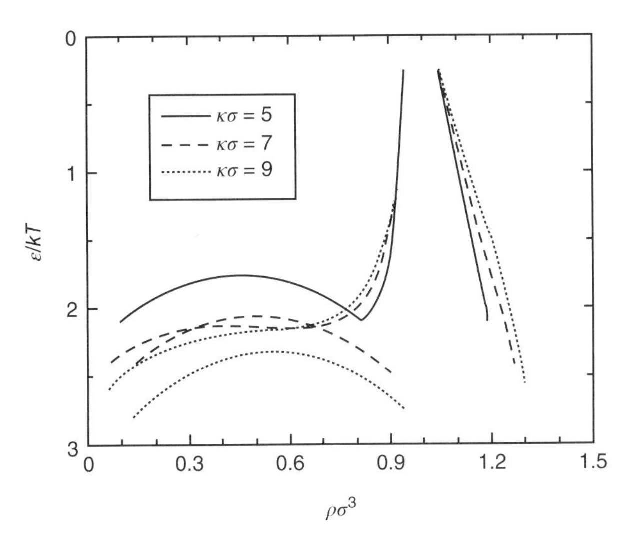

Figure 12.37 Phase diagrams for attractive Yukawa fluids for different extents of the attractive well. When the attraction is short range, as in protein systems, equilibrium between gas and crystal is found, but the liquid–liquid transition becomes metastable.

particle number density) for the hardcore attractive Yukawa potential:

$$V_c(r) = \begin{cases} \infty & r \leq \sigma, \\ -\varepsilon \dfrac{e^{[-\kappa\sigma(r/\sigma - 1)]}}{r/\sigma} & r > \sigma. \end{cases}$$

The *range of the attractive interaction* determines whether or not a *stable liquid phase* exists. The liquid–vapor coexistence curve disappears when $\kappa\sigma \approx 6$ [214]. The critical point (highest point of the curve that curving toward the horizontal axis) drops below the solubility curve (the monotonic increasing curve in the left side of Fig. 12.37) for $\kappa\sigma > 7$, becoming metastable. That is to say, the gas–liquid transition becomes metastable with respect to fluid–solid transition. For sufficiently narrow well widths (i.e., $\kappa\sigma$ is small), the solubility becomes a strong function of the strength of attraction. In this region, a change in ε of a few tens of a k_BT greatly alters protein (colloid) solubility. The value of ε/k_BT at the metastable critical point increases in magnitude as well widths shrink (i.e., $\kappa\sigma$ decreases). For solutions of globular proteins, it has been known for more than 30 years that a liquid–liquid phase separation occurs below the fluid–solid transition [215]. The possible explanation of this metastable liquid–liquid phase transition is again the fact that the range of attraction is small compared to the size of the protein [216]. The presence of a metastable liquid–liquid phase separation drastically changes the crystallization pathway in solutions of globular proteins [6, 217]. Within the metastable region, the *Ostwald rule* (Ostwald proposed that the solid first formed on crystallization of a melt or a solution would be the least stable polymorph) implies that the fluid–fluid phase separation would precede the fluid–crystal transition. Whether crystals are best grown inside or outside the metastable region is a matter of debate [6]. If fluid–fluid phase separation occurs first, nucleation may be enhanced but growth will be determined by the now-high volume fraction. As the crystal grows, a covering film of fluid acts like a buffer between the original fluid and the growing crystal. In protein systems, this fluid layer aids growth because it means monomers are always available for the growth step and they have ample time within the film to find the proper orientation for bonding. Thus, the nucleation is increased because of the effect of metastable states on the dynamics, rather than by an increase in the thermodynamic driving force.

12.3.2 Crystallization Kinetics

Experiments on colloidal crystallization consistently show that it is much easier to crystallize charged colloids than uncharged "hard-sphere" colloids. Clearly, long-ranged repulsions have a large effect on the crystal-nucleation rate. This may even be true for colloidal suspensions of particles that are only weakly charged. Only a few nucleation experiments on charged colloids have been reported, though. The crystallization process of slightly charged spherical TPM-silica (γ-methacryloxypropyltrimethoxysilane grafted on silica) particles, with zeta potential 55 ± 10 mV and surface charge $260 \pm 60e$ at $\kappa^{-1} = 60 \pm 12$ nm and very low ϕ. $\kappa^{-1} = 43 \pm 5$ nm and $\phi = 0.2$, was studied using time-resolved static light scattering [140]. The induction time, the crystallization rate, the scattered intensity after completion of the crystallization process, and the width of the Bragg peaks were found to be strongly dependent on the concentration of the initially metastable colloidal fluid. Assuming a simple crystal geometry, quantities such as the size of the crystallites, the number concentration of the crystallites, and both nucleation and crystallite growth rates, as functions of the concentration, were calculated from the measured quantities.

With the aid of nearly index-matched perfluorinated particles of diameter 90 nm and a surface charge of $520 \pm 50e$, as measured by conductivity measurements, the measurements were extended into the regime of large metastability and nucleation-dominated solidification. With increasing particle number density, n, the solidification time, determined by the appearance of a finite shear modulus, decreased from minutes to milliseconds. Nucleation rate densities, J, were derived from the width of the principal peak in the static structure factor, as measured by means of light scattering after complete solidification. J was observed to increase approximately exponentially with n, as expected from CNT in the absence of a kinetic glass transition (Fig. 12.38). This is somewhat surprising since it implies that both the pair interaction and the surface tension are constant. Additional measurements of the elastic and dynamic behavior, however, showed that for the largest concentrations, the sample properties were consistent with formation of a glass.

Computer simulations were performed to study crystal nucleation in hardcore Yukawa systems as a function of the amplitude of the Yukawa repulsion and the magnitude of the screening length [218]. It was found that the *charge had a strong direct effect on the nucleation* barrier, due to the decrease in surface tension. This effect is strongest for small charges. It was also found that the functional dependence of the barrier height as a function of supersaturation does not change significantly for different values of the charge (Fig. 12.39). The results were in contrast to the experiment, where only a slight dependence of the nucleation rate on supersaturation was observed. In addition, at the same volume fraction, the nucleation barrier is much lower for weakly charged spheres than it is for hard spheres. This is partly because the fluid–solid coexistence of charged spheres occurred at lower volume fractions, implying a higher supersaturation.

Figure 12.39 compares the dimensionless nucleation rate of FCC crystals obtained in simulations [218] and experiments [52] on. The experiments used slightly charged PMMA spheres that freeze at $\phi = 0.38$. It is therefore natural to describe them by a Yukawa model with a freezing point at that volume fraction. However, this condition is not sufficient to fix the values of both $\kappa\sigma$ and $\beta\varepsilon$. Figure 12.39 shows comparison of the reduced nucleation rates from experiments with the simulation results for some combinations of $\kappa\sigma$ and $\beta\varepsilon$ that yield a freezing point near $\phi = 0.38$. The figure

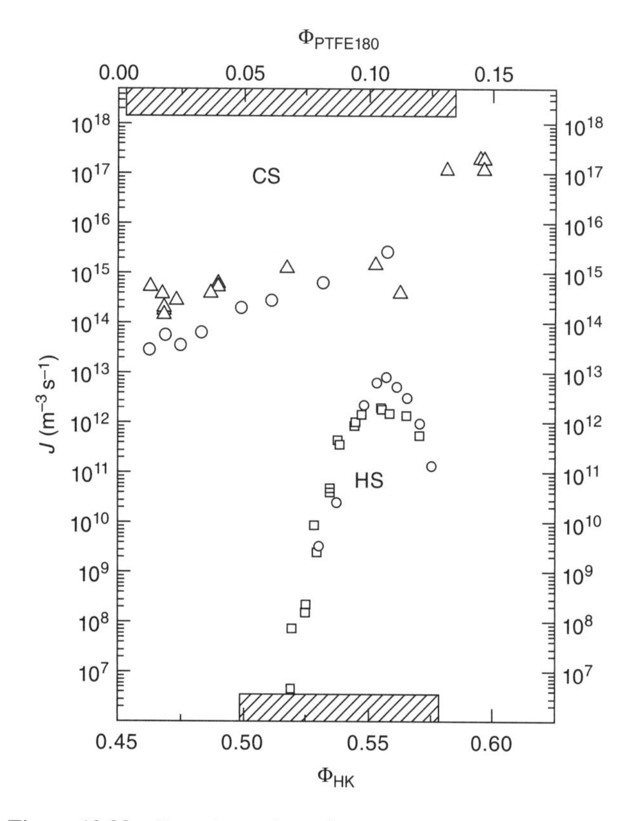

Figure 12.38 *Experimental nucleation rate densities J versus particle volume fraction ϕ.* Upper scale and curves: charged spheres (PTFE180); lower scale and curves: hard spheres (PMMA). The nucleation rates are significantly larger for charged spheres than that of hard spheres. It increases continuously over the range of volume fractions investigated. The density dependence is, however, less pronounced than that of hard spheres.

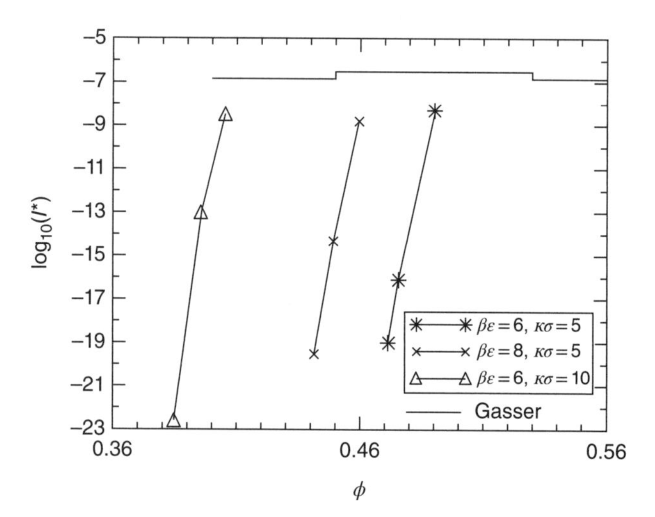

Figure 12.39 *Comparison of slightly charged sphere nucleation rates between the experiment and simulation.* The dimensionless nucleation rate is $I^* = I\sigma^5/D_0$ with I is the nucleation rate is σ the hardcore diameter and D_0 is the self-diffusion coefficient at infinite dilution. The experiment data is the horizontal curve without symbols. Only the data sets that match the freezing density of the experimental system are plotted.

shows that different $\kappa\sigma - \beta\varepsilon$ combinations yield very different nucleation rates. However, slopes of the different curves are all similar. Thus, the main effect of varying $\kappa\sigma$ and $\beta\varepsilon$ is to shift the nucleation curves horizontally. Comparison of the computed nucleation rates with the experimental data results in two observations: First, the experimental rates tend to be much higher than the computed rates; Gasser *et al.* found $-6.9 \leq \log[I^*] \leq -6.5$ for volume fractions between 0.45 and 0.53. Second, and more importantly, the experiments suggest that the nucleation rate barely varies with volume fraction, which was not reproduced at all with the Yukawa model studied in the simulation. Therefore, *quantitative difference exists between experiments and simulations.*

Large variation of nucleation rates with particle density was indeed observed in recent experiments (Fig. 12.40) [219] conducted at constant salt concentration $c_s = 2 \times 10^{-7}$ mol/l and varying particle number densities, $18\,\mu m^{-3} \leq n \leq 66.3\,\mu m^{-3}$. The diameter of the particles is $\sigma = 68 \pm 3$ nm, and the effective charge is $Z^* = 331 \pm 3$. The samples exhibit a BCC structure at low density and an FCC structure at high density. Direct video microscopy observations of individual nucleation events were employed to obtain time-resolved nucleation rate densities. Polarization microscopy and static light scattering on the resulting solids, in combination with the *Avrami* phase transformation theory, were used to determine the steady-state nucleation rate at high undercooling. The final nucleation rate densities J from different methods are observed to be consistent with each other. By increasing the difference in the chemical potential between the melt and the crystal about one order of magnitude, J increases from 10^9 to $10^{17}\,m^{-3}\,s^{-1}$. The data can be well analyzed and interpreted using CNT leading to a linearly increasing melt-crystal surface tension. Surprisingly, the reduced surface tension is about one order of magnitude larger compared to other systems such as metals and hard-sphere colloids. The critical radius of the crystal nuclei decreases down to a very small value of 1.5σ. The determined kinetic prefactors are up to 10 orders of magnitude smaller than the prefactor calculated by CNT.

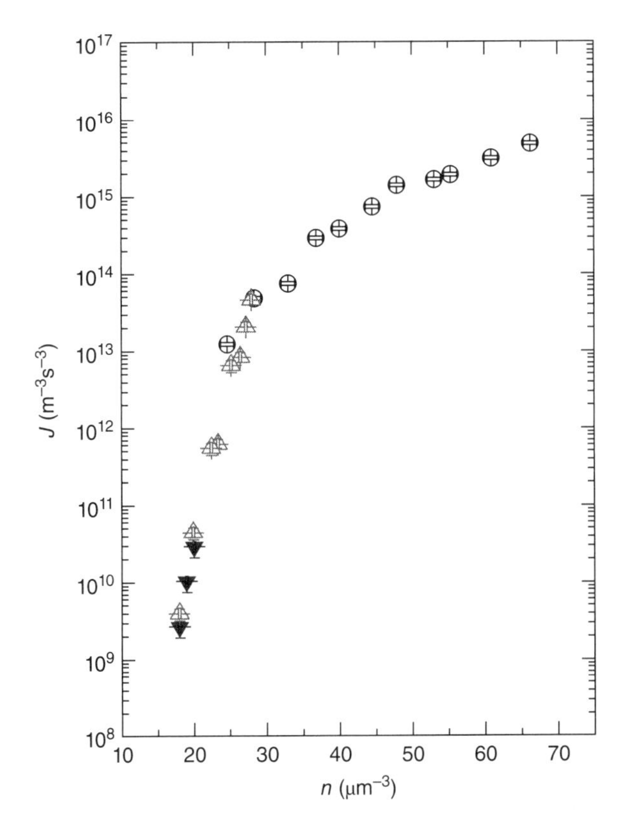

Figure 12.40 *Nucleation rate densities for deionized samples of charge particles.* Data from different experiments show quantitative agreement in the range of equal particle number densities. Filled triangles, J from video microscopy; open triangles, J from microscopic analysis of solidified samples; circles, J from light scattering data obtained from solidified samples. Note the 6.5 orders of magnitude increase in J as n is increased by a factor of 4.

As we have already seen, charged spheres that can be described via hardcore repulsive Yukawa interactions have two *polymorphs* (*polymorphism* is the ability of a solid material to exist in more than one form or crystal structure), BCC and FCC (Fig. 12.33). The relative stability of these phases can be adjusted by changing the potential parameters. This offers the unique opportunity to study polymorph selection, which is important because the structure of a crystal is critical to its properties. Seeded molecular-dynamics simulations were used to follow the evolution of the metastable BCC structure in the FCC stable region [210]. It was found that the degree of undercooling affected the development of the BCC seeds. At small undercoolings, the BCC seed became a BCC crystal, but at large undercoolings the BCC seeds became FCC crystals (Fig. 12.36). Recent experiments demonstrated such a liquid to BCC to FCC crystal transition [220, 221].

Homogenously nucleated simulations were also compared to colloidal experiments [210]. The simulations yielded nuclei of similar shape to the experimentally observed nuclei and a similar minimum size for growth (Fig. 12.41) [52]. The simulated nuclei had more BCC signatures than the experimental nuclei, but upon further growth the BCC signature faded, indicating that the difference between simulations and experiments may be due to the experimental nuclei were observed at later time.

12.4 CRYSTALLIZATION OF MICROGEL PARTICLES

Due to the athermal nature of hard spheres and the near athermal behavior of colloidal charged spheres, we could perhaps say that colloidal crystallization is hard to manipulate. *Smart* colloids, such as PNIPAM microgels, are responsive to external stimuli, such as temperature and pressure [222], and pH and salt concentration, if copolymerized with other

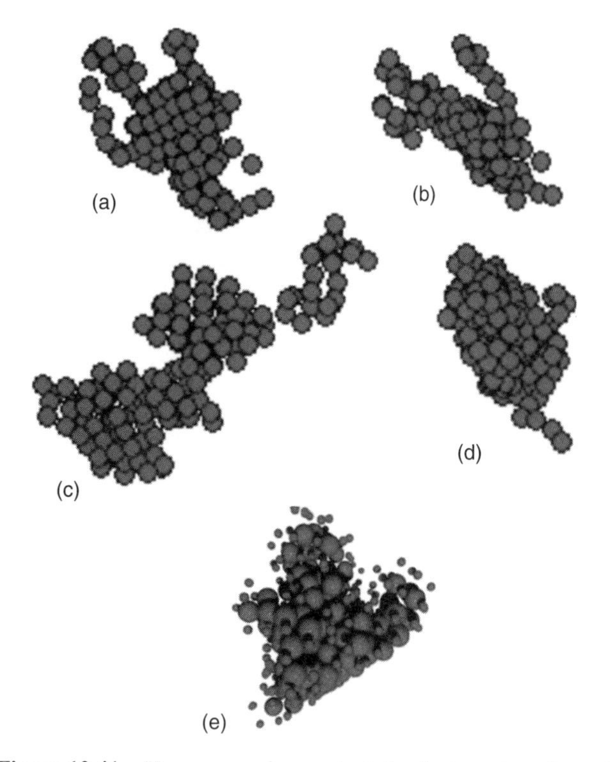

Figure 12.41 *Homogenously nucleated clusters in charged spheres with hardcore Yukawa interaction.* Grown at $\phi = 0.47$ (a–c) and $\phi = 0.45$ (d). Particles are shown smaller than actual size for clarity. (a) and (b) are the same cluster shown from two angles. An experimental nucleus by Gasser *et al.* is shown in (e) for comparison.

monomers such as acrylic acid [223, 224]. Therefore, it is possible to use temperature to melt crystals [158, 225, 226] and eliminate clusters in the metastable fluid that might accelerate nucleation after shear melting of the crystals.

12.4.1 Phase Behavior

Under certain conditions, microgel spheres behave as hard spheres, hence they are often referred to as *thermosensitive hard spheres* [86]. A UV–Vis spectroscopy method was used to determine the phase boundaries for thermosensitive colloids as an alternative to the time-consuming sedimentation method (it will take a long time since hydrogels are almost density matched with the solvent). The Bragg attenuation peak from colloidal crystallites was monitored during quasistatic colloidal crystal melting. The melting and freezing boundaries of the coexistence region were determined via the blueshift of the Bragg peak and the decrease in peak area.

As we can see from Fig. 12.42, the particles behave as thermosensitive hard spheres at low pH. At high pH, the acrylic acid dissociation will result in *charged microgels* [227]. Interestingly, charged microgels were observed to have a much narrow fluid–fluid separation region in polymer concentration than the fluid–fluid separation region of neutral microgels (Fig. 12.42c). Only calculations for neutral, dense microgels with a core-shell structure exist so far [228].

The soft-sphere interparticle interaction typical for these colloidal particles is often described in terms of an inverse power law. $V(r) = \varepsilon\left(\frac{\sigma}{r}\right)^n$, where the exponent n controls the stiffness of the potential. When the interaction gets softer (decreasing n), the fluid–crystal transition moves to higher volume fractions (Fig. 12.43) [229, 230]. It was noticed that when microgels are concentrated at $\phi > 0.3$, the soft-sphere potential is appropriate [231–233], though this is still a matter of debate [234–236].

In many cases, the experimental behavior of microgel suspensions is close to hard-sphere behavior (Fig. 12.42). For example, the crystalline structure of charged microgels is still random hexagonally close packed, even for very dense packing [237]. Currently, there is no experiment confirming the theoretically predicted phase diagram for the (ultra-)soft-ionic microgels (Fig. 12.44) [238].

A more realistic model could perhaps be the *interpenetrating Hertzian sphere model*:

$$V(r) = \begin{cases} \varepsilon(1 - r/\sigma)^{5/2} & r < \sigma, \\ 0 & r \geq \sigma, \end{cases}$$

where σ and ε set the length and energy scales and r is the distance between the centers of the spheres. The phase diagram is shown in Figure 12.45. It shows that the fluid phase freezes into an FCC crystal that, at higher densities, turns into a BCC structure. However, the BCC packing is favored over the FCC at high temperatures because particles in the former have higher vibrational entropy. The fluid reenters at densities larger than the maximum freezing point at $k_BT/\varepsilon = 8.84 \times 10^{-3}$ and $\rho\sigma^3 = 2.40$. At lower temperatures and with increasing density, one finds multiple reentrant melting and four more structures: hexagonal, simple cubic, body-centered-tetragonal, and trigonal. To date, however, only FCC or RHCP structures have been experimentally found for neutral and charged microgel suspensions [237, 239–241].

12.4.2 Crystallization and Melting Kinetics

The importance of melting in nature can hardly be overestimated, and yet a detailed understanding of the mechanisms that drive this transformation is still evolving [242]. The preliminary data for nucleation of hard-sphere crystals using *thermosensitive hard spheres* (Fig. 12.42) [86] was presented in Figure 12.25, which indeed show a good agreement with computer simulation results. The emulsion crystallization method recently introduced showed its promise for the determination of the nucleation rate of hard spheres in the coexistence region. The particles involved in these studies,

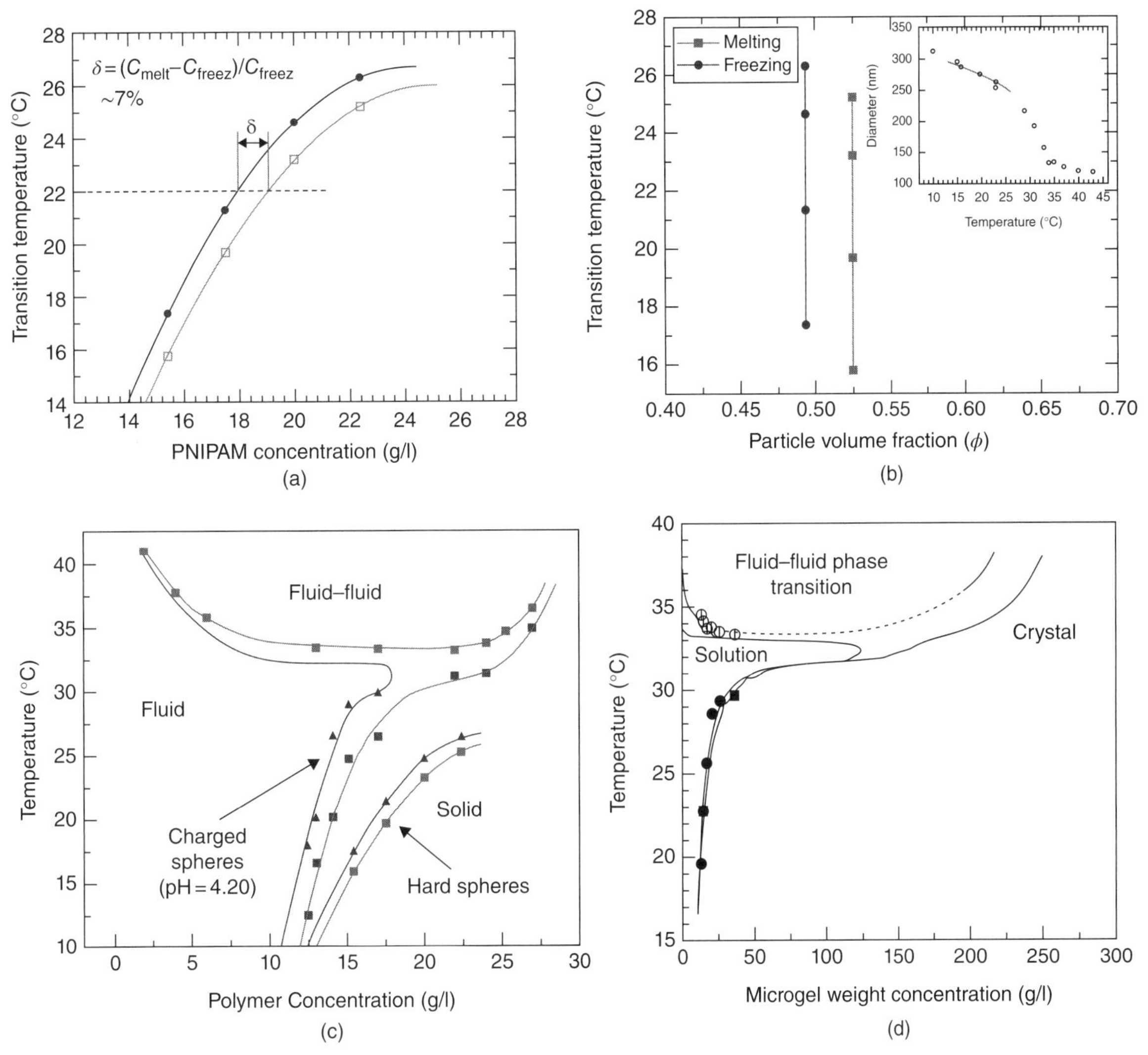

Figure 12.42 *Thermosensitive hard spheres and charged spheres.* (a) Phase diagram of the PNIPAM-*co*-acrylic acid microgel dispersions at pH 2.8 in the temperature–concentration representation. At room temperature, the fluid–crystal transition width normalized to the freezing transition concentration is approximately 7%, consistent with hard spheres with 5% polydispersity. (b) Temperature–volume fraction representation. Inset: Particle diameter with temperature (open circles are DLS). Solid line indicates the diameter of the particles used to transfer from temperature–concentration representation to temperature–volume fraction representation. (c) Charged sphere behavior at pH = 4.20. (d) Theoretical phase diagram for neutral microgels.

when large enough, could also be observed directly under optical microscope [242, 243].

Microgels thus offer opportunities to considerably advance our current knowledge of crystal nucleation. They have also shown promise as model systems for studying melting. Indeed, the melting kinetics of three-dimensional colloidal crystals consisting of poly(*N*-isopropylacrylamide) microgels was investigated by using UV–visible transmission spectroscopy [225]. It was found that the melting was initiated with a decrease in the interplanar spacing of the crystals (cf. the color shifting in Fig. 12.46) and that crystallites broke into smaller pieces at large overheating temperatures. The crystallites reach a minimum burst size before completely broken apart. The complete melting occurs as the average thermal fluctuation of the particles reach around 19%. The experimental results corroborate recent computer simulations that conciliate the Lindemann criterion for melting with Born's mechanical instability criterion.

Premelting is the localized loss of crystalline order at surfaces and defects at temperatures below the bulk melting transition. It can be thought of as the nucleation of the melting process. Premelting has been observed at the surfaces of crystals but not within. Observations of premelting have now been reported at grain boundaries and dislocations within bulk colloidal crystals using real-time

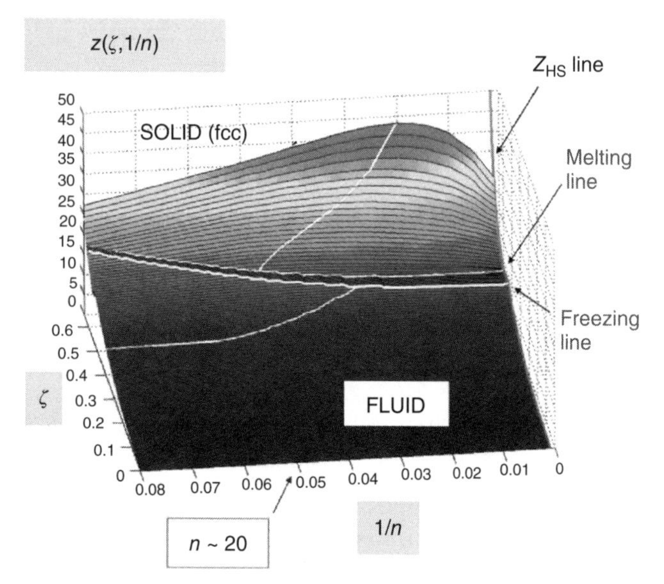

Figure 12.43 Phase diagram of the soft spheres with inverse power-law interparticle potential.

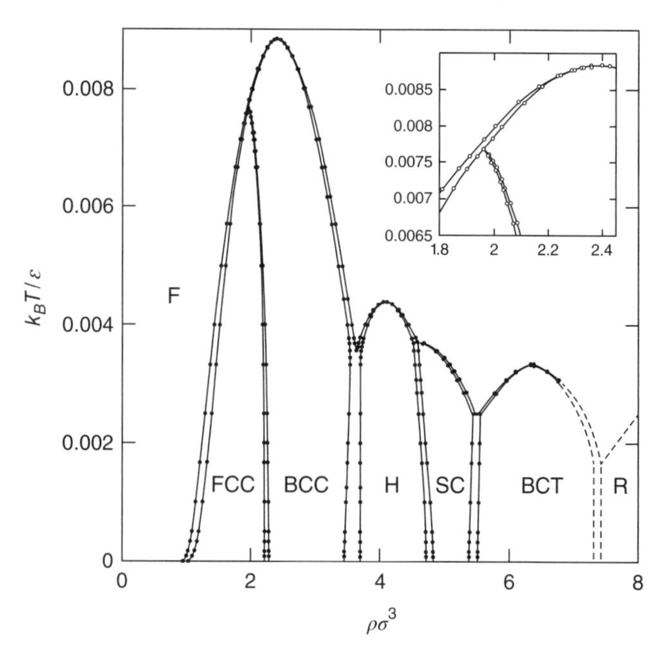

Figure 12.45 *Phase diagram of interpenetrating Hertzian spheres.* Errors are smaller than the size of the circles. Solid lines are a guide to the eye. Broken lines thus indicate approximate phase boundaries. The inset zooms in the region around the F–FCC–BCC triple point.

video microscopy (Fig. 12.47) [242]. The crystals are equilibrium close-packed, three-dimensional colloidal structures made from thermally responsive microgel spheres. Particle tracking revealed increased disorder in crystalline regions bordering defects, the amount of which depends on the type of defect, distance from the defect, and particle volume fraction. The observations suggest that interfacial free energy is the crucial parameter for premelting in colloidal and atomic scale crystals.

Melting mechanisms were also recently investigated in single and polycrystalline colloidal films composed of size-tunable microgel spheres with short-ranged repulsive interactions and confined between two glass walls [244]. Thick films (>4 layers), thin films (≤4 layers), and monolayers exhibit different melting behaviors. Thick films melt from grain boundaries in polycrystalline solid films and from film-wall interfaces in single crystal films; a

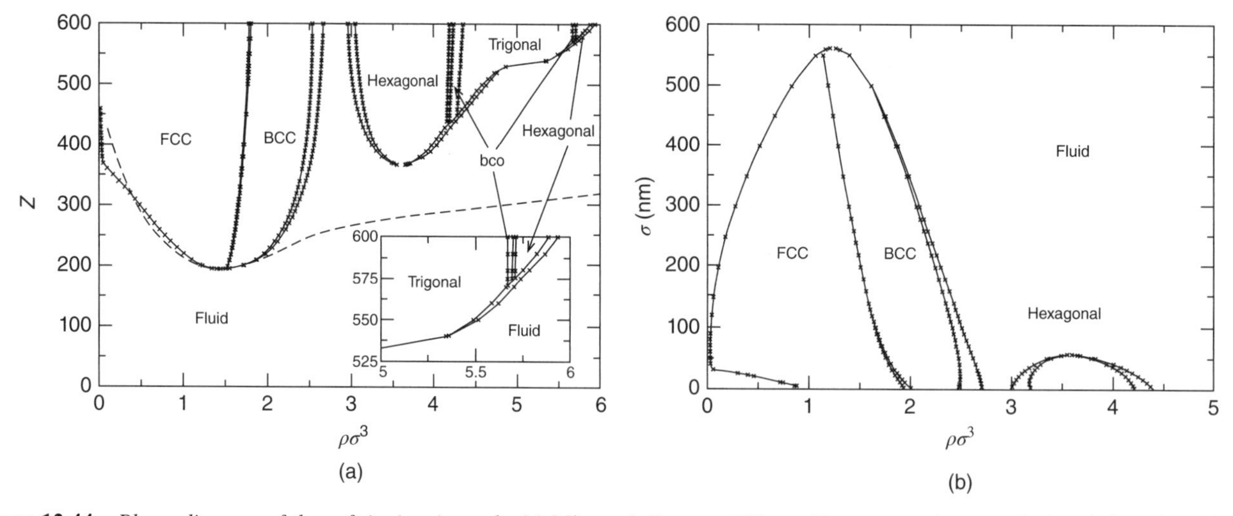

Figure 12.44 *Phase diagram of the soft-ionic-microgels.* (a) Microgel diameter 100 nm. The crosses denote calculated phase boundaries, and the lines are guides to the eye connecting these points. The unlabeled narrow regions between phases denote domains of phase coexistence (density jumps), whereas the dashed line is the locus of points in which the highest peak of $S(k)$ attains the Hansen–Verlet value 2.85. (b) Microgels for fixed charge $Z = 300$ and varying particle diameter σ.

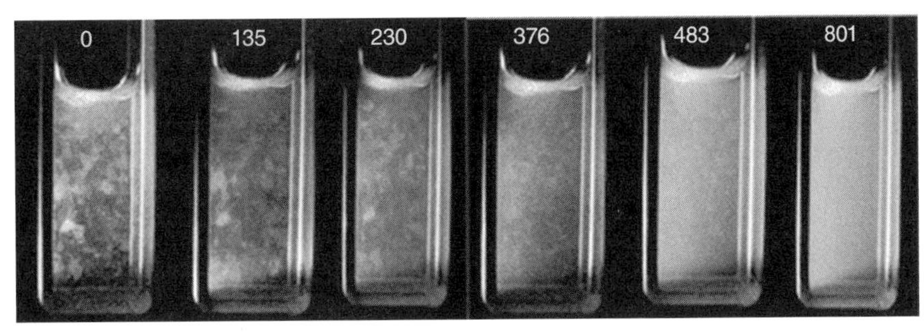

Figure 12.46 The melting of a poly(*N*-isopropylacrylamide) microgel suspension. Numbers on each micrograph indicate the time lapse in seconds after a temperature jump.

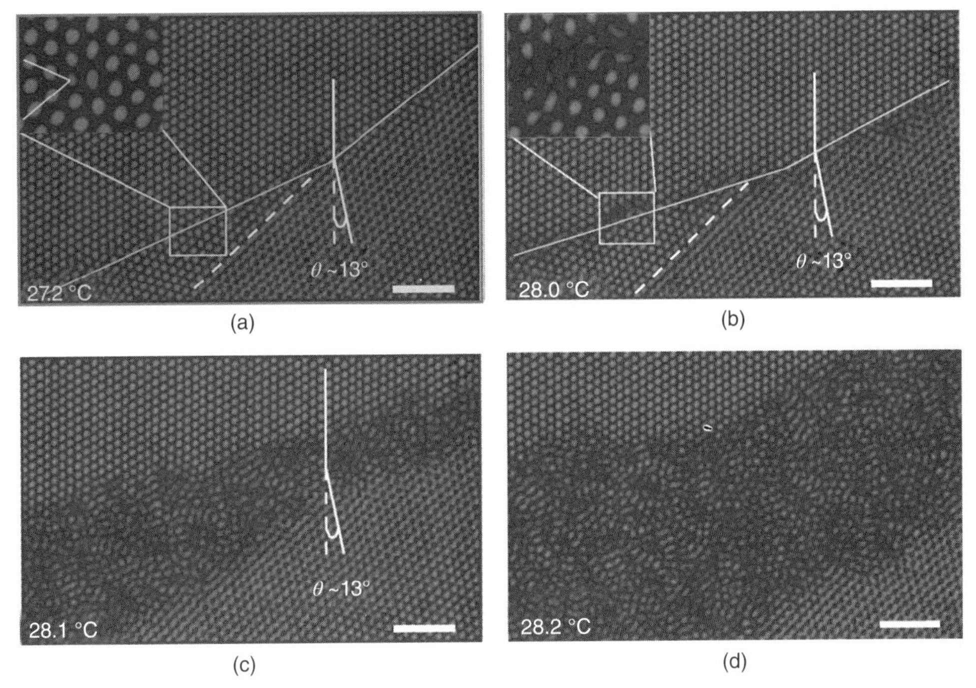

Figure 12.47 Premelting of the colloidal crystal at a grain boundary. The figure shows bright-field images at different temperatures (i.e., particle volume fractions) of two crystallites separated by a grain boundary (crystallites tilted at an angle $\theta \sim 13°$ with respect to one another). (a) Sample at 27.2 °C. The solid and dashed lines show the grain boundary and a partial dislocation, respectively. The grain boundary cuts the two crystals along two different planes (the yellow line has two slopes). It is composed of an array of dislocations; the two extra planes are indicated by lines in the inset. (b) Sample at 28.0 °C. The grain boundary starts to premelt; nearby particles undergo liquid-like diffusion (inset). The partial dislocation, denoted by the dashed line, is not affected. (c and d) The same sample at 28.1 and 28.2 °C, respectively. The width of the premelt region near the grain boundary increases. Scale bars: 5 mm. (*See color plate section for the color representation of this figure.*)

liquid–solid coexistence regime is observed in thick films but vanishes at a critical thickness of four layers. Thin solid films (2–4 layers) melt into the liquid phase in one step from both grain boundaries and from *within crystalline domains*. Monolayers melt in two steps with a middle hexatic phase.

12.5 CONCLUSIONS AND NEW DIRECTIONS

The key advantages using colloids for crystallization (phase transition) study are accessibility to individual "atoms" by optical methods in space and at convenient timescale in dynamics, adjustability of morphology of colloids (size, shape, and their distributions), interparticle interactions (by chemical grating, electric, and magnetic fields), and concentration (by gravity, temperature, and temperature gradient). This research field of colloidal crystallization is making progress along several frontiers: (i) Capture the morphology of the nuclei, demanding refine or new develop in nucleation theory; (ii) quantitatively understand the sensitivity of nucleation rate to the detail of interparticle interactions, especially the softness of near-hard spheres; (iii) incorporate novel techniques to measure the crystallization kinetics, such as droplet microfluidics for emulsion crystallization investigation, and microgravity environment for long-term observation; (iv) bridge to close fields, such as melting, crystal–crystal transformation, phonons, kinetics of defects in crystals and amorphous solids; (v) develop applications of colloidal crystals, such as the fabrication of photonic crystals and chemical sensors.

ACKNOWLEDGMENTS

The author sincerely thanks Prof. Paul M. Chaikin and William B. Russel for the opportunity to participate in CDOT-1, CDOT-2, and PHaSE microgravity experiments for colloidal crystallization in space. He also acknowledges support from NSF (DMR-1006870), NASA (NASA-NNX13AQ60G), and Guangdong Provincial Key Laboratory on Functional Soft Condensed Matter, Guangdong University of Technology, Guangzhou, China.

REFERENCES

[1] Barlow W. A mechanical cause of homogeneity of structure and symmetry geometrically investigated; With special application to crystals and to chemical combination. Sci Proc Roy Soc Dublin 1897;8:527–690.

[2] Lekkerkerker HNW. Ordering in supramolecular fluids. Physica A 1991;176:1–15.

[3] Haymet ADJ. Theory of the equilibrium liquid–solid transition. Annu Rev Phys Chem 1987;38:89–108.

[4] Alder BJ, Wainwright TE. Phase transition in elastic disks. Phys Rev 1962;127:359–361.

[5] Russel WB, Saville DA, Showalter WR. *Colloidal Dispersions*. Cambridge: Cambridge University Press; 1989.

[6] Anderson VJ, Lekkerkerker HNW. Insights into phase transition kinetics from colloid science. Nature 2002; 416:811–815.

[7] Wood WW, Jacobsen JD. Preliminary results from a recalculation of the Monte Carlo equation of state of hard spheres. J Chem Phys 1957;27:1207–1208.

[8] Alder BJ, Wainwright TE. Phase transition for hard sphere fluid. J Chem Phys 1957;27:1208–1209.

[9] Alder BJ, Wainwright TE. Studies in molecular dynamics. J Chem Phys 1959;31:459–466.

[10] Hachisu S, Kobayashi Y. Kirkwood–Alder transition in monodisperse latexes II. Aqueous latexes of high electrolyte concentration. J Colloid Interface Sci 1974;46:470–476.

[11] Hachisu S, Takano K. *Ordering and Organization in Ionic Solutions*. Singapore: World Scientific Publishing Co.; 1988. 376.

[12] Pusey PN, van_Megen W. Phase-behavior of concentrated suspensions of nearly hard colloidal spheres. Nature 1986;320:340–342.

[13] Phan SE, Russel WB, Cheng ZD, Zhu JX, Chaikin PM, Dunsmuir JH, Ottewill RH. Phase transition, equation of state, and limiting shear viscosities of hard sphere dispersions. Phys Rev E 1996;54:6633–6645.

[14] Zhu JX, Li M, Rogers R, Meyer W, Ottewill RH, Russell WB, Chaikin PM. Crystallization of hard-sphere colloids in microgravity. Nature 1997;387:883–885.

[15] Cheng Z, Chaikin PM, Russel WB, Meyer WV, Zhu J, Rogers RB, Ottewill RH. Phase diagram of hard spheres. Mat Des 2001;22:529–534.

[16] Russel WB, Chaikin PM, Zhu J, Meyer WV, Rogers R. Dendritic growth of hard sphere crystals. Langmuir 1997; 13:3871–3881.

[17] Cheng ZD, Zhu JX, Russel WB, Meyer WV, Chaikin PM. Colloidal hard-sphere crystallization kinetics in microgravity and normal gravity. Appl Optics 2001;40:4146–4151.

[18] Pusey PN, van Megen W. Observation of a glass-transition in suspensions of spherical colloidal particles. Phys Rev Lett 1987;59:2083–2086.

[19] Hansen JP, McDonald IR. *Theory of Simple Liquid*. New York: Academic Press; 1976.

[20] Chandler D, Weeks JD, Andersen HC. van der Waals picture of liquids. Science 1983;220:787–794.

[21] Mahanty J, Ninham BW. *Dispersion Forces*. New York/ London: Academic Press; 1976.

[22] Israelachvili JN. van der Waals dispersion force contribution to works of adhesion and contact angles on basis of macroscopic theory. J Chem Soc – Faraday Trans 2 1973;69: 1729–1738.

[23] Likos CN. Effective interactions in soft condensed matter physics. Phys Rep – Rev Sect Phys Lett 2001;348:267–439.

[24] Helden AK, Jansen JW, Vrij A. Preparation and characterization of spherical monodisperse silica dispersions in nonaqueous solvents. J Colloid Interface Sci 1981;81:354–368.

[25] Philipse AP, Vrij A. Preparation and properties of nonaqueous model dispersions of chemically modified, charged silica spheres. J Colloid Interface Sci 1989;128:121–136.

[26] Antl L, Goodwin JW, Hill RD, Ottewill RH, Owens SM, Papworth S, Waters JA. The preparation of poly(methyl methacrylate) lattices in nonaqueous media. Colloids Surf 1986;17:67–78.

[27] Costello BAD, Luckham PF, Tadros TF. Investigation of the interaction forces of polymer-coated surfaces using force balance, rheology, and osmotic-pressure results. Langmuir 1992;8:464–468.

[28] Costello BAD, Luckham PF, Tadros TF. Forces between adsorbed low-molecular-weight graft-copolymers. J Colloid Interface Sci 1993;156:72–77.

[29] Bryant G, Williams SR, Qian L, Snook IK, Perez E, Pincet F. How hard is a colloidal "hard-sphere" interaction? Phys Rev E 2002;66:060501.

[30] Alexander S. Adsorption of chain molecules with a polar head a-scaling description. J Phys 1977;38:983–987.

[31] de Gennes PG. Polymers at an interface – a simplified view. Adv Colloid Interface Sci 1987;27:189–209.

[32] Verwey EJW, Overbeek JTG. *Theory of the Stability of Lyophobic Colloids*. Amsterdam: Elsevier; 1948.

[33] Hynninen AP, Dijkstra M. Phase diagrams of hard-core repulsive Yukawa particles. Phys Rev E 2003;68:021407.

[34] S. Hamaguchi, R. T. F., D.H.E. Dubin, Triple point of Yukawa systems. Phys Rev E 1997, 56: 4671–4682.

[35] Bell GM, Levine S, McCartney LN. Approximate methods of determining the double layer free energy of interaction between two charged colloidal spheres. J Colloid Interface Sci 1970;93:335–359.

[36] Glendinning AB, Russel WB. The electrostatic repulsion from charged spheres from exact solutions to the linearized Poisson–Boltzmann equation. J Colloid Interface Sci 1983; 93:95–104.

[37] Monovoukas Y, Gast AP. The experimental phase-diagram of charged colloidal suspensions. J Colloid Interface Sci 1989;128:533–548.

[38] Smallenburg F, Boon N, Kater M, Dijkstra M, van Roij R. Phase diagrams of colloidal spheres with a constant zeta-potential. J Chem Phys 2011;134:074505.

[39] Toyotama A, Yamanaka J. Heating-induced freezing and melting transitions in charged colloids. Langmuir 2011; 27:1569–1572.

[40] Loeb AL, Overbeek JTG, Wiersma PH. *The Electrical Double-Layer Around a Spherical Colloidal Particle*. Cambridge, Massachusetts: MIT Press; 1961.

[41] Hsu MF, Dufresne ER, Weitz DA. Charge stabilization in nonpolar solvents. Langmuir 2005;21:4881–4887.

[42] Kornbrekke RE, Morrison ID, Oja T. Electrophoretic mobility measurements in low conductivity media. Langmuir 1992;8:1211–1217.

[43] Briscoe WH, Horn RG. Direct measurement of surface forces due to charging of solids immersed in a nonpolar liquid. Langmuir 2002;18:3945–3956.

[44] Keir RI, Suparno , Thomas JC. Charging behavior in the silica/Aerosol OT/decane system. Langmuir 2002;18: 1463–1465.

[45] Thomas JC, Crosby BJ, Keir RI, Hanton KL. Observation of field-dependent electrophoretic mobility with phase analysis light scattering (PALS). Langmuir 2002;18:4243–4247.

[46] McNamee CE, Tsujii Y, Matsumoto M. Interaction forces between two silica surfaces in an apolar solvent containing an anionic surfactant. Langmuir 2004;20:1791–1798.

[47] Smith PG, Ryoo W, Johnston KP. Electrostatically stabilized metal oxide particle dispersions in carbon dioxide. J Phys Chem B 2005;109:20155–20165.

[48] Morrison ID. Electrical charges in nonaqueous media. Colloid Surf A – Physicochem Eng Asp 1993;71:1–37.

[49] Cheng ZD, Chaikin PM, Zhu JX, Russel WB, Meyer WV. Crystallization kinetics of hard spheres in microgravity in the coexistence regime: interactions between growing crystallites. Phys Rev Lett 2002;88:015501.

[50] Clarke SM, Ottewill RH, Rennie AR. Light-scattering-studies of dispersions under shear. Adv Colloid Interface Sci 1995;60:95–118.

[51] Cheng ZD, Russell WB, Chaikin PM. Controlled growth of hard-sphere colloidal crystals. Nature 1999;401:893–895.

[52] Gasser U, Weeks ER, Schofield A, Pusey PN, Weitz DA. Real-space imaging of nucleation and growth in colloidal crystallization. Science 2001;292:258–262.

[53] de Hoog EHA, Kegel WK, van Blaaderen A, Lekkerkerker HNW. Direct observation of crystallization and aggregation in a phase-separating colloid–polymer suspension. Phys Rev E 2001;64:021407.

[54] Yethiraj A, van Blaaderen A. A colloidal model system with an interaction tunable from hard sphere to soft and dipolar. Nature 2003;421:513–517.

[55] Glotzer SC, Solomon MJ. Anisotropy of building blocks and their assembly into complex structures. Nat Mater 2007;6:557–562.

[56] Sugimoto T. Preparation of monodispersed colloidal particles. Adv Colloid Interface Sci 1987;28:65–108.

[57] Matijevic E, Hsu WP. Preparation and properties of monodispersed colloidal particles of lanthanide compounds: 1. Gadolinium, europium, terbium, samarium, and cerium(III). J Colloid Interface Sci 1987;118:506–523.

[58] Matijevic E. Preparation and properties of uniform size colloids. Chem Mater 1993;5:412–426.

[59] Eastoe J, Hollamby MJ, Hudson L. Recent advances in nanoparticle synthesis with reversed micelles. Adv Colloid Interface Sci 2006;128:5–15.

[60] Xia Y, Xiong YJ, Lim B, Skrabalak SE. Shape-controlled synthesis of metal nanocrystals: simple chemistry meets complex physics? Angew Chem – Int Ed 2009;48:60–103.

[61] Gorshkov V, Privman V. Models of synthesis of uniform colloids and nanocrystals. Phys E – Low-Dimens Syst Nanostruct 2010;43: 1–12.

[62] Sugimoto T. The theory of the nucleation of monodisperse particles in open systems and its application to AgBr systems. J Colloid Interface Sci 1992;150:208–225.

[63] Vanblaaderen A, Vangeest J, Vrij A. Monodisperse colloidal silica spheres from tetraalkoxysilanes – particle formation and growth mechanism. J Colloid Interface Sci 1992;154:481–501.

[64] Sollich P. Predicting phase equilibria in polydisperse systems. J Phys – Condens Matter 2002;14:R79–R117.

[65] Germain V, Pileni MP. Size distribution of cobalt nanocrystals: a key parameter in formation of columns and labyrinths in mesoscopic structures. Adv Mater 2005;17:1424.

[66] Kofke DA, Bolhuis PG. Freezing of polydisperse hard spheres. Phys Rev E 1999;59:618–622.

[67] Auer S, Frenkel D. Suppression of crystal nucleation in polydisperse colloids due to increase of the surface free energy. Nature 2001;413:711–713.

[68] Bartlett P. Freezing in polydisperse colloidal suspensions. J Phys – Condens Matter 2000;12:A275–A280.

[69] Schope HJ, Bryant G, van Megen W. Effect of polydispersity on the crystallization kinetics of suspensions of colloidal hard spheres when approaching the glass transition. J Chem Phys 2007;127:084505.

[70] Martin S, Bryant G, van Megen W. Crystallization kinetics of polydisperse colloidal hard spheres: experimental evidence for local fractionation. Phys Rev E 2003;67:061405.

[71] Martin S, Bryant G, van Megen W. Observation of a smectic like crystalline structure in polydisperse colloids. Phys Rev Lett 2003;90:255702.

[72] Martin S, Bryant G, van Megen W. Crystallization kinetics of polydisperse colloidal hard spheres. II. Binary mixtures. Physi Rev E 2005;71:021404.

[73] Schope HJ, Bryant G, van Megen W. Small changes in particle-size distribution dramatically delay and enhance nucleation in hard sphere colloidal suspensions. Phys Rev E 2006;74:060401.

[74] Schall P, Spaepen F. Dislocation imaging in fcc colloidal single crystals. Int J Mater Res 2006;97:958–962.

[75] Schope HJ. Formation of dried colloidal monolayers and multilayers under the influence of electric fields. J Phys – Condens Matter 2003;15:L533–L540.

[76] Schope HJ, Marnette O, van Megen W, Bryant G. Preparation and characterization of particles with small differences in polydispersity. Langmuir 2007;23:11534–11539.

[77] Chaikin PM, Donev A, Man WN, Stillinger FH, Torquato S. Some observations on the random packing of hard ellipsoids. Ind Eng Chem Res 2006;45:6960–6965.

[78] Barker JA, Henderson D. What is liquid – understanding states of matter. Rev Mod Phys 1976;48:587–671.

[79] Ree FH, Hoover WG. 5th + 6th Virial coefficients for hard spheres + hard disks. J Chem Phys 1964;40:939–950.

[80] Alder BJ, Wainwright TE. Studies in molecular dynamics. 2. Behavior of a small number of elastic spheres. J Chem Phys 1960;33:1439–1451.

[81] Hoover WG, Ree FH. Melting transition and communal entropy for hard spheres. J Chem Phys 1968;49:3609–17.

[82] Alder BJ, Hoover WG, Young DA. Studies in molecular dynamics. V. High-density equation of state and entropy for hard disks and spheres. J Chem Phys 1968;49:3688–3696.

[83] Tonks L. The complete equation of state of one, two and three-dimensional gases of hard elastic spheres. Phys Rev 1936;50:955–963.

[84] Wang XZ. Mean-field cage theory for the freezing of hard-sphere fluids. J Chem Phys 2005;122:044515.

[85] Frenkel D, Smit B. *Understanding Molecular Simulation: From Algorithms to Applications*. Boston: Academic Press; 1996.

[86] Clements M, Pullela SR, Mejia AF, Shen J, Gong T, Cheng Z. Thermosensitive hard spheres. J Colloid Interface Sci 2008;317:96–100.

[87] Carnahan NF, Starling KE. Equation of state for nonattracting rigid spheres. J Chem Phys 1969;51:635–636.

[88] Hall KR. Another hard-sphere equation of state. J Chem Phys 1972;57:2252–2254.

[89] Phan SE. *Thermodynamic and Rheological Properties of Hard Sphere Dispersions*. Princeton: Princeton University; 1998.

[90] Hales TC. A proof of the Kepler conjecture. Ann Math 2005;162:1065–1185.

[91] Hales TC, Ferguson SP. A formulation of the Kepler conjecture. Discrete Comput Geom 2006;36:21–69.

[92] Hales TC. Historical overview of the Kepler conjecture. Discrete Comput Geom 2006;36:5–20.

[93] Berryman JG. Random close packing of hard-spheres and disks. Phys Rev A 1983;27:1053–1061.

[94] Torquato S. Mean nearest-neighbor distance in random packings of hard D-dimensional spheres. Phys Rev Lett 1995;74:2156–2159.

[95] Torquato S. Nearest-neighbor statistics for packings of hard-spheres and disks. Phys Rev E 1995;51:3170–3182.

[96] Aste T, Saadatfar M, Senden TJ. Geometrical structure of disordered sphere packings. Phys Rev E 2005;71:061302.

[97] Song C, Wang P, Makse HA. A phase diagram for jammed matter. Nature 2008;453:629–632.

[98] Rintoul MD, Torquato S. Metastability and crystallization in hard-sphere systems. Phys Rev Lett 1996;77:4198–4201.

[99] Torquato S, Truskett TM, Debenedetti PG. Is random close packing of spheres well defined? Phys Rev Lett 2000;84:2064–2067.

[100] Torquato S, Stillinger FH. Multiplicity of generation, selection, and classification procedures for jammed hard-particle packings. J Phys Chem B 2001;105:11849–11853.

[101] Kansal AR, Torquato S, Stillinger FH. Diversity of order and densities in jammed hard-particle packings. Phys Rev E 2002;66:041109.

[102] O'Hern CS, Langer SA, Liu AJ, Nagel SR. Random packings of frictionless particles. Phys Rev Lett 2002;88:075507.

[103] Kamien RD, Liu AJ. Why is random close packing reproducible? Phys Rev Lett 2007;99:155501.

[104] Parisi G, Zamponi F. Mean-field theory of hard sphere glasses and jamming. Giorgio Parisi and Francesco Zamponi, Rev. Mod. Phys. 82: 789 – Published 16 March 2010.

[105] Mezard M, Parisi G, Tarzia M, Zamponi F. On the solution of a 'solvable' model of an ideal glass of hard spheres displaying a jamming transition. J. Stat. Mech. (2011) P03002 (http://iopscience.iop.org/1742-5468/2011/03/P03002).

[106] Jin YL, Makse HA. A first-order phase transition defines the random close packing of hard spheres. Phys A – Stat Mech Appl 2010;389:5362–5379.

[107] Wang P, Song CM, Jin YL, Makse HA. Jamming II: Edwards' statistical mechanics of random packings of hard spheres. Phys A – Stat Mech Appl 2010;390:427–455.

[108] Torquato S, Jiao Y. Dense packings of the Platonic and Archimedean solids. Nature 2009;460:876–879.

[109] Torquato S, Jiao Y. Dense packings of polyhedra: Platonic and Archimedean solids. Phys Rev E 2009;80:041104.

[110] Torquato S, Stillinger FH. Jammed hard-particle packings: from Kepler to Bernal and beyond. Rev Mod Phys 2010;82:2633–2672.

[111] Donev A, Stillinger FH, Chaikin PM, Torquato S. Unusually dense crystal packings of ellipsoids. Phys Rev Lett 2004;92:255506.

[112] Man WN, Donev A, Stillinger FH, Sullivan MT, Russel WB, Heeger D, Inati S, Torquato S, Chaikin PM. Experiments on random packings of ellipsoids. Phys Rev Lett 2005;94:198001.

[113] Donev A, Connelly R, Stillinger FH, Torquato S. Underconstrained jammed packings of nonspherical hard particles: ellipses and ellipsoids. Phys Rev E 2007;75:051304.

[114] Alexander S. Amorphous solids: their structure, lattice dynamics and elasticity. Phys Rep – Rev Sect Phys Lett 1998;296:65–236.

[115] Batten RD, Stillinger FH, Torquato S. Phase behavior of colloidal superballs: shape interpolation from spheres to cubes. Phys Rev E 2010;81:061105.

[116] Alder BJ, Carter BP, Young DA. Crystal transformation for hard spheres. Phys Rev 1969;183:831–833.

[117] Young DA, Alder BJ. Studies in molecular-dynamics. 13. Singlet and pair distribution functions for hard-disk and hard-sphere solids. J Chem Phys 1974;60:1254–1267.

[118] Alder BJ, Young DA, Mansigh MR, Salsburg ZW. Hard sphere equation of state in close-packed limit. J Comput Phys 1971;7:361.

[119] Frenkel D, Ladd AJC. New Monte-Carlo method to compute the free-energy of arbitrary solids – application to the fcc and hcp phases of hard-spheres. J Chem Phys 1984;81:3188–3193.

[120] Bolhuis PG, Frenkel D, Mau SC, Huse DA. Entropy difference between crystal phases. Nature 1997;388:235–236.

[121] Berg BA, Neuhaus T. Multicanonical algorithms for 1st order phase-transitions. Phys Lett B 1991;267:249–253.

[122] Berg BA, Neuhaus T. Multicanonical ensemble – a new approach to simulate 1st-order phase-transitions. Phys Rev Lett 1992;68:9–12.

[123] Woodcock LV. Entropy difference between the face-centred cubic and hexagonal close-packed crystal structures. Nature 1997;385:141–143.

[124] Bruce AD, Wilding NB, Ackland GJ. Free energy of crystalline solids: a lattice-switch Monte Carlo method. Phys Rev Lett 1997;79:3002–3005.

[125] Pusey PN, van Megen W, Bartlett P, Ackerson BJ, Rarity JG, Underwood SM. Structure of crystals of hard colloidal spheres. Phys Rev Lett 1989;63:2753–2756.

[126] Gasser U, Sierra-Martin B, Fernandez-Nieves A. Crystal structure of highly concentrated, ionic microgel suspensions studied by neutron scattering. Phys Rev E 2009;79:051403.

[127] Wilson AJC. Imperfections in the structure of cobalt. II. Mathematical treatment of proposed structure. Proc R Soc Lond A – Math Phys Sci 1942;180:0277–0285.

[128] Hendricks S, Teller E. X-ray interference in partially ordered layer lattices. J Chem Phys 1942;10:147–167.

[129] Loose W, Ackerson BJ. Model-calculations for the analysis of scattering data from layered structures. J Chem Phys 1994;101:7211–7220.

[130] Brindley GW, Mering J. Diffractions des rayons X par les structures en couches desordonnees. 1. Acta Crystallogr 1951;4:441–447.

[131] Guinier A. *X-ray Diffraction*. San Francisco: Freeman; 1963.

[132] Baxter RJ. Ornstein–Zernike relation for a disordered fluid. Aust J Phys 1968;21:563–569.

[133] Verlet L, Weis JJ. Equilibrium theory of simple liquids. Phys Rev A 1972;5:939–952.

[134] Turnbull D. Kinetics of solidification of supercooled liquid mercury droplets. J Chem Phys 1952;20:411–424.

[135] Russel WB. On the dynamics of the disorder order transition. Phase Transitions 1990;21:127–137.

[136] Ackerson BJ, Schatzel K. Classical growth of hard-sphere colloidal crystals. Phys Rev E 1995;52:6448–6460.

[137] Lekkerkerker HNW, van Duijneveldt JS. *Science and Technology of Crystal Growth*. Kluwer; 1995.

[138] He Y, Ackerson BJ, van Megen W, Underwood SM, Schätzel K. Dynamics of crystallization in hard-sphere suspensions. Phys Rev E 1996;54:5286–5297.

[139] Schatzel K, Ackerson BJ. Computer simulations of dense hard-sphere systems. J Chem Phys 1996;105:9258–9265.

[140] Dhont JKG, Smits C, Lekkerkerker HNW. A time resolved static light-scattering study on nucleation and crystallization in a colloidal system. J Colloid Interface Sci 1992;152:386–401.

[141] Gilmer GH. *Handbook of Crystal Growth*. Amsterdam: Elsevier; 1993. Vol. 1.

[142] Wurth M, Schwarz J, Culis F, Leiderer P, Palberg T. Growth-kinetics of body-centered-cubic colloidal crystals. Phys Rev E 1995;52:6415–6423.

[143] Harland JL, vanMegen W. Crystallization kinetics of suspensions of hard colloidal spheres. Phys Rev E 1997;55: 3054–3067.

[144] Harland JL, Henderson SI, Underwood SM, van Megen W. Observation of accelerated nucleation in dense colloidal fluids of hard sphere particles. Phys Rev Lett 1995, 75: 3572–3575.

[145] Tokuyama M, Enomoto Y. Dynamics of crossover phenomenon in phase-separating systems. Phys Rev Lett 1992;69:312–315.

[146] Sagui C, Ogorman DS, Grant M. Nucleation and growth: decay of a metastable state. Phys Rev E 1997;56:R21–R24.

[147] Gast AP, Monovoukas Y. A new growth instability in colloidal crystallization. Nature 1991;351:553–555.

[148] Langer JS. Instabilities and pattern-formation in crystal-growth. Rev Mod Phys 1980;52:1–28.

[149] Gibbs JW. *Collected Works*. New York: Yale University Press; 1948.

[150] Izmailov AF, Myerson AS. Theory of metastable state relaxation in a gravitational-field for noncritical binary-systems with nonconserved order-parameter. J Phys A – Math Gen 1993;26:2709–2725.

[151] Auer S, Frenkel D. Prediction of absolute crystal-nucleation rate in hard-sphere colloids. Nature 2001;409:1020–1023.

[152] Cheng Z. *Colloidal Hard Sphere Crystallization and Glass Transition*. Princeton: Princeton University; 1998.

[153] Kawasaki T, Tanaka H. Formation of a crystal nucleus from liquid. Proc Natl Acad Sci U S A 2010;107:14036–14041.

[154] Tanaka H. Bond orientational ordering in a metastable supercooled liquid: a shadow of crystallization and liquid–liquid transition. J. Stat. Mech. (2010) P12001 (http://iopscience.iop.org/1742-5468/2010/12/P12001).

[155] Tang SJ, Hu ZB, Cheng ZD, Wu JZ. Crystallization kinetics of thermosensitive colloids probed by transmission spectroscopy. Langmuir 2004;20:8858–8864.

[156] Filion L, Ni R, Frenkel D, Dijkstra M. Simulation of nucleation in almost hard-sphere colloids: the discrepancy between experiment and simulation persists. J Chem Phys 2011;134:134901.

[157] Filion L, Hermes M, Ni R, Dijkstra M. Crystal nucleation of hard spheres using molecular dynamics, umbrella sampling, and forward flux sampling: a comparison of simulation techniques. J. Chem. Phys. 133, 244115 (2010); http://dx.doi.org/10.1063/1.3506838

[158] Gong TY, Shen JY, Hu ZB, Marquez M, Cheng ZD. Nucleation rate measurement of colloidal crystallization using microfluidic emulsion droplets. Langmuir 2007;23:2919–2923.

[159] Coupland JN. Crystallization in emulsions. Curr Opin Colloid Interface Sci 2002;7:445–450.

[160] Laval P, Salmon JB, Joanicot M. A microfluidic device for investigating crystal nucleation kinetics. J Cryst Growth 2007;303:622–628.

[161] Kashchiev D, Kaneko N, Sato K. Kinetics of crystallization in polydisperse emulsions. J Colloid Interface Sci 1998;208:167–177.

[162] Velev OD, Lenhoff AM, Kaler EW. A class of microstructured particles through colloidal crystallization. Science 2000;287:2240–2243.

[163] Yi GR, Thorsen T, Manoharan VN, Hwang MJ, Jeon SJ, Pine DJ, Quake SR, Yang SM. Generation of uniform colloidal assemblies in soft microfluidic devices. Adv Mater 2003;15:1300–1304.

[164] Zheng B, Ismagilov RF. A microfluidic approach for screening submicroliter volumes against multiple reagents by using preformed arrays of nanoliter plugs in a three-phase liquid/liquid/gas flow. Angew Chem – Int Ed 2005;44: 2520–2523.

[165] Zheng B, Roach LS, Ismagilov RF. Screening of protein crystallization conditions on a microfluidic chip using nanoliter-size droplets. J Am Chem Soc 2003;125: 11170–11171.

[166] Zheng B, Tice JD, Ismagilov RF. Formation of arrayed droplets of soft lithography and two-phase fluid flow, and application in protein crystallization. Adv Mater 2004; 16:1365–1368.

[167] Zheng B, Gerdts CJ, Ismagilov RF. Using nanoliter plugs in microfluidics to facilitate and understand protein crystallization. Curr Opin Struct Biol 2005;15:548–555.

[168] Cheng ZD, Zhu JX, Russel WB, Chaikin PM. Phonons in an entropic crystal. Phys Rev Lett 2000;85:1460–1463.

[169] Hunter RJ. *Foundations of Colloid Sciences*. 2nd ed. New York: Oxford University Press; 2001.

[170] Herlach DM, Klassen I, Wette P, Holland-Moritz D. Colloids as model systems for metals and alloys: a case study of crystallization. J Phys – Condens Matter 2010;22:153101.

[171] Chaikin PM, Pincus P, Alexander S, Hone D. BCC–FCC, melting and reentrant transitions in colloidal crystals. J Colloid Interface Sci 1982;89:555–562.

[172] Sirota EB, Ouyang HD, Sinha SK, Chaikin PM, Axe JD, Fujii Y. Complete phase-diagram of a charged colloidal system – a synchrotron X-ray-scattering study. Phys Rev Lett 1989;62:1524–1527.

[173] Leunissen ME, Christova CG, Hynninen AP, Royall CP, Campbell AI, Imhof A, Dijkstra M, van Roij R, van Blaaderen A. Ionic colloidal crystals of oppositely charged particles. Nature 2005;437:235–240.

[174] Hiltner PA, Krieger IM. Diffraction of light by ordered suspensions. J Phys Chem 1969;73:2386–2389.

[175] Hachisu S, Kobayash Y, Kose A. Phase separation in monodisperse latexes. J Colloid Interface Sci 1973; 42:342–348.

[176] Pieranski P. Colloidal crystals. Contemp Phys 1983; 24:25–73.

[177] Vanmegen W, Snook I. Ordered states in systems of macroscopic-particles. Nature 1976;262:571–572.

[178] Lindsay HM, Chaikin PM. Elastic properties of colloidal crystals and glasses. J Chem Phys 1982;76:3774–3781.

[179] Medeirosesilva J, Mokross BJ. Solid-like phase-transitions in a screened Wigner lattice – statics. Phys Rev B 1980; 21:2972–2976.

[180] Tata BVR. Colloidal dispersions and phase transitions in charged colloids. Curr Sci 2001;80:948–958.

[181] Crocker JC, Grier DG. Microscopic measurement of the pair interaction potential of charge-stabilized colloid. Phys Rev Lett 1994;73:352–355.

[182] Sainis SK, Germain V, Dufresne ER. Statistics of particle trajectories at short time intervals reveal fN-scale colloidal forces. Phys Rev Lett 2007;99:018303.

[183] Sainis SK, Germain V, Mejean CO, Dufresne ER. Electrostatic interactions of colloidal particles in nonpolar solvents: role of surface chemistry and charge control agents. Langmuir 2008;24:1160–1164.

[184] Pugh RJ, Matsunaga T, Fowkes FM. The dispersibility and stability of carbon-black in media of low dielectric-constant. 1. Electrostatic and steric contributions to colloidal stability. Colloids Surf 1983;7:183–207.

[185] Comiskey B, Albert JD, Yoshizawa H, Jacobson J. An electrophoretic ink for all-printed reflective electronic displays. Nature 1998;394:253–255.

[186] Roberts GS, Wood TA, Frith WJ, Bartlett P. Direct measurement of the effective charge in nonpolar suspensions by optical tracking of single particles. J Chem Phys 2007;126:194503.

[187] Wette P, Schope HJ, Palberg T. Comparison of colloidal effective charges from different experiments. J Chem Phys 2002;116:10981–10988.

[188] Wette P, Schope HJ, Palberg T. Experimental determination of effective charges in aqueous suspensions of colloidal spheres. Colloid Surf A – Physicochem Eng Asp 2003;222:311–321.

[189] Wette P, Klassen I, Holland-Moritz D, Herlach DM, Schope HJ, Lorenz N, Reiber H, Palberg T, Roth SV. Communications: complete description of re-entrant phase behavior in a charge variable colloidal model system. J Chem Phys 2010;132:131102.

[190] Hessinger D, Evers M, Palberg T. Independent ion migration in suspensions of strongly interacting charged colloidal spheres. Phys Rev E 2000;61:5493–5506.

[191] Alexander S, Chaikin PM, Grant P, Morales GJ, Pincus P, Hone D. Charge renormalization, osmotic-pressure, and bulk modulus of colloidal crystals – theory. J Chem Phys 1984;80:5776–5781.

[192] Russ C, von Grunberg HH, Dijkstra M, van Roij R. Three-body forces between charged colloidal particles. Phys Rev E 2002;66:011402.

[193] Biesheuvel PM. Evidence for charge regulation in the sedimentation of charged colloids. J Phys – Condens Matter 2004;16:L499–L504.

[194] van Roij R. Defying gravity with entropy and electrostatics: sedimentation of charged colloids. J Phys – Condens Matter 2003;15:S3569–S3580.

[195] Royall CP, Leunissen ME, Hynninen AP, Dijkstra M, van Blaaderen A. Re-entrant melting and freezing in a model system of charged colloids. J Chem Phys 2006;124:244706.

[196] Trizac E, Bocquet L, Aubouy M. Simple approach for charge renormalization in highly charged macroions. Phys Rev Lett 2002;89:248301.

[197] Ninham, B.W., Parsegia, V.A. Electrostatic potential between surfaces bearing ionizable groups in ionic equilibrium with physiologic saline solution. J Theor Biol. 1971, 31: 405–428.

[198] Popa I, Sinha P, Finessi M, Maroni P, Papastavrou G, Borkovec M. Importance of charge regulation in attractive double-layer forces between dissimilar surfaces. Phys Rev Lett 2010;104:228301.

[199] Royall CP, Leunissen ME, van Blaaderen A. A new colloidal model system to study long-range interactions quantitatively in real space. J Phys – Condens Matter 2003;15:S3581–S3596.

[200] Bevan MA, Eichmann SL. Optical microscopy measurements of kT-scale colloidal interactions. *Current Opinion in Colloid and Interface Science* 2011;16: 149–157.

[201] Larsen AE, Grier DG. Like-charge attractions in metastable colloidal crystallites. Nature 1997;385:230–233.

[202] Schmitz KS. Volume-term theories, Sogami-Ise potential, and the Langmuir model for phase separation in macroion systems: a resolution. Phys Rev E 2002;65:061402.

[203] Warren PB. A theory of void formation in charge-stabilized colloidal suspensions at low ionic strength. J Chem Phys 2000;112:4683–4698.

[204] Go D, Kodger TE, Sprakel J, Kuehne AJC. Programmable co-assembly of oppositely charged microgels. Soft Matter 2014;10:8060–8065.

[205] Vermolen ECM, Kuijk A, Filion LC, Hermes M, Thijssen JHJ, Dijkstra M, van Blaaderen A. Fabrication of large binary colloidal crystals with a NaCl structure. Proc Natl Acad Sci U S A 2009;106:16063–16067.

[206] Bartlett P, Campbell AI. Three-dimensional binary superlattices of oppositely charged colloids. Phys Rev Lett 2005;95:128302.

[207] Sharma V, Yan QF, Wong CC, Cartera WC, Chiang YM. Controlled and rapid ordering of oppositely charged colloidal particles. J Colloid Interface Sci 2009;333:230–236.

[208] Walker DA, Kowalczyk B, de la Cruz MO, Grzybowski BA. Electrostatics at the nanoscale. Nanoscale 2011;3: 1316–1344.

[209] Kalsin AM, Fialkowski M, Paszewski M, Smoukov SK, Bishop KJM, Grzybowski BA. Electrostatic self-assembly of binary nanoparticle crystals with a diamond-like lattice. Science 2006;312:420–424.

[210] Browning AR, Doherty MF, Fredrickson GH. Nucleation and polymorph selection in a model colloidal fluid. Phys Rev E 2008;77:041604.

[211] Rosenbaum DF, Kulkarni A, Ramakrishnan S, Zukoski CF. Protein interactions and phase behavior: sensitivity to the form of the pair potential. J Chem Phys 1999;111:9882–9890.

[212] Berland CR, Thurston GM, Kondo M, Broide ML, Pande J, Ogun O, Benedek GB. Solid liquid-phase boundaries of lens protein solutions. Proc Natl Acad Sci U S A 1992; 89:1214–1218.

[213] Orea P, Tapia-Medina C, Pini D, Reiner A. Thermodynamic properties of short-range attractive Yukawa fluid: simulation and theory. J Chem Phys 2010;132:114108.

[214] Hagen H, Frenkel D. Determination of phase diagrams for the hard-core attractive Yukawa system. J Chem Phys 1994;101:4093–4097.

[215] Ishimoto C, Tanaka T. Critical behavior of a binary mixture of protein and salt-water. Phys Rev Lett 1977;39:474–477.

[216] Vliegenthart GA, Lekkerkerker HNW. Predicting the gas–liquid critical point from the second virial coefficient. J Chem Phys 2000;112:5364–5369.

[217] tenWolde PR, Frenkel D. Enhancement of protein crystal nucleation by critical density fluctuations. Science 1997;277:1975–1978.

[218] Auer S, Frenkel D. Crystallization of weakly charged colloidal spheres: a numerical study. J Phys – Condens Matter 2002;14:7667–7680.

[219] Wette P, Schope HJ. Nucleation kinetics in deionized charged colloidal model systems: a quantitative study by means of classical nucleation theory. Phys Rev E 2007;75:051405.

[220] Xu SH, Zhou HW, Sun ZW, Xie JC. Formation of an fcc phase through a bcc metastable state in crystallization of charged colloidal particles. Phys Rev E 2010;82:010401 (R).

[221] Zhou H, Xu S, Sun Z, Du X, Liu L. Kinetics study of crystallization with the disorder-bcc–fcc phase transition of charged colloidal dispersions. Langmuir 2011;27:7439–7445.

[222] Lietor-Santos JJ, Gasser U, Vavrin R, Hu ZB, Fernandez-Nieves A. Structural changes of poly(*N*-isopropylacrylamide)-based microgels induced by hydrostatic pressure and temperature studied by small angle neutron scattering. J Chem Phys 2010;133:034901.

[223] Pelton R. Temperature-sensitive aqueous microgels. Adv Colloid Interface Sci 2000;85:1–33.

[224] Lu Y, Ballauff M. Thermosensitive core-shell microgels: from colloidal model systems to nanoreactors. Prog Polym Sci 2011;36:767–792.

[225] Tang SJ, Hu ZB, Zhou B, Cheng ZD, Wu JZ, Marquez M. Melting kinetics of thermally responsive microgel crystals. Macromolecules 2007;40:9544–9548.

[226] Peng Y, Wang ZR, Alsayed AM, Yodh AG, Han Y. Melting of multilayer colloidal crystals confined between two walls. Phys Rev E 2011;83:011404.

[227] Cho JK, Meng ZY, Lyon LA, Breedveld V. Tunable attractive and repulsive interactions between pH-responsive microgels. Soft Matter 2009;5:3599–3602.

[228] Wu JZ, Zhou B, Hu ZB. Phase behavior of thermally responsive microgel colloids. Phys Rev Lett 2003;90:048304.

[229] Heyes DM, Branka AC. Interactions between microgel particles. Soft Matter 2009;5:2681–2685.

[230] Agrawal R, Kofke DA. Solid–fluid coexistence for inverse-power potentials. Phys Rev Lett 1995;74:122–125.

[231] Pyett S, Richtering W. Structures and dynamics of thermosensitive microgel suspensions studied with three-dimensional cross-correlated light scattering. J Chem Phys 2005;122:034709.

[232] St John AN, Breedveld V, Lyon LA. Phase behavior in highly concentrated assemblies of microgels with soft repulsive interaction potentials. J Phys Chem B 2007;111:7796–7801.

[233] Eckert T, Richtering W. Thermodynamic and hydrodynamic interaction in concentrated microgel suspensions: hard or soft sphere behavior? J Chem Phys 2008;129:124902.

[234] Paloli D, Mohanty PS, Crassous JJ, Zaccarelli E, Schurtenberger P. Fluid–solid transitions in soft-repulsive colloids. Soft Matter 2013;9:3000–3004.

[235] Mohanty PS, Paloli D, Crassous JJ, Zaccarelli E, Schurtenberger P. Effective interactions between soft-repulsive colloids: experiments, theory, and simulations. J Chem Phys 2014;140:094901.

[236] Hashmi SM, Dufresne ER. Mechanical properties of individual microgel particles through the deswelling transition. Soft Matter 2009;5:3682–3688.

[237] Gasser U, Fernandez-Nieves A. Crystal structure of highly concentrated, ionic microgel suspensions studied by small-angle X-ray scattering. Phys Rev E 2010;81:052401.

[238] Gottwald D, Likos CN, Kahl G, Lowen H. Phase behavior of ionic microgels. Phys Rev Lett 2004;92:068301.

[239] Hellweg T, Dewhurst CD, Bruckner E, Kratz K, Eimer W. Colloidal crystals made of poly(*N*-isopropylacrylamide) microgel particles. Colloid Polym Sci 2000;278:972–978.

[240] Mohanty PS, Richtering W. Structural ordering and phase behavior of charged microgels. J Phys Chem B 2008;112:14692–14697.

[241] Sierra Martin B, Fernandez Nieves A. Phase and non-equilibrium behaviour of microgel suspensions as a function of particle stiffness. Soft Matter 2012;8:4141–4150.

[242] Alsayed AM, Islam MF, Zhang J, Collings PJ, Yodh AG. Premelting at defects within bulk colloidal crystals. Science 2005;309:1207–1210.

[243] Lyon LA, Debord JD, Debord SB, Jones CD, McGrath JG, Serpe MJ. Microgel colloidal crystals. J Phys Chem B 2004;108:19099–19108.

[244] Peng Y, Wang Z, Alsayed AM, Yodh AG, Han Y. Melting of colloidal crystal films. Phys Rev Lett 2010;104:205703.

13

THE GLASS TRANSITION

JOHAN MATTSSON
School of Physics and Astronomy, University of Leeds, Leeds, United Kingdom

13.1 INTRODUCTION

A glass is a solid that lacks an ordered structure of its building blocks. Common glasses are found in windows, computer screens, watches, lenses, building materials or in artistic products such as vases or sculptures. Many polymeric materials either contain glassy parts or are fully glassy; optical fibers are made from silica glass and the preservation

Fluids, Colloids and Soft Materials: An Introduction to Soft Matter Physics, First Edition. Edited by Alberto Fernandez Nieves and Antonio Manuel Puertas.

of proteins, biological cells, or foods is often achieved using glassy states formed from aqueous sugar solutions. Another important example is the rapid development of glassy pharmaceuticals, which is driven by advantages in terms of dissolution and processing of the amorphous state compared with the crystalline state. Glasses also appear in nature; you find glassy states in volcanic lava, in some plants, and animals; and aerosols, in particular types of clouds, can exist in glassy states that might affect a range of important atmospheric processes.

The building blocks of a glass do not have to be molecular or atomic. Instead a glass can be made from crowded colloidal particles dispersed in a fluid, where the particles might be hard or soft, or have different shapes or interactions. In fact, many similarities to glassy behavior are observed for an even wider range of materials, including emulsions, pastes, foams, granular systems, or even the motion of people or vehicles in crowded environments. The common denominator is that as the environment gets crowded, the movement of the fundamental units needs to be coordinated for any motion to take place and at high enough level of crowding, motions on longer length scales come to a halt.

Given the variety of materials that display some elements of glass formation, it is evident that a book chapter requires a narrowed scope. Usually, this is achieved by discussing glass formation in either molecular/atomic or colloidal materials, respectively, but not in both. To some extent, this division reflects that there are obvious differences between the systems, but a less fortunate side effect is the lack of opportunity to highlight interesting and illuminating analogies. To address this, we have here chosen to provide an introduction to both molecular and colloidal glass formation, with a particular focus on comparison between the two. By including both classes of glass-forming materials, we are by necessity forced to compromise on some system-dependent details. However, whenever possible, we have tried to provide the reader with suitable references for further reading. Thus, we hope that this chapter will serve as a broad background to glass formation and stimulate further reading and study across the different subfields, as we are convinced that such diversions will be both interesting and productive. Finally, it is worth pointing out that there are hardly any topics concerning glass formation that are not vividly debated. This is generally a healthy state of affairs since it demonstrates both the importance and vitality of the field as well as the excitement and strong engagement of all of us interested in it. However, it does mean that it is nearly impossible to write a chapter that covers and satisfies all views. We have here tried to strike a balance between providing a coherent view, yet emphasizing particularly debated areas, and providing some key literature references to assist the reader.

13.2 BASICS OF GLASS FORMATION

13.2.1 Basics of Glass Formation in Molecular Systems

For a molecular liquid that is cooled below its melting point, the ordered crystalline state corresponds to the minimum free energy and is thus thermodynamically preferred. Crystallization starts with thermal fluctuations leading to the formation of a crystalline nucleus. The nucleus formation can be facilitated by the presence of specific nucleation sites such as interfaces or heterogeneities in the sample (heterogeneous nucleation) or occur spontaneously without preferential nucleation sites (homogeneous nucleation). The growth of a crystal from the nucleus requires molecular mobility, but as the liquid is cooled below its melting point into its supercooled state, the molecular motions slow down dramatically. Thus, for high enough cooling rates, there is not enough time for crystallization to occur before the molecular mobility becomes so restricted that no long-range motions take place on experimental timescales. The result is a solid material with a disordered liquid-like structure – a glass. In some molecular or polymeric systems, glasses are also easily formed without quenching since crystallization is hindered by the presence of molecular structural motifs that prevent the formation of an ordered state. One example of such a material is atactic polystyrene, which is found in plastic cups or Styrofoam. Its molecular structure includes phenyl rings that are randomly arranged along the polymeric backbone. This arrangement prevents order and makes polystyrene an excellent glass-former. The experimental glass transition is thus simply a crossing of the characteristic molecular and experimental timescales and is in itself no mystery. The mystery lies instead in the origin and evolution of the slowing down of molecular motions as liquids become supercooled and approach their glassy state.

The experimental glass-transition temperature T_g is usually defined as the temperature at which the shear viscosity reaches a value of 10^{12} P. The shear viscosity η is the link between shear stress σ and shear strain rate $\dot{\gamma}$, where $\dot{\gamma} = d\gamma/dt$ and γ is the shear strain. For Newtonian fluids [1] where the viscosity is independent of the strain rate, the simple relation $\sigma = \eta\dot{\gamma}$ holds. The viscosity is thus a measure of a liquid's resistance to flow. The viscosity is also related [2] to the timescale on which the liquid structure relaxes, the so-called structural relaxation or α-relaxation time τ_α, as $\tau_\alpha = \eta / G_\infty$. Here, G_∞ denotes the "instantaneous" or "infinite frequency" shear modulus, which typically has a value in the range of 10^9–10^{10} Pa for solids [2]. Thus, a T_g defined through $\eta = 10^{12}$ P corresponds to the temperature where the structural relaxation time $\tau_\alpha \approx 100$ s, which is a commonly used definition of T_g based on structural relaxation data.

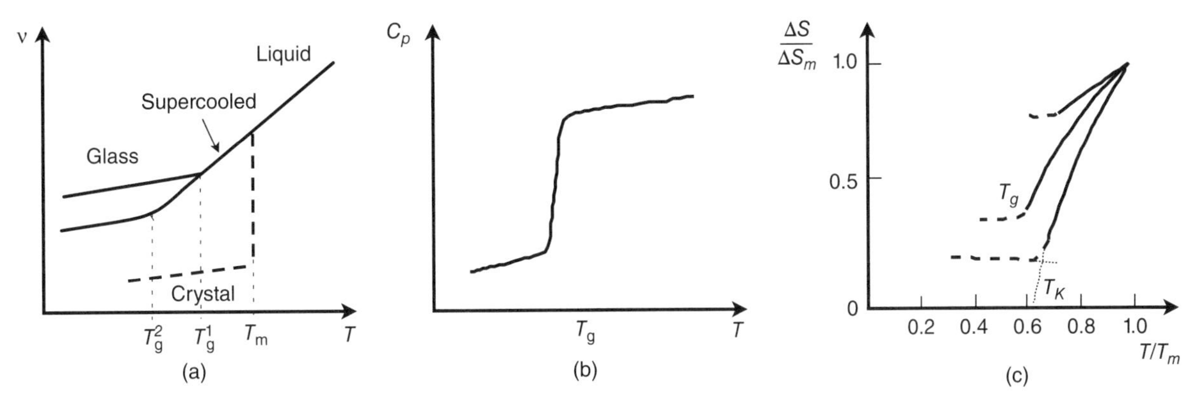

Figure 13.1 (a) Sketch of the specific volume as a function of temperature. The dashed line shows the path for crystallization where T_m denotes the melting point. A high cooling rate results in a glass transition at T_g^1, whereas a low cooling rate yields a transition at T_g^2; (b) sketch of the specific heat as a function of temperature showing the step observed at T_g; and (c) sketch of the temperature-dependent excess entropy for liquids with different fragility (see Section 13.5.3). Source: Adapted from Debenedetti 2001 [3], figure 4. Reproduced with permission of [3]. Copyright 2001 by Macmillan Publishers Ltd.

Crystallization is a so-called first-order thermodynamic phase transition. This means that thermodynamic properties such as volume or entropy, which are obtained as first derivatives of the free energy with regard to a thermodynamic variable, are discontinuous across the transition. This behavior is illustrated for the specific volume by the dashed line in Figure 13.1a. An experimental glass transition, on the other hand, is not a thermodynamic phase transition and is instead characterized by a continuous transition from the liquid state to the glassy state, as shown in Figure 13.1a. The glass-transition is also observed as a quite sharp step in the specific heat (a second order thermodynamic property), as shown in Figure 13.1b. This change is due to the increased molecular mobility corresponding to the degrees of freedom (DOF) that are released as the material goes from glass to liquid. Moreover, the location of the glass transition depends on the cooling/heating rate, as shown in Figure 13.1a, where the glass transition T_g^1 is observed by the use of a higher cooling rate than T_g^2. This rate dependence is a natural consequence of the fact that the glass transition arises from a crossing of the molecular and experimental timescales.

Experimentally, the standard method to determine the specific heat and thus T_g is by the use of differential scanning calorimetry (DSC). The sample is placed in a small pan, and the difference between the heat flows fed to the sample pan and an empty reference pan, respectively, is measured as the temperature is either increased or decreased through the transition at a constant rate. The difference between the sample and reference heat flows divided by the heating rate gives the heat capacity and division by the sample mass yields the specific heat c_p. A simple relationship exists between the applied rate in a DSC experiment and the probed structural relaxation timescale, and it can be shown that for typical glass-forming liquids, the specific heat step occurs at $T_g = T(\tau_\alpha = 100\,\text{s})$ when the measurement is performed at a rate of ca 10 K/min [4].

Another important property to consider with regard to glass formation is the entropy. The entropy of a liquid can be divided into two contributions, one due to vibrational DOF S_v and a second due to configurational DOF S_c [3, 5]. Since the entropy of a liquid $S = S_v + S_c$ is directly related to its heat capacity through $S(T) = \int c_p(T')/T'\,dT'$, it can be determined using calorimetry. In terms of glass formation, one is normally interested in S_c within the supercooled state, and to achieve this and determine it from S, S_v is often approximated by extrapolation of the crystal entropy into the liquid temperature range. Note that the vibrational properties of a crystal are not identical to those of the corresponding liquid [2], but for the main argument here this approximation is sufficient. Figure 13.1c shows the so-called excess entropy ΔS_{exc} defined as the entropy of the liquid minus the extrapolated crystal entropy, $\Delta S_{\text{exc}} = S - S_{\text{crystal}}$ where ΔS_{exc} can be viewed as an estimate of S_c. As the temperature is reduced below the melting temperature T_m, ΔS_{exc} decreases dramatically and this decrease can be directly linked to the slowing down of dynamics as the glass-transition is approached, as described in the paragraph on entropy-based models of glass-formation in Section 8.2. Liquids with different fragility (see Section 5.3) will show ΔS_{exc} decreasing at different rates, as shown in the sketch in Figure 13.1c. The strong decrease of ΔS_{exc} also implies that it extrapolates to zero at some temperature T_K, which means that further cooling at equilibrium would imply a negative ΔS_{exc} and eventually a negative entropy all together. This is clearly unphysical and is often termed an "entropy crisis" [3, 6]. To avoid this "crisis," it has been speculated that a thermodynamic phase transition, an "ideal" glass transition, might occur close to T_K, which would bring the system from the supercooled liquid state into an "ideal" thermodynamically preferred glassy state, thus avoiding the crisis. However, criticism has been raised against the underlying extrapolation [2, 3], and from a general perspective it is not at all clear what properties a unique "ideal" disordered glassy state would have. The question of how an equilibrium liquid behaves at temperatures significantly below T_g continues to be a much discussed topic [3, 6, 7].

13.2.2 Basics of Glass Formation in Colloidal Systems

A colloidal system consists of small nanometer- to micron-sized particles suspended in a liquid. Just as atomic or molecular liquids, the particles of a colloidal suspension can undergo a transition from a fluid to a disordered solid glassy state. In a colloidal system, however, the glass is formed as the colloid volume fraction ϕ is increased enough to induce dynamic arrest by particle crowding. The phase behavior for a simple colloidal system consisting of "hard spheres" is illustrated in Figure 13.2 [8–10]. For a hard-sphere colloid, the interparticle potential goes from zero to infinite as the particles touch and temperature thus plays no role in the interparticle interaction. The phase diagram in Figure 13.2 is thus controlled solely by entropy, where crystallization is driven by the fact that the formation of the ordered crystal state actually increases the DOF available for motion and thus the entropy. As illustrated in Figure 13.2, for $\phi < 0.494$, a hard-sphere colloidal suspension is in a disordered fluid state and the particles can move easily. In the ϕ-range between 0.494 and 0.545, the fluid and crystalline states coexist but for higher ϕ, the preferred state is the crystal. In the ordered crystalline state, volume fractions up to $\phi = 0.74$ can be reached for hexagonally close packed cubic crystal structures. Moreover, above $\phi = 0.494$, the suspension can exist in a metastable "supercooled" state if crystallization is avoided, in analogy to the behavior of molecular liquids. In an experiment, supercooling can be achieved by using colloids with a significant size polydispersity or by quickly changing ϕ for instance by centrifugation to values where the colloid motions are slowed down enough for crystallization to be inhibited. At $\phi_g \geq 0.58$, a near monodisperse hard-sphere suspension forms a glass; the exact value of ϕ_g, however, depends on the size polydispersity. A glassy suspension can be further packed up to the so-called random close packing at $\phi_{rcp} \approx 0.64$, which corresponds to the highest packing fraction a disordered arrangement of monodisperse spheres can reach. As for molecular systems, glass-formation in colloidal systems can be described in terms of the evolution of the configurational entropy [11, 12].

13.3 STRUCTURE OF MOLECULAR OR COLLOIDAL GLASS-FORMING SYSTEMS

In contrast to the crystalline state, the glassy state is not characterized by any long-range ordered arrangements of the building blocks. Figure 13.3a–c shows optical microscopy images of colloidal hard-sphere suspensions at different colloid volume fractions: Figure 13.3a shows a fluid suspension at a relatively low volume fraction ($\phi \approx 0.4$), where the lack of order characteristic of a liquid is found; Figure 13.3b is the corresponding image for a suspension above its glass transition ($\phi \geq 0.58$); and even though the volume fraction and thus the particle density is higher, the lack of any obvious order is clear and stands in contrast to the ordered nature of the colloidal crystal shown in Figure 13.3c. Thus, except for an increase in particle density, no significant structural changes are directly observed during glass formation. We note, however, that detailed investigations of 'particle' coordinations both in molecular and colloidal systems have found evidence for growing bond-orientational order as the glass-transition is approached [15a, 16a].

For molecular systems or for colloids of small size, direct space studies are difficult. Scattering techniques can then alternatively be used to study the arrangement of building blocks. In a scattering experiment, a beam of coherent radiation, such as laser light, is focused onto a sample, and the scattered intensity is recorded as a function of both time and the angle between the incoming and outgoing light, the scattering angle θ. The incoming beam is characterized by a wave vector $\vec{k}_i$ and the outgoing beam by $\vec{k}_f$. Since momentum $\hbar\vec{k}$ has to be conserved, $\vec{k}_f - \vec{k}_i = \vec{Q}$, where $\vec{Q}$ is the so-called scattering vector and

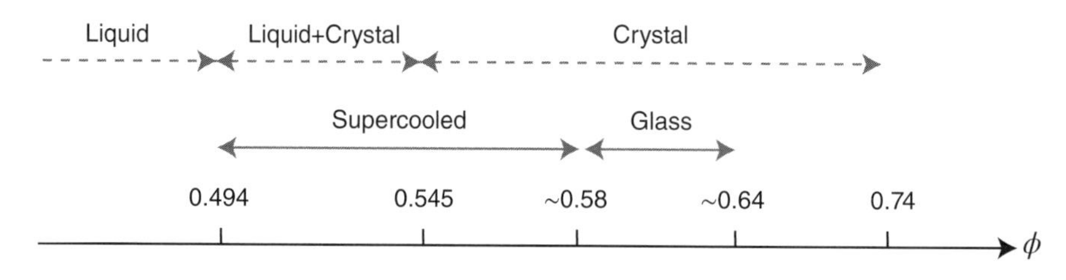

Figure 13.2 Sketch of the equilibrium and nonequilibrium states of a suspension of colloidal hard spheres.

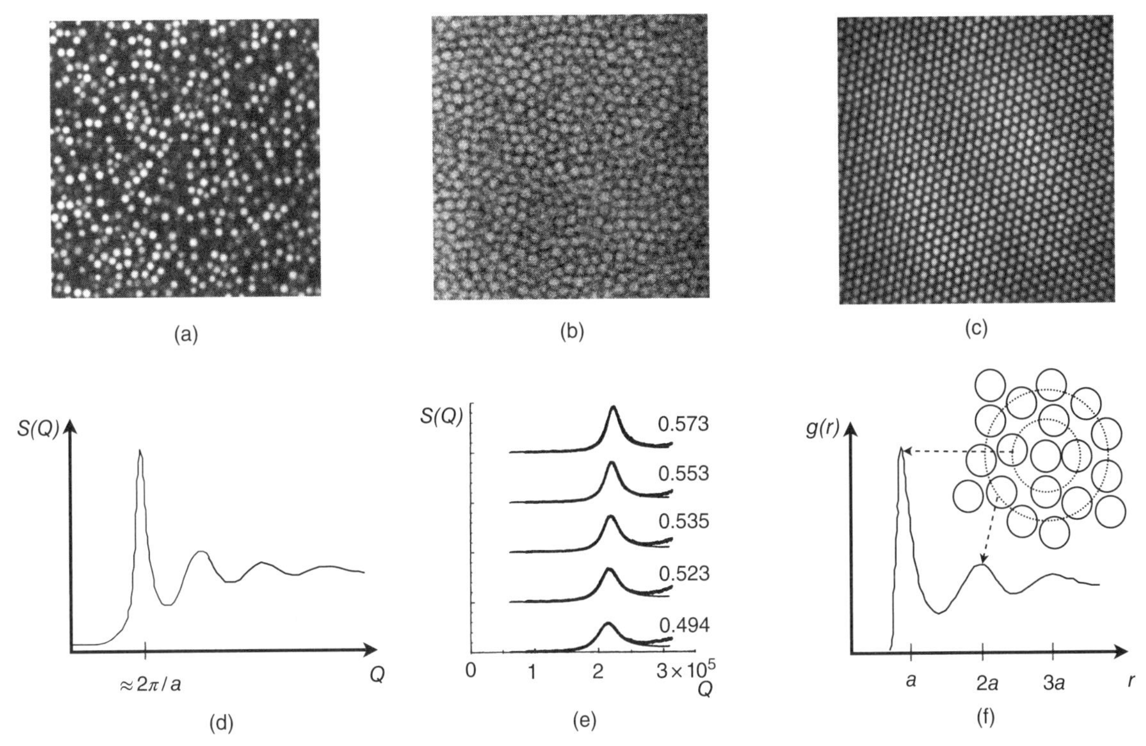

Figure 13.3 Confocal microscopy image of a colloidal hard sphere: (a) liquid [13], (b) glass, and (c) crystal [13]; (d) sketch of a typical structure factor $S(Q)$ for a dense liquid or glass; (e) experimental $S(Q)$ measured using light scattering for a hard-sphere suspension at the indicated volume fractions [14]; (f) sketch of a typical pair distribution function for a dense liquid or glass, indicating the coordination shells surrounding any particle. (d) and (f): Source: Adapted from Reichman 2005 [15b], figure 2. SISSA Medialab Srl. Reproduced with permission of IOP Publishing. All rights reserved. (e): Source: Adapted from van Megen 1991 [14], figure 1. Reproduced with permission of the American Physical Society. Copyright 1991 by the American Physical Society.

$\hbar\vec{Q}$ is the momentum exchanged in the scattering event. In a light-scattering experiment, the energy change involved in the scattering is generally negligible so that $|\vec{k}_f| = |\vec{k}_i| = k$, which leads to a simple relation between $Q = |\vec{Q}|$, k and θ: $Q = 4\pi n/\lambda_0 \cdot \sin(\theta/2)$, where n is the index of refraction of the scattering medium and λ_0 is the vacuum wave length of the laser light. Thus, by setting θ you set the probed Q, and the inverse of Q is directly related to the length scale probed in the experiment.

The real space lattice spacing d of a crystalline material corresponds to a spacing in reciprocal space of $2\pi/d$, and a scattering experiment thus shows strongly scattered intensity at Bragg peaks located at $Q_p = 2\pi/d$. For disordered materials, on the other hand, the reciprocal space is not quantized, but the structural information is still there in the Q-dependent scattering. Here, we introduce a few key concepts that are used throughout this chapter in the discussion on scattering from either molecular or colloidal systems. The number density of "particles" (here meaning colloids, atoms, or molecules) in a liquid can be expressed as $\rho(\vec{r},t) = \sum_i \delta(\vec{r} - \vec{r}_i(t))$, where δ is the Kronecker delta. A spatial Fourier transform of this particle density gives [15b, 16b]

$$\rho_k(t) = \sum_i \int d\vec{r}\, e^{i\vec{k}\cdot\vec{r}}\, \delta(\vec{r} - \vec{r}_i(t)) = \sum_i e^{i\vec{k}\cdot\vec{r}_i(t)},$$

where $\vec{k}$ denotes a particular Fourier component. In a time-dependent scattering experiment, one probes the so-called intermediate scattering function (ISF),

$$F(k,t) = \frac{1}{N}\langle \rho_{-k}(0)\, \rho_k(t)\rangle = \frac{1}{N}\sum_{ij}\left\langle e^{-i\vec{k}\cdot\vec{r}_i(0)} e^{i\vec{k}\cdot\vec{r}_j(t)}\right\rangle,$$

where N is the number of "particles" contributing to the scattering and $\langle\rangle$ denotes an ensemble average. The ISF correlates a particular Fourier component of the density (as chosen by the scattering vector, $k = Q$) at two different times separated by the lag-time t (the initial time is here set to 0 for simplicity, but when the ISF is calculated it is averaged over the initial times). If we are only interested in the static structural arrangement of the "particles," we can study a snapshot of the structure by setting $t = 0$ in the ISF. The resulting spatial correlation function $S(k) = F(k, t = 0)$ is called the static structure factor

$$F(k,\ t = 0) = \frac{1}{N}\langle \rho_{-k}(0)\rho_k(0)\rangle = S(k).$$

The static structure factor can be determined from a scattering experiment by measuring the scattered intensity as a function of Q. A sketch of the typical behavior of $S(Q)$ for a dense one-component liquid or glass is shown in Figure 13.3d. Since the structure is disordered, there are no lattice planes and thus no Bragg peaks. However, the spatial Fourier components probed through the scattering vectors Q still characterize the structure, and the first peak in $S(Q)$ is approximately related to $2\pi/a$ where a is the particle diameter, since for a dense system a is approximately the interparticle distance. Figure 13.3e shows $S(Q)$ as determined using light scattering for a hard-sphere colloidal suspension approaching the glass transition. It is clear that no obvious significant structural evolution is observed except for a small shift of the main peak toward larger Q corresponding to shorter length scales as the particles become more highly packed.

The structure factor $S(k)$ is related through a spatial Fourier transform to a real space function $g(r)$ called the pair distribution function. For a spatially isotropic system, $S(k) = 1 + \rho \int d\vec{r} e^{-i\vec{k}\vec{r}} g(r)$, where ρ is the average particle density [15b]. The function $g(r)$ is a probability density and is directly proportional to the probability of finding a particle a distance r away, given that there is a particle at the origin $r = 0$. Figure 13.3f shows the typical behavior of $g(r)$ for a dense single-component system. Due to packing, $g(r)$ shows oscillations at small distances (small r) from a given particle, resulting from shells with higher than average density of particles. The packing thus gives rise to a local structure also in a disordered system, even though the long-range order is missing.

13.4 DYNAMICS OF GLASS-FORMING MOLECULAR SYSTEMS

13.4.1 Relaxation Dynamics as Manifested in the Time Domain

The dynamics of supercooled liquids and glasses are characterized by a range of relaxation processes spanning a wide range of timescales covering ~15 orders of magnitude ~10^{-12} to 10^3 s. The large dynamic range makes strong requirements on the experimental methods that are suitable. Since the focus of this chapter is not on experimental techniques, we concentrate on data from two widely used techniques, dielectric broadband spectroscopy and dynamic light scattering, both of which can cover the relevant dynamic range.

The relaxation processes observed for a glass-former are often denoted using Greek letters $\alpha, \beta, \gamma, \delta$, and so on starting with the relaxation occurring on the longest timescales. I here restrict the discussion to the α- and β-relaxations since they both have a strongly intermolecular nature and are generic to all or at least most glass-forming liquids [3, 6, 7], whereas the faster relaxation mechanisms can often be related to more localized largely intramolecular motion. Importantly, there is evidence for direct links between the α- and β-relaxations [17, 18] suggesting that they are both fundamental elements of glass-formation.

The structural or α-relaxation takes place on a timescale τ_α that corresponds to the timescale on which the structural correlations observed in $S(Q)$ decay. It is the relation between τ_α and the experimental timescale of observation that determines if a material appears liquid or solid. If you probe a sample on timescales longer than τ_α it behaves like a liquid, whereas on shorter times it appears solid. For an atomic or molecular liquid, τ_α is related to the viscosity, as discussed in Section 13.2.1. However, note that this is not true for polymeric glass-formers where the viscosity is strongly influenced by the polymer chain length, whereas the α-relaxation is independent of chain length for long enough chains [19].

In a dynamic light-scattering experiment performed in time domain, you measure the scattered intensity $I(Q, t)$ as a function of time for a chosen scattering vector, Q. You then calculate the intensity correlation function by correlating pairs of scattered intensities separated by a lag-time t as $\langle I(0)I(t)\rangle$, where $\langle\rangle$ denotes an average over the starting times (the starting time is here set to 0 for simplicity). If the scatterers have not moved during the lag-time t, the correlation function will just show a fixed finite value. However, as the scatterers move, the two intensities will decorrelate and the value of the correlation function will decrease. Moreover, from the intensity correlation function, a corresponding correlation function of the scattered fields can be calculated, $\langle E(0)E^*(t)\rangle$ (the asterisk denotes the complex conjugate) [16b]. One can show that this field correlation function is directly related to the ISF $F(Q, t)$, which, as described earlier, encodes how the scatters are arranged and how they move [16b]. A decay in the ISF, characterized by a timescale τ, essentially tells you that it takes the scatterers a time τ to move the probed length scale as set by the particular choice of Q. Note that $F(Q, t)$ is normalized so that it has a value of 1 if the scatters are fully correlated and a value of 0 if they are fully uncorrelated. A sketch of a typical $F(Q,\ t)$ for a glass-forming liquid is shown in Figure 13.4a [20]. Note that as a system falls out of equilibrium, a scattering experiment becomes more complex since the time-averaged intensity correlation function discussed earlier ceases to correspond to that of

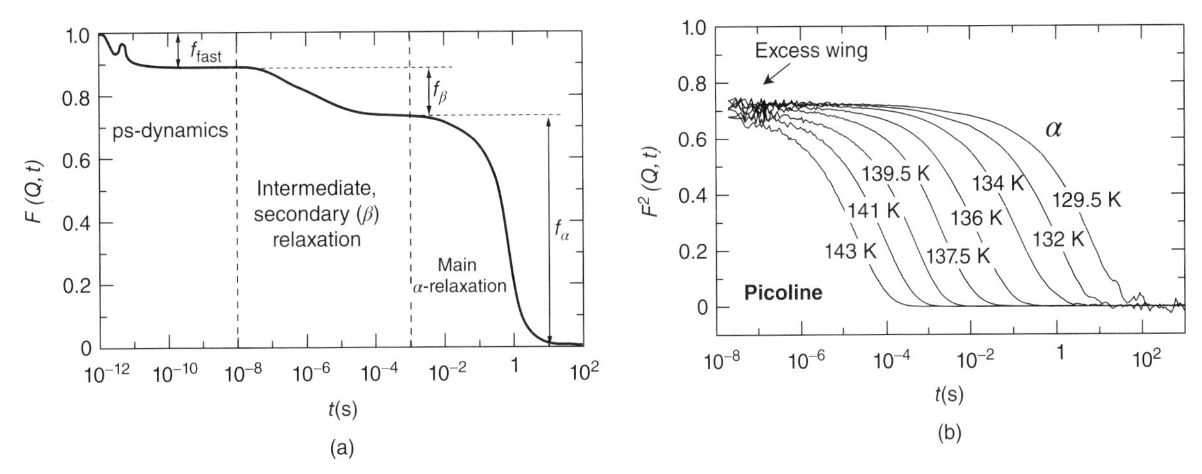

Figure 13.4 (a) Sketch of a typical intermediate scattering function (ISF) for a molecular glass-former. (b) Data showing the square of the ISF for the molecular liquid picoline as it approaches its glassy state. The data are recorded using dynamic light scattering in depolarized geometry. Two relaxation contributions are observed: the structural or α-relaxation at long times and an indication of the so-called excess wing at short times (see text for discussion). (a): Source: Adapted from Brodin 2003 [20], figure 1. Reproduced with permission of Springer (b): Source: Adapted from Brodin 2003 [20], figure 5. Reproduced with permission of Springer.

an ensemble average and care has to be taken to obtain the proper ensemble averaging [21].

The molecular relaxation processes are observed as decays in the ISF, each corresponding to a loss of correlation. The fastest decay for a molecular liquid occurs at picosecond timescales and can be of complex origin [6]; the exact nature of the behavior in this regime can also be quite system specific [6]. These timescales are too fast for standard light scattering performed in the time domain, but they can be probed using, for example, time-domain neutron-scattering techniques [22–24]. In the same time regime around timescales of $\sim 10^{-12} - 10^{-11}$ s, a vibrational contribution is observed as a small oscillation; this is the manifestation of the so-called boson peak, usually studied in the frequency domain and described in detail in Section 13.9. In the dynamic range between the boson peak and the α-relaxation, secondary relaxations (β, γ, δ, etc.) might be observed. However, these are generally more difficult to observe in light scattering compared with dielectric spectroscopy, so I defer the discussion of these to the following discussion about dielectric relaxation. The final long-time decay is due to the α-relaxation.

An example demonstrating the ISF (the figure shows F^2), as measured using light scattering, is shown in Figure 13.4b for the glass-forming liquid picoline [20]. This particular experiment was performed using the depolarized component of the scattered light, which means that this experiment is not directly probing density fluctuations but could correspond more directly to what is measured in a dielectric broadband experiment, as described in the following. To determine the exact contributions to depolarized light scattering constitutes a research field in itself [20], but for identifying the main features in the ISF and for the discussion here, these details are not important. The dominating feature is the α-relaxation decay, which moves to longer times for decreasing temperatures as the molecular motions slow down. Note that for picoline the short-time intercept is not 1, which demonstrates that other faster relaxation processes have occurred outside the experimental window at short times. Within the experimental window, evidence is also found for an additional secondary relaxation, a so-called excess wing [20, 25]. It is, however, very weak and can be much more clearly observed in dielectric spectroscopy, as is generally the case for secondary contributions, as discussed in the following.

In summary, the full decay of the ISF for a glass-forming liquid is complex and a detailed physical model is generally not available. Thus, a simpler approach is often taken where the decays are described using empirical expressions. The simplest empirical description for a relaxation decay is a single exponential, $F \propto \exp(-t/\tau)$, where τ is the relaxation time characterizing the decay. Most relaxation decays for glass-formers can, however, not be described using this simple expression and this is certainly the case for the α-decay, which shows considerable stretching. To account for this, a generalized so-called stretched exponential or Kohlrausch–Williams–Watts (KWW) function is often used instead, $F \propto \exp[-(t/\tau)^{\beta}]$ [26, 27]. The stretching parameter β ranges from 0 to 1 and determines how stretched, or nonexponential, the decay is. Figure 13.5a shows a KWW function for values of β ranging from 0.3 to 1. As described in more detail in Section 13.5.4, the α-relaxation is generally highly stretched with typical values for molecular glass-formers of $\sim$0.3–0.8 near T_g [6, 7, 28].

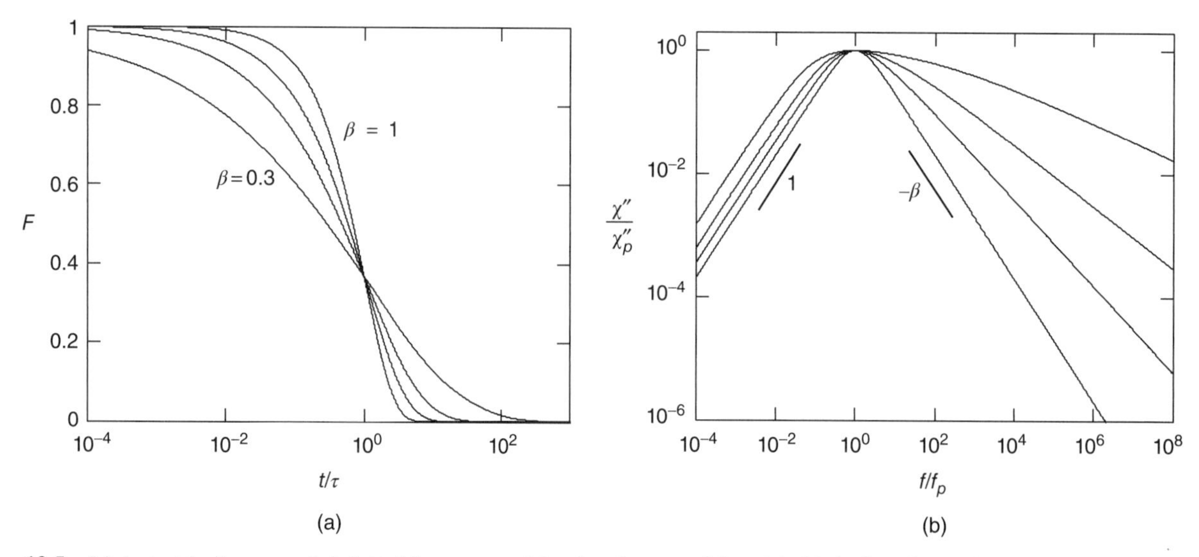

Figure 13.5 (a) A stretched exponential plotted for a range of β-values between 0.3 and 1; (b) the imaginary (loss) part of the corresponding frequency-dependent response function (susceptibility) for the same β-values. All peaks are normalized by the peak height and peak frequency. The power law flanks on each side of the peak are shown with the relevant exponents 1 at low and $-\beta$ at high frequencies.

13.4.2 Relaxation Dynamics as Manifested in the Frequency Domain

In the frequency domain, the ISF corresponds to a frequency-dependent complex response function $\chi^*(f)$ [20]. A time-dependent decay in the ISF corresponds to a frequency-dependent decay in the real part $\chi'(f)$ and a peak in the imaginary part $\chi''(f)$; the real and imaginary parts are linked through a so-called Kramers–Kronig transformation [29]. As an illustration, the imaginary part of the response function corresponding to a KWW decay in the ISF is shown in Figure 13.5b for a range of values of the stretching parameter.

A commonly used frequency-domain technique is broadband dielectric relaxation spectroscopy, which can be used to measure dynamics over a wide frequency range covering ~8–12 orders of magnitude [29]. For dielectric relaxation measurements ranging from frequencies ~10^{-3}–10^{10} Hz the sample is typically placed between two metal electrodes across which an oscillating electric field is applied. The permanent molecular dipoles within the sample respond to the field and this response is monitored through a frequency-dependent response function, the complex permittivity $\varepsilon^*(f)$. When the applied field strength is low, the experiment is performed in the linear response regime and a fluctuation–dissipation relation holds, which means that the response to the applied field probes molecular spontaneous fluctuations [29].

In analogy to the behavior observed in the ISF, a frequency-dependent measurement can in general include the contributions shown in the sketch in Figure 13.6b: (i) an α-relaxation; (ii) one or several secondary relaxations including the β-relaxation discussed in more detail below; (iii) a microscopic relaxation often termed β_{fast} observed in the GHz–THz frequency range; and (iv) the so-called boson peak, described in more detail in Section 13.6.3. In addition, some molecular glass-formers show the so-called excess wing, as shown in Figure 13.6a, where the high-frequency power law flank of the α-relaxation response peak turns into another power law (wing) with a lower characteristic exponent. It has been argued that the excess wing is a generic feature of the α-response, and the fact that it is often not observed is just due to its low strength in many materials [30, 31]. Contrasting this, however, a range of studies have demonstrated that if the α-response is shifted toward lower frequencies, for example, by the application of pressure or by confining the liquid within a restricted geometry, the excess wing turns into a secondary relaxation peak [32–34]. The latter studies suggest that the excess wing is due to a secondary relaxation submerged under the high-frequency flank of the α-relaxation. Thus, the scenario with an α-relaxation and at least one secondary relaxation appears to be a generic feature of glass formation in molecular systems. Figure 13.6c and d shows examples of typical experimental data on two molecular glass-formers: the molecular liquid tripropylene glycol in Figure 13.6d where both α- and β-relaxations are clearly observed and propylene glycol in Figure 13.6c for which the α-relaxation and an excess wing are instead observed.

As for the analysis of time-domain data, data acquired in the frequency domain are also usually analyzed using empirical expressions. As described above the real and imaginary part of the response function are directly linked which means that one can fit either the full complex response or either of its two parts. For the following discussion we focus mainly

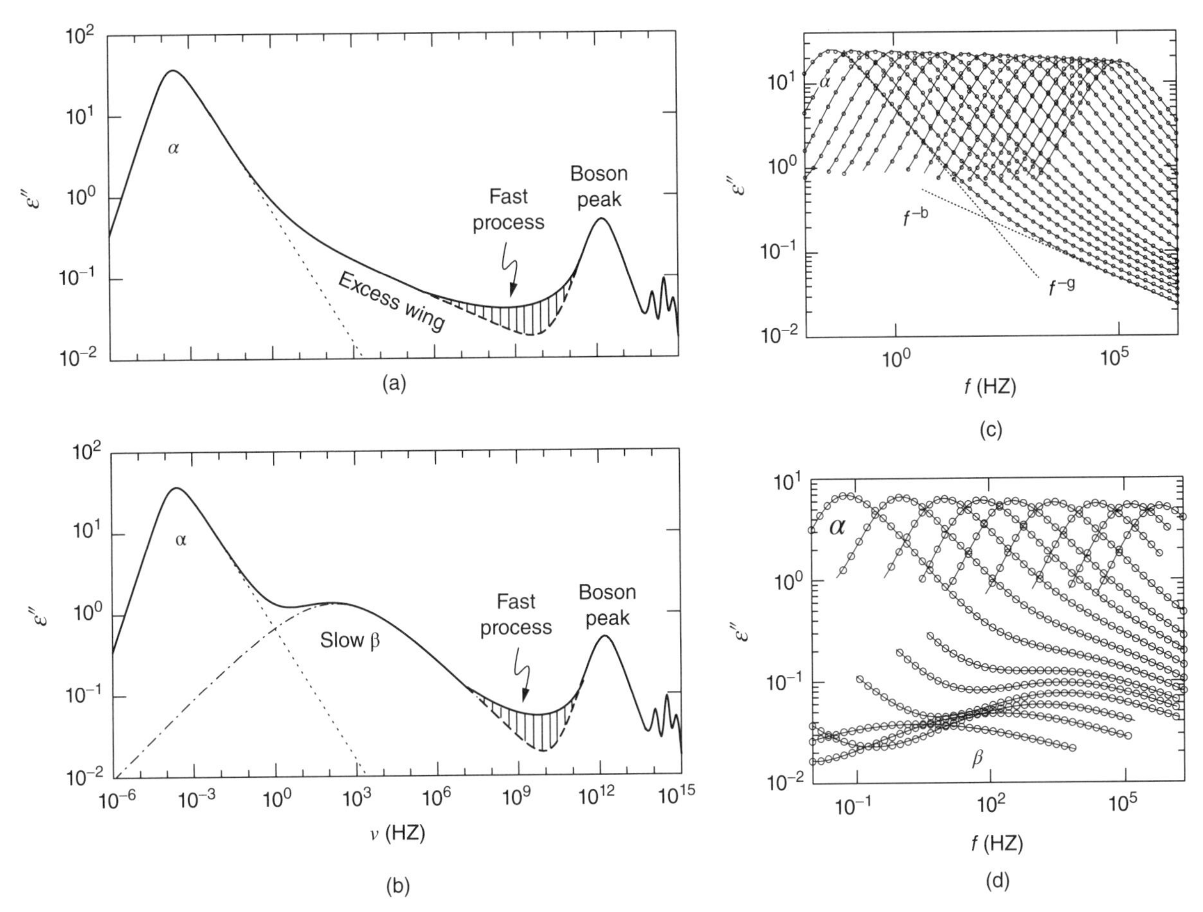

Figure 13.6 (a) Sketch of the frequency-dependent dielectric loss for a glass-former that shows the so-called excess wing. (b) The frequency-dependent loss for a glass-former that shows a β-response peak. (c) Dielectric spectroscopy data for the molecular liquid propylene glycol, showing an α-relaxation and an excess wing. Data are shown for a range of temperatures and the α-relaxation moves to higher frequencies for higher temperatures. (d) Dielectric data for the molecular liquid tripropylene glycol showing both α- and β-relaxations. (a) and (b): Source: Adapted from Lunkenheimer 2001 [25], figure 1. Reproduced with permission of Elsevier.

on the loss peak data of the imaginary part of the response function. To fit relaxation data in the frequency domain a KWW expression transformed into the frequency domain is sometimes used, and Figure 13.5b shows the imaginary (loss) part of the response function corresponding to the stretched exponential decay in Figure 13.5a. In a double logarithmic representation, this response peak consists of a power law flank with exponent 1 at low frequencies and a power law flank with exponent β at high frequencies [35]. Since the use of a KWW expression requires a transformation to the frequency domain and since the bluntness of the resulting loss peak is often too large to fit the data well, at least for most small-molecule glass-formers, other empirical expressions are more commonly used. One often used example is the so-called Havriliak–Negami (HN) expression [35],

$$\chi^*(\omega) = \chi_\infty + \left(\chi_s - \chi_\infty\right) \frac{1}{\left[1 + (i\omega\tau)^\alpha\right]^\gamma},$$

where χ_s and χ_∞ and are the low- and high-frequency limiting values of the response, ω is the angular frequency ($\omega = 2\pi f$), and α and γ are constants characterizing the shape of the response. The HN expression has the advantage that it is defined directly in the frequency domain. It also offers more flexibility since both the low- and high-frequency power law flanks of the loss peak observed in $\chi''(\omega)$ can be varied; α is the power law exponent at low frequencies; and $-\alpha \cdot \gamma$ at high frequencies.

Note that for an α-relaxation, the low-frequency exponent is usually 1 so that $\alpha = 1$ and the HN expression with $\alpha = 1$ is sometimes called a Cole–Davidson expression. Even though this is the most commonly used expression for fitting α-relaxations, it often turns out that exact shape around the loss peak cannot be well fit with this expression either. Other empirical expressions that add extra flexibility in terms of additional fitting parameters that control the shape around the peak have thus been introduced [35]. Moreover, for secondary β-relaxations, the loss peaks generally have a symmetric shape on a logarithmic frequency axis and here low- and high-frequency power laws are thus characterized

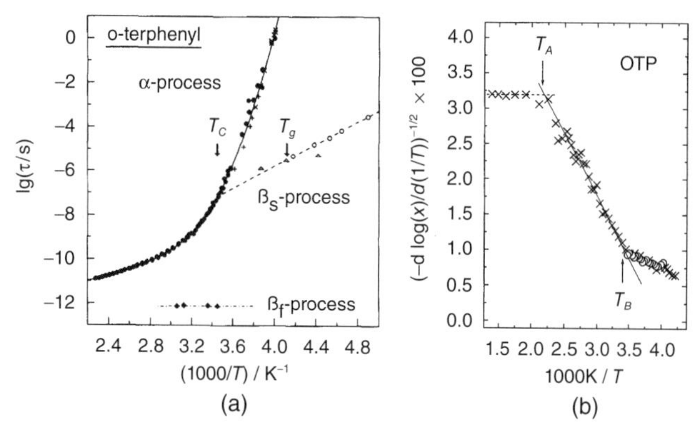

Figure 13.7 (a) Arrhenius plot for the glass-forming liquid OTP, showing the temperature evolution of the α-, β-, and β_{fast}-relaxations. The glass-transition temperature T_g is marked together with the mode-coupling theory temperature T_c, which corresponds to the temperature of the dynamic crossover T^* discussed in Section 13.5.5 (called T_B in panel b). Source: Adapted from Rössler 1994 [22], figure 8. Reproduced with permission of Elsevier. (b) Plot of the α-relaxation data of OTP in such a way that a straight line indicates a VFT behavior and a horizontal line an Arrhenius behavior, see the text for details. Source: Adapted from Hansen 1997 [36], figure 6. Reproduced with permission of [36]. Copyright 1997, AIP Publishing LLC.

by the same exponent. This is often accounted for within the HN expression by setting $\gamma = 1$, and α then gives the power law exponent for the symmetric stretching; the HN expression with $\gamma = 1$ is often called a Cole–Cole expression.

13.4.3 The Structural Relaxation Time

By fitting the relaxation decay in a time-domain experiment or the response peak in a frequency-domain experiment, the characteristic timescales of a particular relaxation process are determined. The typical T-dependence of the α, the β, and the fast microscopic relaxation β_{fast} are shown in Figure 13.7a in the so-called Arrhenius plot, meaning that the logarithm of the timescale is plotted versus inverse temperature. Molecular relaxations can often be characterized by a fixed energy barrier, the so-called activation energy E_A, that needs to be overcome for the particular motions to take place. The characteristic timescale for barrier-activated motion can be expressed as $\tau = \tau_0 \exp(E_A / k_B T)$ where τ_0 is a vibrational timescale that is approached at high T where the thermal energy $k_B T$ is so high that the energy barrier lacks importance. Secondary relaxations within the glassy state are typically characterized by T-independent activation energies; in an Arrhenius plot, they will thus be represented by straight lines with slopes set by E_A, as shown in Figure 13.7a for the β-relaxation (called β_s in the figure). The microscopic picosecond β_f-process has a very weak T-dependence as shown in Figure 13.7a. In contrast, the temperature dependence of the α-relaxation cannot generally be described as due to activation over a fixed energy barrier. Instead, the characteristic timescale often grows more dramatically as the temperature is decreased. One can view this stronger T-dependence as due to a T-dependent activation energy $E_A(T)$, but it is not generally clear what functional shape to use. Many more or less empirical expressions have been proposed to account for the observed temperature dependence, and the most common is the Vogel–Fulcher–Tamman (VFT) expression, $\tau_\alpha = \tau_0 \exp[DT_0/(T - T_0)]$, where D controls how the timescale varies with T as T_0 is approached and T_0 is the temperature where the relaxation time would diverge [6].

Experimentally, a glass-forming system will always fall out of equilibrium before T_0 is reached and there is thus no consensus whether it makes sense to interpret the implied divergence as indicative of the physics underlying the dynamic arrest. The interest in this question is strongly driven by the fact that a divergence could imply the presence of a thermodynamic phase transition, and even though the experimental glass transition just denotes a falling out of equilibrium, there might be an underlying thermodynamic transition at lower temperatures, an "ideal glass transition" that drives the slowing down in the supercooled state. Speculations about an underlying thermodynamic transition have been spurred by the experimental observation that T_0 values obtained from fitting data to a VFT expression often lie close to T_K (the Kauzmann temperature) values determined from thermodynamic data [6], even though the correspondence is not general [37]. It is important to realize, however, that even though it is clear that the VFT expression can describe the data well, it is also clear that one can devise alternative parameterizations that describe the data equally well, but do not imply a low-T divergence [38]. Thus, the question of whether a low-T "ideal glass transition" exists remains open.

Figure 13.8a shows $\tau_\alpha(T)$ in an Arrhenius representation for a range of glass-formers of different molecular character. All data can be described near T_g using a VFT expression except for silica, which is better described using an Arrhenius expression. It is clear from Figure 13.8a that different glass-formers approach their glass-transition temperatures in different ways; the glass-transition temperature is here defined as $T_g = T(\tau_\alpha = 100\,\text{s})$. This is further highlighted by rescaling the abscissa by T_g, as shown in Figure 13.8b. A glass-former for which a small temperature change near T_g leads to a large change in τ_α is termed fragile, whereas a glass-former characterized by an Arrhenius-type dependence where T_g is approached more slowly is called strong [6]. Different metrics are used for fragility, and include the D parameter obtained from a VFT fit and the slope of τ_α at T_g in the Arrhenius plot, from which a fragility index is defined as $m = d\log(\tau_\alpha)/d(T_g/T)$. The latter definition is more direct and is model independent, whereas the D parameter is coupled to the other two VFT parameters, τ_0 and T_0. The fact that the latter parameters are always outside the range of the experimental data generally makes this a less suitable

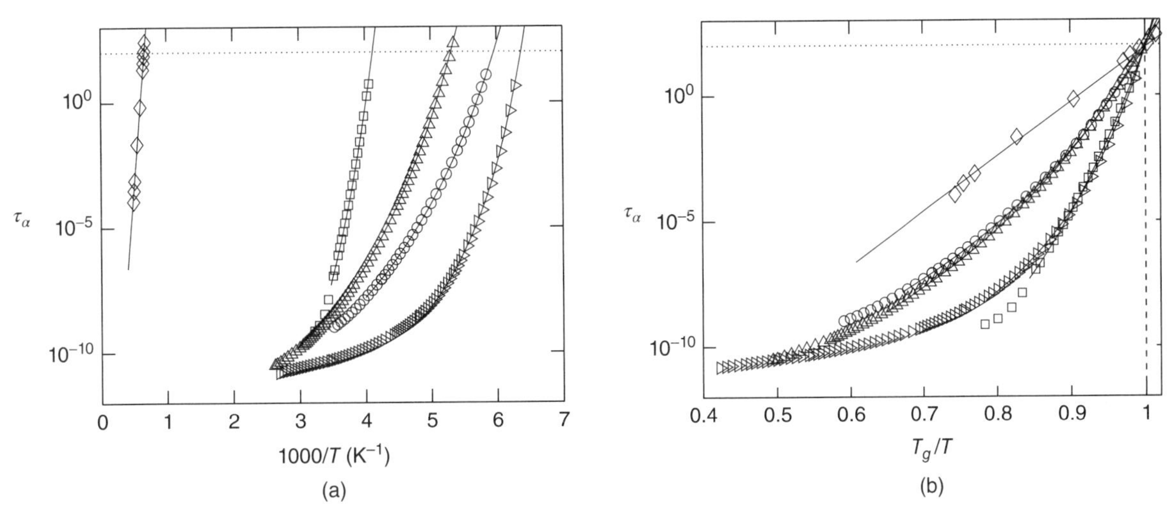

Figure 13.8 (a) Arrhenius plot of the α-relaxation data for a range of glass-formers ranging from strong to fragile: (diamonds) silica [39], (squares) OTP [40], (triangles) glycerol [40, 41], (circles) propylene glycol, (right triangles) propylene carbonate [40]. (b) The same data in a T_g-scaled Arrhenius plot to emphasize the behavior close to T_g and thus the variation in fragility.

choice for characterizing fragility. Typically, small-molecule glass-formers characterized by van der Waals or ionic bonding such as toluene, orthoterphenyl (OTP), or calcium potassium nitrate (CKN) are fragile, whereas those characterized by the formation of intermolecular networks through hydrogen bonding such as glycerol are intermediately fragile, and those characterized by strong covalent networks such as silica are strong. On a quantitative level, however, there is presently no established link between molecular or intermolecular structure, interactions, and fragility.

13.4.4 The Stretching of the Structural Relaxation

As discussed earlier, the α-relaxation can be described in time domain using a stretched exponential with a β-value often significantly below 1 near T_g. In the frequency domain, this stretching is manifested by a high-frequency flank of the α-relaxation peak characterized by a power law exponent larger than -1. Different explanations have been proposed to account for this stretching. The most common is that the stretching is due to the existence of dynamic heterogeneity where the relaxation within a particular spatial region is essentially single exponential with a single associated timescale. This timescale is however different for different regions and the stretching then arises from averaging over the distribution of regions [7]. Alternatively, there are models that link β to the degree of cooperative motions involved in the α-relaxation, where $\beta = 1$ corresponds to independent noncooperative motions and $\beta < 1$ reflects the degree of cooperativity [42]. All in all, there are a wide range of views, ideas, and models on this topic and a clear consensus is still lacking [2, 6, 7, 43].

Two direct experimental observations are important regarding the stretching of the α-relaxation. The first is that for some systems, the stretching stays relatively independent of temperature near T_g (see Fig. 13.9a), which means that the shape of the α-relaxation response is essentially fixed and the response for different T can be superimposed; this is called time–temperature superposition or TTS. At higher T within a dynamic crossover regime (discussed in more detail in Section 13.5.5), β does however change and at high T it approaches values near unity. This behavior is shown in Figure 13.9a where β is plotted as a function of the relaxation frequency and the crossover regime is marked with a gray box [44]. Also, as demonstrated in Figure 13.9a, different glass-formers show different values of β near T_g. In fact, a correlation has been found [28] between the β-value and the fragility, where fragile glass-formers generally show the lowest β-values near T_g, as shown in Figure 13.9b.

13.4.5 The Dynamic Crossover

A VFT function can generally describe the T-dependence of the α-relaxation well over a significant T-range above T_g. However, at a temperature T^* typically situated at $1.2T_g - 1.6T_g$, the dynamics change and undergo a crossover into a different behavior. This crossover is manifested in a range of properties, including (i) a change in the T-dependence of the α-relaxation [36]; (ii) a separate β-relaxation develops for $T < T^*$, as shown in Figure 13.7a; (iii) a change in the T-dependence of the specific heat [45]; (iv) a decoupling of translational and rotational diffusion takes place for $T < T^*$ [46]; (v) a change in the T-dependence of the α-relaxation strength [47]; (vi) the high-T behavior is generally quite well described by the so-called mode-coupling theory (MCT) for $T > T^*$, but standard MCT breaks down for temperatures near T^* [48].

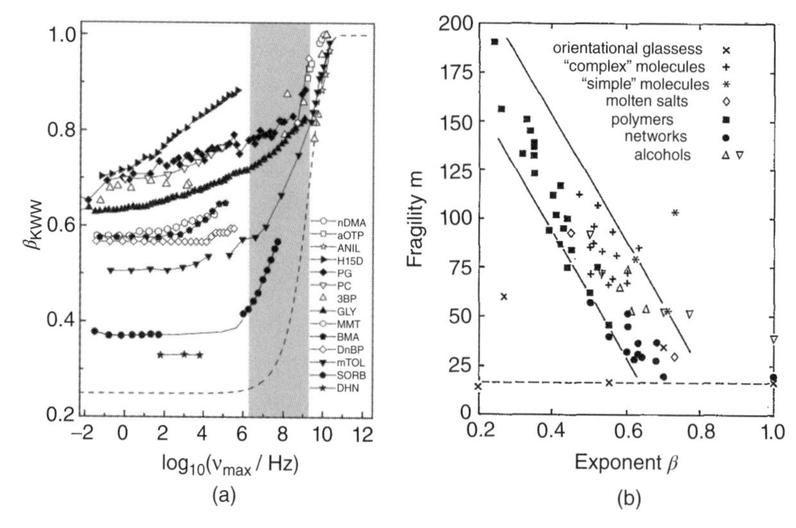

Figure 13.9 (a) Stretching parameter β versus α-relaxation peak frequency $\nu_\alpha = \tau_\alpha/2\pi$ for a range of glass-formers of varying fragility. The dynamic crossover regime is marked with a gray box. Source: Wang 2007 [44], figure 1. Reproduced with permission of the American Physical Society. Copyright 2007 by the American Physical Society. (b) Correlation between the fragility quantified by the m-parameter and the stretching parameter β; highly fragile liquids show lower β-values than strong liquids. Source: Böhmer 1993 [28], figure 2. Reproduced with permission of [28]. Copyright 1993, AIP Publishing LLC.

The change in translational/rotational motion occurring at T^* is discussed in more detail in Section 13.7 and MCT in Section 13.8. Here, we focus on the observations for the $\tau_\alpha(T)$ behavior.

Experimentally, $\tau_\alpha(T)$ can be well described using a VFT expression both above and below T^*, but the T-dependence and thus the fitting parameters change at the transition. A direct way to see this is to linearize the $\tau_\alpha(T)$ data so that it follows a straight line if it is well described by a VFT behavior [36]. Figure 13.7b shows such a linearization where the parameter $Z = 100[d\log(f_p)/d(1/T)]^{-1/2}$ ($f_p = \tau_\alpha/2\pi$) is plotted versus inverse temperature. A horizontal straight line with a zero slope corresponds to an Arrhenius behavior, whereas a non-Arrhenius VFT behavior is observed as a straight line with a slope that is a function of the VFT parameters, T_0 and D; note that τ_0 is irrelevant since it disappears due to the derivative in the definition of Z. The absolute value of the slope in this plot can thus be viewed as a measure of how non-Arrhenius the relaxation behavior is. The linearized plot for the fragile glass-former OTP is shown in Figure 13.7b. The VFT-behavior corresponding to a straight line at low-T changes at the crossover temperature T^* into a second VFT-behavior. At high temperatures, above the temperature T_A in the figure, the behavior for OTP undergoes a second transition into an Arrhenius T-dependence. We will here focus only on the two lower temperature regimes since these are generally observed and are of direct relevance for supercooled liquids and the glass transition.

Since the α-relaxation behavior changes significantly at T^*, and since the value of T^* is different for different glass-formers, it is natural to rescale the relaxation behavior by T^*, as shown in Figure 13.10a. This rescaling places the emphasis on the development around the crossover, which is a natural point of reference. We see from Figure 13.10a that the behavior for $T > T^*$ is quite similar for widely different glass-formers, but below T^* the differences are significant; the more fragile glass-formers reach T_g closer to T^*, as compared with the stronger glass-formers. It is clear that a correlation between the T^*/T_g ratio and fragility exists where the ratio is larger for stronger liquids [49].

Even though the fragility is different for different glass-formers, they can all be described by VFT or Arrhenius expressions for $T < T^*$. This is important since it suggests that despite the differences in fragility, a unified physical mechanism might still be at play. To further demonstrate that the T-evolution below the crossover can be described using the same functional shape, Figure 13.10b shows the results of plotting the dynamics on a rescaled abscissa: $(T_g/T - 1)K$, where $K = T^*/(T^* - T_g)$ (note that in the plot in Figure 13.10b, the crossover temperature is taken as the critical temperature T_c from an MCT analysis; T_c closely coincides with the T^* defined according to the criteria above). This scaling [49, 50] is another way of expanding the data around T^*, but here the data are further collapsed by compensating also for the variation in T_g, which leads to a complete collapse. Note that the fragility is directly related to the parameter K, which is approximately proportional to the fragility metric m [49].

The existence of a dynamic crossover and thus a change of transport properties in deeply supercooled liquids were predicted by Goldstein in 1969 [51]. It was argued that at low temperatures, where molecules become more packed, a change of transport mechanism should occur where crossing of potential energy barriers larger than thermal energies

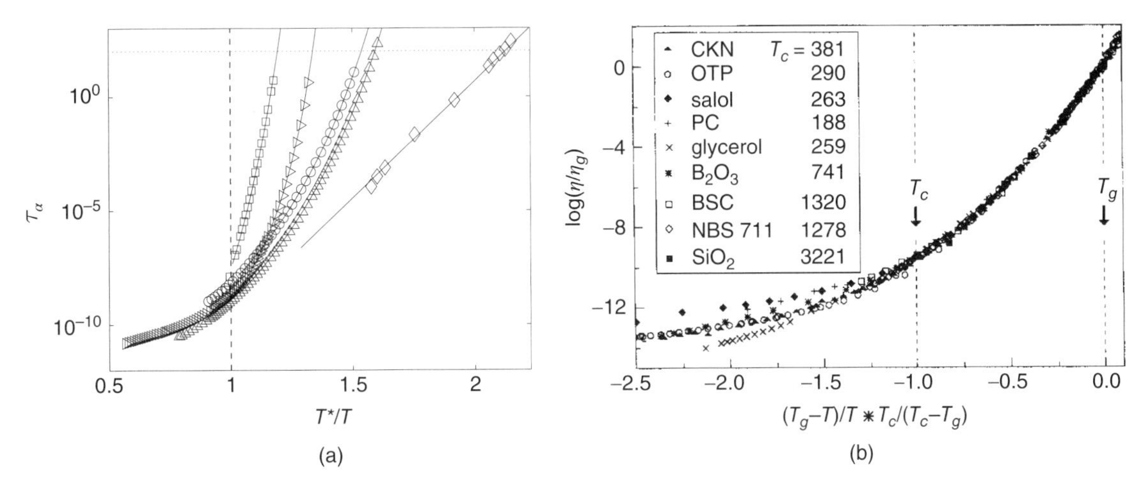

Figure 13.10 (a) The same data as in Figure 13.8, plotted in a T^*-scaled Arrhenius plot to stress the importance of the dynamic crossover. (b) Viscosity data for a range of glass-formers plotted using a scaling that takes account of both the dynamic crossover T^* and T_g to demonstrate that a wide range of liquids can be collapsed onto one master curve, demonstrating the similarity in their temperature evolution [49, 50]. Source: Adapted from Rössler 1996 [50], figure 13. Reproduced with permission of Elsevier.

start to dominate. A convenient way to visualize such a change is to think of the molecular motions in terms of a potential energy landscape (or more accurately a free energy landscape since entropic contributions could play very important roles) [2, 3, 52]. For a system consisting of N molecules, $3N$ coordinates are needed to denote the coordinates of all molecules and each of these configurations is assigned a potential energy, meaning that a configurational energy landscape has $(3N+1)$ dimensions. To visualize this landscape, the $3N$ coordinates are mapped onto one generalized coordinate in Figure 13.11.

At high T, where the thermal energy is significantly higher than the characteristic landscape energies, the landscape's role in the dynamics is small. However, as the thermal energy is reduced, the behavior becomes increasingly influenced by the landscape and this regime can be quite well described using MCT (see Section 13.8.1). As the temperature is further reduced into the deeply supercooled range ($T < T^*$), the experienced landscape becomes rugged and now consists of large minima or basins separated by significant energy barriers, in turn containing smaller minima or subbasins separated by smaller energy barriers. Note that only for low enough temperatures, $T < T^*$ will the subbasins be resolved and the energy landscape picture can thus conceptualize the "bifurcation" at $T < T^*$ into two relaxations, the α- and the β-relaxations, as shown in Figure 13.7a. The system spends a significant amount of time exploring the configurations

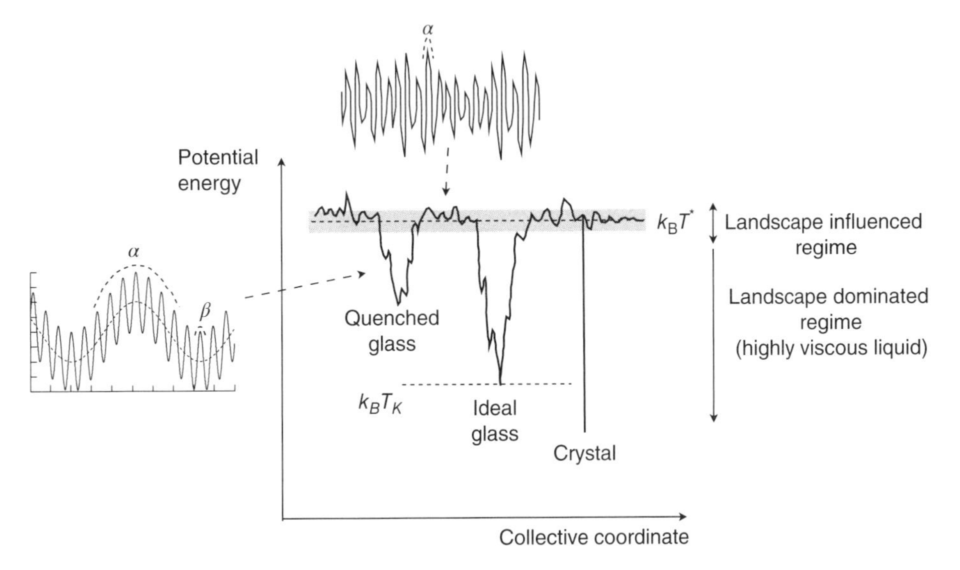

Figure 13.11 Schematic view of the potential energy landscape for a liquid. See the text for a detailed description.

within one of the basins (via β-relaxations) before significant thermal fluctuations make it possible to overcome the barrier into a neighboring basin (α-relaxation), as sketched in Figure 13.11. The excursions that take place within one of the basins are sometimes considered relatively limited so that given enough time the system might return to the same configuration thus yielding a "reversible" character to the exploration (β-relaxation), whereas the vastness of the full landscape effectively makes the crossing from one basin to another an "irreversible" process (α-relaxation) [53, 54].

Finally, the landscape picture is a convenient way to visualize the effects of kinetics on glass formation. If a supercooled liquid is cooled slowly, the system has a long time to explore the landscape, which results in the system ending up in a deeper minimum compared with the result of a fast temperature quench. However, if the liquid could be cooled infinitely slowly, without ending up in the crystalline global minimum, the whole landscape would thus be investigated and the lowest glassy energy minimum found; this would correspond to an "ideal" glass forming at a temperature $T \approx T_K$.

13.5 DYNAMICS OF GLASS-FORMING COLLOIDAL SYSTEMS

13.5.1 General Behavior

Colloidal particles dispersed in a liquid undergo Brownian motion resulting from random collisions with solvent molecules. Figure 13.12b shows typical particle trajectories for a low volume fraction suspension where the random walk character of the particle motions is clearly seen. The random collisions lead to a zero average displacement $\langle \Delta r \rangle$, but a nonzero mean square displacement (MSD), which in 3D is $\langle \Delta r^2 \rangle = \langle [\vec{r}(t+\Delta t) - \vec{r}(t)]^2 \rangle = 6D_t \Delta t$, where Δt is the lag-time, the angular brackets denote an average over all particles and initial times t, and D_t is the translational diffusion coefficient. For a single colloidal particle of radius a, D_t is given by the Stokes–Einstein (SE) relation $D_t = k_B T / 6\pi\eta a$, where η is the solvent viscosity. Thus, in dilute suspensions, the characteristic timescale of diffusive motion is given by the time it takes for a particle to diffuse its own size, the Brownian time, $\tau_B = a^2/6D_t$. Figure 13.12a shows the MSD as a function of Δt for a glass-forming suspension of colloidal microgel particles [55]. Since colloidal microgel particles consist of highly swollen chemically cross-linked polymer networks, they are deformable and as the particles pack together the volume fraction eventually becomes irrelevant as a means of characterizing the suspensions. Instead, for soft colloids, an effective volume fraction ζ is often defined as $\zeta = nV_0$, where $V_0 = 4\pi a^3/3$ is the volume of the undeformed particle of radius a, as measured in a dilute suspension. For low concentrations where the particles are undeformed, ζ equals the volume fraction and for all concentrations, ζ is proportional to the number density of microgels. The use of ζ instead of volume fraction is thus a convenient way to compare data for different batches of deformable colloids.

At short times, for the suspensions that are fluid, the MSD behavior in Figure 13.12a shows a near-unity slope in a log–log plot, which indicates diffusive motion. This behavior persists even for quite concentrated suspensions even though hydrodynamic interactions mean that D_t can vary significantly from that of dilute suspensions [8, 9]. Hydrodynamic interactions arise since colloidal particles are suspended in a fluid, which means that specific interactions occur that are mediated by the fluid and which can be of

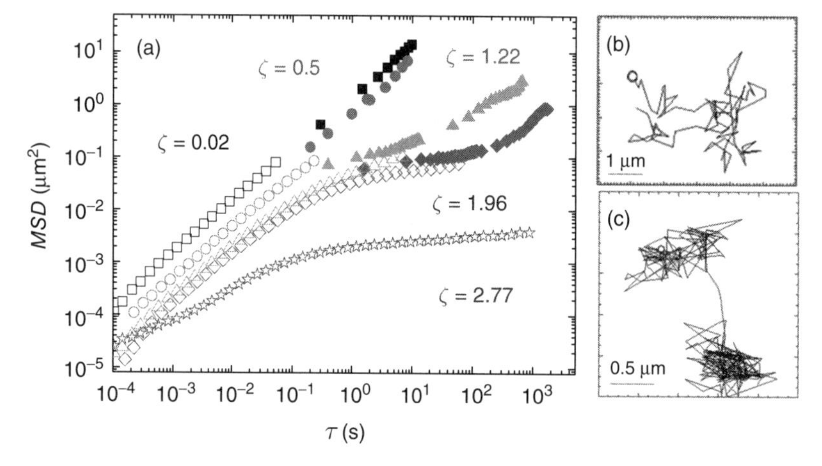

Figure 13.12 (a) Mean square displacement (MSD) versus lag-time τ for a colloidal suspension of microgels. $\xi = 0.02$ (squares), 0.5 (circles), 1.22 (triangles), 1.96 (diamonds), and 2.77 (stars) [55]. (b) Particle trajectories for the low-concentration suspension $\xi = 0.02$, showing a random walk character [55]. (c) Particle trajectories for a high-concentration suspension $\xi = 1.96$, showing the intermittent nature of the motions with long periods characterized only by local restricted motions, interrupted by bursts of long-range excursions [55]. Source: Adapted from Romeo 2012 [55], figure 8. Reproduced with permission of [55]. Copyright 2012, AIP Publishing LLC.

significant importance. However, for glassy dynamics, most studies demonstrate that hydrodynamic effects do not play a significant role [8] and we will thus not discuss these here in detail, but refer to Ref. [8] for a collection of good references on this topic.

As the colloid concentration is increased, the motions of the colloidal particles become restricted due to crowding by neighboring particles. This leads to the development of a plateau in the MSD, as shown in Figure 13.12a. In this dynamic regime, the particle motions become strongly intermittent and the colloid motions are highly restricted for most of the time and only every now and then do significant excursions occur, as shown from a typical particle trace in Figure 13.12c. Note that the features observed here for deformable colloids are very similar to those of hard-sphere colloids [8, 9, 56], where the plateau develops at volume fractions ~52%. For the suspension of deformable microgels described in Figure 13.12, however, the onset of the plateau occurs as the microgels fill space at $\zeta \approx 1$. At long times, the MSD transitions into a second diffusive regime and a long-time diffusion coefficient can thus be defined as $D = \langle \Delta r^2 \rangle / 6 \langle \Delta t \rangle$. The onset of this diffusive regime as observed from an upturn in the MSD corresponds to the α-relaxation and moves to longer times as the glass transition is approached. As the glassy state is finally reached, the colloids are trapped on experimental timescales and the plateau persists to the longest measured timescales.

13.5.2 The Structural Relaxation

Laser scanning confocal microscopy (LSCM) is an excellent experimental technique for studying the motions of colloids directly in real space, as outlined in Chapter 10. To track colloids effectively using LSCM, particles with sizes in the micron-range are usually used, which corresponds to relatively long Brownian timescales and thus a small probed dynamic range. Thus, a good alternative is to use light scattering, where the motion of smaller colloids can easily be studied enabling a wider dynamic range to be probed. Colloidal particles with $a \approx$ 100–200 nm for which the main peak of $S(Q)$ is positioned within the range of standard light-scattering setups can easily be synthesized. For molecular glass-formers, the disadvantage of light scattering is that the probed length scales are large compared with molecular sizes.

Glass formation has been studied extensively in model hard-sphere colloidal systems [8, 9, 14, 57–59]. A commonly used system consists of polymeric colloids made from poly(methyl methacrylate) (PMMA) coated with a brush of poly(12-hydroxystearic acid) [8, 56]. The brush prevents the particles from getting close enough for attractive van der Waals interactions to induce particle–particle attractions that can lead to aggregation. The brush provides a certain "softness" to the interparticle potential, but for the discussion here we can still view the potential as essentially hard sphere. The PMMA colloids are typically suspended in a mixture of organic liquids that is chosen so that both the density and index of refraction difference between the colloids and the liquid are well matched. This (i) minimizes the risk of sedimentation or creaming due to a density mismatch, (ii) minimizes van der Waals interactions and thus the risk of aggregation, and (iii) leads to optically clear nonturbid solutions where light scattering can accurately be performed even at high-colloid volume fractions using standard light-scattering setups. To avoid interference from crystallization in studies of glass formation, a small particle size polydispersity is usually introduced (~10%) [8].

The ISF, as measured for suspensions of PMMA-based hard-sphere colloids [58, 59], is shown in Figure 13.13a for a range of volume fractions, ϕ. In dilute suspensions, the decay is essentially single exponential reflecting the Brownian motion of colloids with a narrow size distribution.

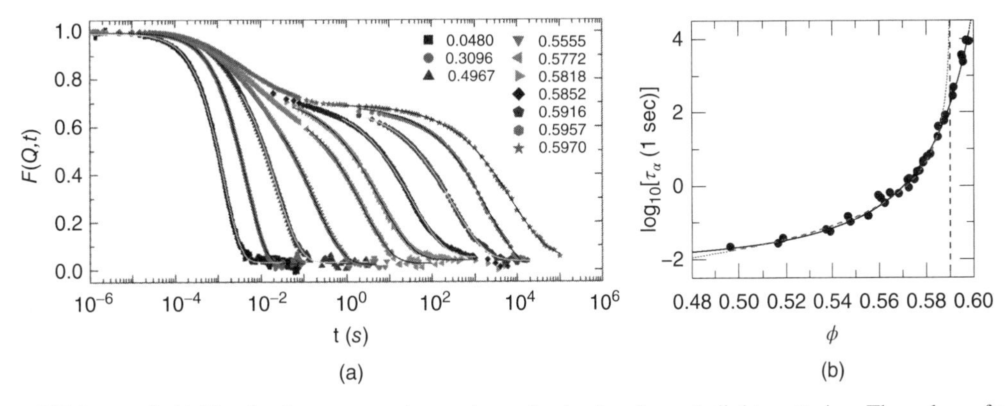

Figure 13.13 (a) ISF for a colloidal hard-sphere suspension as determined using dynamic light scattering. The volume fractions are given in the legend [58]. Source: Adapted from Brambilla 2009 [58], figure 1. Reproduced with permission of the American Physical Society. Copyright 2009 by the American Physical Society. (b) α-Relaxation time as a function of volume fraction for the data shown in panel (a) determined from a KWW fit to the final decay in the ISF [59]. The fit showed with a dashed line is a power law fit as predicted by MCT and the solid line is a fit to an exponential VFT-type expression as described in the text. Source: Adapted from El Masri 2009 [59], figure 11. SISSA Medialab Srl. Reproduced with permission of IOP Publishing. All rights reserved.

As ϕ becomes larger than ~0.5, however, a two-step decay is observed with an intermediate plateau, as expected for a glass-forming system and as discussed earlier for the MSD. The final decay, corresponding to the α-relaxation, is stretched with a stretching parameter $\beta \approx 0.5$, where the exact value depends on the probed length scale (as set by the probed scattering vector Q) in relation to the particle size as well as details of the scattering contributions (coherent vs incoherent scattering [59]). As ϕ is further increased, the final decay shifts toward longer times. By fitting the final decay using a KWW expression, the characteristic α-relaxation timescale $\tau_\alpha(\phi)$ can be determined, as shown in Figure 13.13b.

Most detailed analyses of light-scattering data on hard-sphere colloids have been performed using MCT. As described in more detail in Section 13.8.1, MCT provides detailed predictions on the dynamics, as monitored through the ISF. These predictions have been shown to hold to a significant degree for both hard-sphere colloids and molecular liquids, however, only over a limited dynamic range [48, 58, 59]. Below a crossover temperature for molecular systems and above a crossover volume fraction (or effective volume fraction ξ for deformable colloids) for colloidal systems, activated dynamics not generally included in MCT take over and MCT breaks down. It is important to note that particularly for colloidal systems, the breakdown of MCT is still very debated and different opinions regarding the range of applicability of MCT exist and this continues to be a highly debated area [58, 60–62].

As discussed in more detail in Section 8.1, standard (ideal) MCT predicts that $\tau_\alpha(\phi)$ should follow a power law $\tau_\alpha(\phi) \propto (\phi - \phi_c)^{-\gamma}$, where ϕ_c denotes a critical volume fraction. This behavior can indeed describe the data over a significant dynamic range, but as shown by the dashed line in Figure 13.13b, the power law description breaks down for high volume fractions as the glass transition is approached. Also for molecular glass-formers, a power law expression $\tau_\alpha(T) \propto (T - T_c)^{-\gamma}$ can describe the data quite well over a limited dynamic range for $T > T^*$, but similarly this description breaks down at lower temperatures, where a VFT expression describes the data better. Thus, consistent with the behavior for molecular systems, the hard-sphere data in Figure 13.13b can be fitted with a VFT-type expression, $\tau_\alpha(T) = \tau_0 \cdot \exp[A/(\phi_0 - \phi)^\gamma]$ (for the solid black line in Fig. 13.13b, $\gamma = 2$ was used, but as discussed in [59] $\gamma = 1$ also describes the data well), where ϕ_0 denotes the volume fraction for which τ_α would diverge. A value of $\phi_0 \approx 0.637$ was found in Ref. [59], which coincides quite well with the ϕ_{rcp} of random close packing [63] of hard spheres; this constitutes the highest possible disordered packing of monodisperse spheres. Interestingly, also viscosity measurements on hard-sphere suspensions were best described using a VFT-like approach with $\phi_0 \approx 0.625$ [64]. These results provide an interesting analogy to the behavior of molecular glass-formers where it has been observed that for many systems $T_0 \approx T_K$ and it does raise the question of whether an "ideal" glass transition for colloidal hard spheres would take place near ϕ_{rcp} if it was reached within equilibrium. However, the difficulty in determining colloidal volume fractions with a great degree of accuracy, the effects of size polydispersity, and for the viscosity measurements, the difficulty of measuring at high volume fractions means that it is not easy to reach a consensus regarding how close ϕ_0 is to ϕ_{rcp}.

13.5.3 The Dynamic Crossover

The breakdown of the standard MCT description for colloids is one indication of a dynamic crossover, as discussed earlier. Moreover, as discussed in more detail in Section 13.7.2, both molecular and colloidal systems show a very similar decoupling of translational and rotational diffusion near this dynamic crossover. For thermal systems, the dynamic crossover is also manifested in a change of their potential (or free) energy landscape properties, which have been studied in detail in computer simulations [3, 5, 52, 65]. Typically, the simulated system is equilibrated at a temperature of interest within the liquid state, after which it is subjected to a rapid temperature quench. The quench traps the system in the landscape minimum corresponding to the configurations explored at the time of the quench. The properties of this minimum and its corresponding configurations can then be studied in detail. For thermal particles interacting through soft repulsive interparticle potentials ($U \propto r^{-n}$, where the inverse power law exponent n sets the softness), it has been observed that the properties of the probed minima change at a dynamic crossover in a very similar manner with regard to either ϕ or T-induced glass formation, which suggests a strong link between these two paths toward the glassy state [65]. For a hard-sphere colloidal system, where entropy solely sets the phase behavior, a potential energy landscape model is not useful. However, an analogous technique was developed for quenching hard-sphere systems so that the particle configurations characterizing the system at the time of the quench and at a certain hard-sphere volume fraction can be determined and studied in detail [65]. Using this technique, it was found in computer simulations that both soft-sphere and hard-sphere systems display an analogous change in configurational behavior at a crossover volume fraction $\phi^* \approx 0.53$, which corresponds well with the ϕ for which translational and rotational motions decouple, as determined in both computer simulations [66] and experimental work [67] (see Section 13.7.2).

In conclusion a dynamic crossover is a general feature of glass-forming systems, which manifests itself in a remarkably similar manner for colloidal and molecular systems. For molecular liquids, the ratio between the crossover temperature T^* and T_g varies for different liquids, and this ratio is

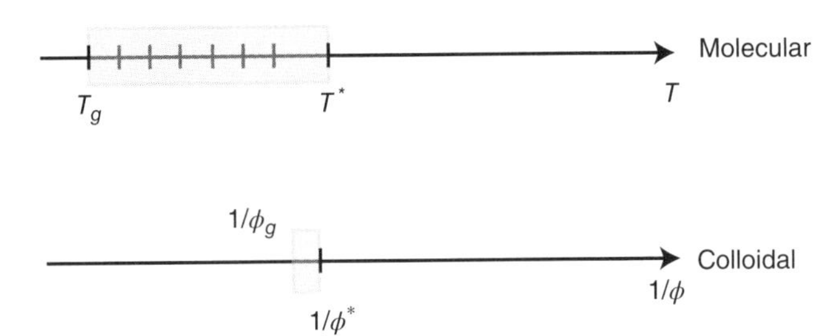

Figure 13.14 Schematic showing the relation between the dynamic crossover and the glass transition for molecular and hard-sphere colloidal systems, respectively. For molecular systems, the relative distance between T^* and T_g varies with the fragility. Hard-sphere colloidal systems, on the other hand, only display a highly "fragile" behavior where ϕ^* and ϕ_g are situated close together.

large for strong and small for fragile liquids, typically spanning a range of 1.2–1.6. For hard-sphere like colloids, the situation is qualitatively similar to that of fragile liquids, where T corresponds to $1/\phi$ and the ratio between ϕ_g and ϕ^* is always small (see Figure 13.14). The wide range of fragilities observed for molecular systems is thus not observed for hard-sphere colloids. An important question is whether the similarities between molecular and colloidal systems cover also the variation in fragility and if so, what the relevant colloidal properties are that lead to this variation.

13.5.4 "Fragility" in Colloidal Systems

A strong to fragile scenario similar to that found for molecular systems has recently been demonstrated also for a colloidal system of deformable colloidal microgels [55, 68], as shown in Figure 13.15 (compare with Figure 13.8b for molecular liquids). The "fragility" for the microgel suspensions is defined as the sensitivity to a change in ζ analogous to the fragility of molecular systems defined as a sensitivity to a T-change. The colloidal fragility is affected by the elastic properties of the colloids and "softer" particles generally show a stronger behavior, whereas stiff hard-sphere like particles are fragile. In fact, it has been demonstrated in several studies on deformable colloids that the viscosity or corresponding structural relaxation time varies much less dramatically with a change in particle number concentration near their glass transition [68, 69]. For deformable colloids such as microgels, as for hard-sphere colloids, the onset of the characteristic glass-transition behavior with a two-step relaxation, a plateau in the ISF, and a final stretched exponential decay are observed as the particles pack together. For hard-sphere systems, the effects of crowding are observed near $\phi^* \sim 0.5$ and for suspensions of deformable microgels near an effective volume fraction $\zeta^* \sim 1$. At higher ζ, the microgel particles will thus be forced to deform and shrink in order to fit into the available volume, as shown in the sketch in Figure 13.15c. For $\zeta > \zeta^*$, the volume is packed with microgels and in order to move, particles have to deform. Thus, the internal elasticity of the particles will start to influence the behavior, and the dynamics should undergo a crossover into a regime dominated by thermally activated behavior where the activation energy is related to the elastic properties of the packed suspensions, which in turn is related to the elastic properties of the particles. In this elasticity-dominated regime for $\zeta > \zeta^*$, τ_α follows either a VFT- or Arrhenius-type behavior depending on the particle properties and a glass transition is reached at ζ_g. Note that microgels that are elastically very soft and/or have particular charge properties might deform below colloid concentrations corresponding to particle packing

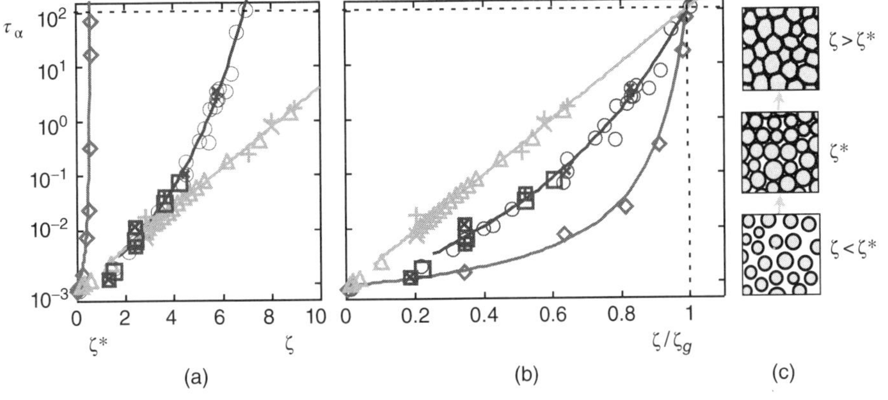

Figure 13.15 (a) Plot of the structural relaxation time τ_α as a function of concentration ζ for microgel suspensions of different "softness" (to account for differences in size between the samples and thus the difference in Brownian timescales, the data were normalized to coincide in the highly dilute state). Note that the more deformable microgels have to be significantly compressed before they reach their glass transition at ζ_g. The dynamic crossover for the suspensions ζ^* occurs close to where the microgels pack. (b) The same plot where the abscissa is rescaled with $\zeta_g = \zeta(\tau_\alpha = 100\,\text{s})$. (c) Schematic illustration of how the microgels pack together near ζ^* and deform for higher concentrations. Source: Adapted from Mattsson 2009 [68], figure 2. Reproduced with permission from [68].

[55]. However, such effects were not observed in the studies discussed here [55, 68] where particle deformation occurred near packing at $\zeta^* \sim 1$.

It is interesting to note the strong similarity between Figure 13.15a for colloids and Figure 13.10a for molecular glass-formers. The core idea of such a comparison is based on the hypothesis that the dynamic crossovers observed for both deformable colloids and molecular glass-formers signify the entering into a thermally activated elasticity-controlled regime where fragility is linked to the elastic properties of the material. For deformable colloids, such a transition should normally occur close to the point where the particles pack as this is the point where the internal elasticity of the particles becomes relevant. Based on this, in Ref. [68], the concentration-dependent α-relaxation timescale $\tau_\alpha(\zeta)$ was viewed as resulting from the ζ-dependent elasticity of the suspensions as probed on timescales relevant for local particle motions. Thus, the ζ-dependent activation energy should approximately relate to thermal energy as $G_p(\zeta)V_c/k_BT$, where G_p is some elastic modulus relevant for local particle motions and V_c is a volume characteristic of the elastic deformations. Assuming to first approximation, a constant characteristic volume V_c, $\tau_\alpha(\zeta)$ would thus be given by the ζ-dependence of the elastic modulus. As discussed in Section 13.8.2, a correlation between $\tau_\alpha(T)$ and T-dependent elastic properties is observed for molecular systems where a T-dependent activation energy relates to thermal energy as $G_\infty(T)V_c/k_BT$ and $G_\infty(T)$ is the high-frequency (short timescale) elastic shear modulus.

In the presented view, for both molecular and colloidal glass-formers, the dynamic crossovers are fundamentally important since they signify the onset of the activated dynamics that takes the system into the glassy state. The fact that the slowing down, both in molecular and colloidal systems, show correlations with the evolution of elastic properties suggests that many significant similarities between the two classes of glass-forming materials might exist.

13.5.5 Glassy "Secondary" Relaxations

Secondary relaxations that persist within the glassy state are generic to molecular glass formation, and it is thus important to ask what glassy relaxation behavior exists in colloidal systems. To date, most studies of the relaxation dynamics of colloidal glasses have focused on hard-sphere-type colloids and/or on dynamics driven by an applied shear. Figure 13.16 shows results from an experimental study on a hard-sphere colloidal glass, where the dynamics are studied using LSCM [70]. The two panels of Figure 13.16a show the same 2D slice of a nonsheared colloidal glass recorded 2.5 min apart. The color coding gives the strength of the local shear strain ε_{yz} (the lateral x- and y-directions are denoted in the figure and the z-direction is perpendicular to the image plane) resulting from thermal fluctuations, where the blue and red colors denote strains of opposite sign; the arrows highlight regions where particles rearrange and one finds that ε_{yz} changes sign at these sites between the two recordings, demonstrating the presence of thermally induced strain fluctuations. At short times, these fluctuations were observed to be reversible, but for longer times, the authors observed also thermally induced irreversible rearrangements.

It is interesting to compare this behavior with the interpretations of secondary β and structural α relaxation processes in molecular systems, as discussed in Section 13.5.5, with regard to the potential (free) energy landscape. The β-processes are here viewed as thermally activated reversible rearrangements corresponding to jumps between small subbasins confined within one of the large basins characterizing the landscape for $T < T^*$. The α-relaxation, on the other hand, corresponds to thermally activated irreversible rearrangements and thus jumps between large landscape basins [52]. Comparing the colloidal results with the picture that has evolved for molecular systems, there are thus interesting similarities. Note that the existence of irreversible rearrangements within a glassy state is not unexpected since a glass by definition is out of equilibrium and its properties will thus slowly evolve, or age, over time [6]. How significant such aging processes are in glasses depends on both their processing history and the distance from their corresponding glass transition.

Moreover, when the colloidal hard-sphere glass shown in Figure 13.16 is sheared (along the yz-plane), significant largely irreversible rearrangements occur. Such regions of concentrated strain occurring in response to mechanical deformation are called shear transformation zones (STZ) [70–73] and play important roles during plastic deformations, somewhat analogous to lattice dislocations in ordered crystalline systems. STZs have been widely studied in metallic glasses [53, 74] and in computational studies [71]. In fact, in Ref. [70], the properties of STZs in colloidal and metallic glasses were reported to be very similar in terms of both the STZ size and activation energy. An STZ has a relatively small core of concentrated strain, but this core is surrounded by a long-range strain field, which means that the spontaneous formation of one STZ can induce the formation of other STZs. This leads to a correlation in both time and space of STZs, which is clearly observed in Figure 13.16b where the accumulated strain is showed in a time series for a slice of the sheared sample. It has been found that the correlations between strains separated by a distance r follow a power law behavior in r, which means that the strain correlations are both long range and scale-free, that is, they are not characterized by any characteristic length scale [72, 75]. The spreading of such rearranging regions will eventually proceed throughout the system leading to macroscopic flow.

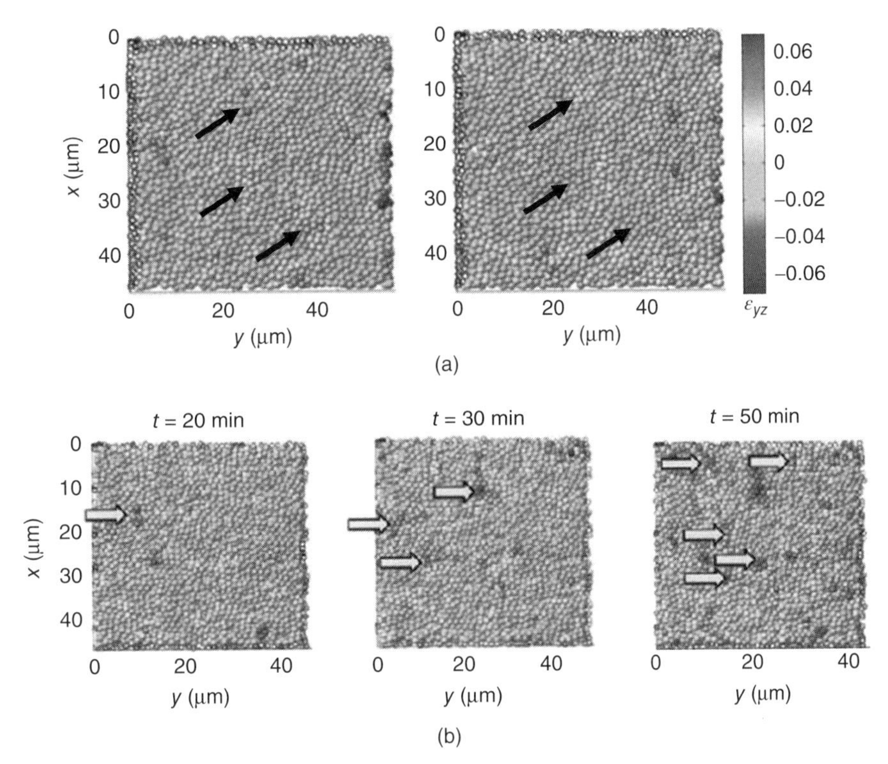

Figure 13.16 (a) Confocal microscopy image of a hard-sphere colloidal suspension showing a 2D slice through the system at two different times, 2.5 min apart. The color coding denotes the local shear strain ε_{yz}, and red and blue strains have opposite signs. The arrows mark regions that change sign of the strain within the interval. Source: Adapted from Schall 2007 [70], figure 1. Reproduced with permission of AAAS. (b) Evolution of cumulative shear strain with time for a sheared hard-sphere suspension. Source: Adapted from Schall 2007 [70], figure 4. Reproduced with permission of AAAS. (*See color plate section for the color representation of this figure.*)

Interestingly, a range of recent publications on metallic glasses have indicated that the activation energy of STZs in sheared glasses and the activation energy of secondary β-relaxations in the same glasses [53, 54, 74] are similar. The emerging picture, at least for metallic glasses, is thus that the properties of STZs observed under shear are related to the properties of glassy β-relaxations, whereas the α-relaxations correspond to the formation of an interconnected network of STZ "hot spots" percolating throughout the system. Although speculative at this stage, the reported similarity between STZ behavior in glassy colloidal and metallic systems together with the suggestion of a link between the behavior of secondary relaxation modes and STZs in metallic systems gives further support for the existence of secondary β-relaxations to be a general feature of molecular, metallic, and colloidal systems.

13.6 FURTHER COMPARISONS BETWEEN MOLECULAR AND COLLOIDAL GLASS FORMATION

13.6.1 Dynamic Heterogeneity

In a liquid far above its glass transition, the "particles," either molecules or colloids, all show the same degree of mobility. However, as the particles slow down close to a glass transition, their motions become increasingly intermittent consisting largely of localized motions that are only occasionally interrupted by long-range movements. On timescales significantly larger than τ_α, the behavior still looks dynamically homogeneous even though on this timescale every particle has undergone periods of both high and low mobilities. For timescales comparable to τ_α, however, many particles have only undergone localized motions, whereas some particles have moved significantly; one observes dynamic heterogeneity [9, 56, 76]. The dynamic homogeneity of a low volume fraction colloidal suspensions, for instance, leads to a Gaussian particle displacement distribution, as shown in Figure 13.17b. The development of dynamic heterogeneity observed in the crowded high volume fraction state corresponds to the development of non-Gaussian tails in the displacement distribution where the tails reflect the displacements of the particles that move the most (see Figure 13.17c). Dynamic heterogeneity can thus be quantified by the deviations from Gaussianity, and the so-called non-Gaussian parameter α_2 is commonly defined from a combination of the second and fourth moments of the displacement

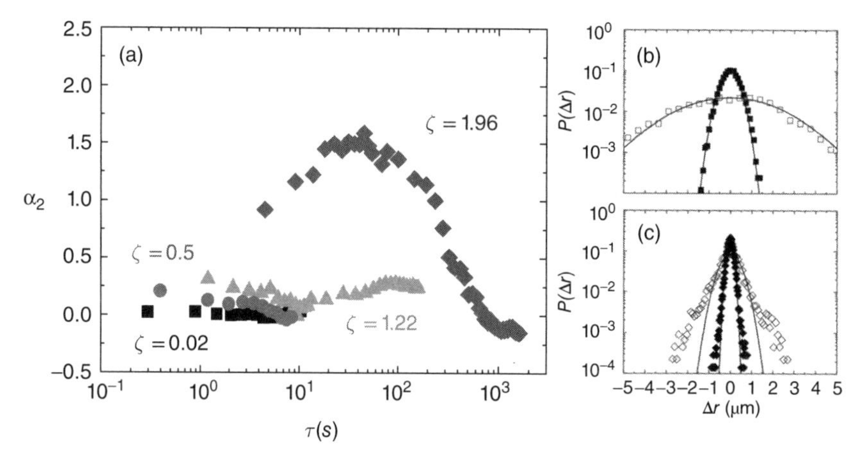

Figure 13.17 (a) Non-Gaussian parameter versus lag-time for the microgel suspensions discussed in Figure 13.12. (b) Displacement distributions for a $\zeta = 0.02$ microgel suspension at short (solid) and long (open) lag-times; the lines are Gaussian fits. (c) Distributions for the $\zeta = 1.96$ suspension. Note that the displacement distributions at these higher concentrations have additional wings reflecting the development of dynamic heterogeneities. Source: Adapted from Romeo 2012 [55], figure 9. Reproduced with permission of [55]. Copyright 2012, AIP Publishing LLC.

distribution as

$$\alpha_2(\tau) = \frac{\langle \Delta r^4 \rangle}{3\langle \Delta r^2 \rangle^2} - 1.$$

For a Gaussian displacement distribution, α_2 is zero and it increases for an increasing degree of non-Gaussianity. The behavior of α_2 observed for a glass-forming microgel suspension [55] is shown in Figure 13.17a. For the most concentrated suspension, α_2 shows a maximum and comparing with the MSD behavior for the same suspension in Figure 13.17a, it is clear that this peak appears at lag-times corresponding to τ_α, which is typical for both colloidal and molecular glass-formers [56].

As an illustration of typical dynamic heterogeneity, confocal microscopy data for a hard-sphere suspension [56] near its glass transition are shown in Figure 13.18a, where fast-moving particles are shown as large spheres and slow-moving particles as small spheres. It is clear that the fast-moving particles are correlated in space and form clusters of high mobility. Consistent with the behavior of α_2, one generally observes that this clustering is most pronounced for timescales $\sim\tau_\alpha$.

To study the development of dynamic heterogeneities in more detail [76–80], it is useful to define an overlap function or mobility (we here follow the approach outlined in [76, 77, 79, 80]):

$$m_i(t) = \exp[-|\vec{r}_i(t) - \vec{r}_i(0)|^2/d^2],$$

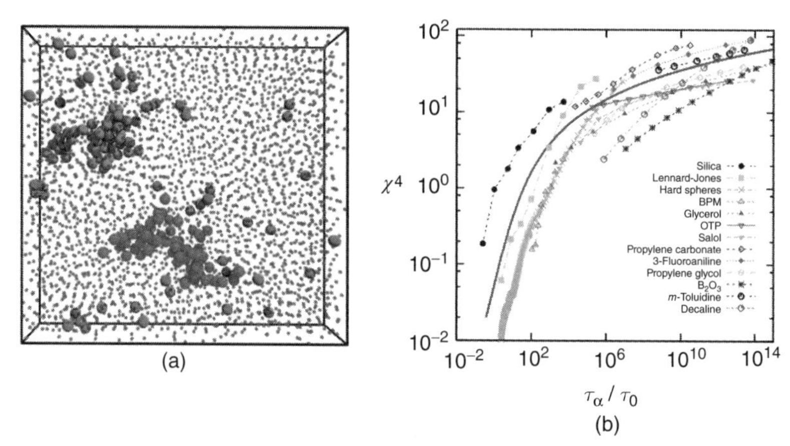

Figure 13.18 (a) Confocal microscopy data for a colloidal hard-sphere suspension with $\phi = 0.56$. The large spheres indicate the fastest moving particles, and it is clear that these form clusters in both time and space. Source: Adapted from Weeks 2000 [56], figure 4. Reproduced with permission of AAAS. (b) The maximum of the dynamic susceptibility $\chi_4^{\max}$, which provides a measure of the size of correlated motions in a liquid, as a function of the normalized α-relaxation timescale for a wide range of molecular liquids and for hard-sphere colloidal suspensions. Source: Adapted from Berthier 2011 [76], figure 5. Reproduced with permission of the American Physical Society. Copyright 2011 by the American Physical Society.

$m_i(t)$ provides a measure of the overlap between the position of the ith particle (molecule or colloid) $\vec{r}_i(t)$ at times 0 and t, measured in terms of the length scale d, which is typically of the order of the particle size. Based on m_i, one can define a mobility that takes all N particles into account,

$$m(\vec{r},t) = \sum_{i=1}^{N} m_i(t)\delta[\vec{r} - \vec{r}_i(0)],$$

where $\delta(\vec{r})$ is the Kronecker delta function. Using m, one can correlate the mobilities of the system with regard to both time and positions by defining the correlation function,

$$G_4(\vec{r},t) = \left\langle m\left(\vec{r},t\right) m(\vec{0},t)\right\rangle - \left\langle m\left(\vec{r},t\right)\right\rangle^2,$$

where the angular brackets denote an ensemble average. $G_4(\vec{r},t)$ correlates the mobilities at two points in space, $\vec{r}$ and $\vec{0}$, and two points in time, t and 0; it is thus often called a four-point correlation function. For systems in equilibrium that are translationally invariant [80], $G_4(r,t)$ gives the correlation at time t between particles separated by a distance r. Since this function tracks correlations in both space and time, it can be used to track the development of dynamic heterogeneities. The space integral of G_4, the so-called dynamic susceptibility,

$$\chi_4(t) = \int G_4(\vec{r},t)d^3\vec{r},$$

is important since its maximum value $\chi_4^{\max}$, which occurs for $t \approx \tau_\alpha$, can be directly related to the size of the dynamically correlated regions. The dynamic susceptibility can also be determined directly from the fluctuations of the total system mobility,

$$M(t) = \int m(\vec{r},t)d^3\vec{r}, \text{ as } \chi_4(t) = N\left[\left\langle M(t)^2\right\rangle - \langle M(t)\rangle^2\right],$$

where N denotes the total number of particles. G_4 and χ_4 have been directly determined for colloids where it is possible to resolve the particle motions in both time and space with the necessary resolution [78]. However, for molecular systems, this is much more difficult. Approximate theoretical relations have, however, been determined, which provides experimentally simpler ways to determine χ_4. These techniques use the fact that for small perturbations of a system, the induced fluctuations are directly related to the spontaneous fluctuations through a fluctuation–dissipation relation. Based on this, one can define a dynamic susceptibility that is used to provide an approximation of χ_4 and is readily accessed experimentally. By studying how the response, as monitored through the measured two-point correlation function such as the ISF, is affected by the experimental control parameter, either T for a molecular or ϕ for a colloidal glass-former, the corresponding dynamic susceptibility, either χ_T or χ_ϕ, is determined [80]. This dynamical susceptibility can then be related to χ_4 using some additional information that is system specific [80, 81]. Using this route, the spatial resolution necessary for a proper determination of $G_4(\vec{r},\ t)$ is not needed, which has made it possible to reach information about χ_4 for a wide range of glass-forming systems [80–82].

Figure 13.18b shows $\chi_4^{\max}$ determined for glass-forming liquids together with data for hard-sphere colloids; the data are here plotted as a function of the structural relaxation time τ_α to emphasize the strong generalities in the observed behavior. τ_α is normalized by a microscopic timescale for liquids (here fixed to 1 ps) and for colloids by a typical Brownian timescale (set to 1 ms); this renormalization facilitates the comparison between different systems. As τ_α grows, either due to a reduced T for molecular or due to an increased ϕ for colloidal glass-formers, $\chi_4^{\max}$ increases, which means that the characteristic length scale of correlated motions increases as the glass transition is approached. Note that the growth near the glass transition is relatively weak and direct estimates based on these data give length scales in the nanometer range for molecular systems, corresponding to motions involving ~10–100 molecules at the glass transition [82].

Recently, χ_4 was determined also for suspensions of deformable colloidal microgels [78]. The results identified dynamic correlations that appeared more long range than what has been observed for hard-sphere colloids [78]. Long-range correlated motions have been observed for a wide range of systems characterized by packed soft objects, and it has been suggested that such long-range correlations are characteristic of systems where elasticity plays a significant role in the rearrangements [78, 83–85]. Similar long-range correlations have been observed also for hard-sphere colloidal systems under shear [72] for which it was suggested that the energy supplied through shear can play a role analogous to the energy stored in a systems consisting of highly packed deformable building blocks. It is presently not clear what the exact distinction between the behavior of dense suspensions of "hard" and "soft" colloids is with regard to dynamic heterogeneities neither with nor without applied shear.

13.6.2 Decoupling of Translational and Rotational Diffusion

The translational diffusion of a tracer molecule/particle in a molecular liquid or a tracer particle in a colloidal suspension is characterized by a translational diffusion coefficient:

$$D_t \equiv \lim_{\Delta t \to \infty} \frac{1}{6\Delta t}\left\langle r^2\left(\Delta t\right)\right\rangle,$$

where $\langle r^2(\Delta t)\rangle$ is the translational MSD and r is the distance moved during the time, Δt. Correspondingly, the rotational diffusion coefficient is as follows:

$$D_r \equiv \lim_{\Delta t\to\infty} \frac{1}{4\Delta t}\langle \varphi^2(\Delta t)\rangle,$$

where $\langle \varphi^2(\Delta t)\rangle$ is the rotational MSD and $\vec{\varphi}(\Delta t)$ is the vectorial rotational displacement [86–88]. The SE relation for translational diffusion and the Stokes-Einstein-Debye (SED) relation for rotational diffusion relate the diffusion coefficient to the ratio of thermal energy (that drives the diffusion) over a friction factor f, $D_{t,r} = kT/f_{t,r}$. The friction factor for a sphere is $f_t = 6\pi\eta R$ for translational and $f_r = 8\pi\eta R^3$ for rotational motion, where η is the viscosity and R is the sphere radius [86]. Both D_t and D_r thus scale with T/η, and the ratio D_t/D_r is a constant as long as the SE/SED relations are valid. This coupling between translational and rotational diffusion generally holds well for liquids at temperatures above the crossover temperature T^*, as shown for the molecular liquid OTP for $T > T^* \approx 290$ K in Figure 13.19a. For $T < T^*$, however, the SE relation usually breaks down and the temperature dependence of translational and rotational diffusion generally differ [46]. Here, rotational diffusion often continues to be strongly coupled to the viscosity as $D_r \propto \eta^{-1}$, whereas translational diffusion shows a weaker viscosity dependence $D_t \propto \eta^{-\alpha}$, where $\alpha < 1$ [67]. An explanation for the decoupling at $T < T^*$ is that translational and rotational motions of a probe couple differently to the underlying distribution of molecular motions [6, 86, 89]. A decoupling of rotational and translational diffusion could thus occur due to the presence of dynamic heterogeneities, even though other explanations for the behavior exist [53, 90]. Note that the size of the probe plays a role for the observed behavior [86, 89]. If the cause of the SE breakdown is linked to the existence of dynamic heterogeneities, then this can readily be conceptualized since a probe that is large compared with the length scales of the dynamic heterogeneities will effectively average out these effects [6, 7].

An analogous breakdown of the SE behavior, which coincides with the onset of dynamic heterogeneities, has been observed in computer simulations on a hard-sphere fluid at $\phi^* \approx 0.52$ [66]. This suggests that a similar behavior should be found also for colloidal suspensions. Indeed, this was recently demonstrated for suspensions of spherical hard-sphere colloids [67] to which a small number of nonspherical tetrahedral tracer clusters were added. By adding asymmetric particles, both the translational and rotational diffusion of the tracers could be investigated using LSCM. The results are shown in Figure 13.19b, where the translational diffusion coefficients are shown in triangles and the rotational diffusion coefficients in circles; the squares is the result from an alternative way of determining D_r; see Ref. [67]. It is clear from Figure 13.19b that D_t and D_r begin to separate at a volume fraction $\phi^* \approx 0.52$, which correspond to the crossover volume fraction where dynamic heterogeneities start to develop [56, 65, 66].

13.6.3 The Vibrational Properties and the Boson Peak

For an ordered crystalline system, the periodic real space structure results in Brillouin zones (Bz) in reciprocal space and enables the use of Bloch formalism [91]. Each normal mode or phonon is characterized by its dispersion relation $E(k)$, where k is the wave vector characterizing the mode. Due to the crystal symmetry, it suffices to describe each mode within the first Brillouin zone (Bz) [91]. The vibrational density of states (DOS) $g(E)$ can be determined using the Debye

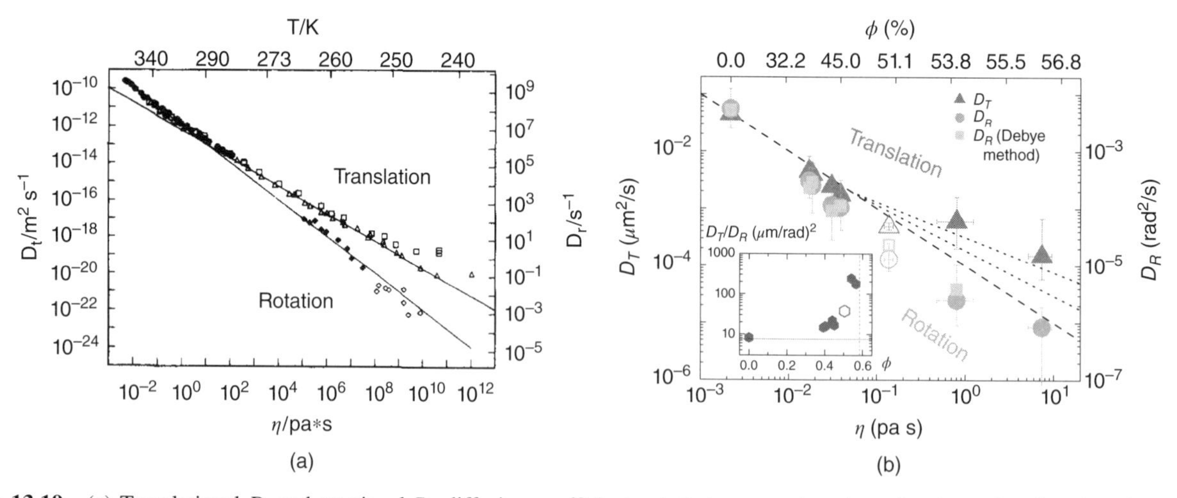

Figure 13.19 (a) Translational D_t and rotational D_r diffusion coefficients plotted versus viscosity η for the molecular glass-former OTP [46, 87]. For temperatures below $T^* = 290$ K, translational and rotational diffusion decouple. Source: Adapted from Chang 1994 [46], figure 6. Reproduced with permission of Elsevier. (b) The analogous plot for a hard-sphere colloidal suspension [67]. Source: Edmond 2012 [67], figure 4. Reprinted with permission.

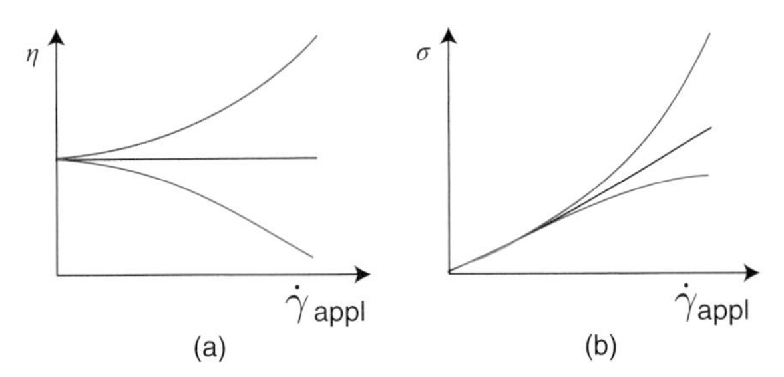

Figure 6.3 Cartoon of three different flow curves in viscosity versus shear rate representation and stress versus shear rate representation: Newtonian (black); shear thinning (red); shear thickening (blue).

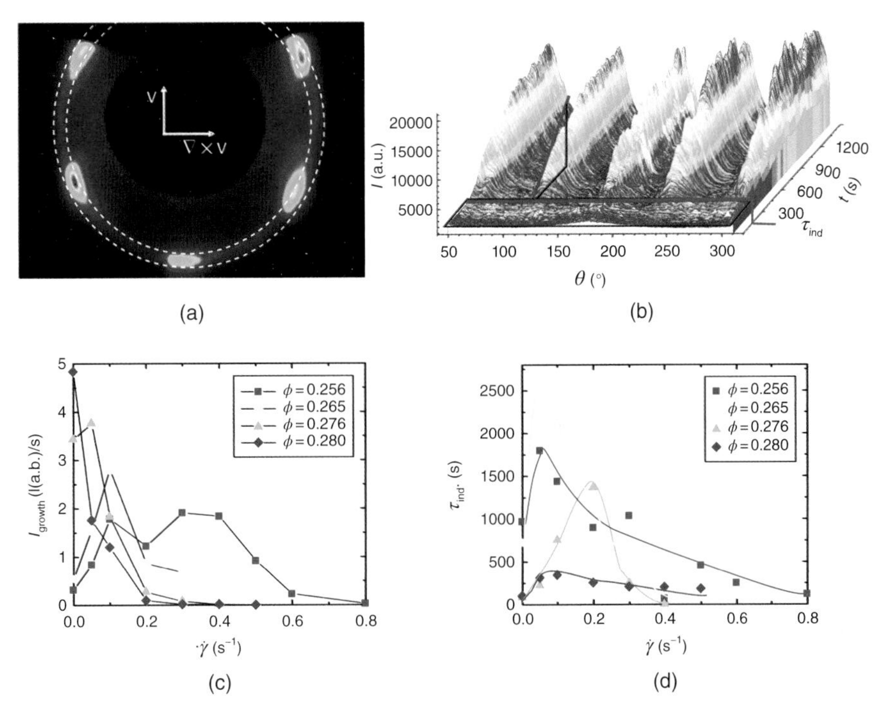

Figure 6.8 Crystallization kinetics of charged colloidal spheres after a quench from a high shear rate where the crystal structure is melted to a final shear rate. (a) Light scattering pattern using a geometry as shown in Figure 6.6a, probing the flow-vorticity plane. The pattern is taken 1500 s after a quench to a shear rate of 0.05 s^{-1}. (b) Azimuthal profiles taken at the scatter angle of the first-order Bragg peaks, dashed lines in (a), as a function of time after the quench to $\dot{\gamma} = 0.05\ s^{-1}$. The growth rate I_{growth} is obtained from the slope of the red line and induction time τ_{ind} is obtained from the start of growth. τ_{ind} and I_{growth} are plotted in (c) and (d), respectively, as a function shear rate ε for various volume fractions. Source: Adapted with permission from Ref. [63]. Copyright 2005 American Chemical Society.

Fluids, Colloids and Soft Materials: An Introduction to Soft Matter Physics, First Edition. Edited by Alberto Fernandez Nieves and Antonio Manuel Puertas.

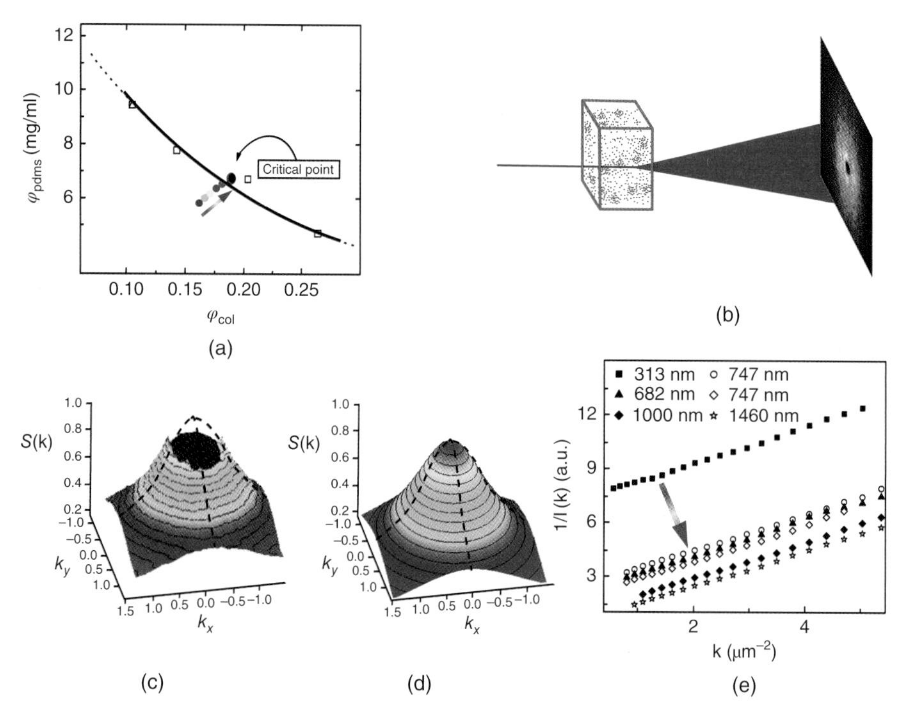

Figure 6.10 (a) The equilibrium phase diagram of mixtures of silica spheres (102 nm diameter, the *x*-axis gives the volume fraction ε_{col}) and polydimethylsiloxane (molecular weight of 206 kg/mol, concentration is given on the *y*-axis in mg/ml). The range of the interaction potential $R_V \approx 25$ nm, since the radius of gyration is 23 nm. Source: The figure is adapted with permission from Ref. [89]. (b) Schematic small-angle light scattering setup to determine the critical structure (c), which can also be calculated using Equation 6.24 (d). The correlation lengths are calculated from the angular averaged structure factors (d), which are schematically shown in (e). Source: The figures (c,d,e) are adapted from Ref. [90] ©IOP Publishing. Reproduced by permission of IOP Publishing. All rights reserved.

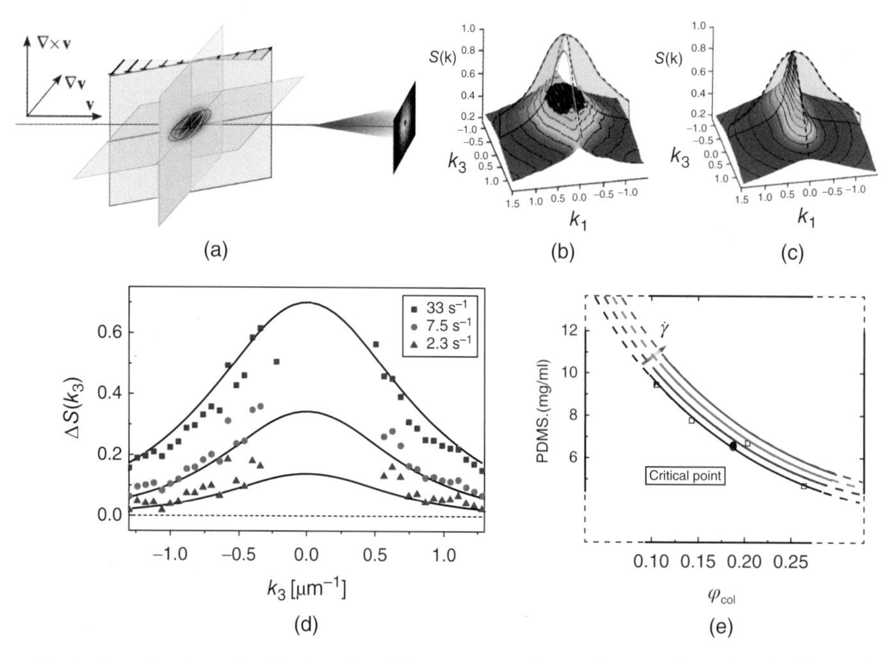

Figure 6.11 (a) Geometrical setup of a sheared critical system. The same experimental system is used as in Figure 6.6. Experimental (b) and theoretically calculated (c) sheared critical structure factor detected in the flow (1,**v**) and vorticity (3, $\nabla \times \mathbf{v}$) scattering plane. The equilibrium correlation length is 750 nm and the dressed Péclet number is $\lambda = 4$. The shaded surfaces show the difference with the equilibrium structure factor. (d) The difference between the equilibrium and sheared structure factor for different shear rates, taken at the axis where no distortion is expected ($k_1 = 0$). The solid lines are fits to the theory, using Equation 6.38. The dashed line indicates the theoretically predicted absence of distortion in this direction. Source: Figure adapted from Ref. [90] ©IOP Publishing. Reproduced by permission from IOP Publishing. All rights reserved. (e) Sketch of the effect of shear flow on the phase boundary. Source: Adapted with permission from Ref. [89].

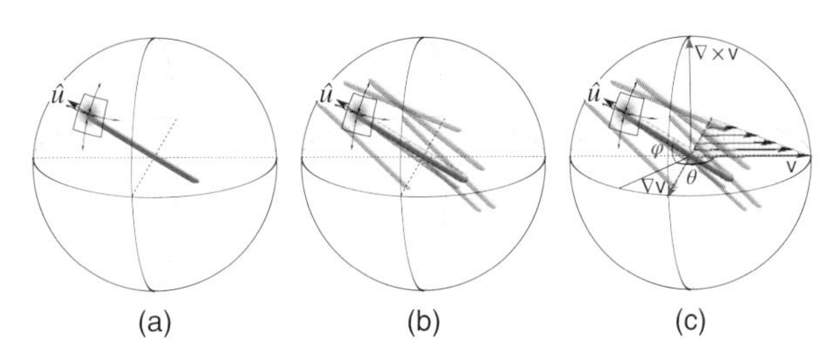

Figure 6.13 (a) Definition of the unit vector **û**. The arrows out of the little surface area indicate the rotational Brownian diffusion given by the first term in Equations 6.42 and 6.50. In concentrated solutions, (b) neighboring rods exert a torque on the rod influencing the rotational diffusion given by the second term in Equations 6.42 and 6.50. Shear flow (c) adds another torque term given by the third term in Equation 6.50.

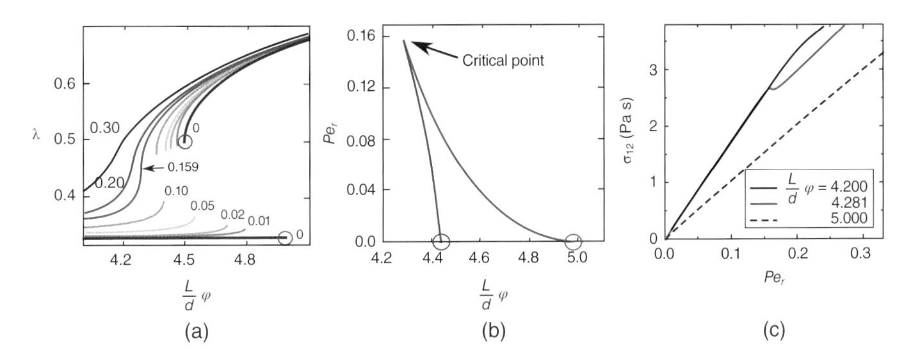

Figure 6.15 (a) Bifurcation diagram of the nematic order parameter λ, the highest eigen value of stationary solutions of **S**, versus the scaled volume fraction $\frac{L}{d}\varphi$, as obtained from the equation of motion (Eq. 6.51) hard rods in shear flow. The numbers refer to values of the bare rotational Péclet number. (b) The nonequilibrium spinodal lines as obtained from (a) by taking the nonequilibrium spinodal points. The critical concentration is defined by the point where both spinodal lines meet and given by a critical concentration of $\frac{L}{d}_{\text{crit}}\varphi = 4.281$ and critical Péclet number of $Pe_r^{\text{crit}} = 0.159$. The circles indicate the equilibrium I–N (red) and N–I (blue) spinodal points and are per definition the end of the isotropic and nematic branch as calculated by Kayser and Raveché [113]. (c) Theoretical flow curves plotting the shear stress σ_{12} versus the bare Péclet number Pe_r. At the critical concentration of $\frac{L}{d}\varphi_r$ the flow curve contains an unstable part where the slope in σ_{12}. Source: Reprinted and adapted from Ref. [107]. Copyright 2003 with permission from Elsevier.

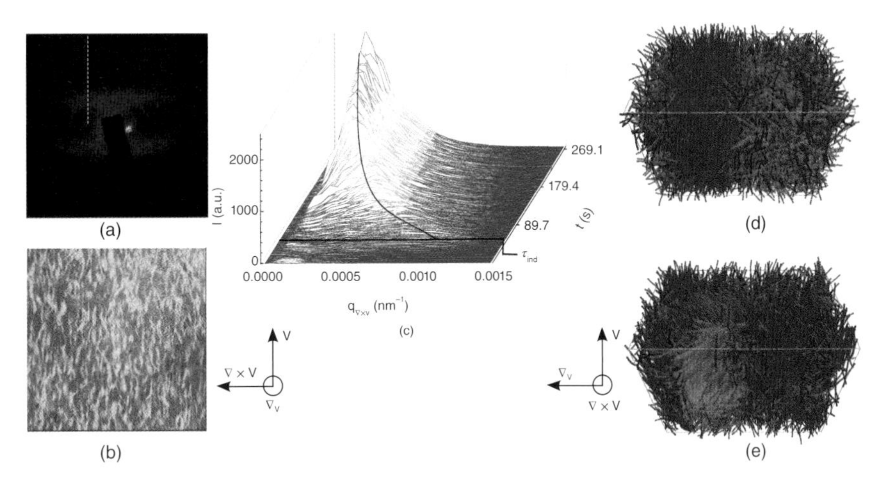

Figure 6.17 I–N phase coexisting *fd* suspension in shear flow probed by SALS (a) and polarization confocal microscopy (b). (c) Cross section of the scatter pattern (taken at the dashed line in (a)) as a function of time. The size of the form structure can be deduced from this plot as well as the induction time τ_{ind}. (d) MPCD simulation of a coexisting isotropic–nematic system of rods. The color indicates the orientation. (e) As (d) but now under shear flow. The original nematic phase is undergoing a tumbling motion while the isotropic flow aligns into a para-nematic phase.

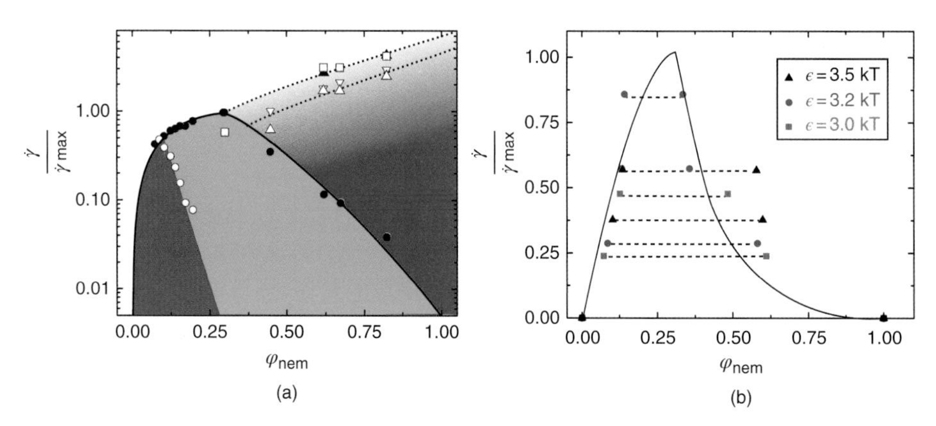

Figure 6.18 (a) Nonequilibrium experimental phase diagram for *fd* virus with 18 mg/ml dextran: binodal (•, solid line is a fit) and N–I spinodal (○) points obtained from SALS; tumbling-to-aligning transition, given by the points where all structures disappear (■) and where the response of the stress after flow reversal is overdamped (□); a wagging motion causes a local maximum in the viscosity (▽) and a minimum damping constant (△). The dotted lines are guides to the eye. The quenched nematic phase is unstable in the blue region and metastable in the green region. In the red region, the full nematic phase is in a tumbling state. (b) Nonequilibrium binodal as obtained from MPCD simulations. The symbols refer to simulations with different depth of the Lennard-Jones potential ϵ causing attractions between the rods. The full line is a fit to the experimental data, as plotted in (a). The dashed lines are the tie lines, connecting binodal points. The shear rate is scaled by the shear rate at the maximum of the binodal. The concentrations in experiment and simulations are given in terms of the fraction of the nematic phase in the quiescent sample.

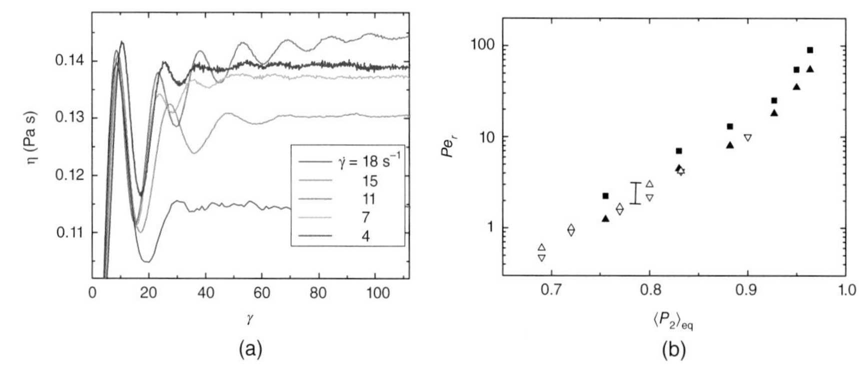

Figure 6.19 (a) Response of the viscosity after a flow reversal for various shear rates at a concentration nem = 0.67. The damping constant first decreases and then increases going toward flow alignment. The time is scaled with the applied shear rate, so the *x*-axis is given in strain units. (b) Dynamic phase diagram of the tumbling nematic phase as a function of the orientational order parameter at equilibrium. The experimental points indicate the Péclet numbers, where the viscosity shows a local maximum (▽) and a minimum damping constant (△). The theoretical points indicate tumbling to wagging (▲) and wagging to flow aligning (■) transitions. The region of the transition as inferred from simulations [119] is indicated by the vertical line. Source: Reprinted (adapted) with permission from Ref. [115]. Copyright 2005 American Chemical Society.

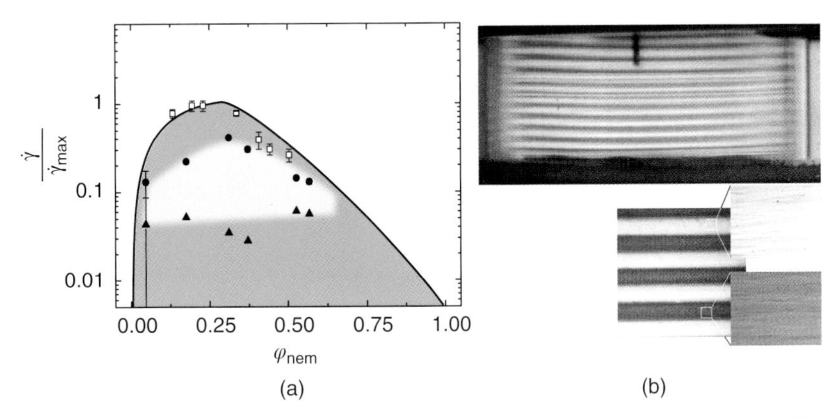

Figure 6.20 (a) Nonequilibrium binodal for the I–N transition of *fd* virus in shear flow, as Figure 6.18a, including the region where vorticity banding is observed. The dextran concentration was 10.6 mg/ml. The binodal line is the same as in Figure 6.18a. (b) Macro- and micrographs of vorticity bands all taken with almost crossed polarizers. The top picture shows that bands are formed throughout the Couette cell. The height of the cell being 3 cm. The lower figure shows a banded state for the same, where the bandwidth is about 2 mm. The two enlargements on the right show the inhomogeneities that are present within the bands.

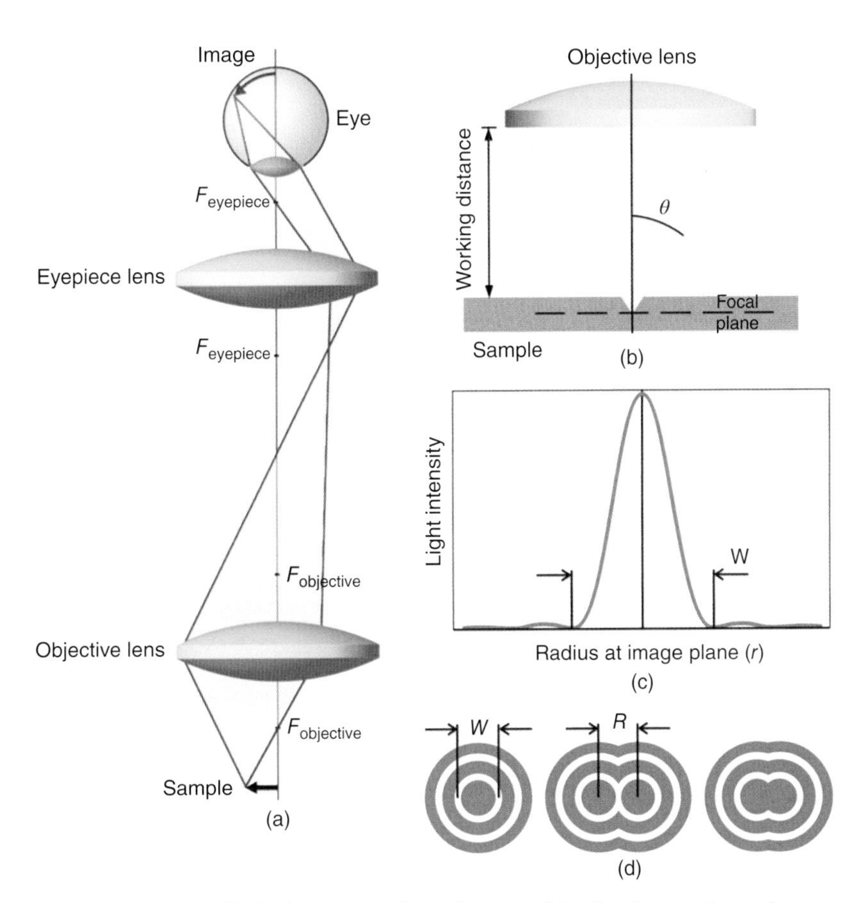

Figure 10.1 Principles of imaging with an optical microscope: (a) ray diagram of the simplest two-lens microscope; (b) definition of parameters of an objective lens; (c) point spread function with the first minimum at $r = W/2$; (d) Airy patterns in the image plane. "$F_{\text{objective}}$" and "F_{eyepiece}" mark the foci of objective and eyepiece lenses, respectively. "R" shows the separation distance between the centers of Airy disks.

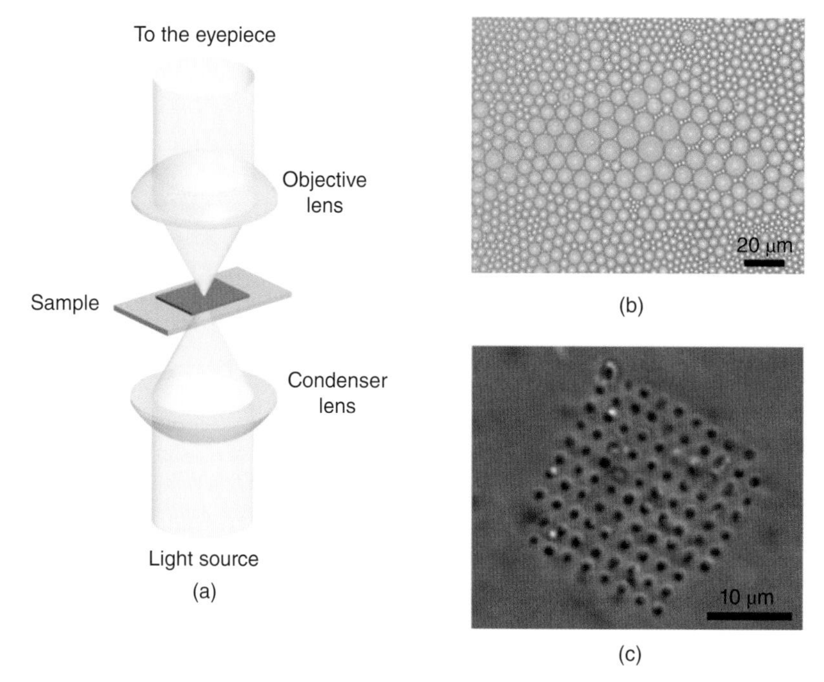

Figure 10.2 Bright field microscopy diagram (a) and textures: (b) toroidal focal conic domains in a smectic liquid crystal (8CB) film spread over a glycerol surface; (c) an array of *P. aeruginosa* cells trapped by an array of infrared laser beams.

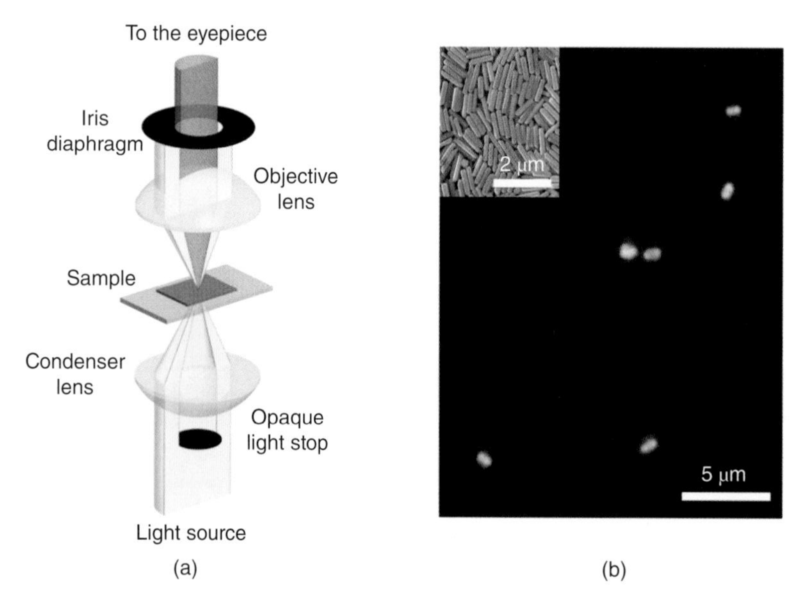

Figure 10.3 Dark field microscopy diagram (a) and texture of gold rod-like colloids (b) dispersed in a smectic (8CB) homeotropic liquid crystal cell. Inset in (b) shows the transmission electron microscopy image of gold rod-like colloids. Source: TEM image courtesy of Nanopartz Inc.

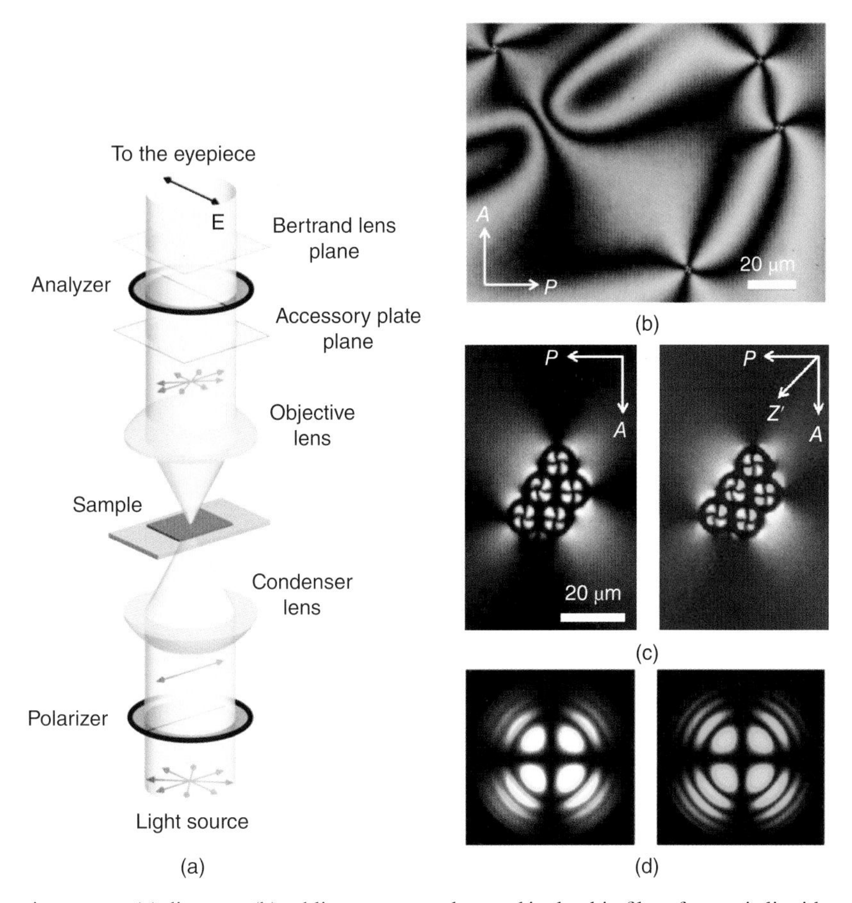

Figure 10.4 Polarizing microscopy: (a) diagram; (b) schlieren texture observed in the thin film of nematic liquid crystal spread over glycerol surface; (c) texture of nematic liquid crystal around a colloidal particle of nonzero genus [2] observed without (left) and with (right) the red plate inserted; (d) conoscopic images obtained for a homeotropic uniaxial smectic A (8CB) sample with a Bertrand lens inserted under the white (left) and green (right) light. Arrows *P*, *A*, and Z′ show a direction of polarizer, analyzer, and "slow axis" of the red plate, respectively.

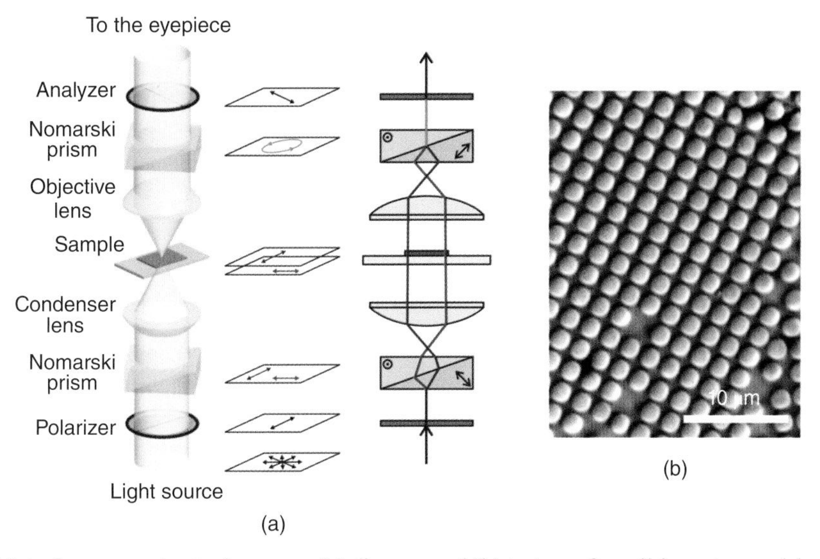

Figure 10.5 Differential interference contrast microscopy (a) diagram and (b) texture of an oil-in-water emulsion. Source: Image courtesy of V.N. Manoharan; see Ref. [26] for detailed description of the image in (b).

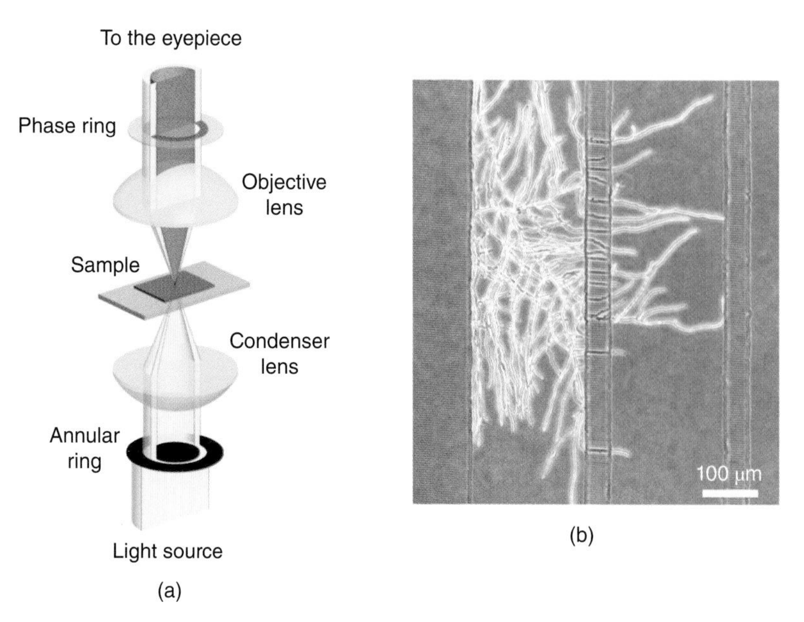

Figure 10.6 Phase contrast microscopy: (a) diagram and (b) the image of elongating mycelium hyphae of a common *Aspergillus* mold confined within polydimethylsiloxane-based microfluidic channels. Source: Image courtesy of L. Millet; see Ref. [27] for detailed description of the image in (b).

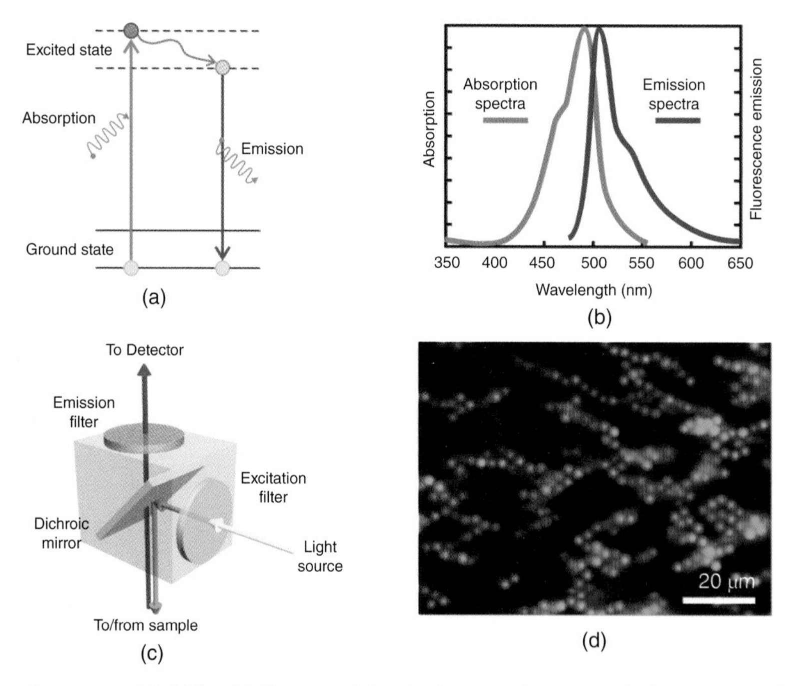

Figure 10.7 Fluorescence microscopy: (a) Jablonski diagram; (b) typical absorption and emission spectra of a fluorescent dye; (c) filters cube; (d) fluorescence texture of spherical microparticles tagged with a dye and dispersed in a nematic liquid crystal.

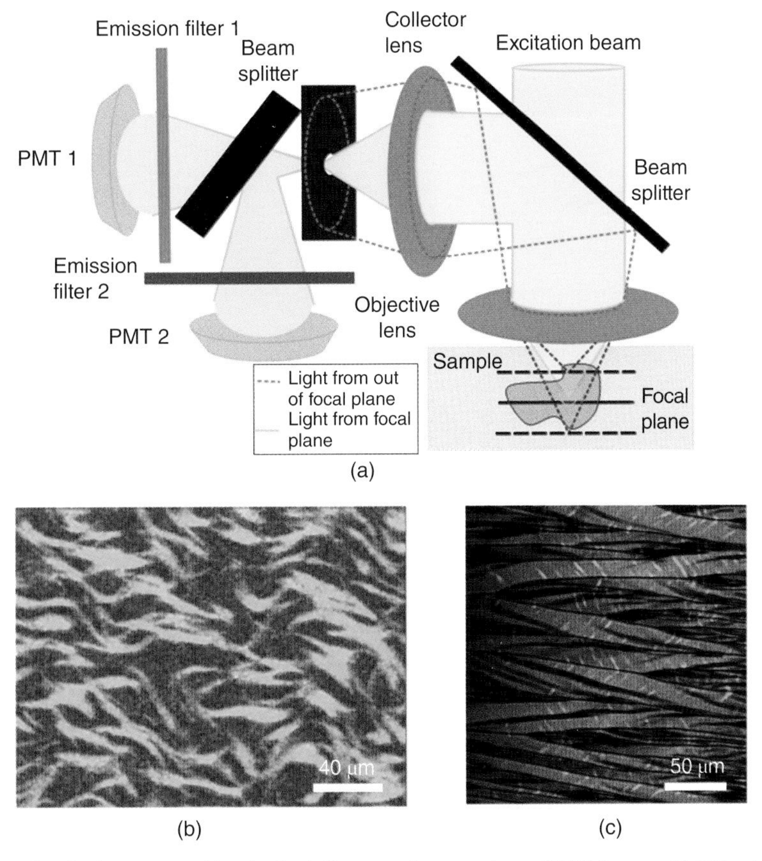

Figure 10.8 Fluorescence confocal microscopy: (a) principal diagram of a two-channel FCM setup: PMT is photomultiplier tube; (b) image of phase-separated domains of F-actin (green) and DNA (red) labeled by two different dyes [5]; (c) polarizing microscope texture of the pattern formed in a drying droplet of aqueous DNA and colocalized with the fluorescence confocal signal from a small number of molecules marked by fluorescent dye (green); see Refs [5, 6] for the detailed description of samples.

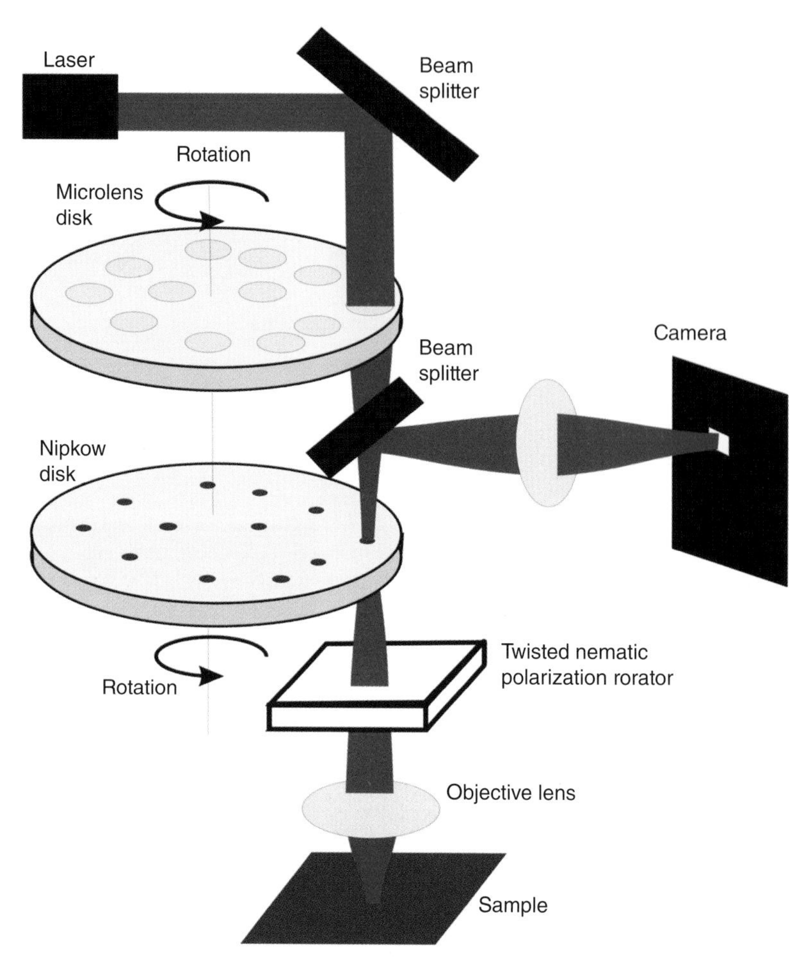

Figure 10.9 A schematic of Nipkow disk fluorescence confocal microscopy. Twisted nematic polarization rotator is typically used only in studies of samples with orientational ordering.

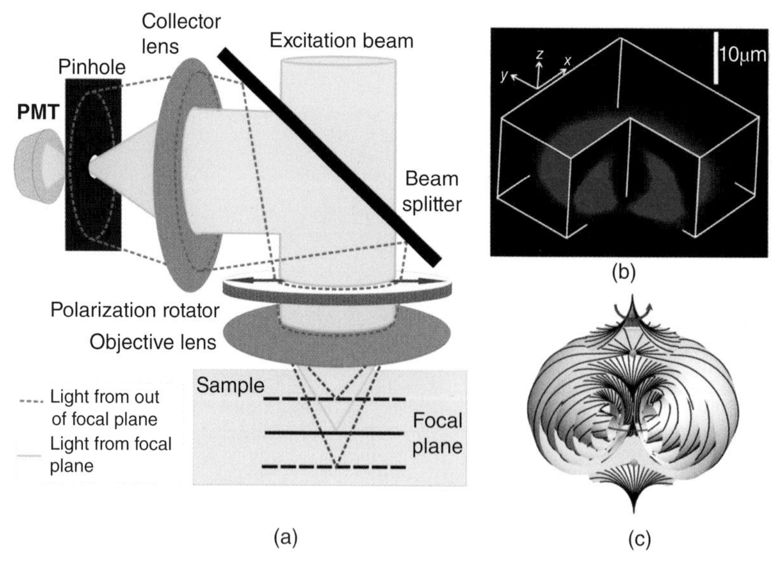

Figure 10.10 Fluorescence confocal polarizing microscopy: (a) a schematic diagram of FCPM with a polarization rotator; (b) 3D image of the T3-1 toron director configuration obtained using FCPM with circularly polarized probing light (see Ref. [7] for details); (c) a schematic representation of T3-1 configuration with the double-twist cylinder looped on itself and accompanied by two hyperbolic point defects (blue dots).

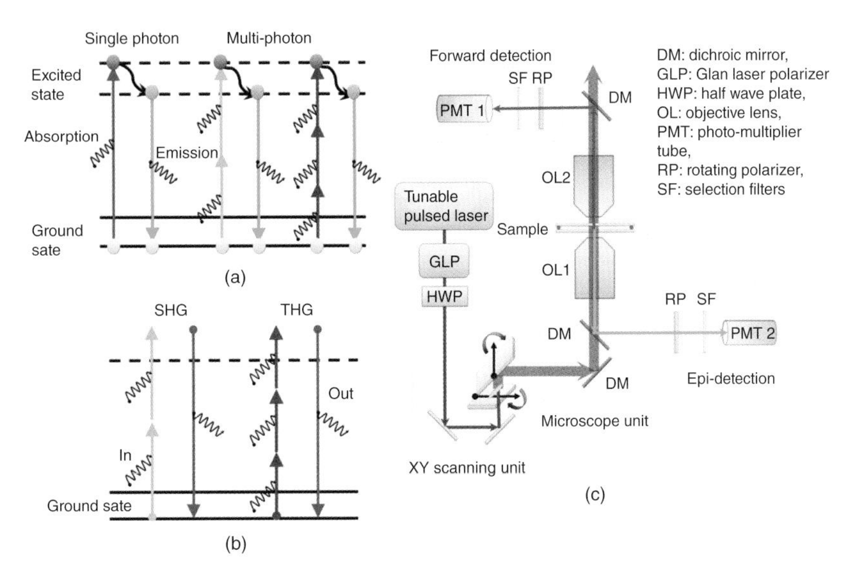

Figure 10.11 Multiphoton excitation fluorescence and multiharmonic generation microscopy. Energy diagram of (a) single and multiphoton processes and (b) second and third harmonic generation processes. (c) A schematic diagram of nonlinear optical imaging setup based on a tunable pulsed laser, *xy* scanning mirror, and inverted microscope with multiple detection channels. GLP, HWP, and RP are used to control light polarization. DM, dichroic mirror; GLP, Glan laser polarizer; HWP, half wave plate; OL, objective lens; PMT, photomultiplier tube; RP, rotating polarizer; SF, selection filters.

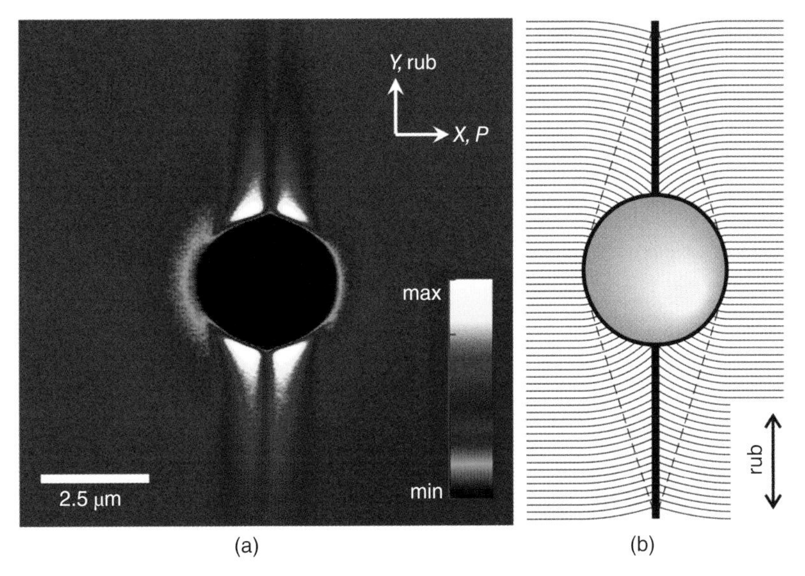

Figure 10.12 Application of multiphoton excitation fluorescence microscopy to imaging of a colloidal particle in a smectic liquid crystal (see Ref. [10] for detailed description). (a) Three-photon excitation fluorescence image of smectic liquid crystal layer deformations around a melamine resin sphere obtained with excitation at 870 nm and detection within 390–450 nm. (b) Reconstructed deformations of smectic layers (thin solid lines) around a spherical inclusion. Liquid crystal molecules are anchored tangentially at the surface of the particle. Arrow marked with "rub" shows the direction of substrates rubbing. Thick solid line shows the singular defect line, and dashed lines enclose the regions of strong layer deformations corresponding to the areas of maximum signal in (a).

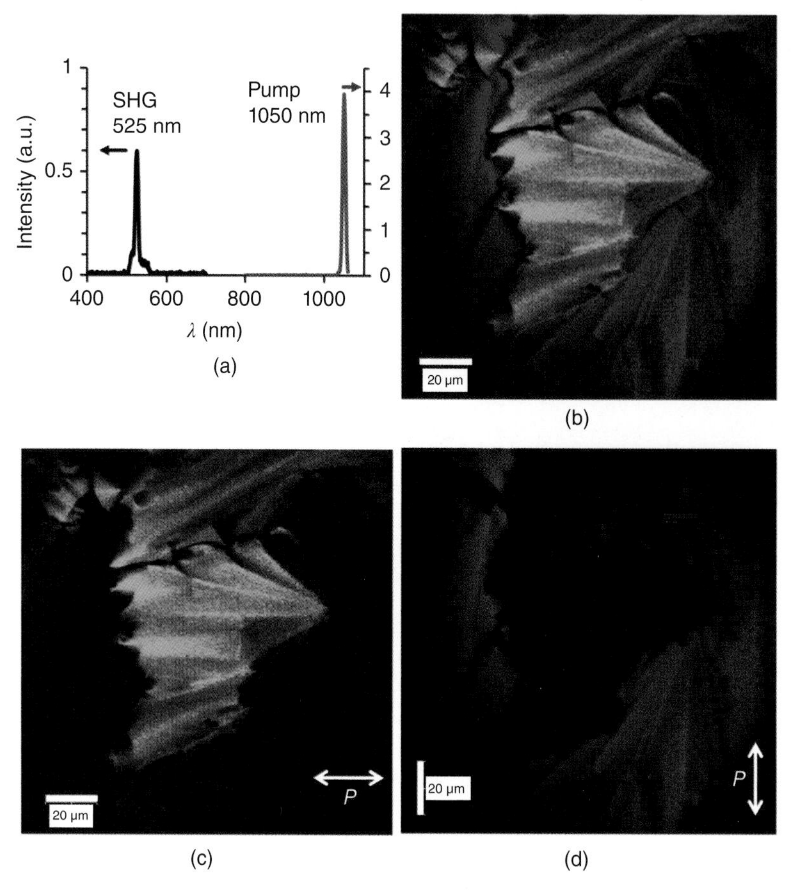

Figure 10.13 Second harmonic generation imaging of a smectic C* liquid crystal. (a) A spectrum showing the excitation pulse and the generated SHG signal. (b) In-plane colocalized superimposed texture of two SHG images of (c) and (d) that were obtained separately for two orthogonal polarizations (see Ref. [10] for a detailed description).

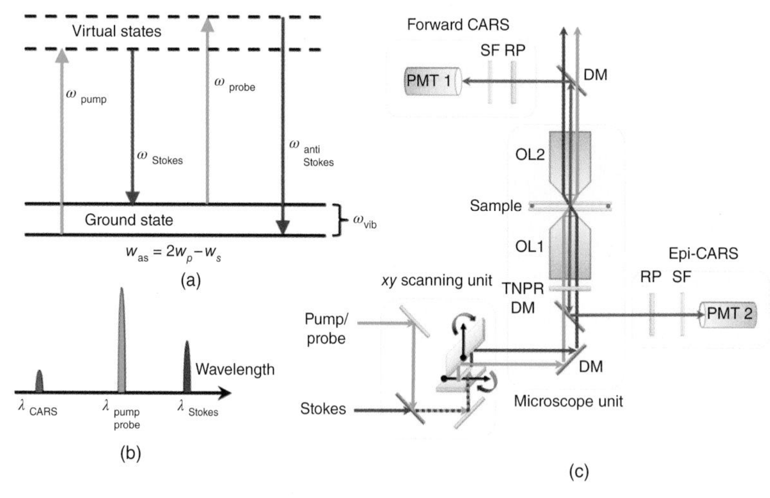

Figure 10.14 Coherent anti-Stokes Raman scattering microscopy. (a) The energy diagrams of CARS at $\omega_{as} = 2\omega_p - \omega_s$ when $\omega_{\text{vib}} = \omega_p - \omega_s$. (b) CARS signal generated at shorter wavelength than the pump and Stokes wavelengths. (c) A schematic diagram of CARS polarizing microscopy setup utilizing the synchronized pump/probe and Stokes pulses, *xy* scanning galvano-mirrors, and inverted microscope with forward and epi-detection channels. DM, dichroic mirror; OL, objective lens; PMT, photomultiplier tube; RP, rotating polarizer; TNPR, twisted nematic polarization rotator; SF, selection filter.

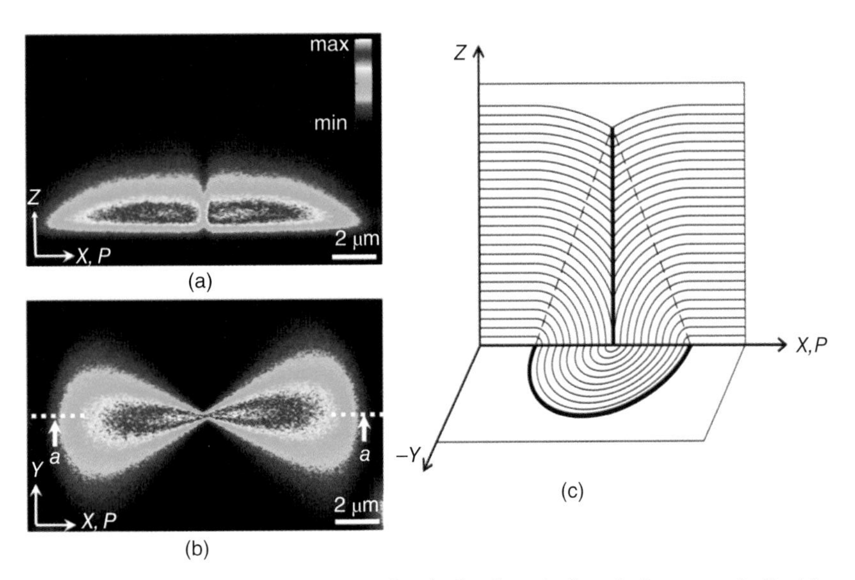

Figure 10.15 Coherent anti-Stokes Raman scattering images of toric focal conic domain in a smectic liquid crystal thin film on a solid substrate. (a) Vertical cross-sectional and (b) in-plane images. (c) Reconstructed structure of smectic layers (thin solid lines) in the toric focal conic domain.

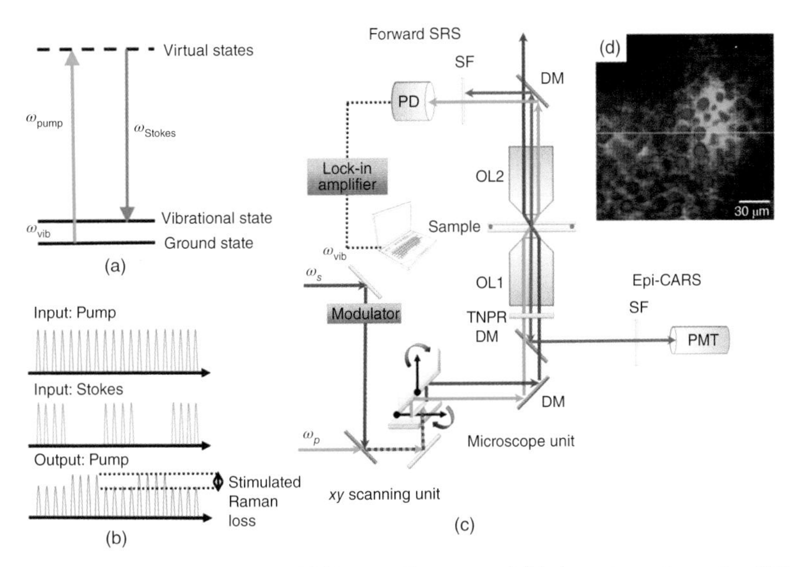

Figure 10.16 Stimulated Raman scattering microscopy. (a) Energy diagram and (b) detection scheme for SRS. Stokes beam is modulated at high frequency, and the resulting amplitude modulation of the pump pulse (stimulated Raman loss) can be detected. (c) A schematic diagram for SRS microscope with forward SRS and epi-CARS detection. (d) Two-color SRS image of DMSO (green) and lipid (red) in the subcutaneous fat layer. DM, dichroic mirror; OL, objective lens; PD, photodiode; PMT, photomultiplier tube; SF, selection filter. Source: Image reprinted from [34] with permission from AAAS.

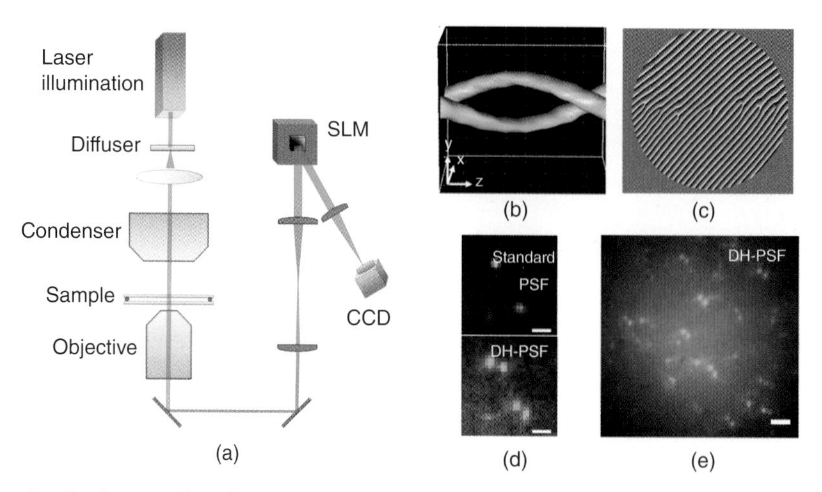

Figure 10.17 Optical imaging by the use of engineered PSF. (a) Optical setup for imaging with pre-engineered PSF. (b) Schematic representation of double-helix PSF. (c) Typical phase mask used to generate double-helix PSF. (d–e) Single molecule detection in 3D using double-helix PSF technique. Source: Images from [36] copyright 2009 National Academy of Sciences, USA.

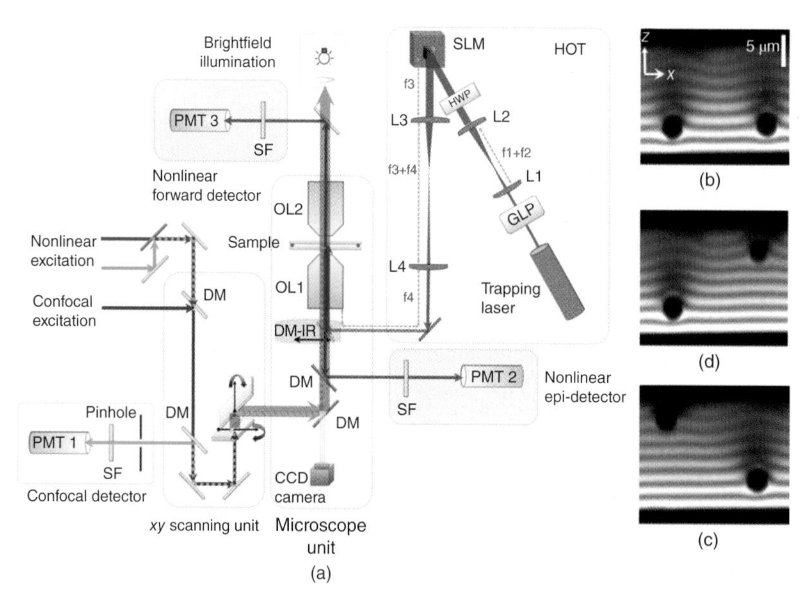

Figure 10.18 Integrated setup of 3D imaging and holographic optical trapping. (a) A schematic diagram of 3D optical manipulation with holographic optical trapping (HOT) and simultaneous imaging with FCPM and nonlinear optical microscopy. (b–d) FCPM vertical cross sections showing spherical particles in cholesteric liquid crystal moved axially and imaged using the integrated setup. The periodicity of bright stripes roughly corresponds to the half pitch of the cholesteric liquid crystal. DM, dichroic mirror; DM-IR, dichroic mirror for trapping; OL, objective lens; PMT, photomultiplier tube; SF, selection filters.

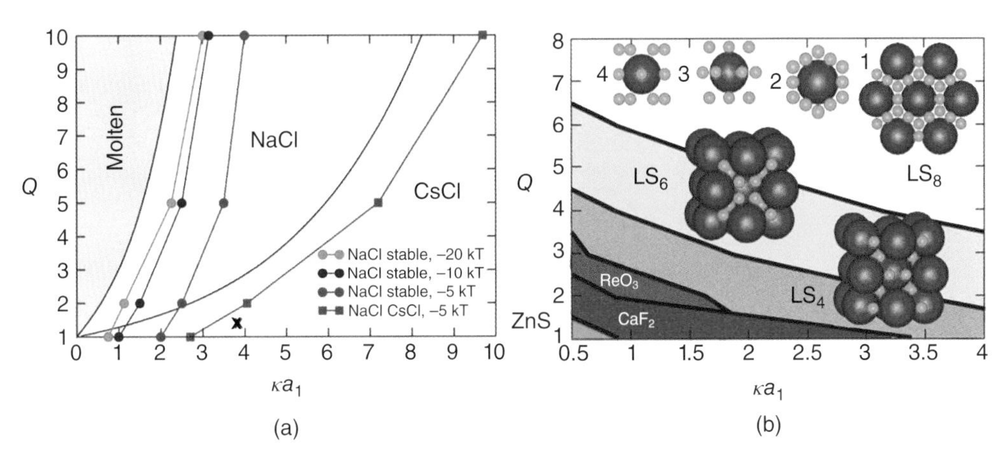

Figure 12.35 *Phase diagram of oppositely charged spheres.* Charge ratio $Q = -Z_1Z_2$ and size ratio a_2a_1 of 1 (a) and 0.31 (b). Full curves indicate crystal–crystal transitions from Madelung energy calculations (zero-temperature; black) and finite-temperature, zero-pressure MC simulations with a contact energy V_{12} of $-20k_BT$ (green), $-10k_BT$ (blue) and $-5k_BT$ (red). (a) The NaCl-structure melting line and NaCl–CsCl phase boundary. (b) Crystal phases coexisting with a dilute gas of small colloids, from Madelung energies. Insets: 1, projection of the three basic planes of LS_8; 2, configuration around a large colloid; 3, side view; 4, side view rotated by 90°.

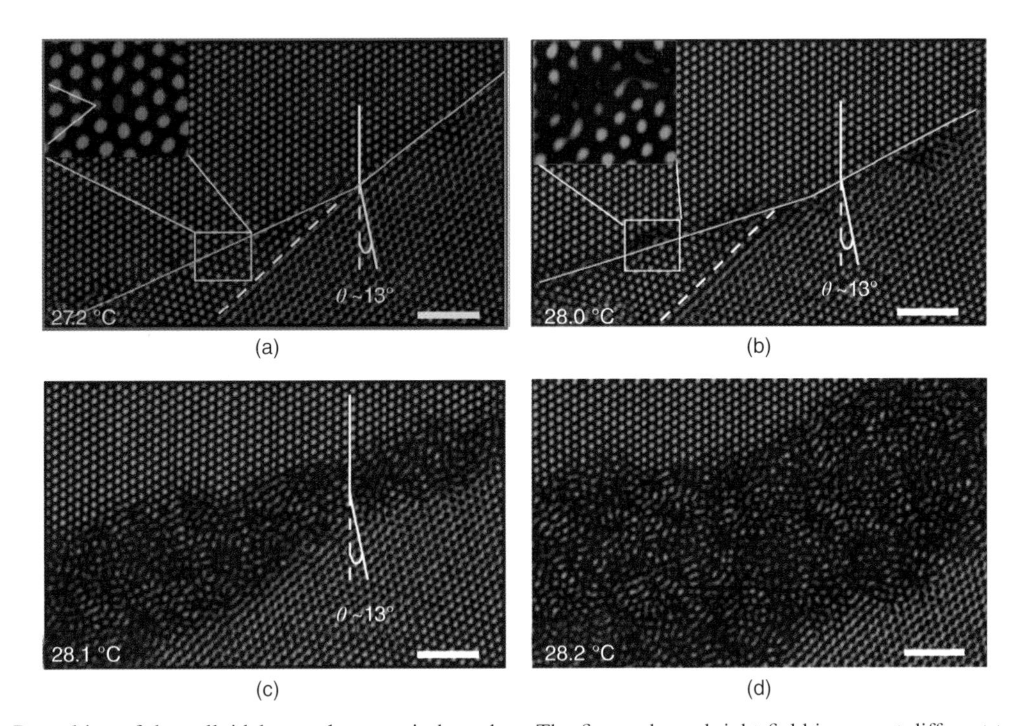

Figure 12.47 Premelting of the colloidal crystal at a grain boundary. The figure shows bright-field images at different temperatures (i.e., particle volume fractions) of two crystallites separated by a grain boundary (crystallites tilted at an angle $\theta \sim 13°$ with respect to one another). (a) Sample at 27.2 °C. The solid and dashed lines show the grain boundary and a partial dislocation, respectively. The grain boundary cuts the two crystals along two different planes (the yellow line has two slopes). It is composed of an array of dislocations; the two extra planes are indicated by lines in the inset. (b) Sample at 28.0 °C. The grain boundary starts to premelt; nearby particles undergo liquid-like diffusion (inset). The partial dislocation, denoted by the dashed line, is not affected. (c and d) The same sample at 28.1 and 28.2 °C, respectively. The width of the premelt region near the grain boundary increases. Scale bars: 5 mm.

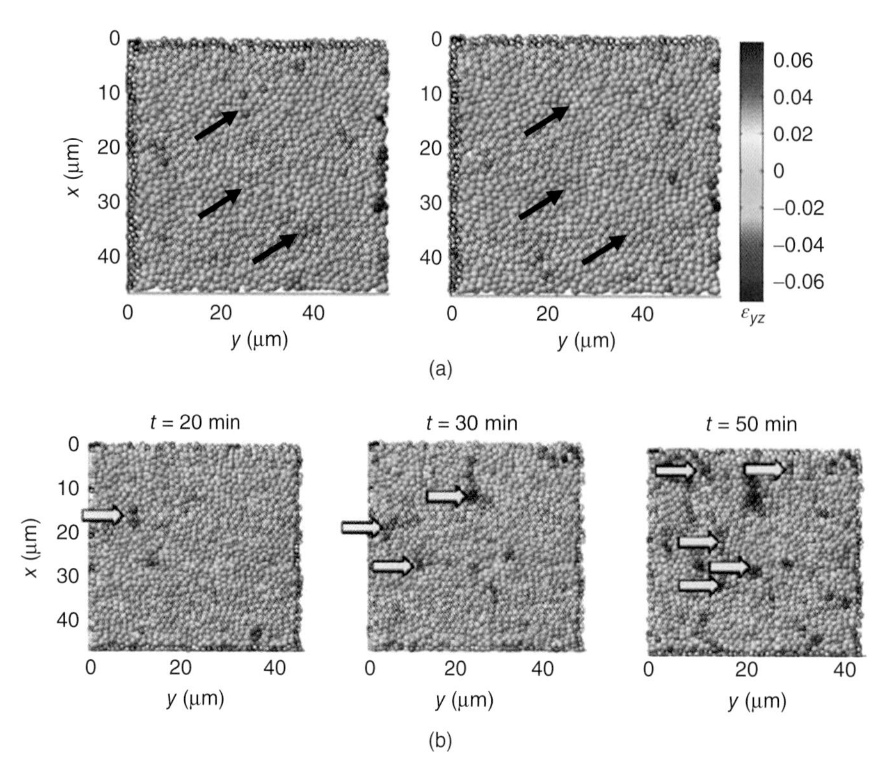

Figure 13.16 (a) Confocal microscopy image of a hard-sphere colloidal suspension showing a 2D slice through the system at two different times, 2.5 min apart. The color coding denotes the local shear strain ϵ_{yz}, and red and blue strains have opposite signs. The arrows mark regions that change sign of the strain within the interval. (b) Evolution of cumulative shear strain with time for a sheared hard-sphere suspension. Source: Adapted from Ref. [70].

model by summing the acoustic phonon contributions, which leads to $g(E) \propto E^2$ in 3D and for a system that obeys the Debye behavior $g(E)/E^2$ thus equals a constant value. At the Bz edge $g(E)$ shows the so-called van Hove singularity, which corresponds to the maximum in $E(k)$ where the group velocity goes to zero [91].

For a disordered system, the lack of order means that a description in terms of acoustic modes described by unique wave vectors is not possible. Still, many similarities remain, and one can approximately consider the first sharp peak in the structure factor $S(k)$ (see Fig. 13.3d) at k_p as a broadened reciprocal lattice point, and $k_p/2$ as a pseudo-Bz edge, in analogy to the behavior of crystals. For small k (long distances), the acoustic modes can be viewed as plane waves, but as k increases toward $k_p/2$, the modes will be modified by the local structure, the plane-wave description breaks down and the modes will have a more complex nature [92]. In analogy with a van Hove singularity of crystals, approaching the pseudo-Bz edge results in a piling up of vibrational modes, which can result in an excess in the DOS compared with the Debye prediction. It has been suggested that this is the origin of much discussed and commonly observed peak, the so-called boson peak in $g(E)/E^2$ [92]. The boson peak is generally observed for disordered materials. Note, however, that the exact origin of the boson peak is still strongly debated, and many alternative views that are not directly related to acoustic modes exist, see Refs [92–94].

As an illustration, a typical behavior for $g(E)/E^2$ is shown for an atomic glass $Na_2FeSi_3O_{8.5}$ in Figure 13.20a for varying pressures [92]. It is clear that an excess in DOS compared with the Debye prediction, a boson peak, is observed. For comparison, the data for a polycrystalline material, aegirine, of the same composition as the glass are also shown, and the van Hove singularity resulting from its transverse acoustic branch is marked as TA. Note the strong similarities between the glass and the crystal, particularly at high pressures. To further compare between the glass and the crystal, the integrated area under the boson peak for the glass and the van Hove singularity for the polycrystal, respectively, are estimated and marked as shaded areas in Figure 13.20b and c. The number of vibrational modes corresponding to the areas is comparable which the authors interpreted as suggesting that the origin of the boson peak might be related to the piling up of acoustic states near the boundary of the pseudo-Brillouin zone.

For suspensions of both hard-sphere [96] and soft-sphere colloids [95], the vibrational DOS have been determined using LSCM. In analogy with the behavior for molecular systems, an excess of low-frequency collective vibrational modes are observed. A difficulty with studying vibrational modes in colloidal systems is that the liquid of the suspension dampens vibrations and induces hydrodynamic effects, and great care has to be taken to minimize or circumvent these effects [95, 96]. The DOS, here denoted $D(\omega)$, for a 2D suspension of colloidal microgels is shown in Figure 13.20d, as normalized by the Debye prediction in 2D ($\propto \omega$) and a boson peak is clearly observed. The boson peak dependence on the volume fraction ϕ is also shown in the inset; the boson peak decreases in amplitude and shifts to higher frequencies for larger ϕ. This behavior is reminiscent of that observed for molecular systems, as shown in Figure 13.20a. The advantage compared with studying molecular systems is that for colloids the vibrational modes can be directly visualized. Figure 13.20e shows the vibrational normal modes observed at the boson peak frequency ω^*, demonstrating the quasi-localized nature of the modes observed for ω^*. Interestingly, it has recently been demonstrated both in computer simulations [97] (Fig. 13.21a) and in experimental studies on hard-sphere colloids [96] (Figure 13.21b) that the locations of these quasi-localized low-frequency modes are strongly correlated with irreversible structural rearrangements and thus with the α-relaxation. These low-energy "soft" modes effectively act as "hot spots," which nucleate structural relaxations. In Figure 13.21a and b, the vibrational mode amplitudes have been overlaid with the regions of high mobility (marked by white dots) to demonstrate the strong correlation.

13.7 THEORETICAL APPROACHES TO UNDERSTAND GLASS FORMATION

A wide range of theoretical approaches, often conceptually very different, have been applied to address the glass-transition phenomenology. Even though these efforts have resulted in a lot of understanding of many of the pieces of the puzzle, the complete picture is arguably still missing. Due to the range of concepts that have been invoked, it is impossible to give a comprehensive review here. Thus, we will only discuss a few key approaches that illustrate common ways of making sense of the observed behavior. For good overviews of the models and theories used to describe glass formation, see for instance [2, 3, 5–7, 17, 43, 98, 99].

13.7.1 Above the Dynamic Crossover: Mode Coupling Theory

The MCT [98, 99] is a highly influential theory since it provides a very detailed set of predictions that can be directly tested in experiments. It has been highly successful at describing the behavior of both supercooled liquids and colloidal suspensions near and above the dynamic crossover. MCT starts from the equations of motions of the system building blocks. The evolution of the density–density autocorrelation function, $\phi_k(t) = \rho_k(t)\rho_k(0)/\left\langle|\rho_k|^2\right\rangle$, is

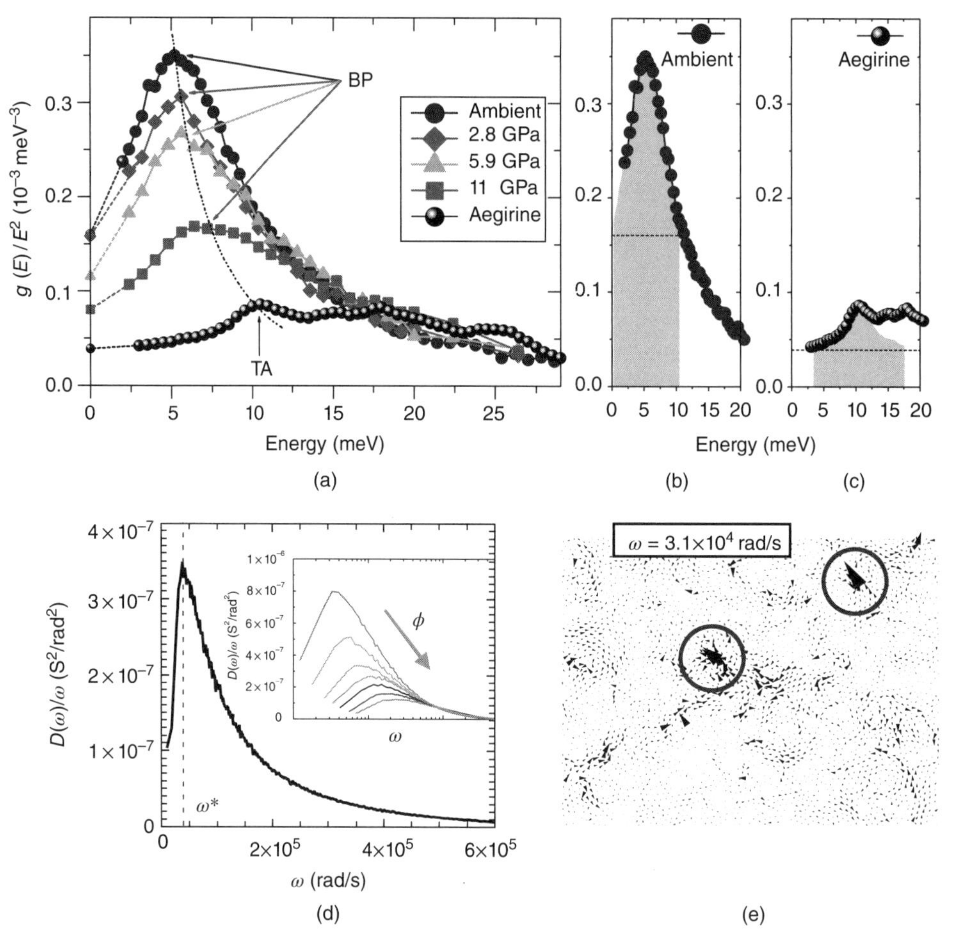

Figure 13.20 (a) The vibrational DOS $g(E)$ normalized by the Debye prediction for an atomic glass $Na_2FeSi_3O_{8.5}$ and its crystalline counterpart aegirine. TA denotes the van Hove singularity due to the transverse acoustic phonon branch. For higher pressures, the boson peak of the glass moves toward higher frequencies and lower amplitudes approaching the crystalline data [92]. (b) Ambient pressure data with the shaded region indicating the area under the boson peak [92]. (c) Data for the crystalline counterpart aegirine with the shaded region marking the estimate of the peak area under its van Hove singularity [92]. (a)–(c): Source: Adapted from Chumakov 2011 [92], figure 1. Reproduced with permission of the American Physical Society. Copyright 2011 by the American Physical Society. (d) The boson peak determined experimentally for a 2D suspension of colloidal microgel beads [95]. Source: Adapted from Chen 2010 [95], figure 3. Reproduced with permission of the American Physical Society. Copyright 2010 by the American Physical Society. The inset shows the boson peak for different volume fractions ranging from 0.84 to 0.885 where it is observed that the peak moves toward higher frequencies and lower amplitudes at higher volume fractions. Source: Adapted from Chen 2010 [95], figure 4. Reproduced with permission of the American Physical Society. Copyright 2010 by the American Physical Society. (e) The normal modes at the boson peak frequency ω^* as visualized using confocal microscopy [95]. Source: Adapted from Chen 2010 [95], figure 3. Reproduced with permission of the American Physical Society. Copyright 2010 by the American Physical Society.

described through the equation of motion:

$$\ddot{\phi}_k(t) + \nu_k\dot{\phi}_k(t) + \Omega_k{}^2\phi(t) + \Omega_k{}^2\int_0^t m_k(t-t')\dot{\phi}_k(t')dt',$$

where the subscript k indicates the wave-vector dependence, ν_k is a friction term, Ω_k is a k-dependent frequency of microscopic excitations, and $m_k(t)$ is the so-called memory function. The memory function is a key ingredient in MCT and describes the feedback mechanism arising due to particle crowding. If the feedback mechanism is strong enough, it results in dynamic arrest at a critical temperature T_c (or volume fraction ϕ_c for an MCT description of a colloidal system). It has been found that even very simple choices of approximating the memory function will induce a dynamic arrest, thus predicting a glass transition. One common example of such a memory function choice is $m(t) = \upsilon_1\phi(t) + \upsilon_2\phi^2(t)$, where υ_1 and υ_2 are model parameters and we here assume no k-dependence. Application of MCT to experimental data on molecular systems have however clearly demonstrated that T_c is situated significantly above T_g for standard implementations of MCT and corresponds quite well with the crossover temperature T^*. Thus, even though MCT can describe data for molecular

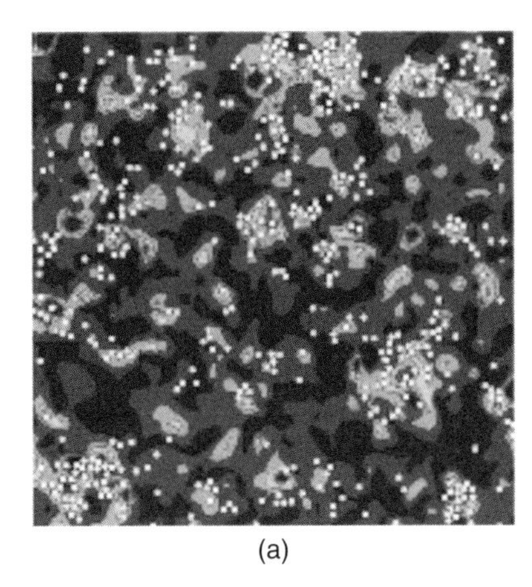
(a)

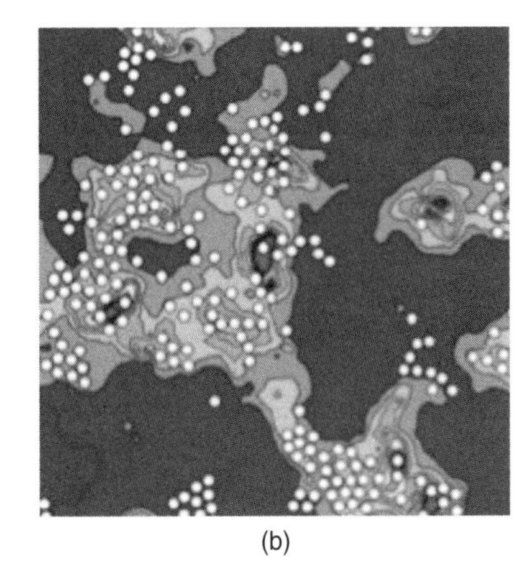
(b)

Figure 13.21 Comparison of the distribution of irreversible structural rearrangements (white dots) and contour plots of the low-frequency mode participation from (a) experiments using confocal microscopy [96]. Source: Adapted from Ghosh 2011 [96], figure 5. Reproduced with permission of the American Physical Society. Copyright 2011 by the American Physical Society. (b) Computer simulations [97]. Source: Adapted from Widmer-Cooper 2008 [97], figure 3. Reproduced with permission from [97]. Copyright 2008 by Macmillan Publishers Ltd.

systems well above T^*, the standard MCT description breaks down at lower temperatures [6, 48].

MCT predicts a two-step decay of the ISF $F(k,t)$ with an intermediate plateau, as sketched in Figure 13.22a. The plateau regime is predicted to be composed of three distinct subregimes marked IIa, II, and IIb in Figure 13.22a, where the first is a power law decay onto the plateau, $F = f_k + At^{-a}$ (IIa), the second is the plateau itself with $F = f_k$ (II), and the third is a power law decay off the plateau, $F = f_k - Bt^b$ (IIb). MCT predicts that near T_c, the power law exponents a and b are linked through a mathematical relationship, $\Gamma(1-a)^2/\Gamma(1-2a) = \Gamma(1+b)^2/\Gamma(1+2b)$, where Γ is the Gamma function. The long-time decay in F, the α-relaxation, can be approximately described as a stretched exponential with a characteristic timescale that follows a power law with the distance to T_c, as $\tau_\alpha \propto (T - T_c)^{-\gamma}$ where γ is a critical exponent; γ is in turn linked to the exponents a and b as $\gamma = 1/2a + 1/2b$. Thus, if one of the exponents a, b, or γ is known, the others are predicted by the theory and these predictions can be directly tested on experimental data. Moreover, in the standard version of the theory, time–temperature superposition is predicted near T_c.

As an illustration of the use of MCT to describe glass formation, Figure 13.22b shows the ISF F as determined for a hard-sphere colloidal suspension for a range of volume fractions together with MCT fits [100]. In this volume fraction range, MCT can describe the data well even though as discussed in Section 13.6.2, there is recent evidence for a breakdown of the applicability of MCT at higher volume fractions, consistent with the observation in molecular systems [6, 48].

Figure 13.22c shows the frequency-dependent response χ'' for a molecular glass-former, as determined using light scattering [48]. In frequency domain, the spectrum corresponding to the complex decay in the ISF constitutes a minimum between two power law flanks characterized by exponents a and b,

$$\chi''(\nu) = \chi''_{\min}\left[\frac{b\left(\nu/\nu_{\min}\right)^a + a(\nu/\nu_{\min})^{-b}}{a+b}\right],$$

where a and b are the two power law exponents corresponding to regimes (IIa) and (IIb) in the ISF. $\chi''_{\min}$ is the amplitude of the minimum and $\nu_{\min}$ is its frequency. Moreover, MCT predicts that the temperature dependence for $T > T_c$ is $\chi''_{\min}(T) \propto \sigma^{1/2}$, $\nu_{\min}(T) \propto \sigma^{1/2a}$, and $\tau_\alpha(T) \propto \sigma^{-\gamma}$, where $\sigma = (T - T_c)/T_c$, the normalized relative distance from the critical temperature T_c. Through the temperature dependencies of these parameters, T_c can be determined. At low temperatures, below ~180 K for the liquid data shown in Figure 13.22d, the MCT scaling behavior breaks down.

13.7.2 Below the Dynamic Crossover: Activated Dynamics

For temperatures below T^* in molecular systems or above ϕ^* in colloidal systems, one can view the dynamics as thermally activated and the T or ϕ dependent increase in τ_α is then due to a T or ϕ dependent activation energy, E_A. The physics entering E_A is different in different classes of models. Here, we only discuss three common choices, where the first links E_A to the so-called "free volume" available for rearrangements, the second to configurational entropy, and the third to an elastic energy. For more thorough discussion about theories applied to the glass-transition phenomenology, see Refs [2, 6–8, 11, 43, 77, 98].

Free-volume models: The essential idea in these models is that the sample volume is divided into the occupied volume and a volume available for motions, the so-called "free

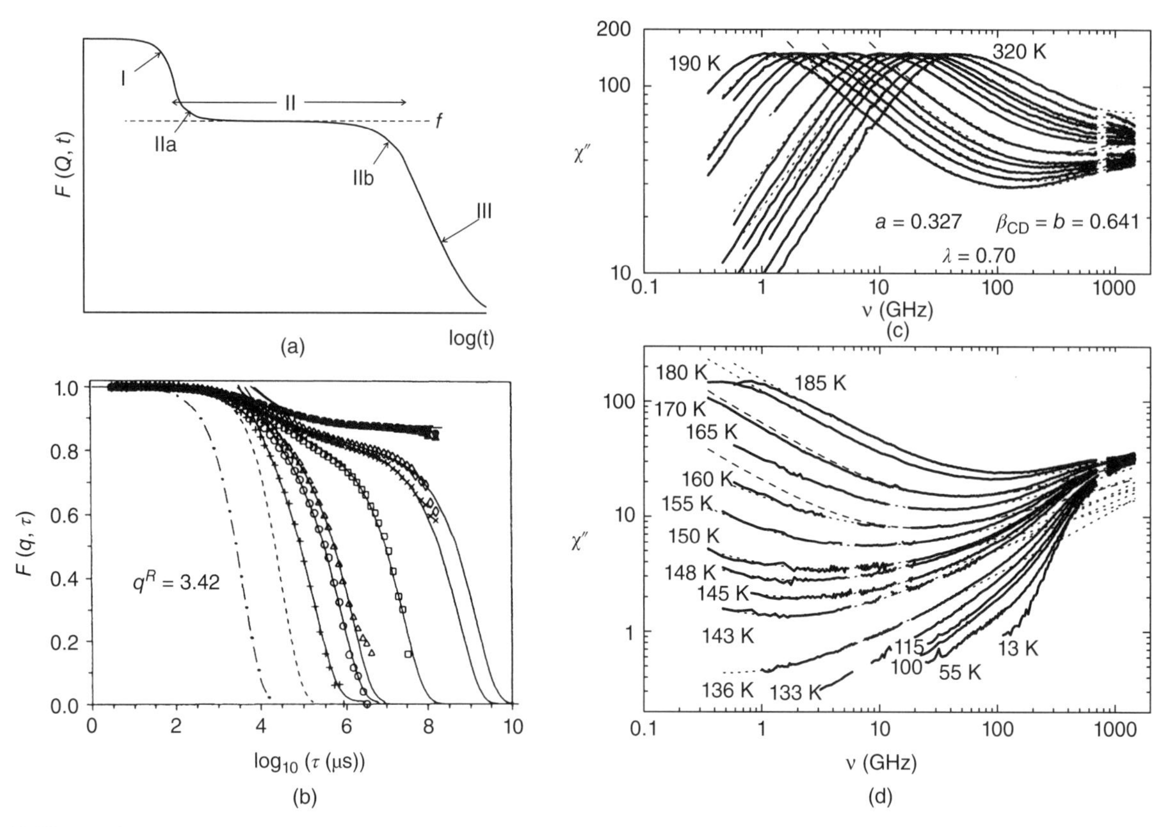

Figure 13.22 (a) Schematic picture of an ISF with several dynamic regions marked. These are discussed in the context of MCT in the text [15b]. Source: Adapted from Reichman 2005 [15b], figure 3. SISSA Medialab Srl. Reproduced with permission of IOP Publishing. All rights reserved. (b) ISF data from light scattering for a hard-sphere suspension at different volume fractions. The lines are MCT fits [57]. Source: Adapted from van Megen 1994 [57], figure 5. Reproduced with permission of the American Physical Society. Copyright 1994 by the American Physical Society. (c) Frequency-dependent loss for a molecular glass-former picoline as determined by dynamic light scattering in the T-range between 190 and 320 K [48]. (d) Data on the molecular liquid picoline for temperatures between 13 and 185 K. At low temperatures, the MCT description breaks down, as shown by the mismatch of the dashed lines and the data [48]. (c)–(d): Source: Adapted from Adichtchev 2002 [48], figure 1. Reproduced with permission of the American Physical Society. Copyright 2002 by the American Physical Society.

volume" [101, 102]. There are many different ways in which the latter can be defined, but without going into details the underlying concept is always the same, that is, the activation energy controlling structural relaxation is parameterized using a T or ϕ dependent "free volume," which is reduced as the glass transition is approached. The glass is formed when there is not enough free volume left for motions to take place. The fact that for molecular liquids τ_α is not solely a function of volume demonstrates that these models have to be used with caution from a fundamental perspective [2].

Entropy-based models: The general idea is here to parameterize E_A in terms of the configurational part of the entropy. Many different ways to achieve this exist, but the most influential model based on configurational entropy is that of Adam and Gibbs [2, 6, 11, 103]. They argued that the α-relaxation generally requires cooperative motions between neighboring molecules within the so-called cooperatively rearranging region, CRR. They made the assumption that E_A is proportional to the CRR volume and based on this derived an expression for E_A in terms of the configurational entropy, which in turn can be approximated by the excess entropy, as discussed in Section 13.2. As shown in Figure 13.1d, the excess entropy decreases dramatically as the glass transition is approached, which will thus be connected to the slowing down of the dynamics. Even though the details of the assumptions going into this model and the experimental verifications are somewhat problematic, the model is highly influential since it incorporates the concept of cooperative motion and implies a growing length scale as the glass transition is approached, both of which have been confirmed even though the detailed numerical predictions do not hold in general [103].

Models based on elasticity: If a liquid is probed on short enough timescales, there is no time for the molecules to rearrange and it behaves elastically as a solid. Even though the structural relaxation time might be long, a local rearrangement takes place on a fast timescale and elasticity is thus often viewed as a key factor in determining E_A [2]. This notion is supported by a range of identified links between properties related to "slow" and "fast" DOF [2], which include:

(i) A correlation between properties of the ISF plateau within the glass, which is related to "fast" timescales, and fragility, which in turn is set by the "slow" final decay of the ISF [104].
(ii) A correlation between the ratio of the high frequency ("fast") bulk and shear moduli K_∞/G_∞ and fragility [105].
(iii) A correlation between glassy moduli ("fast") and T_g ("slow") [2, 105].
(iv) For a range of molecular liquids, the viscosity η ("slow") can be collapsed onto a master curve by plotting $\log(\eta)$ versus $G_\infty(T)/T$, where G_∞ is the high-frequency shear modulus ("fast"), as shown in Figure 13.23 [2, 106].

Behavior (iv) suggests that the viscosity or structural relaxation time can be described as $\eta \propto \tau_\alpha \propto \exp[V_c G_\infty(T)/k_B T]$, where V_c is a volume characteristic of the relaxation event. The implication of this is that the "slow" timescale behavior captured in $\tau_\alpha(T)$ and thus in the fragility is encoded in the T-dependence of the short-time elastic modulus. These observations have resulted in the use of elasticity as the key entity in a range of glass-transition models. One illustrative such model is the so-called "shoving model," which is based on the idea that a molecular rearrangement happens on a short timescale so that the barrier to overcome should be set by the short-time elastic properties. The inset to Figure 13.23 shows a set of molecules where the dark ones take part in a rearrangement. The argument is that the dominating part of the energy cost is due to the "shoving" aside of the surrounding molecules resulting from a thermal fluctuation; this gives room for the rearrangement. Due to the fast timescales involved in the actual rearrangement, the material surrounding the rearranging particles is treated as elastic and the energy needed to increase the volume of the spherical "hole" in which the dark molecules can be shown to be approximately $G_\infty(T)V_c$, where V_c is the characteristic volume of the deformation [2]. This simple reasoning leads to the experimentally observed expression (iii).

13.8 CONCLUSIONS

As this chapter has hopefully demonstrated, the research area focusing on the formation and behavior of out-of-equilibrium glassy materials is exciting in terms of the challenges involved, the phenomenological richness, and the wide range of materials for which similar concepts and ideas come into play. Moreover, there are currently many promising developments that are bound to bring the field forward at a good rate. One area that is steadily getting more powerful is that of computer simulations and as simulation techniques become increasingly efficient, more and more of the characteristic phenomenology of glass formation, for ever more realistic systems, can be studied in detail using computers. In addition to stand-alone computer simulations, we also expect that computer experiments become more important as integral parts of the analysis and interpretation of experimental data. One important example is the structure determination of disordered materials, where neutron diffraction techniques are often used in conjunction with empirical potential structure refinement (EPSR) simulation

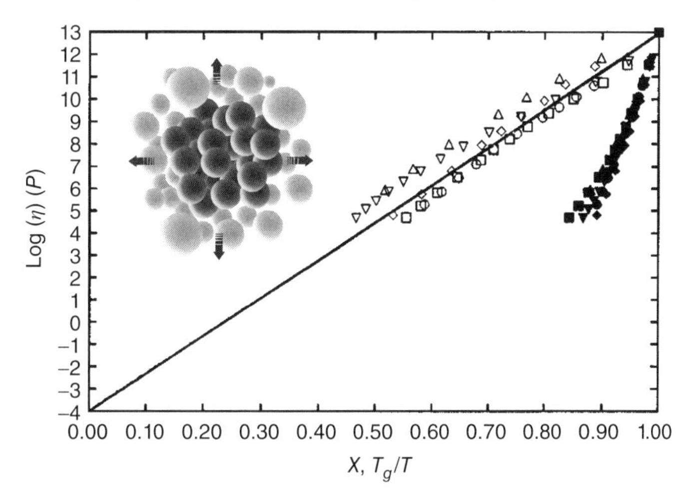

Figure 13.23 The solid symbols show viscosity for a range of molecular glass-formers in a T_g-scaled Arrhenius plot. The unfilled symbols show the data plotted versus $x = G_\infty(T)/T$, which linearizes the behavior. Source: Adapted from Dyre 1996 [106], figure 1. Reproduced with permission of the American Physical Society. Copyright 1996 by the American Physical Society. The upper left inset is a sketch of a relaxation event where the black particles are the ones taking part in the event. The arrows indicate the expansion of the surrounding matrix, which is needed for a relaxation to take place [2, 106]. Source: Adapted from Dyre 2006 [2], figure 8. Reproduced with permission of the American Physical Society. Copyright 2006 by the American Physical Society.

techniques; a realistic atomistic model is simulated and the model simulation parameters are optimized by matching with experimental data. Another interesting example is the combined use of computer simulations and experiments to circumvent the strong hydrodynamic effects and viscous damping of vibrations in the study of colloidal suspensions [95]. An experimental development that should bring significant impact in the years to come is the synchrotron-based X-ray photon correlation spectroscopy, which combines access to the long timescales relevant near the glass transition with atomic length-scale resolution. For colloids, most systematic studies of glass formation have still been performed on hard-sphere-type systems. The opportunity to detailed design and control of colloids with regard to size, shape, elastic properties, and interactions, however, provides an amazingly rich playing field to significantly expand these studies in the future. Colloidal particles in the "real" world are rarely perfectly round and hard and have often broad or complex size distributions. More systematic studies of this larger parameter range promise to be important and reveal a lot of new phenomena. Finally, more and more interactions are taking place between the previously quite separate research fields focusing on molecular, metallic, and colloidal glass formation. The increased flow of ideas and information between these subfields should not only speed up our learning about glassy systems, in general, but certainly also make this research area even more exciting.

REFERENCES

[1] Larson RG. *The Structure and Rheology of Complex Fluids*. Oxford University Press; 1999.

[2] Dyre JC. Colloquium: the glass transition and elastic models of glass-forming liquids. Rev Mod Phys 2006;78:953–972.

[3] Debenedetti PG, Stillinger FH. Supercooled liquids and the glass transition. Nature 2001;410:259–267.

[4] Hensel A, Schick C. Relation between freezing-in due to linear cooling and the dynamic glass transition temperature by temperature-modulated DSC. J Non-Cryst Solids 1998;235–237:510–516.

[5] Debenedetti PG. *Metastable Liquids: Concepts and Principles*. Princeton, New Jersey: Princeton University Press; 1996.

[6] Angell CA et al. Relaxation in glassforming liquids and amorphous solids. J Appl Phys 2000;88(6):3113–3157.

[7] Ediger MD, Angell CA, Nagel SR. Supercooled liquids and glasses. J Phys Chem 1996;100:13200–13212.

[8] Hunter GL, Weeks ER. The physics of the colloidal glass transition. Rep Prog Phys 2012;75:066501.

[9] Cipelletti L, Weeks ER. *"Glassy Dynamics and Dynamical Heterogeneity" in Colloids in "Dynamical Heterogeneities in Glasses, Colloids, and Granular Media"*. Oxford, New York: Oxford University Press; 2010. arXiv:1009.6089.

[10] Pusey PN, van Megen W. Phase behaviour of concentrated suspensions of nearly hard colloidal spheres. Nature 1986;320:340.

[11] Schweizer KS. Entropic barriers, activated hopping, and the glass transition in colloidal suspensions. J Chem Phys 2003;119:1181.

[12] Mirigian S, Schweizer KS. Elastically cooperative activated barrier hopping theory of relaxation in viscous fluids. I. General formulation and application to hard spheres. J Chem Phys 2014;140:194506.

[13] Kim, The image was provided with permission from Chanjoong Kim.

[14] van Megen A, Pusey PN. Dynamic light-scattering study of the glass transition in a colloidal suspension. Phys Rev A 1991;43(10):5429–5441.

[15] (a) Leochmach M, Tanaka H. Roles of icosahedral and crystal-like order in the hard spheres glass transition. Nature Communications 2012;3:974 (b) Reichman DR, Charbonneau P. Mode-coupling theory. J Stat Mech: Theory Exp 2005:P05013.

[16] (a) Tanaka H, Kawasaki T, Shintani H, Watanabe K. Critical-like behaviour of glass-forming liquids. Nat. Mat. 2010;9:324–331. (b) Berne BJ, Pecora R. *Dynamic Light Scattering*. Dover Publications Inc.; 2000.

[17] Ngai KL. *Relaxation and Diffusion in Complex Systems (partially ordered systems)*. New York, Dordrecht, Heidelberg, London: Springer; 2011.

[18] Kudlik A, Tschirwitz C, Blochowicz T, Benkhof S, Rössler E. Slow secondary relaxation in simple glass formers. J. Non-Cryst. Solids 1998;235–237:406–411.

[19] Rubinstein M, Colby R. *Polymer Physics*. Oxford University Press; 2003.

[20] Brodin A et al. Light scattering and dielectric manifestations of secondary relaxations in molecular glassformers. Eur Phys J B 2003;36:349–357.

[21] Pusey PN. Dynamic light scattering by non-ergodic media. Physica A 1989;157:705–741.

[22] Rössler EA et al. Indications for a change of transport mechanism in supercooled liquids and the dynamics close and below T_g. J Non-Cryst Solids 1994;172–174:113–125.

[23] Richter D et al. Decoupling of time scales of motion in polybutadiene close to the glass transition. Phys Rev Lett 1992;68:71.

[24] Petry W et al. Dynamic anomaly in the glass transition region of orthoterphenyl. Zeitschrift für Physik B Condens Matter 1991;83:175–184.

[25] Lunkenheimer P, Loidl A. Dielectric spectroscopy of glass-forming materials: alpha-relaxation and excess wing. Chem Phys 2001;284:205–219.

[26] Williams G, Watts DC. Non-symmetrical dielectric relaxation behaviour arising from a simple empirical decay function. Trans Faraday Soc 1970;66:80.

[27] Kohlrausch R. Theorie des elektrischen Ruckstandes in der Leidener Flasche. Pogg Ann Phys 1854;91:179.

[28] Böhmer R et al. Nonexponential relaxations in strong and fragile glass formers. J Chem Phys 1993;99:4201–4209.

[29] Kremer F, Schönhals A. *Broadband Dielectric Spectroscopy*. Berlin, New York: Springer; 2003.

[30] Petzold N et al. Evolution of the dynamic susceptibility in molecular glass formers: results from light scattering, dielectric spectroscopy, and NMR. J Chem Phys 2013;138:12A510.

[31] Dixon PK et al. Scaling in the relaxation of supercooled liquids. Phys Rev Lett 1990;65:1108–1111.

[32] Schneider U et al. Excess wing in the dielectric loss of glass formers: a Johari–Goldstein beta relaxation? Phys Rev Lett 2000;84(24):5560–5563.

[33] Mattsson J et al. Chain-length-dependent relaxation scenarios in an oligomeric glass-forming system: from merged to well-separated alpha and beta loss peaks. Phys Rev Lett 2003;90(7):075702.

[34] Bergman R et al. Confinement effects on the excess wing in the dielectric loss of glass-formers. Europhys Lett 2003;64(5):675–681.

[35] Bergman R. General susceptibility functions for relaxations in disordered systems. J Appl Phys 2000;88(3):1356.

[36] Hansen C et al. Dynamics of glass-forming liquids. III. Comparing the dielectric alpha- and beta-relaxation of 1-propanol and *o*-terphenyl. J Chem Phys 1997;107:1086–1093.

[37] Tanaka H. Relation between thermodynamics and kinetics of glass-forming liquids. Phys Rev Lett 2003;90:055701.

[38] Hecksher T et al. Little evidence for dynamic divergences in ultraviscous molecular liquids. Nat Phys 2008;4: 737–741.

[39] Bucaro JA, Dardy HD. High-temperature strain relaxation in silica by optical correlation spectroscopy. J Non-Cryst Solids 1977;121–129:121.

[40] Stickel F-J. Untersuchung der Dynamik in niedermolekularen Flussigkeiten mit Dielektrischer Spektroskopie. PhD Thesis. Mainz University, Shaker, Aachen; 1995.

[41] Brodin A, Rössler EA. Depolarized light scattering study of glycerol. Eur Phys J B 2005;44:3–14.

[42] Ngai KL. Short-time and long-time relaxation dynamics of glass-forming substances: a coupling model perspective. J Phys Condens Matter 2000;12:6437–6451.

[43] Donth E. *The Glass Transition; Relaxation Dynamics in Liquids and Disordered Materials*. Berlin, Heidelberg: Springer-Verlag; 2010.

[44] Wang L-M, Richert R. Primary and secondary relaxation time dispersions in fragile supercooled liquids. Phys Rev B 2007;76:064201.

[45] Kisliuk A, Mathers RT, Sokolov AP. Crossover in dynamics of polymeric liquids: back to T? J Polym Sci Part B 2000;38:2785–2790.

[46] Chang I et al. Translational and rotational molecular motion in supercooled liquids studied by NMR and forced Rayleigh scattering. J Non-Cryst Solids 1994;172–174:248–255.

[47] Schönhals A. Evidence for a universal crossover behaviour of the dynamic glass transition. Europhys Lett 2001;56(6):815–821.

[48] Adichtchev SV et al. Anomaly of the nonergodicity parameter and crossover to white noise in the fast relaxation spectrum of a simple glass former. Phys Rev Lett 2002;88:051101.

[49] Rössler E, Hess K-U, Novikov VN. Universal representation of viscosity in glass forming liquids. J Non-Cryst Solids 1998;223:207.

[50] Rössler EA, Sokolov AP. The dynamics of strong and fragile glass formers. Chem Geol 1996;128:143–153.

[51] Goldstein M. Viscous liquids and the glass transition: a potential energy barrier picture. J Chem Phys 1969;51:3728–3739.

[52] Stillinger F. A topographic view of supercooled liquids and glass formation. Science 1995;267:1935.

[53] Yu HB et al. The beta-relaxation in metallic glasses. Nat Sci Rev 2014;1:429–461.

[54] Harmon JS et al. Anelastic to plastic transition in metallic glass-forming liquids. Phys Rev Lett 2007;99:135502.

[55] Romeo G et al. Origin of de-swelling and dynamics of dense ionic microgel suspensions. J Chem Phys 2012;136:124905.

[56] Weeks ER et al. Three-dimensional direct imaging of structural relaxation near the colloidal glass transition. Science 2000;287:627–631.

[57] van Megen W, Underwood SM. Glass transition in colloidal hard spheres: measurements and mode-coupling theory analysis of the coherent intermediate scattering function. Phys Rev E 1994;49(5):4206–4220.

[58] Brambilla G et al. Probing the equilibrium dynamics of colloidal hard spheres above the mode-coupling glass transition. Phys Rev Lett 2009;102:085703.

[59] El Masri D et al. Dynamic light scattering measurements in the activated regime of dense colloidal hard spheres. J Stat Mech: Theory Exp 2009.

[60] Brambilla G et al. Brambilla *et al.* Reply: Reinhardt *et al.* Phys Rev Lett 2010;105:199605.

[61] Reinhardt J, Weysser F, Fuchs M. Comment on "Probing the equilibrium dynamics of colloidal hard spheres above the mode-coupling glass-transition". Phys Rev Lett 2010;105:199604.

[62] van Megen W. Comment on "Probing the equilibrium dynamics of colloidal hard spheres above the mode-coupling glass transition". Phys Rev Lett 2010;104:169601.

[63] Bernal JD, Mason J. Packing of spheres: co-ordination of randomly packed spheres. Nature 1960;188:910.

[64] Cheng Z. Nature of the divergence in low shear viscosity of colloidal hard-sphere dispersions. Phys Rev E 2002;65(4):041405.

[65] Brumer Y, Reichman DR. Mean-field theory, mode-coupling theory, and the onset temperature in supercooled liquids. Phys Rev E 2004;69:041202.

[66] Kumar SK, Szamel G, Douglas JF. Nature of the breakdown in the Stokes–Einstein relationship in a hard-sphere fluid. J Chem Phys 2006;124:214501.

[67] Edmond KV et al. Decoupling of rotational and translational diffusion in supercooled colloidal fluids. Proc Natl Acad Sci 2012;109:17891–17896.

[68] Mattsson J et al. Soft colloids make strong glasses. Nature 2009;462(7269):83–86.

[69] Vlassopoulos D, Fytas G, Pakula T, Roovers J. Multi-arm star polymers dynamics. J. Phys. Condens. Matter 2001;13:R855–R876.

[70] Schall P, Weitz DA, Spaepen F. Structural rearrangements that govern flow in colloidal glasses. Science 2007;318:1895–1899.

[71] Falk ML, Langer JS. Dynamics of viscoplastic deformation in amorphous solids. Phys Rev E 1997;57(6):7192.

[72] Chikkadi V et al. Long-range strain correlations in sheared colloidal glasses. Phys Rev Lett 2011;107:198303.

[73] Johnson WL, Samwer K. A universal criterion for plastic yielding of metallic glasses with a $(T/T_g)2/3$ temperature dependence. Phys Rev Lett 2005;95:195501.

[74] Yu HB et al. Relating activation of shear transformation zones to beta relaxations in metallic glasses. Phys Rev B 2010;81:220201.

[75] Chikkadi V, Schall P. Nonaffine measures of particle displacements in sheared colloidal glasses. Phys Rev B 2012;85:031402.

[76] Berthier L. Dynamic heterogeneity in amorphous materials. Physics 2011;4:42.

[77] Berthier L, Biroli G. Theoretical perspective on the glass transition and amorphous materials. Rev Mod Phys 2011;83:587–645.

[78] Rahmani Y et al. Dynamic heterogeneity in hard and soft sphere colloidal glasses. Soft Matter 2012;8:4264–4270.

[79] Candelier R, Dauchot O, Biroli G. Building blocks of dynamical heterogeneities in dense granular media. Phys Rev Lett 2009;102:088001.

[80] Berthier L et al. Characterizations of dynamic heterogeneity. In: *Dynamic Heterogeneities in Glasses, Colloids and Granular Media*. Oxford University Press; 2011.

[81] Berthier L et al. Direct experimental evidence of a growing length scale accompanying the glass transition. Science 2005;310:1797.

[82] Dalle-Ferrier C et al. Spatial correlations in the dynamics of glassforming liquids: experimental determination of their temperature dependence. Phys Rev E 2007;2007:041510.

[83] Duri A et al. Resolving long-range spatial correlations in jammed colloidal systems using photon correlation imaging. Phys Rev Lett 2009;102:085702.

[84] Sessoms DA et al. Multiple dynamic regimes in concentrated microgel systems. Philos Trans R Soc A 2009;367:5013.

[85] Maccarrone S et al. Ultra-long range correlations of the dynamics of jammed soft matter. Soft Matter 2010;6:5514.

[86] Mazza MG et al. Connection of translational and rotational dynamical heterogeneities with the breakdown of the Stokes–Einstein and Stokes–Einstein–Debye relations in water. Phys Rev E 2007;76:031203.

[87] Debye P. *Polar Molecules*. New York: Dover; 1929.

[88] Einstein A. *Theory of Brownian Motion*. New York: Dover; 1956.

[89] Ediger MD. Spatially heterogeneous dynamics in supercooled liquids. Ann Rev Phys Chem 2000;51:99.

[90] Ngai KL, Capaccioli S. An explanation of the differences in diffusivity of the components of the metallic glass Pd43Cu27Ni10P20. J Chem Phys 2013;138:094504.

[91] Ashcroft NW, Mermin ND. *Solid State Physics*. New York, London: Holt, Rinehart and Winston; 1976.

[92] Chumakov AI et al. Equivalence of the boson peak in glasses to the transverse acoustic van Hove singularity in crystals. Phys Rev Lett 2011;106:225501.

[93] Kob W, Binder K. *Glassy Materials and Disordered Solids: An Introduction*. London: World Scientific; 2011.

[94] Marruzzo A et al. Heterogeneous shear elasticity of glasses: the origin of the boson peak. Sci Rep 2013;3:1407.

[95] Chen K et al. Low-frequency vibrations of soft colloidal glasses. Phys Rev Lett 2010;105:025501.

[96] Ghosh A et al. Connecting structural relaxation with the low frequency modes in a hard-sphere colloidal glass. Phys Rev Lett 2011;107:188303.

[97] Widmer-Cooper A et al. Irreversible reorganization in a supercooled liquid originates from localized soft modes. Nat Phys 2008;4:711–715.

[98] Gótze W. *Complex Dynamics of Glass-Forming Liquids*. Oxford Science Publications; 2012.

[99] Götze W, Sjögren L. Relaxation processes in supercooled liquids. Rep Prog Phys 1992;55:241.

[100] van Megen W, Underwood SM. Glass transition in colloidal hard spheres: measurement and mode-coupling-theory analysis of the coherent intermediate scattering function. Phys Rev E 1994;49:4206.

[101] Cohen MH, Turnbull D. Molecular transport in liquids and glasses. J Chem Phys 1959;31:1164.

[102] Cohen MH, Grest GS. Liquid-glass transition, a free-volume approach. Phys Rev B 1979;B20:1077.

[103] Adam G, Gibbs JH. On the temperature dependence of cooperative relaxation properties in glass-forming liquids. J Chem Phys 1965;43:139.

[104] Scopigno T et al. Is the fragility of a liquid embedded in the properties of its glass? Science 2003;302:849.

[105] Novikov VN, Sokolov AP. Poisson's ratio and the fragility of glass-forming liquids. Nature 2004;431:961.

[106] Dyre JC, Olsen NB, Christensen T. Local elastic expansion model for viscous-flow activation energies of glass-forming molecular liquids. Phys Rev B 1996;53:2171.

14

COLLOIDAL GELATION

Emanuela Del Gado[1], Davide Fiocco[2], Giuseppe Foffi[3], Suliana Manley[4], Veronique Trappe[5], & Alessio Zaccone[6]

[1]*Department of Physics and Institute for Soft Matter Synthesis and Metrology, Georgetown University, Washington, DC, USA*
[2]*Institute of Theoretical Physics, EPFL, Lausanne, Switzerland*
[3]*Laboratoire de Physique des Solides, CNRS UMR 8502, Université Paris-Sud XI, Orsay, France*
[4]*Laboratory of Physics of Biological Systems, EPFL, Lausanne, Switzerland*
[5]*Department of Physics, University of Fribourg, Fribourg, Switzerland*
[6]*Department of Chemical Engineering and Biotechnology, University of Cambridge, Cambridge, United Kingdom*

14.1 INTRODUCTION: WHAT IS A GEL?

Colloidal suspensions constitute interesting model systems and useful materials in their equilibrium states as fluid or crystals. However, any colloidal particle that is suspended in a liquid intrinsically interacts with other particles by short-ranged attractive van der Waals interactions, which implies that any suspension is inherently unstable. Stabilization is necessary because colloids are constantly diffusing due to thermal energy, so in the absence of a barrier, they would collide and stick together. Colloidal suspensions can be stabilized, electrostatically or sterically, to remain dispersed for long periods of time. Upon destabilization, interparticle interactions can drive them to aggregate and locally arrest. In such *nonequilibrium* states, particles are trapped in local energy minima, preventing a full exploration of phase space.

Nonequilibrium states are both useful and fascinating, possessing complex material properties. Among these nonequilibrium states, colloidal glasses can be defined as systems where the local arrest of the particles is sufficient to provide stress-bearing properties. Local arrest is necessary

Fluids, Colloids and Soft Materials: An Introduction to Soft Matter Physics, First Edition. Edited by Alberto Fernandez Nieves and Antonio Manuel Puertas.

but not always sufficient for a system to exhibit solid-like properties. Indeed, at low volume fractions ϕ, cluster phases may form, where particles are locally trapped in a nonequilibrium configuration, but the clusters do not fill space and thus are still free to diffuse. In this case, the overall properties of the system are that of a fluid. To exhibit solid-like properties, the clusters must either become trapped in a cage of neighboring clusters or directly connect to form a space-spanning network, where the particles are unable to reconfigure not only with respect to their nearest neighbors but also relative to any other particle within the network. *Colloidal gels* are defined as systems that, to exhibit solid-like properties, must fulfill both conditions, local arrest of the particles and formation of space spanning networks. Many common substances are colloidal gels, from foods such as yogurt to personal care products such as deodorants.

14.1.1 An Experimental Summary: How Is a Gel Made?

Charge-stabilized colloids experience an electrostatic repulsion that prevents them from aggregating. Aggregation is typically induced by adding salt or by changing the pH of the solution, which screens the electrostatic repulsion, allowing particles to come together and stick to form clusters. Because upon screening, the resulting interaction is based on van der Waals forces, one generally refers here to "strongly attractive" systems, where the van der Waals forces are much greater than thermal energy, $V \gg kT$. Gelation occurs when the clusters fill all of space, or equivalently, when the volume fraction of clusters, ϕ_c, is 1. The experimental signature for this transition is an arrest in particle dynamics and the onset of elasticity.

Sterically stabilized suspensions typically include a surfactant or polymer layer grafted to the surface of the colloids, and this prevents them from getting close enough to interact via van der Waals attraction. Here, the most common method to create attractions between particles is through a polymer-induced depletion interaction. There is a depletion zone around each colloid of thickness equal to the polymer radius of gyration R_g, from which polymer is excluded. As a result, there is a net gain in free volume for the depletant when these zones overlap, increasing the entropy of the system and reducing the total free energy. This results in a "weak attraction" between particles, $V_{ss}(r) > k_B T$, whose range and strength are set by, respectively, the size and concentration of the depletant. Gelation occurs when clusters of colloidal particles pack to fill the sample container or when the space-spanning colloid-rich phase of the phase-separating suspension becomes dense enough to form a glass.

This qualitative description forms the basis for our in-depth discussion of gels. In this chapter, we focus on three key concepts related to colloidal gels: the interactions that allow gels to form, the routes to gelation and their relationship to equilibrium phases, and gel elasticity.

14.2 COLLOID INTERACTIONS: TWO IMPORTANT CASES

An essential component of colloidal gelation is attractive interactions. In this section, we introduce two mechanisms of attraction between colloids: van der Waals and depletion forces. Both mechanisms can induce gelation; however, they have very different intensities and for this reason we refer to them as "strong" and "weak" interactions. Such a description is intended to give an intuitive idea of the intensity of the forces that these mechanisms induce. We explain here the physical origins of these mechanisms and derive the forces between colloids that result from each. This allows us to compare the forces in a more rigorous way and to understand how each force can be tuned using external parameters.

14.2.1 "Strong" Interactions: van der Waals Forces

van der Waals forces, also known as dispersion forces, are interaction forces that act between atoms and molecules. Compared to other forces of chemical (covalent, metallic, etc.) or physical (hydrogen bonds, electrostatic, etc.) origin, they are much weaker and are also more ubiquitous since they pertain to any atom or molecule even when they are uncharged and nonpolar. These forces, resulting from different electrostatic effects, are due to the interactions between two rotating dipoles (*Keesom forces*), between a permanent dipole and an induced one (*Debye forces*), and a third interaction of purely quantum mechanical origin due to the instantaneous values of dipole moments (*London dispersion forces*). van der Waals forces are the sum of these three contributions. They all share the same functional form: they are attractive and decay as $\sim 1/r^6$. Consequently, we can write the general form:

$$V_{\mathrm{vdW}}(r) = -\frac{C_{\mathrm{Keesom}}}{r^6} - \frac{C_{\mathrm{Debye}}}{r^6} - \frac{C_{\mathrm{London}}}{r^6} = -\frac{C_{\mathrm{vdW}}}{r^6}, \tag{14.1}$$

where the constants C_X give the contribution of the different interactions. Note that while the Keesom and Debye forces can be absent in the case of neutral and nonpolar atoms and molecules, London forces are always present and, in fact, they represent in most situations the largest contribution to the van der Waals forces.

The forces presented in Equation 14.1 are restricted to the microscopic domain. However, acting between the atoms and molecules within a colloidal particle, they can induce an attraction at the mesoscopic scale. The mesoscopic description is obtained by integrating out the molecular/atomistic degrees of freedom to obtain an effective interaction potential. In the next section, we derive the van der Waals force between two spherical particles. To this end, we first compute the interaction between a molecule and an infinite plane, and then extend it to the interaction between two planes. The interactions between two spherical objects will be obtained introducing the Derjaguin approximation (DA). This very useful approximation relates the potential energy per unit area between two planes with that between two arbitrarily curved surfaces. With this approximation, it will be straightforward to obtain the interaction between two spheres.

14.2.1.1 vdW Interaction Between One Molecule and One Half-Space or Between Two Half-Spaces We start with a derivation of the simplified case of the interaction potential between a molecule and a half-space filled with molecules. We assume here that two molecules interact with a purely attractive potential depending on their distance of the form:

$$v(x) \sim -\frac{C}{x^n}, \quad n > 3, \tag{14.2}$$

where C is a constant. For $n = 6$, we have the functional form for dispersion forces that we have discussed earlier, but let us start with the case of a general power law. Figure 14.1 illustrates the geometry for the derivation of the molecule–half-space interaction potential. The molecule is placed in the origin $\mathbf{0}$, and we can work in cylindrical coordinates due to the symmetry of the problem.

We need here a further fundamental assumption: interactions are pairwise additive. This means that the interaction between the molecule and the half-space is evaluated as the sum of the interactions of the molecule with each molecule contained in the half-space. With this fundamental hypothesis, we consider an infinitesimal volume in the bulk of the half-space. The number of molecules contained in this volume is $dN = \rho dV = 2\pi\rho r\, dr\, dz$, where we have integrated out the angular degree of freedom to take advantage of the symmetry of the problem. The plane defining the half-space is taken to be $z = L$. The infinitesimal volume is placed at a distance $x = (r^2 + z^2)^{\frac{1}{2}}$ from the molecule, and it is straightforward to write down the interaction potential by integration:

$$\begin{aligned} V(x) &= -C \int dN \frac{1}{x^n} = -2\pi\rho C \int_L^\infty dz \int_0^\infty dr \frac{r}{(r^2 + z^2)^{n/2}} \\ &= -\frac{2\pi C\rho}{(n-2)} \int_L^\infty dz \frac{1}{z^{n-2}} = -\frac{2\pi\rho C}{(n-2)(n-3)} \frac{1}{L^{n-3}}. \end{aligned} \tag{14.3}$$

For the specific intermolecular potential that we are considering here, that is, $n = 6$, the potential in Equation 14.3 reduces to

$$V(x) \underset{n=6}{=} -\frac{\pi C}{6} \frac{1}{L^3}. \tag{14.4}$$

Similarly, the effective interactions between two half-spaces can be simply derived. In this case, we obtain for the interaction potential per unit area:

$$V(x) = -\frac{2\pi\rho^2 C}{(n-2)(n-3)(n-4)} \frac{1}{L^{n-4}}, \tag{14.5}$$

which for $n = 6$ reduces to

$$V(x) = -\frac{A_H}{12\pi} \frac{1}{L^2}, \tag{14.6}$$

where we have introduced a material-dependent constant A_H:

$$A_H = \pi^2 \rho^2 C. \tag{14.7}$$

The constant A_H is known as the Hamaker constant, and it depends on the dielectric constant and refractive indices. When measured, it allows the estimation of the van der Waals pair potential constant C.

14.2.1.2 Derjaguin Approximation In this section, we evaluate the interaction potential between two spheres as a result of van der Waals forces. To this end, we introduce the Derjaguin approximation (DA), which relates the potential energy per unit surface area of two half-spaces to the interaction of two objects with an arbitrary shape. We limit our calculation to the case of two spherical surfaces since we are most interested in spherical colloidal particles. In this particular condition, we can obtain the force between the two spheres by integrating over circular domains around the separating direction. The Derjaguin approximation, which takes these circular surfaces to be planar, holds in the case in

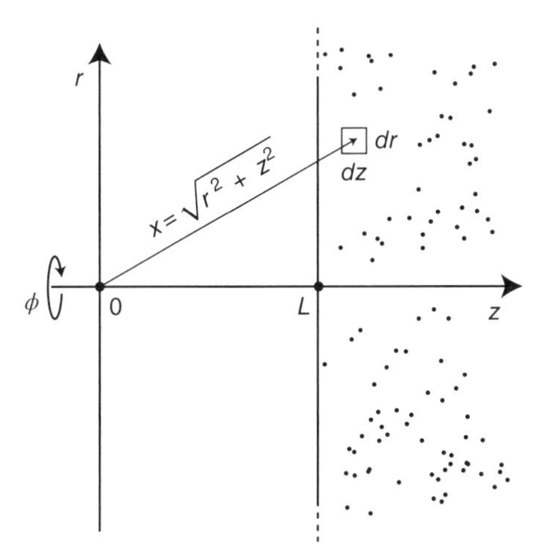

Figure 14.1 Sketch of the geometry used to integrate the potential due to the presence of a half-space.

which the radii of the spherical surfaces R_1 and R_2 are much larger than the separation distance between the surfaces, that is, $R_{1,2} \gg L$. It is possible to show that in this case the force along the direction joining the centers of the two spheres F_{ss} is given by

$$F_{ss}(L) \simeq 2\pi \frac{R_1 R_2}{R_1 + R_2} W_{pp}(L), \tag{14.8}$$

where $W_{pp}(L)$ is the potential energy per unit area between two planar surfaces. We can use the result of Equation 14.6 in the Derjaguin approximation of Equation 14.8 to find the force between two spheres

$$F_{ss}(L) = -\frac{R_1 R_2}{R_1 + R_2} \frac{A_H}{6} \frac{1}{L^2}, \tag{14.9}$$

which can be integrated to obtain directly the potential energy between the two spheres

$$V_{ss}(L) = -\frac{R_1 R_2}{R_1 + R_2} \frac{A_H}{6} \frac{1}{L}. \tag{14.10}$$

For spheres with same radius, that is, $R_1 = R_2 = a$, we obtain the interaction potential

$$V_{ss}(L) = -\frac{A_H}{12} \frac{a}{L}. \tag{14.11}$$

Finally, to estimate the validity of the DA, we calculate the exact van der Waals potential [1] between two spheres of equal diameter V_{full} by performing an integration similar to the one we have performed in Equation 14.3 . The result is

$$V_{ss}(L) = -\frac{A_H}{6} \left(\frac{2a^2}{(4a+L)L} + \frac{2a^2}{(2a+L)^2} + \ln \frac{(4a+L)L}{(2a+L)^2} \right). \tag{14.12}$$

In Figure 14.2, we present the ratio between the approximated (Eq. 14.11) and exact potentials (Eq. 14.12) for two spheres of radius $a = 1$. As expected, the approximation is valid for $L < a$, and it continues to hold up to distances L of the order of 1% of the radius.

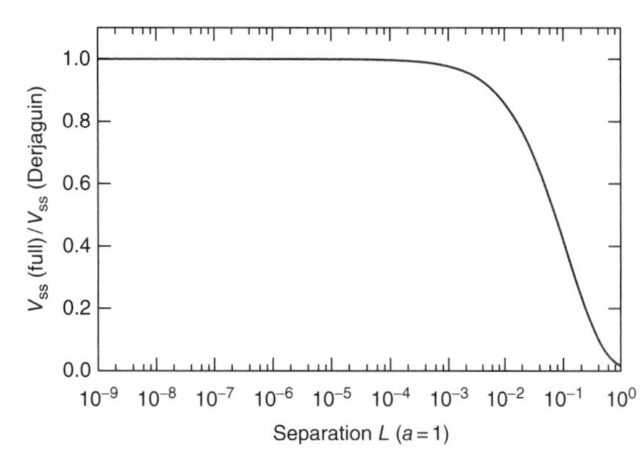

Figure 14.2 Plot of the reliability of the DA in the case of two unit spheres for different values of L.

*14.2.1.3 **What We Have Not Said*** The main approximation in the treatment of vdW forces lies in the assumption of the pairwise additivity of the interactions. As a matter of fact, the fluctuation of a dipole influences and is influenced by more than one dipole at the same time so this approximation is not exact. Another limitation is due to the fact that the propagation of the field generated by dipoles is assumed as instantaneous while it propagates at the speed of light. This is a fair assumption only when the distance between the two interacting surfaces is small enough, while when this does not happen *retardation* effects become significant. A more exact approach is *Lifshitz theory* where the interacting bodies are treated as continuous media. Such an approach requires, at least in its original formulation, methods typical of quantum field theory, and its discussion is beyond the scope of the present exposition. A more detailed discussion can be found in Refs [2, 3].

14.2.2 "Weak" Interactions: Depletion Interactions

A completely different mechanism of colloidal attraction is given by the presence of additional, small solute particles, or "depletants" in the suspension of colloidal particles. In what follows, we explore the effect of such depletants on the behavior of the colloidal particles, focusing in particular on the *Asakura–Oosawa* approximation, which allows the analytical calculation of the forces that arise between colloidal particles as a consequence of the additional component.

For simplicity, let us consider a pair of colloidal particles in a continuous medium with N additional depletant particles in it. Let one of the two colloids sit at the origin of the reference frame and the other at position $\mathbf{R}$. Their pair distribution function is given by

$$g_{cc}(\mathbf{R}) = \frac{V \int e^{-\beta U(\mathbf{R}, \mathbf{r_1}, \dots, \mathbf{r_N})} d\mathbf{r_1} \cdots d\mathbf{r_N}}{\int e^{-\beta U(\mathbf{R}, \mathbf{r_1}, \dots, \mathbf{r_N})} d\mathbf{R}\, d\mathbf{r_1} \cdots d\mathbf{r_N}}, \tag{14.13}$$

where U represents the potential energy associated with the choice of positions of the depletants and colloids $(\mathbf{R}, \mathbf{r_1}, \dots, \mathbf{r_N})$. What does Equation 14.13 tell us? If there are no additional solute particles, g_{cc} is proportional to the Boltzmann factor of the colloid–colloid interaction:

$$g_{cc}(\mathbf{R}) = \frac{V e^{-\beta U(\mathbf{R})}}{\int e^{-\beta U(\mathbf{R})} d\mathbf{R}}. \tag{14.14}$$

When depletant particles are added, the numerator of Equation 14.13 is equal to the partition function of the additional solute particles constrained to move in the field generated by the pair of colloidal particles held fixed at a distance $\mathbf{R}$. The consequence of this is that g_{cc} will be larger at the values $\mathbf{R}$ that allow the partition function of the additional particles to be larger. This means, in other terms, that the colloidal particles will preferentially sit at positions that allow the free energy of the depletant particles to be lower. Intuitively, the free energy of the system of colloids and depletants is dominated by the entropy of the more numerous solute particles, and thus the colloids should minimize the total free energy by sticking together to increase their accessible volume. We derive this result more rigorously below.

Let us assume for a moment that we have no way to detect the positions of the depletant particles. This actually happens in experimental setups, which are able to measure the coordinates of the bigger colloidal particles and thus estimate their pair distribution function, but cannot directly visualize the smaller depletant particles. In practice, polymer coils are often used as depletants. In the case seen earlier, we had been watching a pair of colloids that behave (at least judging from the pair distribution function) as if they interacted with a pair potential $w_{cc}(r)$ such that

$$g_{cc}(\mathbf{R}) = e^{-\beta w_{cc}(\mathbf{R})}. \tag{14.15}$$

w_{cc} is called the *potential of mean force* because if we take its gradient with respect to separation of the colloidal particles, using Equation 14.13, we arrive at

$$-\nabla w_{cc}(\mathbf{R}) = \frac{1}{\beta}\nabla \ln g_{cc}(\mathbf{R}), \tag{14.16}$$

$$= -\nabla F_s(\mathbf{R}), \tag{14.17}$$

$$= \frac{\int -\nabla U\, e^{-\beta U(\mathbf{R},\mathbf{r}_1,\ldots,\mathbf{r}_N)}\, d\mathbf{r}_1\cdots d\mathbf{r}_N}{\int e^{-\beta U(\mathbf{R},\mathbf{r}_1,\ldots,\mathbf{r}_N)}\, d\mathbf{r}_1\cdots d\mathbf{r}_N}, \tag{14.18}$$

$$= \mathbf{f}_{cc} + \left\langle \sum_{i=1}^{N} \mathbf{f}_i \right\rangle, \tag{14.19}$$

where $F_s = \int e^{-\beta U(\mathbf{R},\mathbf{r}_1,\ldots,\mathbf{r}_N)} d\mathbf{r}_1\cdots d\mathbf{r}_N$ is the free energy of the depletant particles moving in the field generated by the fixed pair of colloidal particles, $\mathbf{f}_{cc}$ is the bare colloid–colloid force exerted on a colloid by the other, and the f_i are the forces exerted on such a colloid by the depletant particles while the angle brackets $\langle\cdots\rangle$ denote the canonical ensemble average over all the depletant particle positions.

The meaning of Equation 14.19 can be elucidated with a simple, but relevant, example. Let our pair of colloidal particles interact with a *hard-sphere potential* so that their centers cannot be closer than $2a$ (which is the diameter of the spherical colloidal particles), but do not exert any direct force on each other for $r > 2a$.

$$v_{cc}(r) = \begin{cases} \infty & \text{if } r < 2a, \\ 0 & \text{if } r \geq 2a. \end{cases} \tag{14.20}$$

Let the depletants be particles of size R_g, which do not interact at all with each other, that is,

$$v_{ss} = 0, \tag{14.21}$$

and whose interaction with the colloids is also hard-sphere like:

$$v_{cs}(r) = \begin{cases} \infty & \text{if } r < a + R_g = \sigma_{cs}, \\ 0 & \text{if } r \geq a + R_g = \sigma_{cs}, \end{cases} \tag{14.22}$$

so that the distance r of their centers from the centers of the colloidal particles cannot be lower than some distance σ_{cs}. This is a fair description of mixtures of colloidal particles and polymer coils [4] and is the basis for the *Asakura–Oosawa model* of depletion interactions. A nice feature is that the quantities from Eqations 14.17 to 14.19 can be evaluated analytically without resorting to approximations.

As the total energy can be written as

$$U(\mathbf{R}, \mathbf{r}_1, \ldots, \mathbf{r}_N) = v_{cc}(R) + \sum_i v_{cs}(r_i) + \sum_i v_{cs}(|\mathbf{R} - \mathbf{r}_i|), \tag{14.23}$$

we have that using the definitions of v_{cc} (Eq. 14.20) and v_{cs} (Eq. 14.22) in the calculation of the free energy F_s

$$F_s(\mathbf{R}) = -\frac{1}{\beta}\ln \int e^{-\beta U(\mathbf{R},\mathbf{r}_1,\ldots,\mathbf{r}_N)} d\mathbf{r}_1\cdots d\mathbf{r}_N. \tag{14.24}$$

The Boltzmann factor in the integrand is equal to 1 except for the configurations in which the two colloids overlap or in which some depletant particle overlaps with one of the two colloids (in that case, as $U = \infty$, the integrand is equal to zero). So we end up with

$$F_s(\mathbf{R}) = -\frac{1}{\beta}\ln \int_{V_{acc}} 1\, d\mathbf{r}_1\cdots d\mathbf{r}_N, \tag{14.25}$$

$$= -\frac{N}{\beta}\ln \int_{V_{acc}} 1\, d\mathbf{r}_1, \tag{14.26}$$

where the domain of integration V_{acc} is the region of space where the centers of the depletant particles are free to roam without overlapping with the colloidal particles.

According to Equation 14.26, the evaluation of F_s reduces to the evaluation of the volume V_{acc} accessible to the centers of the additional solute particles, which in the end boils down to the computation of the volume of the lens-shaped region of overlap of the forbidden volumes due to the presence of the colloids (see Fig. 14.3). Once that F_s has been calculated as a function of $\mathbf{R}$ (the derivation is straightforward! We will not show it here, but the reader can refer for instance to Ref. [5]), we can plug it in Equation 14.17 (taking into account that $\mathbf{f}_{cc} = -\nabla v_{cc} = \mathbf{0}$) to get the expression of the mean force (valid in the range $[2a, 2\sigma_{cs}]$)

$$\left\langle \sum_{i=1}^{N} \mathbf{f}_i \right\rangle = \frac{\pi\rho}{\beta}\left(\frac{R^2}{4} - \sigma_{cs}^2\right), \tag{14.27}$$

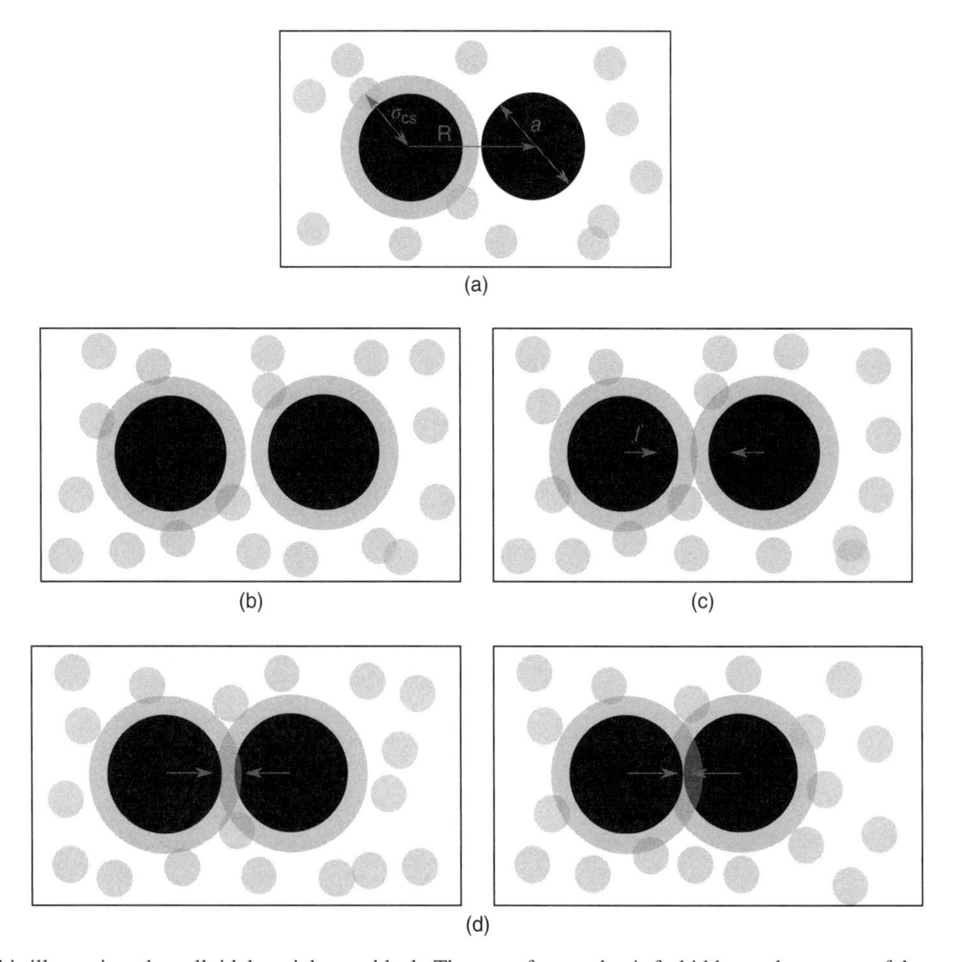

Figure 14.3 In this illustration, the colloidal particles are black. The part of space that is forbidden to the centers of the small solute particles (small gray spheres) includes the black particles as well as a shell around them shown in gray. As the distance between the two colloids is lowered, the forbidden regions fuse and shrink (as they overlap on a lens-shaped domain) and the smaller particles have more space to move, increasing the domain of integration V_{acc} in Equation 14.26 and thus reducing their free energy F_s. (a) Geometrical parameters of the system composed by the two colloidal particles and the smaller solvent particles. The region around the black colloid (left) shaded in gray denotes its associated forbidden zone. (b) If the particles are far apart, the volume that they exclude to the centers of the solute particles is always the same, and no depletion force is present. (c) If the particles are closer than $2\sigma_{cs}$ the volumes of exclusions overlap, and particles are effectively pushed by the smaller particles as they want the excluded region to shrink more. (d) The force (in the Asakura–Oosawa approximation) becomes even stronger at short distances up to the distance when the two colloidal particles touch each other ($2a$).

which is an *attractive* force whose range goes up to $R = 2\sigma_{cs}$ (higher values of R do not change V_{acc} in Equation 14.26, and thus the mean force at such distances is zero), which depends on the number density of additional particles ρ. This force is completely due to the variation of phase space available to the depletant particles as a function of R: that is why in the literature, forces of this kind are named *entropic forces*.

The resulting potential has the form:

$$V_{ss}(R) = -\frac{\pi\rho}{\beta}\left(\frac{R^3}{12} - \sigma_{cs}^2 R\right). \qquad (14.28)$$

14.2.2.1 What We Have Not Said The model potential used earlier does not describe well all binary mixtures: for instance, v_{ss} in Equation 14.21 and v_{cs} in Equation 14.22 are the simplest approximations, assuming no interactions between depletant particles, and a hard-sphere interaction between depletants and colloids. In the case that a more exact expression is used, the effective force would not be obtainable analytically [6, 7]. In addition, one typically deals with systems containing more than two colloids, and in that case the mean force potential capable of reproducing structural data is, in general, a many-body potential and thus cannot be exactly decomposed into a sum of pair potentials (except in the case of short-ranged interactions resulting from a high colloid–polymer size ratio [8]). For a more in-depth treatment of these issues, we refer the reader to the reviews on the subject [4, 9].

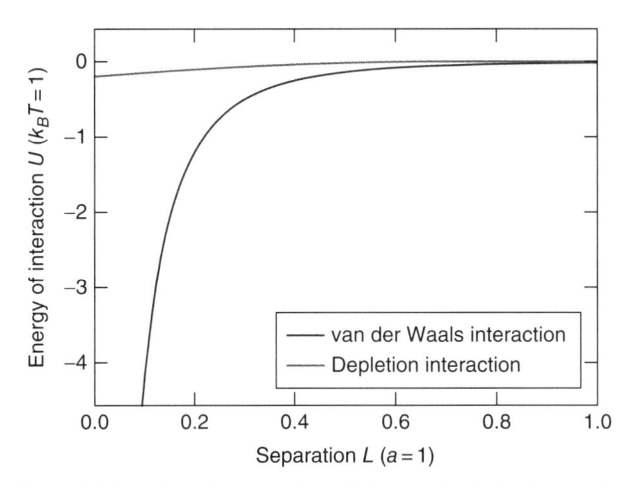

Figure 14.4 Plot of the van der Waals and depletion interaction for the values of the parameters reported in the text.

14.2.3 Putting It All Together

Let us work out a realistic example to see what the typical magnitudes of the forces acting between two colloidal particles are: take two spherical polystyrene beads of diameter $2a = 100\ nm$ in water. In a solution containing hydroxyethyl cellulose (HEC) polymers, the centers of mass of colloids and polymers remain separated by more than $\sigma_{cs} = 85\ nm$. Suppose the volume fraction of HEC polymers is $\eta = 0.064$. The Hamaker constant of polystyrene in water is about 10^{-19} J, and we can suppose the system is at room temperature, 25 °C.

From the above-mentioned data, one can determine that the polymer "radius" $R_g = 35\ nm$. So, the number density is

$$\rho = \frac{3\eta}{4\pi R_g^3} \approx 3.5 \times 10^{20}\ m^{-3}. \tag{14.29}$$

At this stage, we have all the information that we need to calculate the expressions for "strong" (Eq. 14.11) and "weak" (Eq. 14.28) interactions. A plot of the Asakura–Oosawa and the exact van der Waals interaction is shown in Figure 14.4, where energy and distance values are reported in units of $k_BT = 4.11 \times 10^{-21}\ J$ and particle diameter $2a = 100\ nm$.

We will see in the next section that the magnitude of attractive forces and the particle volume fraction can be used to define a phase diagram for gelation. This concept is widely used as a universal framework for understanding gelation under a variety of conditions.

14.3 ROUTES TO GELATION

Amazingly enough, the formation of space spanning structures occurs naturally in attractive colloidal systems, leading to solid-like materials formed from very dilute systems. Short-range attractions lead to a variety of disordered arrested states, where the nature of the arrest is governed by both the volume fraction ϕ and the strength of the attraction U [10, 11]. The lower the volume fraction, the higher U must be to permit the formation of a stress-bearing network. At sufficiently high U, gelation occurs via diffusion-limited cluster aggregation (DLCA), where particles meet and stick irreversibly to each other. These aggregates are fractal in structure and eventually grow to fill all space, forming a network in which the particles are permanently trapped.

At lower U, particles do not simply stick each time they meet. However, gelation is still possible because the system will tend to separate into colloid-rich and colloid-poor phases once it is above some critical volume fraction. This is depicted as a fluid–fluid phase boundary in Figure 14.5. It has been shown in a number of works that the phase separation process will be arrested when the density of the colloid-rich phase reaches that of a colloidal glass [12–17]. Under conditions where the colloid-rich phase forms an interconnected network, the glassy arrest of the particles within the network naturally entails gelation. Interconnected colloidal-rich phases are generally formed around and above the critical composition denoted by the circle in Figure 14.5. By contrast, well below this critical composition, the phase separation process leads to the formation of discontinuous high density phases. In this range, the glassy arrest of the colloidal-rich phase is not sufficient to lead to gelation, and the vitrified droplets must aggregate into a space-spanning network to provide rigidity to the system if it is to form a gel.

The routes to gelation can thus be divided into two classes: one where the constituents of the system, colloidal particles or vitrified droplets, aggregate to form space-spanning structures, and the other where the network is formed first and then becomes vitrified.

14.3.1 Dynamic Scaling

Both routes exhibit similar kinetics since they are both diffusion-limited processes. Because the diffusion coefficient is inversely proportional to the size a of the object, $D = k_BT/6\pi\eta a$, growth points are nearly stationary in space and progressively adsorb all faster diffusing units around them. As a result of this, there is a rather well-defined distance between these accretion points, which is evidenced experimentally by the appearance of a well-defined peak in the low q-range of the scattering vector dependence of the scattered intensity, $I(q)$. In Figure Figure 14.6, we show a schematic of the temporal evolution of the q-dependence of $I(q)$ typically observed for both DLCA [18] and phase separation via spinodal decomposition [19], where q is the magnitude of the scattering vector. In both cases, the low-q peak systematically shifts to lower values and the peak intensity increases as time proceeds, reflecting the

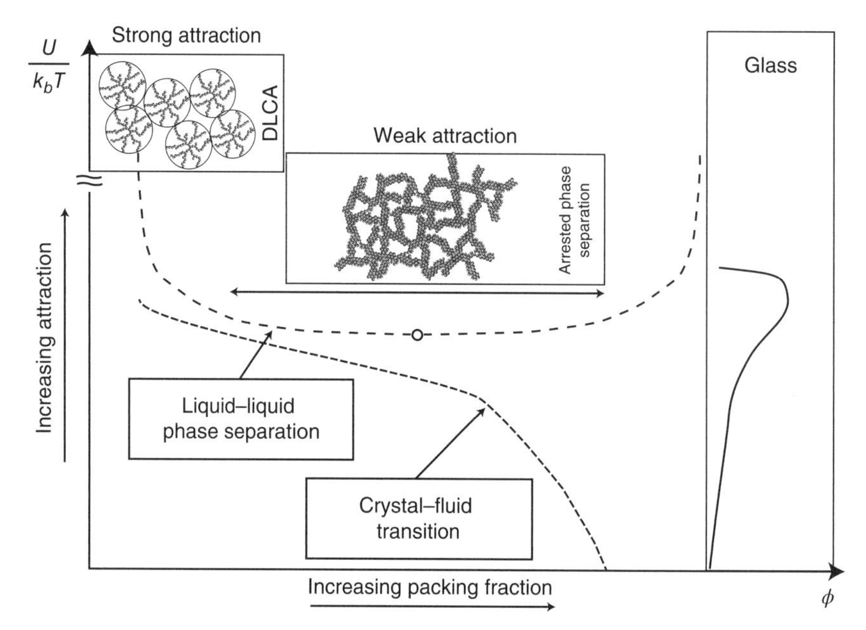

Figure 14.5 Schematic state diagram depicting different routes to arrested solid-like states for suspensions of colloidal particles interacting by short-range attractions. The dashed line denotes the fluid–fluid phase separation boundary, where the critical composition marked with a circle is about $\phi \sim 0.25$. Around this composition and at energies above the boundary, spinodal decomposition leads to the formation of an interconnected dense phase. Gels will form here when the dense phase reaches the critical density of a glass; at this point the arrested dynamics of the particles will prevent further coarsening. The phase separation is arrested and the system exhibits stress-bearing properties, since the particles are both trapped in a local configuration and part of a space-spanning network. At very large attractions, particles and clusters will stick to each other each time they encounter; here, the aggregation process is solely diffusion limited and termed diffusion-limited cluster aggregation (DLCA). The gel point is reached once the clusters fill all space and connect to form a space-spanning network. At lower interaction energies and intermediate concentrations, a fluid–crystal transition (dashed line) occurs, although this is suppressed for polydisperse systems. At larger concentrations, the formation of space-spanning networks is no longer a requirement for rigidity. The solid line depicts the colloidal glass transition beyond which the constituent of the system is permanently trapped in cages of nearest neighbors.

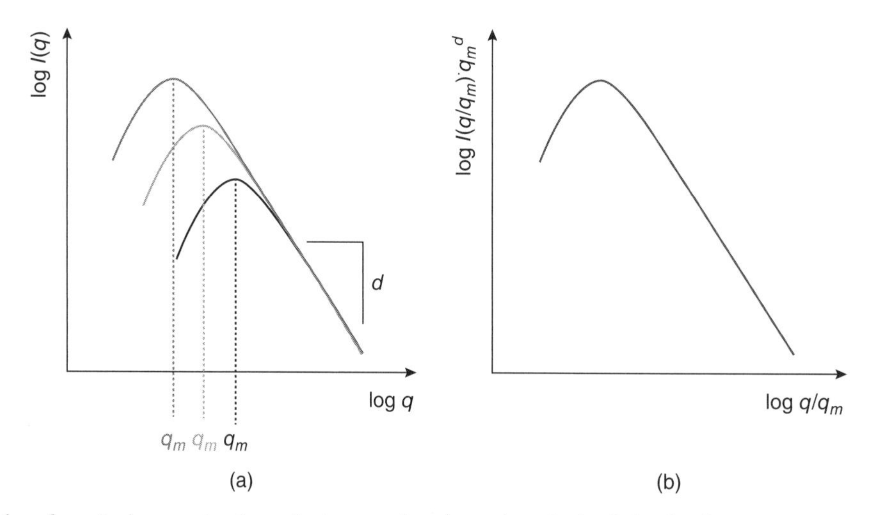

Figure 14.6 Schematic of scattering vector-dependent scattering intensity obtained in the low q-range at different times during diffusion-limited cluster aggregation and/or spinodal decomposition (a). With time, the peak position shifts toward the left to lower scattering vectors q and the intensity at the peak $I(q_m)$ increases. In the higher q-range, the scattering intensity decreases as $I(q) \sim q^{-d}$. Normalization of the q-axis with q_m and the I-axis with q_m^{-d} scales all scattering profiles onto a single master curve (b).

increase in, respectively, the size and mass of the growing structures within the system. At larger q, the scattered intensity decreases as a power law, $I(q) \sim q^{-d}$, with $d = 1.8$ for DLCA and $d = 3$ for spinodal decomposition. In this q-range, the magnitude of $I(q)$ remains stationary during the process, such that a simple normalization of q with the value denoting the position of the peak, q_m, and of $I(q)$ with q_m^{-d} will lead to the collapse of all scattering profiles onto a single master curve, as shown schematically (Fig. 14.6). Such scaling behavior is termed *dynamic scaling* and reflects the self-similarity of the growing units; the unit found at later times is just an upscaled version of that found at earlier times. In the case of spinodal decomposition, this unit is a thick strand for conditions generating interconnected networks and more or less spherical droplets for conditions generating disconnected dense phases. In the case of DLCA, the units are fractal aggregates.

The properties of the coarse gels obtained by arrested spinodal decomposition are to date not completely understood and the interplay between phase separation and dynamics arrest still requires exploration. By contrast, the structure–function relation of fractal gels is rather well established. Indeed, fractal aggregation and the properties of the resulting gels were extensively studied in the 1980s [20–24]. We summarize a few main results in the next section.

14.3.2 Fractal Aggregation

14.3.2.1 Diffusion-Limited versus Reaction-Limited Aggregation The effective interactions between charge stabilized colloids can be conveniently controlled and exploited to investigate their role in aggregation kinetics and aggregate structure. In particular, the repulsive barrier in the effective potential between charged colloids decreases and finally disappears altogether upon increase in the ion concentration [18].

Accordingly, one distinguishes between two limiting cases: reaction-limited cluster aggregation (RLCA) and DLCA. In the former case, the colloids still possess a small but nonnegligible repulsive barrier that needs to be overcome before the colloids irreversibly stick to one another. In the latter case, repulsion is essentially nonexistent, such that the colloids irreversibly stick to each other each time they encounter. These differences in the aggregation kinetics result in differences in the growth laws, the cluster size distribution, and compactness of the fractal objects. Intuitively, in the RLCA case, free colloids can diffuse further within a growing aggregate since they do not immediately stick. This leads to more compact objects.

The net result of both DLCA and RLCA processes is the formation of fractal gels, which can theoretically be formed at essentially infinitely low volume fractions. To understand this, we simply need to consider that the mass of a fractal aggregate, $M \sim R_m^d$, with $d_f \sim 2$. This implies that the density of the aggregate, $\rho \sim M/R_m^3 = R_m^{d-3}$, decreases as its size increases. Thus, no matter how dilute a colloidal suspension you start with, a fractal will always be able to grow large enough to fill all space.

14.3.2.2 Estimating the DLCA Cluster Size and Gelation Time As we mentioned in the previous section, there is a characteristic length scale R_m that emerges during the DLCA aggregation process. As a consequence, we can approximate the cluster size at gelation by estimating the point at which clusters fill all space. To estimate this size, we assume a number of clusters N. Since the clusters are fractals, the number of particles within a cluster is $n = M/m = (R_m/a)^{d_f}$ for particles of size a and mass m. With Nn the total number of particles, we now can write the volume fraction of particles as $\phi_0 \approx Nna^3/V$, while the volume fraction of the clusters is $\phi_{\text{eff}} \approx NR_m^3/V$. Since the clusters, whose size has grown to $R_m = R_c$, fill all space at gelation, we impose the condition $\phi_{\text{eff}} = 1$. We can substitute $N/V = R_c^{-3}$ to find

$$\frac{R_c}{a} = \phi_0^{1/(d_f-3)}, \tag{14.30}$$

R_c thus decreases as ϕ_0 increases, or conversely, as the system becomes more dilute, the clusters will grow to larger sizes before they touch. But no matter how low ϕ_0 is, there is a theoretical cluster size that will yield a gel. In practice, limits to gelation are reached if the sample size is smaller than the cluster size, or if there are thermal or mechanical stresses that break the gel. The characteristic size of the clusters and their fractal dimension determine the dynamical [25] and the mechanical properties of fractal gels, as we see in the following sections.

To estimate the gelation time, one can simplify matters by considering DLCA as a doubling process, whereby monomers become dimers, which then stick together to become tetramers, and so forth. The timescale for doubling also increases as clusters grow and diffuse more slowly. In this scenario, the cluster mass will increase linearly with time, $dM \sim M$; together with the Brownian collision rate $dt \sim 1/(8\pi R_m cD)$, where c is the number concentration of clusters and D is their diffusion coefficient. We can re-express the concentration in terms of the time-dependent cluster mass, $c \approx (m\phi_0)/(Ma^3)$. This gives us enough information to calculate the growth rate of the cluster size using the simplified form

$$dt \approx \frac{1}{8\pi R_m cDM} dM \approx \frac{\eta a^3}{k_B T m \phi_0} dM. \tag{14.31}$$

We integrate and then simplify using $M = m(R_m/a)^{d_f}$ to find

$$R_m(t) \approx a\left(\frac{\phi_0 k_B T}{a^3 \eta} t\right)^{1/d_f}. \tag{14.32}$$

To find the gelation time, we then substitute the expression we found for the cluster size at gelation (Eq. 14.30) and solve

$$t_g \approx \left(\frac{a^3\eta}{k_B T}\right)\phi_0^{\frac{3}{(d_f-3)}}. \tag{14.33}$$

14.3.2.3 Dynamics of Fractal Gels The characteristic size of the clusters at gelation and the fractal dimension determine the internal dynamics and as we will see in the next section, the mechanical properties of fractal gels. Because gels are solid-like systems, their dynamics are limited to internal fluctuations. The larger the cluster radius R_c, the floppier the gel, and the larger the amplitude of the largest possible fluctuation.

To describe such dynamics, a model was developed that considers the short-time fluctuations of fractal gels as a superposition of overdamped elastic modes associated with segments of the gel [25]. Up to R_c, the motion of each segment results not only from direct thermal fluctuations but also from fluctuations of any larger segments within the same cluster. The fluctuation of each mode is constrained to a maximum amplitude and relaxes with a timescale that depends on the size of the mode. This relaxation time is determined by viscous relaxation and is set by $\tau(s) = 6\pi\eta s/\kappa(s)$ for a mode of size s, where η is the viscosity of the fluid and $\kappa(s)$ is the spring constant. The slowest mode is set by the characteristic length of the system, R_c. The timescale of fluctuations of the gel at the cluster size is thus $\tau_c = 6\pi\eta R_c/\kappa(R_c)$, inversely proportional to the elastic modulus $G = \kappa(R_c)/R_c$ as discussed in the following section. The amplitude of the slowest mode sets the maximum mean square displacement δ^2.

By considering the sum over all fluctuations, it was shown that time dependence of the mean square displacement can be approximated as

$$\langle \Delta r^2(\tau)\rangle = \delta^2\left[1 - e^{(-(\tau/\tau_1)^p)}\right] \tag{14.34}$$

with $p = 0.7$. The characteristic relaxation time $\tau_1 = 0.35\tau_c$ and the maximum mean square displacement, therefore, describe the full spectrum of displacements. Note that both τ_1 and δ^2 are determined by the cluster size R_c; this means that the floppiest and slowest modes dominate the internal fluctuations and characterize the dynamics of the gel.

14.4 ELASTICITY OF COLLOIDAL GELS

The variety of soft solids formed through colloidal gelation can combine efficient flow and transport with cohesion, strength, and flexibility. The origin of their unique mechanical properties is their disordered, heterogeneous microscopic structure, where stress transmission is far from trivial. In principle, one could suitably adjust interactions and/or the route to gelation to produce "smart materials" with specifically designed and tunable mechanics. To achieve this, one should rationalize their rigidity and understand how the elastic behavior of a gel depends on its microscopic features such as the interparticle interactions or the specific gel structure. In fact, since gelation does not correspond to a crystalline-ordered structure, this is an open challenge [26, 27].

14.4.1 Elasticity of Fractal Gels

Fractal gels are sparse, tenuous structures, with low connectivity. Kantor and Webman considered that in such random elastic solids, only a small fraction of the network (a stress-bearing backbone) is mainly responsible for its macroscopic elastic response [28, 29]. This is because dangling ends cannot transmit stress, and thus do not contribute to the elasticity. There are two modes of deformation contributing to elasticity of the stress-bearing backbone: bond bending and bond stretching. In two dimensions, the elastic energy of a random chain of N elastic bonds, where the bond vectors $\mathbf{b}$ are of length a, is

$$H = \frac{Q_1}{2}\sum_i \delta\Theta_i^2 + \frac{Q_2}{2a^2}\sum_i \delta b_i^2, \tag{14.35}$$

where Q_1 and Q_2 are local elastic constants, $\delta\Theta_i$ is the change in the angle Θ_i between bond vectors $i-1$ and i (Fig. 14.7), and δb is the change in length of bond i. Thus, the first term is related to bending, while the second term is related to stretching. If one applies a force to one end of the chain $\mathbf{R}_N$ with the other end $\mathbf{R}_0$ held fixed, one can obtain $\delta\Theta_i$ and δb_i by minimizing the total energy, in terms of the magnitude of the applied force and the in-plane projections of the chain parallel and perpendicular to the direction of the force.

In particular, $\delta\Theta_i$ depends on the bond vector component perpendicular to the force direction in the plane. Kantor and Webman therefore rewrote the bending contribution to the elastic energy as $F^2NS_\perp^2/2Q_1$ where F is magnitude of the force and $S_\perp^2$ is the squared radius of gyration of the in-plane projection of the chain along the direction perpendicular to the force. For sufficiently long chains, this term dominates the elastic energy as compared to the bond stretching term. Hence, one can obtain the bending force constant of the chain by relating the elastic energy to the squared displacement of the end of the chain as

$$\kappa = \frac{Q_1}{NS_\perp}. \tag{14.36}$$

It is important to note that this effective force constant depends not only on its length but also on the direction of the applied force and the shape of the chain.

DLCA fractal gels are formed by very large clusters, typically less rigid than the connections between them. Hence,

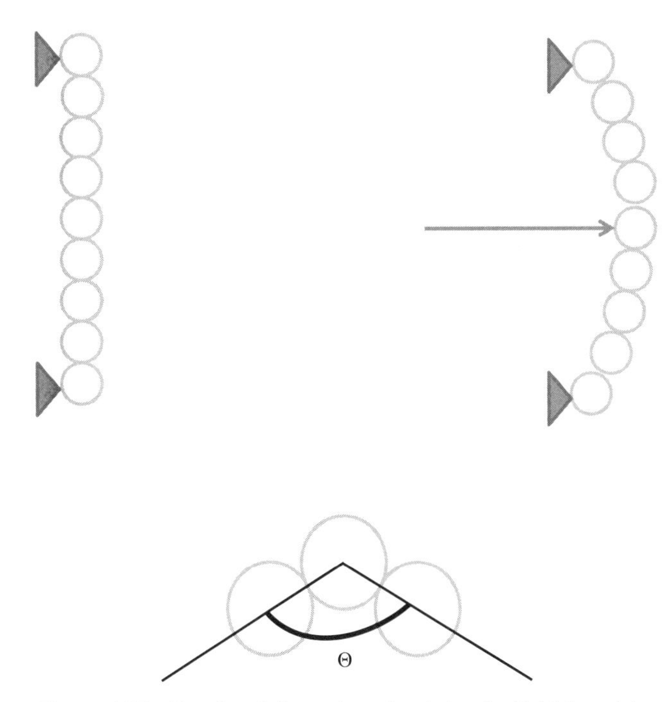

Figure 14.7 Bending deformation of a chain of colloidal particles in $d = 2$. The microscopic bending angle involving three particles of the chain is also shown.

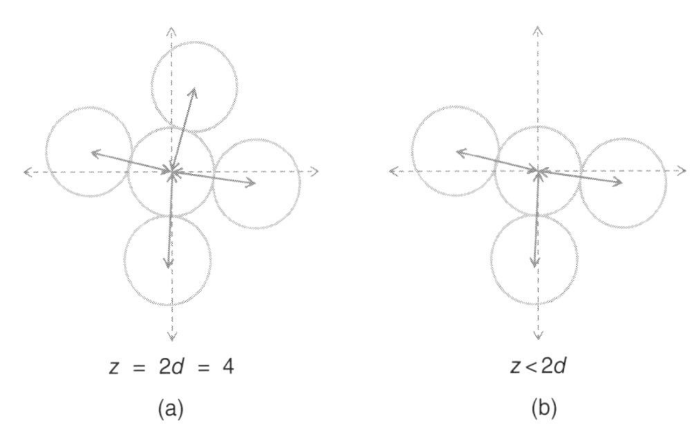

Figure 14.8 A sketch of the Maxwell criterion for mechanical stability in two dimensions ($d = 2$). In an overall rigid system, the average local particle configuration is represented by the figure on the left ($z = 2d$, also called the *isostatic* condition). Dashed arrows indicate the degrees of freedom of the tagged particle in the center, while solid arrows indicate the forces transmitted by its contact neighbors. If the average connectivity of the system is like the one on the right, Maxwell's criterion is not satisfied and the system is overall not rigid.

the approach of Kantor and Webman is expected to give a good description of their elasticity, in terms of the elastic backbone of the cluster size R_c: $S_\perp \sim R_c \sim \phi_0^{1/(d_f-3)}$ where $\phi_0 = \frac{4}{3}\pi a^3 \rho$ is the volume fraction. The number of particles in the stress-bearing backbone is $N \sim R_c^{d_B}$ with the "bond dimension" $d_B \simeq 1.1$ because of the relatively linear backbone of DLCA gels. From this, one obtains the elastic constant of the gel as a function of the volume fraction: $\kappa \sim Q_1/\phi_0^{(2+d_B)/(d_f-3)}$. The elastic modulus of the gel G should, therefore, scale as $G \sim \kappa/R_c \sim \phi_0^{\frac{3+d_B}{3-d_f}}$. This scaling has been confirmed by a number of experiments [30, 31].

As we have seen in the previous section, different types of interactions and arrest conditions can give rise to gels with very different microscopic structures. Also, upon increasing the initial volume fraction, the mesoscopic organization of the gel network goes from the fractal thin structures just discussed to networks of thicker strands and to dense interconnected assemblies of bulkier flocs, typically produced by an incipient phase separation (Fig. 14.5). These changes in the structure may correspond to significant changes in the elastic response of the material [31–36].

14.4.2 Deformations and Connectivity

In crystalline solids, there is a direct link between the specific crystal structure and the elastic constants measured in a mechanical test. This relationship is described in Cauchy Born theory based on the assumption that upon deformation, all particles in the microscopic structure follow the macroscopic strain (*affine deformation*) [37, 38]. In the case of amorphous solids, such as gels, this type of theoretical framework is still missing [26]. One important point is that, for example, the specific microscopic configuration in which the system has been arrested is mechanically but not necessarily thermodynamically stable [27]. Internal stresses and deformations generated due to the gelation process or due to unbalanced osmotic pressure exerted by the solvent phase can dramatically improve the rigidity of the strand connecting two points in a gel network, similar to the manner in which the stretching of a violin string makes it significantly more capable of supporting an applied stress. Hence, internal stresses are able to confer mechanical stability to structures [39], which according to the Maxwell criterion of rigidity would never support a finite stress because their microscopic connectivity (i.e., number of nearest neighbors or mechanical contacts between particles) is too low (see Fig. 14.8).

Another peculiarity of gels or other soft solids, which makes their elastic response intrinsically different from that of crystals, is the *nonaffinity* of the particle displacements under strain [27, 40, 41, 42]. Qualitatively, this means that the particles do not just move proportionally to the global strain but undergo additional displacements, thus implying additional energy dissipation. A schematic view of nonaffine displacements is shown in Figure 14.9. Particles bonded on a tagged particle, initially undergoing affine displacements, do exert a force on the tagged particle. In a crystal, the resultant of these forces would be zero, but in a disordered

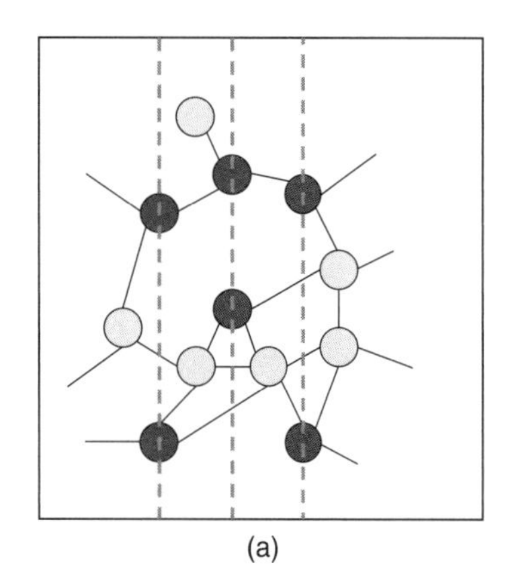

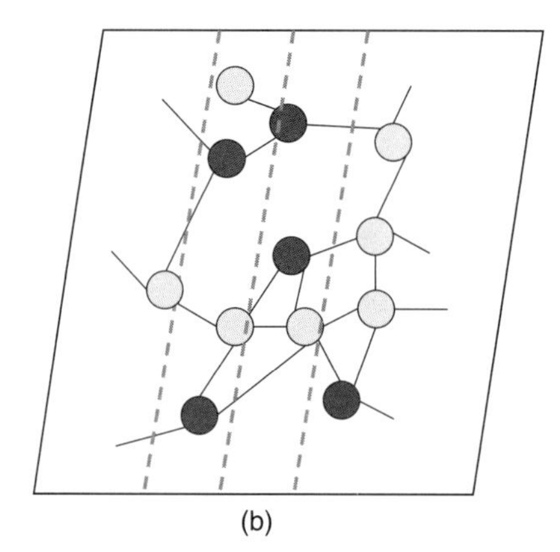

Figure 14.9 Schematic sketch of nonaffine displacements under shear deformation in a colloidal gel. Dark-shaded particles lying on the vertical dashed lines at rest would still lie on the dashed lines under an affine distortion. As the deformation is nonaffine in disordered solids, they do not.

structure such a cancelation does not occur: the resultant force is finite and induces an additional displacement of the tagged particle, in order to maintain mechanical equilibrium. Hence, nonaffine displacements are fundamental to preserving the mechanical equilibrium [42, 43].

14.5 CONCLUSIONS

In this chapter, we have highlighted the essential components and properties of colloidal gelation. In summary, attractive interactions are the basis for aggregation, but whether a gel is formed depends on where the system is located in the phase space described by the strength of the attraction and the particle volume fraction. The state diagram provides a guide to understanding the routes to gelation, which comprise two major classes: aggregation into clusters with rigid bonds, which over time grow to form a space-spanning network, or phase separation to form a network that subsequently vitrifies. Aggregation gives rise to fractal clusters, which allow us to make estimations on key properties, such as cluster size, gelation time, and elasticity. The elasticity in this case arises from bending modes of the stress-bearing backbone of the gel network.

Gelation is an active area of research, and indeed, many fundamental questions remain unanswered. For example, what impact does the addition of anisotropy in the interaction potentials, via "patchy" colloids, or the introduction of nonadditive interactions have on network structure? In the area of elasticity, how do gels break down, collapse, and fracture? How do nonaffine deformations contribute to elastic behavior? How do the ideas developed here apply to biological and gels in biological systems where *active* processes act continuously on their structure? These fascinating questions reflect the interdisciplinary nature of colloid research, requiring ideas from physics, chemistry, and biology to bring depth of understanding and new materials to reality.

REFERENCES

[1] Hamaker HC. The London-van der Waals attraction between spherical particles. Physica 1937;4(10):1058–1072.

[2] Israelachvili JN. *Intermolecular and Surface Forces*. London: Academic Press; 2011.

[3] Russel WB, Russel WB, Saville DA, Schowalter WR. *Colloidal Dispersions*. Cambridge: Cambridge University Press; 1992.

[4] Likos CN. Effective interactions in soft condensed matter physics. Phys Rep 2001;348(4-5):267–439.

[5] Attard P. Spherically inhomogeneous fluids. II. Hard-sphere solute in a hard-sphere solvent. J Chem Phys 1989;91:3083.

[6] Louis AA, Allahyarov E, Löwen H, Roth R. Effective forces in colloidal mixtures: from depletion attraction to accumulation repulsion. Phys Rev E 2002;65(6):061407.

[7] Fiocco D, Pastore G, Foffi G. Effective forces in square well and square shoulder fluids. J Phys Chem B 2010;114(37):12085.

[8] Dijkstra M, Brader JM, Evans R. Phase behaviour and structure of model colloid-polymer mixtures. J Phys Condens Matter 1999;11(50):10079.

[9] Belloni L. Colloidal interactions. J Phys Condens Matter 2000;12:R549.

[10] Cates ME, Fuchs M, Kroy K, Poon WCK, Puertas AM. Theory and simulation of gelation, arrest and yielding in attracting colloids. J Phys Condens Matter 2004;16(42):S4861–S4875. cited By (since 1996)51.

[11] Zaccarelli E. Colloidal gels: equilibrium and non-equilibrium routes. J Phys Condens Matter 2007;19(32):1–50–. cited By (since 1996)193.

[12] Pusey PN, Pirie AD, Poon WCK. Dynamics of colloid-polymer mixtures. Physica A 1993;201(1-3):322–331. cited By (since 1996)68.

[13] Poon WCK, Pirie AD, Pusey PN. Gelation in colloid-polymer mixtures. Faraday Discuss 1995;101:65–76. cited By (since 1996)143.

[14] Verhaegh NAM, Asnaghi D, Lekkerkerker HNW, Giglio M, Cipelletti L. Transient gelation by spinodal decomposition in colloid-polymer mixtures. Physica A 1997;242(1-2):104–118. cited By (since 1996)85.

[15] Manley S, Wyss HM, Miyazaki K, Conrad JC, Trappe V, Kaufman LJ, Reichman DR, Weitz DA. Glasslike arrest in spinodal decomposition as a route to colloidal gelation. Phys Rev Lett 2005;95(23):238302. cited By (since 1996)72.

[16] Cardinaux F, Gibaud T, Stradner A, Schurtenberger P. Interplay between spinodal decomposition and glass formation in proteins exhibiting short-range attractions. Phys Rev Lett 2007;99(11):118301. cited By (since 1996)55.

[17] Lu PJ, Zaccarelli E, Ciulla F, Schofield AB, Sciortino F, Weitz DA. Gelation of particles with short-range attraction. Nature 2008;453(7194):499–503. cited By (since 1996)293.

[18] Carpineti M, Giglio M. Spinodal-type dynamics in fractal aggregation of colloidal clusters. Phys Rev Lett 1992;68(22):3327–3330. cited By (since 1996)179.

[19] Siggia ED. Late stages of spinodal decomposition in binary mixtures. Phys Rev A 1979;20(2):595–605. cited By (since 1996)606.

[20] Witten TA, Sander LM. Diffusion-limited aggregation, a kinetic critical phenomenon. Phys Rev Lett 1981;47(19):1400–1403. cited By (since 1996)2216.

[21] Meakin P. Formation of fractal clusters and networks by irreversible diffusion-limited aggregation. Phys Rev Lett 1983;51(13):1119–1122. cited By (since 1996)772.

[22] Weitz DA, Oliveria M. Fractal structures formed by kinetic aggregation of aqueous gold colloids. Phys Rev Lett 1984;52(16):1433–1436. cited By (since 1996)405.

[23] Ball RC, Weitz DA, Witten TA, Leyvraz F. Universal kinetics in reaction-limited aggregation. Phys Rev Lett 1987;58(3):274–277. cited By (since 1996)121.

[24] Lin MY, Lindsay HM, Weitz DA, Ball RC, Klein R, Meakin P. Universality in colloid aggregation. Nature 1989;339(6223):360–362. cited By (since 1996)392.

[25] Krall AH, Weitz DA. Internal dynamics and elasticity of fractal colloidal gels. Phys Rev Lett 1998;80(4):778–781. cited By (since 1996)167.

[26] Anderson PW. In: Balian R, Maynard R, Toulouse G, editors. *Ill-Condensed Matter: Les Houches 1978, Session XXXI*. New York: North-Holland; 1979.

[27] Alexander S. Amorphous solids: their structure, lattice dynamics and elasticity. Phys Rep 1998;296:65–236.

[28] Kantor Y, Webman I. Elastic properties of random percolating systems. Phys Rev Lett 1984;52:1891–1894.

[29] de Gennes PG. *Scaling Concepts in Polymer Physics*. Ithaca (NY): Cornell University Press; 1979.

[30] Krall AH, Weitz DA. Internal dynamics and elasticity of fractal colloidal gels. Phys Rev Lett 1998;80:778–781.

[31] Laurati M, Petekidis G, Koumakis N, Cardinaux F, Schofield AB, Brader JM, Fuchs M, Egelhaaf SU. Structure, dynamics, and rheology of colloid-polymer mixtures: from liquids to gels. J Chem Phys 2009;130:134907.

[32] Zaccone A, Wu H, Del Gado E. Elasticity of arrested short-ranged attractive colloids: homogeneous and heterogeneous glasses. Phys Rev Lett 2009;103:208301.

[33] Kroy K, Cates ME, Poon WCK. Cluster mode-coupling approach to weak gelation in attractive colloids. Phys Rev Lett 2004;92:148302.

[34] Koumakis N, Petekidis G. Two step yielding in attractive colloids: transition from gels to attractive glasses. Soft Matter 2011;7(6):2456.

[35] Ramakrishnan S, Chen Y-L, Schweizer KS, Zukoski CF. Elasticity and clustering in concentrated depletion gels. Phys Rev E 2004;70:040401.

[36] Zaccone A. The shear modulus of metastable amorphous solids with strong central and bond-bending interactions. J Phys Condens Matter 2009;21(28):285103.

[37] Ladau LD, Lifshitz EM. *Theory of Elasticity*. Pergamon: Oxford; 1981.

[38] Born M, Huang H. *The Dynamical Theory of Crystal Lattices*. Oxford: Oxford University Press; 1954.

[39] Wyart M. On the rigidity of amorphous solids. Ann Phys 2005;30(3):1.

[40] Tanguy A, Wittmer JP, Leonforte F, Barrat J-L. Continuum limit of amorphous elastic bodies: a finite-size study of low-frequency harmonic vibrations. Phys Rev B 2002;66:174205.

[41] DiDonna BA, Lubensky TC. Nonaffine correlations in random elastic media. Phys Rev E 2005;72:066619.

[42] Zaccone A, Scossa-Romano E. Approximate analytical description of the nonaffine response of amorphous solids. Phys Rev B 2011;83:184205.

[43] A. Zaccone, Mod. Phys. Lett. B 27, 1330002 (2013)

SECTION V

OTHER SOFT MATERIALS

15

EMULSIONS

SUDEEP K. DUTTA, ELIZABETH KNOWLTON, & DANIEL L. BLAIR

Department of Physics and The Institute for Soft Matter Synthesis and Metrology, Georgetown University, Washington, DC, USA

15.1 INTRODUCTION

In this chapter, we discuss the basic and applied science relating to emulsions. Our primary scientific focus will be on the mechanical properties of high-volume-fraction systems with a review of the methods for their creation and methods of stabilization, quantifying the flow properties (i.e., rheology), and their applications in industrial settings and the basic sciences. The layout of the chapter is first an overview of emulsions and a brief history of their use. This will be followed by a discussion of the methods and science of creating and stabilizing fluid–fluid interfaces, followed by an overview of the rheological properties of dense emulsions.

The simplest emulsions are thermodynamically metastable fluid droplets suspended within a secondary fluid, more simply, liquid colloids. This *mixture* is very different from that of two miscible fluids. In particular, the creation of emulsions requires a substantial input of energy. Maintaining that *mixture* requires a significant understanding of the physicochemical properties of diverse macromolecular components, fluid–fluid surface interactions, and the thermodynamics of partially miscible species. Our everyday experience exemplifies the difficulty of creating stable emulsions. Indeed, the most familiar example is a salad dressing composed of vinegar (continuous phase) and olive oil (dispersed phase). To generate most emulsions, we are compelled to invoke a violent input of energy through shearing that is mediated by vigorous shaking or using high shear mixing devices.

15.1.1 Background

Emulsions are routinely utilized as a means to transport and deliver value-added materials that may be comprised of and

Fluids, Colloids and Soft Materials: An Introduction to Soft Matter Physics, First Edition. Edited by Alberto Fernandez Nieves and Antonio Manuel Puertas.

potentially volatile materials that are safely encased within a surrounding solvent. The applications of emulsions are widely varied and range from the largest industrial scale production down to integrating the contents of individual nanoliter- to picoliter-sized droplets.

As with many engineered materials, nature often provides the basic framework for controlling and optimizing their design and functionality. Some of the earliest types of emulsion science, most of which are still found today, come in the form of foods. For millennia, butter has been created through the processing of milk solids into semisolid phases that are comprised of water droplets dispersed in milk fats. Another specific example originates in 16th-century French cooking where protein-stabilized suspensions consisting of oil or fat droplets suspended in aqueous phases, such as vinegar and water, were developed and refined. Through the simple process of shearing these fluids, mayonnaise and sauces that have a distinctive rheology and mouth feel are created. Today, there are thousands of variations of emulsion-based sauces found in cooking. In the world of personal care products, emulsions are routinely utilized for their specific ability to deliver materials to the skin or hair, such as conditioning oils, that are transported to surfaces by fluids that can easily evaporate or can be washed away. Emulsions provide versatility in many different applications that require a deep and fundamental understanding of their creation, stability, and rheology, making them materials that we will continue to investigate in basic and applied science.

Emulsion science is a subset of the larger field of colloid science, which is focused on the synthesis and use of particulate laden fluids. As discussed previously within this book, colloidal dispersions are collections of mesoscale particles that have complex surface chemistries that mediate physical interactions through the suspending solvent. In colloidal dispersions with short-range electrostatic interactions, particle concentration determines the bulk-phase behavior [1]. Simple emulsions are categorized by the nature of the suspended phase to the continuous. Oil-in-water (O/W) emulsions are *Direct*, while water-in-oil (W/O) are commonly described as *Invert*. If the concentration of dispersed droplets is sufficiently low, the fluid properties are still classified as Newtonian and many body interactions between droplets can be largely ignored. As with many colloidal suspensions, if the particles are sufficiently small, $a \leq 5$ μm where a is the particle radius, the interactions are infrequent and dominated by Brownian motion and electrostatics. However, as the volume fraction is raised, the system will undergo a rheological (mechanical) transition to a material that rapidly develops a nonzero elasticity. All of this is however predicated on the stability of the particles to flocculation, aggregation, aging, and coarsening, all of which will be described in detail below. For example, stabilized emulsions may still vary in time if the continuous and dispersed phases are near their lower/upper critical solution temperature. This will provide a mechanism for transport of the dispersed fluid through the continuous phase known as Ostwald ripening [2]. Surface instability and partial miscibility can be mitigated with the use of **surf**ace-**act**ive ag**ents** or surfactants. The underlying physics of surface (in)stability originates in the elasticity of surfaces on molecular length scales arising from Marangoni effects [3, 4]. Moreover, surfactants can provide an effective steric barrier to coalescence while also contributing to a thermodynamically relevant energy barrier for the reduction of coarsening. However, in many cases, the timescale for coarsening is determined by the solubility of the dispersed phase in the continuous phase.

15.2 PROCESSING AND PURIFICATION

15.2.1 Creation and Stability

Emulsions that are simply made of two liquid components will eventually phase separate. If the emulsion is going to persist for any significant amount of time, an emulsifier must be added to the system. Surfactants are surface-active compounds, which promote the formation of emulsions. There are a variety of natural and synthetic materials that promote the long-term stability of emulsions [5, 6]; for example, powdered silica, minerals, lattices, soaps, detergents, protein (denatured or globular), and block copolymers. Fundamentally, a good surfactant has a finite solubility (or wettability in the case of solid particles) in both liquids making up the emulsion. It is this ability to dissolve in both liquids that facilitates emulsification. The phase in which the surfactant is most soluble is usually the continuous phase, known as the Bancroft rule [7].

Because emulsions are not in thermodynamic equilibrium, their creation requires the addition of energy, usually through shaking, stirring, vibration, or some other mechanical processing. As the mechanical work generates shear forces on the liquid–liquid interface that exceed the cohesive forces of the dispersing liquid, drops form. The size of the droplets is a complex function of the local shear forces in the system and surface tension between the fluids, neither of which are uniformly distributed since emulsions usually form under turbulent flows. The amount of reversible work W, "permanently" brought into the emulsion system, is stored as the surface area of the newly formed drops. This can be quantified in terms of the surface tension σ between the two fluid phases and the surface area of the newly formed droplets $A \rightarrow dW = \sigma\, dA$.

In an emulsion, surfactants play two important roles. Their first role is to lower surface tension σ, between the two fluids, reducing the work necessary to create droplets. The second role is to coat the outside of the droplets. Once the droplets have formed, the surfactant acts as a physical barrier between the different phases of the system. The surfactant isolates individual droplets from the continuous phase and from each other. For this barrier to be effective, a minimal amount of surfactant, enough to coat every particle in the systems, needs to be on available.

If there is not enough surfactant, the system is described as "emulsifier-poor." If an emulsion forms under these conditions, the mean drop size is inversely proportional to emulsifier concentration [8]. Without enough surfactant to stabilize the droplets, they will readily coalesce until the surface area of each of the remaining particles are fully coated. In emulsifier-rich systems, the size of the emulsion drops is set by the capillary pressure and viscous dissipation associated the forming drops [9]. In batch emulsification, droplet size increases with the viscosity of the dispersed phase and interfacial tension; decreases with rate of agitation; and is not greatly affected by the viscosity of the continuous phase and the volume fraction of the dispersed phase [10–13].

No matter how effective the surfactant is at stabilizing the emulsion, given enough time, the properties of an emulsion will evolve. Depending on the final use of the end product, instabilities can be mitigated long enough to result in an effective stable system. The creaming of fresh milk is unacceptable for many applications, even if the emulsion can be simply recovered by simple agitation. Alternatively, many emulsion applications depend on the instability of the system. Asphalt emulsions cure when exposed to weather because the material that keeps the separated components separate begins to breaks down.

Emulsions can also be broken through chemical, physical, and/or microbiological mechanisms. Chemical changes can include anything from alterations of pH that influence the relative balance of the surfactant between each phase to oxidation reactions that chemically alter the surfactant. Microbial stresses are significant because emulsions generally contain a significant amount of water. Microbial populations including bacteria and fungi can degrade the components of an emulsion, particularly the surfactant; this is especially relevant for emulsions found in foods, cosmetics, and pharmaceuticals.

Physical stresses include instabilities that directly result from the fact that emulsions are generally at the colloidal scale, and have a very high surface to volume ratio. This leads to a variety of potential ways that an emulsion can be destabilized, including aggregation, ripening, phase inversion, and phase separation. Phase separation includes the complete partitioning of the two fluids back to their bulk phases. Another type of phase separation is known as creaming where the emulsion droplets remain intact, but the system separates into two emulsions with drastically different volume fractions. Aggregation (flocculation) is the attraction of distinct particles into clusters that do not coalesce. Ripening is the growth of larger droplets at the expense of smaller ones. Phase inversion occurs when the dispersed phase becomes the continuous phase and the continuous phase becomes the dispersed. Often, as one instability alters the state of the emulsion, other instabilities can begin to dominate. For example, creaming generates stresses on the droplets that can bring particles much closer together, leading to coalescence and eventual phase separation.

Emulsions cream due to density differences between the continuous phase and dispersed phases, provided that the particles are large enough; if emulsion droplets are small (homogenized), thermal fluctuations dominate and creaming is mitigated. Brownian forces prevent creaming of smaller droplets: when $4/3\pi a^3 \Delta\rho g H \leq k_B T$, where a is the droplet radius, $\Delta\rho$ is the mass density difference between the continuous and dispersed phases, H is the droplet height, and k_B is Boltzmann's constant. Homogenization breaks larger droplets into small enough droplets to prevent the formation of a cream layer. Sedimentation occurs if the dispersed phase is denser than the continuous phase, for instance in an O/W emulsion made with PTFE.

The rate at which emulsions cream is a function of particle size, density differences, and viscosity of the continuous phase: the stationary velocity of particles under laminar conditions is $v = 4a^2 g \Delta\rho / 18\eta$. If the viscosity of the continuous phase is high enough, emulsions will cream so slowly that the effect is insignificant on all relevant timescales. Centrifugation speeds up the creaming process and is very commonly used to concentrate emulsions.

Depletion forces also can encourage an emulsion to phase separate. Depletion forces are created by adding free nonadsorbing polymer to an emulsion or through an increase in surfactant above the critical micelle concentration [14, 15]. The concentration of depletents determine the osmotic pressure difference between the droplets, thus setting the strength of the attraction. When the diameter of the polymer (twice its radius of gyration) is larger than approximately one-third of the colloid hard-core diameter and the osmotic pressure is great enough, the emulsion droplets will flocculate [16–18]. If there is no density difference between the two fluids in the emulsion, creaming will not occur, but the system will still phase separate into a droplet-rich phase and a droplet-poor phase. Because depletion forces are dependent on the relative sizes of the components of the system, creaming can be exploited to reduce the polydispersity of an emulsion [19].

Creaming is easily reversed by agitation, whereas phase separation caused by depletion can be reversed by removing the added polymer or excess surfactant. Coalescence involves the combination of individual droplets and leads to complete phase separation of the two liquids. This is a very complex process, involving the approach of two or more droplets with each other, the draining of the thin film of continuous phase between the droplets, the distortion of the droplet shape, and finally the rupture of the droplet. For example, the dynamics of droplet coalescence is an extremely rich field of study by itself.

15.2.2 Destabilization and Aggregation

Ultimately, coalescence (commonly referred to as breaking) requires droplet deformations on a scale large enough to overcome the elasticity of the droplet interface. The ability of the droplet interface to recover from deformation is intimately connected to its Laplace pressure of the interface. For a droplet with a thin interface, the Laplace pressure is $\Delta P = 2\sigma/a$.

Stresses provide a route to the destabilization of droplet interfaces. Surface tension gradients lead to pressure gradients that can exceed the magnitude of the Laplace pressure on a length scale of the droplet size, when $\Delta P > 2\sigma/a^2$ [20, 21]. The dimensionless number that balances viscous forces and the surface tension is the capillary number that is used to characterize whether an emulsion is susceptible coalescence at a shear rate of $\dot{\gamma}$, $Ca = a\dot{\gamma}\eta/2\sigma$. Taylor predicted that the critical capillary number for droplet breakup is $Ca = 0.5$ when the viscosities of two component fluids are roughly equal and Newtonian. This prediction has been experimentally verified for the breakup of single droplets [22–24], although if the dispersed phase viscosity exceeds four times the continuous phase, no breakup occurs at all.

Much of the work in this area is restricted to particular situations, especially with respect to the rheological properties of the overall system or any of its constituents. Predictions are immensely more difficult in systems in which either of the fluids is non-Newtonian, where the surfactant affects the rheology of continuous phase, or in dense emulsions, where the volume fraction causes permanent facets in the faces of the droplets. To further complicate matters, in some cases the critical capillary number is influenced by the history of the flow [25]. However, there are important conclusions that can be made from this kind of analysis. Coalescence occurs more readily when the continuous phase viscosity is high, or when deformations exceed the ability of the surfactant to hold the interface together, or when the system is compressed (droplet facets inevitably cause large surface tension gradients). When droplets associate with each other for an extended period of time, they will flocculate or aggregate. Although a distinct instability, flocculation often precedes rupture during coalescence. Aggregation depends on the nature of the particle–particle interactions. Most generally, if a pair of particles can overcome any electrostatic repulsion they may have for each other, van der Waals' forces will hold them together (DLVO theory) [26, 27]. van der Waals' interactions are attractive forces between two permanent dipoles (Keesom forces), those between a permanent dipole and an induced dipole (Debye forces), and forces between two instantaneously induced dipoles (London dispersion force). Because London forces always exist between different molecules, there is always an attractive element to the net potential. If the droplets overcome the repulsive forces due to electrostatic screening, attractive forces can dominate. If the repulsion dominates the potential during a collision, the emulsion will not flocculate or coalesce. Once particles have come close enough that London dispersion forces dominate the interparticle interaction, the flocculation is essentially permanent. The interplay between van der Waals forces and electrostatics oversimplifies the enormous number of colloidal interactions that can come into play when emulsions flocculate. However, the success of DLVO theory to explain the instabilities in colloidal systems under high salt concentrations is profound. Furthermore, DLVO theory continues to be used to explain a variety of phenomena including those involving such disparate systems such as dispersion of nanoparticles and cells as well as more traditional colloids. Based on DLVO, the best way to stabilize an emulsion to flocculation is to use added charge to the surface of the particle. Often, this can be facilitated by using a charged surfactant. To encourage flocculation, electrostatic repulsion can be reduced by using a conductive continuous phase. Although not formal part of the DLVO theory, the ideas behind the theory can be used to explain how things such as steric repulsions can hinder flocculation. A thick layer of polymer can adsorb to the droplet surface effectively preventing flocculation. The polymer creates a steric barrier that prevents the droplets from getting close enough for the attractive London forces to exert much influence.

15.2.3 Coarsening

Unlike flocculation, instabilities caused by Ostwald ripening do not require direct particle–particle interactions. Instead, Ostwald ripening is the growth of larger particles at the expense of smaller ones through mass diffusion. Finite amounts of the dispersed phase are always dynamically exchanged between the emulsion droplets and the continuous phase. When there are particles of different sizes in the system, smaller particles experience a net loss of mass while larger ones experience a net gain. Larger particles are more energetically favored than smaller ones [2], as illustrated in the relationship between Laplace pressure and particle size. Furthermore, the energy cost of removing one molecule from a small particle is exceeded by the energy gain of adding a molecule to a large particle. During coarsening, the total number of particles does change. However, the average size of droplets can grow with a square root dependence, implying a self-similar scaling [28]. Eventually, the largest particles will become big enough that they are more susceptible to rupture. Since no two fluids are absolutely immiscible or perfectly monodisperse, there is always some degree of ripening in an emulsion. However, the process can be dramatically slowed down so that the size distribution changes imperceptibly. Solid particles and surfactants seem to almost completely arrest Ostwald ripening [8]. Adding a third, less soluble, component to the dispersed phase can also arrest the ripening process [29–31].

Phase inversion involves a profound system-wide change in the emulsion. In this instability, the dispersed phase becomes continuous and the continuous phase becomes dispersed. As mentioned earlier, the Bancroft rule says that the surfactant should be more soluble in the continuous phase. Under certain conditions, the preferential solubility of the surfactant can change from one liquid to the other. The easiest way for this to happen is to heat the system. At the phase inversion temperature (PIT), the preference for one phase or the other is equal. When this happens, the emulsion can invert. To avoid this, emulsions should be kept well away, 10–20$^c irc$C, from their surfactant's PIT. If an emulsion inverted because of heating, it should return to its previous state cooling it. However, the original size and size distribution are rarely recaptured after re-inverting an emulsion. As usually with emulsions, this redistribution of sizes depends on the nature of each of the components of the system and can be exploited when creating an emulsion with specific properties [32]. In addition to changing the preferential solubility of the surfactant by heating the system, there are other mechanisms leading to inversion. For instance, pH, ionic strength, or polarity of the continuous phase can be altered to affect the relative solubility of the surfactant.

15.2.4 Purification: Creaming and Depletion

The fact that emulsions are easy to generate partially explains why they are so widespread. In the kitchen, mayonnaise can be made by hand with a whisk. In the lab, a high shear rate mixer can easily produce submicron droplets. A potential issue is that many applications require a high degree of monodispersity, that is, a narrow distribution of droplet sizes. This is true in industry where the size might control the delivery rate of a drug or the texture of a cream. For basic physics studies, a polydisperse emulsion may complicate the interpretation of results due to a broad range of Laplace pressures and complex packing structure.

Several techniques have been developed to generate monodisperse emulsions directly. These range from microfluidic devices, which can make multiple emulsions with extreme precision, to membrane cross-flow, which is used in industrial applications when large volumes are required. Nonetheless, there are instances when a polydisperse emulsion needs to be fractionated by size. In this section, an iterative purification technique is described along with two interactions important to emulsion science.

The technique takes advantage of the mass density difference between the continuous and disperse phases of a typical oil-in-water emulsion, where the oil might be 5% lighter [33]. Due to this mismatch, an oil droplet experiences a buoyant force $F_b = \rho_w g V$, where ρ_w is the mass density of water, V is the droplet volume, and g is the acceleration due to gravity. As this causes the droplet to rise toward the top of the solution, motion commonly referred to as creaming, it also experiences a viscous drag. At low Reynolds number, this force is given by Stokes' law as $F_d = 3\pi\mu v d$, where μ is the viscosity of water, v is the droplet's velocity, and d is its diameter. Balancing the buoyant, drag, and gravitational forces yields a terminal velocity

$$v = \frac{(\rho_w - \rho_o) g d^2}{18\mu}, \tag{15.1}$$

where ρ_o is the oil mass density [34, 35].

To take advantage of the strong dependence of this velocity on d, the initial step of the purification procedure is to dilute the emulsion to a low oil volume fraction on the order of 10%. After some time, a concentrated layer forms at the top of the emulsion, which will preferentially consist of the largest oil droplets. Next, the cream and the dilute emulsion below are separated. After the cream is diluted back to 10%, both fractions are allowed to cream again. This process of creaming, separating, and diluting is repeated many times to yield a series of creams, each with a different mean droplet diameter and a monodispersity that improves with every iteration. In practice, the procedure is most convenient for droplets that cream in roughly a day, which for typical oil-in-water emulsions occurs for $d \approx 5$ μm.

The technique can be applied to much smaller droplets by making a simple modification, with the added benefit of better size discrimination. The idea is to force small droplets, which would cream very slowly on their own, to aggregate, or flocculate, into clusters that cream much more rapidly.

This is accomplished by increasing the surfactant concentration in the continuous phase above the critical micelle concentration, where the surfactant molecules organize into small structures in order to minimize their free energy. For example, a solution of the common surfactant sodium dodecyl sulfate forms 4 nm spheres, with the anionic hydrophilic head groups on the surface, above a concentration of 8 mM in pure water. These micelles and the emulsion droplets will both be regarded as hard spheres for this discussion.

Figure 15.1a shows a single emulsion droplet surrounded by micelles; the shaded region indicates the volume from which the centers of the micelles are excluded. The micelles collide with the droplet uniformly and thus exert an isotropic osmotic pressure. Under the assumption that the micelles do not interact, this pressure is given by the ideal gas law as $P_{\text{osm}} = n_m k_B T$, where n_m is the number density of micelles, k_B is Boltzmann's constant, and T is the temperature.

In Figure 15.1b, two droplets have come close enough for their excluded volumes to overlap. This results in an attraction, known as the depletion force, between the droplets. It can be viewed as entropic in nature, as the total volume excluded to the micelles decreases with increasing overlap.

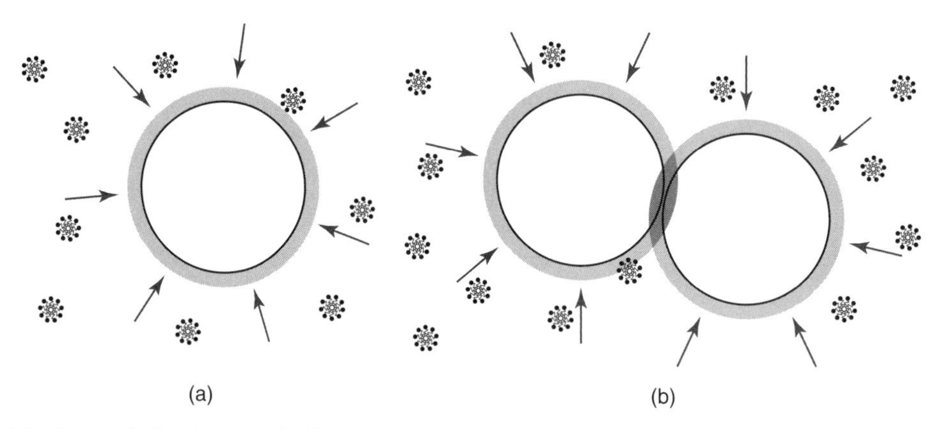

Figure 15.1 A background of surfactant micelles creates short-range attraction between oil droplets through the depletion force.

The magnitude of the force depends on the imbalance of the osmotic pressure due to the overlap region and thus can be estimated geometrically. Once the droplets are in contact, the corresponding potential energy is

$$U = -\frac{3}{2}k_B T \phi_m \frac{d}{d_m}, \quad (15.2)$$

where ϕ_m and d_m are the volume fraction and diameter of the micelles [36]. While the range of the attraction is roughly fixed at d_m, the strength depends on d and can be tuned with ϕ_m.

For the purification process, the surfactant concentration sets the minimum droplet size that will experience a large enough depletion force to stay flocculated. So for a low value of ϕ_m, only the largest droplets will form the cream layer. On the subsequent iteration, ϕ_m can be increased in the dilute fraction to force smaller droplets to cream. Continuing in this way yields droplets monodisperse enough to crystallize in as few as five steps [33].

15.3 EMULSION SCIENCE

15.3.1 Microfluidics: Emulsions on a Chip

Although not central to our scientific discussion, a brief introduction to microfluidic-based emulsion production, with a few interesting applications, is certainly warranted. Microfluidics have recently become the subject of tremendous interest for their potential to produce new materials based on emulsions, while simultaneously revolutionizing the field of microbiology and genetics [37]. Microfluidic devices, produced using soft lithography methods [38–41] or with pulled microcapillaries [42], are being utilized to create droplet-on-demand, *designer* emulsions for microencapsulation of materials. In this environment, each drop is largely independent from their surroundings and from other drops. Moreover, by controlling the relative ratios of the dispersed phase constituents, exquisite control of each droplet's contents is attainable on pico- to nanoliter volumes.

Microfluidic emulsion generators rely on hydrodynamic flow focusing as the physical mechanism for creating droplets. The production of emulsions in microfluidics occurs typically in the regime of low Reynolds number flow, $Re \sim \mathcal{O}(10^{-6})$–$\mathcal{O}(10^{1})$, indicating that viscous stresses dominate over inertia. Moreover, the physics of fluids at small lengths provides a set of physical constraints on the stresses within the system. From a comparison of the stress scales in a focusing geometry, one can directly determine the size R of the droplets produced. Balancing the capillary stresses on a drop, σ/R, and the viscous stresses of the continuous phase, $\eta U/h$, where η is the viscosity of the continuous phase, h is the size of the channel, and U is the fluid velocity, one can immediately determine that $R \sim h/Ca$. This simple analysis indicates that the capillary number Ca and the characteristic length scale of the system h uniquely determine the size of each droplet.

Microfluidics provide a very robust platform to produce drops-on-demand with a high degree of potential functionality. Automated control of reagent content and ratios is a valuable tool for the production of libraries of proteins and enzymes. In fact, microfluidic emulsion manipulation technology has produced a new paradigm for high-throughput analysis of protein crystallization that may lead to breakthroughs in drug discovery [43–45].

15.3.2 Dense Emulsions and Jamming

A disordered suspension of hard-sphere colloidal particles can display a dramatic range of mechanical properties, where the jamming transition separates liquid-like behavior and solid-like behavior [46, 47]. Assuming an athermal system, this transition is controlled by the volume fraction ϕ of the particles and the magnitude of an applied shear. In the absence of shear, neighbors are forced into contact once a critical volume fraction ϕ_c has been reached, which requires that the system be confined by a positive pressure. This results in near structural arrest and the sudden appearance of a bulk modulus. The system will unjam, and flow like a fluid, if a sufficiently large shear is applied.

Consisting of only liquids, a dense collection of emulsion drops displays remarkably similar behavior, where the structural integrity arises from the surface tension of the interfaces. To see how jamming is desirable for many applications, one needs to look no further than mayonnaise. This compressed emulsion is stiff enough to be scooped from a jar, yet flows easily when spread with a knife.

The constituent particles of both emulsions and hard-sphere colloids interact through repulsive contact potentials roughly of the form $V \sim \delta^{\alpha}$, where δ is the distance a sphere is compressed along its radius. While colloids can be reasonably thought of as undeformable hard spheres, some elasticity can be incorporated with $\alpha = 2.5$. This Hertzian model is equivalent to a nonlinear spring that becomes increasingly stiff with compression. For an emulsion droplet, the exponent depends on the degree of deformation and the arrangement of neighbors [48, 49], with typical situations described well for α between the harmonic and Hertzian values of 2 and 2.5. More than the functional form of the potential, its softness is what distinguishes emulsions. For instance, micron-sized droplets of a typical oil-in-water emulsion will deform when they simply come into contact after creaming under gravity. Higher pressures, obtained through centrifugation for example, result in significant deformation and faceting, with ϕ potentially reaching values of 0.9 or higher. Another significant difference between the two systems is that emulsion droplets are essentially frictionless, which removes the history dependence that is common for granular packings. Interestingly, aqueous foams do exhibit shear-induced normal stress differences, similar to the Poynting relation seen in elastic solids [50, 51].

Studying a jammed system above ϕ_c could be quite revealing, as many unusual properties appear in this regime. In an ideal elastic material, an applied shear results in a uniform deformation, as the global strain dictates the motion in any small neighborhood. A hallmark of jamming is that shear introduces nonaffine deformation and rearrangement, as disorder results in isolated regions that can most easily accommodate droplet motion. Because of this, various properties that characterize the local packing and bulk mechanical behavior are expected to scale nontrivially with the compression $\Delta\phi \equiv \phi - \phi_c$. In fact, recent theoretical work suggests that the onset of jamming shares many features with the critical point of a second-order phase transition [52, 53]. In order to test these findings experimentally, it is essential to fine-tune ϕ above its critical value, which is essentially impossible for hard-sphere colloids. Nonetheless, the complex deformations that accompany large $\Delta\phi$ in soft particles may introduce their own complications.

Emulsions are well suited to study with many of the tools of soft matter science, such as rheology, light scattering, and confocal microscopy. For example, it is not difficult to generate monodisperse droplets in the laboratory, using a host of different liquids and surfactants. The continuous and disperse phases can be density-matched to minimize the effects of gravity, refractive index-matched to yield a transparent sample, and fluorescently dyed for imaging. While there is an extensive literature on the jamming of soft particles, the remainder of this section will highlight some of the experimental results that have been obtained with compressed emulsions in both jammed and flowing states.

15.3.3 The Jammed State

The solid-like behavior of the jammed state has long been studied with bulk rheology [54]. Figure 15.2a shows the results of an oscillatory strain sweep for an oil-in-water emulsion compressed to a volume fraction $\phi = 0.72$. The shear stress τ increases linearly with the strain amplitude γ

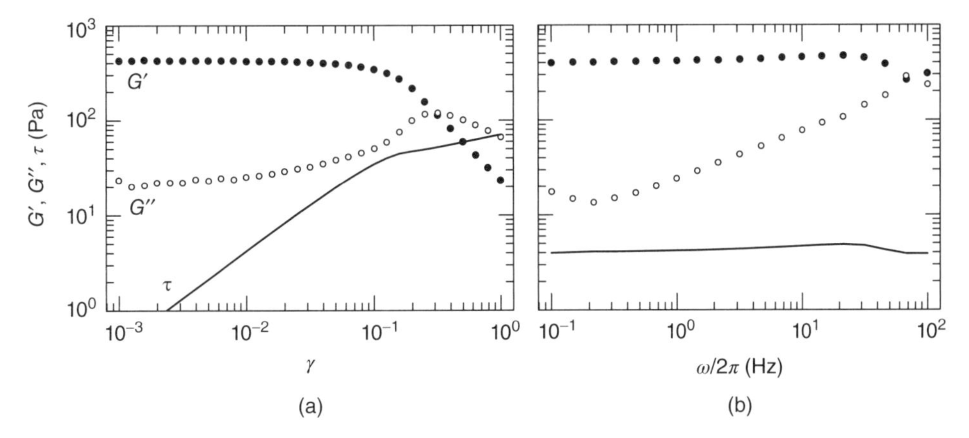

Figure 15.2 Oscillatory rheology of a compressed emulsion [55]. (a) The storage modulus G' (solid circles), loss modulus G'' (open circles), and stress τ (line) are plotted as a function of the strain amplitude γ for data taken at a frequency $\omega/2\pi = 1$ Hz. (b) The same quantities are plotted as a function of ω for $\gamma = 0.01$.

below a value of 0.1, as one would expect from an elastic medium. This characterization is confirmed by the nearly constant value of the dynamic storage modulus G' (whose magnitude is set by the Laplace pressure of the droplets), which is significantly larger than that of the loss modulus G''. In this linear regime, the energy supplied by the applied strain is reversibly stored in deformations of the emulsion interfaces.

The frequency dependence of the moduli in the linear regime is shown in Figure 15.2b. G' is nearly constant for most of the frequency range, consistent with elasticity. One might expect that G'' would increase linearly with frequency for such a medium, but it is instead fairly constant at low frequency and then increases as $\omega^{1/2}$ over the range displayed. This anomalous dissipation can be attributed to the nonaffine dynamics at work in this system [56]. Small pockets of rearranging droplets, rather than uniform deformation across the system, are also responsible for a lower-than-expected G'. Light-scattering measurements confirm that $G'' \sim \omega$ at much higher frequencies above the regime where one may expect glassy dynamics to dominate the material properties [56].

The static elastic moduli are also defining properties of a jammed system. In a system with homogeneous dynamics, the ratio between the shear and bulk moduli G/K is a constant with respect to $\Delta\phi$. The unusual relaxation present in disordered materials modifies both quantities, but the nonaffine motion especially softens the system to shear near the jamming transition. While the particular scaling of G and K depends on the form of the interaction potential, simulations predict that $G/K \sim \Delta\phi^{1/2}$ [57, 58]. Although this relation has yet to be confirmed experimentally, measurements on emulsions have shown $G' \sim \phi\Delta\phi$, with a similar scaling for the osmotic pressure Π [59]. As $K = \phi\, d\Pi/d\phi$, G/K should increase quickly for small $\Delta\phi$.

For $\gamma > 0.2$ in Figure 15.2a, the stress increases sublinearly with strain, a signature of shear thinning, although it still follows a power law relationship. In this range, G' declines rapidly, falling below G'', indicating a loss of elasticity. In this nonlinear regime, the applied strain provides enough energy to create large-scale flow and rearrangement of droplets, rather than additional deformation; this viscous motion accounts for the large values of the loss modulus.

The system is said to flow as it crosses over from linear to nonlinear behavior. A rheologically defined yield stress and strain can be identified as the intersection of power law fits in the two regimes. The results of this sort of analysis for a wide range of compressions are plotted in Figure 15.3 [60]. As indicated by the different symbols, measurements were performed on four monodisperse emulsions, where the average droplet diameter varied by a factor of 3.

The yield strain γ_y has a minimum of just a few percent at a critical volume fraction ϕ_c. Below this, γ_y decreases with increasing ϕ due to a smaller free volume seen by the droplets. A similar effect is seen with hard spheres, but emulsions allow for measurements at higher compression, where another mechanism becomes the dominant yield process. Here, γ_y increases again, as fewer slip planes are available as the droplet deformations become stronger. Another interesting feature of the data is that γ_y is independent of the droplet size, as only the geometry of the packing dictates the onset of flow.

As shown in Figure 15.3b, the yield stress τ_y increases with packing as $\Delta\phi^2$. This further supports the picture of fewer slip planes being available for yield with increasing ϕ. In this case, agreement in the curves for the different-sized emulsions occurs only if the stresses are scaled by the Laplace pressure, emphasizing the importance of this quantity.

The forms of both γ_y and τ_y suggest that the critical volume fraction ϕ_c for jamming is near the monodisperse random close-packing value of 0.63, as expected for frictionless spheres. In addition to ϕ, a packing can also be characterized by the average number of contacts a droplet makes with its neighbors, a quantity known as the coordination number z. For spheres in three dimensions,

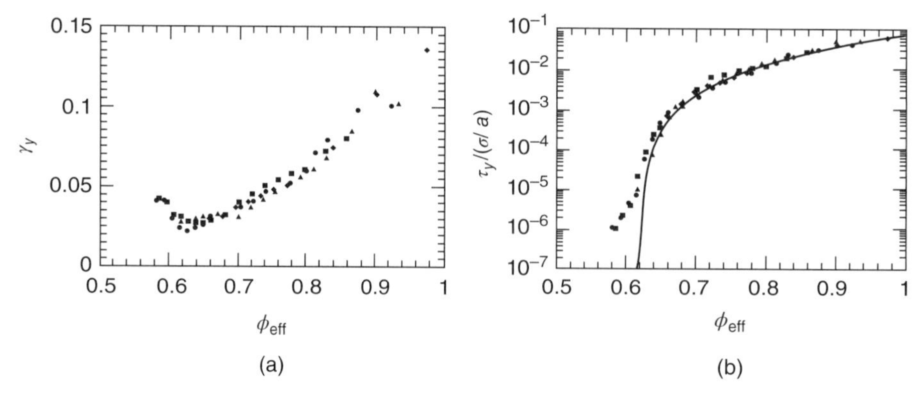

Figure 15.3 Oscillatory measurements of yield. The (a) yield strain and (b) scaled yield stress vary strongly with volume fraction, but are independent of the diameter of the droplets: 0.25 μm (circles), 0.37 μm (triangles), 0.53 μm (squares), and 0.74 μm (diamonds). Source: Mason [60], p. 439. Reproduced with permission of Elsevier.

force balance dictates a minimum coordination number $z_c = 6$ for mechanical stability. This so-called isostatic limit is also thought to mark the transition to jamming, as z jumps discontinuously from 0 to z_c at ϕ_c, consistent with a first-order phase transition. For a uniform system, z would grow linearly with increased compression, but disorder gives rise to the scaling $z - z_c \sim \Delta\phi^{1/2}$ [46, 52].

Isostaticity and other local particle-level properties can be investigated with a confocal microscope, which makes it possible to image throughout the bulk of a compressed emulsion. With a careful choice of fluorescent dyes, the facets that form when droplets come into contact can be directly visualized [61, 62]. Studies of lightly compressed emulsions confirm a coordination number near 6. While similar measurements have not been performed at higher compression in emulsions, the predicted scaling of z with $\Delta\phi$ has been observed in foams.

Extending this type of analysis, recent experimental results on a polydisperse emulsion guided the development of a statistical model that examines packing from the point of view of an individual particle [63]. A key feature of the model is that it distinguishes between the collection of droplets that are neighbors with a central droplet and the subset of those that are actually in contact. Being able to directly image contact patches was essential to confirming that the ratio of the contacts to total neighbors is independent of droplet radius. Another parameter in the model specifies the maximum space around a droplet that is available for placing neighbors, which is also experimentally accessible. The ability of the model to predict the probability distributions for the number of neighbors and contacts, and the local packing fraction, suggests that it may offer some insight into how polydispersity determines the global packing fraction and, ultimately, mechanical properties.

In the jammed state, a colloidal system can support an applied force. Due to the disordered arrangement of the constituent particles, the force is distributed inhomogeneously throughout the material. The form of the force probability distribution is characterized by several features: a limiting value for small forces, a peak near the mean force, and a tail that vanishes exponentially or faster. These contact forces have been directly visualized at the boundary surface of 3D granular systems and in the bulk of 2D photoelastic disks, with distributions measured for static and flowing configurations.

With emulsions, it is possible to identify not just contacts but also to quantify the corresponding magnitude and orientation of the interparticle forces. Rather than using the force model described earlier, which requires precise knowledge of the compression δ, another approach is better suited to the analysis of confocal images. When two droplets are brought into contact, they deform, creating a roughly circular contact patch. A patch of area A is created by a normal force $f = (2\sigma/\tilde{d})A$ between the droplets, where $\tilde{d}$ is the geometric mean of their diameters.

As shown in Figure 15.4, force measurements were performed on two emulsion systems by directly analyzing contact patches [61, 62]. The distributions show a peak near the mean force and an exponential tail, as many simulations have predicted. It remains to be seen how these features change as the systems leave the jammed state.

Direct force measurements also provide information about the spatial distribution of forces. This is of great interest, as forces in a jammed system tend to be roughly localized along one-dimensional paths, known as stress chains, which may be closely tied to the rich nonlinear behavior of these materials. For the emulsion studied in Ref. [62], correlations in the force orientation pointed to

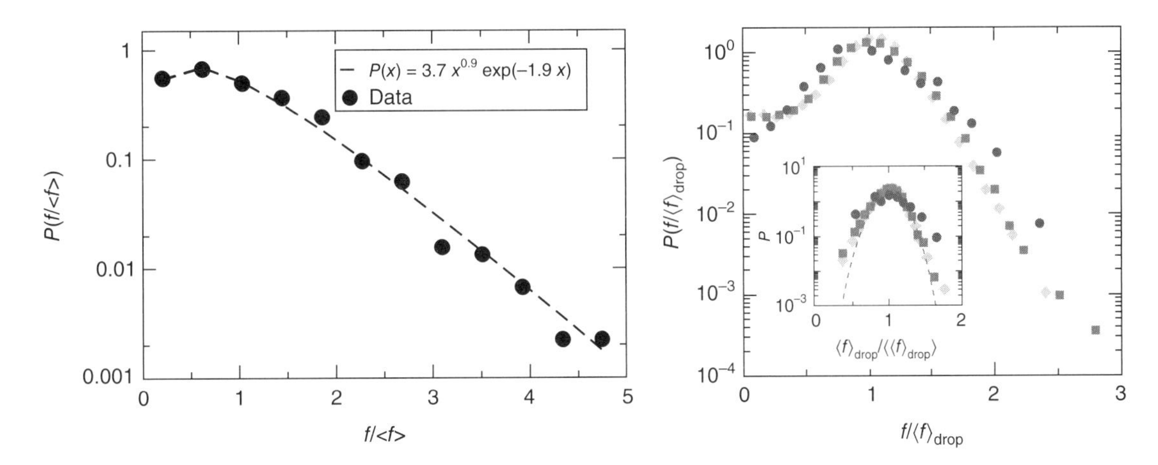

Figure 15.4 Experimentally measured interparticle force distributions in the bulk of compressed emulsions. (a) Data (circles) are well described by a model (dashed line) that has a peak and exponential tail. Source: Brujić *et al.* [61], p. 207. Reproduced with permission of the Royal Society of Chemistry. (b) Data for three emulsions, where each force on a droplet is normalized by the mean force on that droplet, show a well-defined peak. Source: Zhou *et al.* [62], p. 1631. Reproduced with permission of AAAS.

a chain persistence length of 10 droplet diameters. This value dropped to seven when a stress was applied to the emulsion, possibly reflecting a lowering of a correlation length as the system moved further from the jamming threshold, reminiscent of a second-order phase transition. Friction also affects the length of the stress chains, as the presence of tangential forces allow for abrupt changes in the force transmission. Again, direct measurement has revealed that near-perfect force balance is achieved when only repulsive normal forces are taken into account, confirming the minimal role of friction.

15.3.4 The Flowing State

While the oscillatory measurements in Figure 15.2 probed the transition out of the jammed state, a continuous shear provides additional insight into the nature of the flowing state. The flow curves in Figure 15.5a show the shear stress τ required to produce steady shear rates $\dot{\gamma}$ spanning seven orders of magnitude. This measurement is limited to stresses less than the Laplace pressure, which for these 5 μm droplets is roughly 5000 Pa, above which the droplets are susceptible to shear rupture. In the plots, the yield stress is easily identified as the value required to produce even the smallest strain rates. At the largest values of $\dot{\gamma}$, τ increases sublinearly, consistent with the shear thinning seen in the strain sweep of Figure 15.2a. The solid lines show that the data are well described by the empirical constitutive relation known as the Herschel–Bulkley model; $\tau = \tau_y + k\dot{\gamma}^n$, where k is known as the consistency and n is a power law exponent. For emulsions, n is roughly 0.5 for $\phi > \phi_c$. Using a micromechanical constitutive model, based on the elastohydrodynamics of soft spheres, in conjunction with numerical simulations and experiments, Seth *et al.* provide a simple framework that capture the nonlinear rheological behavior of flowing soft glasses [64].

In recent years, confocal microscopy has revealed that the dynamics of flowing emulsions are more complicated than that the rheology might suggest. While tracking the motion of individual droplets is possible, finding average velocities of layers of droplets using standard particle imaging velocimetry provides many interesting insights. For instance, Figure 15.5b shows the average velocity in the strain direction as a function of the location between the parallel plates imposing the shear. The flow profiles vary substantially with the strain rate, just as the rheology does. At high strain rates, the profile is nearly linear, which is the expected situation for a geometry where the stress is constant. At lower strain rates where the stress is near its yield value, however, there is a large gradient in the strain near the bottom plate. This sort of shear localization is commonly seen in a variety of structured fluids [65–67]. Our results indicate that shear bands do exist for purely repulsive systems, whereas attractive interactions were indicated as necessary for strain localization [65, 67]. At the very least, it is clear that the nominal strain applied by the shearing plates does not accurately describe the behavior throughout the emulsion.

The unexpected flow profiles raise the question of whether the Herschel–Bulkley model is capable of providing a full description of the flow behavior. This issue was investigated with pressure-driven flows in narrow microchannels [68]. It was found that there was no unique relationship between the theoretically calculated local stress and the experimentally extracted local strain rate that held for all values of the channel pressure. However, the flow profiles could be reproduced by introducing both a fluidity that defines the local rate of rearrangement and a cooperativity length that provides the scale on which the fluidity can vary. The cooperativity length was found to increase, perhaps linearly, with $\Delta\phi$. In the language of second-order phase transitions, it is expected that a correlation length characterizing the

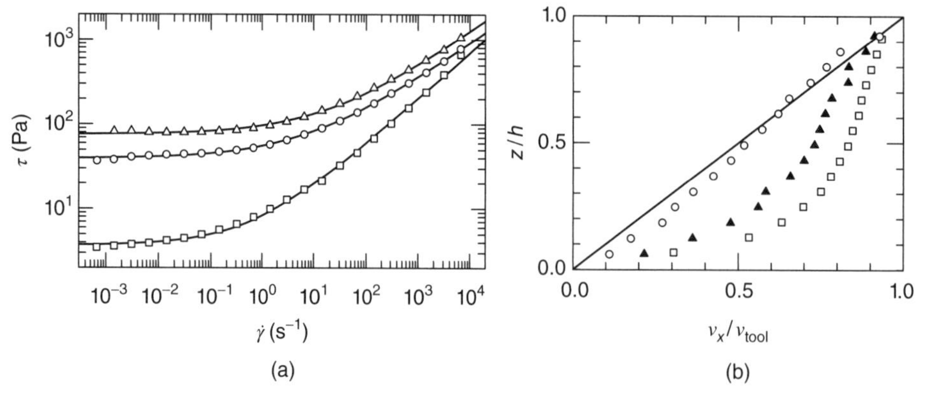

Figure 15.5 Bulk flow curves and local velocity profiles of a yielded emulsion [55]. (a) The measured stress versus strain rate curves (symbols) closely follow a Herschel–Bulkley form (solid lines). For emulsions with $\phi = 0.64$ (squares), 0.76 (circles), and 0.85 (triangles), the power law exponents are 0.54, 0.43, and 0.44, respectively. (b) For $\phi = 0.72$, the normalized flow profiles for strain rates of 0.004 s^{-1} (squares), 0.02 s^{-1} (triangles), and 200 s^{-1} circles) show increasingly uniform flow.

size of fluctuations diverges at a critical point; the length scale found in these experiments is distinct and gives the size of the region over which a rearrangement event is felt. The heterogeneous dynamics that the nonlocal flow rule describes are what lead to the anomalous dissipation in Figure 15.2.

While much has been learned about these seemingly simple systems, there remain a great number of open questions on the nature of the flowing state and how it relates to the jamming transition. Further experiments, ranging from measurements of the bulk viscosity to direct characterization of nonaffine motion at the droplet scale, are needed to develop a deeper understanding.

15.4 CONCLUSIONS

Controlling the production and physical properties of liquid colloids is a historically important and broad subject that is in and of itself the topic of a number of volumes. We hope that this brief introduction provides some context for scope of the required physics and physical chemistry that is required to produce, purify, and utilize emulsions.

REFERENCES

[1] Pusey PN, van Megen W. Phase behaviour of concentrated suspensions of nearly hard colloidal spheres. Nature 1986;320:340.

[2] Ratke L, Voorhees PW. *Growth and Coarsening: Ostwald Ripening in Material Processing*, Engineering Materials. Berlin Heidelberg and New York: Springer-Verlag; 2002, ISBN: 3-540-4563-2.

[3] Valkovska DS, Danov KD, Ivanov IB. Surfactants role on the deformation of colliding small bubbles. Colloids Surf, A 1999;156:547–566.

[4] Danov KD, Valkovska DS, Ivanov IB. Effect of surfactants on the film drainage. J Colloid Interface Sci 1999;211:291–303.

[5] Ramsden W. Separation of solids in the surface-layers of solutions and 'suspensions' (observations on surface-membranes, bubbles, emulsions, and mechanical coagulation).—Preliminary account. Proc R Soc Lond 1903;72:156–164.

[6] Pickering SU. Cxcvi.-Emulsions. J Chem Soc, Trans 1907;91:2001–2021.

[7] Bancroft WD. The theory of emulsification, V. J Phys Chem 1912;17:501–519.

[8] Tcholakova S, Denkov ND, Lips A. Comparison of solid particles, globular proteins and surfactants as emulsifiers. Phys Chem Chem Phys 2008;10:1608.

[9] Davies J. Drop sizes of emulsions related to turbulent energy dissipation rates. Chem Eng Sci 1985;40:839–842.

[10] Binks BP, Whitby CP. Silica particle-stabilized emulsions of silicone oil and water: aspects of emulsification. Langmuir 2004;20:1130–1137.

[11] Golemanov KK, Tcholakova SS, Kralchevsky PAP, Ananthapadmanabhan KPK, Lips AA. Latex-particle-stabilized emulsions of anti-Bancroft type. Langmuir 2006;22:4968–4977.

[12] Tcholakova S, Denkov ND, Sidzhakova D, Ivanov IB, Campbell B. Effects of electrolyte concentration and pH on the coalescence stability of beta-lactoglobulin emulsions: experiment and interpretation. Langmuir 2005;21:4842–4855.

[13] Vankova NN, Tcholakova SS, Denkov NDN, Vulchev VDV, Danner TT. Emulsification in turbulent flow. J Colloid Interface Sci 2007;313:612–629.

[14] Bibette J, Roux D, Pouligny B. Creaming of emulsions: the role of depletion forces induced by surfactant. J Phys II France 1992;2:401–424.

[15] Bibette J, Roux D, Nallet F. Depletion interactions and fluid-solid equilibrium in emulsions. Phys Rev Lett 1990;65: 2470–2473.

[16] Asakura S, Oosawa F. On interaction between two bodies immersed in a solution of macromolecules. J Chem Phys 1954;22:1255.

[17] Vrij A. Polymers at interfaces and the interactions in colloidal dispersions. Pure Appl Chem 1976;48:471–483.

[18] Lekkerkerker HNW, Poon WC-K, Pusey PN, Stroobants A, Warren PB. Phase behaviour of colloid + polymer mixtures. Europhys Lett 1992;20:559.

[19] Bibette J. Depletion interactions and fractionated crystallization for polydisperse emulsion purification. J Colloid Interface Sci 1991;147:474–478.

[20] Taylor GI. The viscosity of a fluid containing small drops of another fluid. Proc R Soc Lond Ser A 1932;138:41–48.

[21] Taylor GI. The formation of emulsions in definable fields of flow. Proc R Soc Lond Ser A 1934;146:501–523.

[22] Rumscheidt F, Mason S. Particle motions in sheared suspensions XI. Internal circulation in fluid droplets (experimental). J Colloid Sci 1961;16:210–237.

[23] Bentley BJ, Leal LG. A computer-controlled four-roll mill for investigations of particle and drop dynamics in two-dimensional linear shear flows. J Fluid Mech 1986;167: 219–240.

[24] Guido S, Villone M. Measurement of interfacial tension by drop retraction analysis. J Colloid Interface Sci 1999;209:247–250.

[25] Stone HA. A simple derivation of the time-dependent convective-diffusion equation for surfactant transport along a deforming interface. Phys Fluids A Fluid Dyn 1990;2: 111–112.

[26] Derjaguin BV, Landau L. Theory of the stability of strongly charged lyophobic sols and of the adhesion of strongly charged particles in solutions of electrolytes. Acta Phys Chim URSS 1941;14:633–662.

[27] Verwey E, Overbeek J. *Theory of the Stability of Lyophobic Colloids*. Amsterdam: Elsevier; 1948.

[28] Lambert J, Mokso R, Cantat I, Cloetens P, Glazier JA, Graner F, Delannay R. Coarsening foams robustly reach a self-similar growth regime. Phys Rev Lett 2010;104:248304.

[29] Higuchi WI, Misra J. Physical degradation of emulsions via the molecular diffusion route and the possible prevention thereof. J Pharm Sci 1962;51:459–466.

[30] Kabal'nov A, Pertzov A, Shchukin E. Ostwald ripening in two-component disperse phase systems: application to emulsion stability. Colloids Surf 1987;24:19–32.

[31] Webster AJ, Cates ME. Stabilization of emulsions by trapped species. Langmuir 1998;14:2068–2079.

[32] Shinoda K, Kuneida H. Phase properties of emulsions. In: *Encyclopedia of Emulsion Technology*. Marcel Dekker; 1983.

[33] Bibette J. Depletion interactions and fractionated crystallization for polydisperse emulsion purification. J Colloid Interface Sci 1991;147:474.

[34] Chanamai R, McClements DJ. Creaming stability of flocculated monodisperse oil-in-water emulsions. J Colloid Interface Sci 2000;225:214.

[35] Derkach SR. Rheology of emulsions. Adv Colloid Interface Sci 2009;151:1.

[36] Vrij A. Polymers at interfaces and the interactions in colloidal dispersions. Pure Appl Chem 1976;48:471.

[37] Squires T, Quake S. Microfluidics: fluid physics at the nanoliter scale. Rev Mod Phys 2005;77:977–1026.

[38] Xia Y, Whitesides GM. Soft lithography. Angew Chem Int Ed 1998;37:550–575.

[39] Thorsen T, Roberts RW, Arnold FH, Quake SR. Dynamic pattern formation in a vesicle-generating microfluidic device. Phys Rev Lett 2001;86:4163–4166.

[40] Anna SL, Bontoux N, Stone HA. Formation of dispersions using "flow focusing" in microchannels. Appl Phys Lett 2003;82:364.

[41] Link DRD, Anna SLS, Weitz DAD, Stone HAH. Geometrically mediated breakup of drops in microfluidic devices. Phys Rev Lett 2004;92:054503.

[42] Utada ASA, Lorenceau EE, Link DRD, Kaplan PDP, Stone HAH, Weitz DAD. Monodisperse double emulsions generated from a microcapillary device. Science 2005;308:537–541.

[43] Unger MA, Chou HP, Thorsen T, Scherer A, Quake SR. Monolithic microfabricated valves and pumps by multilayer soft lithography. Science 2000;288:113–116.

[44] Hansen CC, Quake SRS. Microfluidics in structural biology: smaller, faster… better. Curr Opin Struct Biol 2003;13: 538–544.

[45] Selimović S, Gobeaux F, Fraden S. Mapping and manipulating temperature-concentration phase diagrams using microfluidics. Lab Chip 2010;10:1696–1699.

[46] van Hecke M. Jamming of soft particles: geometry, mechanics, scaling and isostaticity. J Phys Condens Matter 2010;22: 033101.

[47] Liu AJ, Nagel SR. The jamming transition and the marginally jammed solid. Annu Rev Condens Matter Phys 2010;1:347.

[48] Morse DC, Witten TA. Droplet elasticity in weakly compressed emulsions. Europhys Lett 1993;22:549.

[49] Lacasse MD, Grest GS, Levine D, Mason TG, Weitz DA. Model for the elasticity of compressed emulsions. Phys Rev Lett 1996;76:3448.

[50] Labiausse V, Höhler R, Cohen-Addad S. Shear induced normal stress differences in aqueous foams. J Rheol 2007;51:479.

[51] JSTOR: Proceedings of the Royal Society of London. Series A, Containing Papers of a Mathematical and Physical Character, Vol. 82, No. 557 (Jul. 31, 1909), pp. 546–559.

[52] O'Hern CS, Silbert LE, Liu AJ, Nagel SR. Jamming at zero temperature and zero applied stress: the epitome of disorder. Phys Rev E 2003;68:011306.

[53] Olsson P, Teitel S. Critical scaling of shear viscosity at the jamming transition. Phys Rev Lett 2007;99:178001.

[54] Princen HM, Kiss AD. Rheology of foams and highly concentrated emulsions. J Colloid Interface Sci 1986;112:427.

[55] Dutta SK, Mbi A, Arevalo RC, Blair DL. Development of a confocal rheometer for soft and biological materials. Rev Sci Instrum 2013;84:063702.

[56] Liu AJ, Ramaswamy S, Mason TG, Gang H, Weitz DA. Anomalous viscous loss in emulsions. Phys Rev Lett 1996;76: 3017.

[57] O'Hern CS, Silbert LE, Liu AJ, Nagel SR. Jamming at zero temperature and zero applied stress: the epitome of disorder. Phys Rev E 2003;68:011306. DOI: 10.1103/PhysRevE.68.011306.

[58] Liu AJ, Nagel SR. The jamming transition and the marginally jammed solid. Annu Rev Condens Matter Phys 2010;1:347–369.

[59] Mason TG, Bibette J, Weitz DA. Elasticity of compressed emulsions. Phys Rev Lett 1995;75:2051.

[60] Mason TG, Bibette J, Weitz DA. Yielding and flow of monodisperse emulsions. J Colloid Interface Sci 1996;179:439.

[61] Brujić J, Edwards SF, Grinev DV, Hopkinson I, Brujić D, Makse HA. 3D bulk measurements of the force distribution in a compressed emulsion system. Faraday Discuss 2003;123:207.

[62] Zhou J, Long S, Wang Q, Dinsmore AD. Measurement of forces inside a three-dimensional pile of frictionless droplets. Science 2006;312:1631.

[63] Clusel M, Corwin EI, Siemens AON, Brujić J. A 'granocentric' model for random packing of jammed emulsions. Nature 2009;460:611.

[64] Seth JR, Mohan L, Locatelli-Champagne C, Cloitre M, Bonnecaze RT. A micromechanical model to predict the flow of soft particle glasses. Nat Mater 2011;10:838–843.

[65] Coussot P, Ovarlez G. Physical origin of shear-banding in jammed systems. Eur Phys J E 2010;33:183–188.

[66] Schall P, van Hecke M. Shear bands in matter with granularity. Annu Rev Fluid Mech 2010;42:67.

[67] Paredes J, Shahidzadeh-Bonn N, Bonn D. Shear banding in thixotropic and normal emulsions. J Phys Condens Matter 2011;23:284116.

[68] Goyon J, Colin A, Ovarlez G, Ajdari A, Bocquet L. Spatial cooperativity in soft glassy flows. Nature 2008;454:84.

16

AN INTRODUCTION TO THE PHYSICS OF LIQUID CRYSTALS

JAN P. F. LAGERWALL
Physics and Materials Science Research Unit, University of Luxembourg, Luxembourg

16.1 OVERVIEW OF THIS CHAPTER

This chapter aims to introduce the reader with an interest in general soft matter to liquid crystals, assuming very little prior knowledge in the field. We treat the two main classes of liquid crystals, thermotropics and lyotropics, on equal footage, discussing them in parallel as much as possible and highlighting their similarities as well as their differences. We start with the introduction of the simplest liquid crystal phase, the nematic, explaining fundamental concepts such

Fluids, Colloids and Soft Materials: An Introduction to Soft Matter Physics, First Edition. Edited by Alberto Fernandez Nieves and Antonio Manuel Puertas.

as long-range order and the director in this context. The hallmarks of thermo- and lyotropic liquid crystals are then described, whereafter the most fundamental liquid crystal phases with 1D and 2D positional orders are introduced. A few words on the effect of chirality on the structures of different liquid crystal phases follow, and then we briefly discuss liquid crystal phase sequence notation.

An overview of the anisotropic physical properties of liquid crystals is provided in Section 16.3. After introducing the orientational order parameter, we consider optical, dielectric, diamagnetic, conductivity, and viscosity anisotropy before we start discussing distortions and defects in the director fields and the related unique liquid crystal elasticity. We then return to chiral liquid crystals with a very brief summary of which special physical properties chirality can give rise to in liquid crystals, with a slightly deeper discussion of the peculiar optical properties. The chapter closes with a few selected examples of the most exciting subtopics of current research on liquid crystalline soft matter.

16.2 LIQUID CRYSTAL CLASSES AND PHASES

16.2.1 The Foundations: Long-Range Order, the Nematic Phase, and the Director Concept

What makes liquid crystals unique is that they in spite of their fluidity (a property that we usually connect to disordered, isotropic, liquids, and gases) exhibit long-range order (a property that we usually connect to solid crystals, exhibiting a lattice and strict arrangement of the building blocks). Either the orientation or position, or both, of the entities building up the phase (building blocks) is correlated over a long distance. In the simplest and best studied liquid crystal phase, the ***nematic***[1] (abbreviated N), the long-range order is of purely orientational type: the anisometric building blocks (often rods or disks) are roughly oriented along one and the same direction. In contrast, the positions are as dynamic and disordered as in ordinary isotropic liquids; hence, a lattice cannot be defined. Instead, the orientational order is described via the ***director*** (abbreviated **n**), indicating the local average direction of the principal symmetry axis of the building blocks, as illustrated in Figure 16.1. Because liquid crystal phases generally exhibit head–tail symmetry, **n** is equivalent to −**n**. In the absence of aligning forces, the director varies smoothly throughout a nematic sample without discrete changes (except in defects, cf. Section 16.4), hence one often speaks of the ***director field*** **n**(**r**), where **r** is the space coordinate.

[1]The name is derived from the Greek word for thread, νῆμα, referring to the threadlike defects that are typical of the nematic phase when observed through a polarizing microscope (cf. Figure 16.14).

Figure 16.1 Cartoon of a nematic phase built from rod-shaped building blocks, drawn with a vertical director **n**.

16.2.2 Thermotropics and Lyotropics: The Two Liquid Crystal Classes

The reason for discussing phase building blocks rather than molecules is that two main classes of liquid crystal exist, differing in the nature of their smallest unit. ***Thermotropic*** liquid crystals are built up by single molecules and no further molecular species (specifically solvent molecules) are required for the liquid crystal phase formation. The name thermotropic reflects the fact that temperature is the fundamental thermodynamic control parameter determining the phase. Two temperatures are particularly important, namely those defining the beginning and end of liquid crystalline order (cf. Fig. 16.2). At the ***melting point*** T_m the solid crystal melts into a liquid crystalline state. On further heating, it may or may not be followed by interliquid crystal phase transitions, but eventually the ***clearing point*** T_c is reached, at which the material turns into an ordinary isotropic liquid. It owes its name to the fact that a bulk liquid crystal scatters light, whereas the isotropic liquid is clear.

Being intermediate between the crystalline solid and isotropic liquid states, liquid crystal phases are sometimes collectively referred to as ***mesomorphic***[2] and molecules

[2]The term is derived from the Greek words *μέσος= intermediate*, and *μορφή = shape*.

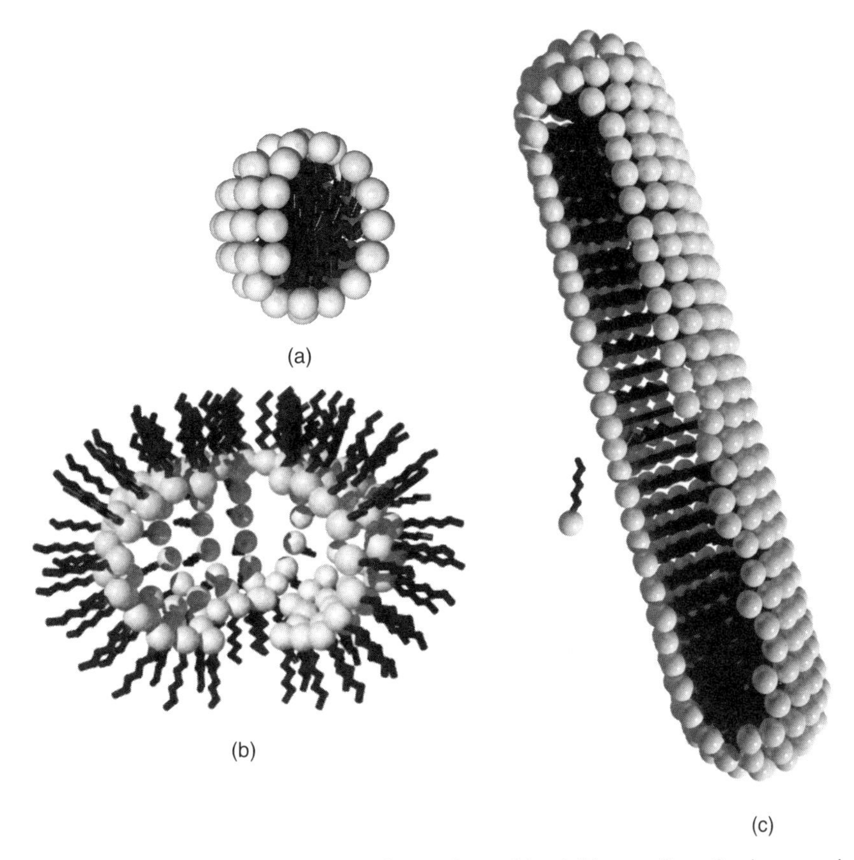

Figure 16.4 Cartoons of a spherical and a rod-shaped micelle of normal type (a)and (b), as well as of an inverse micelle (c). In order to reveal the micelle interior, a sector of amphiphiles has been left out in each case. In reality, the hydrated polar head groups (depicted as white balls) constitute a complete shell around (normal) or within (inverse) the nonpolar chains (black zigzags). Note that the order is greatly exaggerated in all three drawings.

crystals [9], where the constituent molecules (often dyes) stack up into columns of varying lengths.

16.2.3 The Smectic and Lamellar Phases

The nematic is the simplest liquid crystal phase, and it therefore often serves as a first model for describing what is meant by liquid crystallinity. Its orientational order without any long-range positional order makes it unique as the only anisotropic 3D liquid, but there are many other mesomorphic phases, exhibiting long-range positional order in one, two, or three dimensions. In our brief discussion of the most important positionally ordered phases, we start with the case of 1D positional order, found in the ***smectic*** (thermotropics) and ***lamellar*** (lyotropics) phases, both with a layered structure.

In a lamellar phase (L), we have a periodic stack of parallel molecular bilayers separated by water (cf. Figure 16.5a and b). Within each bilayer, the hydrophobic chains of the amphiphiles can be in a flexible liquid-like state with many *gauche* conformers and strong orientational dynamics (a), in which case the phase is denoted L_α. The chain fluctuations are unbiased with respect to directions in the lamellar plane; hence, their average direction is simply along the bilayer normal **k**. The L_α phase thus has the same uniaxial (cylindrical) symmetry as the nematic, with the director **n** parallel to **k**. This phase is of great biological importance, because it is the fundamental state of our cell membranes, of which the main component is phospholipids. If our body would fail to keep the right temperature, allowing the cell temperature to cool down a few degrees, the membrane could undergo a transition to a state with stiff chains and positional order within the bilayer, referred to as L_β if the cylindrical symmetry is retained or $L_{\beta'}$ for the more common case of uniformly tilted chains that locally (single bilayer) break the rotational symmetry (Fig. 16.5b). Globally also the $L_{\beta'}$ phase can have cylindrical symmetry because there may be no correlation in tilt direction between different bilayers. The L_β and $L_{\beta'}$ phases are often referred to as ***gel phases***, reflecting their low degree of fluidity. A cell membrane in this state would be fatal to the cell function and consequently to life; hence, the gel phases do not occur in living organisms. While many surfactants do not develop gel phases, their sole lamellar phase being L_α, amphiphilic lipids normally form both phases (and sometimes also others), the transition between

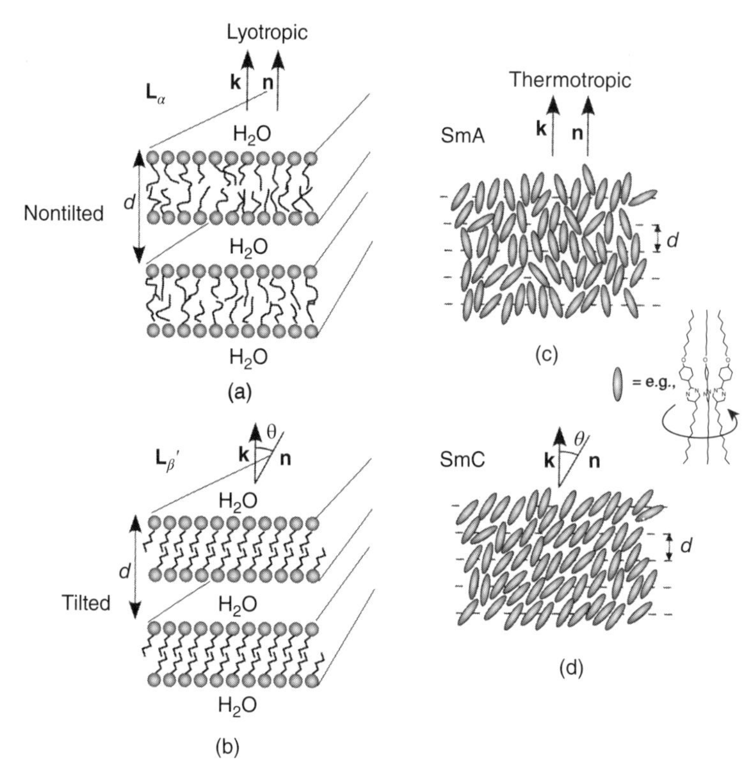

Figure 16.5 Schematic drawings of lamellar and smectic phases. Amphiphiles are drawn with a ball for the polar head group and smectic mesogens as ellipsoids. The distance d along the layer normal **k** is the repeat distance in each phase.

them largely being thermally controlled. The transition temperature is then often called the "main transition."

In thermotropics, the analog to L_α is the smectic-A phase,[9] abbreviated SmA. It (generally) has the same cylindrical symmetry with the director **n** along the layer normal **k**. As the molecules (usually rodlike) are not amphiphilic and as there is no solvent, we here do not have bilayers but monolayers. Furthermore, they are usually not as well defined as in the lyotropic case (where the solvent "seals off" adjacent bilayers from each other), cf. Figure 16.5c. While the SmA phase is almost always uniaxial, a few examples of biaxial SmA phases have recently been reported [10, 11]. Much more important as biaxial smectic is, however, the family of smectic-C-type phases, characterized by a tilted director and in this respect thus a thermotropic relative of the lyotropic $L_{\beta'}$ phase, cf. Figure 16.5d. The tilt angle (θ) is usually temperature dependent, often approaching zero on heating toward the SmA phase, while it normally saturates around 30° at lower temperatures. An important difference from the lyotropic $L_{\beta'}$ analog is that the tilt direction in smectics is always correlated over macroscopic distances in all three dimensions, thereby removing global cylindrical symmetry and giving rise to the biaxiality of the phase.

[9] The letters A, B, C, and so on in the smectic nomenclature simply signify the historical order of discovery.

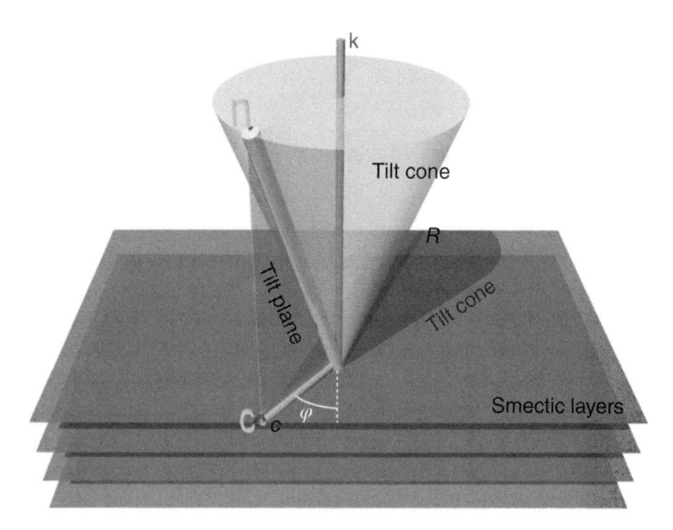

Figure 16.6 The geometry of the SmC phase calls for the definition of many new auxiliary concepts, such as the tilt plane, the tilt cone, and the C-director **c**.

There are several versions of the basic smectic-C-type structure, distinguished by different interlayer tilting direction correlation schemes. In the fundamental SmC phase, the tilt in adjacent layers is in the same direction. One often defines the ***SmC tilt plane*** as the plane spanned by the director **n** and the layer normal **k**, cf. Figure 16.6 where also

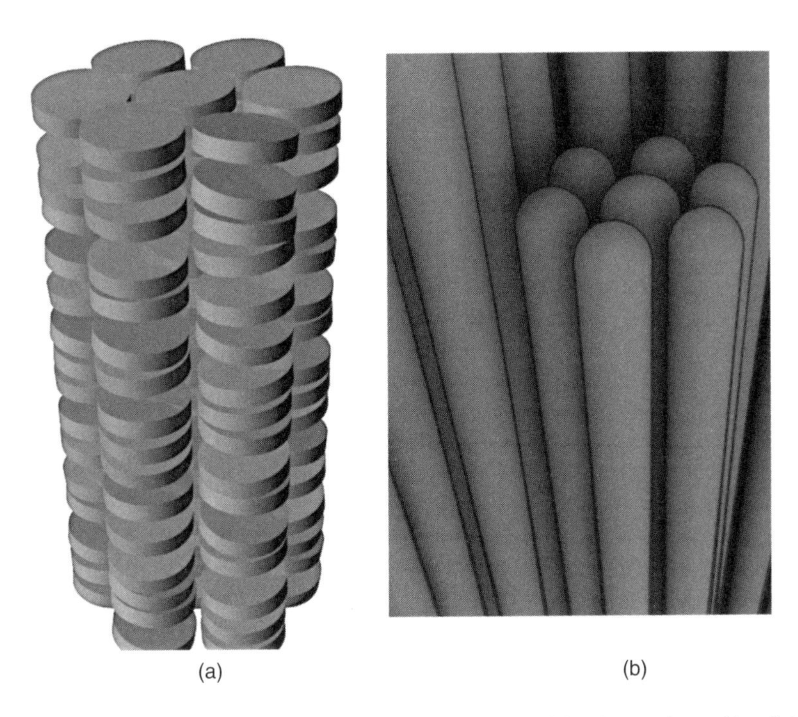

Figure 16.7 Cartoons of columnar phases. Part (a) depicts the thermotropic case, with columns formed by disk-shaped mesogens stacking with their cores on top of each other. The lyotropic hexagonal columnar phase (b) is formed by uniformly aligned rod micelles (here depicted as smooth cylinders) positionally organized on a hexagonal lattice in the plane perpendicular to the micelle long axis (defining **n**). To emphasize the hexagonal ordering, seven micelles are shown with their top ends at the same level, within the picture frame.

the SmC "tilt cone," comprising all possible tilting directions, is drawn. In addition, when considering the physics of the SmC phase PIERRE-GILLES DE GENNES realized [12] that it is convenient to define an additional vector field, closely related to the director. Reflecting the phase in which it finds use, it is called the ***C-director***, abbreviated **c**, and it is defined as the projection of the actual director **n** onto the smectic layer plane (cf. Figure 16.6). The C-director is very helpful in the description of the structure and behavior of SmC and analog phases. For a more detailed overview of the present understanding of smectic-C-type phases, in particular chiral versions, see Ref. [13].

In lamellar and smectic phases, the (bi-)layer thickness is a function, on the one hand, of the effective average length of the amphiphiles or mesogens and, on the other, of the average inclination that the molecules exhibit with respect to the layer normal (see Ref. [13] for a more detailed discussion). Basically, the smectic layer thickness is the projection of the average mesogen length on **k**, whereas the thickness of the lamellar bilayer is twice the corresponding projection of the average amphiphile length. Finally, we should point out that the bilayer thickness is not the periodicity of the lamellar phase: its period is the sum of the bilayer thickness and the thickness of the water layer separating two adjacent bilayers. This means that the translational periodicity in lyotropic lamellar phases is a function of water content.

In the smectic and lamellar phases, the quasi-long-range positional order is 1D along **k**. Within the layers of L_α and SmA/C phases, the molecule positions have only the randomly fluctuating short-range order of isotropic liquids. The layers of these phases can thus be regarded as 2D liquids. In the case of the $L_{\beta'}$ lamellar phase [14] and in the so-called higher ordered smectic phases (e.g., SmB, SmI, and SmF), the so-called hexatic order prevails within the layers [15]. This means, first, that a somewhat extended but still relatively short-range (correlation length on the order of 100 molecules) hexagonal close-packed positional order exists. Second, the orientation of the corresponding hexagonal unit cell is long-range correlated within the layer, a special type of orientational order that is referred to as bond orientational order.[10]

16.2.4 The Columnar Phases

Adding one more dimension of long-range positional order, we encounter the columnar liquid crystal phases. Now the building blocks are positioned on a 2D lattice, but in the third direction there is no long-range positional correlation (cf. Figure 16.7). In the thermotropic case the columnar order is typically formed by discotic mesogens, the aromatic cores stacking up as in piles of coins, each pile being located on a point of a lattice that is basically

[10]The term is misleading as it has nothing to do with chemical bonds. The interested reader is referred to standard liquid crystal textbooks or handbooks [15] for more information.

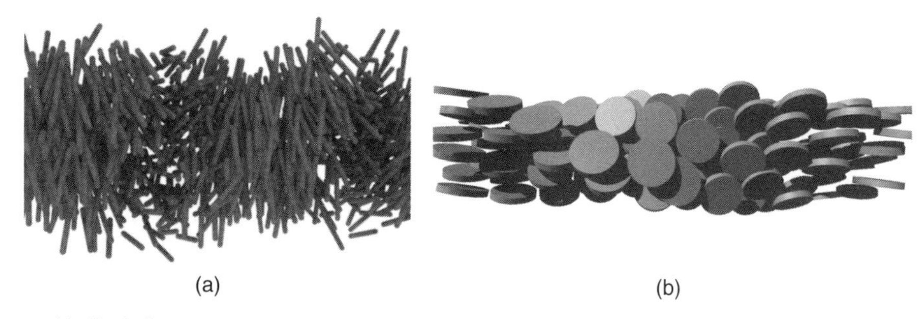

Figure 16.8 Cartoons of helical director modulation in the chiral nematic (cholesteric) phase with rod- (a) and disk-shaped (b) building blocks, respectively. Each rod or disk represents either a mesogen (thermotropics) or micelle (lyotropics) with the corresponding shape. Note that, for clarity, the pitch has been drawn orders of magnitude smaller than in reality and the local degree of order is much too high.

hexagonal (Fig. 16.7a). Within a pile, the molecule positions are usually not long-range correlated (the stack is internally disordered), hence, the structures can be regarded as 1D liquids. Moreover, the peripheral chains of each mesogen are rather flexible and disordered, putting the interstitial volume "between the columns" in a state with very low degree of order, an aspect that has been ignored in Figure 16.7a. The average orientation of the disk is often perpendicular to the column, but it can also be inclined in a common direction, distorting the column arrangement from a truly hexagonal one and thus reducing the symmetry of the phase. For more details on columnar thermotropic phases, see, for example, Ref. [16].

In the lyotropic columnar phases, we have "infinitely" long rod-shaped micelles positionally correlated on a 2D lattice in a plane perpendicular to the rods, hence also perpendicular to **n** (cf. Figure 16.7b). The disorder of the system is mainly found in the water (or other solvent) found between the columns, just like the disordered peripheral chains of discotic mesogens in thermotropic columnar phases. In addition, the amphiphilic molecules diffuse within each column (=micelle) much better than between columns, leading also here to a lesser degree of positional order along **n**. Most often the packing of the rod micelles in the plane perpendicular to **n** is hexagonal and the normal columnar phase is then labeled H_α. The inverted version (water inside, oil between the micelles) gets the label H^i_α. Also rectangular and square lyotropic columnar phases occur; see Ref. [17].

16.2.5 Chiral Liquid Crystal Phases

With the introduction of chirality into liquid crystals, another fascinating and useful aspect of this class of ordered soft matter arises. In most cases, chiral liquid crystals are a result of chiral molecules being present, either as chiral dopants in an achiral host phase or by the use of mesogens that are themselves chiral. The presence of stereoisomerically[11] enriched chiral molecules, rendering the phase unichiral on the macroscopic scale, is for most thermotropic liquid crystals indicated by adding a star to the shorthand of the corresponding achiral phase, for example, N^* for a chiral nematic.[12] In lyotropics, the same indication is used for chiral nematics, but the convention has not been adopted for lamellar and columnar phases. Usually no indication of phase chirality is used here, although the amphiphiles forming lamellar phases are often unichiral (e.g., in naturally occurring phospholipids).

The chiral nematic phase (often referred to as the ***cholesteric*** phase[13]) is the best studied chiral liquid crystal, a result of its simplicity, ubiquity, beauty as well as applicability. The molecular chirality here expresses itself macroscopically in the phase as a helical modulation of the director along an axis perpendicular to **n** (cf. Figure 16.8). If the periodicity (pitch) of the helix is short, the physical properties change dramatically compared to the nonchiral nematic. The mechanical properties along the helix can start resembling those of a weak solid and spectacular optical phenomena occur, to be discussed further in Section 16.5.

The cholesteric helix is a result of the interaction between chiral molecules that confers a certain twist in molecule orientation along an axis perpendicular to **n**. With this arrangement, the material can continuously fill space without any defects. The tendency to twist, however, applies along *all* directions perpendicular to any local director, so the single-helix situation of a cholesteric is not the most favored state if we consider the very local scale. Instead, the minimizing structure would be one where the director twists along any radius perpendicular to a certain central director, as schematically drawn in Figure 16.9a. Although this

[11]Stereoisomers are the different versions of a chiral molecule. In a stereoisomerically enriched sample, a particular stereoisomer is present at much greater concentration than the others. The opposite is a racemate, where all stereoisomers are present at equal concentrations, leading to a cancelling out of chirality effects.

[12]Racemates, where all stereoisomers of a chiral species are present at equal amounts, are effectively nonchiral in their physical properties. The star is thus used only when one stereoisomer dominates over the other(s).

[13]The term "cholesteric" has historical origin: the phase was first discovered in a sample of cholesteryl benzoate, exhibiting BP* and $N*$ phases [18].

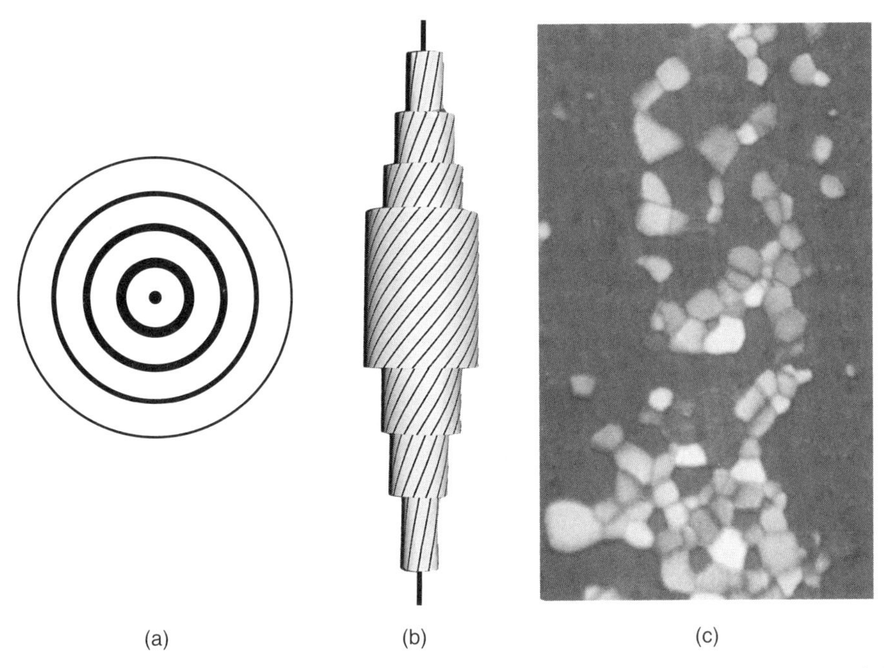

Figure 16.9 (a) The interactions between adjacent chiral molecules in a phase with nematic-like order promote twisting along all directions perpendicular to a certain starting director (here out of the paper plane as represented by the central black point). The director rotation around axes in the paper plane is indicated with a line thickness coding, the thickest (the filled circle in the center) for **n** close to perpendicular, the thinnest (the outermost circle) for **n** 45° out of the paper plane. (b) A model of the double-twist cylinder structure that is the building block of BP I* and BP II* blue phases. Lines represent the director field. The drawing in (a) corresponds to a horizontal cut through the center of the model. (c) Example of the multicolored platelet polarizing microscopy texture characteristic of BP I* and BP II*.

configuration, referred to as double-twist, locally minimizes the elastic energy in the chiral system, it cannot fill 3D space without generating a network of defects or voids. The concept "local scale" here means within a distance from the central director where the twist has reached about 45° as illustrated in Figure 16.9b where the multiply twisting director field in *a* has been extended along the central director into what is called a *double-twist cylinder* unit.

At high temperatures, in the vicinity of the clearing point, the elastic energy of a local defect-free regime can be markedly lower for double-twist than single-twist, while the energetic cost of defects is much less than at low temperatures where the nematic order is better developed. In this case, three new liquid crystal states, related to but not equal to the chiral nematic phase, may become energetically stable. In these ***blue phases*** [19–21], abbreviated[14] BP*, double-twist cylinder units, indeed develop in a thermodynamically stable state. Three chiral nematic blue phases have been identified, two of which (BP I* and BP II*) have body-centered cubic and simple cubic, respectively, arrangements of the double-twist cylinders, combined with a regular network of defect lines (disclinations). The third (BP III*) is amorphous, and its local structure is still not conclusively elucidated. The temperature range of blue phases is usually so small (often less than 1 K) that they are difficult to study and applications of them long seemed unrealistic. Recently, important advances have, however, been made in expanding the temperature range of BP* phases, in particular by polymer stabilization of the defect network [22]. This has kindled a new interest in applying blue phases in devices, including displays.

In smectic phases, a helical director modulation perpendicular to **n** would break the layers and is thus not admitted.[15] The SmA* phase therefore has the same structure as an achiral SmA phase. Its physical properties are, however, quite different from its achiral analog [13], and it is thus important to recognize that the phase is still chiral despite the absence of a helix. In contrast to SmA*, the chiral SmC* phase does allow a helical superstructure, but it applies to the C-director rather than the normal director. In other words, **n** in SmC* does not twist as in N^*, but **c** does, that is, the tilting *direction* processes helically along the smectic layer normal, as illustrated in Figure 16.10.

[14] In the literature, one finds the blue phase acronyms either with or without the star for chirality.

[15] This nevertheless happens in the so-called *twist grain boundary* (TGB*) phases, sometimes appearing with chiral mesogens with very high twisting power. Their structure can be described as finite blocks of smectic order, each block rotated a specific angle around an axis in the smectic layer plane with respect to its neighbor blocks, thereby forming a discrete helical structure.

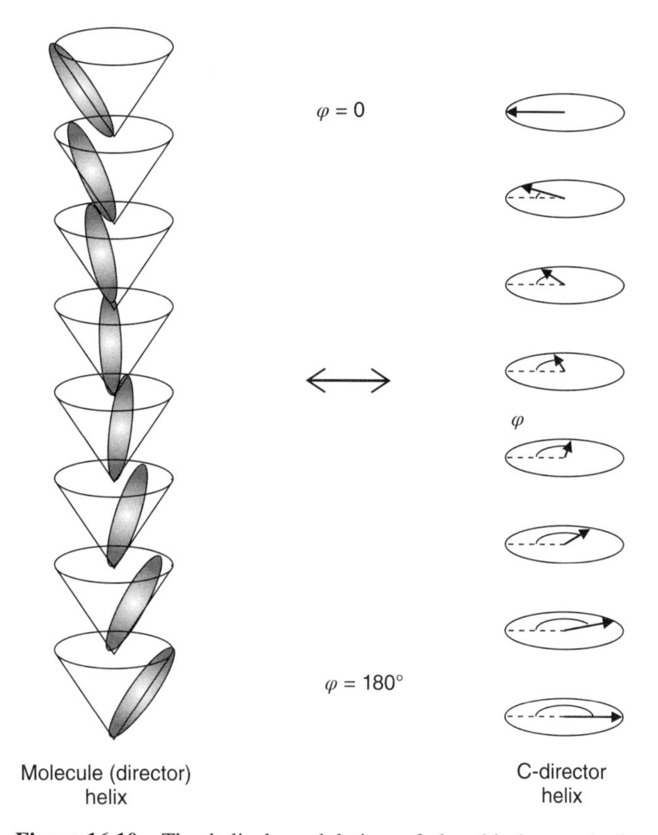

Figure 16.10 The helical modulation of the chiral smectic-C (SmC*) phase applies to the director tilting direction φ or, equivalently, to the C-director **c**. To the left, a cartoon of the molecular order is shown, each layer being represented by a tilt cone on which the actual director is indicated with an ellipse. The right drawing shows the corresponding modulation in **c** and φ. The twist is greatly exaggerated for reasons of clarity. A real short-pitch SmC* phase has a pitch on the order of 100 smectic layers.

The helical pitch p can be anywhere from a few hundred nanometers to infinity, for cholesterics as well as SmC*-type phases. It depends, on the one hand, on the so-called "helical twisting power" (HTP) of the chiral molecule (in other words, it is structure-specific) and, on the other, on its enantiomeric enrichment ("enantiomeric excess," *ee*),[16] and there is also generally a rather strong temperature dependence. This leads to a color that changes on heating or cooling (see Section 16.5), forming the basis for the common use of N^* liquid crystals in temperature sensors. The handedness of the helix depends on which stereoisomer builds up the phase or is added to the achiral host phase. If more than one stereoisomer is present, the surplus stereoisomer will determine the handedness, and the pitch will be longer than for a unichiral sample. Quantitatively, the *inverse* pitch depends linearly on *ee*, going from 0 (infinite pitch) at the racemate to p_0^{-1} at 100% *ee*, where p_0 is the pitch of the unichiral material.

[16]Enantiomers are two stereoisomers that are each other's mirror reflections, such as the right and left hands. Apart from the handedness and its physical consequences (most commonly rotation of plane-polarized light), the physical properties of the two enantiomers are identical. This is not necessarily true for other types of stereoisomerism.

In a fairly recently discovered class of chiral smectic liquid crystals, all molecules forming the phases are actually achiral [23–25]. It can arise in smectic phases formed by molecules with a strongly bent core because such molecules can only pack efficiently if the bend direction is uniform over large distances. If in addition the director is now tilted (as is often the case in these systems), we have three nonparallel directions describing quasi-long-range order: the layer normal, the tilting direction, and the bend direction. Together, these three vectors span a Cartesian coordinate system, thus exhibiting a handedness. In other words, although no chiral molecules are present, the phase is locally chiral, with a handedness L or R that is randomly selected [26]. These phases currently attract much attention, largely due to their unusual type of chirality [27] and the special properties that it gives rise to, for example, in terms of helical superstructures [28] and polar ordering phenomena [29–31].

In columnar phases, the requirement for a macroscopic chiral superstructure reflecting molecular chirality is, just like in smectics, a nonzero director tilt of mesogens with respect to the column axis. If the director is along the columns, a chiral superstructure would require a breaking of the hexagonal 2D lattice, a situation that has not yet been reported. If **n**, on the other hand, is tilted, we can have a helical modulation of the tilt direction, this time along the column axis.

16.2.6 Liquid Crystal Polymorphism

Figure 16.2 summarizes what has been stated above about rod-shaped thermotropic and lyotropic liquid crystals, respectively, in highly simplified generic phase diagrams. In reality, the phase diagrams are generally more complex and they often include several more phases. Moreover, the details of the typical multiphase regions occurring in lyotropic systems are ignored in the right diagram. The polymorphism of liquid crystals is incredibly rich and encompasses many more phases, but these are outside the scope of this brief introduction.

Some thermotropic substances show a greater number of liquid crystalline phases on cooling than on heating, the reason being that some of them may occur only as metastable supercooled phases when cooling the substance past its melting point. If crystallization does not set in immediately, a phase transition into a new ***monotropic*** liquid crystalline state may take place. Sooner or later, the supercooled phase will however be replaced by the crystalline state corresponding to the global energy minimum. When heating the substance from the crystalline state, the monotropic phase will not develop, but the substance will melt directly into the first ***enantiotropic*** liquid crystal phase of the compound, that is, the first thermodynamically stable liquid crystal phase, forming on heating as well as on cooling.

When stating the phase sequence of a substance with monotropic phases, these are enclosed in parentheses, together with the transition temperatures (which must obviously refer to a cooling run). A compound with a monotropic SmC phase underneath an enantiotropic SmA phase could thus have the phase sequence: Cr. 31 (SmC 28) SmA 40 Iso. Note that the melting point (31 °C) is higher than the SmA–SmC transition (28 °C). Furthermore, all enantiotropic transition temperatures are generally given on heating, since transition temperatures can be determined with greater accuracy in this direction.

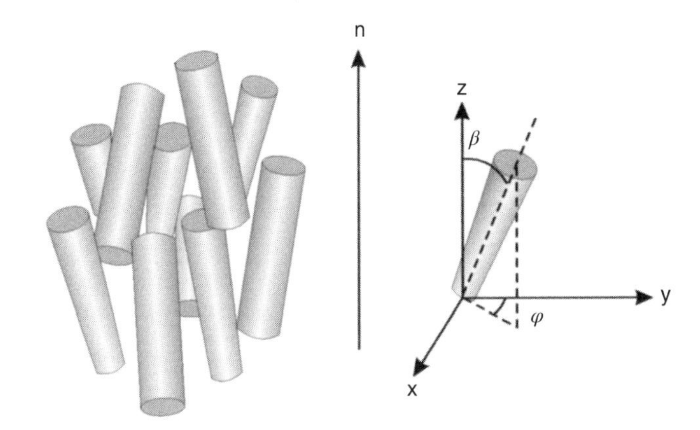

Figure 16.11 Graphical depiction of the concepts required for defining the orientational order parameter. The building blocks of the liquid crystal phase are depicted as rods, on average oriented along **n** but each deviating by some angle β in a direction φ.

16.3 THE ANISOTROPIC PHYSICAL PROPERTIES OF LIQUID CRYSTALS

The macroscopic anisotropy of liquid crystals, a property shared with no other fluid, is the basis for their successful commercial exploitation. The anisotropy takes on several facets, the most important ones of which will be discussed in the following. In this section, we restrict ourselves to achiral phases, the special properties arising as a result of chirality being the topic of Section 16.5.

16.3.1 The Orientational Order Parameter

The definition of the director provides a first qualitative description of the long-range orientational order of a nematic phase and enters as the natural variable in what is referred to as the *elastic continuum description* of the free energy of a nematic, in the context of which it was introduced.[17] But apart from the *direction* of the anisotropy axis in space, one often desires also a quantitative measure of the anisotropy or degree of orientational order. In other words, we are looking for a number that tells us to what extent the building blocks align along **n** or how large the fluctuations around this direction are. The parameter filling our needs is nowadays often called the ***nematic order parameter*** S (sometimes written as P_2 for reasons that will become clear below), even when applied to non-liquid crystalline systems and despite the fact that it was originally developed not for nematics but for describing the chain orientation in polymer fibers. For 3D systems, this parameter, introduced in 1939 by Hermanns and Platzek [32] and adopted for liquid crystals in 1942 by Tsvetkov [33], has the form:

$$S = \frac{1}{2}\langle 3\cos^2\beta - 1\rangle \tag{16.1}$$

[17]The concept of the director was introduced by the Swede CARL-WILHELM OSEEN while the name was introduced by FRANK LESLIE and JERALD ERICKSEN of British and American nationality, respectively [18].

Here, we have introduced the angle β, describing the deviation in orientation away from **n** of a particular building block at a particular moment in time (cf. Figure 16.11). The deviation is averaged over the ensemble of building blocks constituting the phase considered, as indicated by the pointed brackets. In the figure we again use the angle φ to define the tilting direction. This time the tilt is that of the molecule with respect to **n**. While not directly entering the definition of the order parameter, it is important to be aware of the degeneracy in φ (all *directions* of deviation from **n** can occur with equal probability), affecting any statistical description of liquid crystal order.

The order parameter (16.1) has the desirable property of being zero for total disorder, since in three dimensions $\langle\cos^2\beta\rangle = 1/3$ for a random orientational distribution, and 1 for perfect order (in which case $\langle\cos^2\beta\rangle = 1$).[18] You may recognize the expression as the second Legendre polynomial, and the order parameter is often referred to as $\langle P_2\rangle$ to emphasize this equivalence, but in this chapter we use the simpler shorthand S. In fact, the full scalar orientational order parameter[19] is a series of Legendre polynomials, the first term of which is $S = \langle P_2\rangle$, but for many considerations it suffices to use this first term.

It has been empirically as well as theoretically established that the thermotropic isotropic–nematic transition is first order, that is, the order parameter jumps discretely at the phase transition from zero to a nonzero value, typically around $S = 0.4$. On cooling a thermotropic nematic, S increases, initially fast but rapidly approaching saturation

[18]The minimum value of Equation 16.1 is not zero but −1/3. However, the negative values describe a physically irrelevant state of building blocks preferentially aligning *perpendicular* to the director (in conflict with its definition), hence this regime can be ignored.

[19]A complete description of the orientationally ordered state requires a tensorial order parameter that includes both **n** and S [12].

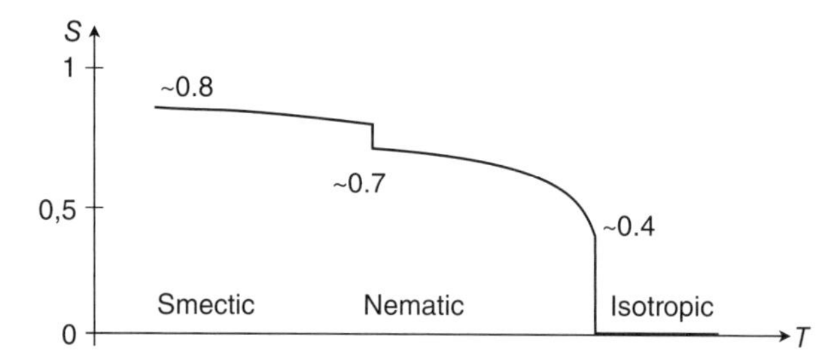

Figure 16.12 Typical temperature dependence of the orientational order parameter in a thermotropic material exhibiting a nematic–smectic phase sequence.

around $S = 0.6$–0.7 (cf. Figure 16.12). If smectic phases develop on further cooling, the orientational order may increase discretely by a small amount at that phase transition as well, but intersmectic transitions such as SmA–SmC generally do not affect S markedly. In thermotropic liquid crystals, the orientational order rarely increases beyond $S \approx 0.8$, columnar phases being a possible exception.

In lyotropic liquid crystals, one can basically distinguish two subclasses as far as the orientational order is concerned. Among surfactant-based lyotropics, the hexagonal phases have very high-order parameters in the range of $S \approx 0.9$ [34]. The same holds for nematic phases formed not by surfactant micelles but by colloidal suspensions of more or less hard anisometric particles (viruses, carbon nanotubes, bentonite platelets, etc.). In both cases, the order parameter is largely insensitive to temperature, that is, these lyotropic systems are in many respects athermal. Micellar nematic phases, on the other hand—much more rarely observed as they generally require a smart mixture of surfactant and cosurfactant—generally have a strong temperature dependence of the order parameter, often looking qualitatively similar to Figure 16.12. Moreover, the value S_{NI} at the transition point (which can here be defined either as a temperature or as a concentration) can here be very low, sometimes substantially lower than S_{NI} of thermotropic liquid crystals. It is typically $S_{NI} \approx 0.3$ [35–38], but some experimental works report values as low as $S_{NI} \approx 0.1$ [36, 37].

16.3.2 Optical Anisotropy

The anisometry of the building blocks and the presence of long-range orientational order together generally render liquid crystals optically anisotropic, that is, their refractive index $n_{\parallel}$ parallel to **n** is different from that perpendicular to the director, $n_{\perp}$. For a uniaxial material, these two indices are the extreme refractive index values and the director is equivalent to the optic axis of the liquid crystal. The magnitude of anisotropy, the ***birefringence***, is defined as $\Delta n = n_{\parallel} - n_{\perp}$. Consequently, if $n_{\parallel}$ is the maximum refractive index, $\Delta n > 0$ and we call the liquid crystal positive uniaxial (this is the general case for rod-shaped mesogens), whereas if it is the minimum the phase is negative uniaxial (this is the standard case for discotics).

A birefringent material can affect the polarization of light, changing the ellipticity continuously between linear and circular polarizations as well as changing the main oscillation direction or reverse the rotation handedness. This is the basis for the use of liquid crystals in most electrooptic devices such as displays (liquid crystal displays are well described in numerous books, and we thus refer the reader to other sources for an explanation of their operation). But the effective birefringence depends on the orientation of **n**, going down to zero for a director, and consequently optic axis, aligned parallel to the direction of light propagation. This is the case for normal light incidence (the light ray is perpendicular to the sample plane) onto a sample in the so-called ***homeotropic***[20] director geometry (Fig. 16.13a), a term used when **n** is perpendicular to the sample plane. The

[20]The term *homeotropic* may be understood as a condensed form of *homeoisotropic*, meaning *like isotropic*, referring to the absence of birefringence for light propagating along the optic axis.

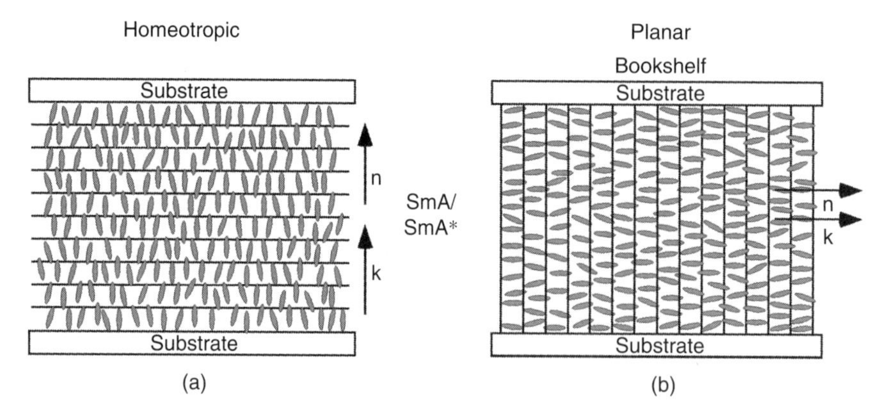

Figure 16.13 Schematic drawing of homeotropic and planar alignment, respectively, here illustrated for the case of an SmA phase confined between parallel flat substrates. Normal light incidence means that the light beam is propagating vertically in the picture. Note that the drawing is not to scale: a smectic layer is some three orders of magnitude thinner than the distance between the sample substrates.

electric field vector of the light will then be perpendicular to the optic axis, regardless of the light's particular polarization state; hence, it experiences solely the refractive index $n_{\perp}$. In contrast, the maximum effective birefringence is experienced when the propagation direction is perpendicular to the optic axis. Again considering the case of normal incidence, this corresponds to **n** being in the sample plane, horizontal as in Figure 16.13. For standard liquid crystal cells, this corresponds to what is referred to as ***planar*** alignment,[21] although the more general meaning of this term is that **n** is in the plane of the interface between liquid crystal and surrounding medium (the difference becomes important, e.g., for drops or shells of liquid crystal).

In Figure 16.13, the two principal alignment types are illustrated for the case of a smectic-A phase between flat substrates. Note that homeotropic alignment corresponds to smectic layers in the plane of the sample, whereas the planar alignment is equivalent to layers standing up. The latter geometry is therefore often called *bookshelf geometry*, referring to the layers standing as books in a shelf. The nematic case is identical to the SmA case apart from the absence of layers, but the SmC case is somewhat different. Here, one usually maintains the *layer* geometries in Figure 16.13 in the definition of the concepts homeotropic and planar, although **n** is then not along the normal incidence direction in the former case (it is therefore better referred to as ***quasi-homeotropic***) and not necessarily in the sample plane in the latter.

In a planar-aligned nematic sample where **n** has no restriction to any particular direction in the sample plane (we refer to this geometry as ***degenerate planar*** alignment), the director and thus the optic axis vary randomly throughout the sample area. When observing the sample between crossed polarizers, all locations where **n** is parallel to one of the polarizers will appear black. This is because the light is here polarized exactly along or perpendicular to the optic axis, hence the effect of birefringence vanishes (if this is unclear, please refer to a basic optics book). Since these locations are continuously connected, forming randomly curving "threads" or "streaks," we get a texture that is very characteristic for the nematic phase (cf. the example in Figure 16.14). This texture is called a ***schlieren*** texture after the German word *Schliere*, meaning streak.

The magnitude of the birefringence of a phase with uniform director orientation depends on the degree of orientational order S and the polarizability anisotropy $\Delta\alpha$ of the building blocks of the phase according to

$$\Delta n \propto S\Delta\alpha, \tag{16.2}$$

[21] An older term sometimes encountered is *homogeneous alignment*. There are many reasons to avoid this term that is not only obsolete but also misleading: *homeotropic* alignment is *homogeneous* in the proper sense of the word and a *planar* sample without a well-defined direction of **n** in the sample plane is an example of a highly *inhomogeneous* director field.

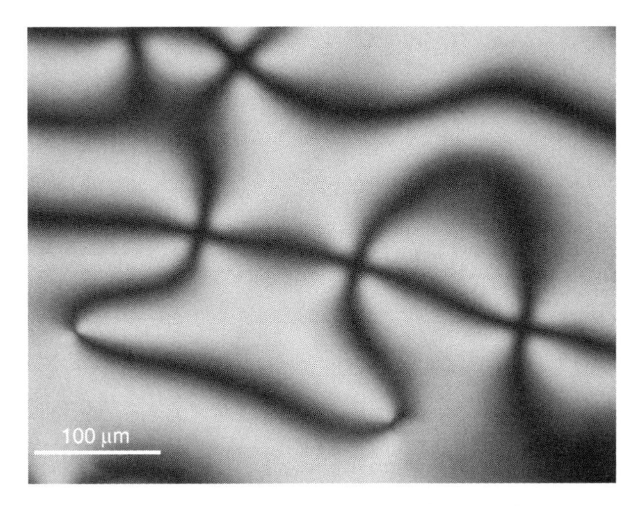

Figure 16.14 An example of the characteristic schlieren texture exhibited by nematic liquid crystals in degenerate planar alignment. The micrograph is of a lyotropic nematic formed by disk-shaped micelles, but the same type of texture can be formed by any type of achiral nematic in degenerate planar geometry.

that is, it is a linear function of the nematic order parameter and the difference in polarizability along and perpendicular to the principal symmetry axis of the building block. While $\Delta\alpha$ is often almost temperature independent, the order parameter generally changes strongly at temperatures close to the transition to the isotropic phase, as discussed earlier, hence Δn decreases rapidly as a thermotropic liquid crystal is heated toward its clearing point (cf. Figure 16.15).

The linear aromatic core commonly found in rod-shaped thermotropic mesogens results in a high value of $\Delta\alpha$; hence, thermotropic rods belong to some of the strongest birefringent materials known. The circular core structure of discotic mesogens yields somewhat lower $\Delta\alpha$ although its high degree of aromaticity still allows for reasonably high values. Lyotropic liquid crystals, in contrast, are most often formed by molecules without aromatic moieties, making $\Delta\alpha$ rather small. Consequently, a lyotropic nematic has orders of magnitude smaller birefringence than its thermotropic counterpart, although $S(T)$ is similar for the two cases.

If we cool a nematic-forming material down from its isotropic into its nematic phase, the isotropic refractive index will split up in $n_{\parallel}$ and $n_{\perp}$ at the phase transition (cf. Figure 16.15). At that temperature, these three refractive indices are related as follows:

$$n_{\text{iso}}(T_{NI}) = \frac{n_{\parallel} + 2n_{\perp}}{3} \tag{16.3}$$

a result of the geometrical averaging—weighted for the dual axes perpendicular to the director—taking place in the isotropic phase. This holds regardless of whether the material has positive (Fig. 16.15a) or negative (b) optical anisotropy. As a result, the isotropic refractive index at the

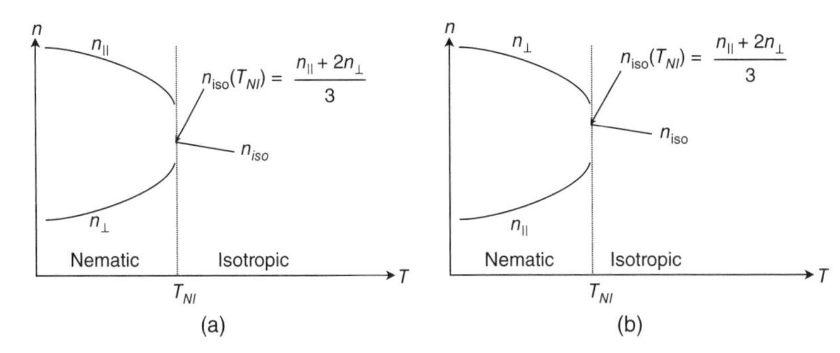

Figure 16.15 Typical temperature dependencies of the refractive indices of a positive (a) and negative (b) uniaxial nematic, below and above the clearing point.

phase transition is closer to the greater nematic refractive index for a negative uniaxial material, whereas it is closer to the smaller one for a positive material. The decrease in n_{iso} upon heating through the isotropic phase is a result of the dependence of the refractive index on the density of the phase, which decreases on further heating.

The visible effect of birefringence of a material between crossed polarizers is generally not simply one of bright or dark, but one of color. The difference in optical path length, $\Delta\Lambda$, between light polarized along the director (experiencing $n_{\|}$) and light polarized perpendicular to it (experiencing $n_{\perp}$) depends on birefringence Δn and sample thickness d as follows:

$$\Delta\Lambda = \Delta n \cdot d. \tag{16.4}$$

If $\Delta\Lambda$ is an odd multiple of $\lambda/2$, where λ is the wavelength of the light (defining its color), the polarization after the liquid crystal is again linearly polarized but its polarization plane is mirrored in **n**, giving this light a maximum in intensity when viewed through crossed polarizers. In contrast, if $\Delta\Lambda$ is a multiple of λ, the birefringence has no effect and the polarization is the same as before the liquid crystal; hence, the light is extinguished by the second polarizer. Thus, a 4-μm-thick sample with $\Delta n = 0.1$ (typical for thermotropic liquid crystals) viewed between crossed polarizers and illuminated by white light (this is the typical situation when studying liquid crystals in a polarizing microscope) will not let through any blue light ($\lambda \approx 400$ nm $= \Delta\Lambda$), whereas the intensity is greater the more red the light is (the maximum intensity, given for $\Delta\Lambda = \lambda/2$, is for $\lambda = 800$ nm, thus just into the infrared regime). The sample will appear with an orange-red color. For a sample thickness of around 2 μm, $\Delta\Lambda$ is about 200 nm, much less than λ of any visible light and close enough to $\lambda/2$ for the whole visible spectrum that all colors get transmitted well through the second polarizer. The sample then appears quite neutral in color.

The color variations with Δn and d are graphically summarized in a *Michel–Lévy* chart, available from many microscope manufacturers. In the chart, one can immediately find the color that a sample of a certain thickness and birefringence will have when viewed between crossed polarizers. Conversely, if we know the sample thickness, we can use the chart to estimate Δn or at least track changes in birefringence. For $\Delta n \approx 0.1$, one finds that a regime of strong colors (varying quasi-periodically over two to three orders) for sample thicknesses in the range $d \approx 3$μm to $d \approx 20$μm is surrounded by a gray-white regime for very thin cells ($d \approx 1$–2μm) and a regime with an ill-defined and uninformative pallid appearance for very thick cells. The latter is one of the reasons why thermotropic liquid crystal samples are rarely much thicker than some tens of microns.

The reason for the indistinct color of thick samples is that the distance between wavelengths that are maximally transmitted gets increasingly smaller as the sample gets thicker. Two consecutive wavelengths λ_+ and λ_- that both experience maximal transmission are related by

$$\Delta\Lambda = \Delta n d = (2k+1)\frac{\lambda_+}{2} = (2k+3)\frac{\lambda_-}{2}, \tag{16.5}$$

where k is an integer. The distance between λ_+ and λ_- is thus

$$\Delta\lambda = 2\Delta n d\left(\frac{1}{2k+1} - \frac{1}{2k+3}\right), \tag{16.6}$$

which, after rewriting the integer k in terms of λ_+, becomes

$$\Delta\lambda = \Delta n d \lambda_+ \left(\frac{1}{\Delta n d} - \frac{1}{\Delta n d + \lambda_+}\right). \tag{16.7}$$

Plotting $\Delta\lambda$ as a function of cell gap d, for $\Delta n = 0.1$ and $\lambda_+ = 500$ nm, we get the curve shown in Figure 16.16. We see that $\Delta\lambda$ quickly gets rather small as the sample thickness reaches a few tens of microns, meaning that so many wavelengths are transmitted at high intensity that the distinction in color is lost. Of course, we have an equally frequent set of wavelengths that are blocked by the second polarizer, as well as intermediate wavelengths that are transmitted with various elliptical polarizations. The end result is a mix of wavelengths that makes thick samples appear with an unsaturated color tone that varies between weak green and pink.

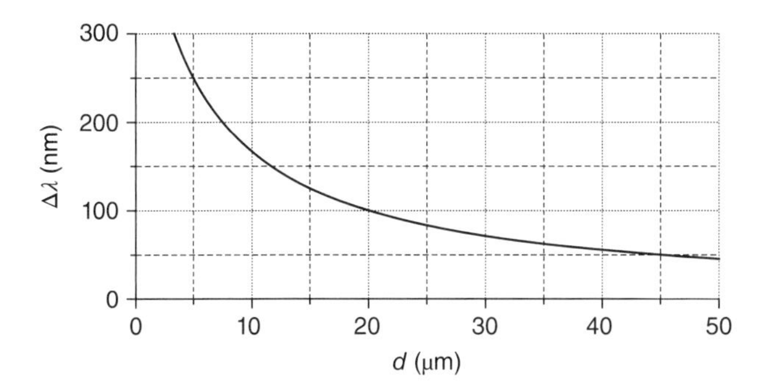

Figure 16.16 The wavelength offset $\Delta\lambda$ between adjacent wavelengths that are maximally transmitted by a birefringent sample observed between crossed polarizers rapidly decreases as the sample thickness d increases, leading to loss of distinct colors for thick samples.

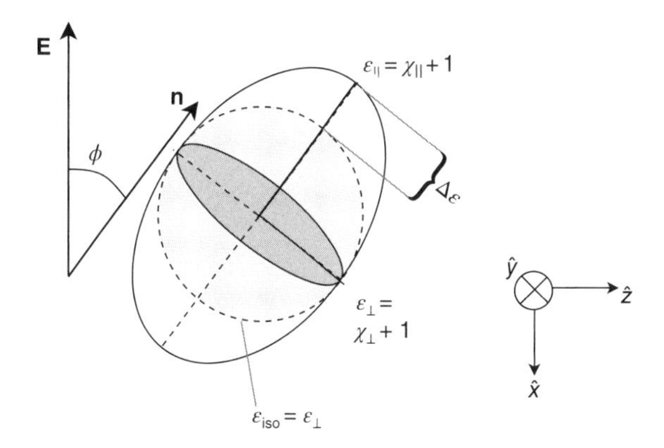

Figure 16.17 When an electric field is applied over an anisotropic dielectric medium such as a liquid crystal, a polarization is induced that may be nonparallel to the field. This results in a torque which, in case of the highly fluid nematic, reorients the director until the induced polarization is along the electric field.

This brief discussion of the optical anisotropy of liquid crystals has been done using the simplest case of uniaxial birefringence. The optical situation gets more complicated in the case of a biaxial phase, which has two optic axes and three extreme values of refractive index n_1, n_2, and n_3. However, when working with liquid crystals we often deal with uniaxial phases (N, SmA, L_α, ...) and even the biaxial phases (SmC, ...) are often very weakly biaxial, such that we in a first approximation can neglect the biaxiality.

16.3.3 Dielectric, Conductive, and Magnetic Anisotropy and the Response to Electric and Magnetic Fields

The ability to modulate a liquid crystal display element between dark and bright is normally due to the dielectric anisotropy. By applying an electric field, this gives us a handle for switching the director orientation between different defined geometries, giving different transmission. When an electric field **E** is applied over a dielectric medium such as a liquid crystal, the field will polarize the medium, that is, an electric dipole moment will be induced, proportional to the field strength and to the dimensionless electrical susceptibility χ^e of the medium:

$$\mathbf{P} = \epsilon_0 \chi^e \mathbf{E}. \tag{16.8}$$

Here, the vacuum permittivity $\epsilon_0 \approx 8.85 \times 10^{-12}$ C/Vm is required to get the dimension of a polarization. In an isotropic medium, χ^e is independent of direction and can be represented by a scalar. The resulting polarization will thus always be along **E**. In anisotropic media, in contrast, the susceptibility depends on the direction of the field, and its full description requires a tensorial χ^e. We here consider only the simplest uniaxial nematic case, for which we require the two scalar susceptibility values $\chi^e_\parallel$ parallel and $\chi^e_\perp$ perpendicular to **n**. Analogous to the optical anisotropy, we can now define the susceptibility anisotropy as $\Delta\chi^e = \chi^e_\parallel - \chi^e_\perp$.

The important consequence of the susceptibility being anisotropic is that the induced polarization **P** is not necessarily directed along **E**. Field application thus generally results in a local torque $\mathbf{\Gamma}_\epsilon = \mathbf{P} \times \mathbf{E}$ on the director, acting to align **n** along ($\Delta\chi^e > 0$) or perpendicular ($\Delta\chi^e < 0$) to the field. The electrical susceptibility is rarely considered in practice, as the quantity that is typically known (or measured) is the dielectric permittivity $\epsilon = \chi^e + 1$. However, as the one quantity is just the other rescaled by the addition of a constant, their anisotropies are numerically identical, $\Delta\epsilon = \epsilon_\parallel - \epsilon_\perp = \Delta\chi^e$.

Consider the situation depicted in Figure 16.17: an electric field **E** is applied vertically (in the $-\hat{x}$-direction as defined in the figure) over a uniaxial nematic with positive $\Delta\epsilon$, the director of which is inclined at an angle ϕ from the vertical direction. The anisotropic permittivity is represented graphically by a rotationally symmetric ellipsoid, the distance between its center and its surface in each point being equal to ϵ in that particular radial direction. It is convenient to separate the permittivity in an isotropic component $\epsilon_{\text{iso}} = \epsilon_\perp$ (graphically represented by the dotted sphere) and an anisotropic addition $\Delta\epsilon$. The former component can be analyzed as for any isotropic system, that is, the resulting polarization is parallel to the field and has the magnitude $P_{\text{iso}} = \epsilon_0 \chi^e_\perp E = \epsilon_0(\epsilon_\perp - 1)E$. More interesting is the anisotropic component, which generates a polarization along the director but is driven only by the component of the field along **n**, that is, $E_\parallel = E \cos \phi$.

We can now calculate the dielectric torque $\mathbf{\Gamma}_\epsilon = \mathbf{P} \times \mathbf{E}$ imposed on the nematic director by the electric field, writing it as the sum of the torques corresponding to the isotropic permittivity and the anisotropic addition, respectively (we remind that the anisotropic addition of the susceptibility is $\Delta\chi^e = \Delta\epsilon$):

$$\begin{aligned}\boldsymbol{\Gamma}_\epsilon &= \epsilon_0(\epsilon_\perp - 1)\mathbf{E}\times\mathbf{E} + \epsilon_0\Delta\epsilon E\cos\phi(\mathbf{n}\times\mathbf{E})\\ &= 0 - \epsilon_0\Delta\epsilon E^2\cos\phi\sin\phi\hat{y}.\end{aligned} \tag{16.9}$$

In the last step, we have specifically indicated that the cross product of $\mathbf{E}$ with itself is zero, resulting in a vanishing torque related to the isotropic permittivity component (as was perhaps obvious from the beginning), and we have evaluated the cross product between the director and field, amounting to $|\mathbf{n}\times\mathbf{E}| = E\sin\phi$ and directed out of the paper, that is, in the $-\hat{y}$ direction. We can finally rewrite the product of cosine and sine as sine for the double angle to get a compact final expression:

$$\boldsymbol{\Gamma}_\epsilon = -\frac{1}{2}\epsilon_0\Delta\epsilon E^2\sin 2\phi\hat{y}. \tag{16.10}$$

Applying this expression to the situation, we see that the torque turns the director into the field direction, that is, the field switches $\mathbf{n}$ until it lines up with the electric field. At that point, $\Phi = 0$ and the torque disappears. While the same torque would arise in the corresponding situation with an anisotropic crystalline solid instead of a nematic, the only mechanical result that can be expected is that the whole crystal, in one piece, rotates until the maximum permittivity is along the field direction, a freedom that the sample does not necessarily have. In the nematic liquid crystal, on the other hand, the sample as such is not rotated at all since its fluidity allows the *internal* structure to reorient; only $\mathbf{n}$ aligns up with the field. This switching process is fundamental for the use of liquid crystals in the displays of your computer, mobile phone, handheld electronic devices, and so on.

In display devices, the starting alignment is typically planar, that is, $\mathbf{n}||\hat{z}$ in Figure 16.17 and $\phi = 90°$. A curious thing is that if you enter that angle into Equation 16.10, you will obtain the result that there is no torque at all on the director and one might at first think that $\mathbf{n}$ would stay planar despite the field. The $\phi = 90°$ situation is, however, an unstable state of maximum energy in the presence of the field. Thermal fluctuations constantly turn the local director out of the display plane, by randomly varying small amounts. The angle ϕ is thus not simply 90°, but it fluctuates around this value, any deviation resulting in a nonzero dielectric torque from the applied field trying to enhance the fluctuation and turn $\mathbf{n}$ away from the electrostatically energy-maximizing planar state.

In a device or in a typical research sample cell, there is however also a strong influence from the sample substrates, imposing the planar starting configuration. Because of the elastic properties of the nematic (to be discussed below), this surface-imposed alignment is propagated through the liquid crystal in order to minimize deformations in $\mathbf{n}(\mathbf{r})$, which would increase the energy. When an electric field is applied, we thus get a competition between the dielectric contribution to the free energy, promoting a reorientation of the director and the elastic contribution, counteracting any change from the initial planar-aligned state. At weak fields, the elastic contribution "wins" and the director remains unaffected by the field. At strong fields, it is the other way around and the sample is switched. The borderline case is called the ***Frederiks threshold*** after the Polish-Russian physicist VSEVOLOD FREDERIKS, who was the first to study the switching phenomenon in detail [18]. The switching process is nowadays called the ***Frederiks transition***.

The anisotropy of liquid crystals is also reflected in their magnetic and conductive properties. As for the former, the diamagnetic anisotropy $\Delta\chi^m = \chi^m_{\|} - \chi^m_{\perp}$ of a thermotropic liquid crystal most often has the same sign as its dielectric anisotropy, that is, it is positive for rod-shaped and negative for discotic mesogens. In lyotropics, on the other hand, this relation as well as the reverse one occurs and lyotropic nematic phases are therefore often denoted with one of the shorthands N^+_C, N^-_C, N^+_D, or N^-_D. The indices C and D refer to cylindrical and discotic micelle shape, respectively, and the + or − superscript to the sign of the diamagnetic anisotropy.

As with dielectric anisotropy and electric fields, the nonzero diamagnetic anisotropy means that we can exert a torque on the director by applying a magnetic field. The same type of analysis as above yields analogous results, including the definition of a (magnetic) Frederiks threshold. The influence of magnetic fields on $\mathbf{n}(\mathbf{r})$ can be quite important in the research on lyotropics, since the water usually used as solvent complicates the use of electric fields for reorienting the director. The diamagnetic anisotropy is, however, often not large enough to allow a reorientation of higher ordered lyotropic phases such as the lamellar and columnar phases; hence, field alignment in lyotropics is a good option only when working with nematic phases.

Just like the optical anisotropy, the dielectric and diamagnetic anisotropies depend on the orientational order and anisotropy in the molecular origins of the susceptibilities, and the type of behavior sketched in Figure 16.15 for Δn is seen also for these parameters. The numerical relations are, however, somewhat more complex than Equation 16.2 since we are here considering electric and magnetic fields of frequencies much lower than those of light, extending down to the static case. At optical frequencies, the electronic polarizability (the field-induced temporary distortion of the electron cloud within the molecule) is the only contribution to the polarization of the material; hence, it is the only molecular property appearing in Equation 16.2. But if the field is static, or oscillating but with a relatively low frequency (kHz–MHz), permanent dipoles in the molecules contribute strongly, as they are (partially) aligned by the applied field, resulting in a much stronger induced macroscopic dipole than what would result from the electronic polarization alone (corresponding to α). Thus, while the dielectric and diamagnetic susceptibility anisotropies will still be linearly proportional to the order parameter, becoming exactly zero for $S = 0$ in the isotropic phase, the various low-frequency contributions to the macroscopic polarization or magnetization must also be considered.

The molecular origin of the macroscopic dielectric susceptibility anisotropy is thus the molecular polarizability anisotropy and the molecular dipole moment. Both are of course present whether or not we have a liquid crystal phase, so we might expect that we can align the molecules by applying an electric field over a sample even if it is in its isotropic state. A typical mesogen dipole moment is on the order of 5 D or about 17×10^{-30} C m so the energy of interaction with a typical applied electric field (say 10 V/μm = 10 MV/m) is on the order of $-\mathbf{p} \cdot \mathbf{E} \approx 2 \times 10^{-22}$ J. This should now be compared to the thermal energy per molecule, $k_B T$, which at room temperature is about $4{\times}10^{-21}$ J, thus about an order of magnitude larger than the electrical interaction energy. Hence, it is clear that the alignment induced by a reasonable electric field applied over a mesogenic compound in its isotropic phase is negligible. Only because of the long-range order prevailing in the liquid crystalline phases, resulting in some 10^{23} molecules responding together in unison to applied fields, do we get the strong response to relatively weak electric fields that is the hallmark of liquid crystals. While this example was worked out for the case of electric fields, an analogous estimation for the magnetic field response yields the same results, see, for example, Ref. [12].

While the dielectric and diamagnetic anisotropies are not too much affected by an inter-liquid crystal phase transition, the situation can be radically different for the electrical conductivity anisotropy $\Delta\sigma$, which typically changes sign at the transition between a nematic (having maximum σ along **n**) and a smectic phase. This is because the appearance of long-range positional order greatly influences the charge carrier mobility μ, the smectic structure increasing μ (and thus σ) along the layers but decreasing it perpendicular to them. Also in columnar phases, the translational order greatly affects the conductivity σ, the greatest value appearing along the columns. Their high conductivity in this direction has rendered these phases particularly interesting for organic electronics. However, the orientational and positional fluctuations in columnar liquid crystal phases are still sufficiently large that a non-negligible scattering of charge carriers arises. This prevents the conductivity in the liquid crystal state from reaching values that are truly competitive with alternative materials. When considering discotics for organic electronics, the typical desired route is to use the liquid crystal phase for achieving a macroscopically aligned columnar geometry without the differently oriented domains and the boundaries between them that pester the corresponding crystalline phases, and then vitrifying the system (it should undergo a glass transition) to get rid of the fluctuations without losing the desired boundary-free columnar geometry [16].

16.3.4 The Viscous Properties of Liquid Crystals

In contrast to isotropic liquids, a liquid crystal has not one but several shear viscosities, reflecting the importance of the direction in which the flow occurs relative to the director. Moreover, a liquid crystal exhibits a type of viscosity that has no counterpart in isotropic fluids, namely, reorientational viscosity. We will here give only a very simplified description of the most important aspects of liquid crystal hydrodynamics, considering first the viscous response to shear flow of a nematic, and then looking at the reorientational viscosities of nematics and smectics.

Figure 16.18a shows a classic setup for defining shear viscosity: a fluid is encapsulated between a stationary bottom plate and a top plate that moves with velocity v, as indicated by the long thick arrow. Since the fluid at each plate is at rest with respect to that plate, the fluid velocity as measured in the external reference frame increases continuously from zero at the bottom plate to v at the top plate, as illustrated by the arrows increasing in length from bottom to top. The resulting shear flow is counteracted by dissipative forces described by the shear viscosity of the fluid. So far everything holds for any fluid experiencing shear flow. But we are here

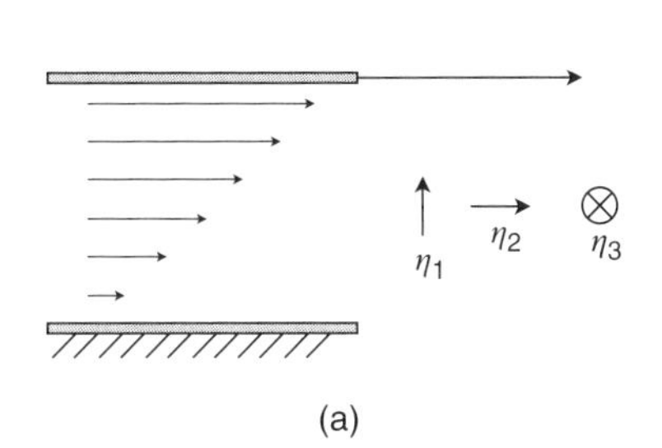

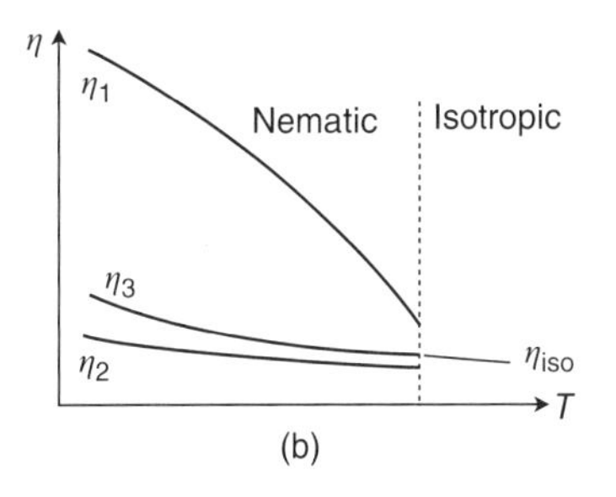

Figure 16.18 (a) The three Mięsowicz viscosities are defined by considering a classic shear flow experiment, except that **n** is blocked by an external magnetic field, giving three possible orthogonal director orientations with respect to the flow direction. (b) The single isotropic viscosity splits up below the transition to the nematic phase, η_3 often essentially following the isotropic viscosity value, whereas η_1 is higher and η_2 lower than the viscosity in the isotropic phase.

considering a liquid crystal, the orientational order of which must also be taken into account. It makes a great difference if the flow is along or perpendicular to **n**, and for the latter case also the direction of the flow velocity gradient with respect to **n** is important. In the 1930s, these three situations, depicted by three orthogonal director orientations in Figure 16.18a, were for the first time successfully realized for measuring the three corresponding viscosities, η_1, η_2, and η_3, by the Polish physicist MARIAN MIĘSOWICZ. In his classic experiments, described in two Nature papers [39, 40] published before and after the war, respectively, the flow cell was oriented in three orthogonal directions relative to a strong external magnetic field, controlling the direction of **n**. Consequently, these three special viscosity values of a nematic are nowadays referred to as the ***Męsowicz viscosities***.

It is probably intuitively obvious that the lowest of the three shear viscosities is η_2, corresponding to flow along the director. The largest viscous resistance is obtained when the director is perpendicular to the flow but parallel to the velocity gradient, that is, η_1 is the largest Męsowicz viscosity. The η_3 value, finally, is of intermediate magnitude, generally considerably closer to η_2 than η_1. A typical temperature dependence is sketched in Figure 16.18b.

In a very approximate way, the Męsowicz viscosities are sufficient for understanding the basic viscous behavior of liquid crystals. For instance, if we make no particular effort to control the director orientation, flow will align the director for minimum dissipation, that is, **n** will be more or less along the flow. This also means that if one attempts to measure the viscosity of a nematic with a standard viscometer, one will generally measure a viscosity not far from η_2, obtaining a value on the order of a Poise (0.1 Pa s) for thermotropics. Flow alignment can be very useful, sometimes even essential, for achieving uniformly aligned samples of liquid crystals that are difficult to align by means of field application or surface action, for example, when dealing with lyotropic columnar or lamellar liquid crystals (however, for liquid crystals with positional order, the minimum dissipation occurs for flow along the layers or columns, i.e., not necessarily along **n**).

While the Męsowicz viscosities bear a close resemblance to the scalar viscosity η in an isotropic liquid, there are in a nematic also new types of viscosity that have no correspondence in an isotropic liquid. If the director is free to reorient, two such viscosity coefficients appear. Consider, for instance, that a magnetic field **B** has aligned the sample homogeneously with an initial director $\mathbf{n}_0$ along **B** in the xy-plane. This is the minimum energy situation and there is no flow. If we now rotate the field slowly some angle Φ around the z-axis, the director will reorient, following the field with a certain phase lag such that the magnetic torque balances a viscous torque that counteracts the reorientation. The phase building blocks simply rotate everywhere around an axis perpendicular to **n** without any bulk flow. The viscous torque is

$$\Gamma = -\gamma_1 \cdot \frac{d\phi}{dt}, \tag{16.11}$$

where the change in azimuthal direction ϕ is counteracted by a viscosity γ_1. This viscosity is called the ***rotational viscosity***. An analog case occurs if we have a uniformly planar-aligned nematic phase ($\mathbf{n}_0$ along one direction in the xy-plane) with positive dielectric anisotropy over which we apply an electric field in the z-direction. As described earlier, the director will now tend to reorient until it is aligned along $\hat{z}$ (cf. Figure 16.19a). This rotation of the director around a perpendicular axis is again counteracted by the same rotational viscosity γ_1. This viscosity also determines the speed of back-relaxation of the director when we take the field away (assuming that the surfaces promote the original planar alignment $\mathbf{n}_0$). The second coefficient of the same nature, γ_2, regards the rotation of the director in a shear velocity gradient, that is, the reorientation that would be the result if we took the magnetic field away during the Męsowicz experiment. It describes the coupling between the orientation of **n** and the shear flow.

We have now introduced five independent viscosity coefficients for the nematic phase. Indeed, the hydrodynamics of nematics requires five independent viscosities for its general description [12] as worked out by Ericksen and Leslie. We

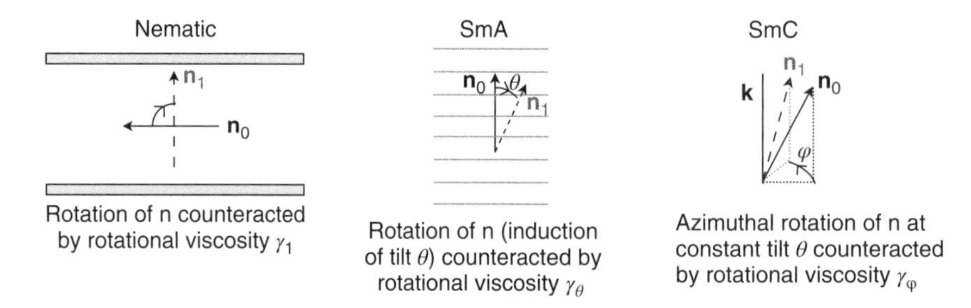

Figure 16.19 Rotational viscosities in the nematic, SmA and SmC phases. See main text for further explanations.

will not discuss this description but only mention that the fundamental five viscosity coefficients $\alpha_1, \ldots, \alpha_5$ in this theory are very different from the viscosities introduced earlier. However, $\eta_1, \eta_2, \eta_3, \gamma_1$, and γ_2 can be expressed in these coefficients and vice versa when convenient.

In smectics, the viscosities are essentially different in character. A rotation of **n** that induces a director tilt in SmA or changes the SmC tilt angle θ is usually coupled to a layer contraction [41]. Such a motion is thus very limited, but for small tilts it is useful to define the so-called "soft mode" viscosity γ_θ for this motion (cf. Figure 16.19b). In the SmC phase, we have another characteristic rotational viscosity different from γ_1 and γ_2 of the nematic. Because the director is naturally tilted with respect to the layer normal, we now have the possibility of reorientation of the tilting direction φ around the tilt cone (cf. Figure 16.19c). The corresponding viscosity is denoted γ_φ and in contrast to γ_θ it is generally very low. This is because the reorientation in question has no impact on the smectic structure, and, at least in the absence of external boundary conditions, it corresponds to a change between energetically equivalent states.

16.4 DEFORMATIONS AND SINGULARITIES IN THE DIRECTOR FIELD

16.4.1 Liquid Crystal Elasticity

While it is probably no surprise that liquid crystal phases with long-range positional order can exhibit solid-like elasticity as a response to mechanically induced changes in a lamellar, smectic, or columnar structure, it may to the novice be more mysterious at first that elastic deformation and restoration exist also in nematics. However, this is not the standard solid-like elasticity related to shape change, but an orientational elasticity connected to deformations in the director field. We will, in this introductory chapter, discuss only this type of elasticity, which is quite unique to liquid crystals. It is the only type of elasticity existing in nematics, but it is just as important in smectics and lamellar phases. In columnar phases, the 2D lattice makes director field distortions more complicated, but also there they play a role.

The undeformed elastic ground state of an achiral nematic is one where the director field is totally uniform, that is, **n** points in the same direction everywhere throughout the sample. An elastic deformation in the case of a nematic results from a *local* reorientation $\delta\mathbf{n}$ of the director, leading to a director field distortion transmitted over a larger scale. Since distortions in the director field cost energy, the nematic tries to reestablish the uniform director field, not by a force as in solid elasticity but by an elastic *torque*. The strength of the torque can be calculated by differentiating the elastic free energy. Our starting point for analyzing elastic deformations and their consequences in a nematic is thus to establish the mathematical expression for the elastic contributions to the free energy. This was first done successfully by the Swede CARL-WILHELM OSEEN in 1928, followed by the Englishman Sir CHARLES FRANK who in 1958 introduced the notation and terminology that is still used today [18]. Consequently, the result is nowadays referred to as the ***Oseen–Frank theory.***

We cannot go into the details of the work of Oseen and Frank here but simply state the key results and discuss the meaning. The main conclusion of Oseen's work was that, neglecting surface effects, *any* generic deformation of the nematic director field can be described as a linear combination of three elementary elastic deformations: ***splay***, ***twist***, and ***bend*** (the English names were coined by Frank). Oseen also introduced a fourth type of deformation (named saddle-splay by Frank), the mathematical description of which contains a complete divergence. This means that it can be transformed into a surface integral over the volume boundaries via Gauss' theorem. The consequence is that we generally can neglect this term when considering bulk nematics, but it becomes important close to boundaries, thus, for example, in vesicles and other membranes. The three fundamental distortions, defined graphically in Figure 16.20, are independent of one another, that is, a twist can only be compensated by a twist of opposite handedness, not by any combination of splay and bend, and analogously for the other deformations. As a result, the full elastic free energy of a bulk nematic can be written as a sum of three terms, each giving the total elastic free energy contribution due to splay, twist, and bend deformations in the sample, respectively:

$$G_{\text{deform.}} = \frac{1}{2}K_1(\nabla\cdot\mathbf{n})^2 + \frac{1}{2}K_2(\mathbf{n}\cdot(\nabla\times\mathbf{n}))^2 + \frac{1}{2}K_3(\mathbf{n}\times(\nabla\times\mathbf{n}))^2. \quad (16.12)$$

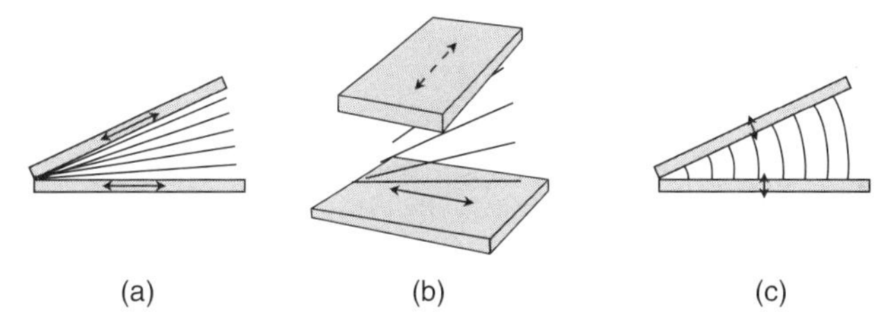

Figure 16.20 Graphical definitions of the three elementary director field deformations (a) splay ($K_1(\nabla\cdot\mathbf{n})^2 \neq 0$), (b) twist $K_2(\mathbf{n}\cdot(\nabla\times\mathbf{n}))^2 \neq 0$, and (c) bend $K_2(\mathbf{n}\cdot(\nabla\times\mathbf{n}))^2 \neq 0$, together with their corresponding terms in the elastic free energy and their respective elastic constants K_1, K_2, and K_3. The drawings illustrate how each deformation can be produced in practice by encapsulating a nematic between flat substrates, the insides of which are prepared so as to ensure uniform planar or homeotropic director orientation at the substrate (the preferred director is indicated with a thick double-headed arrow). Source: Adapted from de Gennes and Prost [12].

Each deformation is described by the appropriate combination of the vector operators divergence ($\nabla \cdot \mathbf{n}$) and curl ($\nabla \times \mathbf{n}$) squared (the handedness or direction of a deformation cannot play a role in an achiral system, hence linear terms are ruled out) and the energy cost for a certain deformation is proportional to the respective elastic constant, K_1 for splay, K_2 for twist, and K_3 for bend.[22] These constants are very important material parameters, critically dictating the behavior of a nematic. They have the dimension of a force, the magnitude typically in the pN range, and they must be positive (required for stability). Often K_1 and K_2 have about the same value, whereas K_3 is about twice that value. It can be convenient to approximate them all equal to a single value K, the so-called "one-constant approximation."

Some aspects of the meaning of Equation 16.12 can be made clearer by making some simplifications. First, since divergence and curl essentially are two types of spatial derivatives of a vector field, we may see each term in (16.12) as proportional to $\delta\mathbf{n}^2$, rendering an analogy with Hooke's law for solid elasticity obvious: for solid and nematic elasticity alike, the energy is proportional to the square of the deformation magnitude. Second, since a spatial derivative gives us an inverse length, we see that each term in (16.12) is proportional to $1/R^2$, where R is the characteristic length of each particular deformation. The characteristic length is the pitch in case of a twist deformation, the radius of curvature of a bend deformation, and the distance to the "source" of a splay deformation. Note that the former is independent of location, that is, the energy cost of a twist deformation depends on how tightly twisted $\mathbf{n}(\mathbf{r})$ is but not on which point in the sample we consider. In contrast, the characteristic length of bend and splay is only defined once we have chosen a location, and it is not uniform. This means that the energy cost of a bend or splay deformation varies throughout the sample (which is natural, since a bend or splay locally gets less and less apparent the further away from the "source" we are).

The $1/R^2$ dependence and the small magnitude of the elastic constants means that elastic deformations cost very little energy if their characteristic length is not very small, so little in fact that large-scale deformations are all the time thermally generated in a nematic liquid crystal. It is these constantly generated random large-scale deformations that yield the nonuniformity in $\mathbf{n}(\mathbf{r})$ of a bulk nematic, giving rise to its characteristic turbidity: since $\mathbf{n}(\mathbf{r})$ changes randomly over distances that are large (but not too large) on an optical scale, so does the effective refractive index for a light beam going through the sample, and strong scattering results.

The expression of the elastic free energy (16.12) is a most important result, and it is also very helpful in practice. Probably the most common use is to find the equilibrium state under a certain set of conditions by minimizing the free energy, which of course must contain any other relevant terms, for instance, electric field terms. If only the elastic energy terms are present, it is clear that the ground state will be one of a uniform, or at least, minimally deformed director field (compatible with the boundary conditions) throughout the sample. This is the foundation for our ability to control the ground-state director orientation in a simple manner, an absolute prerequisite for using nematics in devices. By using substrates prepared for inducing planar or homeotropic anchoring, in the former case with one direction in the substrate plane being the preferred direction of $\mathbf{n}$, we can control the director configuration throughout the sample. By assembling the substrates with their preferred direction nonparallel, we can even induce a well-defined twist, as is the case in the classic *Twisted Nematic* display configuration.

As mentioned in Section 16.2.5 the chiral nematic, or cholesteric, phase features a helical director modulation spontaneously, that is, without any external influence the director field twists along an axis perpendicular to $\mathbf{n}$. A similar thing holds for chiral smectic-C-type phases, with the important difference that the helix axis is now perpendicular to the C-director $\mathbf{c}$, that is, along the smectic layer normal. Obviously, both helical structures are incompatible with Equation 16.12, which predicts a totally uniform ground-state director field. As was pointed out by Frank in 1958, the expression for the elastic contribution in the free energy must be slightly modified for the case of a chiral nematic. For a cholesteric phase, the standard achiral twist term is replaced by a slightly extended term that expresses the spontaneous twist:

$$G_{twist}^{N*} = \frac{1}{2}K_2(\mathbf{n} \cdot \nabla \times \mathbf{n} + q_0)^2. \qquad (16.13)$$

The added term $q_0 = 2\pi/p_0$ is the wave vector of the cholesteric helix, p_0 being its natural pitch. It is a straight-forward exercise to show that the twist term in (16.13) is minimized not by a uniform director field but by a twisted one, for which q is equal to the natural wave vector q_0.

When the natural twist is very weak, q_0 is small and the term in Equation 16.13 does not add significantly in case of a nontwisted director field to the energy Equation 16.12 for a nonchiral nematic with the same $\mathbf{n}(\mathbf{r})$. For strong natural twist, however, typically with the pitch on the same order as the wavelength of visible light or smaller, the added term in Equation 16.13 means a new contribution to the elastic energy that can be dominant. If we compress or dilate the helix, the elastic energy is a quadratic function of the deviation from the natural wave vector. This means that the cholesteric helix acts as an elastic spring trying to keep the pitch equal to the natural (spontaneous) pitch p_0, in analogy with Hooke's law for springs in basic mechanics.

[22]Because the three elastic constants in a way represent diagonal elements in the elastic energy (they are actually components of a tensor of fourth rank), they are sometimes written K_{11}, K_{22}, and K_{33}. The saddle-splay constant is K_{24} in this notation.

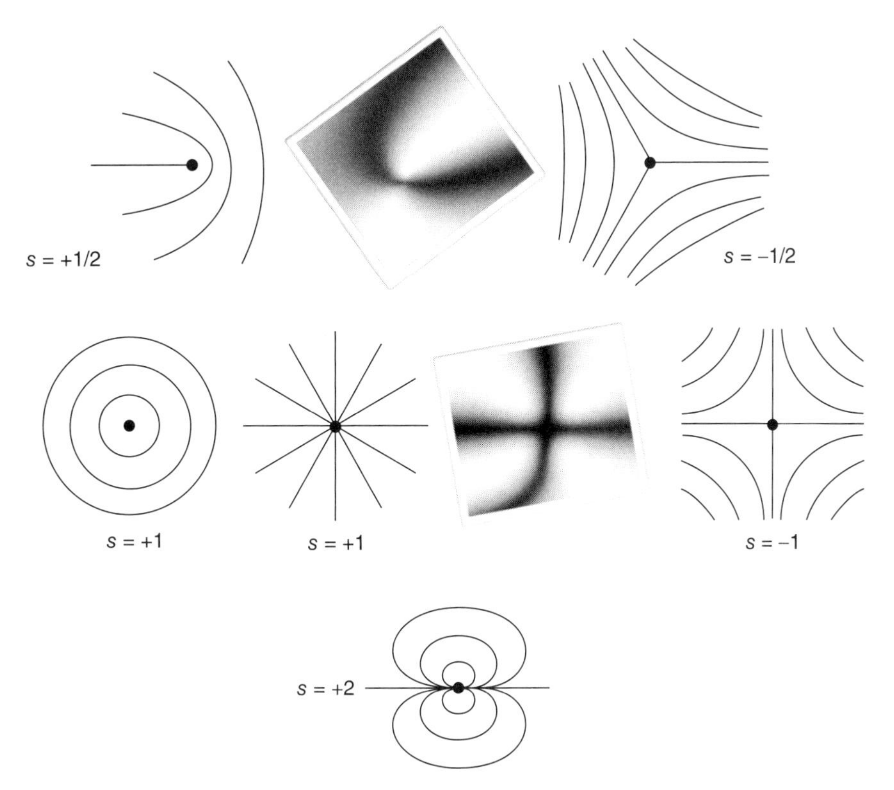

Figure 16.21 Examples of disclinations of magnitude $\frac{1}{2}$, 1, and 2, of positive and negative signs, in a planar-aligned nematic. For the $s = \pm\frac{1}{2}$ and $s = \pm 1$ cases, polarizing micrographs are provided as illustrations of how these disclinations can be recognized in the polarizing microscope. Note that the sign of the disclination cannot easily be determined from a single micrograph.

This very much resembles the elasticity of smectics, keeping the layer distance constant. Although in the cholesteric case, this elasticity is much weaker, it is of the same nature as in a smectic or a solid and is responsible for the fact that short-pitch cholesterics show very similar textures (e.g., focal-conic defects) to those of smectics.

16.4.2 The Characteristic Topological Defects of Liquid Crystals

The director fields shown in Figure 16.20 are all smooth and locally defect-free despite their deformed state. If we prepare a nematic sample with degenerate planar alignment, we will however almost always have defects in the director field, that is, points where $\mathbf{n}(\mathbf{r})$ is not defined. Such a defect is called a ***disclination***[23]. Some examples of disclinations are shown in Figure 16.21 together with exemplifying polarizing micrographs of how the most common defects may appear visually. The drawings are 2D as is the plane of observation in the microscope. The lines illustrate $\mathbf{n}(\mathbf{r})$ and the black dot indicates the disclination itself (the point where the director is undefined).

You will notice that each disclination has been given a magnitude and a sign in the figure, and disclinations are indeed found with different strengths. In fact, when two or more disclinations are present in a sample, these parameters are of fundamental importance for determining the interactions between the defects. The strength is denoted by s and a simple way to determine s is illustrated in Figure 16.22. Take some kind of linear object with a distinct head–tail asymmetry (in the figure we use a match) and place it along the director at some starting point of the director field near the disclination. The starting point (**1** in the figure) is arbitrary, as is the direction of the head of the object. Now move the object one full turn around the disclination (the rotation sense is also arbitrary), always making sure that the object orientation follows the variations of the director field exactly and stopping first when the object is back where you started. In the left example in Figure 16.22, the match has made one turn anticlockwise around the disclination, and as a result the match itself has also rotated one full turn anticlockwise. The director thus rotates in the *same sense* (we attribute a positive sign to this) and by the *same number of turns* (the magnitude of the disclination is the ratio of the

[23]The term was coined by Frank who first called them *disinclinations*, in analogy to dislocations in solids: while a dislocation is a defect in a crystal lattice where the *location* is ill-defined, a disclination is a defect in a liquid crystal where the *inclination* of the director is ill-defined. Over time, the shorter term disclination has, however, been generally accepted [18].

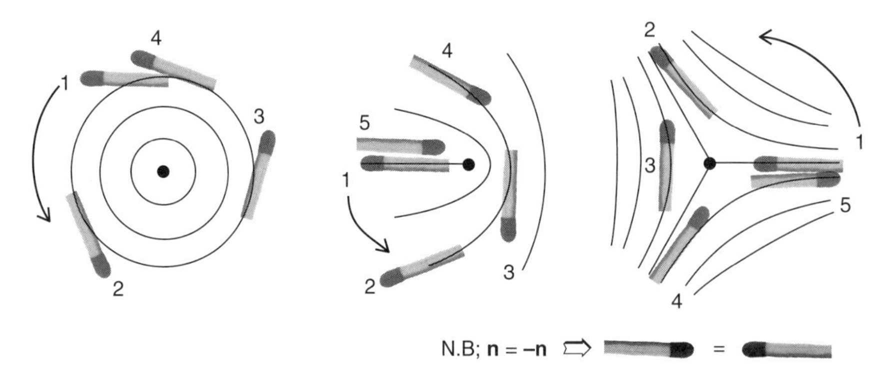

Figure 16.22 A simple way to determine the magnitude and sign of disclinations (see main text for explanation).

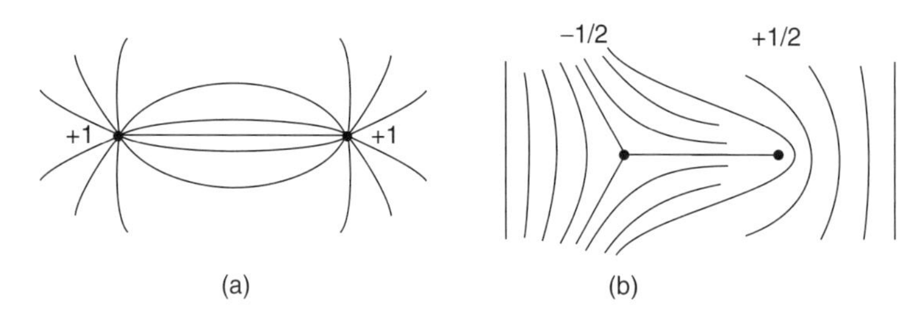

Figure 16.23 Two like-signed disclinations repel each other (a), effectively creating a distortion to the director field that at a distance much longer than the defect separation corresponds to that of a single disclination with strength equal to the sum of the actual disclinations. If the disclinations have opposite signs (b), they tend to compensate each other, reducing the total director field distortion and thus effectively being attracted by each other. If they are kept from merging and thus annihilating each other, they constitute an elastic dipole.

number of turns) as we go around the disclination, and we thus characterize this disclination by $s = +1$.

When doing the same thing around the second disclination, we see that the director again rotates anticlockwise when we follow $\mathbf{n}(\mathbf{r})$ in an anticlockwise loop around the disclination; hence, its sign must be positive. But the match has not come to its original orientation: it has only rotated half a turn. While a half-turn rotation is no symmetry operation for our match, it is for the director ($\mathbf{n} \Longleftrightarrow -\mathbf{n}$); hence, this disclination is perfectly allowed. Its magnitude is 1/2, since the director rotated only half a turn as a result of the full circumvention of the disclination. When following the procedure for the third defect in Figure 16.22, finally, we again end up with a director that has been rotated only half a turn, but in addition it has been rotated in the opposite sense: while we went around the disclination anticlockwise, the match rotated clockwise. Hence, this defect is an $s = -\frac{1}{2}$ disclination. Disclinations with $|s| > 1$ are rarely observed since the energy contained in the surrounding field of a singularity increases as s^2. Energy is thus gained by splitting an $|s| = 2$ disclination into two $|s| = 1$, or one $|s| = 1$ into two $|s| = \frac{1}{2}$.

In the three dimensions of our real world, the 2D drawings discussed here correspond either to a defect line through the bulk or to a point defect stuck at the surface. It turns out that line defects of integer strength are thermodynamically unstable, because the liquid crystal can decrease its energy by smoothly inclining the director around the defect toward one of the bounding surfaces, thereby removing the line discontinuity and transferring it to a point defect at the surface. Thus, integer defects are always point defects stuck at one of the bounding substrates of our sample. This way of smoothing out a line defect (called *escape in the third dimension*) is not available for half-integer defects; hence, these are always line discontinuities. For a richer discussion of these issues, see, for example, the textbook by de Gennes and Prost [12].

What happens if two $s = +1$ disclinations approach each other? In Figure 16.23a, we have sketched this situation. The director field gets strongly distorted between the disclinations, and the distortion gets stronger the closer the singularities. If we were to push them closer to each other, the effect on the director field would be like tightening a spring. We would have to do work on the system to achieve this, whereas, on the other hand, the energy stored in the strongly distorted director field would allow the system to push the defects away from each other if nothing keeps them in place, that is, *two disclinations of the same sign repel each other*. If we nevertheless were to push the defects closer and closer until the point that they overlap, we would get an $s = +2$ pattern (cf. Figure 16.21). Thus, *merging two disclinations* simply creates a new disclination with sign and strength given by the *sum of the original disclinations*.

If we apply this new result to two disclinations of the same strength but opposite sign, as in Figure 16.23b, we notice

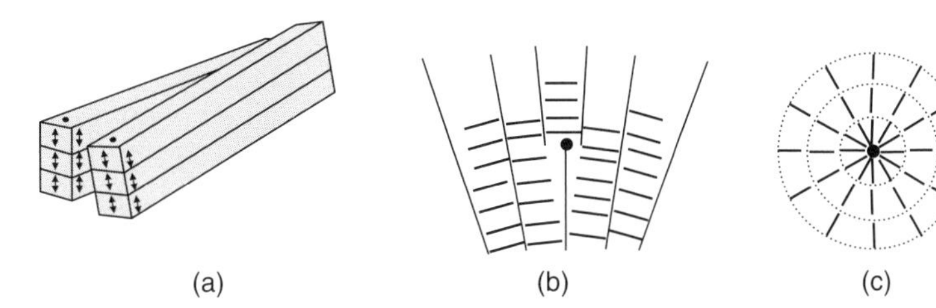

Figure 16.24 The geometry of SmA does not allow twist (a) and bend (b) in the director field but splay (c) occurs. The director is to the left illustrated with double-headed arrows, in the middle, and to the right with lines thicker than the lines indicating the smectic layer boundaries.

that the net result should be no disclination at all. Indeed, we see in the figure that the director field is uniformly vertical already quite close to the two disclinations, which in a configuration like this, close to each other but not merged, constitute an *elastic dipole*. As the distortions of the director field get smaller, the closer the disclinations, they are effectively attracted and it is not too difficult to realize that if they eventually are allowed to merge they will annihilate each other, leaving behind a deformation-free director field. With disclinations in a pure nematic, such a dipole structure is thus practically difficult to realize, but, as we will see in Section 16.6.1, they can be prepared relatively easily by adding particles, drops, or bubbles.

The elasticity theory described earlier was developed for nematics but it can be extended and modified to phases with partial positional order. For each particular phase, one must however take into account the special restrictions imposed by its symmetry, an issue we will now demonstrate using the examples of SmA and SmC phases. The results that we will obtain in principle apply also to all lyotropic lamellar phases discussed here, L_α, L_β, and $L_{\beta'}$, since they have the corresponding symmetries. However, the greater rigidity of the gel phases in practice often prevents strong bending of the lamellae in these phases.

If we first consider SmA, it is relatively easy to realize that two of the three elementary deformations are actually not allowed by the structure, as illustrated in Figure 16.24. Considering twist first (a), we see that, since the director also defines the layers through its double role as layer normal, a twist in the director field also entails a twist in the layer structure, something that would require a breaking up of the layers. This cannot happen in ordinary SmA phases (but it occurs in the already mentioned Twist Grain Boundary, TGB*, phases[15]). As for director bend, drawing (b) illustrates that also this gives problems with the layer structure. Since a bend in $\mathbf{n}(\mathbf{r})$ effectively corresponds to a splay in the layers, the layer thickness would have to change constantly, a situation that does not occur in reality since the layer thickness is defined by the effective length and inclination of the molecules (cf. Section 16.2.3). Therefore, if a director bend were to appear, it would induce a network of dislocations in the layered structure.

While director twist and bend are not allowed the *layers* are allowed to bend, for instance, into a cylindrical roll (Fig. 16.24c). They can also easily undulate with a layer thickness essentially constant. This corresponds to mainly splay in the director field and the deformation energy is equal to

$$G^{SmA} = \frac{1}{2}K_1(\nabla \cdot \mathbf{n})^2 + \frac{1}{2}B\delta d^2, \tag{16.14}$$

where the second term represents the ordinary (solid-like) elasticity due to variations δd in layer thickness. In the case of a cylindrical (or spherical) shape with perfectly circular cross section, the layer thickness is unchanged and only the splay term prevails.

When smectic layers form out of a nematic phase or from the isotropic liquid, they often do so in torus-like structures referred to as *Dupin cyclides*. In the simplest case, the half cross section of the cyclide looks like the drawing in Figure 16.24c, that is, we here have a combination of director splay and layers bent into tubes. The discontinuities in $\mathbf{n}(\mathbf{r})$ that must appear as smectic layers rolled into Dupin cyclides are to fill space, and the resulting hyperbolic and elliptical defect lines are the origin of the characteristic *focal-conic* defects, which are often seen in the textures of smectics in the microscope [12, 21]. Again, space is too limited here to go into these issues, and we must refer the reader wishing to learn more about defects to the standard textbooks.

Although SmC is less symmetric and thus in some sense more ordered than SmA (occurring at lower temperatures), it actually allows more fluctuations in the director field than SmA. This is because the addition of a tilting direction gives a new freedom for deformations, which do not interfere with the layered structure. In fact, the projection of $\mathbf{n}$ onto the layer plane, the C-director $\mathbf{c}$, behaves in many respects similar to the normal director in the nematic. The best illustration is perhaps the helical structure of a chiral SmC* phase, where $\mathbf{c}$ mimics the twisted structure that $\mathbf{n}$ shows in a chiral nematic phase (cf. Figure 16.10). But also $\mathbf{n}$ itself has gained new degrees of freedom. By making a drawing of the helical SmC* structure, as in Figure 16.10 (a 3D model is even better), it is not too difficult to convince oneself that $\mathbf{n}$ actually undergoes a combination of bend and twist deformation in the helix. If we instead have an achiral SmC phase, such deformations can obviously be induced thermally as temporary fluctuations if their characteristic length is large.

The sequence of strong→weak→strong thermal fluctuations in an N→SmA→SmC material can easily be observed at a polarizing microscope. By placing a homeotropically aligned achiral nematic sample, exhibiting SmA and SmC phases at lower temperatures, in the microscope and observing it on cooling, we see that the N phase fluctuates relatively strongly, and there are many defects in the basically black texture. At the transition to SmA, most defects disappear and the fluctuations suddenly stop, reflecting the great reduction in degrees of freedom for director fluctuations. When the temperature reaches the transition to SmC, on the other hand, the characteristic schlieren texture of a *quasi-homeotropic* SmC sample (the changing curvature of the schlieren reflecting the variations in **c**) appears, the texture gets brighter and fluctuates quite substantially. This is certainly an experiment worth doing if you have access to a suitable substance and the required lab equipment, as it gives a direct hands-on illustration of the issues we have discussed earlier. The quasi-homeotropic SmC schlieren texture can look very similar to the schlieren texture of a planar-aligned nematic, but there is one important difference: there are only integer disclinations. Because the sign invariance of the director, $\mathbf{n} \Longleftrightarrow -\mathbf{n}$, does not apply to the C-director ($\mathbf{c} \not\Leftrightarrow -\mathbf{c}$ because $\mathbf{c}$ and $-\mathbf{c}$ correspond to opposite director tilting directions; hence, physically different states) half-integer disclinations as in the top row of Figure 16.21 are not possible in an SmC phase.

We have so far used mainly 2D drawings and stated that the corresponding 3D situation is either a point defect stuck at one substrate or a uniform extension of the 2D pattern between the sample substrates. However, very interesting things happen when we allow also curved surfaces in 3D, for example, a drop or a shell. As you can easily convince yourself of by looking at a globe, the topology of a sphere requires a total defect sum of $s = +2$. The south pole and north pole each constitute an $s = +1$ disclination in the meridians, but the rest of the earth's surface is free of defects in this "director field." There are many other ways that one could draw a continuous director field on the globe, but regardless of choice, a summation of all defects will always end with a total of $s = +2$. The same holds for a spherical planar-aligned liquid crystal sample.

We end this brief introduction to liquid crystal elasticity by pointing out that this is a *continuum* description: it relies fundamentally on the assumption that the liquid crystal's true discrete molecular nature can be neglected, the fluid being treated as a continuum instead. This assumption holds as long as the objects and phenomena under discussion have sizes that are very large on a molecular scale. Since this is indeed the case in most practical uses of liquid crystals, the elasticity theory is most often valid and extremely useful. There are however situations where the theory may break down, in particular in the research on nanoparticles in liquid crystals.

16.5 THE SPECIAL PHYSICAL PROPERTIES OF CHIRAL LIQUID CRYSTALS

Chirality has several important consequences for the physical properties of liquid crystal phases: sometimes related to the helical modulation and sometimes simply to the loss of mirror symmetry. The chiral SmA* phase, for instance, shows a totally different response to electric fields than an achiral SmA phase (SmA* can be considered paraelectric), although structurally the two phases are identical since a helical modulation is not allowed. And in the chiral SmC* phase, the chirality leads to an even more dramatic effect: the existence of an internal spontaneous polarization density, giving rise to ferro- and antiferroelectric behavior [13]. There are several important helix-related effects in nematics, such as strong optical activity, an apparent viscosity that for short-pitch cholesterics may be several orders of magnitude higher than that of an achiral nematic, and striking characteristic textures and defects that are very different from those of non-chiral nematics. If the helix pitch is short, on the order of 300 nm, then the optical anisotropy of the phase reverses compared to the nonhelical analog, the helix axis becoming the effective optic axis. Moreover, the substance may show so-called ***selective reflection*** of visible light, a striking phenomenon that is both useful and beautiful. In this brief introduction, we can unfortunately not go into all these interesting and useful phenomena. We will discuss only the optical activity and selective reflection, as these are perhaps the most commonly occurring.

16.5.1 Optical Activity and Selective Reflection

Few liquid crystals are as immediately fascinating as a short-pitch cholesteric. Placed on a dark background and observed by the naked eye, it can exhibit striking iridescent colors, ranging from deep violet to dark red. If the observer tilts the sample, he or she will notice that the color changes with the angle at which the sample is observed, shifting toward shorter wavelengths (more toward the violet) for more oblique angles. Moreover, the color will often also change as a function of temperature, generally toward longer wavelengths on cooling. The origin of these striking colors is selective reflection. A somewhat more elaborate examination will reveal that the selectively reflected light is circularly polarized if the sample is viewed at normal incidence, elliptical if it is tilted. Sometimes, the cholesteric will not be colored because it is so strongly twisted that the reflected wavelength is below the visible region, but a careful investigator might then find that plane-polarized light sent through the sample along the helix axis gets

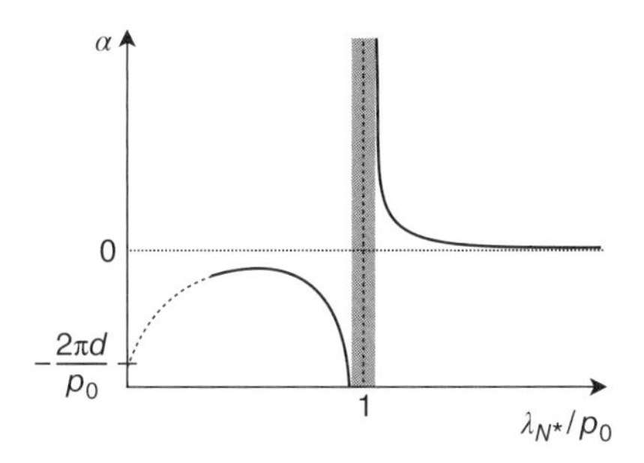

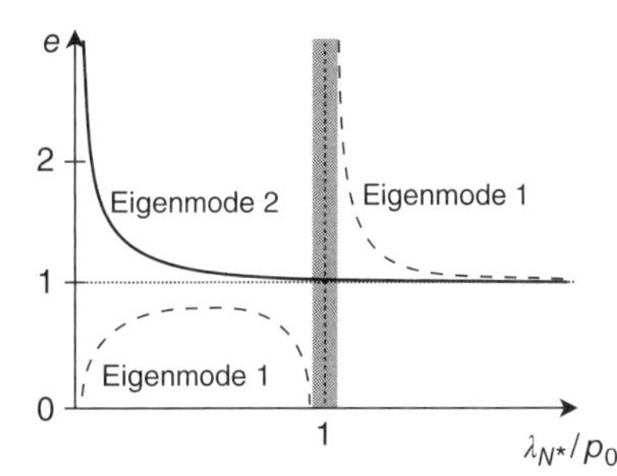

Figure 16.25 Generic sketches of the optical rotatory power α (a) ($\alpha > 0$ defined as right-handed rotation) and the ellipticity e of the eigenmodes (b) of a left-handed cholesteric liquid crystal along its helix axis, as a function of the ratio of the light wavelength in the cholesteric medium to its helix pitch, λ_{N*}/p_0. The handedness in eigenmode 1 is the same as in the helix, in mode 2 it is opposite.

its polarization plane strongly rotated: the cholesteric is optically active, in some cases extremely strongly so.

The complete derivation of the optical properties of cholesterics is a rather advanced optics exercise, involving the solution of Maxwell's equations in the helically modulated medium. As it is outside the scope of this chapter, we refer the interested reader to, for example, Chapter 11 of the *Introduction to Liquid Crystals* by Preistley and Wojtowicz [42], where a very readable account of the procedure can be found. We will here simply state the results of this derivation, beginning with the fact that we must distinguish three regimes of different behavior, depending on the ratio of the light wavelength inside the cholesteric, λ_{N*}, and the pitch of the cholesteric helix, p_0:

- $\mathbf{p_0 \gg \lambda_{N*}}$ ***(Mauguin limit)*:** For very long helix pitch, light propagates along the helix of the chiral nematic essentially as two orthogonal linearly polarized eigenmodes (they are slightly elliptic), their planes of polarization following the rotation of the local optical indicatrix in the structure. In the $\lambda_{N*}/p_0 \to 0$ limit, the formal optical "rotatory power" $|\alpha|$ (the magnitude of the polarization plane rotation) for each mode is simply equal to $2\pi d/p_0$ (see Fig. 16.25a), where d is the sample thickness, that is, for a sample of thickness $d = p_0$ the polarization plane is rotated one full turn. This is not optical activity in the normal sense, the rotation is just given by the helix and is independent of λ. It is thus *achromatic* but requires that the incident light is polarized according to one of the eigenmodes of the cholesteric structure (Fig. 16.25b) in order to work. This long-pitch regime is often called the Mauguin limit, in recognition of the first analysis of the situation by Charles Mauguin [18].
- $\mathbf{p_0 \approx \lambda_{N*}}$: As the light wavelength in the cholesteric approaches the helix pitch from the short-wavelength side, the ellipticity e of eigenmode 1 grows from zero (vertical linear polarization) to near 1 (circular polarization) and then back to zero at the band edge. At the same time, eigenmode 2 goes from horizontal linear polarization (infinite ellipticity) to circular. The medium becomes truly optically active, with a rotatory power that diverges, $|\alpha|$ becoming extremely high in the direct vicinity of the selective reflection band. The edges of the reflection band are defined by $\lambda_{N*} = p_0 \pm \frac{1}{2}\Delta n p_0$, where Δn is the birefringence the sample would have had if the helix were unwound. Within the reflection band, the light is truly separated into two circularly polarized components, the one with the same handedness as the cholesteric helix no longer being accepted by the medium and thus fully reflected, the other one fully transmitted. The attribute "selective" refers to the rather narrow width of the reflection band. In modern terminology, the cholesteric liquid crystal has a natural *photonic band gap*. On the long-wavelength side of the selective reflection band, the cholesteric is again strongly optically active, just as on the short-wavelength side, but the sign of the optical rotation has changed. While the rotation of the polarization plane was in the same sense as the helix for $\lambda_{N*} \ll p_0$, it is now in the opposite sense.
- $\mathbf{p_0 \ll \lambda_{N*}}$: All light is again transmitted through the sample but both eigenmodes are now very nearly circularly polarized. In this regime, the optical rotatory power decreases with increasing wavelength as $\alpha \sim \lambda_{N*}^{-2}$, like in other common optically active media, for example, quartz.

These results are summarized graphically in Figure 16.25.

The origin of the selective reflection is the optically periodic structure resulting from the helical modulation. Although the periodicity of the helix is p_0 the optical periodicity is half of this, since already a 180° turn of the director takes us back to an optically equivalent situation. Based on this periodicity, we can apply the standard geometrical method well known from the analysis of X-ray diffraction from crystals to, in a simple way, obtain an expression for the relation between the selectively reflected color and the viewing angle.

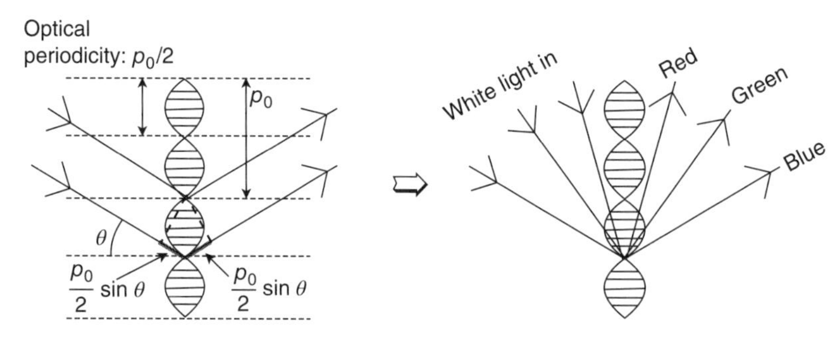

Figure 16.26 The viewing angle dependence of the color of a cholesteric liquid crystal, summarized to the right, can be understood following a simple graphical procedure, sketched to the left. See text for further explanations.

Consider the schematic cholesteric structure drawn in the left part of Figure 16.26. The sample is aligned with its helix axis vertical, and it is illuminated by light coming from the left in the picture at some arbitrary angle θ. Just as in the analysis of Bragg scattering in crystals, the rays entering the cholesteric are reflected at a periodic set of planes, the spacing here being the optical period $p_0/2$. These virtual planes are defined by the local optic axis having a certain angle with respect to the plane of incidence (the paper plane in Fig. 16.26). Basic geometrical optics tells us that the reflection angle is equal to the angle of incidence. Looking at the figure, we see that the lower of the two depicted reflected rays has gone a slightly longer path than the upper one, the difference being exactly $2 \cdot (p_0/2) \sin \theta = p_0 \sin \theta$. These two rays will interfere to produce the effective reflected light and we know (again from basic optics) that the interference will be constructive if the path difference is equal to a multiple of a full wavelength. In other words, we obtain for the reflected light the result:

$$p_0 \sin \theta = m\lambda_{N^*}, \qquad (16.15)$$

where m is an integer. We recognize this from crystallography as Bragg's law. In cholesterics, Equation 16.15 is however fully correct only for oblique incidence ($\theta \neq 90°$). For normal incidence (along the helix), all higher order reflections are absent, that is, only the fundamental reflection ($m = 1$) is present. Basically, the reason is the perfectly sinusoidal modulation in the N^* phase: higher order reflections represent higher harmonics of the periodic structure but a sinusoidal modulation has no harmonics, only the fundamental. For oblique incidence the modulation is no longer perfectly sinusoidal since we are not following the helix; hence, higher harmonics are present in the structure and we get higher order reflections (albeit very weak).

For the normal incidence case, $\theta = 90°$ and $m = 1$, Equation 16.15 gives us the result stated previously, namely, that the selectively reflected light has a wavelength in the liquid crystal equal to the cholesteric pitch. The light wavelength in the cholesteric is shortened with respect to the wavelength in air by a factor equal to the average refractive index. For typical chiral thermotropics, this is about 1.5, so we can immediately estimate the helix pitch of a cholesteric that under observation straight on looks, for instance, green (~550 nm wavelength in air) as roughly 370 nm. A decrease in the angle of incidence θ of the light leads, according to Equation 16.15, to a decreased wavelength of the selectively reflected light. This immediately explains the observation that the color changes toward shorter wavelengths (in the direction from red to blue) if we tilt a cholesteric sample away from us, as schematically summarized in the right part of Figure 16.26.

Since the helical modulation imposed on **n** in cholesterics applies to **c** in chiral smectic-C-type liquid crystals, also these chiral phases exhibit the special optical properties described above, including the optical activity (with sign and magnitude depending on the relation between light wavelength and helix pitch) and selective reflection. For the basic SmC* phase, the quantitative results derived earlier regarding the selective reflection—Equation 16.15 complemented with the restriction that $m = 1$ for normal incidence—are directly valid. For more complex chiral smectic-C-type structures, the situation can be slightly different.

Having cubic symmetry or being amorphous, all blue phases are optically isotropic in that they show no linear birefringence, but they show optical activity and reflect circularly polarized light. For BP I* and BP II*, the reflected wavelength is not unique (and not necessarily blue) but depends on the orientation in space of the cubic lattice. These phases show a characteristic platelet texture (Fig. 16.9c), each platelet corresponding to a certain lattice orientation. The color arises from Bragg reflection from the "crystal planes" of the structure, different orientations yielding constructive interference for different wavelengths, hence the multitude of colors.

16.6 SOME EXAMPLES FROM PRESENT-DAY LIQUID CRYSTAL RESEARCH

We end this introduction to liquid crystals by giving five examples of the exciting directions in which soft matter research dealing with liquid crystals is heading today. The choice of examples obviously reflects the interests of the author of the chapter, but they do in fact belong to the topics in liquid crystal research getting the most attention on an international scale. They are thus a representative selection, although other authors might have emphasized certain aspects less and others more. The topics will only be introduced very briefly, with references provided for further reading.

16.6.1 Colloid Particles in Liquid Crystals and Liquid Crystalline Colloid Particles

The last 15 years have seen a strongly increasing interest in the combination of liquid crystals and colloids, with two quite different approaches playing the lead roles. On the one hand, colloidal particles can be introduced into a liquid crystalline host, thereby ordering the particles in various ways via the interaction between topological defects, mediated through the elasticity of the nematic director field. The resulting arrangements depend on particle size, shape, and how the particle surfaces interact with the liquid crystal. On the other hand, one can prepare colloidal particles that contain a liquid crystal in their core or shell and disperse these in an isotropic dispersion medium such as water.

Although beautiful experiments on the interaction of colloidal scale particles with the nematic director field were conducted by Pieranski and co-workers already in the 1970s [43], the large-scale interest in this new class of responsive colloid took off after a seminal article published in *Science* by Poulin *et al.* [44]. Their elegant study revealed how surfactant-loaded water droplets dispersed in a thermotropic nematic are organized in chains, thanks to the fact that every droplet constitutes an $s = +1$ disclination in the nematic director field, each of which in a flat sample for topological reasons must be accompanied by an $s = -1$ disclination. A number of intriguing modes of self-assembly were observed and explained as a result of the strong impact of topology in nematic liquids crystals. Several researchers around the world were inspired by these results, the most active group perhaps being the team around Igor Musevic and Slobodan Zumer in Ljubljana, Slovenia, which has produced a long series of fascinating papers on colloidal crystals and smaller scale regular structures stabilized by forces arising from the nematic elasticity [45–52]. Other very active teams are, for example, that around Oleg Lavrentovich at the Kent State Liquid Crystal Institute, which has studied in particular the dynamics of liquid crystal colloids [53–56] and the group of Ralf Stannarius in Magdeburg, which specialized on colloids in smectic films [57–59]. The self-assembled colloidal structures arising in nematic host phases are fascinating from an academic point of view, but they are also technologically interesting, for example, for the generation of photonic crystals and metamaterials.

Going down in particle size to the nanometer scale, many researchers are studying colloids of carbon nanotubes or metallic or semiconducting nanorods in liquid crystals, thermotropic as well as lyotropic, where the liquid crystal so far has primarily acted as an aligning agent [60–65]. This concept can work very well, a lyotropic nematic host phase being able to hold a sufficient concentration of well-dispersed carbon nanotubes, while aligning them along **n**, that the colloid turns into a fluid linear polarizer [66]. The resulting phase also acquires interesting viscoelastic properties, allowing thin and very long filaments with uniformly aligned nanotubes to be extracted rapidly from the bulk phase [67]. Other interesting results in this field are, for instance, the report that carbon nanotubes can act as chiral dopants for thermotropic liquid crystals [68] (a truly surprising result considering that carbon nanotube samples are racemates), that carbon nanotube doping enhances the performance of liquid crystals in display devices [61, 69, 70], or that actuators can be made using nematic elastomers doped with carbon nanotubes [71]. Also spherical nanoparticles are interesting to disperse in liquid crystal hosts, several interesting new effects, for example, on alignment switching and chirality transfer being observed in case of alkane thiol or alkane amine-capped metal or semiconducting nanoparticles in nematics [72–74]. These are only a few examples from the very active field focusing on colloids of nano- and microparticles in liquid crystals.

The reverse situation, where the liquid crystal is inside the colloidal particle, dispersed in an isotropic host, has been studied mainly along two tracks. On the one hand, lyotropic liquid crystals have turned out to hold promise for drug or nutrient delivery purposes [75–78]. Here, a cubic or hexagonal liquid crystal phase formed by food-grade amphiphiles in water and possibly also a (food-grade) oil are contained in a thin shell of amphiphilic block copolymer. These soft matter nanoparticles, generally prepared via ultrasonication giving diameters in the 100 nm range, are referred to as cubosomes and hexosomes, respectively, and they can be heavily loaded with drugs or nutrients for oral delivery. On the other hand, thermotropic liquid crystals are being studied as colloidal drops [79–81] or shells [81–87], stabilized against the aqueous continuous phase by a layer of surfactant or amphiphilic polymer. In this case, it is primarily the intriguing and potentially useful defect arrangements forming in the colloidal liquid crystal particles that trigger the interest. This fascinating research field is discussed in depth in the chapter by Alberto Fernandez-Nieves.

16.6.2 Biodetection with Liquid Crystals

The sensitive response of nematics in terms of director reorientation as a result of a change in environment, together with the strong effect on the optical properties of such a reorientation, form the basis for employing liquid crystals as biodetectors, a research field that was pioneered mainly by Nicholas Abbott at the University of Wisconsin-Madison [88–94]. The basic idea is to prepare a liquid crystal at an interface with an aqueous phase with a particular alignment, either uniformly planar or homeotropic, giving it a corresponding fingerprint in polarizing microscopy. The analyte is then added to the aqueous phase and if a biomolecular binding event such as antigen–antibody recognition takes place at the interface, it modifies the liquid crystal alignment, for example, by rearranging amphiphiles at the interface, giving rise to a distinct change in polarizing microscopy texture. The liquid crystal thus functions primarily as an amplifier, making the biomolecular binding event easily visible on microscopic scale. The concept is most often applied to planar liquid crystal samples, but liquid crystal droplets were also successfully used for detecting bacteria and viruses [88]. Since the liquid crystal–based method makes fluorescent tagging or labeling of molecules unnecessary, it is often referred to as label-free biodetection.

Recently, other researchers extended this concept to detect the presence of single-stranded (ss) DNA using liquid crystals [95, 96]. Price and Schwartz [96] prepared a liquid crystal at an interface with an aqueous solution of a cationic surfactant, the latter inducing homeotropic alignment of the liquid crystal, which thus appears dark in polarizing microscopy. If ssDNA is added electrostatic complexation between anionic DNA and the cationic surfactant occurs,[24] changing the liquid crystal alignment and thus the texture. If complementary DNA strands are added, the two strands form a duplex, altering the electrostatically complexed DNA–surfactant structure and thus again changing the liquid crystal orientation, with a consequent change in optical appearance. A similar approach was followed by Chen and Yang [95], and they even succeeded in quantitative determination of the DNA concentration based on the liquid crystal realignment.

16.6.3 Templating and Nano-/Microstructuring Using Liquid Crystals

The self-assembled long-range ordered yet nanostructured arrangements inherent of liquid crystals are very attractive from a broad materials science point of view. Several approaches are therefore being followed to take advantage of the liquid crystalline order for other materials, either by making the liquid crystal-generated order permanent via polymerization/gelation or by preparing a liquid crystalline sample such that a regular array of defects forms, which can then be used, for example, for positioning nano- or microparticles over large areas.

Lyotropic hexagonal and cubic structures are often employed as templates for making mesoporous silica, of immense use in catalysis applications [97]. An organosilicate compound such as tetraethylorthosilicate (TEOS) is added to the lyotropic phase as a silica precursor, and this is then hydrolyzed into SiO_2. As the reaction takes place in the liquid crystal environment, the produced silica adopts the structure of the aqueous channels of the lyotropic template. The surfactant is then removed via thermal decomposition (calcination), leaving only the nanostructured silica, which now has the hexagonal or cubic structure given by the liquid crystal. Alternatively, it may sometimes be advantageous to instead replace the ordinary surfactants with surfactants with reactive headgroups [98–100], allowing the polymerization of the surfactant component of the lyotropic liquid crystal. The resulting structures have potential, for example, for water filtration applications or as synthetic biodegradable materials for medical applications.

Also thermotropic liquid crystals are highly attractive as templates of various kinds. An important example is the production of helical polyacetylene through polymerization in a chiral nematic reaction field, pioneered by Hideki Shirakawa and Kazuo Akagi and co-workers at Kyoto University, Japan [101]. The polyacetylene chains twist into fibers that exhibit physical properties that can be quite unique, in particular in terms of their magnetoresistance. The chiral liquid crystal reaction field is absolutely essential for the success in the preparation of this new material. Another very interesting example is the case of liquid crystalline elastomers, weakly cross-linked nematic or smectic side- or main-chain polymers. Because the polymer backbone is stretched out in the liquid crystal phase but adopts a random coil conformation in the isotropic phase, these materials can dramatically change their shape as a result of a phase transition between liquid crystalline and isotropic. Several groups have demonstrated impressive actuators based on the effect [102–106], triggered either by temperature change or by light irradiation, the latter option made possible by the addition of photoisomerizable groups such as azobenzene.

But polymerization is not necessary to take advantage of the liquid crystalline order in this way. Jung and co-workers at KAIST, Korea, have developed a concept where SmA liquid crystals are confined in microstructured channels such that they develop regular arrays of focal-conic domains [107]. These can be used as anchor points for positioning microparticles [108], but the liquid crystal with its regular defect array can also be used in general as a soft lithographic template [109] or as a basis for fabricating superhydrophobic surfaces [110].

16.6.4 Liquid Crystals for Photovoltaic and Electromechanical Energy Conversion

Another area where the spontaneous long-range order of liquid crystals and the ease of controlling their alignment

[24] Chemists frequently use the terms cation and anion for positively and negatively charged ions, respectively. In the field of soft matter also, physicists frequently adopt this terminology.

over large areas are very attractive is organic electronics. In particular, the use of liquid crystals for generating organic photovoltaic devices is currently attracting much attention. Although a large number of aromatic molecules that are potentially useful for organic semiconductors or solar cells have been designed as a result of strong efforts in this direction over many years on an international scale, the single-molecule properties are only one factor in determining the performance of an organic electronic device. In order to ensure high charge carrier mobility on a device scale, the molecules must be packed in a well-ordered arrangement providing good overlap of the π orbitals of the aromatic moieties, allowing hopping conduction between adjacent molecules. This precludes amorphous materials since the molecules are disordered already on a small scale, while crystalline structures might at first seem to be ideal. However, the arrangement must extend uniformly over distances corresponding to the interelectrode distances of the device, typically ranging from some hundreds of nanometers to several microns, as otherwise the good conduction path is broken wherever there is a grain boundary, and this renders crystalline materials difficult to work with since single-crystal growth over the distances required is challenging.

This is where liquid crystals enter the stage since the molecules in a liquid crystalline state are orientationally ordered and the structure can often be uniformly aligned over very large distances without much trouble. Importantly, there are many ways of designing molecules such that they are both mesogenic and useful for photovoltaics/organic electronics [111, 112]. A drawback is that the molecule mobility in the liquid crystalline state is still rather high, such that the π–π stacking required for good hopping conduction is considerably poorer than in a crystalline state. The problem tends to be greater for the phases that are easy to align, that is, nematics and even SmA and SmC phases. Therefore, the approach typically aimed for is to prepare the large-scale ordered arrangement in the liquid crystalline phase and then cool the system down in a way that precludes crystallization, instead giving a glassy state in which the liquid crystalline order is frozen in, but now without the rather strong molecular fluctuations of the true liquid crystalline state.

The fundamental structure in an organic solar cell is a heterojunction between an electron-donating and an electron-accepting material. When a photon is absorbed its energy leads to the creation of an exciton, that is, a bound electron–hole pair, diffusing toward the interface where it dissociates into free charge carriers in the two respective materials. Key parameters for the photoelectrical conversion efficiency is the exciton diffusion length, that is, how long an exciton can travel within the material before the electron and hole recombine (the distance to the heterojunction must thus be smaller than the exciton diffusion length in order to harvest the energy) and the charge carrier mobility. Both parameters are particularly high in discotic columnar liquid crystals, and this is why considerable attention has been devoted to using these liquid crystals for photovoltaics [16, 113]. While promising results have been obtained, there are still some important challenges that need to be addressed. It is for instance not trivial to create a uniformly aligned heterostructure of two different discotic columnar phases designed for being electron-donating and electron-accepting, respectively. Some recent reports, however, provide good indications that the hurdles can be overcome by careful molecule design and optimization of the device preparation strategy, see, for example, Ref. [114].

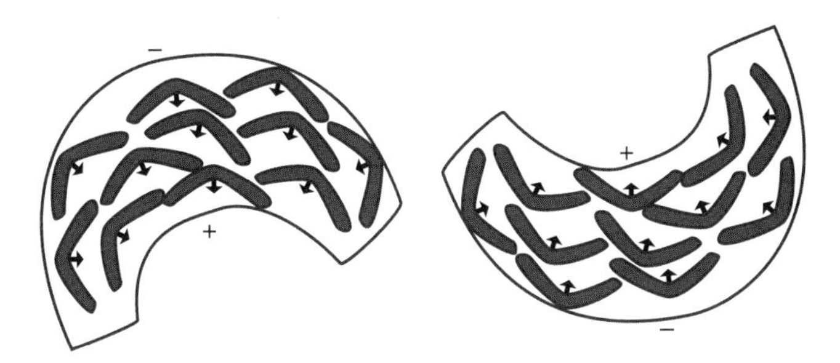

Figure 16.27 Cartoon of how the flexoelectric effect of a bent-core nematic liquid crystal can be used for generating an alternating electrical current from mechanical motion. By flexing the sheet containing the liquid crystal between concave and convex the bend direction reverses, and with that the polarization direction also reverses. This reverses the surface charge, giving a current peak for each reversal of the curvature.

Another highly interesting approach to harvesting energy using liquid crystals has recently been proposed by the group around Antal Jakli at Kent State University, USA. In this case, the basis is the very high flexo-/piezoelectric coefficients of some nematic phases formed by bent-shaped mesogens, which allows electrical current to be extracted from mechanical flexing of the liquid crystal material [31, 115, 116]. If a uniformly aligned nematic phase of bent-shaped mesogens is contained between flexible substrates, or inside an elastomer, the macroscopic sample shape can be changed between convex and concave, as illustrated in Figure 16.27. In the bent sample, the bent molecules pack best if their bend direction coincides with that of the macroscopic sample, hence the molecule bend reverses upon flexing. Due to the structure of the molecules, they exhibit a very large molecular dipole moment in the bend direction, thus the bend reversal also leads to a dipole reversal throughout the sample, eventually giving rise to a reversal of the surface charge at the sample boundaries. By attaching electrodes to the sample, an alternating electrical current can thus be extracted from the flexing liquid crystal sample. This is a very recently introduced concept and we can certainly expect strong development in the next few years.

16.6.5 Lipidomics and the Liquid Crystal Phases of Cell Membranes

As mentioned earlier, the L_α lamellar liquid crystalline phase is the natural state of cell membranes. In recent years, it has become increasingly clear, however, that one actually needs to distinguish two subclasses of this phase and the concepts *liquid-ordered* (Lo) and *liquid-disordered* (Ld) states were introduced [117–120]. The term "liquid" here refers to the lack of long-range positional order (and thus lack of a lattice) of the lipids within the bilayer plane in contrast to the gel phases that, in this terminology, are considered as "solid" (they do have positional order within the membrane). The additions "ordered" or "disordered" refer to the state of the acyl chains. The classic L_α picture fits the Ld phase best because there are many *gauche* conformers among the chains and they are very flexible and dynamic. In the end of the 1980s, it was realized, however, that a state can develop where the chains are mainly in all-*trans* conformation, yet without developing the in-plane positional order of the gel phases, and this is the Lo state.

The Lo and Ld phases coexist laterally in model membranes as a result of phase separation and the formation of the so-called "lipid rafts." These are dynamic fluctuating nanoscale regimes enriched in sphingolipids, cholesterol, and transmembrane proteins and in them the lipids adopt an Lo state. These "rafts" are distributed in a membrane "sea" in an Ld state, enriched in ordinary glycerophospholipid. While the Lo and Ld states do not necessarily constitute the states occurring in the live cell membrane, as living systems are far from equilibrium, the lipid raft paradigm with local increases in sphingolipid and cholesterol content has now been generally accepted, also for membranes of living cells. The tendency of this lipid composition to promote increased chain order certainly plays a role in the process, most likely in providing connectivity, which is then further modulated by the specificity of the membrane proteins as well as their antibodies [120].

The investigation of lipid rafts and Lo–Ld phase coexistence, and the consequences for cell signaling and trafficking, is a very active research area that forms an important component of the emerging field of "lipidomics." We can expect very interesting results in the future and as the (still relatively new) liquid-ordered phase becomes better understood, this will certainly be a good subject for study from a classic liquid crystal physics point of view. At present, the research is conducted primarily following a biophysical chemistry approach.

16.6.6 Active Nematics

A very recent and highly dynamic area of soft matter research, with a strong component of liquid crystalline ordering phenomena, is the focus on active soft matter [121], or active nematics, more specifically. We are here dealing with energy dissipation systems composed of anisotropic objects of relatively large scale (at least in one dimension), such that thermal motion of the particles becomes unimportant in the context of the large-scale ordering phenomena that are observed, making a granular matter approach more appropriate than the colloid science approach typical of lyotropic liquid crystals [122]. While conventional lyotropic liquid crystalline ordering results from entropy maximization (or, equivalently, minimization of excluded volume) driven by thermal motion, as originally explained by Onsager [123], the physics of active nematics goes beyond equilibrium thermodynamics and requires energy dissipation to be explicitly considered. The systems considered span several orders of magnitude in size, ranging from actin and microtubuli [124–128], which are filamentous proteins normally found in the cytoskeleton (stabilizing the cell membrane shape), to animals like birds and fish, well known to form ordered structures like flocks and schools when present at large numbers [121].

Although the origin of ordering is different, the organized states of active nematics can resemble typical liquid crystal textures to a large extent. However, there are subtle differences that give tangible effects. For instance, in contrast to regular nematics, active nematics may be polar [127, 128], breaking the director sign invariance. This prohibits topological defects of half-fold strength; hence, only integer defects should be observed in this active liquid crystal systems [124]. There is also a temporal difference in the sense that the concentration of particles frequently varies with time in active nematics. Liquid crystal-like ordering can thus come and go, for instance, as the concentration of bacteria in a microchannel increases as the bacteria divide, eventually pushing them to self-organize with very high nematic-like order along the channel due to packing constraints [129].

REFERENCES

[1] Chandler D. Interfaces and the driving force of hydrophobic assembly. Nature 2005;437(7059):640–647.

[2] Israelachvili JN. *Intermolecular and Surface Forces*. 3rd ed. Burlington (MA): Academic Press; 2010.

[3] Fennell Evans D, Wennerström H. *The Colloidal Domain: Where Physics, Chemistry, Biology, and Technology Meet*, Advances in Interfacial Engineering. New York: Wiley-VCH; 1999.

[4] Barry E, Dogic Z. Entropy driven self-assembly of nonamphiphilic colloidal membranes. Proc Natl Acad Sci U S A 2010;107(23):10 348–10 353.

[5] Mourad MC, Wijnhoven JE, Van't Zand DD, van der Beek D, Lekkerkerker HN. Gelation versus liquid crystal phase transitions in suspensions of plate-like particles. Philos Trans R Soc London, Ser A 2006;364(1847):2807–2816.

[6] Ao GY, Nepal D, Aono M, Davis VA. Cholesteric and nematic liquid crystalline phase behavior of double-stranded

DNA stabilized single-walled carbon nanotube dispersions. ACS Nano 2011;5(2):1450–1458.

[7] Lu L, Chen W. Large-scale aligned carbon nanotubes from their purified, highly concentrated suspension. ACS Nano 2010;4(2):1042–1048.

[8] Zhang SJ, Kinloch IA, Windle AH. Mesogenicity drives fractionation in lyotropic aqueous suspensions of multiwall carbon nanotubes. Nano Lett 2006;6(3):568–572.

[9] Lydon J. Chromonic mesophases. Curr Opin Colloid Interface Sci 2004;8(6):480–490.

[10] Cheng XH, Das MK, Diele S, Tschierske C. Novel liquid-crystalline phases with layerlike organization. Angew Chem 2002;114(21):4203–4207.

[11] Tschierske C. Micro-segregation, molecular shape and molecular topology partners for the design of liquid crystalline materials with complex mesophase morphologies. J Mater Chem 2001;11(11):2647–2671.

[12] de Gennes P-G, Prost J. *The Physics of Liquid Crystals*. Oxford: Clarendon Press; 1993.

[13] Lagerwall JPF, Giesselmann F. Current topics in smectic liquid crystal research. ChemPhysChem 2006;7(1):20–45.

[14] Smith GS, Sirota EB, Safinya CR, Clark NA. Structure of the L_β phases in a hydrated phosphatidylcholine multimembrane. Phys Rev Lett 1988;60(9):813–816.

[15] Demus D, Goodby JW, Gray G, Spiess H-W, Vill V, editors. *Handbook of Liquid Crystals*. Weinheim: Wiley-VCH; 1998.

[16] Laschat S, Baro A, Steinke N, Giesselmann F, Hägele C, Scalia G, Judele R, Kapatsina E, Sauer S, Schreivogel A, Tosoni M. Discotic liquid crystals: from tailor-made synthesis to plastic electronics. Angew Chem Int Ed 2007;46(26):4832–4887.

[17] Figueiredo Neto AM, Salinas SRA. *The Physics of Lyotropic Liquid Crystals: Phase Transitions and Structural Properties*, Monographs on the Physics and Chemistry of Materials. New York: Oxford University Press; 2005.

[18] Sluckin TJ, Dunmur DA, Stegemeyer H. *Crystals That Flow: Classic Papers from the History of Liquid Crystals*. London: Taylor and Francis; 2004.

[19] Crooker PP, Kitzerow HS. Blue phases. Condens Matter News 1992;1(3):6–10.

[20] Kitzerow HS. Blue phases at work! ChemPhysChem 2006;7(1):63–66.

[21] Dierking I. *Textures of Liquid Crystals*. Weinheim: Wiley-VCH; 2003.

[22] Kikuchi H, Yokota M, Hisakado Y, Yang H, Kajiyama T. Polymer-stabilized liquid crystal blue phases. Nat Mater 2002;1(1):64–68.

[23] Reddy RA, Tschierske C. Bent-core liquid crystals: polar order, superstructural chirality and spontaneous desymmetrisation in soft matter systems. J Mater Chem 2006;16(10):907–961.

[24] Takezoe H, Takanishi Y. Bent-core liquid crystals: their mysterious and attractive world. Jpn J Appl Phys, Part 1 2006;45(2A):597–625.

[25] Pelzl G, Diele S, Weissflog W. Banana-shaped compounds - a new field of liquid crystals. Adv Mater 1999;11(9):707–724.

[26] Link DR, Natale G, Shao R, Maclennan JE, Clark NA, Körblova E, Walba DM. Spontaneous formation of macroscopic chiral domains in a fluid smectic phase of achiral molecules. Science 1997;278(5345):1924–1927.

[27] Hough LE, Spannuth M, Nakata M, Coleman DA, Jones CD, Dantlgraber G, Tschierske C, Watanabe J, Koerblova E, Walba DM, Maclennan JE, Glaser MA, Clark NA. Chiral isotropic liquids from achiral molecules. Science 2009;325(5939):452–456.

[28] Hough LE, Jung HT, Krueerke D, Heberling MS, Nakata M, Jones CD, Chen D, Link DR, Zasadzinski J, Heppke G, Rabe JP, Stocker W, Koerblova E, Walba DM, Glaser MA, Clark NA. Helical nanofilament phases. Science 2009;325(5939):456–460.

[29] Reddy RA, Zhu C, Shao R, Korblova E, Gong T, Shen Y, Garcia E, Glaser MA, Maclennan JE, Walba DM, Clark NA. Spontaneous ferroelectric order in a bent-core smectic liquid crystal of fluid orthorhombic layers. Science 2011;332(6025):72–77.

[30] Keith C, Prehm M, Panarin YP, Vij JK, Tschierske C. Development of polar order in liquid crystalline phases of a banana compound with a unique sequence of three orthogonal phases. Chem Commun 2010;46(21):3702–3704.

[31] Harden J, Chambers M, Verduzco R, Luchette P, Gleeson JT, Sprunt S, Jakli A. Giant flexoelectricity in bent-core nematic liquid crystal elastomers. Appl Phys Lett 2010;96(10):102 907–.

[32] Hermans PH, Platzek P. Beitr ge zur Kenntnis des Deformationsmechanismus und der Feinstruktur der Hydratzellulose. Kolloid Z 1939;88:68–72.

[33] Tsvetkov V. Acta Physiochim (USSR) 1942;16:132–147.

[34] Quist PO, Halle B, Furo I. Nuclear-spin relaxation in a hexagonal lyotropic liquid-crystal. J Chem Phys 1991;95(9):6945–6961.

[35] Beica T, Moldovan R, Tintaru M, Enache I, Frunza S. Measurements of optical anisotropy of a calamitic lyotropic liquid crystal. Cryst Res Technol 2004;39(2):151–156.

[36] Johannesson H, Furo I, Halle B. Orientational order and micelle size in the nematic phase of the cesium pentadecafluorooctanoate-water system from the anisotropic self-diffusion of water. Phys Rev E 1996;53(5):4904–4917.

[37] Boden N, Clements J, Dawson KA, Jolley KW, Parker D. Universal nature of the nematic-to-isotropic transition in solutions of discotic micelles. Phys Rev Lett 1991;66(22):2883–2886.

[38] Mcclymer JP, Labes MM. Simultaneous measurement of guest and host ordering in a nematic lyophase via fluorescence spectroscopy. Mol Cryst Liq Cryst 1991;195:39–44.

[39] Miesowicz M. Influence of a magnetic field on the viscosity of para-azoxyanisole. Nature 1935;136:261.

[40] Miesowicz M. The three coefficients of viscosity of anisotropic liquids. Nature 1946;158:27.

[41] Krueger M, Giesselmann F. Dielectric spectroscopy of 'de vries' type smectic A* - C* transitions. Phys Rev E 2005;71(4):041 704.

[42] Preistley EB, Wojtowicz PJ. *Introduction to Liquid Crystals*. New York: Plenum Publishing Corporation; 1976.

[43] Cladis PE, Kleman M, Pieranski P. New method for decoration of mesomorphic phase of para methoxybenzilidene para betylaniline. C R Acad Sci B Phys 1971;273(6):275–279.

[44] Poulin P, Stark H, Lubensky TC, Weitz DA. Novel colloidal interactions in anisotropic fluids. Science 1997;275(5307):1770–1773.

[45] Tkalec U, Ravnik M, Zumer S, Musevic I. Vortexlike topological defects in nematic colloids: chiral colloidal dimers and 2D crystals. Phys Rev Lett 2009;103(12):127 801.

[46] Ognysta U, Nych A, Nazarenko V, Skarabot M, Musevic I. Design of 2D binary colloidal crystals in a nematic liquid crystal. Langmuir 2009;25(20):12 092–12 100.

[47] Conradi M, Ravnik M, Bele M, Zorko M, Zumer S, Musevic I. Janus nematic colloids. Soft Matter 2009;5(20):3905–3912.

[48] Musevic I, Skarabot M. Self-assembly of nematic colloids. Soft Matter 2008;4(2):195–199.

[49] Skarabot M, Ravnik M, Zumer S, Tkalec U, Poberaj I, Babic D, Musevic I. Hierarchical self-assembly of nematic colloidal superstructures. Phys Rev E 2008;77(6):061 706.

[50] Tkalec U, Skarabot M, Musevic I. Interactions of micro-rods in a thin layer of a nematic liquid crystal. Soft Matter 2008;4(12):2402–2409.

[51] Ravnik M, Skarabot M, Zumer S, Tkalec U, Poberaj I, Babic D, Osterman N, Musevic I. Entangled nematic colloidal dimers and wires. Phys Rev Lett 2007;99(24): 247 801.

[52] Musevic I, Skarabot M, Tkalec U, Ravnik M, Zumer S. Two-dimensional nematic colloidal crystals self-assembled by topological defects. Science 2006;313(5789):954–958.

[53] Pishnyak OP, Sergij SV, Lavrentovich OD. Inelastic collisions and anisotropic aggregation of particles in a nematic collider driven by backflow. Phys Rev Lett 2011;106(4):047 801.

[54] Pishnyak OP, Tang S, Kelly JR, Shiyanovskii SV, Lavrentovich OD. Levitation, lift, and bidirectional motion of colloidal particles in an electrically driven nematic liquid crystal. Phys Rev Lett 2007;99(12):127 802.

[55] Smalyukh I, Kuzmin AN, Kachynski AV, Prasad PN, Lavrentovich OD. Optical trapping of colloidal particles and measurement of the defect line tension and colloidal forces in a thermotropic nematic liquid crystal. Appl Phys Lett 2005;86(2):021 913.

[56] Liao G, Smalyukh II, Kelly JR, Lavrentovich OD, Jákli A. Electrorotation of colloidal particles in liquid crystals. Phys Rev E 2005;72(3):31 704–.

[57] Bohley C, Stannarius R. Inclusions in free standing smectic liquid crystal films. Soft Matter 2008;4(4):683–702.

[58] Bohley C, Stannarius R. Colloidal inclusions in smectic films with spontaneous bend. Eur Phys J E 2007;23(1):25–30.

[59] Voltz C, Stannarius R. Buckling instability of droplet chains in freely suspended smectic films. Phys Rev E 2005;72(1):011 705.

[60] Scalia G. Alignment of carbon nanotubes in thermotropic and lyotropic liquid crystals. ChemPhysChem 2010;11(2):333–340.

[61] Rahman M, Lee W. Scientific duo of carbon nanotubes and nematic liquid crystals. J Phys D: Appl Phys 2009;42(6):063 001.

[62] Lagerwall JPF, Scalia G. Carbon nanotubes in liquid crystals (Feature article). J Mater Chem 2008;18(25):2890–2898.

[63] Liu Q, Cui Y, Gardner D, Li X, He S, Smalyukh II. Self-alignment of plasmonic gold nanorods in reconfigurable anisotropic fluids for tunable bulk metamaterial applications. Nano Lett 2010;10(4):1347–1353.

[64] Popa-Nita V, Kralj S. Liquid crystal-carbon nanotubes mixtures. J Chem Phys 2010;132(2):024 902.

[65] Schoot P, Popa-Nita V, Kralj S. Alignment of carbon nanotubes in nematic liquid crystals. J Phys Chem B 2008;112(15):4512–4518.

[66] Scalia G, von Bühler C, Hägele C, Roth S, Giesselmann F, Lagerwall JPF. Spontaneous macroscopic carbon nanotube alignment via colloidal suspension in hexagonal columnar lyotropic liquid crystals. Soft Matter 2008;4(3):570–576.

[67] Schymura S, Dölle S, Yamamoto J, Lagerwall J. Filament formation in carbon nanotube-doped lyotropic liquid crystals. Soft Matter 2011;7(6):2663–2667.

[68] Basu R, Boccuzzi KA, Ferjani S, Rosenblatt C. Carbon nanotube-induced chirality in an achiral liquid crystal. Appl Phys Lett 2010;97(12):121 908.

[69] Lee C-W, Shih W-P. Quantification of ion trapping effect of carbon nanomaterials in liquid crystals. Mater Lett 2010;64(3):466–468.

[70] Chen HY, Lee W, Clark NA. Faster electro-optical response characteristics of a carbon-nanotube-nematic suspension. Appl Phys Lett 2007;90(3):033 510.

[71] Courty S, Mine J, Tajbakhsh AR, Terentjev EM. Nematic elastomers with aligned carbon nanotubes: new electromechanical actuators. Europhys Lett 2003; 64(5):654–660.

[72] Kinkead B, Hegmann T. Effects of size, capping agent, and concentration of CdSe and CdTe quantum dots doped into a nematic liquid crystal on the optical and electro-optic properties of the final colloidal liquid crystal mixture. J Mater Chem 2010;20(3):448–458.

[73] Qi H, Hegmann T. Multiple alignment modes for nematic liquid crystals doped with alkylthiol-capped gold nanoparticles. ACS Appl Mater Interfaces 2009;1(8):1731–1738.

[74] Qi H, Neil JO, Hegmann T. Chirality transfer in nematic liquid crystals doped with (S)-naproxen-functionalized gold nanoclusters: an induced circular dichroism study. J Mater Chem 2008;18(4):374–380.

[75] Yaghmur A, Glatter O. Characterization and potential applications of nanostructured aqueous dispersions. Adv Colloid Interface Sci 2009;147-148:333–342.

[76] Sagalowicz L, Leser ME, Watzke HJ, Michel M. Monoglyceride self-assembly structures as delivery vehicles. Trends Food Sci Technol 2006;17(5):204–214.

[77] Barauskas J, Johnsson M, Tiberg F. Self-assembled lipid superstructures: beyond vesicles and liposomes. Nano Lett 2005;5(8):1615–1619.

[78] Spicer PT. Progress in liquid crystalline dispersions: Cubosomes. Curr Opin Colloid Interface Sci 2005;10(5-6):274–279.

[79] Blanc C, Kleman M. The confinement of smectics with a strong anchoring. Eur Phys J E 2001;4(2):241–251.

[80] Humar M, Musevic I. 3D microlasers from self-assembled cholesteric liquid-crystal microdroplets. Opt Express 2010;18(26):26 995–27 003.

[81] Lopez-Leon T, Fernandez-Nieves A. Drops and shells of liquid crystal. Colloid Polym Sci 2011;289(4):345–359.

[82] Lopez-Leon T, Koning V, Devaiah KBS, Vitelli V, Fernandez-Nieves A. Frustrated nematic order in spherical geometries. Nat Phys 2011;7:391–394.

[83] Liang H-L, Schymura S, Rudquist P, Lagerwall J. Nematic-smectic transition under confinement in liquid crystalline colloidal shells. Phys Rev Lett 2011;106(24):247 801.

[84] Lopez-Leon T, Fernandez-Nieves A, Nobili M, Blanc C. Nematic-smectic transition in spherical shells. Phys Rev Lett 2011;106(24):247 802.

[85] Lopez-Leon T, Fernandez-Nieves A. Topological transformations in bipolar shells of nematic liquid crystals. Phys Rev E 2009;79(2):021 707.

[86] Fernandez-Nieves A, Vitelli V, Utada AS, Link DR, Marquez M, Nelson DR, Weitz DA. Novel defect structures in nematic liquid crystal shells. Phys Rev Lett 2007;99(15):157 801.

[87] Skacej G, Zannoni C. Controlling surface defect valence in colloids. Phys Rev Lett 2008;100(19):197 802.

[88] Sivakumar S, Wark KL, Gupta JK, Abbott NL, Caruso F. Liquid crystal emulsions as the basis of biological sensors for the optical detection of bacteria and viruses. Adv Funct Mater 2009;19(14):2260–2265.

[89] Gupta JK, Meli MV, Teren S, Abbott NL. Elastic energy-driven phase separation of phospholipid monolayers at the nematic liquid-crystal-aqueous interface. Phys Rev Lett 2008;1(4):048 301.

[90] Jang CH, Cheng LL, Olsen CW, Abbott NL. Anchoring of nematic liquid crystals on viruses with different envelope structures. Nano Lett 2006;6(5):1053–1058.

[91] Lockwood NA, Mohr JC, Ji L, Murphy CJ, Palecek SP, de Pablo JJ, Abbott NL. Thermotropic liquid crystals as substrates for imaging the reorganization of matrigel by human embryonic stem cells. Adv Funct Mater 2006;16(5):618–624.

[92] Brake JM, Daschner MK, Luk YY, Abbott NL. Biomolecular interactions at phospholipid-decorated surfaces of liquid crystals. Science 2003;302(5653):2094–2097.

[93] Kim SR, Abbott NL. Rubbed films of functionalized bovine serum albumin as substrates for the imaging of protein-receptor interactions using liquid crystals. Adv Mater 2001;13(19):1445.

[94] Gupta VK, Skaife JJ, Dubrovsky TB, Abbott NL. Optical amplification of ligand-receptor binding using liquid crystals. Science 1998;279(5359):2077–2080.

[95] Chen C-H, Yang K-L. Detection and quantification of DNA adsorbed on solid surfaces by using liquid crystals. Langmuir 2010;26(3):1427–1430.

[96] Price AD, Schwartz DK. DNA hybridization-induced reorientation of liquid crystal anchoring at the nematic liquid crystal/aqueous interface. J Am Chem Soc 2008;130(26):8188–8194.

[97] Wan Y, Zhao DY. On the controllable soft-templating approach to mesoporous silicates. Chem Rev 2007;107(7):2821–2860.

[98] Sievens-Figueroa L, Guymon CA. Cross-linking of reactive lyotropic liquid crystals for nanostructure retention. Chem Mater 2009;21(6):1060–1068.

[99] Clapper JD, Iverson SL, Guymon CA. Nanostructured biodegradable polymer networks using lyotropic liquid crystalline templates. Biomacromolecules 2007;8(7):2104–2111.

[100] Hatakeyama ES, Gabriel CJ, Wiesenauer BR, Lohr JL, Zhou M, Noble RD, Gin DL. Water filtration performance of a lyotropic liquid crystal polymer membrane with uniform, sub-1-nm pores. J Membr Sci 2011;366(1-2):62–72.

[101] Goh M, Matsushita S, Akagi K. From helical polyacetylene to helical graphite: synthesis in the chiral nematic liquid crystal field and morphology-retaining carbonisation. Chem Soc Rev 2010;39(7):2466–2476.

[102] Yoshino T, Kondo M, Mamiya J-I, Kinoshita M, Yanlei Y, Ikeda T. Three-dimensional photomobility of crosslinked azobenzene liquid-crystalline polymer fibers. Adv Mater 2010;22(12):1361.

[103] Chambers M, Finkelmann H, Remskar M, Sanchez-Ferrer A, Zalar B, Zumer S. Liquid crystal elastomer-nanoparticle systems for actuation. J Mater Chem 2009;19(11):1524–1531.

[104] Yamada M, Kondo M, Mamiya JI, Yu YL, Kinoshita M, Barrett C, Ikeda T. Photomobile polymer materials: towards light-driven plastic motors. Angew Chem Int Ed 2008;47(27):4986–4988.

[105] Camacho-Lopez M, Finkelmann H, Palffy-Muhoray P, Shelley M. Fast liquid-crystal elastomer swims into the dark. Nat Mater 2004;3(5):307–310.

[106] Yu YL, Nakano M, Ikeda T. Directed bending of a polymer film by light - miniaturizing a simple photomechanical system could expand its range of applications. Nature 2003;425(6954):145.

[107] Kim YH, Yoon DK, Jeong HS, Lavrentovich OD, Jung HT. Smectic liquid crystal defects for self-assembling of building blocks and their lithographic applications. Adv Funct Mater 2011;21(4):610–627.

[108] Yoon DK, Choi MC, Kim YH, Kim MW, Lavrentovich OD, Jung HT. Internal structure visualization and lithographic use of periodic toroidal holes in liquid crystals. Nat Mater 2007;6(11):866–870.

[109] Kim YH, Yoon DK, Jeong HS, Jung HT. Self-assembled periodic liquid crystal defects array for soft lithographic template. 2010;6(7):1426.

[110] Kim YH, Yoon DK, Jeong HS, Kim JH, Yoon EK, Jung H-T. Fabrication of a superhydrophobic surface from a smectic liquid-crystal defect array. Adv Funct Mater 2009;19(18):3008–3013.

[111] Pisula W, Zorn M, Chang JY, Muellen K, Zentel R. Liquid crystalline ordering and charge transport in semiconducting materials. Macromol Rapid Commun 2009;30(14):1179–1202.

[112] Neill MO, Kelly SM. Ordered materials for organic electronics and photonics. Adv Mater 2011;23(5):566–584.

[113] Sergeyev S, Pisula W, Geerts YH. Discotic liquid crystals: a new generation of organic semiconductors. Chem Soc Rev 2007;36(12):1902–1929.

[114] Thiebaut O, Bock H, Grelet E. Face-on oriented bilayer of two discotic columnar liquid crystals for organic donor-acceptor heterojunction. J Am Chem Soc 2010;132(20):6886.

[115] Chambers M, Verduzco R, Gleeson JT, Sprunt S, Jakli A. Calamitic liquid-crystalline elastomers swollen in bent-core liquid-crystal solvents. Adv Mater 2009;21(16):1622.

[116] Chambers M, Verduzco R, Gleeson JT, Sprunt S, Jakli A. Flexoelectricity of a calamitic liquid crystal elastomer swollen with a bent-core liquid crystal. J Mater Chem 2009;19(42):7909–7913.

[117] Mouritsen OG. The liquid-ordered state comes of age. Biochim Biohys Acta 2010;1798(7):1286–1288.

[118] Mouritsen OG, Mouritsen O. *Life - As a Matter of Fat*. Berlin Heidelberg: Springer-Verlag; 2004.

[119] Simons K, Gerl MJ. Revitalizing membrane rafts: new tools and insights. Nat Rev Mol Cell Biol 2010;11(10):688–699.

[120] Lingwood D, Simons K. Lipid rafts as a membrane-organizing principle. Science 2010;327(5961):46–50.

[121] Marchetti MC, Joanny JF, Ramaswamy S, Liverpool TB, Prost J, Rao M, Simha AR. Hydrodynamics of soft active matter. Rev Mod Phys 2013;85(3):1143–1189.

[122] Aranson IS, Tsimring LS. *Granular Patterns*. New York: Oxford University Press; 2009.

[123] Onsager L. The effects of shape on the interaction of colloidal particles. Ann NY Acad Sci 1949;51(4):627–659.

[124] Schaller V, Bausch AR. Topological defects and density fluctuations in collectively moving systems. Proc Natl Acad Sci U S A 2013;110(12):4488–4493.

[125] Sanchez T, Chen DTN, DeCamp SJ, Heymann M, Dogic Z. Spontaneous motion in hierarchically assembled active matter. Nature 2012;491(7424):431.

[126] Sumino Y, Nagai KH, Shitaka Y, Tanaka D, Yoshikawa K, Chate H, Oiwa K. Large-scale vortex lattice emerging from collectively moving microtubules. Nature 2012;483(7390):448–452.

[127] Schaller V, Weber C, Frey E, Bausch AR. Polar pattern formation: hydrodynamic coupling of driven filaments. Soft Matter 2011;7(7):3213.

[128] Schaller V, Weber C, Semmrich C, Frey E, Bausch AR. Polar patterns of driven filaments. Nature 2010;467(7311):73–77.

[129] Volfson D, Cookson S, Hasty J, Tsimring LS. Biomechanical ordering of dense cell populations. Proc Natl Acad Sci U S A 2008;105(40):15 346–15 351.

17

ENTANGLED GRANULAR MEDIA

NICK GRAVISH & DANIEL I. GOLDMAN

School of Physics, Georgia Institute of Technology, Atlanta, GA, USA

Granular materials (GM) are collections of discrete, dissipative, and athermal particles [1–4] and when dry, GM interact through frictional and repulsive forces only. GM are important in industry, engineering, and science, in daily life are commonly encountered as bags of coffee, rice, sugar, etc. Despite the apparent simplicity of particle–particle interaction in dry GM, collections of even simple spherical particles may exhibit complex rheological properties such as transitioning between jammed and flowing states (see Ref. 5 for a comprehensive overview). Study of GM is motivated by industrial and engineering applications in addition to fundamental science research. Experiments on GM often use simple table-top apparatus' in which the forces and motion can be visualized with visible light cameras, thus making GM research very accessible. Lastly, GM can be simulated using the discrete element method (DEM), which enables tandem computational and experimental studies of GM [6–11].

The study of how soft matter flows under external or internal stresses is called rheology [1–4]. The development of principles of granular rheology is a focus of GM research. One area of granular rheology study is to understand how the features of the granular particle—such as roughness, shape, or material stiffness—may influence the material properties (e.g., stiffness or yield strength) of the bulk GM. The majority of such rheological studies have focused on particles of convex shape [2, 12–14], and less research has been focused on concave particles [15–17]. However, recent studies have shown that nonconvex particles display rheological properties that may be desirable for engineered GM, such as nonzero tensile strength, high

Fluids, Colloids and Soft Materials: An Introduction to Soft Matter Physics, First Edition. Edited by Alberto Fernandez Nieves and Antonio Manuel Puertas.

Figure 17.1 Concave particle assemblies from the nanoscale to macroscale. (a) Macromolecular assemblies of rigid oligomers (insets) interpenetrate depending on oligomer shape (top); Source: McKeown and Budd [24] and may form into large entangled networks (bottom); Source: McKeown *et al.* [25]. Particle dimensions are in the nanometer scale. (b) Concave particle suspensions (top); Source: Brown *et al.* [17] and concave colloids (bottom-left); Source: Manna *et al.* [22] and (bottom-right); Chen *et al.* [32]. (c) Concave particle assemblies found in nature, living fire ants (top), and rigid branches of bald eagles nests (bottom). Source: Mlot *et al.* [28].

yield stress, and desirable compaction properties [18, 19]. Thus, a new focus on nonconvex GM will impact scientific and engineering pursuits as diverse as jamming and soft robotics [20], designed granular materials [19], industrial granular processes, and a better understanding of soft matter systems [21].

Examples of nonconvex particle assemblies can be found at all scales of natural and industrial systems (Fig. 17.1). While few nonconvex granular materials studies have been performed, nonconvex particles are currently a focus of engineering efforts at the microscale and nanoscale [22, 23]. For example, macromolecules with concave shapes pack together with large voids in the bulk, which results in materials with a high microporosity [24, 25] (Fig. 17.1). Such high microporosity materials have applications ranging from nanodrug delivery systems to gas trapping [24, 25]. Furthermore, the design of concave colloidal particles have application to self-organizing, smart materials [26]. In addition to engineered systems, concave particles are found in biological systems (Fig. 17.1). The packing of actin or other protein filaments within eukaryotic cells has been modeled as concave granular materials [27]. The bridges and rafts collectively built by ants [28–30] are held together through the entanglement of ant limbs and mandibles, which can be considered concave particles. Structures constructed by animals from branches and twigs such as birds' nests may be considered as concave granular materials [31].

The goal of this chapter is to understand how the shape of a simple nonconvex particle - a "u-particle" influences bulk rheology. In experiment and computer simulation, we systematically study how particle concavity affects bulk properties of a granular material. The fundamental difference between concave and convex particle assemblies is the ability for concave particles to interpenetrate, which we call being entangled. Mechanical entanglement of particles alters the rheology of particle assemblies through their resistance to separate. In Section 17.1, we review previous studies of convex granular materials. Additionally, we introduce experimental techniques used to characterize granular rheology. In Section 17.2, we describe our experiments studying the stability of assemblies of U-shaped concave particles in which we vary concavity. In Section 17.3, we describe theoretical and numerical modeling of u-particle assemblies and their implications for assembly stability. In Section 17.4, we discuss the implications for these results to other systems of concave particles.

17.1 GRANULAR MATERIALS

17.1.1 Dry, Convex Particles

Granular materials are assemblies of macroscopic particles that are typically of size greater than 10 µm [33]. Particles smaller than 10 µm are subject to thermal effects while granular materials are athermal [33]. Granular materials come in many shapes and sizes [14]; however, in the majority of physics studies granular materials are convex (often spherical) particles.

A simple experiment with GM is to pour particles of a known density and mass into a container and measure the volume occupied by the assembly. This experiment measures the fraction of total volume, V, versus the solid particle volume, V_p, and is called the volume fraction $\phi = \frac{V_p}{V}$. Mechanically stable ensembles of spherical, dry granular materials are found in a range of ϕ from random loose-pack $\phi_{\rm rlp} \approx 0.55$ [4, 34] to random close-pack volume fraction $\phi_{\rm rcp} \approx 0.64$ [4]. The maximum ϕ for ordered, uniform, spheres is that of a face-centered-cubic lattice in which $\phi_{\rm fcc} = 0.77$. A loose, ordered, packing is that of a simple-square lattice with $\phi_{\rm ssl} = 0.52$. Although the packing fraction of ordered ensembles (i.e., on a lattice) may be analytically solved for, the calculation of the maximum ϕ of randomly distributed particles of arbitrary shape must be done computationally [13].

A fundamental feature of granular materials is their ability to act like solids, fluids, or gases [2]. The "phase" of the granular material may spontaneously change under an applied load or other perturbation. An avalanche down a granular slope is an example of such a transition: the slope begins as an inert solid until an external perturbation is applied (tilting the slope for instance) after which a section of the slope becomes unstable, and the grains are put into motion. Avalanches typically occur above immobile grains trapped below, an example of the coexistence between the fluid-like and solid-like states of GM.

Dry granular materials consist of particles that interact through frictional and repulsive forces only [1–4]. Forces within a granular material are spatially heterogeneous and are transmitted through filamentary force chains [35, 36]. Because of the absence of particle–particle attraction, dry, spherical granular material cannot support tensile loading [3]. Furthermore, since particle–particle interactions are frictional, the force laws of granular flows are typically rate insensitive at low speeds [3]. At higher flow speeds, however, momentum transfer between particles becomes important and force laws take on a velocity dependence (with a force law dependent on the packing density of the material [37]).

17.1.2 Cohesion through Fluids

When water is added to a granular material, particles can cohere through the formation of capillary bridges (see Refs 38, 39 for comprehensive reviews). Unlike dry GM, wet granular media can support tensile loading because of the particle–particle attraction from capillary bridges [40, 41]. The attractive force from a capillary bridge between two wettable granular spheres is proportional to the surface tension of the fluid and inversely proportional to the radius of curvature. Thus, the strength of the capillary bonds and the rheology of the bulk in general are sensitive to the fluid chemistry, the particle diameter, and the fluid volume.

The presence of even a small amount of water can dramatically alter the rheology of a granular material. In a dry granular material, the maximum slope angle that can be formed is $\approx 30°$. However, wet GM can be formed into piles and structures with much steeper slopes of 90° or above [42, 43]. It is because of this property that sandcastles are able to be built and structurally supported. The angle of repose of a wet GM is a function of the fraction of water present [43] and is a useful metric for determining the cohesion between grains.

Another technique for characterizing the strength of capillary bonds in a wet GM is to study the solid to fluid transition of the ensemble under vertical vibration [44, 45]. When the GM is in the fluid state, capillary bonds are repeatedly broken and reformed. The parameters of the fluid–solid transition for a given granular material are thus related to the strength of the capillary bonds between grains. It has been shown that the vibration energy (peak kinetic energy for sinusoidal vibrations) at which the fluid–solid transition occurs in wet granular materials is linearly proportional to the capillary bond energy [46]. Thus, vibration and relaxation experiments give insight into the particle–particle interactions in wet GM, and we look to these techniques for inspiration in the work we describe next.

17.1.3 Cohesion through Shape

Addition of fluid to dry granular media is not the only way to create cohesive effects. Numerous experiments studying the packing of rods, granular chains, and more complex 3D printed grains may support tensile stresses [19, 21, 47–59]. For a review of the packing and rheology of nonspherical, elongated grains, see Ref. 59.

Rods within a granular assembly lack rotational freedom, and through this frustration of motion, an effective tensile strength develops. Granular rod piles may form with wall angles of 90° or more because of this effective cohesion. The stability and packing of rods is sensitive to the length to diameter ratio of the rods [47, 48, 53]. Longer rods pack together in lower volume fraction ensembles [47] and can be built into taller stable ensembles [53]. Another example of a nonspherical granular system is ensembles of granular chains, solid spheres attached by flexible links. These chains form bulk materials with yield strength that depends on the chain length [60]. Longer chains entangle in the bulk, and through experiment and simulation Brown *et al.* [60] showed that the enhanced yield properties of these ensembles were due to this entanglement.

The addition of bends at the rod ends creates a u-particle. The u-particle shape is arguably the simplest particle shape that possesses concavity. We define a concave particle as a

Figure 17.2 u-Particle assemblies. (a) u-Particles interpenetrate to support tensile loading. (b) A free-standing column of entangled u-particles. (c) Grasping the tower with tweezers illustrates the tensile strength of this granular assembly.

solid body where there may be found a line segment that connects two sections of the particle while not being fully contained within the solid body of the particle. For the purposes of u-particles, line segments connecting points on opposing barbs must pass through the free-space between the barbs, and thus the particle shape is concave. The concavity of a u-particle is defined as the total volume of the concave internal region of the particle. For u-particles, the concavity may be varied by changing the length of the ends compared to the width of the opening. Clusters of u-particles readily form mechanically entangled solids that are easily formed into columns that maintain their shape under gravity and, when pulled from the top by a pair of tweezers, may be lifted as almost a solid plug (see Fig. 17.2).

17.1.4 Characterize the Rheology of Granular Materials

Before we describe our experiments studying the stability and packing of mechanical entangled u-particle ensembles, we briefly review commonly employed experimental methods used to characterize the rheology and packing of granular material.

17.1.4.1 Packing Measurements Mechanically stable ensembles of spherical GM are found in a range of ϕ from random loose-pack $\phi_{\text{rlp}} \approx 0.55$ to random close-pack volume fraction $\phi_{\text{rcp}} \approx 0.64$. Although the packing fraction of ordered ensembles (i.e., on a lattice) may be analytically determined, the prediction of the maximum volume fraction for particles of arbitrary shape is a computationally intensive process [13]. A primary goal of packing studies is to understand how particles arrange in mechanically stable configurations. Furthermore, ϕ may be used as a measurement of the ensembles state and thus used to predict the dynamical response of GM subject to shear [61], impact [37], or intruder drag [62].

Study of the packing of granular media under mechanical [63] or air-fluidized [64] perturbation has revealed that the evolution of ϕ is dependent on the forcing parameters such as peak mechanical acceleration or air pressure. The relaxation dynamics of granular material are complex, exhibiting signatures of multiple timescales [64] or stretched exponential behavior [63]. A common feature of these experiments is that compaction of granular material, and subsequently the increase in ϕ, occurs slowly, over many thousands of iterations.

17.1.4.2 Stress–Strain Rheology Stress–strain experiments are a fundamental tool used to develop and test constitutive equations of material flow. The stress–strain response of granular material subject to a wide variety of tests in various geometries has been studied [1–4, 65]. The prototypical stress–strain experiment is simple shear in which granular material is placed between two (semi-)infinite, horizontal, planes with the top plane translated at constant velocity. An important note about stress–strain experiments is that they effectively take place in a steady-state in which the force along the moving plate fluctuates about a mean value that will persist as long as the experiment proceeds.

17.1.4.3 Vibration Experiments An important experimental tool in understanding the relaxation dynamics and steady-states of granular gases/fluids/solids results from applying uniform vibration [2, 4, 66–68]. Vibro-fluidization has been used to study the properties of wet cohesive beads by exploring the solid–liquid and liquid–gas phase transitions that occur as a function of oscillation amplitude and frequency [45, 46, 69]. A transition from a liquid to gaseous phase occurs in wet cohesive granular media when the injected energy from vibration exceeds the capillary bond energy [69]. In avalanching experiments with dry granular media, vibration has been used to mobilize particles and thus cause the GM to relax under gravity faster or slower depending on the amplitude and frequency of vibration [68, 70]. Relaxation of slopes of granular material follows a Boltzmann-like exponential function with vibration amplitude analogous to a thermal energy [68].

17.2 EXPERIMENT

In this section, we describe a set of experiments to study the packing and relaxation of "u-particle" columns. We formed vertical, free-standing columns from collections of u-particles of varied barb length to width ratio l/w. By varying l/w, we vary the concavity of the particles with $l/w = 0$ being rods with no concave region and large l/w particles possessing a large amount of concavity. We focus on two properties of the u-particle columns: (i) the packing behavior of u-particles studied through measurement of ϕ and (ii) the relaxation of columns under gravity and subject vertical vibration from the floor (see Fig. 17.3).

17.2.1 Experimental Apparatus

U-Particles consisted of steel staples (Duo-fast; Vernon Hills, IL) of constant width, $w = 1.17$ cm, and variable barb length, l ($l/w \in [0.02, 1.125]$). The cross section of all particles was rectangular with thickness of 0.5 mm and width 1.27 mm, which corresponded to a rod-like aspect ratio for $l/w = 0.02$ particles of ≈ 14. We cut particles to size $l/w = 0.02 \pm 0.02, 0.13 \pm 0.02, 0.15 \pm 0.03$, and 0.28 ± 0.04, and other particles were purchased at that size.

Collections of monodisperse particles with fixed l/w were formed into free standing cylindrical columns with column diameter, $d = 4.4$ cm or $d = 5.6$ cm, and height, $h_0 = 3$ cm. Columns were prepared by pouring particles into the cylindrical container followed by a 20 s sinusoidal vibration of the base at a frequency, $f = 30$ Hz, and peak acceleration, $\Gamma = 2$ (in units of gravitational acceleration g). We confirmed that steady-state volume fraction was reached through our preparation protocol in separate experiments conducted over a 60 s time period. Columns occupied a volume $V = \pi h(d/2)^2$, and the volume fraction was calculated as $\phi = \frac{M}{\rho_{st} V}$ where M is total particle mass and $\rho_{st} = 7.85$ g/cm^3 is the density of steel.

Sinusoidal oscillation was generated by an electromagnetic shaker (VTS; Aurora, OH; Fig. 17.1c). The shaker piston was attached to a linear, square-shaft, air bearing, which insured that the motion was primarily vertical. The shaker was mounted to a thick aluminum plate through a collection of springs. This mounting system reduced vibrational coupling, which would occur if the shaker was mounted to the ground. Vibration experiments were performed at a frequency of $f = 30$ Hz and variable peak acceleration $\Gamma \in [1.2, 2.5]$ (in units of gravitational acceleration g). The shaker was controlled by LabVIEW and a Tecron 7550 power amplifier. Acceleration of the vibration table was measured by an accelerometer embedded in the vibration table (PCB Piezotronics; Depew, New York).

Column collapse was monitored using a high-speed camera (Point-Grey; Richmond, BC, Canada). Image capture was triggered externally by a function generator controlled by LabVIEW such that images were captured at a

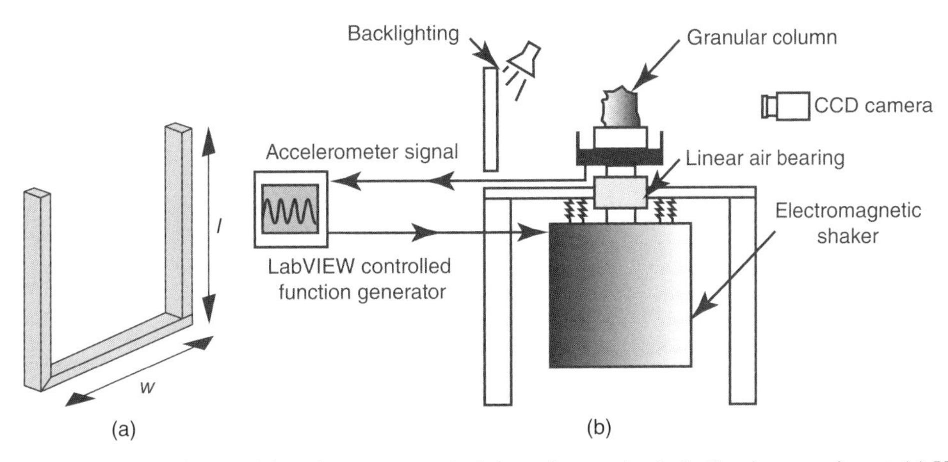

Figure 17.3 Relaxation dynamics of U-particle columns are studied through a mechanical vibration experiment. (a) U-Particle geometry. Width, w, is held constant and length, l, is varied. (b) A computer-controlled shaker table applies sinusoidal forcing to a granular column. A high-speed camera records column collapse from the side.

constant phase of the oscillation cycle, and at frequencies of $f, f/2$, and $f/4$. Images were analyzed in Matlab. Columns appeared black on a white background, and we extracted the foreground column using an image threshold. We then dilated the foreground region to insure that the column was a singly connected region and finally measured the centroid height, $h(t)$, of the column and the projected 2D area $A(t)$.

17.2.2 Packing Experiments

Particles were poured into the cylinder and came to rest at an initial volume fraction ϕ_0, which was dependent on particle packing (Fig. 17.4). Applying vibration for 60 s resulted in a steady-state final volume fraction, ϕ_f, which reached steady-state within approximately 20 s. As can be seen in Figure 17.4b, ϕ_0 and ϕ_f decreased with increasing l/w. Compaction, defined as $\chi = \frac{\phi_f - \phi_0}{\phi_0}$, linearly increased with l/w and was fit by the function $\chi = 0.23(l/w) + 0.12$ ($R^2 = 0.65$). Larger l/w particles likely exhibit a higher compaction because their long barbs cause them to jam in lower ϕ_0 initial states, while their large internal volume also allows packing to high ϕ_f.

For a comparison with similar experiments, the value of $\phi_f = 0.28 \pm 0.01$ we observed for $l/w = 0.02$ particles is close to the range $\phi_f = 0.28$–0.34, observed in cylindrical rod packs with comparable aspect ratio (length/thickness ≈ 14) [48, 52, 71, 72]. The variation in cylinder rod values was due to difference in preparation method. The lower value observed in our $l/w = 0.02$ particles is likely due to the fact that our particle have a rectangular cross section while the values we compare to are from cylindrical cross-sectional rods.

The final volume fraction, ϕ_f, decreased monotonically with increasing l/w. This is consistent with what is observed in rod packing studies in which increasing the length (aspect ratio) of rods decreases the volume fraction [48, 52, 71, 72]. For long rods, the volume fraction scales inversely with rod length, and this behavior is described through a statistical model of particle packing called the random contact model, which we describe in Section 17.3. One way to qualitatively understand this decrease in ϕ_f with increasing l/w is that larger l/w particles have larger internal volumes and thus pack less efficiently.

17.2.3 Collapse Experiments

After we formed cylinders of packed u-particles in the packing experiment, we removed the confining container; this left the column free-standing. During removal of the confining cylinder, the $l/w = 0.02$ particles were marginally stable with partial column collapse occurring approximately 50% of the time, similar to the results reported in Ref. 53. Spontaneous collapse of the $l/w > 0.02$ columns was rarely

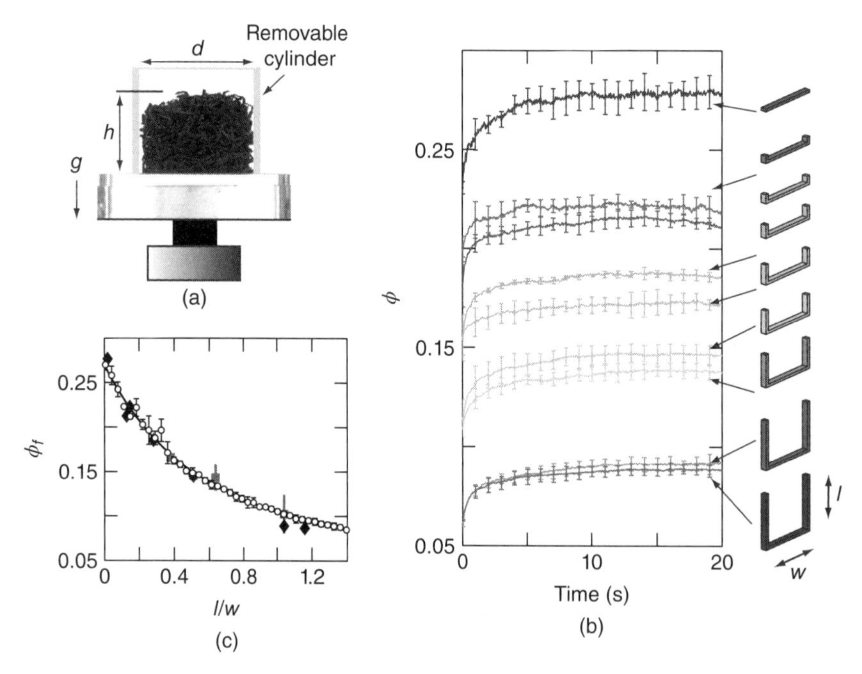

Figure 17.4 Formation and packing of u-particle columns. (a) Experimental setup to form free-standing u-particle columns. Particles were packed within a containing cylinder of diameter $d = 4.4$ cm or $d = 5.6$ cm, which was removed after the packing protocol. (b) $\phi(t)$ during column preparation for various u-particle assemblies. (c) Final packing fraction, ϕ_f, as a function of particle geometry in experiment (column diameter $d = 4.4$ cm diamonds and $d = 5.6$ cm squares) and simulation (white circles). Line is theory prediction from the random contact model.

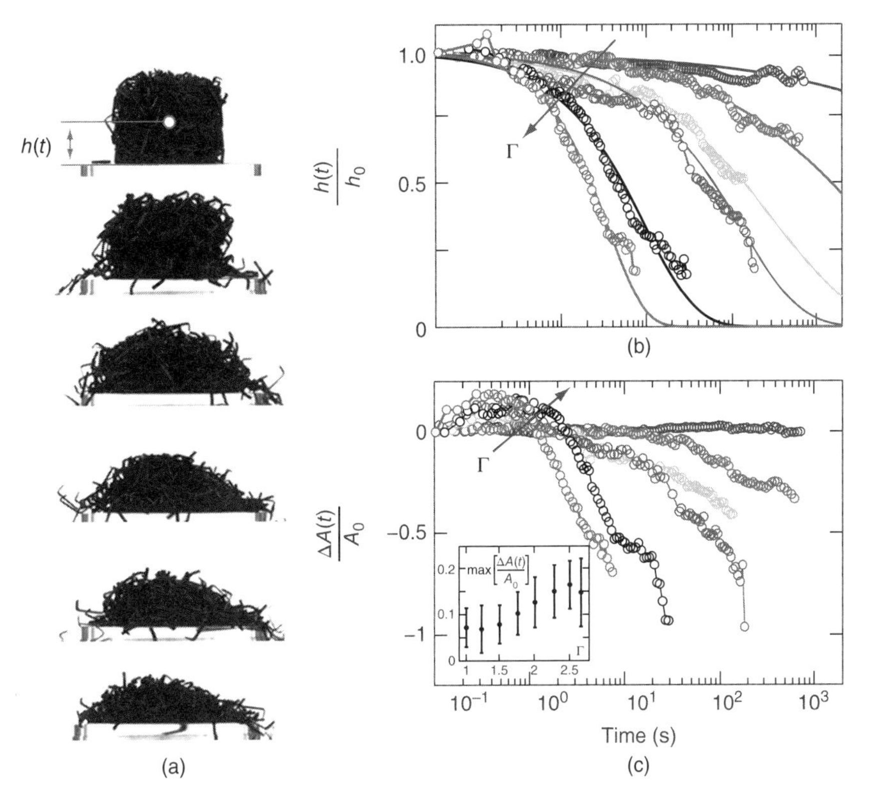

Figure 17.5 Column collapse dynamics. (a) The normalized centroid height, $h(t)/h_0$, of the column during collapse. Vibration parameters are $\Gamma = 2, f = 30$ Hz, and images are separated by 90 oscillation periods. (b) Relaxation of $h(t)/h_0$ as a function of time is shown for $l/w = 0.379$ for $\Gamma = 1.23, 1.48, 1.70, 1.96, 2.20, 2.53$, respectively (arrow denotes increasing Γ). Fit lines are stretched exponentials with equation given in the text. (c) Change in projected area of the column, $\frac{\Delta A(t)}{A_0}$, as a function of time. Γ corresponds to values in (b) with arrow denoting direction of increasing Γ. Inset shows the peak area increase, $\max[\frac{\Delta A(t)}{A_0}]$, as a function of Γ averaged over all l/w (error bars are standard deviation).

observed. To explore the dynamical stability of u-particle columns, we next subjected them to vertical vibration from the base and observed column collapse.

We applied sinusoidal vibration to the base of the free-standing column and observed the collapse process from a lateral view with our camera (Fig. 17.5a). We characterized collapse dynamics by monitoring the centroid height, $h(t)$, and cross-sectional area, $A(t)$, of the column (Fig. 17.5b). The collapse dynamics of $h(t)$ were well described by a phenomenological stretched exponential fit function $\frac{h(t)}{h_0} = e^{[-(\frac{t}{\tau})^{\beta}]}$. The parameter τ is the characteristic collapse time and β is the stretching parameter [73]. Consistent with previous studies [63, 74], β was in the range of 0.5–1 and decreased slightly as Γ increased but was independent of particle geometry. The stretched exponential function is frequently applied to the description of relaxation dynamics of disordered systems [73]; however, a physical interpretation of how it applies to the collapse of geometrically entangled particles is an open question.

For fixed l/w, the collapse time of the column found from the stretched exponential, τ, decreased with increasing Γ. This supports our intuition that larger perturbations cause a more rapid collapse of the column. Furthermore, the logarithm of τ increased linearly with $1/\Gamma$ (Fig. 17.4a) and τ was fit by an exponential $\tau = f^{-1}e^{\Delta/\Gamma}$ with Δ as the single fit parameter ($f = 30$ Hz).

The exponential fit is indicative of an Arrhenius-like process observed in the relaxation of activated systems. The Arrhenius process describes the escape probability of a thermally or mechanically activated particle from a potential well of depth Δ. In thermal systems, the escape time is proportional to one over the Boltzmann factor exp $(-\frac{E}{kT})$, where E is the activation energy required to overcome the potential barrier. In our system, thermal effects are negligible, and instead mechanical excitation plays the role of a thermal energy-like source (Γ analogous to kT) and Δ is analogous to an energy barrier resulting from particle entanglement.

The second quantity we measured during column collapse was the change in projected cross-sectional area, $\frac{\Delta A(t)}{\Delta T}$ of the column. The cross-sectional area displayed an initial increase during the first second of vibration indicating that prior to collapse and particle shedding from the column, the structure initially expands (dilates). The amount of dilation that occurred during collapse, $\max[\frac{\Delta A(t)}{A_0}]$, was an increasing function of Γ for all experiments (see inset Fig. 17.5b).

The stretched exponential fit (Fig. 17.5) suggests that the column collapse process may be qualitatively similar across varied Γ with only the timescale changing. However, the variation in column dilation during collapse suggests that the internal particle processes leading to collapse may differ as a function of Γ. At small Γ, we hypothesize that frictional contacts are mobilized through vibration and thus particles can relax through a sliding process while collisions are not important. At higher Γ, we observe that particles appear highly mobilized and often collide with each other, which likely leads to the dilation we observe during the initial collapse process. Thus, we hypothesize that in different regimes of Γ, the particle scale dynamics of collapse may differ; however, the macroscale collapse time is well described by the stretched exponential.

Column collapse occurred through the separation of entangled particles during vibration. We therefore expected that the hindrance of motion due to particle entanglement – and thus Δ – would increase monotonically with the size of the concave region and thus particle length. Instead we found that Δ was a nonmonotonic function of l/w (Fig. 17.6) with Δ reaching a maximum value at intermediate $l/w = 0.394 \pm 0.045$.[1] Δ appears in an exponential, and, thus, the relaxation time for fixed Γ displays a strong sensitivity to the variation of particle shape (see inset Fig. 17.6). We posit that the maximum in Δ is related to the statistics of particle entanglement within the bulk, and we next study entanglement propensity in theory and simulation.

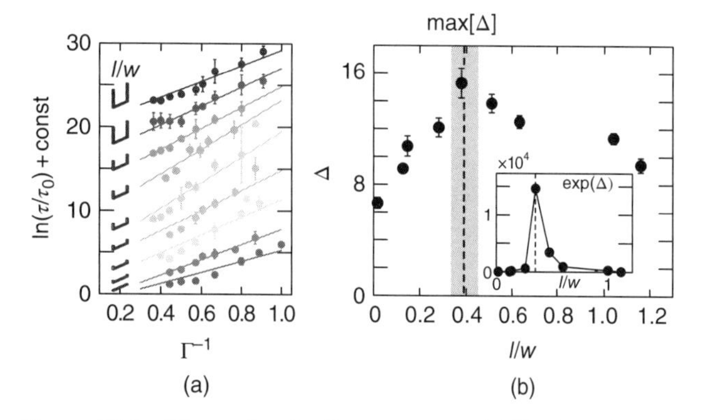

Figure 17.6 Timescale of collapse process. (a) The logarithm of the relaxation time versus inverse acceleration with exponential fit lines $\tau = f^{-1}e^{\Delta/\Gamma}$ ($\tau_0 = 1$ s). Curves are offset vertically for clarity. Error bars are standard deviation of 4 or greater replicates. (b) Δ as a function of l/w. Dashed line indicates estimated maximum of Δ (see Ref. [16]). Error bars represent 95% confidence interval of the best fit lines from (a). Source: Figure reprinted from Ref. [16].

[1] We estimate the maximum and standard deviation of l/w in experiment using a weighted average of points near the peak.

17.3 SIMULATION

In this section, we describe modeling of u-particle packing and examine the ability of particles to mechanically entangle as a function of particle shape. We discuss how excluded volume plays an important role in particle ensembles in which only steric interactions are important, such as our macroscale granular particles. We introduce the random contact model proposed for colloidal rods. We apply this model to our u-particle system and obtain good agreement between the model and our experimental results. Finally, we generate u-particle ensembles in computer simulation and investigate the statistical packing properties of particle arrangements. We show that particles at intermediate l/w display maximum particle entanglement. Lastly, we discuss the relationship between the non-monotonic particle entanglement statistics from simulation, and the non-monotonic relaxation dynamics of columns as measured in experiment.

17.3.1 Random Contact Model of Rods

The random contact model was originally proposed to describe the packing of straight, rod-shaped colloids [72]. This model relates the bulk volume fraction, ϕ, of the ensemble to the particle volume, V_p, and excluded volume, V_e, of the constituent particles. The random contact model assumes only that particles are homogeneously distributed in space and has been shown to work well for rod-shaped particles over a large range of aspect ratios [48, 71, 72, 72]. In the following, we derive the random contact model and explain how it is used in the calculation of "u-particle" packing statistics.

The particle's excluded volume is defined as the volume of space that one particle excludes from another, averaged over all possible particle–particle configurations. Another definition of V_e is in relation to the probability of finding two particles in contact within a larger volume, V. This can be represented as

$$V_e = pV, \tag{17.1}$$

where p is the contact probability. A simple example to consider is a spherical particle of radius r. A spherical particle excludes a volume $V_e = \frac{4}{3}\pi(2r)^3$ from another identical particle. Thus, there is a volume of space, V_e, in which particles cannot be placed without overlapping the original particle. This again is indicative of the probabilistic nature of excluded volume since the probability to randomly place a sphere in a position overlapping the original sphere is $p = \frac{V_e}{V}$.

The relationship between volume fraction and V_e can be determined by the following method. For a volume of space, V with N particles of volume V_p, the solid volume occupied by the particles is $V_o = NV_p$. The volume fraction is defined as $\phi = \frac{V_o}{V}$, which using Equation 17.1 we can rewrite

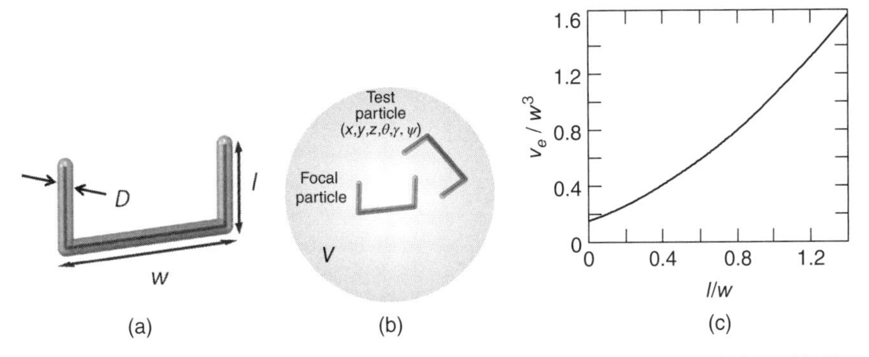

Figure 17.7 Overview of simulation. (a) Three spherocylinders form a u-particle in computer simulation with dimensions given in the text. (b) We compute the excluded volume of u-particles through a Monte Carlo simulation measuring the probability for two randomly placed particles to overlap within a large volume V. (c) Excluded volume of u-particles as a function of l/w.

as $\phi = pN\frac{V_p}{V_e}$. We interpret pN as the average number of contacts per particle within the packing, C, and arrive at the random contact equation[2]

$$\phi = C\frac{V_p}{V_e}. \tag{17.2}$$

The random contact model describes the bulk packing of homogeneously distributed particles in free space, with particle properties V_p and V_e and average contact number C. This model has been tested in experiment and simulation with rod-shaped objects at the microscopic and macroscopic scales [48, 52, 71, 72], and surprisingly all experiments have found a similar contact number $C \approx 10$. We note that this model is applicable for particles that pack with spatially uncorrelated contact points. Applying this model to spheres fails because contact points between particles are always spatially correlated (by definition a distance from the particle location).

We compute V_e numerically for u-particles in a Monte Carlo simulation by using the probabilistic definition of excluded volume. We form u-particles from a combination of three spherocylinders oriented at right angles with each other to form a u-particle. u-Particle dimensions are normalized by the spherocylinder cross-sectional diameter, D. The width of the base spherocylinder is fixed at 14D and the barb lengths are varied from 0 to 16D, consistent with the u-particles used in experiment. In simulation, we randomly place test-particles within a large volume, V, with respect to a focal particle fixed at the center. For each iteration of the computation, we choose a random location (x, y, z) and random orientation (defined by the Euler angles of the particle, θ, γ, ψ) to place the test particle. We then check if the test particle overlaps with the focal particle at this location–orientation combination. To determine the overlap of two u-particles, we must simply compute the pairwise minimum distance between each particle's constituent spherocylinders. If any of these nine pairwise distances are less than the spherocylinder diameter, D, the particles overlap. To detect if particles overlap, we compute the minimum distance between two line segments along the centers of the spherocylinders. We compute the distance between line segments using an algorithm originally developed for computer graphics [75].[3] If particles are found to overlap, we increment a counter N_o. After N iterations of this algorithm, the fraction $\frac{N_c}{N} \to p$ and thus our calculation converges on the excluded volume $V_e = V\frac{N_c}{N}$.

We fit a polynomial to V_e and find that $V_e = 0.460(l/w)^2 + 0.530(l/w) + 0.148$ (in units of W^3, see Fig. 17.7). We approximate the particle volume as $V_p = \pi W(D/2)^2 + 2\pi L(D/2)^2 + \frac{4}{3}\pi(D/2)^3$. With a contact number $C = 9$, the measured volume fraction in experiments and the random contact model prediction are in a good agreement. This value of C is close to the values reported for rod packings of $C = 8.4$–10.8, which depends on preparation [47, 48, 71]; this is surprising given the difference in particle shape between rods and u-particles. We emphasize that the random contact model may only be applied to particles in which the spatial arrangement of contacts is suitably random (i.e., cannot be used for spheres). This process does not work for sphere or for near-sphere packings, which have a lower contact number [76]. Having verified that the random contact model works for u-particles, we may proceed with the calculation of packing statistics for u-particle ensembles.

[2]We note that in the original text of Philipse [72], the random contact model is introduced with the prefactor $2\langle c\rangle$ instead of C. In this case, $\langle c\rangle$ is the ratio of total number of contacts by the number of particles and multiplying this value by two results in, C the average number of contacts per particle. We use this form of the equation in the text and when comparing to studies using the alternate version, we convert reported values of $\langle c\rangle$ to C.

[3]We have uploaded a Matlab implementation of this algorithm to http://www.mathworks.com/matlabcentral/fileexchange/32487-shortest-distance-between-two-line-segments.

17.3.2 Packing Simulations

We study the packing of u-particles in a computer simulation to identify properties of the particle entanglement within the pile. Particle packings were generated through a Monte Carlo simulation. We do not perform molecular dynamics in these simulations; instead we solely enforce the condition that particle configurations which result in an overlap are not allowed. From these packings, we study the statistics of u-particle entanglement.

To generate u-particle packings, we used a brute force packing algorithm to generate close packings of nonoverlapping particles. Packing proceeded in two steps: In the first step, particles were placed at random position and orientation inside a cubic volume of cross-sectional area $(52 \times 52D^2)$ such that the particles did not overlap. If a newly placed particle resulted in an overlap, this particle was removed and a new position was randomly selected. If after 10,000 iterations a suitable particle location was not found, then the algorithm proceeded to step two. In the second step, particles in the volume were selected at random and displaced downward a small random direction and distance $\frac{D}{10}$. If the new location of the particle resulted in particle overlap, the particle was returned to the original location and a new particle chosen. The algorithm was halted after the center of mass height of the ensemble appeared to reach a steady-state. The volume fraction of the simulated packings was determined by measuring the average height of the pile and multiplying it by the areal dimension to obtain the occupied volume and then dividing this by the total volume of particles. A sample packing simulation and packing dynamics are shown in Figure 17.8. In simulation, particles gradually approach a steady-state volume fraction that is consistent with the experimental data (Fig. 17.4c).

We hypothesized that particle entanglement within the column would influence the relaxation time during vertical vibration. Thus, we expected that the maximum in Δ should correspond to a maximum in the density of particle entanglements. In simulation, we defined two particles as entangled when the center line of one particle intersected the internal plane of the neighboring particle (see inset Fig. 17.5a). We measured the number of entanglements per particle, N, for each particle in simulation. The probability distribution function, $P(N)$, was sensitive to l/w (Fig. 17.5a) with mean value $\langle N \rangle$ increasing monotonically with l/w (Fig. 17.5b). The increase was sublinear, indicating that $\langle N \rangle$ grew slower than that of the particle's convex area $(l - D)(w - 2D)$.

The scaling of $\langle N \rangle$ with l/w can be determined by considering the solid volume occupied by the entangled particles in the focal particles convex region (the convex area with infinitesimal thickness δ). Assuming a homogeneous packing, the solid volume in this region is $V_{\text{ent}} = \phi_f(l - D)(w - 2D)\,\delta$. Since each entangled particle contributes only a portion to V_{ent} in the shape of an ellipse of thickness δ, on average $V_{\text{ent}} = \alpha\langle N \rangle \pi\delta\frac{D^2}{4}$ where $\alpha > 1$ accounts for the nonplanar crossings (Fig. 17.9). Solving the above relations yields

$$\langle N \rangle = \frac{4C}{\alpha}\left(\frac{V_p(l - D)(w - 2D)}{\pi V_e D^2}\right). \tag{17.3}$$

With a single fit parameter, $\alpha = 2.648 \pm 0.108$, we find excellent agreement between the predicted number of entanglements per particle and those measured in simulation (Fig. 17.10b).

The spatial density of particle entanglements is $\rho_{\text{ent}} = \langle N \rangle \rho$ where $\rho = \frac{C}{V_e}$ is the number density of particles (Fig. 17.10b). Substitution for $\langle N \rangle$ yields

$$\rho_{\text{ent}} = \frac{4C^2}{\pi\alpha}\left(\frac{V_p(l - D)(w - 2D)}{V_e^2 D^2}\right) \tag{17.4}$$

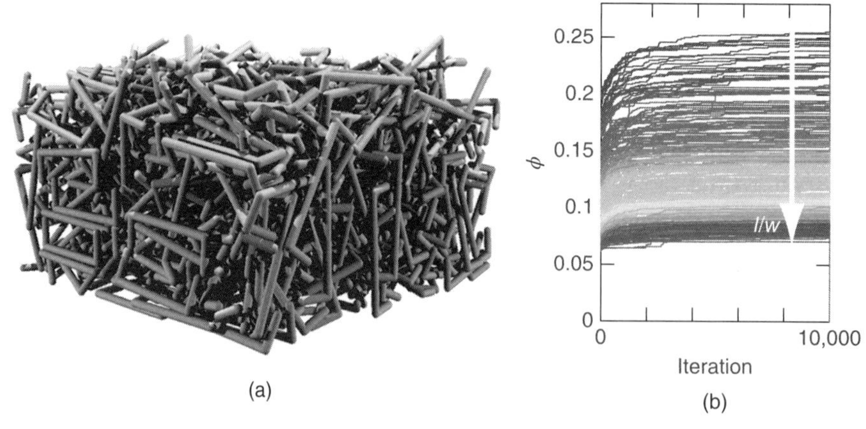

Figure 17.8 Computer simulated u-particle ensembles. (a) A computer-generated ensemble of u-particles of $l/w = 0.35$. (b) Volume fraction of particle ensembles as a function of simulation iteration. Particle ensembles are packed together in a Monte Carlo simulation until the volume fraction reaches a steady-state. Particle l/w is varied from 0 to 1.4 with increasing l/w indicated by arrow.

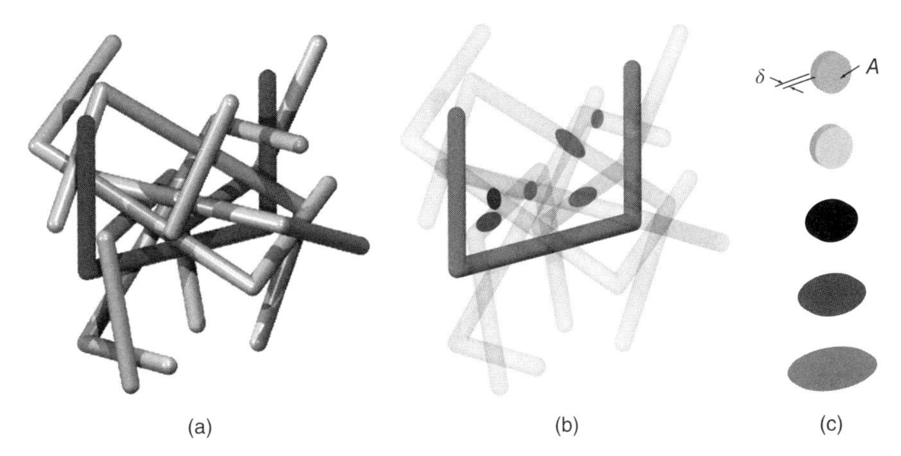

Figure 17.9 Entanglement counting in simulation. (a) A rendering of entangled particles within the ensemble. (b) To predict the number of entanglements within the packing, we consider the infinitesimally thin volume of space within the concave region of the central focal particle (dark gray). The intersection of entangled particles with this plane forms thin ellipses. (c) The cross-sectional area of the intersection region can vary from a circle of diameter D to an ellipse with minor axis $D/2$.

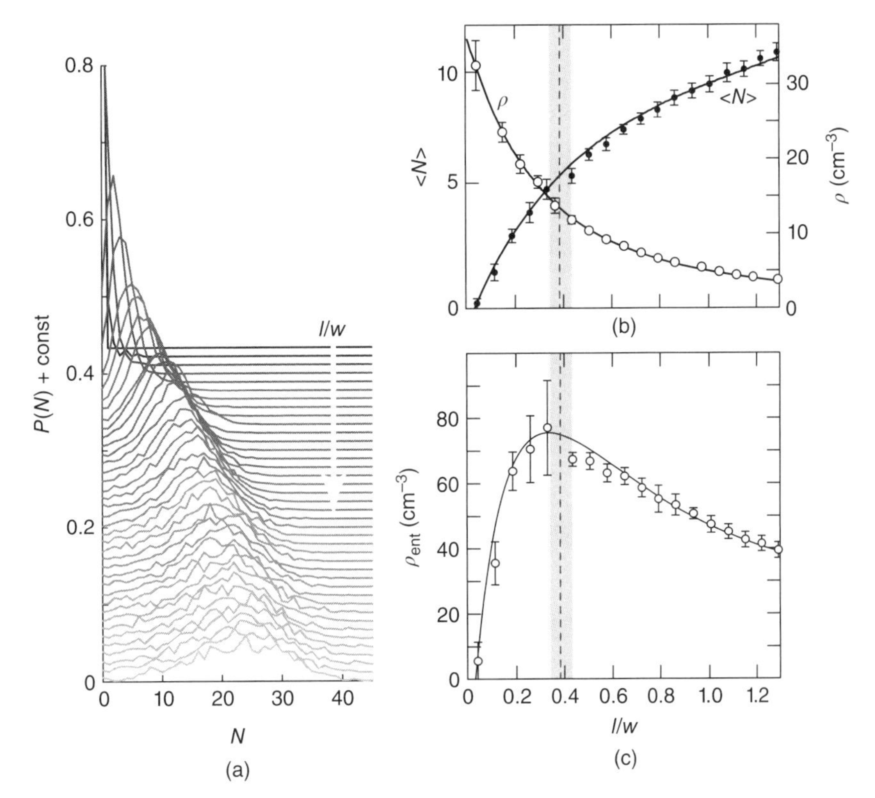

Figure 17.10 Statistics of particle entanglement in simulation. (a) The probability distribution of entanglement number, N, as a function of l/w. Curves are shifted vertically for clarity; $l/w = 0$ at top and increases in increments of 0.036 down. (b) Mean values for N and ρ measured in simulation (circles) and the theoretical fit (black line). (c) Density of entanglements as a function of l/w and the theoretical fit (black line). Vertical dashed line and gray bar correspond to the mean and standard deviation of the estimated maximum of Δ from experiment. Source: Figure reprinted from Ref. [16].

and again the simulation and theory are in a good agreement (Fig. 17.10c) using the previously determined fit parameters C and α. Furthermore, the experimental maximum max[Δ] at $l/w = 0.394 \pm 0.045$ is close to the value obtained in simulation and theory of $l/w = 0.340 \pm 0.015$, suggesting that the large relaxation times for the intermediate u-particle columns are due to the large density of mechanical entanglements.

17.4 CONCLUSIONS

Similar to rod-like particles [48, 52, 71, 72], columns formed from u-particles are stabilized through the inhibition of particle rotation and translation among the entangled particles. The addition of the transverse ends, which form concave u-particles, leads to mechanical entanglement and increases column stability. However, the increase in entanglement with increasing length is offset by the decrease in particle-packing density. These two trends conspire to generate a maximum in the density of mechanical entanglements in collections of nonconvex particles of intermediate l/w—thus columns of these particles most strongly resist separation.

Relaxation processes studied under oscillatory perturbations are found in many soft-matter systems such as oscillatory shear experiments in colloids and suspensions. Vibro-fluidization has been previously used to study the relaxation of piles and columns of dry granular materials under gravity [68, 70, 77]; however, this method has not been applied to characterizing the strength of cohesive GM. We envision that vibration–relaxation experiments similar to those reported here will be useful to explore rheological properties of fluid or electrostatic mediated cohesive GM. Although granular materials in the natural world often posses some interstitial fluids, there is still much to be learned about cohesive granular materials.

Macroscale model systems similar to those described here and elsewhere [60, 78] are useful tools within which to explore how particle shape influences ensemble rheology at other scales. Future study of the particle-scale dynamics of nonconvex particles may provide further insight into the rheology of entangled or crowded particulate systems. For example, model systems such as granular particles may help understand the particle scale dynamics of anomalous diffusion within the crowded cellular environment [79–82]. Furthermore, particles found in nature are often nonspherical [14]; thus, we hope that experiments like those described here will advance the experimental and computational tools used to study nonspherical or even nonconvex particulate systems.

The random contact model utilized to explain the optimum geometry for entanglement of U-shaped particles assumes only uncorrelated particle contacts within the bulk. Thus, we expect the results to apply to rigid nonconvex particulate systems of all scales. A recent study of suspension rheology found that convex particles of differing shape collapsed to a viscosity–stress master curve while concave particle did not collapse to this curve; this difference was attributed to particle entanglement effects [17]. At the microscale, polymers with rigid pendants oriented perpendicular to the polymer chain increase internal molecular free volume and hinder polymer motion, which significantly affects rheology similar to geometric entanglement [83]. At the macroscale, strain-stiffening of model polymers is associated with entanglement [60]. Even organisms can benefit from geometric entanglement. For example, the fire ant *Solenopsis invicta* and the army ant *Eciton burchelli* create waterproof rafts and shelters—which have been described as akin to living chain mail [84]—through the interlocking and entanglement of limbs and mandibles [28, 30].

ACKNOWLEDGMENTS

The authors would like to acknowledge David Hu, Paul Umbanhowar, and Scott Franklin for helpful discussion. Funding support was provided by NSF Physics of Living Systems #0957659, NSF (DMR-0706353), and NSF (PHY-0848894).

REFERENCES

[1] Jaeger HM, Nagel SR. Physics of the granular state. Science 1992;255:1523–1531.

[2] Jaeger HM, Nagel SR, Behringer RP. Granular solids, liquids, and gases. Rev Mod Phys 1996;68(4):1259–1273.

[3] Nedderman RM. *Statics and Kinematics of Granular Materials*. Cambridge: Cambridge University Press; 1992.

[4] Duran J. *Sands, Powders, and Grains: An Introduction to the Physics of Granular Materials*. New York: Springer-Verlag; 2000.

[5] Liu AJ, Nagel SR. *Jamming and Rheology: Constrained Dynamics on Microscopic and Macroscopic Scales*. Boca Raton, FL: CRC Press; 2001.

[6] Cundall PA, Strack ODL. A discrete numerical model for granular assemblies. Geotechnique 1979;29:47.

[7] Silbert LE, Ertas D, Grest GS, Halsey TC, Levine D. Geometry of frictionless and frictional sphere packings. Phys Rev E 2002;65:031304.

[8] Silbert LE, Ertas D, Grest GS, Halsey TC, Levine D, Plimpton SJ. Granular flow down an inclined plane: Bagnold scaling and rheology. Phys Rev E 2001;64:051302.

[9] Silbert LE, Ertas D, Grest GS, Halsey TC, Levine D. Analogies between granular jamming and the liquid-glass transition. Phys Rev E 2002;65:051307.

[10] Ertas D, Grest GS, Halsey TC, Levine D, Silbert LE. Gravity-driven dense granular flows. Europhys Lett 2001;56:214.

[11] Pournin L, Weber M, Tsukahara M, Ferrez JA, Ramaioli M, Libling ThM. Three-dimensional distinct element simulation of spherocylinder crystallization. Granular Matter 2005;7:119.

[12] Azéma E, Estrada N, Radjaï F. Nonlinear effects of particle shape angularity in sheared granular media. Phys Rev E 2012;86:041301.

[13] Damasceno PF, Engel M, Glotzer SC. Predictive self-assembly of polyhedra into complex structures. Science 2012;337(6093):453–457.

[14] Cho GC, Dodds J, Santamarina JC. Particle shape effects on packing density, stiffness, and strength: natural and crushed sands. J Geotech Geoenviron Eng 2006;132(5):591–602.

[15] Meng L, Li S, Lu P, Li T, Jin W. Bending and elongation effects on the random packing of curved spherocylinders. Phys Rev E 2012;86(6):061309.

[16] Gravish N, Franklin SV, Hu DL, Goldman DI. Entangled granular media. Phys Rev Lett 2012;108(20):208001.

[17] Brown E, Zhang H, Forman NA, Maynor BW, Betts DE, DeSimone JM, Jaeger HM. Shear thickening and jamming in densely packed suspensions of different particle shapes. Phys Rev E 2011;84:031408.

[18] Baule A, Makse HA. Fundamental challenges in packing problems: from spherical to non-spherical particles. Soft Matter 2014;10(25):4423.

[19] Miskin MZ, Jaeger HM. Adapting granular materials through artificial evolution. Nat Mater 2013;12(4):326–331.

[20] Brown E, Rodenberg N, Amend J, Mozeika A, Steltz E, Zakin MR, Lipson H, Jaeger HM. Universal robotic gripper based on the jamming of granular material. Proc Natl Acad Sci U S A 2010;107(44):18809–18814.

[21] Stannarius R. Granular materials of anisometric grains. Soft Matter 2013;9(31):7401.

[22] Manna L, Milliron DJ, Meisel A, Scher EC, Alivisatos AP. Controlled growth of tetrapod-branched inorganic nanocrystals. Nat Mater 2003;2(6):382–385.

[23] Glotzer SC, Solomon MJ. Anisotropy of building blocks and their assembly into complex structures. Nat Mater 2007;6(8):557–562.

[24] McKeown NB, Budd PM. Exploitation of intrinsic microporosity in polymer-based materials. Macromolecules 2010;43(12):5163–5176.

[25] McKeown NB, Budd PM, Msayib KJ, Ghanem BS, Kingston HJ, Tattershall CE, Makhseed S, Reynolds KJ, Fritsch D. Polymers of intrinsic microporosity (PIMS): bridging the void between microporous and polymeric materials. Chem Eur J 2005;11(9):2610–2620.

[26] Whitesides GM, Boncheva M. Beyond molecules: self-assembly of mesoscopic and macroscopic components. Proc Natl Acad Sci U S A 2002;99(8):4769–4774.

[27] Miao L, Vanderlinde O, Liu J, Grant RP, Wouterse A, Shimabukuro K, Philipse A, Stewart M, Roberts TM. The role of filament-packing dynamics in powering amoeboid cell motility. Proc Natl Acad Sci U S A 2008;105(14):5390–5395.

[28] Mlot NJ, Tovey CA, Hu DL. Fire ants self-assemble into waterproof rafts to survive floods. Proc Natl Acad Sci U S A 2011;108(19):7669–7673.

[29] Franks NR. Thermoregulation in army ant bivouacs. Physiol Entomol 2008;14(4):397–404.

[30] Anderson C, Theraulaz G, Deneubourg J-L. Self-assemblages in insect societies. Insectes Soc 2002;49(2):99–110.

[31] Elliott JM, Hansell MH. Animal architecture and building behaviour. J Anim Ecol 1984;54(2):676.

[32] Chen S, Wang ZL, Ballato J, Foulger SH, Carroll DL. Monopod, bipod, tripod, and tetrapod gold nanocrystals. J Am Chem Soc 2003;125(52):16186–16187.

[33] de Gennes P-G. Granular matter: a tentative view. Rev Mod Phys 1999;71(2):374–382.

[34] Jerkins M, Schröter M, Swinney HL, Senden TJ, Saadatfar M, Aste T. Onset of mechanical stability in random packings of frictional spheres. Phys Rev Lett 2008;101:018301.

[35] Cates ME, Wittmer JP, Bouchaud JP, Claudin P. Jamming, force chains, and fragile matter. Phys Rev Lett 1998;81(9):1841–1844.

[36] Majmudar TS, Behringer RP. Contact force measurements and stress-induced anisotropy in granular materials. Nature 2005;435(7045):1079–1082.

[37] Umbanhowar P, Goldman DI. Granular impact and the critical packing state. Phys Rev E 2010;82(1):010301.

[38] Mitarai N, Nori F. Wet granular materials. Adv Phys 2006;55(1–2):1–45.

[39] Herminghaus S. Dynamics of wet granular matter. Adv Phys 2005;54(3):221–261.

[40] Richefeu V, Radjaı F, El Youssoufi MS. Stress transmission in wet granular materials. Eur Phys J E 2006;21(4):359–369.

[41] Richefeu V, El Youssoufi MS, Radjaï F. Shear strength properties of wet granular materials. Phys Rev E 2006;73(5):051304.

[42] Nowak S, Samadani A, Kudrolli A. Maximum angle of stability of a wet granular pile. Nat Phys 2005;1(1):50–52.

[43] Tegzes P, Vicsek T, Schiffer P. Avalanche dynamics in wet granular materials. Phys Rev Lett 2002;89(9):94301.

[44] Fournier Z, Geromichalos D, Herminghaus S, Kohonen MM, Mugele F, Scheel M, Schulz M, Schulz B, Schier Ch, Seemann R, Skudelny A. Mechanical properties of wet granular materials. J Phys Condens Matter 2005;17(9):S477.

[45] Scheel M, Geromichalos D, Herminghaus S. Wet granular matter under vertical agitation. J Phys Condens Matter 2004;16(38):S4213.

[46] Fingerle A, Roeller K, Huang K, Herminghaus S. Phase transitions far from equilibrium in wet granular matter. New J Phys 2008;10(5):053020.

[47] Philipse AP. The random contact equation and its implications for (colloidal) rods in packings, suspensions, and anisotropic powders. Langmuir 1996;12:1127.

[48] Blouwolff J, Fraden S. The coordination number of granular cylinders. Europhys Lett 2006;76:1095.

[49] Lumay G, Vandewalle N. Compaction of anisotropic granular materials: experiments and simulations. Phys Rev E 2004;70:051314.

[50] Galanis J, Harries D, Sackett DL, Losert W, Nossal R. Spontaneous patterning of confined granular rods. Phys Rev Lett 2006;96(2):028002.

[51] Stokely K, Diacou A, Franklin SV. Two-dimensional packing in prolate granular materials. Phys Rev E 2003;67(5):051302.

[52] Desmond K, Franklin SV. Jamming of three-dimensional prolate granular materials. Phys Rev E 2006;73(3):031306.

[53] Trepanier M, Franklin SV. Column collapse of granular rods. Phys Rev E 2010;82(1):011308.

[54] Saraf S, Franklin SV. Power-law flow statistics in anisometric (wedge) hoppers. Physical Review E 2011;83(3):030301.

[55] Zou L-N, Cheng X, Rivers ML, Jaeger HM, Nagel SR. The packing of granular polymer chains. Science 2009;326(5951):408–410.

[56] Karayiannis NCh, Foteinopoulou K, Laso M. Contact network in nearly jammed disordered packings of hard-sphere chains. Phys Rev E 2009;80(1):011307.

[57] Lopatina LM, Reichhardt CJO, Reichhardt C. Jamming in granular polymers. Phys Rev E 2011;84(1):011303.

[58] Mohaddespour A, Hill RJ. Granular polymer composites. Soft Matter 2012;8(48):12060–12065.

[59] Borzsonyi T, Stannarius R. Granular materials composed of shape-anisotropic grains. Soft Matter 2013;9: 7401–7418.

[60] Brown E, Nasto A, Athanassiadis AG, Jaeger HM. Strain stiffening in random packings of entangled granular chains. Phys Rev Lett 2012;108(10):108302.

[61] Schofield AN, Wroth P. *Critical State Soil Mechanics*. New York: McGraw-Hill; 1968.

[62] Gravish N, Umbanhowar PB, Goldman DI. Force and flow transition in plowed granular media. Phys Rev Lett 2010;105(12):128301.

[63] Philippe P, Bideau D. Compaction dynamics of a granular medium under vertical tapping. Europhys Lett 2007;60(5):677.

[64] Knight JB, Fandrich CG, Lau CN, Jaeger HM, Nagel SR. Density relaxation in a vibrated granular material. Phys Rev E 1995;51(5):3957.

[65] Oda M, Kazama H. Microstructure of shear bands and its relation to the mechanisms of dilatancy and failure of dense granular soils. Geotechnique 1998;48(4):465–481.

[66] Knight JB, Jaeger HM, Nagel SR. Vibration-induced size separation in granular media: the convection connection. Phys Rev Lett 1993;70(24):3728–3731.

[67] De Bruyn JR, Bizon C, Shattuck MD, Goldman D, Swift JB, Swinney HL. Continuum-type stability balloon in oscillated granular layers. Phys Rev Lett 1998;81(7):1421–1424.

[68] Jaeger HM, Liu C-H, Nagel SR. Relaxation at the angle of repose. Phys Rev Lett 1989;62(1):40–43.

[69] Huang K, Röller K, Herminghaus S. Universal and non-universal aspects of wet granular matter under vertical vibrations. Eur Phys J Spec Top 2009;179(1):25–32.

[70] Rubin D, Goldenson N, Voth GA. Failure and strengthening of granular slopes under horizontal vibration. Phys Rev E 2006;74(5):051307.

[71] Wouterse A, Luding S, Philipse AP. On contact numbers in random rod packings. Granular Matter 2009;11(3):169–177.

[72] Philipse AP. The random contact equation and its implications for (colloidal) rods in packings, suspensions, and anisotropic powders. Langmuir 1996;12(5):1127–1133.

[73] Phillips JC. Stretched exponential relaxation in molecular and electronic glasses. Rep Prog Phys 1999;59(9):1133.

[74] Mattsson J, Wyss HM, Fernandez-Nieves A, Miyazaki K, Hu Z, Reichman DR, Weitz DA. Soft colloids make strong glasses. Nature 2009;462(7269):83–86.

[75] Eberly DH. *3D Game Engine Design: A Practical Approach to Real-Time Computer Graphics*. San Francisco (CA): Morgan Kaufmann Publishers; 2007.

[76] Donev A, Cisse I, Sachs D, Variano EA, Stillinger FH, Connelly R, Torquato S, Chaikin PM. Improving the density of jammed disordered packings using ellipsoids. Science 2004;303(5660):990–993.

[77] Sánchez I, Raynaud F, Lanuza J, Andreotti B, Clément E, Aranson IS. Spreading of a granular droplet. Phys Rev E 2007;76(6):060301.

[78] Raymer DM, Smith DE. Spontaneous knotting of an agitated string. Proc Natl Acad Sci U S A 2007;104(42):16432–16437.

[79] Wong IY, Gardel ML, Reichman DR, Weeks ER, Valentine MT, Bausch AR, Weitz DA. Anomalous diffusion probes microstructure dynamics of entangled F-actin networks. Phys Rev Lett 2004;92(17):178101.

[80] Szymanski J, Weiss M. Elucidating the origin of anomalous diffusion in crowded fluids. Phys Rev Lett 2009;103:038102.

[81] Saxton MJ. Anomalous diffusion due to obstacles: a Monte Carlo study. Biophys J 1994;66(2):394–401.

[82] Tolić Nørrelykke IM, Munteanu E-L, Thon G, Oddershede L, Berg-Sørensen K. Anomalous diffusion in living yeast cells. Phys Rev Lett 2004;93:078102.

[83] Tsui NT, Paraskos AJ, Torun L, Swager TM, Thomas EL. Minimization of internal molecular free volume: a mechanism for the simultaneous enhancement of polymer stiffness, strength, and ductility. Macromolecules 2006;39(9):3350–3358.

[84] Mitchell M. *Complexity: A Guided Tour*. New York: Oxford University Press; 2009.

18

FOAMS

Reinhard Höhler[1,2] & Sylvie Cohen-Addad[1,2]

[1]Institut des NanoSciences de Paris, CNRS-UMR 7588, UPMC Univ Paris 06, Paris, France
[2]Institut Francilien des Sciences Appliquées, Université Paris-Est, Champs-sur-Marne, France

18.1 INTRODUCTION

Liquid foams have characteristic structures on a wide range of length scales. Viewed from a distance, they appear as a continuous material, but a closer look reveals a packing of bubbles. It is generally disordered, but not random: Geometrical and topological laws govern the equilibrium structure [1, 2]. The shape of the individual bubbles depends on the liquid volume fraction. If it is large, the bubbles are approximately spherical and the foam is called "wet." When more and more liquid is extracted, "dry foam" with polyhedral bubbles is obtained (cf. Fig. 18.1). Foam structures cannot only be 3D but also 2D: Examples are bubble rafts floating on a liquid or Langmuir monolayers that are divided into bubble-like cellular domains. To prevent the liquid films that separate neighboring bubbles from rupturing, the gas–liquid interfaces must be stabilized by adsorbed surfactant molecules, proteins, or solid particles.

Liquid foams occur naturally, for instance, on waves in the sea or in volcanic eruptions where magma and gases are ejected simultaneously [4]. Some animals use the properties of foam: Spittlebugs are so named because in the early stages of their development, their nymphs generate a froth covering to protect themselves [5]. The honeycombs made by bees and cell assemblies in fly eyes [6] are foam-like structures because their geometry presents a minimal surface-to-volume ratio. Liquid foams are also used in many industrial applications [7], such as fire-fighting,

Fluids, Colloids and Soft Materials: An Introduction to Soft Matter Physics, First Edition. Edited by Alberto Fernandez Nieves and Antonio Manuel Puertas.

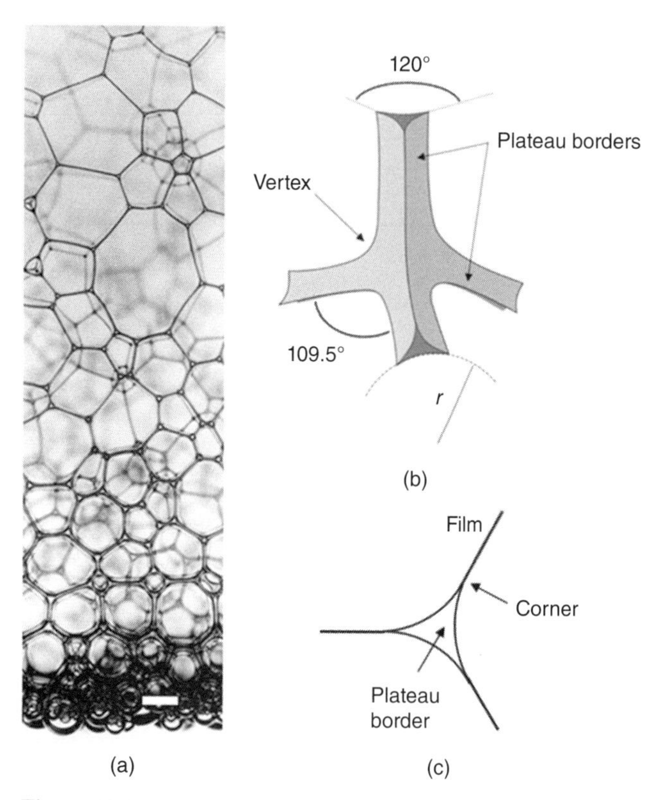

Figure 18.1 (a) Side view of foam floating on a liquid surface (Source: Courtesy of D. Durian). Drainage due to gravity induces a vertical gradient of liquid volume fraction: Near the top the foam is dry and the bubbles are polyhedral, whereas near the bottom the foam is wet and the bubbles are spherical. The white scale bar is 3 mm long. The soap films are not visible themselves, but their junctions, called Plateau borders, appear as dark lines. (b) Connection of Plateau borders at a vertex. The indicated angles illustrate Plateau's rules, a set of geometric constraints valid in dry foams. r is the radius of curvature of the Plateau border interfaces. Source: Reprinted with permission from [3]. Copyright 2010 Belin. (c) Sectional view of a Plateau border, connecting three films.

food, cosmetics, and floatation, a separation process that relies on the ability of hydrophobic particles to attach to gas–liquid interfaces, in contrast to hydrophilic particles. Floatation is widely used in the mining industry for plastics waste recycling as well as for recovering proteins and microorganisms from a cultivation medium [8]. Moreover, many solid cellular materials such as foamed polymers or metals have liquid foam precursors. Further examples are foamed plaster, concrete or ceramics, and also foamed gels, which may be of interest as tissue scaffolds in regenerative medicine [9]. Foams are also used in the oil industry, as a drilling fluid. Such fluids must be injected into the well bore to keep the drill bit cool and to carry out drill cuttings. Often muds are used for this purpose, but if the well is deep, a large hydrostatic pressure builds up at the bottom, which may lead to an invasion of the surrounding porous rock by the fluid. This problem is reduced if foams are injected as a lightweight drilling fluid. Using a yield stress fluid such as foam has the additional advantage that the cuttings are suspended while drilling is paused [7].

From a fundamental point of view, foams are of interest as model materials that help to gain insight into the complex structures and dynamics of soft condensed matter. The structure of wet foams can be described as a packing of spheres without any static friction. The topical question of how such random assemblies may jam or flow is related to the physics of granular materials and glasses [10, 11]. The first studies of ordered 3D foams were theoretical or numerical [2], but since the development of microfluidic devices, small highly monodisperse bubbles can easily be produced and real bubble crystals have been investigated [12, 13] (cf. Chapter 1). Among the fundamental motivations of this work is an unsolved mathematical problem considered by Lord Kelvin [14]: What is the partition of 3D space into equal cells of a given volume that minimizes the interfacial area per cell? Such a partition would be an ideal dry foam structure since it minimizes interfacial energy. The mechanical behavior of liquid foam is governed by coupled processes on a wide range of length and timescales. Subjected to a slowly increasing applied strain, foams first respond elastically. Beyond a yield strain, the bubble packing undergoes intermittent local rearrangements, leading to macroscopic plastic flow [15]. Rearrangement of driven flow is observed in a variety of materials with a local random packing structure such as concentrated emulsions or soft particle pastes, suggesting that a generic mechanism may govern the flow dynamics of a large class of "soft glassy materials" [16]. While the dynamics on scales larger than the individual bubbles, droplets, or particles may indeed present analogous collective behavior, local structures and interactions cannot be expected to be universal: In foams, the rearrangement dynamics depend on the viscosity of the continuous phase, the interfacial tension, and also on the 2D rheology of the surfactant-covered gas–liquid interfaces. The mechanical properties of foams evolve with foam age, defined as the time elapsed since the production. An aging mechanism due to gas diffusion among neighboring bubbles called coarsening relaxes internal stresses that arise when the sample is made or as a consequence of plastic flow [17]. Thus, aging can help to create a well-defined mechanical reference state in foams. In the following, we focus on the structures and dynamics of 3D foams and we compare their behavior to other complex fluids.

18.2 EQUILIBRIUM STRUCTURES

18.2.1 Equilibrium Conditions

As a starting point, we consider a dilute polydisperse suspension of gas bubbles in a liquid, in the absence of gravity.

In static equilibrium the bubbles are spherical as this shape minimizes interfacial energy for a given gas volume. Due to the tension T of the gas–liquid interface, the gas pressure in a bubble of radius R is larger than the liquid pressure outside. This difference ΔP is given by Laplace's law:

$$\Delta P = 2T/R. \tag{18.1}$$

To transform the dilute dispersion into a foam, liquid must be extracted until the average number z of mechanical contacts between a bubble and its neighbors reaches a critical value $z_c = 6$ where jamming occurs in 3D packings [11, 18]. If even more liquid is extracted, bubbles are squeezed against their neighbors so that the coordination number z, the interfacial area, and the energy all increase and an elastic mechanical response of the packing sets in. A foam is characterized by its liquid volume fraction ε, defined as the liquid volume divided by the total foam volume. The critical volume fraction ε_c reached when $z = z_c$ depends on packing structure. For 3D monodisperse random close packings, we have $\varepsilon_c = 0.36$ whereas for ordered face-centered cubic (FCC) or hexagonal close-packed (HCP) structures, $\varepsilon_c = 0.26$. We now analyze the mechanical equilibrium in a foam. The liquid pressure in Plateau borders is smaller than the gas pressure in the bubbles due to the interfacial tension and the radius of curvature r (cf. Fig. 18.1). This difference is called capillary pressure and it is equal to T/r, according to Laplace's law. The pressures in the films and in the gas are almost equal because the film surfaces are much less curved than the Plateau border interfaces. As a consequence, there is a pressure gradient in the liquid that drives it out of the films. As they thin down below about 100 nm, the so-called long-range forces between the opposite gas–liquid interfaces become significant [19]. These forces, due to combined van der Waals, electrostatic, and steric interactions, are often described in terms of a film thickness dependent on disjoining pressure [19]. It pushes the two opposite interfaces apart, thus ensuring a static equilibrium in the film when a thickness typically in the range 10–40 nm is reached.

To discuss the equilibrium of foam structures for $\varepsilon < \varepsilon_c$ on a macroscopic scale, we consider the experiment schematically illustrated in the inset of Figure 18.2 and where gravity is assumed to be negligible: A bubble dispersion is held in a container whose bottom is a membrane. The foaming liquid can freely pass through it, but the bubbles cannot. Extraction of a liquid volume element dV through the membrane requires mechanical work: Indeed, the bubbles are squeezed against each other, their shape becomes increasingly nonspherical, and the interfacial area of the foam increases by dS. The work per extracted volume is called osmotic pressure Π, and it is related to surface tension T and area increase dS by an energy balance [20]:

$$-\Pi \; dV = T\,dS. \tag{18.2}$$

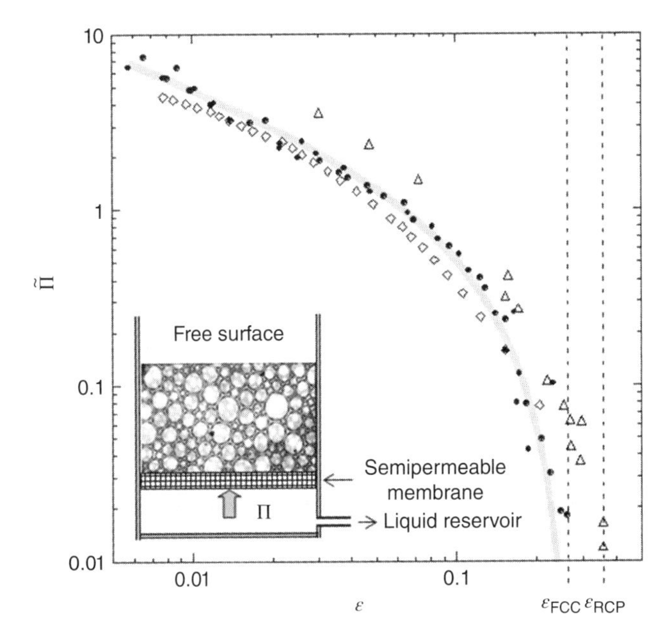

Figure 18.2 Osmotic pressure, normalized by $T/\langle R\rangle$, versus liquid fraction, for disordered monodisperse emulsions (triangle), polydisperse emulsions (squares) and ordered monodisperse foams (filled circles). For the foam samples, $T = 37.5$ mN/m and $\langle R\rangle = 150\,\mu$m, yielding $P = 1400$ Pa for a liquid fraction $\varepsilon = 0.01$. The vertical dashed lines indicate the liquid fractions where the osmotic pressure is expected to go to zero, for face-centered cubic (FCC) and random close-packed (RCP) structures. The full line corresponds to Equation 18.2 with $k = 0.72$. The inset illustrates the concept of osmotic pressure. Source: Reprinted with permission from [12]. Copyright 2008 American Chemical Society.

Equivalently, Π may be defined as the difference between the external pressure exerted on the foam at its free surface and the liquid pressure (cf. Fig. 18.2). Π fundamentally differs from the osmotic pressure of entropic origin considered in the context of molecular solutions. A much closer analogy links Π to the confinement pressure introduced in the physics of granular materials (see Chapter 17). Dimensional arguments show that Π must scale as $T/\langle R\rangle$, where $\langle R\rangle$ is an average bubble radius and, therefore, a dimensionless pressure $\tilde{\Pi} = \Pi/(T/\langle R\rangle)$ is often considered. The variation of osmotic pressure with liquid fraction has so far been analyzed theoretically only close to the wet and dry limits [2, 21]. The following empirical expression has been proposed on the basis of experiments and surface minimization simulations [12], for liquid fractions ranging from the dry limit up to a few percent below ε_c:

$$\tilde{\Pi}(\varepsilon) = k\frac{(\varepsilon - \varepsilon_c)^2}{\sqrt{\varepsilon}}. \tag{18.3}$$

The constant k is equal to 7.3 for ordered foams with FCC or HCP structures and $k = 3.2$ for random foams [22].

The critical behavior close to the jamming transition is hard to observe in foams, but theoretical arguments and experiments with disordered emulsions suggest a scaling $\tilde{\Pi}(\varepsilon) \propto (1-\varepsilon)^2(\varepsilon_c - \varepsilon)$ in this regime [23].

So far, we have ignored gravity. Its effect is illustrated in Figure 18.1a by the vertical liquid fraction gradient established in equilibrium. A hydrostatic force balance yields the following relation between the inverse function of the vertical liquid fraction profile $\varepsilon(Z)$ and the osmotic pressure [20]:

$$Z(\varepsilon) = \frac{T}{\langle R \rangle \rho g} \int_0^{\tilde{\Pi}} \frac{d\tilde{\Pi}(\varepsilon)}{1-\varepsilon}. \tag{18.4}$$

ρ is the liquid density, g is the acceleration due to gravity and the Z-axis is directed upwards. Equation 18.4 shows that the characteristic length scale of the liquid fraction profile is not the usual capillary length $\kappa^{-1} = (T/\rho g)^{1/2}$, but $\kappa^{-2}/\langle R \rangle$.

18.2.2 Geometrical and Topological Properties

Liquid foam equilibrium structures are often disordered, but never totally random [1, 2]. Plateau has shown that in dry foams, liquid films must meet three by three at angles of 120°. Their junctions, the Plateau borders, must meet four by four, at regular tetrahedral angles, as shown in Figure 18.1. Foam structures of very small but finite liquid content can be decorated by liquid channels of constant concave cross section (cf. Fig. 18.1b). Within this approximation, the radius of curvature of the Plateau borders r is related to the liquid fraction and the bubble radius R_V as follows:

$$\varepsilon = c\left(\frac{r}{R_V}\right)^2, \tag{18.5}$$

where R_v is defined as the radius of a sphere that would have the same volume as a bubble and c is a geometrical constant depending on the bubble packing [1, 2]. For Kelvin foam (presented below), $c = 0.333$.

As more and more liquid is added to a foam, its description as a dry foam decorated by Plateau borders fails: Stable vertices of coordination higher than 4 appear, and finally, it becomes hard to distinguish Plateau borders from vertices. Close to the wet limit where ε reaches ε_c, the foam structure is best described as a packing of spheres that are slightly flattened at their contacts. In this regime, 3D simulations of disordered foams as well as confocal microscopy observations of emulsions [24] have shown that the average number of contacts per bubble z scales with liquid volume fraction according to the following relation [1, 21] where $z_c = 6$ and $\varepsilon_c = 0.36$:

$$z - z_c \propto \sqrt{\varepsilon_c - \varepsilon}. \tag{18.6}$$

Crystalline foam structures are much simpler to model than disordered foams since they are entirely specified by their symmetry and the content of the unit cell. Therefore, experimental studies of bubble crystals provide a quantitative testing ground for our understanding of interactions at the scale of bubbles and gas–liquid interfaces. Simulations and experiments have shown that monodisperse bubbles assemble into FCC or HCP crystals when the liquid fraction is reduced to 0.26 [12, 13, 25]. If further liquid is extracted, the bubbles are more and more deformed, giving rise to an increase in interfacial energy. Below a liquid fraction close to 0.07, a transition to a body-centered cubic (BCC) packing is observed as illustrated on Figure 18.3. This latter structure is named after Lord Kelvin who first pointed its relevance as a dry foam structure [14].

Simulations using the Surface Evolver software [26] show that the interfacial energy density of the BCC structure becomes lower than that of the FCC structure at a liquid fraction close to the one where this transition is observed experimentally [12] (Fig. 18.4). The foam therefore gains interfacial energy upon the structural transition. However, to pass from the FCC to the BCC structure via a homogeneous deformation, a perfect bubble crystal would have to overcome an energy barrier and it is not clear where the necessary energy may come from. The answer may be related to the grain boundaries and dislocations that are present in real bubble crystals and that may help them to pass toward the structure of lower energy. The energetically most favorable known dry foam structure is called the Weaire–Phelan structure [2]. It is difficult to produce experimentally because its unit cell is more complex than the Kelvin cell. However, Weaire–Phelan foam can be made in containers with textured walls that act like a template [27].

18.2.3 Static Bubble Interactions

Close to the wet limit, foam can be described as an ordered or disordered assembly of spherical bubbles that are squeezed against each other by the osmotic pressure. The equilibrium of such a structure can be simulated by minimizing the interfacial energy density for fixed bubble volumes and liquid fraction, but this requires very large calculations. Alternatively, bubbles may approximately be described as soft "particles" that interact via a repulsive potential [28, 29]. The force exerted between two neighboring bubbles can be estimated as the capillary pressure times the contact facet area. However, a bubble that is squeezed at one contact must expand to conserve its volume, and this process is constrained by its other neighbors. The effective static interaction between bubbles, therefore, depends on their coordination number z. The following effective interaction potential u between two neighboring bubbles has been deduced from simulations of 3D ordered foams [21, 30]:

$$u(h) = 4\pi R^2 T\, C(z)\ \xi^{\nu(z)}. \tag{18.7}$$

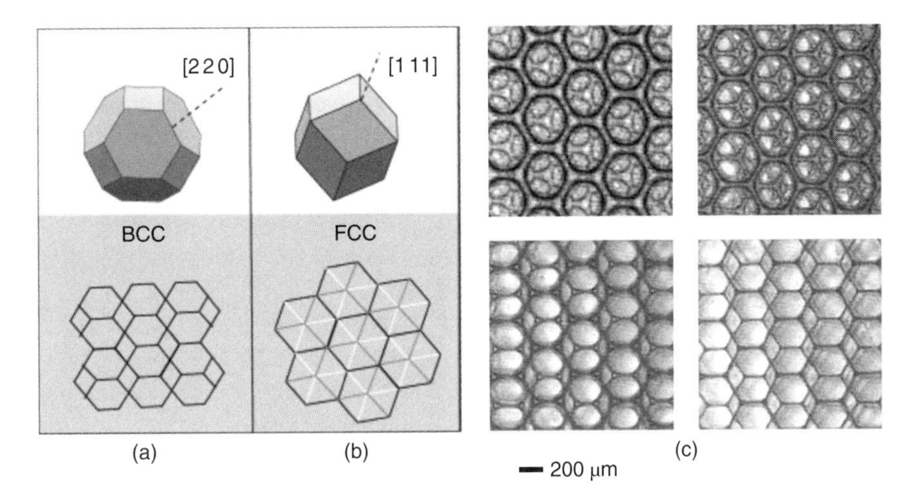

Figure 18.3 Part figures (a) and (b), respectively, show the unit cells of BCC (Kelvin) and FCC bubble crystals. In each case, the dashed lines are perpendicular to the crystal plane that is often observed at a surface of a sample. The patterns below each unit cell show what an observer looking at such a crystal along the dashed line sees. Partfigure (c) shows photographs of experimentally obtained bubbles crystals with decreasing liquid volume fractions 0.186, 0.102, 0.046, and 0.008. Comparison with images (a) and (b) reveals a transition from an FCC to a BCC (Kelvin) structure with decreasing liquid content. Source: Reprinted with permission from [12] Copyright 2008 American Chemical Society.

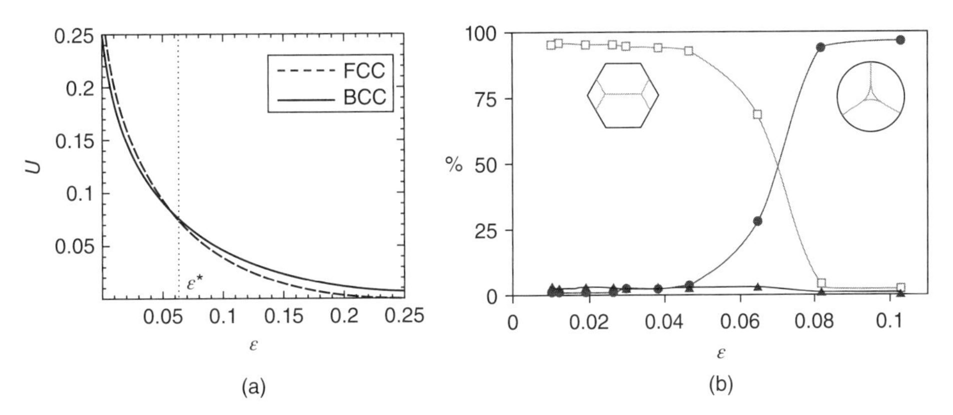

Figure 18.4 (a) Interfacial excess energy density of a foam, normalized by the ratio of surface tension and bubble radius T/R, for either an FCC or a BCC structure, versus liquid volume fraction ε. For a given structure, the excess is defined as the increase in energy density, induced when liquid is extracted from a close packing of spheres, up to a fraction ε. The data are obtained by a simulation using the Surface Evolver software [12]. (b) Percentage of bubbles visible at the surface of a foam that are part of an FCC (circles) or a BCC structure (squares). Examples of such structures are shown on Figure 18.3. Triangles indicate distorted bubbles that cannot be identified with one of these structures. Source: Reprinted with permission from [12] Copyright 2008 American Chemical Society.

$\xi = 1 - d/RZ$ is a dimensionless displacement, where d is half the distance between the bubble centers. Both the exponent ν and the dimensionless coefficient C are increasing functions of z, respectively, in the range $2.2 \leq \upsilon \leq 2.5$ and $0.25 \leq C \leq 0.75$ for $6 \leq z \leq 12$. Thus, bubble interactions in 3D are generally both nonlocal and weakly anharmonic.

18.3 AGING

Foams are intrinsically nonequilibrium materials and their structure ages due to three mechanisms. The first is drainage of the liquid contained in the foam, either in individual films due to capillary pressure or on the scale of the bubbles, due to gravity and capillary pressure gradients (cf. Fig. 18.1). Another aging mechanism is coalescence: As the films separating neighboring bubbles thin down, they may rupture. Even if both drainage and coalescence are suppressed, the foams structure coarsens due to the gas pressure difference between neighboring bubbles. In the following, we describe these three mechanisms in more detail.

18.3.1 Drainage

The network of Plateau borders and vertices in a foam structure can be compared to the voids in porous media such as sponges, ceramics, or sandstone. Darcy's equation describes the rate q at which a pressure gradient ∇p drives liquid through such materials depending on their permeability coefficient α and on the liquid viscosity η:

$$q = -\frac{\alpha}{\eta}\nabla p. \tag{18.8}$$

Foams are different from other porous materials because their permeability increases as liquid penetrates into the bubble packing and opens up the Plateau borders and vertices. Therefore, the permeability is dynamically coupled to the flow rate [31]. In addition, for a given bubble packing and liquid fraction, permeability depends on the hydrodynamic boundary conditions. In ordinary porous materials, the flow velocity must always go to zero at the interface between the liquid and the solid matrix. Even though the continuous phase of foam is a fluid, the same rigid boundary condition may be imposed by the surfactant molecules that cover the gas–liquid interfaces. The resistance of these monolayers to interfacial flow is described by the interfacial shear viscosity η_s. Thus, the coupling between this flow and the bulk flow depends on η_s, the liquid viscosity η, and the size of the Plateau border described by its radius of curvature r. Dimensional arguments show that the impact of interfacial shear viscosity is controlled by the Boussinesq number $Bo = \eta_s/\eta r$, sometimes expressed as an interfacial mobility parameter defined as $M = 1/Bo$. Assuming that the bulk flow velocity goes to zero in the corners of a Plateau border, there is a crossover from rigid to mobile hydrodynamic boundary conditions as a function of the Boussinesq number, as illustrated in Figure 18.5. For small r (i.e., small liquid fraction ε), strong interfacial shear viscosity imposes a rigid boundary condition all over the channel surface, leading to a Poiseuille flow profile inside the Plateau border [1, 2, 32, 33]. In this case, the permeability α is predicted to scale as $R^2\varepsilon^2$. If, on the other hand, the interfacial viscosity is low or the Plateau border perimeter is large, the velocity field in the Plateau border is reminiscent of a plug flow, involving negligible bulk liquid viscous friction. In this case, dissipative flows in the vertices must be considered; their contribution to the permeability is predicted [1, 2, 32, 33] to scale as $R^2\varepsilon^{3/2}$. The most general description of foam permeability takes into account both Plateau border and vertex contributions, in agreement with experimental data [1, 2, 31, 32, 34].

The drainage of foam close to the dry limit, due to the balance of gravity and capillary pressure, is described as the evolution of its liquid content $\varepsilon(Z, t)$ with time t and vertical coordinate Z. It is governed by the following equation obtained using Darcy's law and the conservation of liquid volume [1, 2, 32, 35]:

$$\frac{d\varepsilon}{dt} + \vec{\nabla}\cdot\left(\frac{\alpha\,\varepsilon}{\eta}\vec{g}\right) - \vec{\nabla}\cdot\left(\frac{T\alpha\,\varepsilon^{-3/2}}{3.5R\eta}\vec{\nabla}\varepsilon\right) = 0. \tag{18.9}$$

Depending on the choice of boundary conditions, this equation describes how liquid drains out of an initially homogeneous foam column ("free drainage"), how a small quantity of liquid injected into a foam spreads out with time [36], or how water poured at a constant rate onto a foam penetrates into it ("forced drainage"). This latter experiment is remarkable in that the front separating the wet foam at the top from dry foam below does not spread out with time as it moves downward, due to the coupling between liquid content and permeability modeled by Equation 18.9. To conclude this brief introduction to foam drainage, we note that present theories are most accurate for liquid fractions up to a few percent. The present understanding of drainage in wet foams is much less quantitative.

18.3.2 Coarsening

When the liquid film separating two bubbles is curved, the gas pressures on either side are different due to surface tension. This pressure difference drives a diffusive gas flow across the film depending on its permeability [32]. As a

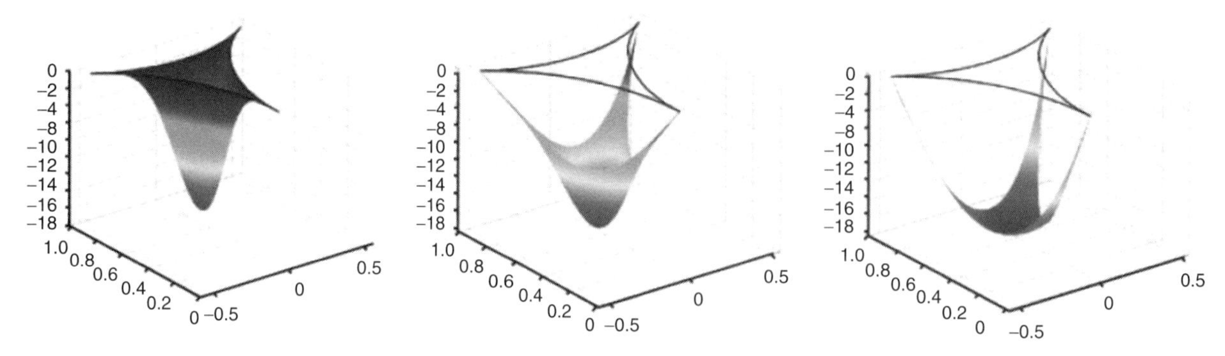

Figure 18.5 Impact of interfacial mobility on the flow profile in a Plateau border. The Boussinesq number Bo in the three illustrated simulations is, from left to right: 10, 1, and 0.1. The vertical axis corresponds to the velocity in the direction along the Plateau border. The flow velocity is plotted as a function of the position in the plane perpendicular to the Plateau border. With decreasing Boussinesq number, the interface is more and more entrained by the bulk flow [32]. Source: Copyright by Wiebke Drenckhan.

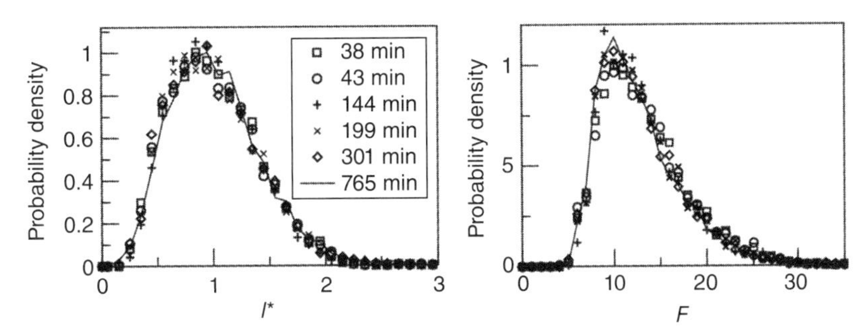

Figure 18.6 Distribution of normalized bubble sizes $l^* = V^{1/3}/\langle V(t)^{1/3}\rangle$ (deduced from the bubble volume V) and facet numbers F in a coarsening foam, measured by X-ray tomography [38]. Data taken over the wide range of indicated foam ages t superpose on a masterplot, illustrating the scaling regime described in the text. The liquid volume fraction is 2%. Source: Reprinted with permission from [38]. Copyright 2010 American Physical Society.

consequence, large bubbles tend to grow at the expense of smaller neighbors; the number of bubbles decreases and the average bubble size $\langle R\rangle$ grows with time. If coarsening is the dominant aging process, a statistically self-similar growth regime is established where the distribution of normalized bubble sizes $R/\langle R\rangle$ no longer changes with time, and the evolution with time is described by the growth of a single independent characteristic length scale [37, 38]. In this scaling state, topological characteristics such as the distribution of facet numbers per bubble become time-invariant [2, 38]. Statistically self-similar growth regimes are also found in other systems where interfacial energy drives mass transfer, such as polycrystals or ceramics [39]. In 3D foams, statistically self-similar growth has been evidenced using X-ray tomography, illustrated on Figure 18.6.

The characteristic length scale grows following a parabolic law (Eq. 18.10), well known in metallurgy [40]:

$$R^2(t) - R^2(t_0) = k_d(t - t_0). \tag{18.10}$$

In foams, the constant k_d is set by the speed of diffusive gas transfer through the liquid film [32] and also by interfacial viscoelasticity [41]. The coarsening induced growth in 2D foams or metal films has been explained theoretically in pioneering work by von Neumann who showed that bubble growth depends on topology. The area of a 2D bubble changes at a rate that scales as the number of its neighbors −6 [2]. An extension of this theory to 3D has been achieved only recently [39].

Coarsening induces an evolution of the foam structure and is also accompanied by characteristic dynamics: As small bubbles shrink and large bubbles grow, they must lose or gain contacts. These topological changes occur as intermittent events where clusters of bubbles reorganize and settle into new equilibrium of smaller interfacial energy [37].

Intermittent dynamics as a material settles into deeper energy minima have been observed [42] and modeled [43] for soft glassy materials. In a material at rest, the rearrangement dynamics progressively slow down, but never stop entirely. When transient flow is applied to such a system, it is "rejuvenated," in the sense that its dynamics are accelerated as the system is driven in into a metastable state of higher energy. However, if transient shear is applied to a coarsening foam, the opposite effect is observed: the rearrangement dynamics are temporarily slowed down [44]. This evidence shows that the aging dynamics in liquid foams are significantly different from those in other soft glassy materials.

18.3.3 Coalescence

In dry foams, the rate of coalescence events rises sharply above a critical gas volume fraction whose value depends on the nature of the surfactant and its concentration, but not on bubble size [45]. Since individual films suspended on porous rings are stable up to disjoining pressures much higher than those where film rupture is observed in foams, it has been proposed that coalescence is triggered by drainage or coarsening-induced bubble rearrangements, which drive rapid film stretching [1, 45].

To conclude the section about foam aging, we note that drainage, coarsening, and coalescence may be evidenced separately only in model systems. In most real foams, these processes coexist and they are strongly coupled among each other [32]. In addition, there is a complex interplay between aging processes and mechanical behavior.

18.4 RHEOLOGY

Even though foams are only made of fluids, they can behave mechanically as either solids or liquids depending on the liquid volume fraction and applied stress as illustrated in Figure 18.7.

18.4.1 Elastic Response

For liquid fractions below the critical value ε_c and in the absence of aging, the static foam response to a small applied

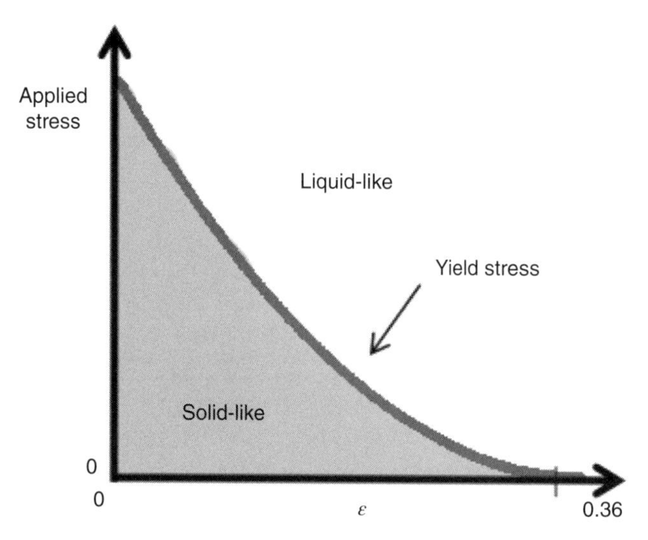

Figure 18.7 Schematic overview, showing domains of solid-like and liquid-like behavior for disordered foams.

strain is elastic: The interfacial energy density scales as $T/\langle R\rangle$, which is the ratio of the surface area and the volume of a typical bubble, multiplied by surface tension. Since the elastic energy increase in response to a shear strain γ must be independent of the sign of γ, the induced change of elastic energy scales to leading order as $\gamma^2 T/\langle R\rangle$. As a consequence, the shear modulus G is expected to scale as $T/\langle R\rangle$. A calculation for 3D disordered dry foams taking into account the nonaffine deformation at the bubble scale provides the estimate [1]: $G = 0.55T/\langle R\rangle$, in reasonable agreement with experimental data [15]. For a typical surface tension of $T = 30$ mN/m and $R = 100$ µm, we obtain $G = 170$ Pa, showing that foams are indeed soft solids. The decrease of elasticity to zero in the wet limit can be analyzed in terms of the repulsive interactions between neighboring bubbles discussed earlier. Two effects are involved [46]: A decrease in the coordination number associated with a softening of the bubble interactions (cf. Eq. 18.7). This mechanism will be discussed in more detail at the end of this chapter where we present the physics of the jamming transition. Experiments with disordered foams are to a good approximation described by the empirical law [15, 31]:

$$G = 1.4\frac{T}{R_{32}}(1-\varepsilon)(\varepsilon_c - \varepsilon), \qquad (18.11)$$

where R_{32} is defined as the ratio of the third to the second moment of the bubble radius distribution, and $\varepsilon_c = 0.36$.

A shear strain applied in the x_1-direction of a Cartesian coordinate system (cf. Fig. 18.8b) induces not only a shear stress σ_{12} but also normal stress differences $N_1 = \sigma_{11} - \sigma_{22}$ and $N_2 = \sigma_{22} - \sigma_{33}$. This nonlinear effect arises because a large amplitude elastic shear deformation tends to align the films in a plane perpendicular to the velocity gradient [15]. The surface tensions induce traction forces tangentially to the films, leading to the observed normal stress differences. The nonlinear response has been explained by a calculation of the strain-induced change of interfacial area and energy, using large deformation kinematics and the simplifying assumption of affine deformations down to the scale of the films [47]. This tensorial constitutive law yields the following relations:

$$\sigma_{12} = G\gamma, \quad N_1 = G\gamma^2, \quad N_2 = -\frac{6}{7}G\gamma^2. \qquad (18.12)$$

Polymeric complex fluids typically have a second normal stress difference, whose magnitude is very small compared to N_1, in contrast to foams. This remarkable difference arises because in contrast to films, the shear aligned polymer strands only pull in the direction of shear x_1 but not in the x_2-direction perpendicular to it. One may therefore distinguish complex fluids into polymer-based "line fluids" and "film fluids," such as emulsions and foams [48].

18.4.2 Linear Viscoelasticity

The linear viscoelastic response of foam can be characterized by measuring either the response to an imposed stress (or strain) step or the response to an imposed oscillatory stress (or strain). If coarsening is the dominant aging process, a viscoelastic relaxation due to coarsening-induced intermittent local bubble rearrangements is observed. Upon such an event, the bubbles may settle into one among several new equilibrium-packing configurations. The outcome of this process is biased toward the configuration, which relaxes the applied strain or stress most strongly. As a consequence, for stresses and liquid fractions in the solid-like regime (cf. Fig. 18.1), a coarsening foam flows as an extremely viscous Newtonian liquid. The creep viscosity describing this steady flow is expected to scale as the number of rearrangements events per unit volume and unit time, multiplied by the characteristic volume of the regions in which the structure is modified upon a rearrangement. This prediction has been validated by experiments where the rearrangement rate was measured by a multiple light scattering technique called diffusing-wave spectroscopy [49] and the mechanical response was probed *in situ* using a rheometer [17]. The creep viscosity divided by the static shear modulus (Eq. 18.11) sets the relaxation time that governs the slow mechanical response of coarsening foams. This response is fundamentally different from the one found in glassy materials where a wide distribution of slow relaxation times or stretched exponential relaxations are often observed.

The elastic and the dissipative contributions to the oscillatory linear foam response at small stress or strain amplitudes are described by the real and imaginary parts of the complex shear modulus $G^* = G' + iG''$. Figure 18.9 shows such data for moderately wet foam as a function of frequency ω. In the low-frequency limit, G' falls to

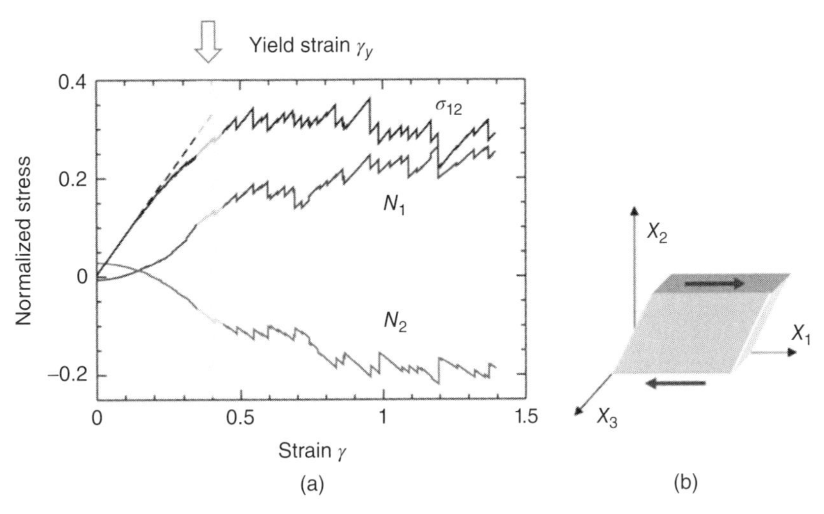

Figure 18.8 (a) Shear stress σ_{12}, first normal stress difference N_1, and second normal stress difference N_2 versus shear strain γ, obtained in a quasistatic simulation of 3D monodisperse disordered dry foam, using the Surface Evolver software. All stresses are normalized by $T/V^{1/3}$, where V is the bubble volume. For strains below the yield strain γ_y, shear stress scales as γ and the normal stress differences with γ^2, in agreement with Equation 18.12. For strains beyond γ_y, plastic flow sets in and the stress growth is saturated. Serrations on the stress versus strain growth curves are a finite sample size effect; they correspond to individual bubble rearrangements. Source: Courtesy A. Kraynik (b) Illustration of the shear geometry.

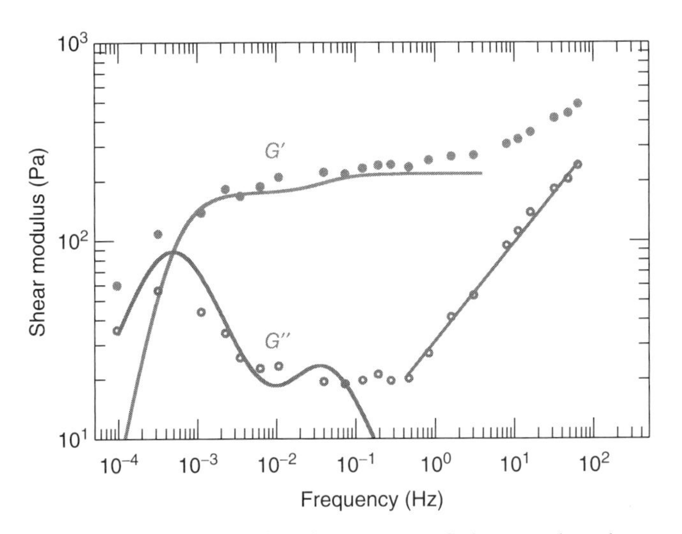

Figure 18.9 Real and imaginary parts of the complex shear modulus G^*, measured for Gillette shaving cream with 8% liquid fraction, at a foam age of 100 min (bubble size, 25 μm; surface tension, 30 mN/m). Data from Refs [17, 50, 51]. Creep flow data have been transformed in the frequency domain using a Laplace transform. The straight line has a slope of ½.

zero whereas G'' goes through a maximum. This is the signature of the coarsening-induced relaxation process whose relaxation time τ has been discussed earlier.

For frequencies ω such that $\omega\tau \gg 1$, the coarsening process is too slow to affect the viscoelastic response, but, after dropping to a minimum value, the experimentally measured loss modulus G'' increases again and scales to a good approximation with frequency as $\omega^{1/2}$. In the same range, the real part of the complex shear modulus starts increasing with frequency, and this behavior can been described as follows [52]:

$$G^*(\omega) = G(1 + \sqrt{i\omega/\omega_c}) + i\eta_\infty\omega, \tag{18.13}$$

where ω_c is a characteristic relaxation frequency, set by processes at the scale of the individual liquid films and Plateau borders, and η_∞ is an effective viscosity due to flow in the continuous phase. The experimentally observed scaling of ω_c with bubble size indicates that for surfactants inducing rigid interfaces the dissipation due to shear flow within the liquid films is dominant, whereas for mobile interfaces dissipation is mainly due to flow at the junctions between the Plateau borders and the films [53]. Behavior similar to the one described by Equation 18.13 has been reported for concentrated emulsions [52] and soft pastes [54], suggesting that a generic mechanism may be involved such as relaxations in regions of the disordered packing of bubbles, droplets, or particles where the local elastic response is weak. However, the complex interfacial viscoelasticity of the gas–liquid interfaces may also contribute to the scaling of G^* with frequency [31].

18.4.3 Yielding and Plastic Flow

When a slowly increasing strain is applied to a foam, the elastic stress first increases. At a characteristic strain denoted as γ_y, the material yields and the stress saturates (cf. Fig. 18.8). At the bubble scale, the packing becomes unstable under these conditions and local plastic rearrangements are induced. In dry foams, they correspond to abrupt topological

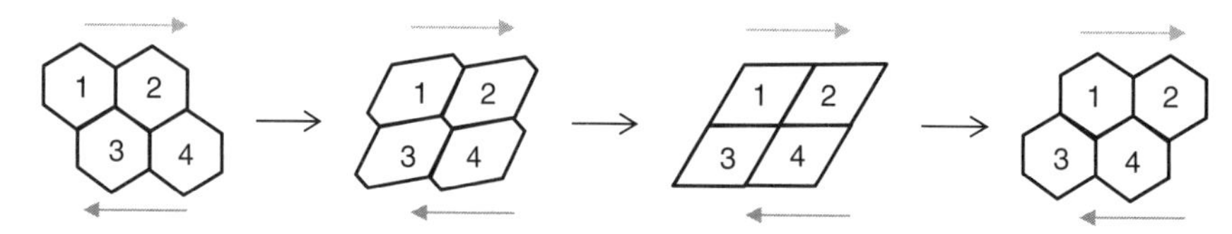

Figure 18.10 Deformation of a 2D foam due to an increasing applied shear strain. A T1 event occurs as bubbles two and three separate and the topology changes.

changes called T1 events, as illustrated schematically for the 2D foam shown on Figure 18.10. The yield stress σ_y is approximately given by the relation $\sigma_y = \gamma_y G$, and 3D experiments [15, 31] show that $\sigma_y \cong 0.5(\varepsilon - \varepsilon_c)^2 T/\langle R\rangle$.

Close to the wet limit, the bubbles are approximately spherical, suggesting that the yielding behavior may be similar to that of a random packing of solid spheres, often considered as a model system for granular materials. In this latter case, the onset of yielding is governed by the ratio of two stress components that have opposite effects: The shear stress that induces flow and the confinement pressure that favors solid-like behavior because it pushes the grains against each other. This stress ratio is equal to the tangent of the largest possible equilibrium slope angle of a granular pile θ^*, called angle of repose (Chapter 17). Simulations show that in the absence of static friction between grains (as expected in foams), the angle of repose is much smaller than that for sand, close to 4°. Remarkably, this is in agreement with experiments [10] where the mechanical stability of wet foam confined below an inclined plate is probed, as a function of its inclination angle. The plate is immersed in a liquid-filled tank, so that buoyancy forces pushing the bubbles upward have an effect that is comparable to the weight of the grains in a sand pile. These findings show that close to the jamming point, there is an analogy between foams and granular materials. However, for liquid fractions ε far below the jamming point, θ^* cannot be expected to remain constant for foams because yield stress and osmotic pressure scale differently with ε.

Homogeneous plastic flows are studied in laboratory experiments and simulations to establish constitutive laws, but in applications, the flow is generally more complex. An example is the Couette geometry where the sample is sheared between concentric cylinders, leading to a shear stress that decreases with the distance from the axis of symmetry of the device. As a consequence, heterogeneous flow is observed: The part of the sample subjected to stress larger than σ_y flows whereas the rest behaves elastically [15]. If the material obeys a constitutive law of the form Equation 18.14 presented in the following paragraph, the strain rate continuously goes to zero at the transition from flowing to elastic regions. This variation is discontinuous in materials that exhibit a phenomenon called shear-banding, such as worm-like micelle solutions. In a recent MRI study with moderately dry 3D foams and a variety of foaming liquids, no evidence for such intrinsic shear-banding was found [55]. In 2D foams confined below a glass plate, localization of plastic flow is induced by wall friction [15]. However, 2D simulations [56] and 3D experiments [57] suggest that shear-banding may exist in dry foams. Such studies are difficult because long-term transient flows and bubble size segregation phenomena need to be taken into account.

18.4.4 Viscous Flow

When 3D foam is subjected to a steady shear flow at a rate $.\gamma$, the stress response generally follows the empirical Herschel–Bulkley law [15]:

$$\sigma(\dot{\gamma}) = \sigma_y + \sigma_V(\dot{\gamma}) = \sigma_y + K\dot{\gamma}^n. \tag{18.14}$$

The viscous increase in stress with strain rate is non-Newtonian, with a power law exponent n depending on the interfacial rheology of the surfactant monolayers covering the gas–liquid interfaces, as shown on Figure 18.11a. If the dilational surface modulus is low, the impact of the liquid viscosity η, surface tension T, and bubble size R_{32} are captured by introducing a dimensionless viscous stress $\sigma_V\ R_{32}/T$ and a dimensionless shear rate, expressed as a capillary number $Ca = \eta\dot{\gamma}\ R_{32}/T$. Represented in this way, data with different bubble sizes and liquid viscosities form a masterplot. If the dilational surface modulus is high, the dimensionless viscous stress is larger for a given capillary number, compared to the previous case. In addition, the exponent n is smaller and the impact of the liquid viscosity is not simply captured by introducing a capillary number.

The behavior shown on Figure 18.11a has been explained by a model on the bubble scale [59] illustrated in Figure 18.11c and d. In a shear flow, bubbles collide and the films separating them thin down, under the effect of capillary pressure. At the same time, the liquid in the films is sheared as the bubbles motion continues, until they finally separate. Dissipation has been predicted to arise mainly due to the shear flow in the bubble contacts [59]. If the shear rate is high, the films do not have time to thin down and the shear friction in their contacts is weaker than for low shear rates. Therefore, the effective friction between bubbles is a nonlinear function of strain rate. Additional contributions to the dissipation are due to flows in the

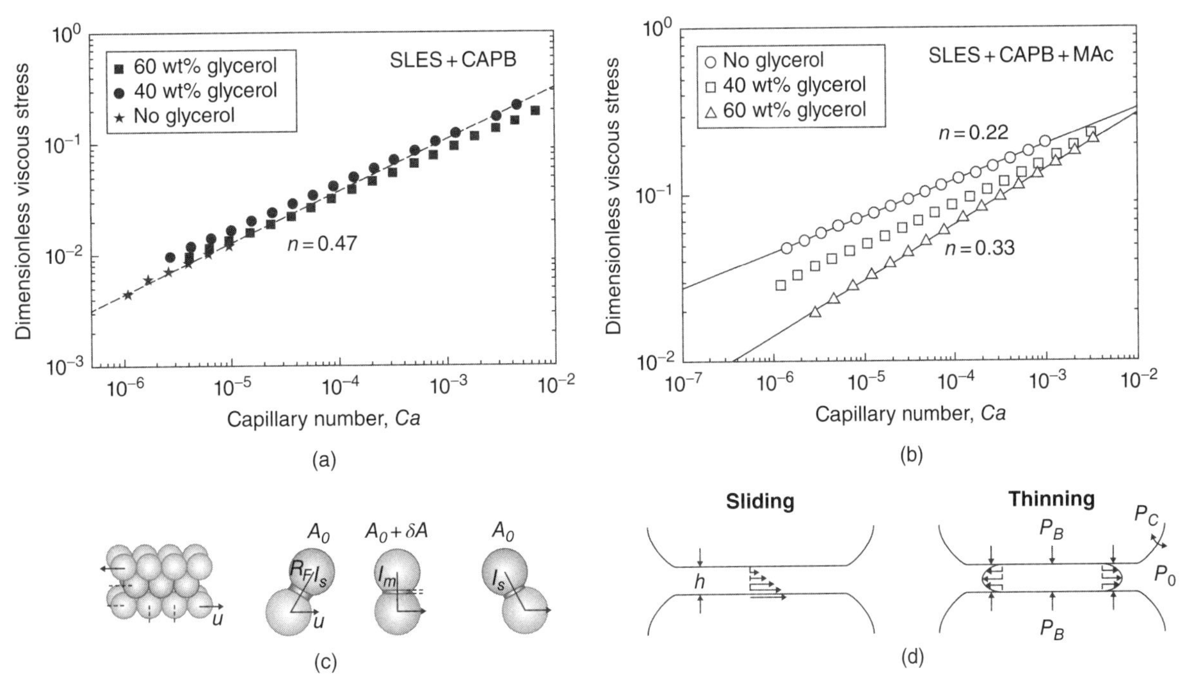

Figure 18.11 Experimentally measured dimensionless viscous stress, plotted versus the capillary number, defined in the text: (a) for a foam stabilized by the surfactants SLES–CAPB, yielding a low surface modulus and (b) for a foam stabilized by the surfactants SLES–CAPB with myristic acid, yielding a high surface modulus [58]. Glycerol is added to the solutions to enhance their bulk viscosity. The liquid volume fraction is 0.1. (c) Shear flow of a FCC bubble packing. As the bubbles are squeezed against each other, their interfacial area increases from its initial value A_0 to $A_0+\delta A$. (d) Illustration of two contributions to the flow in the bubble contact upon the collision shown in (c). Source: Reprinted from [58] with permission. Copyright 2009 by Royal Society of Chemistry .

Plateau borders and vertices as well as due to dilational interfacial viscous friction. This latter effect is induced by the modulation of the bubble surface area during a collision illustrated on Figure 18.11c. A calculation on this basis assuming rigid interfaces and a moderately wet-ordered foam structure predicts a power law exponent n close to 0.5, consistent with experimental data [59]. The variation of the exponent n as a function of surface modulus is not yet fully understood (Fig. 18.11b). Another open issue is the role of disorder in the bubble packing: In simulations of 2D foams where neighboring bubbles interact via Newtonian friction, macroscopic non-Newtonian viscous behavior has been evidenced [60]. This result suggests that, in general, both disorder and nonlinear local bubble interactions may contribute to the experimentally observed macroscopic behavior. Viscoelastic interactions between soft particles in a close packing have also been shown to be crucial for the nonlinear rheology of emulsions and pastes [61].

18.4.5 Rheology near the Jamming Transition

The transition from solid-like to liquid-like behavior at a critical volume fraction of the continuous phase is observed not only in foams but also in soft pastes and emulsions. Simulations of jammed soft disk or soft particle packings have revealed that as the transition is approached, the continuum mechanics description only holds on scales larger than a characteristic length, which diverges as the inverse of the excess number of contacts $1/(z-z_c) \sim (\varepsilon_c-\varepsilon)^{-1/2}$ (cf. Eq. 18.6) [11]. On smaller scales, the spatial distribution of forces induced in the material by a point loading is very heterogeneous, as illustrated in Figure 18.12. This highly nonaffine local response has a strong impact on the macroscopic mechanical behavior. Near the jamming transition, the macroscopic shear modulus has been predicted to scale linearly both with the excess number of contacts $z-z_c$ and with the contact stiffness, deduced from the interaction potential (Eq. 18.7). Since this potential is nonlinear, the contact stiffness for small perturbations is a function of the distance between neighboring bubbles centers. Taking into account the variation of this distance with liquid fraction, an analytical expression of the shear modulus has been deduced which is in good agreement with the experimental evidence, described by the empirical expression (Eq. 18.11) [46]. Close to the jamming transition, the yield stress is predicted to scale with liquid fraction as the shear modulus, multiplied by $(\varepsilon_c-\varepsilon)^{0.7}$, again in agreement with experimental observations. However, a complete theory of the jamming transition is not yet available.

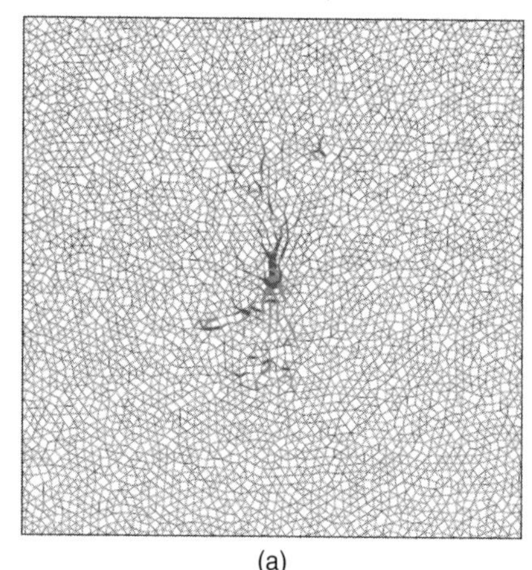

(a)

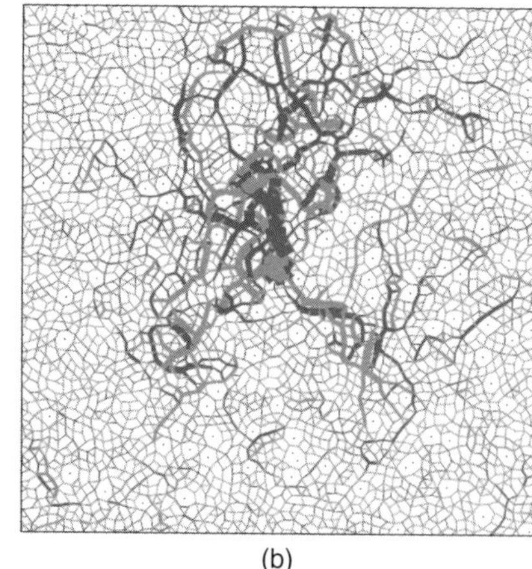

(b)

Figure 18.12 A 2D packing of disks with Hertzian interactions is simulated (a) at a packing fraction far above the jamming transition where the coordination number is $z = 5.41$ and (b) close to the jamming transition, with $z = 4.05$. The same force is applied in both cases to a disk at the center of the packing, pushing it to the right. The induced change of contact forces between neighboring disks throughout the sample is represented by lines linking the disk centers. The disks themselves are not shown. The figure illustrates the highly irregular force distribution on short length scales that appears near the jamming transition. Source: Reprinted from [62] with permission. Copyright 2009 by APS.

REFERENCES

[1] Cantat I, Cohen-Addad S, Elias F, Graner F, Höhler R, Pitois O, Rouyer F, Saint Jalmes A. *Foams: Structure and Dynamics*. Oxford: Oxford University Press; 2013.

[2] Weaire D, Hutzler S. *The Physics of Foams*. Oxford: Oxford University Press; 1999.

[3] Cantat I, Cohen-Addad S, Elias F, Graner F, Höhler R, Pitois O, Rouyer F, Saint Jalmes A. *Les Mousses*. Paris: Belin; 2010.

[4] Slezin YB. *The mechanism of volcanic eruptions (a steady state approach)*. J Volcanol Geotherm Res 2003;122(1–2):7–50.

[5] Mello MLS et al. *Composition and structure of the froth of the spittlebug*. Deois SP Insect Biochemistry 1987;17(3):493–502.

[6] Hilgenfeldt S, Erisken S, Carthew RW. *Physical modeling of cell geometric order in an epithelial tissue*. Proceedings of the National Academy of Sciences of the United States of America 2008;105(3):907–911.

[7] Stevenson P, editor. *Foam Engineering*. Chichester: Wiley-Blackwell; 2012.

[8] Schügerl K. *Recovery of proteins and microorganisms from cultivation media by foam flotation*. In: *New Products and New Areas of Bioprocess Engineering*. Berlin/Heidelberg: Springer; 2000. p 191–233.

[9] Chung KY et al. *Fabricating scaffolds by microfluidics*. Biomicrofluidics 2009;3(2):22403.

[10] Lespiat R, Cohen-Addad S, Höhler R. *Jamming and flow of random-close-packed spherical bubbles: an analogy with granular materials*. Phys Rev Lett 2011;106(14):148302.

[11] Van Hecke M. *Jamming of soft particles: geometry, mechanics, scaling and isostaticity*. J Phys Condens Matter 2010;22(3):033101.

[12] Höhler R et al. *Osmotic pressure and structures of monodisperse ordered foam*. Langmuir 2008;24(2):418–425.

[13] van der Net A et al. *Crystalline arrangements of microbubbles in monodisperse foams*. Colloids and Surfaces A 2007;309(1–3):117–124.

[14] Weaire D. *The Kelvin Problem*. London: Taylor & Francis; 2000.

[15] Höhler R, Cohen-Addad S. *Rheology of liquid foams*. Journal of Physics: Condensed Matter 2005;17:R1041–R1069.

[16] Sollich P et al. *Rheology of soft glassy materials*. Phys Rev Lett 1997;78(10):2020–2023.

[17] Cohen-Addad S, Höhler R, Khidas Y. *Origin of the slow linear viscoelastic response of aqueous foams*. Phys Rev Lett 2004;93(2):028302–028304.

[18] Dennin M. *Discontinuous jamming transitions in soft materials: coexistence of flowing and jammed states*. J Phys Condens Matter 2008;20(28):283103.

[19] Ivanov IB, editor. *Thin liquid films*. In: *Surfactant Science Series*Vol. 29.. New York: CRC Press; 1988.

[20] Princen HM. *Osmotic pressure of foams and highly concentrated emulsions. 1. Theoretical considerations*. Langmuir 1986;2(4):519–524.

[21] Mason TG et al. *Osmotic pressure and viscoelastic shear moduli of concentrated emulsions*. Physical Review E 1997;56(3):3150–3166.

[22] Maestro A et al. *Liquid dispersions under gravity: volume fraction profile and osmotic pressure*. Soft Matter 2013;9(8):2531–2540.

[23] Mason TG, Scheffold F. *Crossover between entropic and interfacial elasticity and osmotic pressure in uniform disordered emulsions*. Soft Matter 2014;10(36):7109–7116.

[24] Jorjadze I, Pontani L-L, Brujic J. *Microscopic approach to the nonlinear elasticity of compressed emulsions*. Phys Rev Lett 2013;110(4):048302.

[25] Heitkam S, Drenckhan W, Frohlich J. *Packing spheres tightly: influence of mechanical stability on close-packed sphere structures*. Phys Rev Lett 2012;108(14):5.

[26] Brakke K. *The surface evolver*. Exp Math 1992;1:141.

[27] Gabbrielli R et al. *An experimental realization of the Weaire–Phelan structure in monodisperse liquid foam*. Philos Mag Lett 2012;92(1):1–6.

[28] Durian DJ. *Bubble-scale model of foam mechanics: melting, nonlinear behavior, and avalanches*. Physical Review E 1997;55(2):1739–1751.

[29] Tighe BP et al. *Model for the scaling of stresses and fluctuations in flows near jamming*. Phys Rev Lett 2010;105(8):088303.

[30] Lacasse M-D, Grest GS, Levine D. *Deformation of small compressed droplets*. Physical Review E 1996;54(5):5436–5446.

[31] Cohen-Addad S, Hohler R, Pitois O. *Flow in foams and flowing foams*. Annual Review of Fluid Mechanics 2013;45:241–267.

[32] Saint-Jalmes A. *Physical chemistry in foam drainage and coarsening*. Soft Matter 2006;2(10):836–849.

[33] Koehler SA, Hilgenfeldt S, Stone HA. *A generalized view of foam drainage: experiment and theory*. Langmuir 2000;16(15):6327–6341.

[34] Pitois O et al. *Node contribution to the permeability of liquid foams*. J Colloid Interface Sci 2008;322(2):675–677.

[35] Verbist G, Weaire D, Kraynik AM. *The foam drainage equation*. J Phys Condens Matter 1996;21:3715–3731.

[36] Saint-Jalmes A et al. *Diffusive liquid propagation in porous and elastic materials: the case of foams under microgravity conditions*. Phys Rev Lett 2007;98(5):058303.

[37] Durian DJ, Weitz DA, Pine DJ. *Multiple light-scattering probes of foam structure and dynamics*. Science 1991;252: 686–688.

[38] Lambert J et al. *Coarsening foams robustly reach a self-similar growth regime*. Phys Rev Lett 2010;104(24): 248304.

[39] MacPherson RD, Srolovitz DJ. *The von Neumann relation generalized to coarsening of three-dimensional microstructures*. Nature 2007;446(7139):1053–1055.

[40] Anderson MP, Grest GS, Srolovitz DJ. *Computer simulation of normal grain growth in three dimensions*. Philos Mag B 1989;59(3):293–329.

[41] Stocco A et al. *Interfacial behavior of catanionic surfactants*. Langmuir 2010;26(13):10663–10669.

[42] Viasnoff V, Lequeux F. *Rejuvenation and overaging in a colloidal glass under shear*. Phys Rev Lett 2002;89(6): 065701–065704.

[43] Sollich P. *Rheological constitutive equation for a model of soft glassy materials*. Physical Review E 1998;58(1):738–759.

[44] Cohen-Addad S, Höhler R. *Bubble dynamics relaxation in aqueous foam probed by multispeckle diffusing-wave spectroscopy*. Phys Rev Lett 2001;86(20):4700–4703.

[45] Carrier V, Colin A. *Coalescence in draining foams*. Langmuir 2003;19(11):4535–4538.

[46] Scheffold F, Cardinaux F, Mason TG. *Linear and nonlinear rheology of dense emulsions across the glass and the jamming regimes*. J Phys Condens Matter 2013;25(50):502101.

[47] Höhler R, Cohen-Addad S, Labiausse V. *A constitutive equation describing the nonlinear elastic response of aqueous foams and concentrated emulsions*. J Rheol 2004;48(3):679–690.

[48] Larson RG. *The elastic stress in "film fluids"*. J Rheol 1997;41(2):365–372.

[49] Höhler R, Cohen-Addad S, Durian DJ. *Multiple light scattering as a probe of foams and emulsions*. Current Opinion in Colloid & Interface Science 2014;19(3):242–252.

[50] Gopal AD, Durian DJ. *Relaxing in foam*. Phys Rev Lett 2003;91(18):188303.

[51] Cohen-Addad S, Hoballah H, Höhler R. *Viscoelastic response of a coarsening foam*. Physical Review E 1998;57(6):6897–6901.

[52] Liu A et al. *Anomalous viscous loss in emulsions*. Phys Rev Lett 1996;76(16):3017–3020.

[53] Krishan K et al. *Fast relaxations in foam*. Physical Review E 2010;82(1):011405.

[54] Cloitre M. *Yielding, flow and slip in microgel suspensions: from microstructure to macroscopic rheology*. In: Fernandez-Nieves HWA, Mattsson J, Weitz DA, editors. *Microgel Suspensions: Fundamentals and Applications*. Wiley; 2011. p. 285–311.

[55] Ovarlez G, Krishan K, Cohen-Addad S. *Investigation of shear banding in three-dimensional foams*. Europhys Lett 2010;91(6):68005.

[56] Wyn A, Davies IT, Cox SJ. *Simulations of two-dimensional foam rheology: localization in linear Couette flow and the interaction of settling discs*. European Physical Journal E 2008;26(1–2):81–89.

[57] Rouyer F, Cohen-Addad S, Höhler R. *Is the yield stress of aqueous foam a well-defined quantity?* Colloids and Surfaces A 2005;263(1–3):111–116.

[58] Denkov ND et al. *The role of surfactant type and bubble surface mobility in foam rheology*. Soft Matter 2009;5(18): 3389–3408.

[59] Tcholakova S et al. *Theoretical model of viscous friction inside steadily sheared foams and concentrated emulsions*. Physical Review E 2008;78(1):011405.

[60] Langlois VJ, Hutzler S, Weaire D. *Rheological properties of the soft-disk model of two-dimensional foams*. Physical Review E 2008;78(2):021401.

[61] Seth JR et al. *A micromechanical model to predict the flow of soft particle glasses*. Nat Mater 2011;10(11):838–843.

[62] Ellenbroek WG, Van Hecke M, Van Saarloos W. *Jammed frictionless disks: connecting local and global response*. Physical Review E – Statistical, Nonlinear, and Soft Matter Physics 2009;80(6):061307.

SECTION VI

ORDERED MATERIALS IN CURVED SPACES

19

CRYSTALS AND LIQUID CRYSTALS CONFINED TO CURVED GEOMETRIES

VINZENZ KONING & VINCENZO VITELLI

Instituut-Lorentz, Universiteit Leiden, 2300 RA Leiden, The Netherlands

19.1 INTRODUCTION

Whether it concerns biological matter such as membranes, DNA, and viruses, or synthesized anisotropic colloidal particles, the deformations inherent to soft matter almost inevitably call for a geometric description. Therefore, the use of geometry has always been essential in our understanding of the physics of soft matter. However, only recently geometry has turned into an instrument for the design and engineering of micron-scaled materials. Key concepts are

Fluids, Colloids and Soft Materials: An Introduction to Soft Matter Physics, First Edition. Edited by Alberto Fernandez Nieves and Antonio Manuel Puertas.

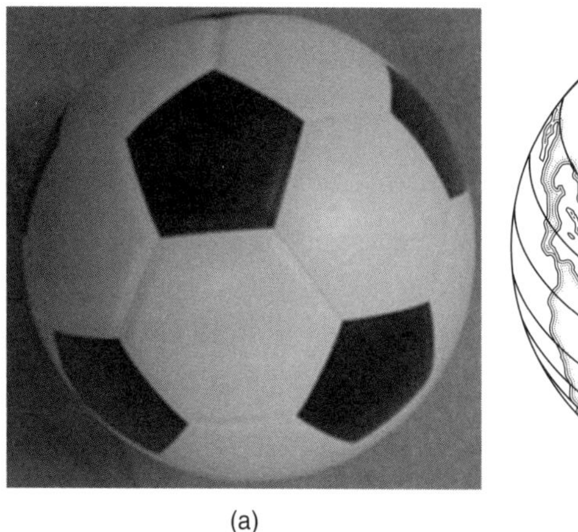
(a)

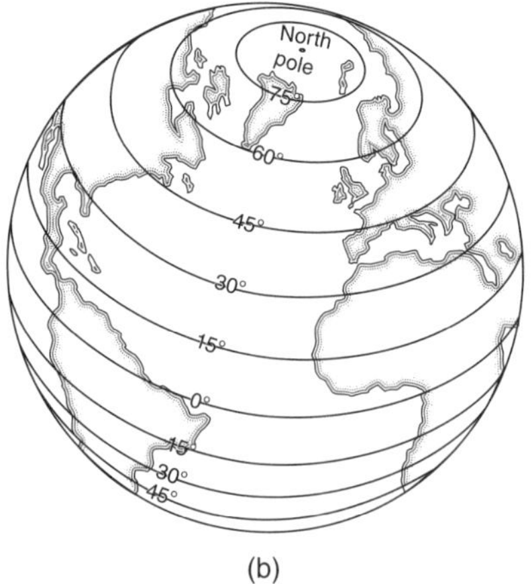

(b)

Figure 19.1 (a) Geometric frustration in a football. A perfect tiling of hexagonal panels cannot be achieved everywhere, resulting in black pentagonal panels (defects). (b) Geometric frustration on the globe. The lines of latitude shrink to a point at the north and south poles (defects). Source: Adapted from http://commons.wikimedia.org/wiki/File:Latitude_lines.svg.

geometrical frustration and the topological defects that are often a consequence of this frustration [1–3].

Geometrical frustration refers to the impossibility of local order to propagate throughout a chosen space. This impossibility is of geometric nature and could, for instance, be due to the topology of the space. Probably your first and most familiar encounter with this phenomenon was while playing (association) football. The mathematically inclined among you may have wandered off during the game and wondered: "Why does the ball contain hexagonal *and* pentagonal panels?" The ball cannot merely contain hexagonal panels: a perfect tiling of hexagons (an example of local order) cannot be achieved on the spherical surface (the space considered). There exists a constraint on the number of faces, F, edges, E, and vertices, V. The constraint is named after Euler and reads [4]

$$F - E + V = \chi, \tag{19.1}$$

where χ is the Euler characteristic. The Euler characteristic is a quantity insensitive to continuous deformations of the surface of the ball such as twisting and bending. We call such quantities topological. Only if one would perform violent operations such as cutting a hole in the sphere and glueing a handle to the hole, a surface of differently topology can be created [4, 5]. For a surface with one handle $\chi = 0$, just as for a torus or a coffee mug. The Euler characteristic χ equals 2 for the spherical surface of the ball. Thus, Euler's polyhedral formula (Eq. 19.1) ensures the need of 12 pentagonal patches besides the hexagonal ones, no matter how well inflated the ball is. To see this, write the number of faces F as the sum of the number of hexagons, H, and pentagons, P, that is, $F = H + P$. One edge is shared by two faces, hence $E = \frac{1}{2}(6H + 5P)$. Moreover, each vertex is shared among three faces, hence $V = \frac{1}{3}(6H + 5P)$. Substituting the expressions for F, E, and V into Equation 19.1 yields $P = 12$. These pentagons are the defects. Similarly, protein shells of spherical viruses that enclose the genetic material consist of pentavalent and hexavalent subunits [6, 7]. Another condensed matter analog of the geometrical frustration in footballs is the "colloidosome." Colloidosomes are spherical colloidal crystals [8–10] that are of considerable interest as microcapsules for delivery and controlled release of drugs [8] (see Fig. 19.1).

The lines of latitude on the surface of a globe are another macroscopic example of geometrical frustration. The points where these lines shrink to a point, that is, the north and south poles, are the defects. Just like the pentagons on the football, the defects on the globe are also required by a topological constraint, namely, the Poincare–Hopf theorem [5]:

$$\sum_a s_a = \chi. \tag{19.2}$$

The lines of latitude circle once around both poles. Hence, there are two defects with an unit winding number, s (see Section 19.5 for a more precise definition). Similar to the lines of longitude and latitude on the globe, a coating of a nanoparticle with a monolayer of ordered tilted molecules also has two polar defects [11–15]. Recently, Stellacci *et al.* have been able to functionalize the defects to assemble linear chains of nanoparticles [15]. A nematic liquid crystal coating possesses four defects at the vertices of a regular tetrahedron in the ground state [12]. Attaching chemical linkers to these defects could result in a three-dimensional diamond structure [13] rather than a one-dimensional chain. This defect arrangement has been recently observed in nematic double emulsion droplets [16], in which a nematic droplet itself contains another smaller water droplet. However, this is only one of the many defect arrangements that are observed, as the size and location of the inner water droplet is varied [16, 17].

Functionalization of the defects, thus resulting ordered structures confined to curved surfaces or shells, offers an intriguing route to directed assembly.

The types of order that we will discuss in this chapter are crystalline and (nematic) liquid crystalline. After introducing mathematical preliminaries, we will discuss the elasticity of crystals and liquid crystals and give a classification of the defects in these phases of matter. We will elucidate the role of geometry in this subject. In particular, we will explicitly show that, in contrast to the two examples given in the introduction, a topological constraint is not necessary for geometrical frustration. After that, we will explore the fascinating coupling between defects and curvature. We will briefly comment on the screening by recently observed charge-neutral pleats in curved colloidal crystals. We will then cross over from a two-dimensional surface to curved films with a finite thickness and variations in this thickness. The particular system we are considering is a spherical nematic shell encapsulated by a nematic double emulsion droplet. We will finish this chapter with a discussion on nematic droplets of toroidal shape. Though topology does not prescribe any defects, there is frustration due to the geometric confinement.

19.2 CRYSTALLINE SOLIDS AND LIQUID CRYSTALS

In addition to the familiar solid, liquid, and gas phases, there exist other fascinating forms of matter, which display phenomena of order intermediate between conventional isotropic fluids and crystalline solids. These are therefore called liquid crystalline or mesomorphic phases [18, 19]. Let us consider the difference between a solid crystal and a liquid crystal. In a solid crystal, all the constituents are located in a periodic manner, such that only specific translations return the same lattice. Moreover, the bonds connecting neighboring crystal sites define a discrete set of vectors, which are the same throughout the system. In a crystal, there is thus both bond-orientational and translational order. In liquid crystals there is orientational order, as the anisotropic constituent molecules define a direction in space, but the translational order is partially or fully lost. The latter phase, in which there is no translational order whatsoever, is called a nematic liquid crystal. The loss of translational order is responsible for the fluidic properties of nematic liquid crystals. A thorough introduction to liquid crystals can be found in the chapter by Lagerwall.

19.3 DIFFERENTIAL GEOMETRY OF SURFACES

19.3.1 Preliminaries

For a thorough introduction to the differential geometry of surfaces, please consult Refs 3, 20, 21. In this section, we introduce the topic briefly and establish the notation. Points on a curved surface embedded in the three-dimensional world we live in can be described by a three-component vector $\mathbf{R}(\mathbf{x})$ as a function of the coordinates $\mathbf{x} = (x^1, x^2)$. Vectors tangent to this surface are given by

$$\mathbf{t}_\alpha = \partial_\alpha \mathbf{R}, \tag{19.3}$$

where $\partial_\alpha = \frac{\partial}{\partial x^\alpha}$ is the partial derivative with respect to x^α. These are in general neither normalized nor orthogonal. However, it does provide a basis to express an arbitrary tangent vector $\mathbf{n}$ in:

$$\mathbf{n} = n^\alpha \mathbf{t}_\alpha. \tag{19.4}$$

Here we have used the Einstein summation convention, that is, an index occurring twice in a single term is summed over, provided that one of the them is a lower (covariant) index and the other is an upper (contravariant) index. We reserve Greek characters $\alpha, \beta, \gamma, \ldots$ as indices for components of vectors and tensors tangent to the surface. The so-called metric tensor reads

$$g_{\alpha\beta} = \mathbf{t}_\alpha \cdot \mathbf{t}_\beta \tag{19.5}$$

and its inverse is defined by

$$g^{\alpha\beta} g_{\beta\gamma} = \delta^\alpha_\gamma, \tag{19.6}$$

where δ^α_γ is equal to one if $\alpha = \gamma$ and zero otherwise. We can lower and raise indices with the metric tensor and inverse metric tensor, respectively, in the usual way, for example,

$$g_{\alpha\beta} n^\alpha = n_\beta. \tag{19.7}$$

It is straightforward to see that the inner product between two vectors $\mathbf{n}$ and $\mathbf{m}$ is

$$\mathbf{n} \cdot \mathbf{m} = n^\alpha \mathbf{t}_\alpha \cdot m^\beta \mathbf{t}_\beta = g_{\alpha\beta} n^\alpha m^\beta = n^\alpha m_\alpha. \tag{19.8}$$

The area of the parallelogram generated by the infinitesimal vectors $dx^1 \mathbf{t}_1$ and $dx^2 \mathbf{t}_2$, given by the magnitude of their cross product, yields the area element

$$\begin{aligned} dS &= \left| dx^1 \mathbf{t}_1 \times dx^2 \mathbf{t}_2 \right| \\ &= \sqrt{(\mathbf{t}_1 \times \mathbf{t}_2)^2}\, dx^1\, dx^2 \\ &= \sqrt{|\mathbf{t}_1|^2 |\mathbf{t}_2|^2 - (\mathbf{t}_1 \cdot \mathbf{t}_2)^2}\, dx^1\, dx^2 \\ &= \sqrt{g_{11} g_{22} - g_{12} g_{21}}\, dx^1\, dx^2 \\ &= \sqrt{g}\, d^2x, \end{aligned} \tag{19.9}$$

where $g = \det(g_{\alpha\beta})$, the determinant of the metric tensor, and d^2x is shorthand for $dx^1\ dx^2$. More generally, the magnitude of the cross product of two vectors **m** and **n** is

$$|\mathbf{m} \times \mathbf{n}| = \left|\gamma_{\alpha\beta} m^\alpha n^\beta\right|, \tag{19.10}$$

which introduces the antisymmetric tensor

$$\gamma_{\alpha\beta} = \sqrt{g}\epsilon_{\alpha\beta}, \tag{19.11}$$

where $\epsilon_{\alpha\beta}$ is the Levi-Civita symbol satisfying $\epsilon_{12} = -\epsilon_{21} = 1$ and is zero otherwise.

Since we will encounter tangent unit vectors, for example, indicating the orientation of some physical quantity, it is convenient to decompose this vector in a set of orthonormal tangent vectors, $\mathbf{e}_1(\mathbf{x})$ and $\mathbf{e}_2(\mathbf{x})$, such that

$$\mathbf{e}_i \cdot \mathbf{e}_j = \delta_{ij} \quad \text{and} \quad \mathbf{N} \cdot \mathbf{e}_i = 0, \tag{19.12}$$

alternative to the basis defined in Equation 19.3. Here, **N** is the vector normal to the surface. We use the Latin letters i, j, and k for the components of vectors expressed in this orthonormal basis. As they are locally Cartesian, they do not require any administration of the position of the index. Besides the area element, we need a generalization of the partial derivative. This generalization is the covariant derivative, D_α, the projection of the derivative onto the surface. The covariant derivative of **n** expressed in the orthonormal basis reads in component form [3]

$$\begin{aligned} D_\alpha n_i &= \mathbf{e}_i \cdot \partial_\alpha \mathbf{n} \\ &= \mathbf{e}_i \cdot \partial_\alpha n_j \mathbf{e}_j + \mathbf{e}_i \cdot \partial_\alpha \mathbf{e}_j n_j \\ &= \partial_\alpha n_i + \epsilon_{ij} A_\alpha n_j, \end{aligned} \tag{19.13}$$

where $\epsilon_{ij} A_\alpha = \mathbf{e}_i \cdot \partial_\alpha \mathbf{e}_j$ is called the spin-connection. The final line is justified because the derivative of any unit vector is perpendicular to this unit vector. More generally, the covariant derivative of the vector **n** along x^α is [21]

$$D_\alpha n^\beta = \partial_\alpha n^\beta + \Gamma^\beta_{\alpha\gamma} n^\gamma, \tag{19.14}$$

where the Christoffel symbols are

$$\Gamma^\alpha_{\beta\gamma} = \frac{1}{2} g^{\alpha\delta} \left(\partial_\gamma g_{\beta\delta} + \partial_\beta g_{\delta\gamma} - \partial_\delta g_{\beta\gamma}\right). \tag{19.15}$$

Finally, with the antisymmetric tensor and the area element in hand, we can state a useful formula in integral calculus, namely, Stokes' theorem

$$\int d^2x\ \sqrt{g}\gamma^{\alpha\beta} D_\alpha n_\beta = \oint dx^\alpha\ n_\alpha. \tag{19.16}$$

19.3.2 Curvature

The curvature is the deviation from flatness and therefore a measure of the rate of change of the tangent vectors along the normal, or, put the other way around, a measure of the rate of change of the normal along the tangent vectors. This can be cast in a curvature tensor defined as

$$K_{\alpha\beta} = \mathbf{N} \cdot \partial_\beta \mathbf{t}_{\alpha} \cdot = -\mathbf{t}_\alpha \cdot \partial_\alpha \mathbf{N}. \tag{19.17}$$

From this tensor, we extract the intrinsic Gaussian curvature

$$G = \det(K^\alpha_\beta) = \frac{1}{2}\gamma^{\alpha\beta}\gamma^{\gamma\delta} K_{\alpha\beta} K_{\gamma\delta} = \kappa_1\kappa_2 \tag{19.18}$$

and extrinsic mean curvature

$$H = \frac{1}{2}\mathrm{Tr}(K^\alpha_\beta) = \frac{1}{2} g^{\alpha\beta} K_{\alpha\beta} = \frac{1}{2}(\kappa_1 + \kappa_2), \tag{19.19}$$

where $\kappa_1 = \mathbf{N} \cdot \partial_1 \tilde{\mathbf{e}}_1$ and $\kappa_2 = \mathbf{N} \cdot \partial_2 \tilde{\mathbf{e}}_2$ are the extremal or principal curvatures, the curvature in the principal directions $\tilde{\mathbf{e}}_1$ and $\tilde{\mathbf{e}}_2$. These eigenvalues and eigenvectors can be obtained by diagonalizing the matrix associated with the curvature tensor. If at a point on a surface κ_1 and κ_2 have the same sign, the Gaussian curvature is positive and from the outsiders' point of view the surface curves away in the same direction whichever way you go, as is the case on tops and in valleys. In contrast, if at a point on a surface κ_1 and κ_2 have opposite signs, the Gaussian curvature is negative, the saddle-like surface curves away in opposite directions. The magnitude of κ_1 and κ_2 is equal to the inverse of the radius of the tangent circle in the principal direction (Fig. 19.2).

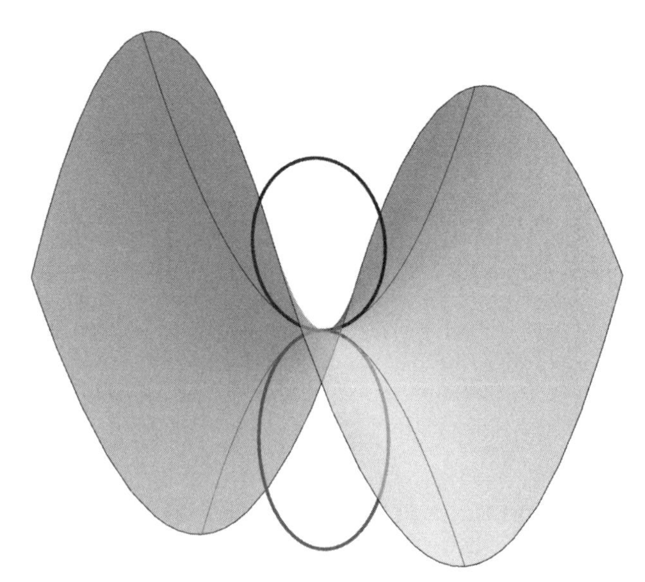

Figure 19.2 Saddle surface has negative Gaussian curvature. κ_1 and κ_2 have different signs. Tangent circles are drawn in the directions of maximum and minimum curvature.

It turns out that the Gaussian curvature and the spin-connection are related. We see how in a moment by considering the normal (third) component of the curl (denoted by $\nabla\times$) of the spin-connection

$$\begin{aligned}(\nabla \times \mathbf{A})_3 &= \epsilon_{3jk}\partial_j(\mathbf{e}_1 \cdot \partial_k \mathbf{e}_2) \\ &= \epsilon_{3jk}\partial_j \mathbf{e}_1 \cdot \partial_k \mathbf{e}_2 \\ &= \epsilon_{3jk}(\mathbf{N} \cdot \partial_j \mathbf{e}_1)(\mathbf{N} \cdot \partial_k \mathbf{e}_2)\end{aligned} \tag{19.20}$$

where we have used the product rule and the antisymmetry of ϵ_{ijk} in the second equality sign. The final line is justified by the fact that the derivative of a unit vector is perpendicular to itself and therefore we have, for example, $\partial_j \mathbf{e}_1 = (\mathbf{N} \cdot \partial_j \mathbf{e}_1)\mathbf{N} + (\mathbf{e}_2 \cdot \partial_j \mathbf{e}_1)\mathbf{e}_2$. If we now with the aid of Equations 19.18 and 19.17 note that

$$G = (\mathbf{N} \cdot \partial_1 \mathbf{e}_1)(\mathbf{N} \cdot \partial_2 \mathbf{e}_2) - (\mathbf{N} \cdot \partial_1 \mathbf{e}_2)(\mathbf{N} \cdot \partial_2 \mathbf{e}_1), \tag{19.21}$$

we easily see that the normal component of the curl of the spin-connection equals the Gaussian curvature:

$$(\nabla \times \mathbf{A}) \cdot \mathbf{N} = G, \tag{19.22}$$

or alternatively [22]

$$\gamma^{\alpha\beta} D_\alpha A_\beta = G. \tag{19.23}$$

This geometrical interpretation of $\mathbf{A}$ shows its importance in Section 19.4, where we comment on its implications on the geometrical frustration in curved nematic liquid crystal films.

19.3.3 Monge Gauge

A popular choice of parametrization of the surface is the Monge gauge or height representation in which $\mathbf{x} = (x, y)$ and $\mathbf{R} = (x, y, f(x, y))$, where $f(x, y)$ is the height of the surface above the xy-plane. In this representation the Gaussian curvature reads

$$G = \frac{\det \partial_\alpha \partial_\beta f}{g}, \tag{19.24}$$

where the determinant of the metric is given by

$$g = 1 + (\partial_x f)^2 + (\partial_y f)^2. \tag{19.25}$$

19.4 ELASTICITY ON CURVED SURFACES AND IN CONFINED GEOMETRIES

19.4.1 Elasticity of a Two-Dimensional Nematic Liquid Crystal

In a nematic liquid crystal, the molecules (assumed to be anisotropic) tend to align parallel to a common axis. The

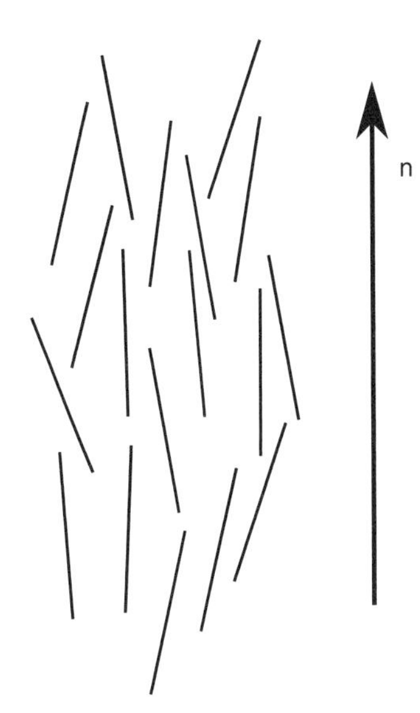

Figure 19.3 The director **n** specifies the average local orientation of the nematic molecules.

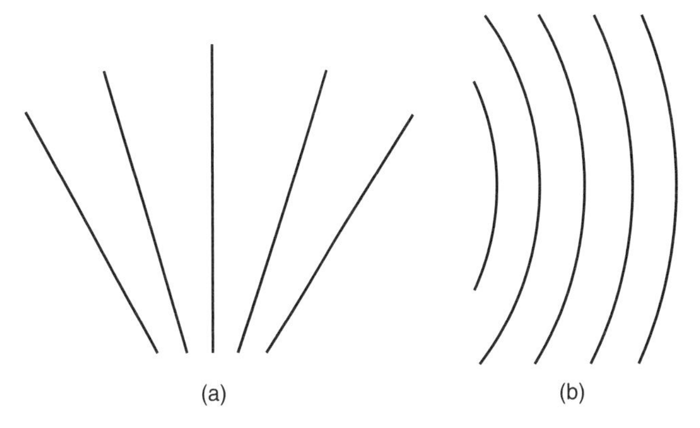

Figure 19.4 Conformations with (a) a nonvanishing divergence of the director and (b) a nonvanishing curl of the director.

direction of this axis is labeled with a unit vector, $\mathbf{n}$, called the director (see Fig. 19.3). The states $\mathbf{n}$ and $-\mathbf{n}$ are equivalent.

Any spatial distortion of a uniform director field costs energy. If we assume that these deformations are small on the molecular length scale, l,

$$|\partial_i n_j| \ll \frac{1}{l}, \tag{19.26}$$

we can construct a phenomenological continuum theory. The resulting Frank free energy F for a two-dimensional flat nematic liquid crystal reads [18, 23, 24]

$$F = \frac{1}{2}\int d^2x \left[k_1\left(\partial_i n_i\right)^2 + k_3\left(\epsilon_{ij}\partial_i n_j\right)^2\right], \tag{19.27}$$

where the splay and bend elastic constants, k_1 and k_3, respectively, measure the energy of the two independent distortions shown in Fig. 19.4.

To simplify the equations, one often makes the assumption of isotropic elasticity. In this approximation, the Frank elastic constants are equal, $k_1 = k_3 = k$, and up to boundary terms the free energy reduces to [23]

$$F = \frac{1}{2}k \int d^2x \; \partial_i n_j \; \partial_i n_j. \tag{19.28}$$

When the coupling of the director to the curvature tensor $K_{\alpha\beta}$ [25–33] is ignored, the elastic free energy on a curved surface generalizes to [11, 13, 14, 34, 35]

$$F = \frac{1}{2}k \int d^2x \; \sqrt{g} D_\alpha n^\beta D^\alpha n_\beta. \tag{19.29}$$

In this equation, the area element has become $dS = d^2x\sqrt{g}$ and partial derivatives have been promoted to covariant derivatives. Because the director is of unit length, we can conveniently specify it in terms of its angle with a local orthonormal reference frame, $\Theta(\mathbf{x})$, as follows:

$$\mathbf{n} = \cos(\Theta)\mathbf{e_1} + \sin(\Theta)\mathbf{e_2}. \tag{19.30}$$

Then, since $\partial_\alpha n_1 = -\sin(\Theta)\partial_\alpha\Theta = -n_2\partial_\alpha\Theta$ and $\partial_\alpha n_2 = \cos(\Theta)\partial_\alpha\Theta = n_1\partial_\alpha\Theta$, we see that

$$\partial_\alpha n_i = -\epsilon_{ij} n_j \partial_\alpha \Theta \tag{19.31}$$

with which we find the covariant derivative to be

$$D_\alpha n_i = -\epsilon_{ij} n_j (\partial_\alpha \Theta - A_\alpha). \tag{19.32}$$

Therefore, we can rewrite the elastic energy as [22]

$$F = \frac{1}{2}k \int d^2x \; \sqrt{g}(\partial_\alpha\Theta - A_\alpha)(\partial^\alpha\Theta - A^\alpha), \tag{19.33}$$

where we have used that $(-\epsilon_{ij}n_j)(-\epsilon_{ij}n_j) = \delta_{jk}n_j n_k = \cos^2(\Theta) + \sin^2(\Theta) = 1$. This form of the free energy[1] clearly shows that nematic order on curved surface is geometrically frustrated. The topological constraints of the introductory section are merely a special example of the frustration of local order due to the geometrical properties of the system.

[1]Note that if we had chosen orthonormal reference frame differing by a local rotation $\Psi(\mathbf{x})$

$$\mathbf{e}_1(\mathbf{x}) \to \cos(\Psi(\mathbf{x}))\mathbf{e}_1(\mathbf{x}) - \sin(\Psi(\mathbf{x}))\mathbf{e}_2(\mathbf{x}), \tag{19.34}$$

$$\mathbf{e}_2(\mathbf{x}) \to \sin(\Psi(\mathbf{x}))\mathbf{e}_1(\mathbf{x}) + \cos(\Psi(\mathbf{x}))\mathbf{e}_2(\mathbf{x}) \tag{19.35}$$

implying

$$\Theta(\mathbf{x}) \to \Theta(\mathbf{x}) + \Psi(\mathbf{x}) \quad A_\alpha(\mathbf{x}) \to A_\alpha(\mathbf{x}) + \partial_\alpha\Psi(\mathbf{x}), \tag{19.36}$$

the free energy, Equation 19.33, remains the same.

Note that for a curved surface without such a topological constraint (e.g., a Gaussian bump), the ground state can be a deformed director field. Since the curl of the spin-connection equals the Gaussian curvature (Eq. 19.23), if the Gaussian curvature is nonzero, the spin-connection is irrotational and cannot be written as the gradient of a scalar field, $A_\alpha \neq \partial_\alpha\Theta$, just like the magnetic field cannot be described by a scalar field either. Therefore, F in Equation 19.33 is nonzero and we can conclude that there is geometrical frustration present in the system.

19.4.2 Elasticity of a Two-Dimensional Solid

Similar to the construction of the continuum elastic energy of a nematic liquid crystal, we can write down the elastic energy of a linear elastic solid as an integral of terms quadratic in the deformations, that is, strain. This strain is found in the following way. Consider a point $\mathbf{x} = (x, y, 0)$ on an initially flat solid. This point is displaced to $\mathbf{x}'(\mathbf{x}) = (x', y', f)$ in the deformed solid, and so we may define a displacement vector $\mathbf{u}(\mathbf{x}) = \mathbf{x}' - \mathbf{x} = u_x\mathbf{e}_x + u_y\mathbf{e}_y + f\mathbf{e}_z$. The square of the line element in the deformed plate is then given by $ds'^2 = (dx + du_x)^2 + (dx + du_x)^2 + df^2$. Noting that $du_x = \partial_i u_x \; dx_i$ with $x_i = x, y$ and similarly for u_y and f, we find [36]

$$ds'^2 = ds^2 + 2u_{ij} \; dx_i \; dx_j. \tag{19.37}$$

Thus, the strain tensor $u_{ij}(\mathbf{x})$ encodes how infinitesimal distances change in the deformed body with respect to the resting state of the solid and reads

$$u_{ij} = \frac{1}{2}(\partial_i u_j + \partial_j u_i + A_{ij}), \tag{19.38}$$

where we have omitted nonlinear terms of second order in $\partial_i u_j$ and where the tensor field $A_{ij}(\mathbf{x})$ is now defined as

$$A_{ij} \equiv \partial_i f \partial_j f. \tag{19.39}$$

We assume that curvature plays its part only through this coupling of gradients of the displacement field to the geometry of the surface, and we therefore adopt the flat-space metric. This is a valid approximation for moderately curved solids, as we comment on at the end of the section [37, 38]. To leading order in gradients of the height function, A_{ij} is related to the curvature as (see Eq. 19.24)

$$-\frac{1}{2}\epsilon_{ik}\epsilon_{jl}\partial_k\partial_l A_{ij} = \det(\partial_i\partial_j f) = G. \tag{19.40}$$

Isotropy of the solid leaves two independent scalar combinations of u_{ij} that contribute to the stretching energy [36]:

$$F = \frac{1}{2}\int dS(2\mu u_{ij}^2 + \lambda u_{ii}^2). \tag{19.41}$$

The elastic constants λ and μ called the Lame coefficients. Minimization of this energy with respect to u_j leads to the force balance equation:

$$\partial_i \sigma_{ij} = 0, \tag{19.42}$$

where the stress tensor $\sigma_{ij}(\mathbf{x})$ is defined by Hooke's law:

$$\sigma_{ij} = 2\mu u_{ij} + \lambda \delta_{ij} u_{kk}. \tag{19.43}$$

The force balance equation can be solved by introducing the Airy stress function, $\chi(\mathbf{x})$, which satisfies

$$\sigma_{ij} = \epsilon_{ik}\epsilon_{jl}\partial_k \partial_l \chi, \tag{19.44}$$

since this automatically gives

$$\partial_i \sigma_{ij} = \epsilon_{jk}\partial_k [\partial_1, \partial_2]\chi = 0 \tag{19.45}$$

by the commutation of the partial derivatives. If one does not adopt the flat-space metric, the covariant generalization of the force balance equation is not satisfied, because the commutator of the covariant derivatives, known as the Riemann curvature tensor, does not vanish. It is actually proportional to the Gaussian curvature and indicates why the range of validity of this approach is limited to moderately curved surfaces [37, 38]. Finally, for small $\partial_i u_j$ the bond angle field, $\Theta(\mathbf{x})$, is given by

$$\Theta = \frac{1}{2}\epsilon_{ij}\partial_i u_j. \tag{19.46}$$

19.4.3 Elasticity of a Three-dimensional Nematic Liquid Crystal

Besides splay and bend, there are two other deformations possible in a three-dimensional nematic liquid crystal. They are twist and saddle-splay, measured by elastic moduli K_2 and K_{24}. The analog of Equation 19.27 reads

$$\begin{aligned} F[\mathbf{n}(\mathbf{x})] = \frac{1}{2}\int & dV(K_1(\nabla \cdot \mathbf{n})^2 + K_2(\mathbf{n}\cdot\nabla\times\mathbf{n})^2 \\ & + K_3(\mathbf{n}\times\nabla\times\mathbf{n})^2) \\ & - K_{24}\int \mathbf{dS}\cdot(\mathbf{n}\nabla\cdot\mathbf{n} + \mathbf{n}\times\nabla\times\mathbf{n}). \end{aligned} \tag{19.47}$$

The integration of the splay, twist, and bend energy density is over the volume to which the nematic is confined. The saddle-splay energy per unit volume is a pure divergence term, hence the saddle-splay energy can be written as the surface integral in Equation 19.47. In addition to the energy in Equation 19.47, there is an energetic contribution coming from the interfacial interactions, often larger in magnitude. Therefore, the anchoring of the nematic molecules at the boundary can be taken as a constraint. In one of the possible anchoring conditions, the director is forced to be tangential to the surface, yet free to rotate in the plane. In this case, the saddle-splay energy reduces to [39]

$$F_{24} = K_{24}\int dS(\kappa_1 n_1^2 + \kappa_2 n_2^2), \tag{19.48}$$

thus coupling the director to the boundary surface. We refer to the chapter by Lagerwall for a more detailed discussion on the origin of Equation 19.47.

19.5 TOPOLOGICAL DEFECTS

Topological defects are characterized by a small region where the order is not defined. Topological defects in translationally ordered media, such as crystals, are called *dislocations*. Defects in the orientational order, such as in nematic liquid crystals and again crystals, are called *disclinations*. The defects are topological when they cannot be removed by a continuous deformation of the order parameter. As we will see momentarily, they are classified according to a topological quantum number or topological charge, a quantity that may only take on a discrete set of values and that can be measured on any circuit surrounding the defect.

19.5.1 Disclinations in a Nematic

Consider for concreteness a two-dimensional nematic liquid crystal. A singularity in the director field is an example of a disclination. Such a point defect can be classified by its winding number, strength, or topological charge, s, which is the number of times the director rotates by 2π, when following one closed loop in counterclockwise direction around the singularity:

$$\oint d\Theta = \oint dx^\alpha\ \partial_\alpha \Theta = 2\pi s. \tag{19.49}$$

We can express Equation 19.49 in differential form by invoking Stokes' theorem:

$$\gamma^{\alpha\beta} D_\alpha \partial_\beta \Theta = q\delta(\mathbf{x} - \mathbf{x}_a), \tag{19.50}$$

where we use an alternative labeling, $q = 2\pi s$, of the charge of the defect, which is located at $\mathbf{x}_a$. The delta-function obeys

$$\delta(\mathbf{x} - \mathbf{x}_a) = \frac{\delta(x^1 - x_a^1)\delta(x^2 - x_a^2)}{\sqrt{g}}, \tag{19.51}$$

such that the integral over the surface yields one. The far field contribution of the defect to the angular director in a flat plane reads

$$\Theta = s\phi + c, \tag{19.52}$$

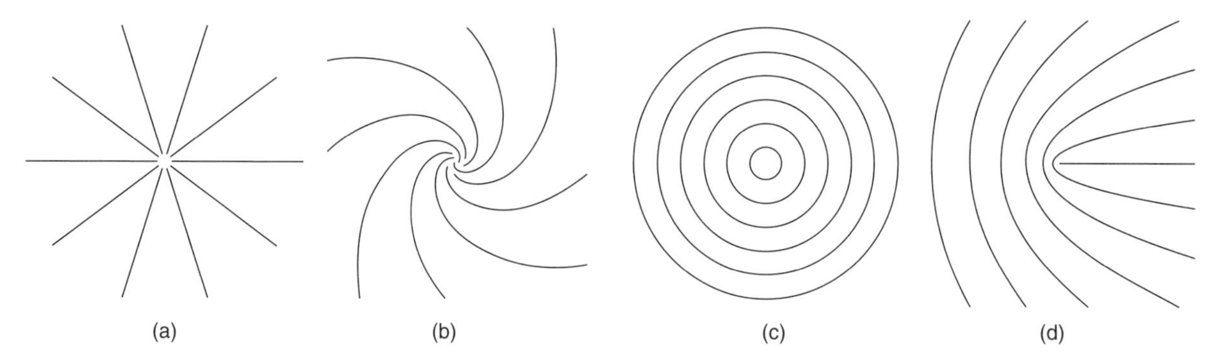

Figure 19.5 Director configurations, $n_1 = \cos\Phi$, $n_2 = \sin\Phi$, for disclinations of strength s and constant c. (a) $s = 1, c = 0$ (b) $s = 1, c = \frac{\pi}{4}$ (c) $s = 1, c = \frac{\pi}{2}$ (d) $s = \frac{1}{2}, c = 0$

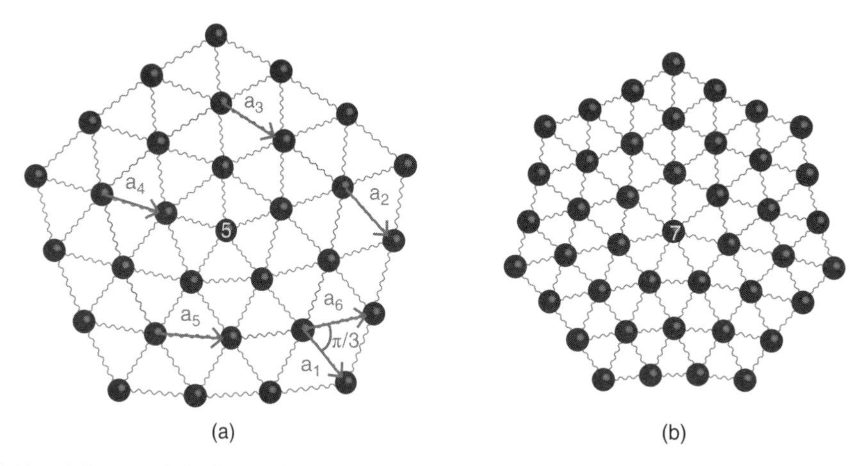

Figure 19.6 (a) Fivefold and (b) sevenfold disclinations. When following a closed counterclockwise loop around the fivefold disclination, the initial lattice vector $\mathbf{a}_1$ rotates via $\mathbf{a}_2$, $\mathbf{a}_3$, $\mathbf{a}_4$, and $\mathbf{a}_5$ over an angle of $\pi/3$ to $\mathbf{a}_6$.

as it forms a solution to the Euler–Lagrange equation of the elastic free energy

$$\partial^2\Theta = 0. \tag{19.53}$$

Here, ϕ is the azimuthal angle and c is just a phase. Examples are presented in Figure 19.5.

Note that since the states $\mathbf{n}$ and $-\mathbf{n}$ are equivalent, defects with half-integer strength are also possible. In fact, it is energetically favorable for an $s = 1$ defect to unbind into two $s = \frac{1}{2}$ defects [13, 40].

19.5.2 Disclinations in a Crystal

Though energetically more costly, disclinations also arise in two-dimensional crystals. At these points, the coordination number deviates from its ordinary value, which is six for a crystal on a triangular lattice. Just like in nematic liquid crystals, disclinations in crystals are labeled by a topological charge, q, which is the angle over which the vectors specifying the lattice directions rotate when following a counterclockwise circuit around the disclination. If we parametrize these lattice direction vectors with $\Theta(\mathbf{x})$, the bond angle field, this condition reads mathematically

$$\oint d\Theta = q. \tag{19.54}$$

Thus, for disclinations in a triangular lattice with fivefold and sevenfold symmetries, as displayed in Figure 19.6, $q = \frac{\pi}{3}$ and $q = -\frac{\pi}{3}$, respectively. Analogous to Equation 19.50, the flat-space differential form of Equation 19.54 for a disclination located at $\mathbf{x}_a$ reads

$$\epsilon_{ij}\partial_i\partial_j\Theta = q\delta(\mathbf{x} - \mathbf{x}_a). \tag{19.55}$$

19.5.3 Dislocations

Besides disclinations, dislocations can occur in crystals. Dislocations are characterized by a Burgers vector $\mathbf{b}$. This vector measures the change in the displacement vector, if we make a counterclockwise loop surrounding the dislocation,

$$\oint d\mathbf{u} = \mathbf{b}. \tag{19.56}$$

Just like the strength of disclinations can only take on a value out of a discrete set, the Burgers vector of a dislocation

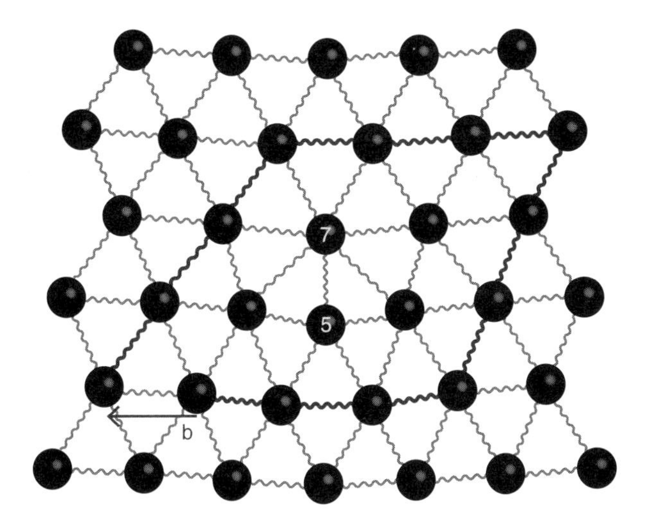

Figure 19.7 Dislocation in a triangular lattice. The Burgers vector specifies by how much a clockwise circuit (marked in bold) around the dislocation fails to close. A dislocation can be viewed as disclination dipole with a moment perpendicular to its Burgers vector.

is equal to some integer multiple of a lattice vector. Also note that a dislocation can be viewed as a pair of closely spaced disclinations of opposite charge [41], as can be seen in Figure 19.7.

The flat-space differential form of Equation (19.56) for a dislocation at $\mathbf{x}_a$ is

$$\epsilon_{ij}\partial_i\partial_j u_k = b_k\delta(\mathbf{x} - \mathbf{x}_a), \tag{19.57}$$

which again can be obtained by using Stokes' theorem.

19.6 INTERACTION BETWEEN CURVATURE AND DEFECTS

19.6.1 Coupling in Liquid Crystals

It is possible to recast the free energy in terms of the locations of the topological defects rather than the director or displacement field if smooth (i.e., nonsingular) deformations are ignored. In this case, the energy in Equation 19.33 is minimized with respect to Θ, leading to

$$D^\alpha(\partial_\alpha\Theta - A_\alpha) = 0. \tag{19.58}$$

This needs to be supplemented with an equation for the effective charge distribution:

$$\gamma^{\alpha\beta}D_\alpha(\partial_\beta\Theta - A_\beta) = \rho - G, \tag{19.59}$$

obtained by combining Equation 19.23 for the curvature and Equation 19.50 for the defect density $\rho(\mathbf{x})$,

$$\rho = \sum_a q_a\delta(\mathbf{x} - \mathbf{x}_a). \tag{19.60}$$

Equation 19.58 is automatically satisfied if one chooses [35]

$$\partial_\alpha\Theta - A_\alpha = \gamma_\alpha^\beta\partial_\beta\chi, \tag{19.61}$$

where $\chi(\mathbf{x})$ is an auxiliary function. At the same time, substituting Equation 19.61 into Equation 19.59 leads to

$$-D^2\chi = \rho - G. \tag{19.62}$$

The source in this Poisson equation contains both topological point charges and the Gaussian curvature with opposite sign. The analog of the electrostatic potential is χ. The role of the electric field is played by $\partial_\alpha\chi$. Indeed, substituting Equation 19.61 in Equation 19.33 shows that the energy density is proportional to the square of the electric field:

$$F = \frac{1}{2}k\int dS\partial_\alpha\chi\partial^\alpha\chi. \tag{19.63}$$

Next, we formally solve Equation 19.62

$$\chi = -\int dS'\Gamma_L(\mathbf{x},\mathbf{x}')(\rho(\mathbf{x}') - G(\mathbf{x}')), \tag{19.64}$$

where $\Gamma_L(\mathbf{x},\mathbf{x}')$ is the Green function of the Laplace–Beltrami operator, $D^2 = D_\alpha D^\alpha$, satisfying

$$D^2\Gamma_L(\mathbf{x},\mathbf{x}') = \delta(\mathbf{x} - \mathbf{x}'). \tag{19.65}$$

Integrating Equation 19.63 by parts and substituting our expressions for χ and the Laplacian of χ (Equations 19.64 and 19.62, respectively) results (up to boundary terms) in

$$F = -\frac{k}{2}\int dS\int dS'(\rho(\mathbf{x}) - G(\mathbf{x}))\Gamma_L(\mathbf{x},\mathbf{x}')(\rho(\mathbf{x}') - G(\mathbf{x}')), \tag{19.66}$$

from which we again deduce the analogy with two-dimensional electrostatics. In this analogy, the defects are electric point sources with their electric charge equal to the topological charge q and the Gaussian curvature with its sign reversed is a background charge distribution. Therefore, the defects will be attracted toward regions of Gaussian curvature with the same sign as the topological charge [11, 30, 35, 42–46]. Such screening will be perfect if $S = \rho$ everywhere since $F = 0$ then. However, unless the surface contains singularities in the Gaussian curvature, like the apex of a cone, perfect screening will be impossible, as the topological charge is quantized, whereas the Gaussian curvature is typically smoothly distributed.

19.6.2 Coupling in Crystals

Note that an *arbitrary* field χ solves Equation 19.45. However, χ must be physically possible and we therefore need to accompany Equation (19.45) with another equation, which we will obtain by considering the inversion of Equation 19.43 [36, 47]:

$$u_{ij} = \frac{1+\nu}{Y}\sigma_{ij} - \frac{\nu}{Y}\sigma_{kk}\delta_{ij}, \tag{19.67}$$

$$= \frac{1+\nu}{Y}\epsilon_{ik}\epsilon_{jl}\partial_k\partial_l\chi - \frac{\nu}{Y}\partial^2\chi\delta_{ij}, \tag{19.68}$$

where the two-dimensional Young's modulus, Y, and Poisson ratio, ν, are given by

$$Y = \frac{4\mu(\mu+\lambda)}{2\mu+\lambda}, \tag{19.69}$$

$$\nu = \frac{\lambda}{2\mu+\lambda}. \tag{19.70}$$

Applying $\epsilon_{ik}\epsilon_{jl}\partial_k\partial_l$ to Equation 19.68 gives

$$\frac{1}{Y}\partial^4\chi = \epsilon_{ik}\epsilon_{jl}\partial_k\partial_l u_{ij}. \tag{19.71}$$

By invoking Equations 19.38 and 19.46, the differential expressions for the defects, namely, Equations 19.57 and 19.55, as well as Equation 19.40 for the curvature, one can rewrite the right-hand side to arrive at the crystalline analog of Equation 19.62:

$$\frac{1}{Y}\partial^4\chi = \rho - G, \tag{19.72}$$

where the defect distribution, ρ, of disclinations with charge q_a and dislocations with Burgers vector $\mathbf{b}^b$ reads

$$\rho = \sum_a q_a\delta(\mathbf{x}-\mathbf{x}_a) + \sum_b \epsilon_{ij}b_i^b\partial_j\delta(\mathbf{x}-\mathbf{x}_b). \tag{19.73}$$

We can also rewrite the free energy (up to boundary terms) in terms of the Airy stress function as follows:

$$F = \frac{1}{2Y}\int dS(\partial^2\chi)^2. \tag{19.74}$$

If we integrate this by parts twice and use Equation 19.72 to eliminate χ and $\partial^4\chi$, we find (up to boundary terms)

$$F = \frac{Y}{2}\int dS \int dS'(\rho(\mathbf{x}) - G(\mathbf{x}))\Gamma_B(\mathbf{x},\mathbf{x}')(\rho(\mathbf{x}') - G(\mathbf{x}')), \tag{19.75}$$

where Γ_B is the Greens function of the biharmonic operator

$$\partial^4\Gamma_B(\mathbf{x},\mathbf{x}') = \delta(\mathbf{x}-\mathbf{x}'). \tag{19.76}$$

Equation 19.75 is the crystalline analog of Equation 19.66. Again, the defects can screen the Gaussian curvature. The interaction, however, is different from the Coulomb interaction in the liquid crystalline case. If the surface is allowed to bend, disclinations will induce buckling, illustrated in

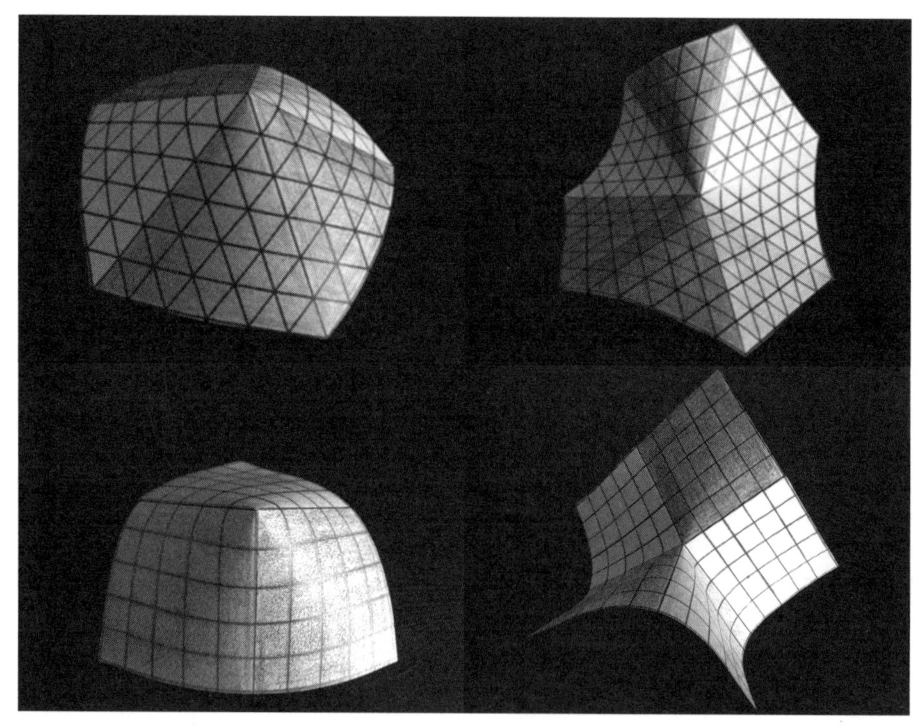

Figure 19.8 Paper models illustrating the coupling between disclinations and curvature. *Left panels:* Positively (*right panels*: negatively) charged disclinations and positive (negative) Gaussian curvature attract. *Top left panel:* Fivefold coordinated particle in a triangular lattice. *Top right panel:* Sevenfold coordinated particle in a triangular lattice. *Bottom left panel:* Threefold coordinated particle in a square lattice. *Bottom right panel:* Fivefold coordinated particle in a square lattice.

Figure 19.8 with paper models. In these cones, the integrated Gaussian curvature is determined by the angular deficit of the disclination

$$\int dSG = q. \tag{19.77}$$

19.6.3 Screening by Dislocations and Pleats

Surprisingly, also charge-neutral dislocations and pleats can screen the curvature [10, 22, 37, 47, 48]. Pleats are formed by arrays of dislocations and allow for an extra piece of crystal, just like their fabric analogs. The opening angle, $\Delta\Theta$, of the pleat (or low-angle grain boundary) is given by

$$\Delta\Theta \approx n_d a, \tag{19.78}$$

where a is the lattice spacing and n_d is the dislocation line density. Since this opening angle can be arbitrarily small, pleats can provide a finer screening than quantized disclinations.

19.6.4 Geometrical Potentials and Forces

The cross terms of Equation 19.75 represent the interaction energy

$$\zeta = -Y \int dS\ \rho(\mathbf{x}) \int dA'\ \Gamma_B(\mathbf{x}, \mathbf{x}')G(\mathbf{x}'). \tag{19.79}$$

By introducing an auxiliary function $V(\mathbf{x})$ satisfying

$$\partial^2 V = G, \tag{19.80}$$

Equation 19.79 can, by integrating by parts twice, be rewritten as

$$\zeta = -Y \int dS\ \rho(\mathbf{x}) \int dS'\ \Gamma_L(\mathbf{x}, \mathbf{x}')V(\mathbf{x}'). \tag{19.81}$$

The field $\zeta(\mathbf{x})$ can be viewed as a geometric potential, that is, the potential experienced by a defect due to the curvature of the crystal [37, 38]. Another more heuristic way to study the interaction of dislocations and curvature is the following. We consider the stress that exist in the monolayer as a result of curvature only, σ_{ij}^G, as the source of a Peach–Koehler force, $\mathbf{f}$, on the dislocation:

$$f_k = \epsilon_{kj} b_i \sigma_{ij}^G. \tag{19.82}$$

Note that, by setting $\rho = 0$, the Airy stress function χ^G satisfies

$$\frac{1}{Y}\partial^4 \chi^G = -G. \tag{19.83}$$

This equation can be solved in two steps. First, we make use of an auxiliary function U obeying

$$\partial^2 U = G. \tag{19.84}$$

This leaves the following equation to be solved:

$$\frac{1}{Y}\partial^2 \chi^G = -U + U_H, \tag{19.85}$$

where U_H is a harmonic function (i.e., $\partial^2 U_H = 0$) introduced to fulfill the boundary conditions [37].

19.7 NEMATICS IN SPHERICAL GEOMETRIES

19.7.1 Nematic Order on the Sphere

As a naive guess for the ground state of a two-dimensional nematic liquid crystal phase on the surface of the sphere, one could imagine the excess of topological charge to be located at the poles, as in the case of tilted molecules on the sphere. However, the order parameter, the director, has the symmetry of a headless arrow instead of a vector. Therefore, this makes it possible for the two $s = 1$ defects to unbind into four $s = \frac{1}{2}$ defects relaxing at the vertices of a regular tetrahedron [12]. The baseball-like nematic texture is illustrated in Figure 19.9. The repulsive nature of defects with like charges can be seen from the free energy, which, as shown in the previous section,

Figure 19.9 The baseball-like ground state of a two-dimensional spherical nematic coating has four $s = \frac{1}{2}$ at the vertices of a tetrahedron in the one-constant approximation. Source: Figure from Ref. 14.

can entirely be reformulated in terms of the defects rather than the director [12, 13]:

$$F = -\frac{\pi k}{2} \sum_{i \neq j} s_i s_j \log\,(1 - \cos\theta_{ij}) + E(R) \sum_j s_j^2. \quad (19.86)$$

Here, θ_{ij} is the angular separation between defects i and j, that is, $\theta_{ij} = \frac{d_{ij}}{R}$, with d_{ij} being the geodesic distance. The first term yields the long-range interaction of the charges. The second term accounts for the defect self-energy

$$E(R) = \pi k \log\left(\frac{R}{b}\right) + E_c, \quad (19.87)$$

where we have imposed a cut-off b representing the defect core size, which has energy E_c. This cut-off needs to be introduced in order to prevent the free energy from diverging. Heuristically, this logarithmically diverging term in the free energy is responsible for the splitting of the two $s = 1$ defects into four $s = \frac{1}{2}$ defects. Two $s = 1$ defects contribute $(2 \times 1^2)\pi k \log\left(\frac{R}{b}\right) = 2\pi k \log\left(\frac{R}{b}\right)$ to the free energy, whereas four $s = \frac{1}{2}$ defects contribute only $\left(4 \times \left(\frac{1}{2}\right)^2\right)\pi k \log\left(\frac{R}{b}\right) = \pi k \log\left(\frac{R}{b}\right)$.

In addition to this ground state, other defect structures have been observed in computer simulations [49–52]. If there is a strong anisotropy in the elastic moduli, the four defects are found to lie on a great circle rather than the vertices of a regular tetrahedron [51, 52].

19.7.2 Beyond Two Dimensions: Spherical Nematic Shells

An experimental model system of spherical nematics is nematic double emulsion droplets [16, 17, 53–59]. These are structures in which a water droplet is captured by a larger nematic liquid crystal droplet, which in turn is dispersed in an outer fluid. There are some crucial differences between a two-dimensional spherical nematic and these systems. Not only is the nematic coating of a finite thickness, this thickness can be inhomogeneous as a result of buoyancy-driven displacement (or other mechanisms) of the inner droplet out of the center of the nematic droplet.

Like point disclinations in two dimensions, there exist disclination *lines* in a three-dimensional nematic liquid crystal, which are categorized in a similar manner. However, charge one lines and integral lines, in general, do not exist. Such lines lose their singular cores [60, 61] by "escaping in the third dimension." In shells, such an escape leads to another type of defects, namely, point defects at the interface, known as boojums (Fig. 19.10).

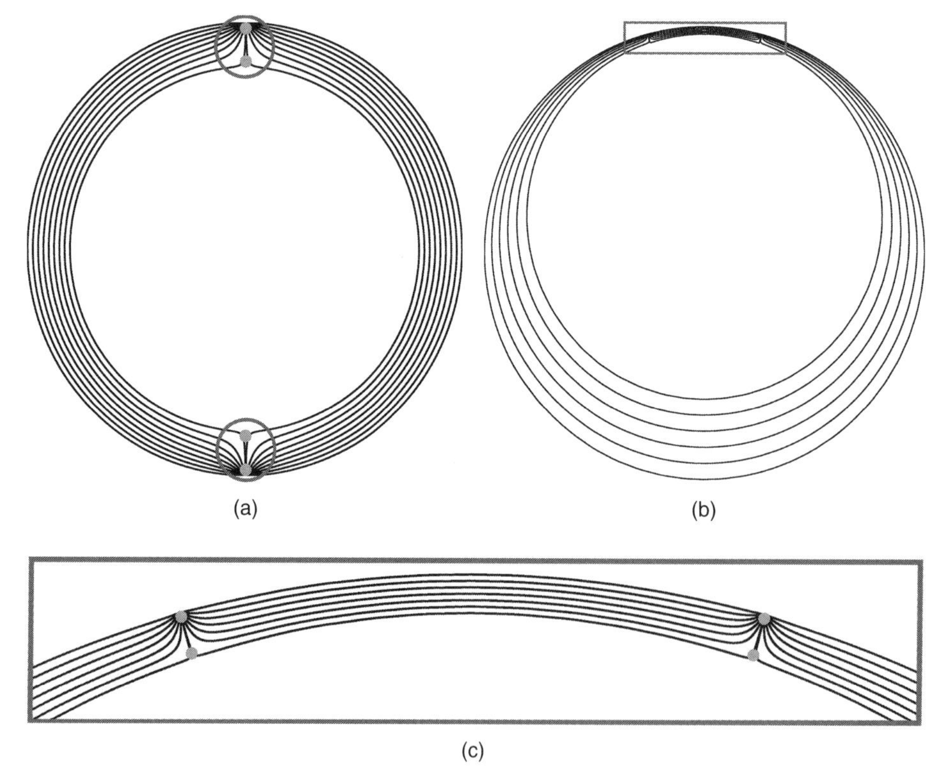

Figure 19.10 (a) Schematic of the deconfined defect configuration in a homogeneous shell. Two pairs (each encircled) of boojums, indicated by dots, are located at the top and bottom of the shell. (b) Schematic of the confined defect configuration in an inhomogeneous shell. All boojums are located at the thinnest, top part of the shell, inside the red rectangle. (c) Zoom of the thinnest section of the inhomogeneous shell in (b). From Ref. 17 – Reproduced by permission of The Royal Society of Chemistry.

In a spherical nematic layer of *finite* thickness, calculations show that the baseball structure with four $s = \frac{1}{2}$ disclination lines spanning the shell becomes energetically less favorable than two antipodal pairs of boojums beyond a critical thickness [14]. Instead of unbinding, the singular lines escape in the third dimension, leaving two pairs of boojums on the bounding surfaces. These two defect configurations are separated by a large energy barrier. As a consequence, both configurations are observed in droplets in the same emulsion. If, in addition, the shell thickness is inhomogeneous, the energy landscape becomes even more complex.

As a consequence of the inhomogeneity, the defects cluster in the thinnest part of the shell, where the lengths of the disclination lines (or distance between boojums forming a pair) are shorter. Since the self-energy of the disclination is proportional to its length, it is attracted toward this region of the shell. One of the intriguing outcomes of the study of inhomogeneous shells is that the pairs of surface defects can make abrupt transitions between the state in which the defects are confined in the thinnest part of the shell and the deconfined state, in which the interdefect repulsion places them diametrically [17]. These confinement and deconfinement transitions occur when the thickness or thickness inhomogeneity is varied. A defect arrangement with a corresponding local minimum in the energy landscape makes the transition to the global minimum when the local minimum loses its stability. This explains both the abruptness of the transitions and the hysteresis between them.

In agreement with this picture, Monte Carlo simulations of nematic shells on uniaxial and biaxial colloidal particles have shown the tendencies for defects to accumulate in the thinnest part and in regions of the highest curvature [62].

19.8 TOROIDAL NEMATICS

The torus has a zero topological charge. Hence, in a nematic droplet of toroidal shape, no defects need to be present. The director field to be expected naively in such a geometry is one that follows the tubular axis, as shown in Figure 19.11. This achiral director configuration contains only bend energy. Simple analytical calculations show, however, that if the toroid becomes too fat, it is favorable to reduce bend deformations by twisting. The price of twisting is screened by saddle-splay deformations, provided that $K_{24} > 0$ [39, 63]. The twisted configuration is chiral. Chirality stems from the Greek word for hand and is indeed in this context easily explained: your right hand cannot be turned into a left hand by moving and rotating it. It is only when viewed in the mirror that your right hand appears to be a left hand and vice versa. Indeed, for small aspect ratios and small values of $(K_2 - K_{24})/K_3$, nematic toroids display either a right- or left-handedness despite the achiral nature of nematics. This phenomenon is recognized as spontaneous chiral symmetry breaking. Typical corresponding plots of the energy as a function of the amount of twist are shown in Figure 19.11.

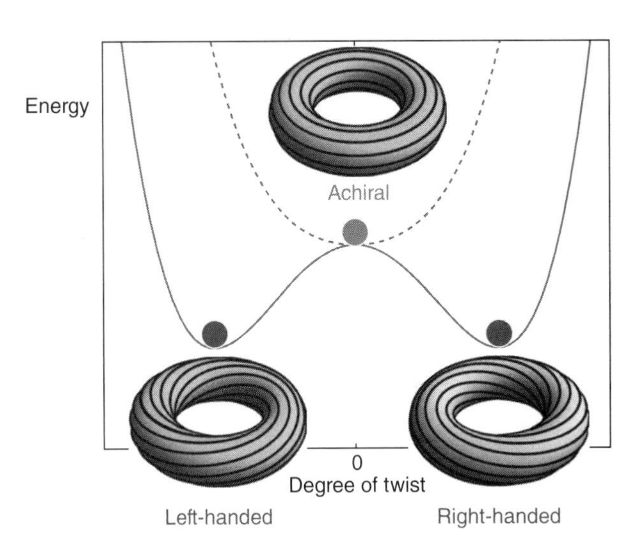

Figure 19.11 Energy as a function of the degree of twist has either a single achiral minimum (dashed) or shows spontaneous chiral symmetry breaking in toroidal nematics (solid) depending on the aspect ratio and elastic constants. The chiral state is favored for fat toroids and small values of $(K_2 - K_{24})/K_3$.

19.9 CONCLUDING REMARKS

We hope to have shared our interest in the rich subject of geometry in soft matter, in particular the interplay of defects and curvature in two-dimensional ordered matter and the confinement of nematic liquid crystals in various geometries. For readers interested in a more detailed treatment, we refer to excellent reviews by Kamien [3], Bowick and Giomi [38], Nelson [2], David [21], and Lopez-Leon and Fernandez-Nieves [64].

REFERENCES

[1] Sadoc J-F, Mosseri R. *Geometrical Frustration*. Cambridge: Cambridge University Press; 2006.

[2] Nelson DR. *Defects and Geometry in Condensed Matter Physics*. Cambridge: Cambridge University Press; 2002.

[3] Kamien RD. Rev Mod Phys 2002;74:953. DOI: 10.1103/RevModPhys.74.953.

[4] Monastyrsky M. *Riemann, Topology, and Physics*. Boston (MA): Birkhäuser Boston; 1999.

[5] Needham T. *Visual Complex Analysis*. Oxford: Oxford University Press; 2000.

[6] Caspar D, Klug A. Cold Spring Harbor Symp Quant Biol 1962;27:1.

[7] Lidmar J, Mirny L, Nelson DR. Phys Rev E 2003;68:051910. DOI: 10.1103/PhysRevE.68.051910.

[8] Dinsmore AD, Hsu MF, Nikolaides MG, Marquez M, Bausch AR, Weitz DA. Science 2002;298:1006. DOI: 10.1126/science.1074868.

[9] Bausch AR, Bowick MJ, Cacciuto A, Dinsmore AD, Hsu MF, Nelson DR, Nikolaides MG, Travesset A, Weitz DA. Science 2003;299:1716, arXiv:cond-mat/0303289. DOI: 10.1126/science.1081160.

[10] Irvine WTM, Vitelli V, Chaikin PM. Nature (London) 2010;468:947. DOI: 10.1038/nature09620.

[11] MacKintosh FC, Lubensky TC. Phys Rev Lett 1991;67:1169. DOI: 10.1103/PhysRevLett.67.1169.

[12] Lubensky TC, Prost J. J Phys II 1992;2:371. DOI: 10.1051/jp2:1992133.

[13] Nelson DR. Nano Lett 2002;2:1125, arXiv:cond-mat/0206552. DOI: 10.1021/nl0202096.

[14] Vitelli V, Nelson DR. Phys Rev E 2006;74:021711, arXiv:cond-mat/0604293. DOI: 10.1103/PhysRevE.74.021711.

[15] DeVries GA, Brunnbauer M, Hu Y, Jackson AM, Long B, Neltner BT, Uzun O, Wunsch BH, Stellacci F. Science 2007;315:358. DOI: 10.1126/science.1133162.

[16] Lopez-Leon T, Koning V, Devaiah KBS, Vitelli V, Fernandez-Nieves A. Nat Phys 2011;7:391. DOI: 10.1038/nphys1920.

[17] Koning V, Lopez-Leon T, Fernandez-Nieves A, Vitelli V. Soft Matter 2013;9:4993. DOI: 10.1039/C3SM27671F.

[18] de Gennes PG, Prost J. *The Physics of Liquid Crystals*. New York: Oxford University Press; 1993.

[19] Stephen MJ, Straley JP. Rev Mod Phys 1974;46:617. DOI: 10.1103/RevModPhys.46.617.

[20] Struik DJ. *Lectures on Classical Differential Geometry*. Addison Wesley Publishing Company; 1988.

[21] David F. Geometry and field theory of random surfaces and membranes. In: Nelson SWD, Piran T, editors. *Statistical Mechanics of Membranes and Surfaces*. World Scientific; 2004.

[22] Nelson DR, Peliti L. J Phys France 1987;48:1085. DOI: 10.1051/jphys:019870048070108500.

[23] Kléman M. *Points, Lines and Walls. In Liquid Crystals, Magnetic Systems and Various Ordered Media*. New York: John Wiley & Sons, Ltd; 1983.

[24] Kleman M, Lavrentovich OD. *Soft Matter Physics: An Introduction*. New York: Springer-Verlag; 2003.

[25] Santangelo CD, Vitelli V, Kamien RD, Nelson DR. Phys Rev Lett 2007;99:017801. DOI: 10.1103/PhysRevLett.99.017801.

[26] Kamien RD, Nelson DR, Santangelo CD, Vitelli V. Phys Rev E 2009;80:051703–. DOI: 10.1103/PhysRevE.80.051703.

[27] Jiang H, Huber G, Pelcovits RA, Powers TR. Phys Rev E 2007;76:031908–. DOI: 10.1103/PhysRevE.76.031908.

[28] Frank JR, Kardar M. Phys Rev E 2008;77:041705. DOI: 10.1103/PhysRevE.77.041705.

[29] Mbanga BL, Grason GM, Santangelo CD. Phys Rev Lett 2012;108:017801. DOI: 10.1103/PhysRevLett.108.017801.

[30] Selinger RLB, Konya A, Travesset A, Selinger JV. J Phys Chem B 2011;115:13989. DOI: 10.1021/jp205128g.

[31] Napoli G, Vergori L. Phys Rev Lett 2012;108:207803. DOI: 10.1103/PhysRevLett.108.207803.

[32] Napoli G, Vergori L. Phys Rev E 2012;85:061701. DOI: 10.1103/PhysRevE.85.061701.

[33] Napoli G, Vergori L. Int J Non Linear Mech 2013;49:66. DOI: 10.1016/j.ijnonlinmec.2012.09.007. ISSN: 0020-7462.

[34] Park J-M, Lubensky TC. Phys Rev E 1996;53:2648. DOI: 10.1103/PhysRevE.53.2648.

[35] Vitelli V, Nelson DR. Phys Rev E 2004;70:051105. DOI: 10.1103/PhysRevE.70.051105.

[36] Landau LD, Lifshitz EM. *Theory of Elasticity*. Volume 7, Course of Theoretical Physics. 3rd ed. Reed Educational and Professional Publishing Ltd.; 1986.

[37] Vitelli V, Lucks JB, Nelson DR. Proc Natl Acad Sci U S A 2006;103:12323. DOI: 10.1073/pnas.0602755103.

[38] Bowick M, Giomi L. Adv Phys 2009;58:449, arXiv:0812.3064 [cond-mat.soft]. DOI: 10.1080/00018730903043166.

[39] Koning V, van Zuiden BC, Kamien RD, Vitelli V. ArXiv e-prints 2013, arXiv:1312.5092 [cond-mat.soft].

[40] Chaikin PM, Lubensky TC. *Principles of Condensed Matter Physics*. Cambridge: Cambridge University Press; 1995.

[41] Nelson DR, Halperin BI. Phys Rev B 1979;19:2457. DOI: 10.1103/PhysRevB.19.2457.

[42] Bowick M, Nelson DR, Travesset A. Phys Rev E 2004;69:041102. DOI: 10.1103/PhysRevE.69.041102.

[43] Vitelli V, Turner AM. Phys Rev Lett 2004;93:215301. DOI: 10.1103/PhysRevLett.93.215301.

[44] Xing X, Shin H, Bowick MJ, Yao Z, Jia L, Li M-H. Proc Natl Acad Sci U S A 2012;109:5202. DOI: 10.1073/pnas.1115684109.

[45] Jesenek D, Perutková vS, Kralj-Iglič V, Kralj S, Iglič A. Cell Calcium 2012;52:277. DOI: /10.1016/j.ceca.2012.04.001. ISSN: 0143-4160, REGULATED EXOCYSTOSIS.

[46] Tie-Yan S, Yi-Shi D. Commun Theor Phys 2006;46:319. http://stacks.iop.org/0253-6102/46/i=2/a=028.

[47] Seung HS, Nelson DR. Phys Rev A 1988;38:1005. DOI: 10.1103/PhysRevA.38.1005.

[48] Hexemer A, Vitelli V, Kramer EJ, Fredrickson GH. Phys Rev E 2007;76:051604. DOI: 10.1103/PhysRevE.76.051604.

[49] Dzubiella J, Schmidt M, Löwen H. Phys Rev E 2000;62:5081. DOI: 10.1103/PhysRevE.62.5081.

[50] Skačej G, Zannoni C. Phys Rev Lett 2008;100:197802. DOI: 10.1103/PhysRevLett.100.197802.

[51] Shin H, Bowick MJ, Xing X. Phys Rev Lett 2008;101:037802, arXiv:0712.4012 [cond-mat.soft]. DOI: 10.1103/PhysRevLett.101.037802.

[52] Bates MA. J Chem Phys 2008;128:104707. DOI: 10.1063/1.2890724.

[53] Fernández-Nieves A, Vitelli V, Utada AS, Link DR, Márquez M, Nelson DR, Weitz DA. Phys Rev Lett 2007;99:157801. DOI: 10.1103/PhysRevLett.99.157801.

[54] Lopez-Leon T, Fernandez-Nieves A. Phys Rev E 2009;79:021707. DOI: 10.1103/PhysRevE.79.021707.

[55] Liang H-L, Schymura S, Rudquist P, Lagerwall J.

Phys Rev Lett 2011;106:247801. DOI: 10.1103/PhysRevLett.106.247801.

[56] Lopez-Leon T, Fernandez-Nieves A, Nobili M, Blanc C. Phys Rev Lett 2011;106:247802. DOI: 10.1103/PhysRevLett.106.247802.

[57] Seč D, Lopez-Leon T, Nobili M, Blanc C, Fernandez-Nieves A, Ravnik M, Žumer S. Phys Rev E 2012;86:020705. DOI: 10.1103/PhysRevE.86.020705.

[58] Liang H-L, Zentel R, Rudquist P, Lagerwall J. Soft Matter 2012;8:5443. Doi: 10.1039/C2SM07415J.

[59] Liang H-L, Noh J, Zentel R, Rudquist P, Lagerwall JP. Philos Trans R Soc A Math Phys Eng Sci 2013;371. DOI: 10.1098/rsta.2012.0258.

[60] Cladis PE. J Phys France 1972;33:591.

[61] Meyer R. Philos Mag 1973;27:405. DOI: 10.1080/14786437308227417.

[62] Bates MA, Skacej G, Zannoni C. Soft Matter 2010;6:655. DOI: 10.1039/B917180K.

[63] Pairam E, Vallamkondu J, Koning V, van Zuiden BC, Ellis PW, Bates MA, Vitelli V, Fernandez-Nieves A. Proc Natl Acad Sci U S A 2013;110:9295. DOI: 10.1073/pnas.1221380110.

[64] Lopez-Leon T, Fernandez-Nieves A. Colloid Polym Sci 2011;289:345. DOI: 10.1007/s00396-010-2367-7, ISSN: 0303-402X.

20

NEMATICS ON CURVED SURFACES – COMPUTER SIMULATIONS OF NEMATIC SHELLS

Martin Bates
Department of Chemistry, University of York, York, UK

20.1 INTRODUCTION

Nematic shells are spherical surface volumes obtained by coating micron-sized objects with a thin layer of nematic liquid crystals [1]. The constraint imposed by such confinement combined with planar anchoring of the liquid crystal at the surface inevitably results in the presence of topological defects [1] in the orientational order of the nematic layer: this inevitability is discussed further in Section 20.2. Nematic liquid crystals in this type of geometry have sparked recent interest in the soft matter community for both theoretical and perhaps more practical reasons. From a theoretical point of view, the defect patterns resulting from the interactions between the defects are interesting in their own right. An understanding of the defect configurations gives us an insight into the origin of the interactions between the defects, which result from the energy-dependent elastic deformations of the nematic order in the shell. This understanding can, in turn, suggest new ways to control the number and type of defects for different shell geometries. Another reason for the interest in this field comes from a seminal idea by Nelson [2] to use nematic-coated particles as "colloidal atoms." A subsequent and more practical chemical challenge is functionalizing the defect cores as attachment points for linkers to act as bonding sites in order to be able to attach the colloidal atoms together, much as bonds do for real molecules. One possible way to do this is to use the fact that the defects are naturally high energy points in the structure as they are disordered regions, where the liquid crystal order breaks down [1]. If a nonliquid-crystalline guest species, which itself should be miscible in at least the isotropic phase of the liquid crystal, is added to the shell, then we might expect this to aggregate at the high energy defect sites. This minimizes the energy as the guest would

Fluids, Colloids and Soft Materials: An Introduction to Soft Matter Physics, First Edition. Edited by Alberto Fernandez Nieves and Antonio Manuel Puertas.

be taking up volume in a region of orientational disorder, displacing the liquid crystal molecules from the regions of disorder. If the guests are designed such that they can be subsequently chemically linked to create a physical chain across the thickness of the shell, from the outer surface of the nematic through to the inner particle, then these linkers could be used to connect neighboring particles in a similar way that chemical bonds connect atoms; the difference is that the lengthscale is orders of magnitude larger for the colloidal systems. This mechanism could thus allow for the construction of colloidal photonic organizations, where the architecture and performance of such superstructures is determined by both the number and the spatial distribution of the linker bonds of each colloidal particle. Of course, this, in turn, depends on the number, type, and position of the defects in the coating layer forming the nematic shell. In other words, if colloids or droplets were regarded as atoms, then the number, type, and location of the defects would determine their valency and the geometry of the bonds formed. This would therefore influence the corresponding micron-scale properties of the assemblies that could possibly be formed. A particularly interesting case would be colloidal atoms with a tetrahedral geometry, similar to an sp_3 carbon atom, which could be used to form diamond lattice photonic crystals for metamaterials.

To make progress in this area, it is first important to understand the origins of the formation of the topological defects in the ground state of nematic shells. Second, it is important to understand how the manipulation of nematic shells, through their thickness, shape, homogeneity, application of external fields, and so on, can allow the locations and the number of defects to be varied. We discuss these aspects from a simulation point of view in this chapter. We start with a brief introduction to the theory in Section 20.2 and experimental evidence in Section 20.3. Simulation methodologies to investigate nematic shells are described in Section 20.4 and results of simulations of nematic shells are discussed in Section 20.5.

20.2 THEORY

We first point out that the theory of the ordering, location, and type of defects formed in nematic shells is discussed in a mathematically rigorous manner in the chapter by Vitelli [3]. Here, we use more simple hand-waving arguments to discuss the origin of defects in nematic shells and their resulting defect structures.

A nematic liquid crystal is a fluid in which there is orientational order but no long-range positional order [1]. Many of the best-known nematogenic molecules are elongated or rod-shaped, although many other nematic forming molecules have distinctly different shapes, from disks to fans to bent cores [4]. However, we shall consider the simple example of rod-shaped molecules and assume that they can be described by uniaxial rods. The molecules in a nematic phase have no long-range translational structure, but they do possess long-range orientational order. We assume that the molecules are aligned, on average, along a common direction known as the director, $\hat{n}$. The director is more formally defined in terms of an eigenvector associated with the largest eigenvalue of the second rank ordering tensor [5]. For our purposes, we can consider it to be an indicator of the local direction of order. As there is no polarity in a nematic phase along the director axis, $\hat{n}$ and $-\hat{n}$ are equivalent. Rotating a nematic in the absence of any aligning potential does not change the energy, and thus the director can adopt any direction if unconstrained.

Defects are spatial regions where the characteristic order of the liquid crystal is disrupted. As we shall see, when the orientations of the molecules in a nematic are confined to the spherical surface of a sphere, the presence of defects is unavoidable [2, 6]. These cannot be removed in any way while retaining the planar anchoring of the molecules at the surface; thus, they are topological defects present in the ground state of the system. The defects are typically characterized by their topological strength, s [1]. The strength is a semi-integer winding number that indicates the rotation that the director undergoes as the defect is encircled; $s = 1/2$ corresponds to a rotation of π in the director field on a ring around the defect. It is also common to refer to the topological strength of a defect as a topological charge. This highlights the fact that the interactions between defects can be either repulsive or attractive. As for electrical charges, topological charges of opposite sign attract, while those of the same sign repel. Indeed, if a system contains multiple topological charges of opposite sign such that their total sum is zero, then the attractive interactions between pairs of oppositely topologically charged defects will eventually cause the defects to approach and annihilate [7]. This is exactly the case for flat liquid crystal cells and the annihilation of opposite topological charges has been well studied for this geometry [8–10]. In contrast, if the total sum of topological charges is nonzero, then the repulsive interaction forces between pairs of like charges leads to a separation of the defects such that the distance between them is maximized [11]. Even if there are initially multiple defects with opposite signs, they cannot all be annihilated if there is a total nonzero charge.

In their seminal theoretical study of ordered liquids on the two-dimensional surface of a sphere, Lubensky and Prost [6] showed that a nematic shell with planar degenerate anchoring should contain four $s = 1/2$ defects. A reasonable question is why four $s = 1/2$ defects are formed rather than two higher charge $s = 1$ boojum-like defects at opposite poles on the sphere (see Fig. 20.1). Both configurations are allowed topologically for a spherical surface and both have the same total defect charge; alternatives such as one $s = 1$ and two $s = 1/2$

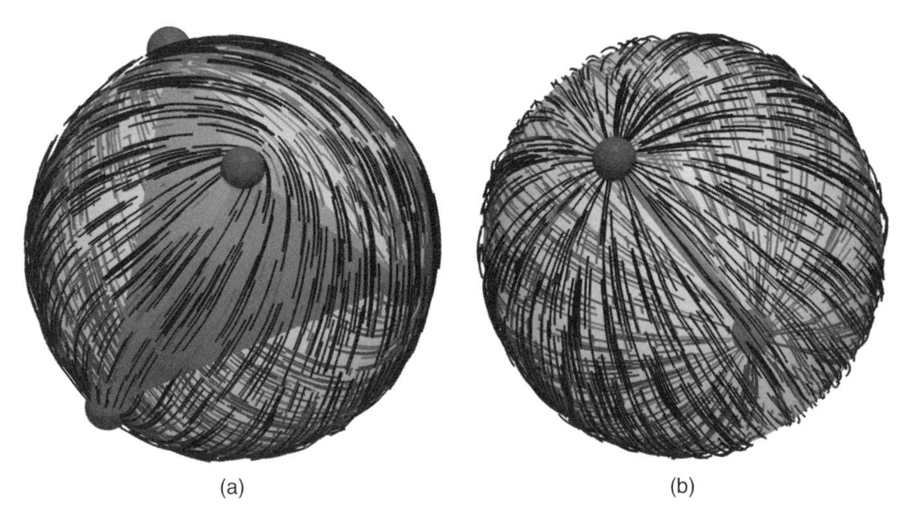

Figure 20.1 Examples of nematic ordering on a spherical shell. The local director is shown by the black streamlines and defects are shown by the small spheres. (a) Four half ($s = 1/2$) defects in a tetrahedral arrangement. (b) Two full ($s = 1$) defects at the poles of a sphere.

defects are also possible. The reason for the stability of the four $s = 1/2$ defects is because the energy of an individual defect is proportional to the square of its strength. Therefore, two $s = 1$ defects are of higher energy and therefore less stable than four $s = 1/2$ defects. This also indicates why topological charges of opposite sign tend to attract and annihilate each other; the energy of the system is minimized by the removal of defects.

Within a one elastic constant assumption, that is, under the assumption that the splay and bend elastic constants [1] (note that the twist elastic constant can be ignored for planar, 2D systems) of the liquid crystal are equal, then these four defects should uniformly repel each other. Given four equally repulsive defects, the defects in the ground state of a 2D nematic shell should be equally spaced on the spherical surface, such that their positions form the vertices of a tetrahedron inside the sphere (see Fig. 20.1a). Thus, a 2D shell formed from a nematic with equivalent splay and bend elastic constants should provide a route to form the tetrahedral geometry necessary to form colloidal atoms with sp_3-like bonding sites [2].

Although possible in theory and simulation, in reality the fabrication of a truly 2D system is difficult and so it is necessary to understand how the defect configuration changes when the surface layer has a finite thickness. It makes sense here to define two limits, thin shells and thick shells. The difference between these comes from the behavior of the nematic director inside the shell. We define a thin shell as one in which the local director defining the nematic orientation inside the shell remains perpendicular to a radial vector (the vector joining the center of the spherical volume to the location of the nematic volume element, which is equivalent to the surface normal at the surface of the shell) at all points inside the shell, except at the defects where the director is undefined. In this sense, thin shells can be regarded as close to being two dimensional. A thick shell is one in which this is no longer true and the director need not remain perpendicular to the radial vector throughout the shell volume; however, the director must be perpendicular to the surface normal at both the inner and outer surfaces.

As the thickness is now finite, the defect points predicted for an infinitely thin surface layer will turn into disclination lines that span from the inner surface of the thin shell to the outer surface. However, as for the infinitely thin layer, the director rotates by π on any closed path inside the nematic layer that encircles the disclination, as shown in Figure 20.2a. Thus, finite thin shells should have similar defect configurations to infinitely thin shells. In contrast, for thick shells the local director can be tilted with respect to the surface normal. This clearly implies that thick shells cannot be viewed in the same way as two-dimensional planar shells. The existence of the third dimension allows the director to escape by turning out of plane and so "escaped" defects can occur, as shown in Figure 20.2b. Vitelli and Nelson [12] have shown that for thick nematic shells this escape mechanism leads to two $s = 1$ defects. As the repulsion here is between just two like charges, these are expected to locate at the poles of a sphere. Thus, theory predicts two types of colloidal atoms: tetrahedral four-valent and linear two-valent.

20.3 EXPERIMENTS ON SPHERICAL SHELLS

20.3.1 Nematics

In principle, nematic shells could be realized experimentally through a number of different types of systems. For example, a solid colloidal object could be coated in a nematic layer and suspended in a solvent. In this case, the shell would have its interior surface imposed on it by the shape of the interior particle. Alternatively, the interior could be liquid

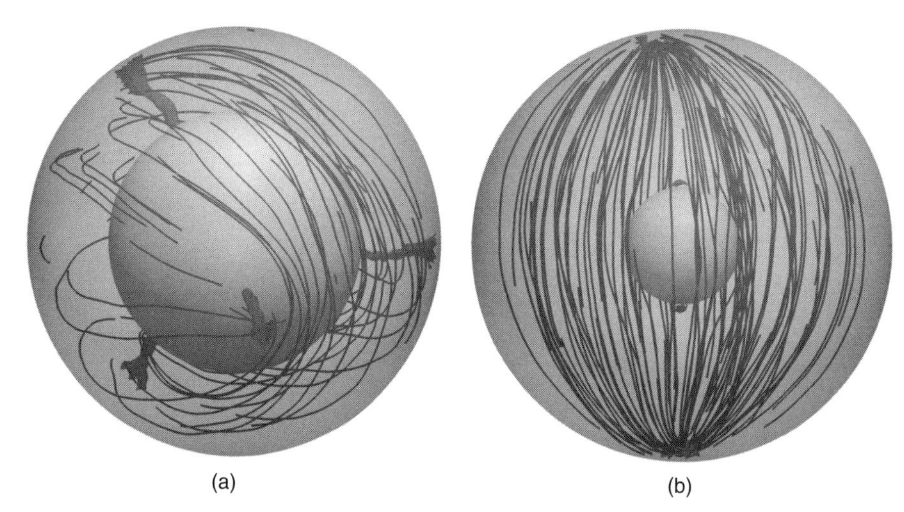

Figure 20.2 "Thin" and "thick" nematic shells. (a) In the thin shell, the director is parallel to the surface (or perpendicular to the surface normal) at all points in the shell. This leads to four $s = 1/2$ defects in the shell. Note here that however a disclination line is encircled, the director rotates by 180°. (b) In the thick shell, the director is able to escape into the third dimension and is no longer parallel to the surface at all points. Note that it escapes along the line joining the poles of the sphere. The defects in this case are not disclination lines but point defects on the inner and outer surfaces of the shell.

(as discussed below). The outer surface when suspended in a solvent is likely to be spherical in the same way that an air bubble in water is spherical, since the energy of the system will be typically minimized when the surface area for the given volume is minimal. This clearly depends on a number of variables, including the thickness of the shell, shape of the inner particle (especially if nonspherical), nature of the interactions between the nematic liquid crystal and the exterior solvent and hence surface tension at the outer surface. However, given that the exterior solvent and the nematic are necessarily immiscible, we can assume that the surface tension will be relatively large and thus the system will try to minimize the surface area leading to a spherical shape so long as the interior particle is not strongly anisometric.

The most widely studied nematic shells so far have not been based on coated solid particles but rather double emulsion droplets formed through microfluidics. A double emulsion is simply a smaller droplet inside a larger one. As already noted, the outer surface of the shell will tend to be spherical, minimizing the surface area. The same argument holds for the inner surface too and thus we expect a sphere inside a sphere geometry for the nematic shell. Most experimental studies so far have concentrated on nematic shells composed of a droplet of water inside a nematic droplet, which itself is suspended in water. Clearly, this is not the only way shells can be formed. For example, the inner droplet need not be chemically equivalent to the outer solvent; so long as the liquid crystal is immiscible in both the inner and outer solvent, the nematic shell should be stable. Indeed, some interesting experiments in which the anchoring at the outer surface is changed during the experiment can be performed by modifying the chemical nature of the exterior solvent once the double emulsion droplet has been formed [13, 14].

The first experiments on nematic shells were performed by Weitz, Fernandez-Nieves, and Lopez-Leon [15–18], described in a series of papers in which the methodology of the production of shells using microfluidics was developed and enhanced. It turns out that the theoretically predicted tetrahedral configuration is not so easy to achieve for two reasons. The first is due to the buoyancy of the inner droplet inside the shell. As noted earlier, we expect the inner and outer boundaries to be spherical. However, this does not guarantee that the nematic shell is homogeneous; this is only true if both spheres are concentric. If the density of the shell and the droplet are not exactly the same, then the inner droplet will float or sink inside the shell depending on the relatively densities, distorting the shell leading to an inhomogeneous thickness. Thus, if the inner solvent is less dense than the nematic phase, then we can imagine that it will tend to float in the nematic and hence narrow the shell at the top [15]. The nonuniform thickness of the shell can also be stabilized in another way, which has its origins in the shells not being truly 2D. Since the shells have finite thickness, the defects are not point defects but disclination lines and the energy of these disclination lines depend not only on the square of their strength but also on the length of the disclination line. In other words, the energy of a defect depends on the thickness of the shell where it resides. If the thickness is not constant, then the defects will tend to congregate in a thinner region of the shell in order to minimize the energy. While the defects repel each other due to the elastic energy associated with deforming the director inside the shell, they will also tend to be attracted to the same region of space.

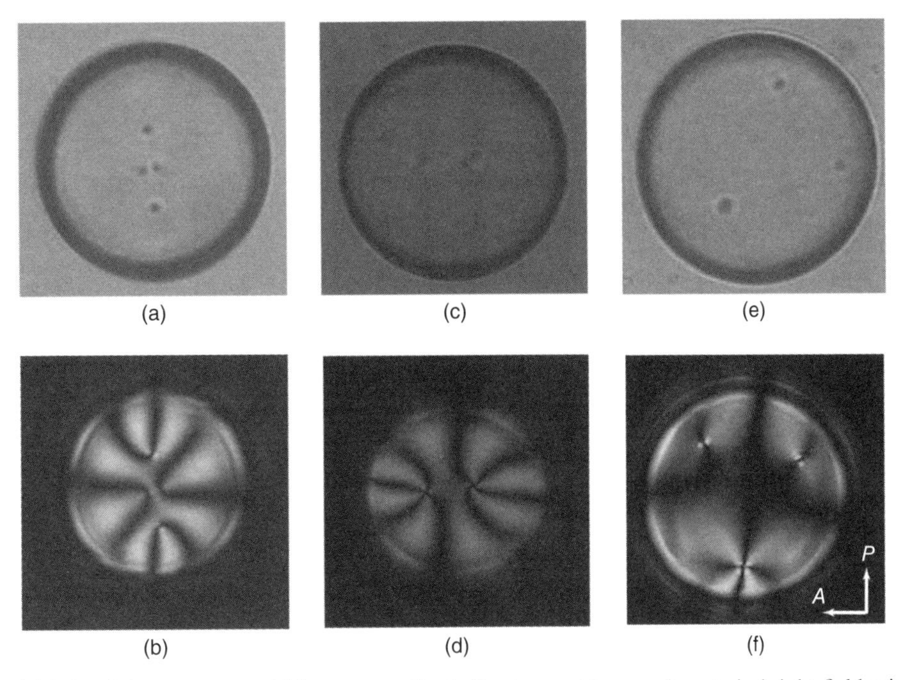

Figure 20.3 Types of thin but inhomogeneous thickness nematic shells observed in experiments in bright field microscopy in which the defects appear as dark regions and under crossed polarizers in which defects appear as points at which the dark brushes meet. The latter allow the strength of the defect to be determined; two brushes meeting at a point correspond to a $s = {}^1\!/_2$ defect and four brushes correspond to a $s = 1$ defect. (a, b) Four $s = {}^1\!/_2$ defects, (c, d) two $s = 1$ defects and (e, f) one $s = 1$ and two $s = {}^1\!/_2$ defects. Source: Fernandez-Nieves [15]. Reproduced with permission of American Physical Society.

Typical defects are shown in Figure 20.3 [15]. Note that numerous arrangements can be achieved for two, three, and four defects depending on the relative thickness of the shell.

An experimental approach to tend toward the true 2D case has shown that this inhomogeneity can be overcome. If the internal droplet is swollen by an osmotically driven flow from the outer solvent phase, then the shell can become relatively thin and approach this limit, appearing to overcome the inhomogeneity. In this case, the tetrahedral geometry of the four $s = {}^1\!/_2$ defects has been observed [18] (see Fig. 20.4a). Similarly, shells with two $s = 1$ defects can similarly be expanded such that the two defects occur on opposite sides of the sphere (Fig. 20.4c) and shells with three defects (two $s = {}^1\!/_2$ defects and one $s = 1$ defect, fulfilling the overall $s = 2$ surface charge) form an isosceles triangle arrangement (Fig. 20.4e). Note here that on swelling the inner water droplet, the defect types do not change. Even though the shell itself tends toward being infinitely thin, the defect types are essentially frozen and do not transform to the four $s = {}^1\!/_2$ defects we might expect to observe for an infinitely thin shell [3]. Thus, there is hysteresis in the defect types when the thickness of the shell is changed. This illustrates that the different defect configurations observed are local minima in the elastic energy and there is presumably a significant energy barrier to overcome for each of the $s = 1$ defects to split into the expected $s = {}^1\!/_2$ defects for infinitely thin shells.

20.3.2 Smectics

Although the majority of experimental work on liquid crystal shells performed so far has been based on the simplest liquid crystal phase, namely the nematic, two recent studies have investigated the behavior of spherical shells in which the enclosed phase is the smectic A [19, 20]. The smectic A phase differs from the nematic in that it has quasi-long-range order in one dimension; the molecules are layered along the director. As the cross-sectional area of the shell increases from the inner to the outer surfaces, combined with the lack of a constant orientation for the director, then frustration must necessarily occur as the layers try to pack in the shell. This leads to buckling instabilities, where the director twists through the shell [20]. Another interesting result of this layering is not so much due to the layered structure, but the divergence of the bend-to-splay elastic constant ratio as the smectic A phase is approached within the nematic phase. As we shall see in Section 20.4, some of the first simulations of nematic shells (predating these experiments) were performed for hard-rod models in which the bend-to-splay elastic constant ratio is large. The simulations for the model systems indicated that in the limit of a large elastic constant ratio, the defects should not form a tetrahedral geometry but rather form on a great circle. Studies of spherical shells for mesogens with a smectic A phase have allowed experiments to test this prediction.

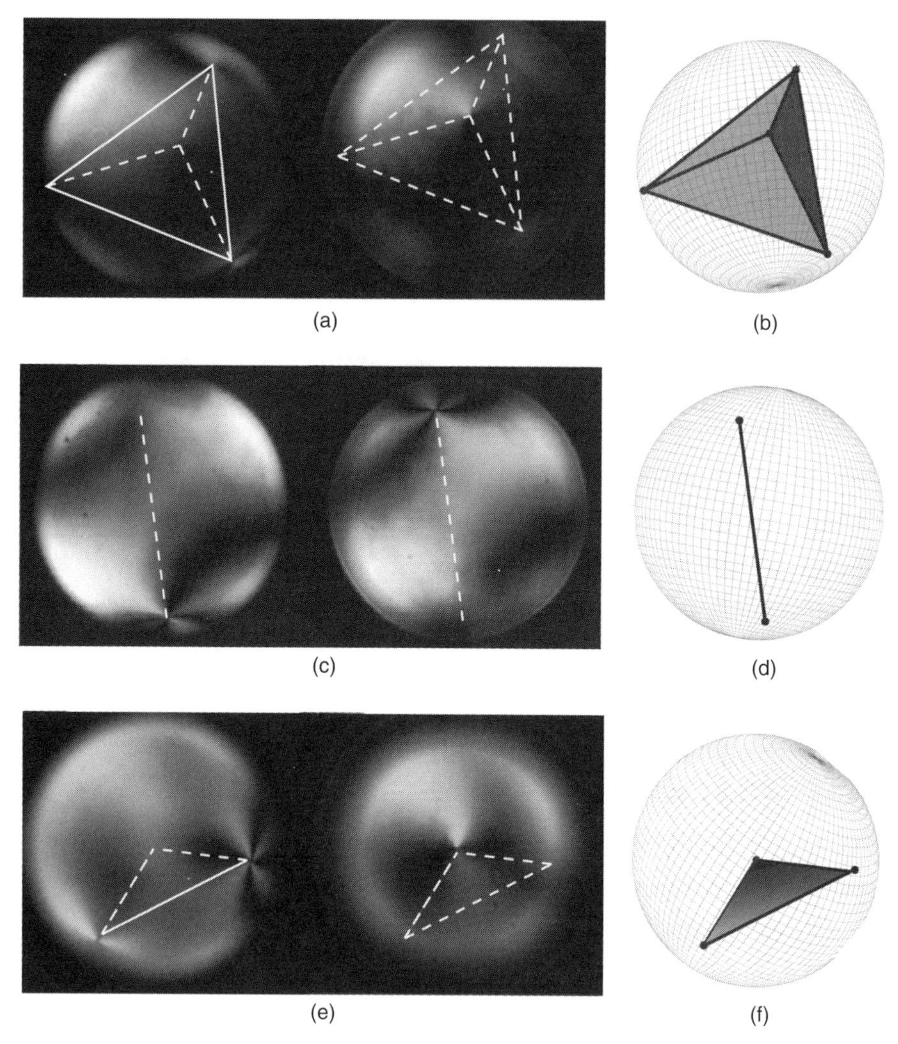

Figure 20.4 Defect configurations in homogeneous thickness osmotically swelled nematic shells. (a) Shell viewed under cross polarizers, with different focus planes through the double emulsion droplet for a shell with four $s = 1/2$ defects, (b) schematic of the defect locations, similar images for (c, d) two $s = 1$ defects and (e, f) one $s = 1$ and two $s = 1/2$ defects. Source: Adapted from Lopez-Leon [18]. Reproduced with permission of Nature Publishing Group.

From a simulation point of view, it is pleasing to note that two series of independent parallel experiments have indeed confirmed that the defect structure starts to deviate from the tetrahedral geometry as the nematic – smectic A transition is approached and confirms the great circle defect configuration observed for nematic phases with a high bend-to-splay elastic constant ratio.

20.4 COMPUTER SIMULATIONS – PRACTICALITIES

Here we discuss some practicalities of using simulations to investigate nematic shells. Typical simulation results are left until Section 20.5.

20.4.1 Introduction

Computer simulation has become a useful technique to model physical systems in many areas of science. If the individual behavior of, or the interactions between, all the components of a system are specified, then a simulation can be used to understand the behavior of the system as a whole by generating representative scenarios based on the values of any imposed system variables. Thus, by comparing simulations with known real-life scenarios, it is possible to test the realism of the behavioral or interaction model and, therefore, extend our knowledge of how the components of the system interact. Alternatively, when the model has been shown to be realistic, simulations can be used as a method for prediction of properties for real-life systems under a given set of circumstances.

Simulation methods for complex liquid systems have been used for many years to understand a range of physical phenomena in physics and chemistry [21, 22]. At one extreme, large systems of small molecules or individual large molecules can be built using an atom building approach familiar to chemists, using an atomic force field such as AMBER [23]. However, for bulk systems of larger molecules, such as liquid crystals and polymers, the models used tend not to consider every atom in the system. Instead, for simplicity and computation tractability, the individual molecules are often modeled as a single, complete unit rather than a collection of individual atoms. This greatly reduces the number of interacting species in a simulation and thus proves more efficient computationally; indeed, simulations of long-range phenomena over distances significantly larger than the length of the molecule in question are rarely tractable if atomic-based models are used. The two common techniques used to investigate bulk systems of complex liquids at finite temperature are Monte Carlo (MC) and molecular dynamics (MD); the former is a stochastic approach, while the latter solves Newton's equations of motion to follow the time dynamics of the system [21, 22]. As we shall see, to simulate nematic shells, we will require the use of external constraints in order to fix the surface anchoring, that is, the orientations of the molecules at the inner and outer surfaces. In principle, these constraints could be applied to the equations of motion in the MD technique. However, the constraints are more straightforward to apply in an MC simulation and thus most simulations of nematic shells have been through MC simulation. Hence, we shall give only a brief introduction to the MC technique.

20.4.2 Monte Carlo Simulations

Monte Carlo (as applied to the simulation of liquids [21, 22]) is an important sampling technique, the idea of which is to sample configurations for the system of interest in accordance with the appropriate Boltzmann distribution. MC simulations use a Markov chain procedure to move from one state to the next in a random and memory-less way; the new trial state depends only on the current state and not on any of the previous states that preceded it in the simulation. The most frequently used procedure for canonical (constant density) simulations of bulk systems is based on that devised by Metropolis *et al.* [24]. In this method, a particle (atom or molecule) is picked at random and given a random displacement or change in orientation or both as appropriate for the model being investigated, the maximum displacement(s) being an adjustable parameter that balances the ratio of accepting too many small moves and rejecting too many large ones. The decision whether to accept or reject this trial state depends on the relative Boltzmann probabilities of the old (ρ_O) and new (ρ_N) states and thus on the energies of the old (U_O) and new (U_N) states. There are two cases to consider. If the energy has gone down, such that the change in energy $\Delta U = U_N - U_O \leq 0$, then the new state is more probable than the old one and the trial move is accepted. If the energy has gone up, then the new state is less probable than the old one. However, this does not mean that the new state should not be visited. The move has to be has accepted with a reduced probability equal to the ratio of the relative Boltzmann probabilities ρ_N/ρ_O. This ratio can be expressed simply as the Boltzmann factor for the energy difference of the two states, $\rho_N/\rho_O = \exp(-\Delta U/kT)$. The decision to accept or reject the move takes place using a random number between 0 and 1. If the random number is less than the ratio, the move is accepted. Otherwise, the trial is rejected and the old state is recounted as a new state in the Markov chain. This is repeated many, many times until the system has reaches equilibrium and statics on whatever properties are of interest can be determined. It is worth noting that this method applies not just to particles that interact via continuous potentials but also to hard body systems. The difference is that the overlap of two hard bodies instantly leads to the trial being rejected, since the energy is infinite. In contrast, if the trial results in no overlapping particles, then it is immediately accepted since the energy difference is zero; all states with no overlaps between the particles have equal Boltzmann factor and are therefore equally probable. The reason that only one molecule or particle is moved at a time is as follows. If we assume that the probability of accepting a trial if we move just one molecule is ½, then the probability of accepting a move involving two molecules will be of the order of $(½)^2$, and so on. Thus, there is a very small chance of accepting a trial involving many molecules at once.

Infinitely Thin Shells Infinitely thin shells are as close to two-dimensional systems as possible. We physically define an infinitely thin nematic shell as a collection of elongated particles arranged on the surface of a sphere such that the center of mass of each particle is a fixed distance from the center of the sphere. The orientation of each particle is tangential to the surface; that is, the vector defining the orientation is fixed perpendicular to the radial vector joining the center of the sphere and the center of mass of the particle. Each particle has rotational freedom in two dimensions in the tangent plane, but the orientation is not free to deviate from this plane. Thus, one type of trial move immediately suggests itself as a rotation in this plane. A second type of trial translates particles on the surface, subject to the orientation being amended such that it remains tangential in the new position. For such a move, some care has to be taken in order to ensure that the choice of the new orientation is not unduly biased. One such method is to translate the particle either within the plane defined by the radial vector and the orientation vector or within a similar plane defined by the radial vector and the vector in the tangent plane that is perpendicular to the orientation vector. These can be considered as moves along the length of the molecule and sideways. These are particularly convenient for highly elongated molecules, since different maximum displacements along these two directions can be applied to sample configuration space more effectively. Trials can consider one type of move (rotation or translation) or a combination of both in either order; translations can be single translations along one direction or combination translations both along the length and sideways on the surface. A third possibility is that particles are just randomly inserted in a random position and orientation (subject to the tangent plane constraint) anywhere on the sphere. However, at the relatively high densities required to form a nematic phase on the surface, these trials are likely to lead to at least one overlap and therefore rejection of the trial. Infinitely thin shells have been investigated for thin hard rods [25, 26], and are discussed in Section 20.5.

Finite Thickness Shells There is nothing, in principle, stopping the simulation of nonlattice models for finite thickness shells. As noted earlier, one of the key reasons for choosing a particular model is often the computational cost and whether or not the problem is computationally tractable for a given model. Simulations of bulk (3D) volumes of liquid crystal systems to determine their phase behavior are typically done on systems of the order of about 4 times the molecular length. Larger systems are of course possible, but doubling the system size in each dimension leads to approximately 8 times the computer power needed for a simulation of the same length. While it is possible to study an infinitely thin surface for simple models as already discussed, it is clearly unfeasible to extrapolate these to finite thickness shells in which the thickness is significantly larger than the length of the rod, let alone orders of magnitude larger. Thus, instead of using off-lattice molecular models, to investigate the effect of thickness for finite shells, a simpler lattice model is often more appropriate. The simplest of all the lattice models for liquid crystals is that devised by Lebwohl and Lasher [27, 28]. This is a simple spin model in which headless rotors are placed on the lattice points of a cubic lattice and interact with their neighbors via the potential $-\varepsilon_{ij}P_2(\boldsymbol{u}_i \cdot \boldsymbol{u}_j)$, where $\boldsymbol{u}_i$ is the orientation of the rotor i, the energy parameter $\varepsilon_{ij} = \varepsilon\ (> 0)$ for nearest neighbors only and zero otherwise and $P_2(x) = (3x^2 - 1)/2$ is the second Legendre function. This choice of potential ensures that the parallel alignment of neighboring spins is favored and thus leads to an orientationally ordered nematic phase as the temperature is lowered past a critical value [28]. Note that as the spins are headless, parallel and antiparallel are the same thing; the high energy state for a pair of spins is perpendicular. Although the rotors could clearly be used to describe molecules, more typically we consider them to represent small volumes of liquid crystal in which the direction of the rotor represents the local director. Although these models have the correct physics, that is, the right form for the interaction potential between neighboring fluid units to produce nematic ordering, this does introduce a lengthscale problem. Namely, since we do not define the size of a fluid element, we cannot say without unambiguity what the lengthscale of any simulation is. However, the reason for using such a model is that simulations using lattice models are very fast in comparison to off-lattice systems and so significantly larger systems can be studied including, of interest to us here, finite thickness nematic shells.

A nematic shell can be defined as the lattice points in the volume between two spherical surfaces; if the bounding spheres are concentric, then a uniform thickness shell will be formed; otherwise, a nonuniform thickness shell will remain. Planar anchoring can be defined by using an additional layer of radially aligned ghost sites with a suitable energy coupling to ensure molecules have tendency to align planar at the surface, that is, perpendicular to the ghost sites [29].

Simulations of lattice systems proceed in a slightly different way to off-lattice systems. Since the particles are at fixed lattice sites, only the orientations of the particles are varied. To generate a trial state, a single particle is selected at random and its orientation is changed. The energies in the old and new states are determined and the Metropolis algorithm applied to decide whether to accept or reject the trial state. Note that ghost particles defining the anchoring should not be rotated at all during the simulations; these have fixed orientations depending on their locations and are only present to determine the anchoring energy for the particles inside the nematic region. It is also worth noting that generating completely random orientations for the trials can work quite well in lattice simulations, especially at high temperatures just below the NI transition and where regions of low order are expected. Since there is no physical volumetric shape, there is no problem with rejections due to overlaps. The trials can, of course, still be rejected if the energy has increased and rejection is more likely than for a small rotation (especially in an orientationally ordered region of space), but since the calculation of the energy is very quick for lattice models, then it may be more optimal to try large rotations (via completely new orientations) and rejecting 90% of those trials rather than solely using relatively small rotations of a few degrees and accepting, for example, roughly 50% of those. In particular, since we know that defects must occur in nematic shells, then we know that there will be volumes of low order, where such moves can possibly explore phase space around the defects more effectively. A compromise can also be used, in which a type of trial can be used a certain percentage of the time; for example, completely random orientations could be used for 50% (on average) of the trials and rotations based on the orientation of the particle of interest in its old state for the other 50%. Simulations of lattice-based models have been used for a number of shell geometries, including shells defined by concentric spheres, nonconcentric spheres, and ellipsoidal volumes. So long as the confining volume of the liquid crystal region can be defined mathematically and the tangent plane identified for the anchoring via ghost particles at the boundary, then this class of model is relatively straightforward to apply. Some results are discussed in Section 20.5.

20.5 COMPUTER SIMULATIONS OF NEMATIC SHELLS

In this section, we consider the results of computer simulations of nematic shells and show how these results can both complement and contrast the existing theoretical and experimental evidence, as discussed in Sections 20.2 and 20.3, and also be used to predict defect topologies and director configurations for novel systems.

20.5.1 Spherical Shells

Computer simulations have been used to investigate defect formation in spherical nematic liquid crystal shells for two distinct types of model, which we shall discuss in turn. The first is the study of infinitely thin shells, using freely translating (nonlattice) models. The second is the study of finite thickness shells, using lattice-based models.

Infinitely Thin Shells The behavior of elongated rods on the (infinitely thin) surface of a sphere has been investigated in two independent series of simulations [25, 26]. Both simulations investigate long hard rods, which interact solely through excluded volume interactions. The rods are located on the surface of a sphere of fixed diameter and the orientations of the rods taken to be tangential to the spherical surface. Starting with disordered systems, the surface coverage (the two-dimensional density) was increased, using grand canonical simulations [25] or by adiabatic compression of the sphere surface [26], such that the coverage was above that where a two-dimensional nematic phase is expected to occur. The relative locations of the defects formed were then investigated. Both simulations gave a similar, if somewhat unexpected, result. Although four $s = 1/2$ defects were observed in the nematic shell as expected from previous theoretical works, their locations were far from the expected tetrahedral result predicted by theory for infinitely thin shells. Instead, they locate on a great circle, or circumference of the sphere, as shown in Figure 20.5. This behavior appears not to be related to the diameter of the sphere. A range of shell diameters from $200D_R \approx 8L_R$ up to $800D_R \approx 32L_R$, where D_R and L_R are the diameter and length of the rod, respectively, were used for rods of length $L_R = 24\,D_R$ in Ref. [25]. Such large shell diameters ensure that the ends of the rods do not pass underneath each other due to the surface curvature and so, locally, the systems are essentially flat. Studies of a shorter rod ($L_R = 15D_R$) on a relatively smaller shell (diameter $75D_R \approx 6L_R$) studied in Ref. [26] resulted in the same behavior. The formation of four defects on a great circle also appears not to be influenced by the coverage density. Simulations at just above, and at 50% larger than, the coverage at which the NI surface transition occurs for a true flat 2D system both show the same behavior [25].

The formation of the defects in a great circle rather than the tetrahedral arrangement has its origins in the elastic constants of the nematic phase of the model investigated. Since the bend elastic constant is significantly larger than the splay elastic constant for long hard rods [30], it costs less energy to splay rather than bend the director. Thus, from an elastic point of view, the system would prefer to be splay-like and so the configuration becomes splay dominated. Note that this does not mean the full splay-like $s = 1$ defects are preferred. Energetically it is still favorable to form the four half-strength $s = 1/2$ defects rather than two full-strength $s = 1$ defects. The repulsion between the defects depends on the orientation of the director field with respect to the locations of the defects. Since the director field will prefer to adopt a splay-like configuration rather than bend, all directions are not equal as assumed in the one elastic constant approach often followed in theoretical models. The result observed in the simulations is that the sphere is essentially split into two hemispheres, in each of which the director field is splay-like. Where the hemispheres join, the director field follows the great circle, as shown in Figure 20.5. This director arrangement means that the hemispheres may be rotated with respect to each other without significantly deforming the director configuration of each hemisphere and thus without change in elastic energy. This has hence been dubbed the "cut and rotate" splay geometry. The interaction between defects on different hemispheres appears to be very weak. The angular distribution for the rotation of one hemisphere against the other is remarkably flat [25], until the pairs are brought so close together that they would form a single $s = 1$ defect, at which point they repel and, in some cases, deviate from the great circle geometry. It is also worth pointing out that if the simulations are started from a perfect splay state configuration in which there are two $s = 1$ defects with perfect nematic order elsewhere, then the defects rapidly break into four $s = 1/2$ defects and eventually give a similar distribution of configurations similar to those started with a disordered surface indicating the stability of the four defect state over the two defect state.

A later theoretical study of the Onsager model of infinitely thin rods confined to the surface of a sphere does not agree with these simulation results [31]. Although the theory correctly predicts splay-like behavior, it predicts that the cut and rotate geometry observed in both series of simulations is not stable with respect to the splay state with two $s = 1$ boojum-type defects. The reason for this is not clear; although the theory uses geodesic rods (curved arcs) rather than straight lines, this should not make any

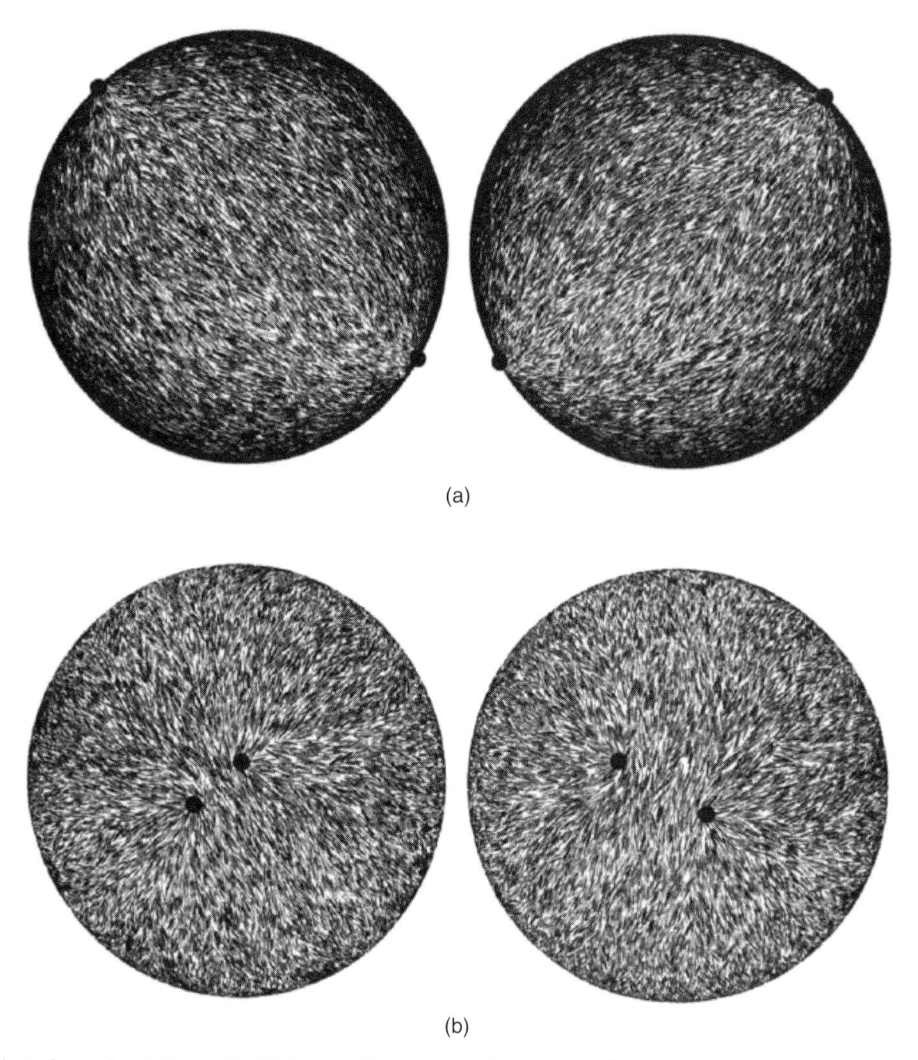

Figure 20.5 Snapshots from simulations of a 2D layer of long hard rods on the surface of a sphere for two common configurations. For clarity, in each case the sphere is cut through its center perpendicular to the viewing direction, and the top and bottom hemispheres shown separately from the same viewing point. The small black spheres indicate the positions of the defects. (a) Four $s = 1/2$ defects locate around a circumference of the sphere, here roughly equally spaced around the circumference. (b) The director configurations of the two hemispheres rotate such that the defects are located close to each other; note in this case, the pairs of defects do not coalesce into $s = 1$ defects, but still repel. Source: Bates [25]. Reproduced with permission of AIP Publishing LLC.

significant difference given the low curvature of the larger systems studied in the simulations.

A more recent simulation study [32] focuses on the effect of elastic constants by using infinitely thin shells of rods composed of a line of soft repulsive spheres. In this type of model, the elastic constants can be continuously varied by modifying the length, coverage, and temperature (or, conversely, softness). For the series of models studied, the defect configuration in the ground state is found to vary from the tetrahedral configuration in the limit that the elastic constants are equal through to the cut and rotate splay geometry when the elastic constants start to differ significantly.

Finite Thickness Shells Skacej and Zannoni [29] have used the Lebwohl–Lasher model to investigate the defects formed in uniform nematic shells, running the simulations at just below the nematic–isotropic transition temperature. Using an elegant algorithm by Callan-Jones *et al.* [33], the local director and a relevant order parameter can be determined to visualize both the director field and the locations of the defects. Due to the simplicity of the interaction potential in the Lebwohl–Lasher model, the elastic constants are equal. Not surprisingly, this leads to four defects being observed in a tetrahedral arrangement for a finite thickness nematic shell. The equal repulsion between all defects maximizes the distances between all unique pairs, resulting in the theoretically predicted tetrahedral arrangement, as shown in Figure 20.2a. Simulations also show how when the thickness of the shell is increased, then the four $s = 1/2$

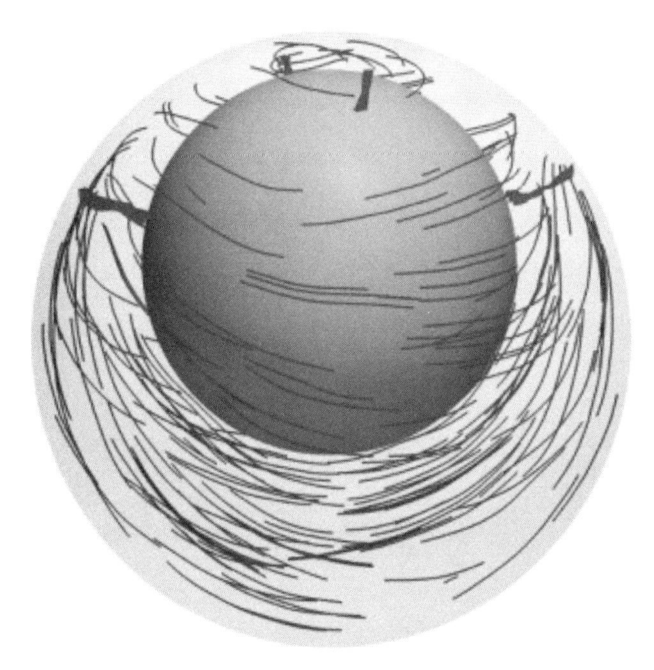

Figure 20.6 Simulation of a nonconcentric, inhomogeneous thickness nematic shell, indicating the relocation of disclination lines from the tetrahedral geometry toward the thinner (top) region of the shell. Source: From Ref. [14].

defect arrangement is no longer stable, replaced instead by two $s = 1$ defects, as shown in Figure 20.2b. One of the advantages of a simulation is that simple extensions can be performed with little effort. For example, the application of a homogeneous field to the nematic shells can be applied to observe what might be expected in a real system under such conditions. In this case, as the field strength is increased, the defects tend toward the top and bottom of the nematic shell, such that the liquid crystal around the waist of the shell aligns parallel to the applied field. The application of a field, therefore, gives one method to remove the tetrahedral symmetry of the ground state. On further increasing the field strength, the defects can be forced to coalesce into two $s = 1$ defects, similar to those for thicker shells [29].

Simulations can also be performed using nonconcentric shells to investigate the location of the disclination lines. By shifting the inner sphere, the simulations can be used to mimic the nonsymmetric cases studied experimentally [15]. Such simulations [14] clearly show that under these conditions, the defects tend toward the thinner regions of the shell minimizing the length of the disclination lines, as shown in Figure 20.6.

20.5.2 Nonspherical Shells

Since most experiments so far have been performed using double emulsion droplets (i.e., liquid-in-liquid-in-liquid droplets), the surfaces have tended to be spherical. The large interfacial tension between immiscible components tends to minimize the surface area between them, thus leading to spherical surfaces. However, if a rigid interior particle or exterior containing volume is used, then it would be possible to have nonspherical shells. Here we examine how the shape might influence the defect configurations in their ground state.

Elongated Infinitely Thin Shells We start with a simple extension of the infinitely thin sphere, by extending the surface to a capsule shape. This is defined as a cylinder, capped by two hemispheres of the same radius. Simulations using long hard rods confined to such a surface have been performed as a function of the cylinder length [34]. We recall that simulations on the surface of a sphere for long hard rods lead to the defects occurring not in the tetrahedral arrangement as predicted by theory using a one-elastic constant approach, but rather around a circumference of the sphere in the cut and rotate splay geometry. Thus, we might expect, and indeed observe, a similar effect for nematic shells on elongated particles when the ratio of the bend to splay elastic constants is large. Defects in nematic shells on the surfaces of elongated capsules are found to be located around the circumference of the hemispherical end-caps, where the hemispheres join the cylinders (see Fig. 20.7). This appears to be largely independent of the length of the cylindrical section. So long as the cylinder is long enough, one pair of defects locates on the common circumference at each end and do not interact with the pair at the other end of the capsule. At each end the repulsions appear to be quite strongly correlated, with the defects forming pairs on opposite sides of the hemisphere. The configurations are remarkably similar to the spherical case, in that the director field is essentially splay-like over the hemispherical cap, which then joins to a uniform director field that encircles the cylindrical section as shown in Figure 20.7. It is clear that the defects tend toward the region of maximum Gaussian curvature; since the hemisphere regions curve in two dimensions (positive Gaussian curvature) and the cylinder region in just one (zero Gaussian curvature), the most curved regions are the hemispheres. The two defects on each hemisphere naturally repel each other in a similar way as if they were an interacting pair on a sphere. They are very rarely located away from the circumference of the hemisphere, since this would lead to a reduction in splay-like character and an increase in bend-like character in the director field and a subsequent increase in elastic energy.

Finite Thickness Ellipsoidal Shells Just as the infinitely thin nematic coating on a sphere can be extended to elongated (and other shaped) particles, so the finite thickness systems can be extended to other shapes. So long as the surface normal can be defined, then any shape imaginable can be simulated. However, it makes sense to focus on

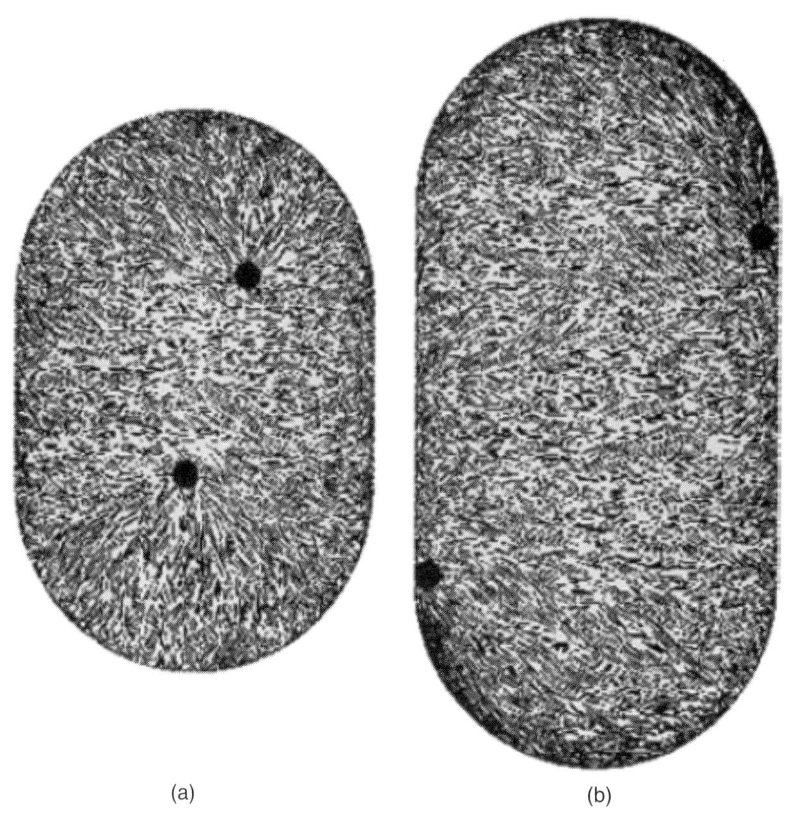

Figure 20.7 Snapshots from simulations of a 2D layer of long hard rods on the surface of capsule-shaped particles. Note that the defects tend to locate on the spherical region of the shell, rather than on the cylindrical region; thus, they tend to be found in the region with larger Gaussian curvature. Here, since the ratio of bend to splay elastic constants is large, the hemispheres tend to form splay-like configurations and thus the defects occur on the circumference of the hemispherical end-caps. Source: Bates [34]. Reproduced with permission of The Royal Society of Chemistry.

relatively simple shapes that could be made experimentally. For example, a polymer matrix such as that in a polymer-dispersed liquid crystal could be used to create the outer surface of the nematic shell. If the matrix is then stretched or compressed, then the spherical shell could be deformed into an elongated or flattened ellipsoid. Similarly, biaxial shapes could be achieved by unequal stretching in two dimensions.

The behavior of finite thickness ellipsoidal shells has been studied by Bates *et al.* [35]. As for spherical shells [29], the Lebwohl–Lasher model was used to study the ellipsoidal cases. Going from spherical to ellipsoidal shells introduces a range of new (shape-related) variables. For example, the shells could be finite and uniform thickness, the shells could be nonhomogeneous in thickness but have the same aspect ratio for the inner and outer surfaces, the inner surface could be spherical (as would be the case for a water or other immiscible fluid droplet in a nematic system) but the outer surface ellipsoidal, the outer surface could be spherical but the inner one ellipsoidal (mimicking a rigid particle inside a nematic droplet in water), one surface could be flattened and the other elongated, and so on.

We start with an elongated uniaxial ellipsoidal shell, with uniform thickness. The defects in the ground state are shown in Figure 20.8a. There are clearly four $s = 1/2$ defects, as have been observed for other thin shells. However, it is just as clear that they are located toward the ends of the elongated particle. This migration of the defects toward the ends of the particle is a similar result to that for the infinitely thin rods on the surface of an elongated capsule, that the defects tend to be located in areas of high Gaussian curvature but still repel as before such that they do not combine into an $s = 1$ defect. There again appears to be little, if any, correlation between the pairs at the opposite ends of the particle. A similar result should be expected for flattened ellipsoidal shells. However, since here the highest curvature region is around the waist of the particle, the defects tend toward the waist. Since all four defects tend toward the waist, each defect interacts strongly with the other three, resulting in four defects equally spaced around the circumference of the disk-shaped object (Fig. 20.8b and c). Biaxial ellipsoids also show a similar behavior with the migration of the defects toward areas of large curvature. In the case of extreme biaxiality, the defects are pinned to the highest curvature regions whereas for less

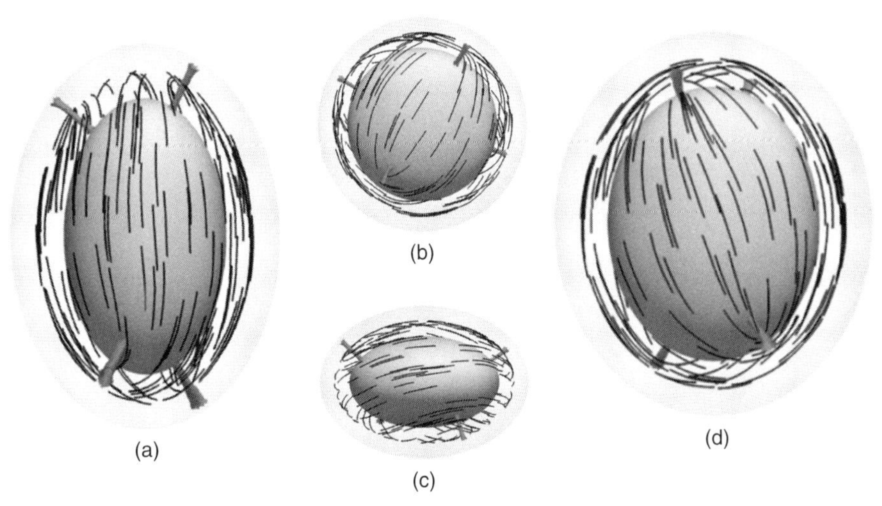

Figure 20.8 Snapshots from simulations of nonspherical colloidal particles covered in a homogeneous layer of nematic liquid crystal, using the Lebwohl–Lasher model. (a) Side view of a rod-shaped ellipsoidal particle, in which the defects tend toward the ends of the particles, (b) top and (c) side views of a disk-shaped particle, in which the defects tend toward the waist, and (d) side view of a biaxial particle that is intermediate between the two. Source: Bates [35]. Reproduced with permission of The Royal Society of Chemistry.

extreme ones, the elastic energy tends to make them repel leading to a configuration in between that of the elongated and flattened ellipsoids (Fig. 20.8d). This behavior has since been confirmed theoretically [36].

As noted earlier, it may be possible to trap an ellipsoidal particle inside a nematic droplet, which itself is suspended inside a solvent, leading to an ellipsoid inside a sphere. Similarly, a polymer matrix could be stretched to leave a spherical water droplet inside an ellipsoidal nematic volume. Thus, understanding the defects occurring in these shapes is useful. In such cases, it is important to remember that elastic energy of a defect is proportional to the length of the disclination line and since this depends on the thickness of the shell, we should expect the defects to occur in the thinner regions of the shell. Figure 20.9 shows the defect structures and director fields when rod-like, biaxial, and disk-like elongated particles are trapped at the center of spherical nematic volumes, and for spherical droplets trapped inside elongated, biaxial and flattened nematic volumes. It is not at all surprising that the defects tend to the narrowest regions of the shells, just as they do when two nonconcentric spherical shells are used to define the nematic volume. There is a particularly pleasing duality observed between these systems. For example, an elongated particle inside an outer spherical volume has defects that tend toward the thin regions at the top and bottom of the particle, one pair at the top, one at the bottom. Similarly, a spherical particle inside a flattened volume also has defects that tend toward the top and bottom. There is also a dual pair of the flattened particle inside a sphere and the sphere inside an elongated volume; in both these cases, the four defects tend toward the waist of the particle.

20.6 CONCLUSIONS

Liquid crystal shells are a relatively new area of research. Although they are clearly still in their infancy, a combination of experiment, theory, and simulation has progressed our understanding of these systems dramatically in only a few years. The fact that the idealized system of a pair of concentric spherical surfaces forming a homogeneous thickness shell is not necessarily the ground state due to the elastic energy, even if the fluids are density matched removing the buoyancy issue, is clearly an important discovery. This shows that the idea that the tetrahedral geometry for the disclinations is not necessarily stable depending on the size and shape of the shell, and thus that the subsequent fabrication of tetrahedral colloidal particles through defect functionalization is not as straightforward as first imagined. The importance of the (relative) elastic constants is also clearly important for understanding the types of defect structure that can be obtained for a particular system. Indeed, since this ratio can be temperature dependent, the arrangement of the disclination lines may also be temperature dependent. The idea that first appeared from simulation evidence that four defects form on a great circle, rather than in a tetrahedral geometry, has been confirmed in experiments where the liquid crystal phase is adapted such that the elastic constant ratio starts to diverge, showing how careful consideration of the material properties of the liquid crystal can be used to tailor the equilibrium defect configuration. The effect of external fields in simulations have shown how the defects can be shifted into a new configuration, or even changed from one type to another. Similarly, different shaped (solid) interior particles or containing outer surfaces

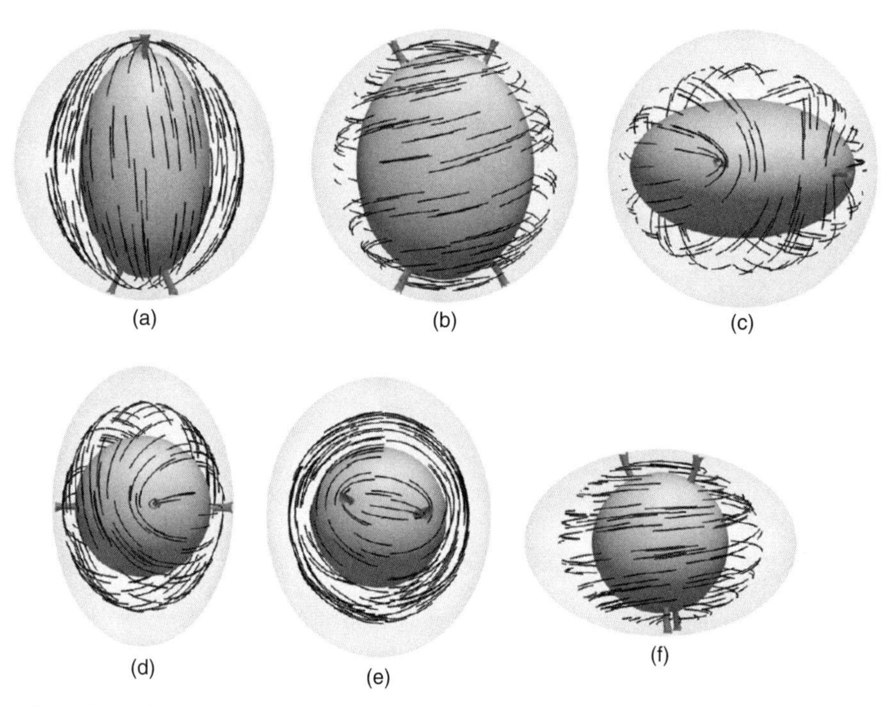

Figure 20.9 Snapshots from simulations of nonspherical particles in spherical cavities and spherical particles in nonspherical cavities. (a) Rod, (b) biaxial, and (c) disk-shaped particles in spherical cavities. Spherical particles in (d) rod, (e) biaxial, and (f) disk-shaped cavities. Note that the defects tending to be located in the thinnest regions of the shell leads to a duality between (a) and (f), (b) and (e), and (c) and (d). Source: Bates [35]. Reproduced with permission of The Royal Society of Chemistry.

can be used to force defects to adopt new configurations. These latter two ideas, explored by simulation and theory, still need to be explored by experiment.

There is still a lot that remains to be discovered. The underlying chemistry of the molecules used in the shells has remained essentially constant, using "standard" nematogens. No systematic studies of nematic mixtures to tailor their materials properties have been studied as yet. Similarly, mixtures of nematics with miscible but nonliquid crystalline dopants that could be used as for future functionalization have yet to be investigated. Other phases, such as the chiral nematic, have not been studied in shells. The fundamental physics investigations so far have been performed on relatively large (50 μm) particles, which are ideal for experiments in the laboratory, but the problem of reducing the size for real applications will need to be overcome for more practical uses. There are also still fundamental problems to be tackled and understood, in experiments, theory, and simulation. These include, for example, whether defects can be transformed from one type to another without the application of external fields or are the defect types fixed once stabilized; for example, we do not expect the $s = 1$ defect to have the same energy as a pair of $s = 1/2$ defects, and thus one of these arrangements should be energetically favored. However, if there is a large barrier between the states, then the energetically favored state may not be reached if the other type forms first. Thus, an important question is if the defects can be transformed from one type to the other, can the typical pathway for this transformation be determined and understood, in the hope of reducing any energy barriers that need to be overcome to favor the transition. Another important consideration is whether the most useful configuration, the tetrahedral geometry, can be stabilized even if the central droplet shifts from the exact center leaving an inhomogeneous thickness shell or can the tetrahedral geometry be induced in some other reproducible way.

Other related curved systems may also prove interesting. In addition to spherical shells, nematics can be contained in more unusual volumes such as simple toroidal geometries or multiple joined tori [37], thus forming an object of higher genus. Combinations of experiments, theory, and simulations can lead to an understanding of the formation of defects and how they arrange themselves to minimize the energy in such systems.

These problems, and no doubt many others unforeseen as yet, mean that there should be an exciting future for this field, generating both interesting science at a fundamental level as well as of a more applied nature that could possibly lead to novel metamaterials. These questions will need experimental, theoretical, and simulation treatments to improve our knowledge of the interactions within the shells, how these depend on the structure and properties of the liquid crystal forming the shell and how the defects might be stabilized and transformed into configurations that are technologically useful.

REFERENCES

[1] Lagerwall J. Chapter in this volume.

[2] Nelson D. Nanoletters 2002;2:1125.

[3] Vitelli V. Chapter in this volume.

[4] Demus D, Goodby JW, Gray G, Spiess H-W, Vill V, editors. *Handbook of Liquid Crystals*. Weinheim: Wiley-VCH; 1998.

[5] Zannoni C. In: Luckhurst GR, Gray GW, editors. *The Molecular Physics of Liquid Crystals*. Academic Press; 1979.

[6] Lubensky TC, Prost J. J de Physique II 1991;2:371.

[7] Chuang I, Yurke B, Pargellis AN, Turok N. Phys Rev E 1993;47:3343.

[8] Bogi A, Martinot-Lagarde P, Dozov I, Nobili M. Phys Rev Lett 2002;89:225501.

[9] Blanc C, Svensek D, Zumer S, Nobili M. Phys Rev Lett 2005;95:097802.

[10] Svetec M, Kralj S, Bradac Z, Zumer S. Eur Phys J E 2006;20:71.

[11] Kleman M, Lavrentovich OD. *Soft Matter Physics: An Introduction*. Springer-Verlag; 2003.

[12] Vitelli V, Nelson DR. Phys Rev E 2006;74:021711.

[13] Lopez-Leon T, Fernandez-Nieves A. Phys Rev E 2009;79:021707.

[14] Lopez-Leon T, Bates MA, Fernandez-Nieves A. Phys Rev E 2012;86:030702.

[15] Fernandez-Nieves A, Vitelli V, Utada AS, Link DR, Marquez M, Nelson DR, Weitz DA. Phys Rev Lett 2007;99:157801.

[16] Shah RK, Shum HC, Rowat AC, Lee D, Agresti JJ, Utada AS, Chu L-Y, Kim J-W, Fernandez-Nieves A, Martinez CJ, Weitz DA. Mater Today 2008;11:18.

[17] Lopez-Leon T, Fernandez-Nieves A. Colloid Polym Sci 2011;289:345.

[18] Lopez-Leon T, Koning V, Devaiah KBS, Vitelli V, Fernandez-Nieves A. Nat Phys 2011;7:391.

[19] Liang H-L, Schymura S, Rudquist P, Lagerwall J. Phys Rev Lett 2011;106:247801.

[20] Lopez-Leon T, Fernandez-Nieves A, Nobili M, Blanc C. Phys Rev Lett 2011;106:247802.

[21] Allen MP, Tildesley DJ. *Computer Simulation of Liquids*. Clarendon Press; 1987.

[22] Frenkel D, Smit B. *Understanding Molecular Simulation: From Algorithms to Applications*. Academic Press; 1996.

[23] Cornell WD, Cieplak P, Bayly CI, Gould IR, Merz KM, Ferguson DM, Spellmeyer DC, Fox T, Caldwell JW, Kollman PA. J Am Chem Soc 1995;117:5179.

[24] Metropolis N, Rosenbluth AW, Rosenbluth MN, Teller AH, Teller E. J Chem Phys 1953;21:1087.

[25] Bates MA. J Chem Phys 2008;128:104707.

[26] Shin H, Bowick MJ, Xing X. Phys Rev Lett 2008;101:037802.

[27] Lebwohl PA, Lasher G. Phys Rev A 1972;6:426.

[28] Fabbri U, Zannoni C. Mol Phys 1986;58:763.

[29] Skacej G, Zannoni C. Phys Rev Lett 2008;100:197802.

[30] Poniewierski A, Holyst R. Phys Rev A 1990;41:6871.

[31] Zhang W-Y, Jiang Y, Chen JZY. Phys Rev Lett 2012;108:057801.

[32] Dhakal S, Solis FJ, Olvera de la Cruz M. Phys Rev E 2012;86:011709.

[33] Callan-Jones AC, Pelcovits RA, Slavin VA, Zhang S, Laidlaw DH, Loriot GB. Phys Rev E 2006;74:061701.

[34] Bates MA. Soft Matter 2008;4:2059.

[35] Bates MA, Skacej G, Zannoni C. Soft Matter 2010;6:655.

[36] Kralj S, Rosso R, Virga E. Soft Matter 2011;7:670.

[37] Pairam E, Vallamkondu J, Koning V, van Zuiden BC, Ellis PW, Bates MA, Vitelli V, Fernandez-Nieves A. Proc Natl Acad Sci 2013;110:9295.

INDEX

Fluids, Colloids and Soft Materials: An Introduction to Soft Matter Physics, First Edition. Edited by Alberto Fernandez Nieves and Antonio Manuel Puertas.